STRING THEORY AND M-THEORY
A MODERN INTRODUCTION

String theory is one of the most exciting and challenging areas of modern theoretical physics. This book guides the reader from the basics of string theory to very recent developments at the frontier of string theory research.

The book begins with the basics of perturbative string theory, world-sheet supersymmetry, space-time supersymmetry, conformal field theory and the heterotic string, and moves on to describe modern developments, including D-branes, string dualities and M-theory. It then covers string geometry (including Calabi–Yau compactifications) and flux compactifications, and applications to cosmology and particle physics. One chapter is dedicated to black holes in string theory and M-theory, and the microscopic origin of black-hole entropy. The book concludes by presenting matrix theory, AdS/CFT duality and its generalizations.

This book is ideal for graduate students studying modern string theory, and it will make an excellent textbook for a 1-year course on string theory. It will also be useful for researchers interested in learning about developments in modern string theory. The book contains about 120 solved exercises, as well as about 200 homework problems, solutions of which are available for lecturers on a password protected website at www.cambridge.org/9780521860697.

KATRIN BECKER is a Professor of physics at Texas A & M University. She was awarded the Radcliffe Fellowship from Harvard University in 2006 and received the Alfred Sloan Fellowship in 2003.

MELANIE BECKER is a Professor of physics at Texas A & M University. In 2006 she was awarded an Edward, Frances and Shirley B. Daniels Fellowship from the Radcliffe Institute for Advanced Studies at Harvard University. In 2001 she received the Alfred Sloan Fellowship.

JOHN H. SCHWARZ is the Harold Brown Professor of Theoretical Physics at the California Institute of Technology. He is a MacArthur Fellow and a member of the National Academy of Sciences.

This is the first comprehensive textbook on string theory to also offer an up-to-date picture of the most important theoretical developments of the last decade, including the AdS/CFT correspondence and flux compactifications, which have played a crucial role in modern efforts to make contact with experiment. An excellent resource for graduate students as well as researchers in high-energy physics and cosmology.

Nima Arkani-Hamed, Harvard University

An exceptional introduction to string theory that contains a comprehensive treatment of all aspects of the theory, including recent developments. The clear pedagogical style and the many excellent exercises should provide the interested student or researcher a straightforward path to the frontiers of current research.

David Gross, Director of the Kavli Institute for Theoretical Physics, University of California, Santa Barbara and winner of the Nobel Prize for Physics in 2004

Masterfully written by pioneers of the subject, comprehensive, up-to-date and replete with illuminating problem sets and their solutions, *String Theory and M-theory: A Modern Introduction* provides an ideal preparation for research on the current forefront of the fundamental laws of nature. It is destined to become the standard textbook in the subject.

Andrew Strominger, Harvard University

This book is a magnificent resource for students and researchers alike in the rapidly evolving field of string theory. It is unique in that it is targeted for students without any knowledge of string theory and at the same time it includes the very latest developments of the field, all presented in a very fluid and simple form. The lucid description is nicely complemented by very instructive problems. I highly recommend this book to all researchers interested in the beautiful field of string theory.

Cumrun Vafa, Harvard University

This elegantly written book will be a valuable resource for students looking for an entry-way to the vast and exciting topic of string theory. The authors have skillfully made a selection of topics aimed at helping the beginner get up to speed. I am sure it will be widely read.

Edward Witten, Institute for Advanced Study, Princeton, winner of the Fields Medal in 1990

STRING THEORY AND M-THEORY
A Modern Introduction

KATRIN BECKER,
Texas A & M University

MELANIE BECKER,
Texas A & M University

and

JOHN H. SCHWARZ
California Institute of Technology

CAMBRIDGE UNIVERSITY PRESS
Cambridge, New York, Melbourne, Madrid, Cape Town, Singapore, São Paulo

Cambridge University Press
The Edinburgh Building, Cambridge CB2 2RU, UK

Published in the United States of America by Cambridge University Press, New York

www.cambridge.org
Information on this title: www.cambridge.org/9780521860697

© K. Becker, M. Becker and J. H. Schwarz 2007

This publication is in copyright. Subject to statutory exception
and to the provisions of relevant collective licensing agreements,
no reproduction of any part may take place without
the written permission of Cambridge University Press.

First published 2007

Printed in the United Kingdom at the University Press, Cambridge

A catalog record for this publication is available from the British Library

ISBN-13 978-0-521-86069-7 hardback
ISBN-10 0-521-86069-5 hardback

Cambridge University Press has no responsibility for
the persistence or accuracy of URLs for external or
third-party internet websites referred to in this publication,
and does not guarantee that any content on such
websites is, or will remain, accurate or appropriate.

To our parents

An Ode to the Unity of Time and Space

Time, ah, time,
how you go off like this!

Physical things, ah, things,
so abundant you are!

The Ruo's waters are three thousand,
how can they not have the same source?

Time and space are one body,
mind and things sustain each other.

Time, o time,
does not time come again?

Heaven, o heaven,
how many are the appearances of heaven!

From ancient days constantly shifting on,
black holes flaring up.

Time and space are one body,
is it without end?

Great indeed
is the riddle of the universe.

Beautiful indeed
is the source of truth.

To quantize space and time
the smartest are nothing.

To measure the Great Universe with a long thin tube
the learning is vast.

Shing-Tung Yau

Contents

Preface		*page* xi
1	**Introduction**	1
1.1	Historical origins	2
1.2	General features	3
1.3	Basic string theory	6
1.4	Modern developments in superstring theory	9
2	**The bosonic string**	17
2.1	p-brane actions	17
2.2	The string action	24
2.3	String sigma-model action: the classical theory	30
2.4	Canonical quantization	36
2.5	Light-cone gauge quantization	48
3	**Conformal field theory and string interactions**	58
3.1	Conformal field theory	58
3.2	BRST quantization	75
3.3	Background fields	81
3.4	Vertex operators	85
3.5	The structure of string perturbation theory	89
3.6	The linear-dilaton vacuum and noncritical strings	98
3.7	Witten's open-string field theory	100
4	**Strings with world-sheet supersymmetry**	109
4.1	Ramond–Neveu–Schwarz strings	110
4.2	Global world-sheet supersymmetry	112
4.3	Constraint equations and conformal invariance	118
4.4	Boundary conditions and mode expansions	122

4.5	Canonical quantization of the RNS string	124
4.6	Light-cone gauge quantization of the RNS string	130
4.7	SCFT and BRST	140
5	**Strings with space-time supersymmetry**	**148**
5.1	The D0-brane action	149
5.2	The supersymmetric string action	155
5.3	Quantization of the GS action	160
5.4	Gauge anomalies and their cancellation	169
6	**T-duality and D-branes**	**187**
6.1	The bosonic string and Dp-branes	188
6.2	D-branes in type II superstring theories	203
6.3	Type I superstring theory	220
6.4	T-duality in the presence of background fields	227
6.5	World-volume actions for D-branes	229
7	**The heterotic string**	**249**
7.1	Nonabelian gauge symmetry in string theory	250
7.2	Fermionic construction of the heterotic string	252
7.3	Toroidal compactification	265
7.4	Bosonic construction of the heterotic string	286
8	**M-theory and string duality**	**296**
8.1	Low-energy effective actions	300
8.2	S-duality	323
8.3	M-theory	329
8.4	M-theory dualities	338
9	**String geometry**	**354**
9.1	Orbifolds	358
9.2	Calabi–Yau manifolds: mathematical properties	363
9.3	Examples of Calabi–Yau manifolds	366
9.4	Calabi–Yau compactifications of the heterotic string	374
9.5	Deformations of Calabi–Yau manifolds	385
9.6	Special geometry	391
9.7	Type IIA and type IIB on Calabi–Yau three-folds	399
9.8	Nonperturbative effects in Calabi–Yau compactifications	403
9.9	Mirror symmetry	411
9.10	Heterotic string theory on Calabi–Yau three-folds	415
9.11	K3 compactifications and more string dualities	418
9.12	Manifolds with G_2 and $Spin(7)$ holonomy	433
10	**Flux compactifications**	**456**
10.1	Flux compactifications and Calabi–Yau four-folds	460
10.2	Flux compactifications of the type IIB theory	480

10.3	Moduli stabilization	499
10.4	Fluxes, torsion and heterotic strings	508
10.5	The strongly coupled heterotic string	518
10.6	The landscape	522
10.7	Fluxes and cosmology	526
11	**Black holes in string theory**	**549**
11.1	Black holes in general relativity	552
11.2	Black-hole thermodynamics	562
11.3	Black holes in string theory	566
11.4	Statistical derivation of the entropy	582
11.5	The attractor mechanism	587
11.6	Small BPS black holes in four dimensions	599
12	**Gauge theory/string theory dualities**	**610**
12.1	Black-brane solutions in string theory and M-theory	613
12.2	Matrix theory	625
12.3	The AdS/CFT correspondence	638
12.4	Gauge/string duality for the conifold and generalizations	669
12.5	Plane-wave space-times and their duals	677
12.6	Geometric transitions	684
Bibliographic discussion		690
Bibliography		700
Index		726

Preface

String theory is one of the most exciting and challenging areas of modern theoretical physics. It was developed in the late 1960s for the purpose of describing the strong nuclear force. Problems were encountered that prevented this program from attaining complete success. In particular, it was realized that the spectrum of a fundamental string contains an undesired massless spin-two particle. Quantum chromodynamics eventually proved to be the correct theory for describing the strong force and the properties of hadrons. New doors opened for string theory when in 1974 it was proposed to identify the massless spin-two particle in the string's spectrum with the graviton, the quantum of gravitation. String theory became then the most promising candidate for a quantum theory of gravity unified with the other forces and has developed into one of the most fascinating theories of high-energy physics.

The understanding of string theory has evolved enormously over the years thanks to the efforts of many very clever people. In some periods progress was much more rapid than in others. In particular, the theory has experienced two major revolutions. The one in the mid-1980s led to the subject achieving widespread acceptance. In the mid-1990s a second superstring revolution took place that featured the discovery of nonperturbative dualities that provided convincing evidence of the uniqueness of the underlying theory. It also led to the recognition of an eleven-dimensional manifestation, called M-theory. Subsequent developments have made the connection between string theory, particle physics phenomenology, cosmology, and pure mathematics closer than ever before. As a result, string theory is becoming a mainstream research field at many universities in the US and elsewhere.

Due to the mathematically challenging nature of the subject and the above-mentioned rapid development of the field, it is often difficult for someone new to the subject to cope with the large amount of material that needs to be learned before doing actual string-theory research. One could spend several years studying the requisite background mathematics and physics, but by the end of that time, much more would have already been developed, and one still wouldn't be up to date. An alternative approach is to shorten the learning process so that the student can jump into research more quickly. In this spirit, the aim of this book is to guide the student through the fascinating subject of string theory in one academic year. This book starts with the basics of string theory in the first few chapters and then introduces the reader to some of the main topics of modern research. Since the subject is enormous, it is only possible to introduce selected topics. Nevertheless, we hope that it will provide a stimulating introduction to this beautiful subject and that the dedicated student will want to explore further.

The reader is assumed to have some familiarity with quantum field theory and general relativity. It is also very useful to have a broad mathematical background. Group theory is essential, and some knowledge of differential geometry and basics concepts of topology is very desirable. Some topics in geometry and topology that are required in the later chapters are summarized in an appendix.

The three main string-theory textbooks that precede this one are by Green, Schwarz and Witten (1987), by Polchinski (1998) and by Zwiebach (2004). Each of these was also published by Cambridge University Press. This book is somewhat shorter and more up-to-date than the first two, and it is more advanced than the third one. By the same token, those books contain much material that is not repeated here, so the serious student will want to refer to them, as well. Another distinguishing feature of this book is that it contains many exercises with worked out solutions. These are intended to be helpful to students who want problems that can be used to practice and assimilate the material.

This book would not have been possible without the assistance of many people. We have received many valuable suggestions and comments about the entire manuscript from Rob Myers, and we have greatly benefited from the assistance of Yu-Chieh Chung and Guangyu Guo, who have worked diligently on many of the exercises and homework problems and have carefully read the whole manuscript. Moreover, we have received extremely useful feedback from many colleagues including Keshav Dasgupta, Andrew Frey, Davide Gaiotto, Sergei Gukov, Michael Haack, Axel Krause, Hong Lu, Juan Maldacena, Lubos Motl, Hirosi Ooguri, Patricia Schwarz, Eric Sharpe, James Sparks, Andy Strominger, Ian Swanson, Xi Yin and especially Cumrun Vafa. We have further received great comments and suggestions from many graduate students at Caltech and Harvard University. We thank Ram Sriharsha for his assistance with some of the homework problems and Ketan Vyas for writing up solutions to the homework problems, which will be made available to instructors. We thank Sharlene Cartier and Carol Silberstein of Caltech for their help in preparing parts of the manuscript, Simon Capelin of Cambridge U. Press, whose help in coordinating the different aspects of the publishing process has been indispensable, Elisabeth Krause for help preparing some of the figures and Kovid Goyal for his assistance with computer-related issues. We thank Steven Owen for translating from Chinese the poem that precedes the preface.

During the preparation of the manuscript KB and MB have enjoyed the warm hospitality of the Radcliffe Institute for Advanced Studies at Harvard University, the physics department at Harvard University and the Perimeter Institute for theoretical physics. They would like to thank the Radcliffe Institute for Advanced Study at Harvard University, which through its Fellowship program made the completion of this project possible. Special thanks go to the Dean of Science, Barbara Grosz. Moreover, KB would also like to thank the University of Utah for awarding a teaching grant to support the work on this book. JHS is grateful to the Rutgers high-energy theory group, the Aspen Center for Physics and the Kavli Institute for Theoretical Physics for hospitality while he was working on the manuscript.

KB and MB would like to give their special thanks to their mother, Ingrid Becker, for her support and encouragement, which has always been invaluable, especially during the long journey of completing this manuscript. Her artistic talents made the design of the cover of this book possible. JHS thanks his wife Patricia for love and support while he was preoccupied with this project.

Katrin Becker
Melanie Becker
John H. Schwarz

NOTATION AND CONVENTIONS

A	area of event horizon
AdS_D	D-dimensional anti-de Sitter space-time
A_3	three-form potential of $D = 11$ supergravity
b, c	fermionic world-sheet ghosts
b_n	Betti numbers
b_r^μ, $r \in \mathbb{Z} + 1/2$	fermionic oscillator modes in NS sector
B_2 or B	NS–NS two-form potential
c	central charge of CFT
$c_1 = [\mathcal{R}/2\pi]$	first Chern class
C_n	R–R n-form potential
d_m^μ, $m \in \mathbb{Z}$	fermionic oscillator modes in R sector
D	number of space-time dimensions
$F = dA + A \wedge A$	Yang–Mills curvature two-form (antihermitian)
$F = dA + iA \wedge A$	Yang–Mills curvature two-form (hermitian)
$F_4 = dA_3$	four-form field strength of $D = 11$ supergravity
F_m, $m \in \mathbb{Z}$	odd super-Virasoro generators in R sector
$F_{n+1} = dC_n$	$(n+1)$-form R–R field strength
$g_s = \langle \exp \Phi \rangle$	closed-string coupling constant
G_r, $r \in \mathbb{Z} + 1/2$	odd super-Virasoro generators in NS sector
G_D	Newton's constant in D dimensions
$H_3 = dB_2$	NS–NS three-form field strength
$h^{p,q}$	Hodge numbers
$j(\tau)$	elliptic modular function
$J = ig_{a\bar{b}} dz^a \wedge d\bar{z}^b$	Kähler form
$\mathcal{J} = J + iB$	complexified Kähler form
k	level of Kac–Moody algebra
K	Kaluza–Klein excitation number
$\mathcal{K}$	Kähler potential
$l_p = 1.6 \times 10^{-33}$ cm	Planck length for $D = 4$
ℓ_p	Planck length for $D = 11$
$l_s = \sqrt{2\alpha'}$, $\ell_s = \sqrt{\alpha'}$	string length scale
L_n, $n \in \mathbb{Z}$	generators of Virasoro algebra
$m_p = 1.2 \times 10^{19}$ GeV$/c^2$	Planck mass for $D = 4$
$M_p = 2.4 \times 10^{18}$ GeV$/c^2$	reduced Planck mass $m_p/\sqrt{8\pi}$
$M, N, \ldots$	space-time indices for $D = 11$
$\mathcal{M}$	moduli space

N_L, N_R	left- and right-moving excitation numbers
Q_B	BRST charge
$R = d\omega + \omega \wedge \omega$	Riemann curvature two-form
$R_{\mu\nu} = R^\lambda{}_{\mu\lambda\nu}$	Ricci tensor
$\mathcal{R} = R_{a\bar{b}} dz^a \wedge d\bar{z}^{\bar{b}}$	Ricci form
S	entropy
S^a	world-sheet fermions in light-cone gauge GS formalism
$T_{\alpha\beta}$	world-sheet energy–momentum tensor
T_p	tension of p-brane
W	winding number
$x^\mu, \mu = 0, 1, \ldots D-1$	space-time coordinates
$X^\mu, \mu = 0, 1, \ldots D-1$	space-time embedding functions of a string
$x^\pm = (x^0 \pm x^{D-1})/\sqrt{2}$	light-cone coordinates in space-time
$x^I, I = 1, 2, \ldots, D-2$	transverse coordinates in space-time
Z	central charge
$\alpha_m^\mu, m \in \mathbb{Z}$	bosonic oscillator modes
α'	Regge-slope parameter
β, γ	bosonic world-sheet ghosts
γ_μ	Dirac matrices in four dimensions
Γ_M	Dirac matrices in 11 dimensions
$\Gamma_{\mu\nu}{}^\rho$	affine connection
$\eta(\tau)$	Dedekind eta function
Θ^{Aa}	world-volume fermions in covariant GS formalism
λ^A	left-moving world-sheet fermions of heterotic string
$\Lambda \sim 10^{-120} M_\mathrm{p}^4$	observed vacuum energy density
$\sigma^\alpha, \alpha = 0, 1, \ldots, p$	world-volume coordinates of a p-brane
$\sigma^0 = \tau, \sigma^1 = \sigma$	world-sheet coordinates of a string
$\sigma^\pm = \tau \pm \sigma$	light-cone coordinates on the world sheet
$\sigma^\mu_{\alpha\dot\beta}$	Dirac matrices in two-component spinor notation
Φ	dilaton field
$\chi(M)$	Euler characteristic of M
ψ^μ	world-sheet fermion in RNS formalism
Ψ_M	gravitino field of $D = 11$ supergravity
$\omega_\mu{}^\alpha{}_\beta$	spin connection
Ω	world-sheet parity transformation
Ω_n	holomorphic n-form

- $\hbar = c = 1$.
- The signature of any metric is 'mostly +', that is, $(-, +, \ldots, +)$.
- The space-time metric is $ds^2 = g_{\mu\nu}dx^\mu dx^\nu$.
- In Minkowski space-time $g_{\mu\nu} = \eta_{\mu\nu}$.
- The world-sheet metric tensor is $h_{\alpha\beta}$.
- A hermitian metric has the form $ds^2 = 2g_{a\bar{b}}dz^a d\bar{z}^{\bar{b}}$.
- The space-time Dirac algebra in $D = d+1$ dimensions is $\{\Gamma_\mu, \Gamma_\nu\} = 2g_{\mu\nu}$.
- $\Gamma^{\mu_1\mu_2\cdots\mu_n} = \Gamma^{[\mu_1}\Gamma^{\mu_2}\cdots\Gamma^{\mu_n]}$.
- The world-sheet Dirac algebra is $\{\rho_\alpha, \rho_\beta\} = 2h_{\alpha\beta}$.
- $|F_n|^2 = \frac{1}{n!} g^{\mu_1\nu_1}\cdots g^{\mu_n\nu_n} F_{\mu_1\ldots\mu_n} F_{\nu_1\ldots\nu_n}$.
- The Levi–Civita tensor $\varepsilon^{\mu_1\cdots\mu_D}$ is totally antisymmetric with $\varepsilon^{01\cdots d} = 1$.

1
Introduction

There were two major breakthroughs that revolutionized theoretical physics in the twentieth century: general relativity and quantum mechanics. General relativity is central to our current understanding of the large-scale expansion of the Universe. It gives small corrections to the predictions of Newtonian gravity for the motion of planets and the deflection of light rays, and it predicts the existence of gravitational radiation and black holes. Its description of the gravitational force in terms of the curvature of space-time has fundamentally changed our view of space and time: they are now viewed as dynamical. Quantum mechanics, on the other hand, is the essential tool for understanding microscopic physics. The evidence continues to build that it is an exact property of Nature. Certainly, its exact validity is a basic assumption in all string theory research.

The understanding of the fundamental laws of Nature is surely incomplete until general relativity and quantum mechanics are successfully reconciled and unified. That this is very challenging can be seen from many different viewpoints. The concepts, observables and types of calculations that characterize the two subjects are strikingly different. Moreover, until about 1980 the two fields developed almost independently of one another. Very few physicists were experts in both. With the goal of unifying both subjects, string theory has dramatically altered the sociology as well as the science.

In relativistic quantum mechanics, called quantum field theory, one requires that two fields that are defined at space-time points with a space-like separation should commute (or anticommute if they are fermionic). In the gravitational context one doesn't know whether or not two space-time points have a space-like separation until the metric has been computed, which is part of the dynamical problem. Worse yet, the metric is subject to quantum fluctuations just like other quantum fields. Clearly, these are rather challenging issues. Another set of challenges is associated with the quantum description of black holes and the description of the Universe in the very early stages of its history.

The most straightforward attempts to combine quantum mechanics and general relativity, in the framework of perturbative quantum field theory, run into problems due to uncontrollable infinities. Ultraviolet divergences are a characteristic feature of radiative corrections to gravitational processes, and they become worse at each order in perturbation theory. Because Newton's constant is proportional to (length)^2 in four dimensions, simple power-counting arguments show that it is not possible to remove these infinities by the conventional renormalization methods of quantum field theory. Detailed calculations demonstrate that there is no miracle that invalidates this simple dimensional analysis.[1]

String theory purports to overcome these difficulties and to provide a consistent quantum theory of gravity. How the theory does this is not yet understood in full detail. As we have learned time and time again, string theory contains many deep truths that are there to be discovered. Gradually a consistent picture is emerging of how this remarkable and fascinating theory deals with the many challenges that need to be addressed for a successful unification of quantum mechanics and general relativity.

1.1 Historical origins

String theory arose in the late 1960s in an attempt to understand the strong nuclear force. This is the force that is responsible for holding protons and neutrons together inside the nucleus of an atom as well as quarks together inside the protons and neutrons. A theory based on fundamental one-dimensional extended objects, called strings, rather than point-like particles, can account qualitatively for various features of the strong nuclear force and the strongly interacting particles (or hadrons).

The basic idea in the string description of the strong interactions is that specific particles correspond to specific oscillation modes (or quantum states) of the string. This proposal gives a very satisfying unified picture in that it postulates a single fundamental object (namely, the string) to explain the myriad of different observed hadrons, as indicated in Fig. 1.1.

In the early 1970s another theory of the strong nuclear force – called quantum chromodynamics (or QCD) – was developed. As a result of this, as well as various technical problems in the string theory approach, string

---

[1] Some physicists believe that perturbative renormalizability is not a fundamental requirement and try to "quantize" pure general relativity despite its nonrenormalizability. Loop quantum gravity is an example of this approach. Whatever one thinks of the logic, it is fair to say that despite a considerable amount of effort such attempts have not yet been very fruitful.

theory fell out of favor. The current viewpoint is that this program made good sense, and so it has again become an active area of research. The concrete string theory that describes the strong interaction is still not known, though one now has a much better understanding of how to approach the problem.

String theory turned out to be well suited for an even more ambitious purpose: the construction of a quantum theory that unifies the description of gravity and the other fundamental forces of nature. In principle, it has the potential to provide a complete understanding of particle physics and of cosmology. Even though this is still a distant dream, it is clear that in this fascinating theory surprises arise over and over.

1.2 General features

Even though string theory is not yet fully formulated, and we cannot yet give a detailed description of how the standard model of elementary particles should emerge at low energies, or how the Universe originated, there are some general features of the theory that have been well understood. These are features that seem to be quite generic irrespective of what the final formulation of string theory might be.

Gravity

The first general feature of string theory, and perhaps the most important, is that general relativity is naturally incorporated in the theory. The theory gets modified at very short distances/high energies but at ordinary distances and energies it is present in exactly the form as proposed by Einstein. This is significant, because general relativity is arising within the framework of a

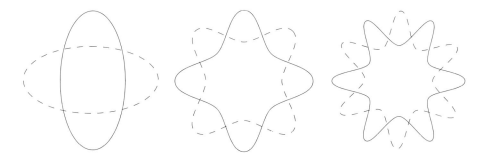

Fig. 1.1. Different particles are different vibrational modes of a string.

consistent quantum theory. Ordinary quantum field theory does not allow gravity to exist; string theory requires it.

Yang–Mills gauge theory

In order to fulfill the goal of describing all of elementary particle physics, the presence of a graviton in the string spectrum is not enough. One also needs to account for the standard model, which is a Yang–Mills theory based on the gauge group $SU(3) \times SU(2) \times U(1)$. The appearance of Yang–Mills gauge theories of the sort that comprise the standard model is a general feature of string theory. Moreover, matter can appear in complex chiral representations, which is an essential feature of the standard model. However, it is not yet understood why the specific $SU(3) \times SU(2) \times U(1)$ gauge theory with three generations of quarks and leptons is singled out in nature.

Supersymmetry

The third general feature of string theory is that its consistency requires supersymmetry, which is a symmetry that relates bosons to fermions is required. There exist nonsupersymmetric bosonic string theories (discussed in Chapters 2 and 3), but lacking fermions, they are completely unrealistic. The mathematical consistency of string theories with fermions depends crucially on local supersymmetry. Supersymmetry is a generic feature of all potentially realistic string theories. The fact that this symmetry has not yet been discovered is an indication that the characteristic energy scale of supersymmetry breaking and the masses of supersymmetry partners of known particles are above experimentally determined lower bounds.

Space-time supersymmetry is one of the major predictions of superstring theory that could be confirmed experimentally at accessible energies. A variety of arguments, not specific to string theory, suggest that the characteristic energy scale associated with supersymmetry breaking should be related to the electroweak scale, in other words in the range 100 GeV to a few TeV. If this is correct, superpartners should be observable at the CERN Large Hadron Collider (LHC), which is scheduled to begin operating in 2007.

Extra dimensions of space

In contrast to many theories in physics, superstring theories are able to predict the dimension of the space-time in which they live. The theory is only consistent in a ten-dimensional space-time and in some cases an eleventh dimension is also possible.

To make contact between string theory and the four-dimensional world of everyday experience, the most straightforward possibility is that six or seven of the dimensions are compactified on an internal manifold, whose size is sufficiently small to have escaped detection. For purposes of particle physics, the other four dimensions should give our four-dimensional space-time. Of course, for purposes of cosmology, other (time-dependent) geometries may also arise.

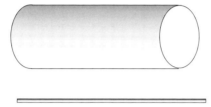

Fig. 1.2. From far away a two-dimensional cylinder looks one-dimensional.

The idea of an extra compact dimension was first discussed by Kaluza and Klein in the 1920s. Their goal was to construct a unified description of electromagnetism and gravity in four dimensions by compactifying five-dimensional general relativity on a circle. Even though we now know that this is not how electromagnetism arises, the essence of this beautiful approach reappears in string theory. The Kaluza–Klein idea, nowadays referred to as *compactification*, can be illustrated in terms of the two cylinders of Fig. 1.2. The surface of the first cylinder is two-dimensional. However, if the radius of the circle becomes extremely small, or equivalently if the cylinder is viewed from a large distance, the cylinder looks effectively one-dimensional. One now imagines that the long dimension of the cylinder is replaced by our four-dimensional space-time and the short dimension by an appropriate six, or seven-dimensional compact manifold. At large distances or low energies the compact internal space cannot be seen and the world looks effectively four-dimensional. As discussed in Chapters 9 and 10, even if the internal manifolds are invisible, their topological properties determine the particle content and structure of the four-dimensional theory. In the mid-1980s Calabi–Yau manifolds were first considered for compactifying six extra dimensions, and they were shown to be phenomenologically rather promising, even though some serious drawbacks (such as the moduli space problem discussed in Chapter 10) posed a problem for the predictive power of string theory. In contrast to the circle, Calabi–Yau manifolds do not have isometries, and part of their role is to break symmetries rather than to make them.

The size of strings

In conventional quantum field theory the elementary particles are mathematical points, whereas in perturbative string theory the fundamental objects are one-dimensional loops (of zero thickness). Strings have a characteristic length scale, denoted l_s, which can be estimated by dimensional analysis. Since string theory is a relativistic quantum theory that includes gravity it must involve the fundamental constants c (the speed of light), $\hbar$ (Planck's constant divided by 2π), and G (Newton's gravitational constant). From these one can form a length, known as the Planck length

$$l_\mathrm{p} = \left(\frac{\hbar G}{c^3}\right)^{1/2} = 1.6 \times 10^{-33}\,\mathrm{cm}.$$

Similarly, the Planck mass is

$$m_\mathrm{p} = \left(\frac{\hbar c}{G}\right)^{1/2} = 1.2 \times 10^{19}\,\mathrm{GeV}/c^2.$$

The Planck scale is the natural first guess for a rough estimate of the fundamental string length scale as well as the characteristic size of compact extra dimensions. Experiments at energies far below the Planck energy cannot resolve distances as short as the Planck length. Thus, at such energies, strings can be accurately approximated by point particles. This explains why quantum field theory has been so successful in describing our world.

1.3 Basic string theory

As a string evolves in time it sweeps out a two-dimensional surface in space-time, which is called the string *world sheet* of the string. This is the string counterpart of the world line for a point particle. In quantum field theory, analyzed in perturbation theory, contributions to amplitudes are associated with Feynman diagrams, which depict possible configurations of world lines. In particular, interactions correspond to junctions of world lines. Similarly, perturbation expansions in string theory involve string world sheets of various topologies.

The existence of interactions in string theory can be understood as a consequence of world-sheet topology rather than of a local singularity on the world sheet. This difference from point-particle theories has two important implications. First, in string theory the structure of interactions is uniquely determined by the free theory. There are no arbitrary interactions to be chosen. Second, since string interactions are not associated with short-distance singularities, string theory amplitudes have no ultraviolet divergences. The string scale $1/l_s$ acts as a UV cutoff.

World-volume actions and the critical dimension

A string can be regarded as a special case of a p-brane, which is an object with p spatial dimensions and tension (or energy density) T_p. In fact, various p-branes do appear in superstring theory as nonperturbative excitations. The classical motion of a p-brane extremizes the $(p+1)$-dimensional volume V that it sweeps out in space-time. Thus there is a p-brane action that is given by $S_p = -T_p V$. In the case of the fundamental string, which has $p = 1$, V is the area of the string world sheet and the action is called the Nambu–Goto action.

Classically, the Nambu–Goto action is equivalent to the string sigma-model action

$$S_\sigma = -\frac{T}{2} \int \sqrt{-h} h^{\alpha\beta} \eta_{\mu\nu} \partial_\alpha X^\mu \partial_\beta X^\nu d\sigma d\tau,$$

where $h_{\alpha\beta}(\sigma, \tau)$ is an auxiliary world-sheet metric, $h = \det h_{\alpha\beta}$, and $h^{\alpha\beta}$ is the inverse of $h_{\alpha\beta}$. The functions $X^\mu(\sigma, \tau)$ describe the space-time embedding of the string world sheet. The Euler–Lagrange equation for $h^{\alpha\beta}$ can be used to eliminate it from the action and recover the Nambu–Goto action.

Quantum mechanically, the story is more subtle. Instead of eliminating h via its classical field equations, one should perform a Feynman path integral, using standard machinery to deal with the local symmetries and gauge fixing. When this is done correctly, one finds that there is a conformal anomaly unless the space-time dimension is $D = 26$. These matters are explored in Chapters 2 and 3. An analogous analysis for superstrings gives the critical dimension $D = 10$.

Closed strings and open strings

The parameter τ in the embedding functions $X^\mu(\sigma, \tau)$ is the world-sheet time coordinate and σ parametrizes the string at a given world-sheet time. For a closed string, which is topologically a circle, one should impose periodicity in the spatial parameter σ. Choosing its range to be π one identifies both ends of the string $X^\mu(\sigma,\tau) = X^\mu(\sigma+\pi,\tau)$. All string theories contain closed strings, and the graviton always appears as a massless mode in the closed-string spectrum of critical string theories.

For an open string, which is topologically a line interval, each end can be required to satisfy either Neumann or Dirichlet boundary conditions (for each value of μ). The Dirichlet condition specifies a space-time hypersurface on which the string ends. The only way this makes sense is if the open string ends on a physical object, which is called a D-brane. (D stands for Dirichlet.) If all the open-string boundary conditions are Neumann, then the ends of the string can be anywhere in the space-time. The modern interpretation is that this means that space-time-filling D-branes are present.

Perturbation theory

Perturbation theory is useful in a quantum theory that has a small dimensionless coupling constant, such as quantum electrodynamics (QED), since it allows one to compute physical quantities as expansions in the small parameter. In QED the small parameter is the fine-structure constant $\alpha \sim 1/137$. For a physical quantity $T(\alpha)$, one computes (using Feynman diagrams)

$$T(\alpha) = T_0 + \alpha T_1 + \alpha^2 T_2 + \ldots$$

Perturbation series are usually asymptotic expansions with zero radius of convergence. Still, they can be useful, if the expansion parameter is small, because the first terms in the expansion provide an accurate approximation.

The heterotic and type II superstring theories contain oriented closed strings only. As a result, the only world sheets in their perturbation expansions are closed oriented Riemann surfaces. There is a unique world-sheet topology at each order of the perturbation expansion, and its contribution is UV finite. The fact that there is just one string theory Feynman diagram at each order in the perturbation expansion is in striking contrast to the large number of Feynman diagrams that appear in quantum field theory. In the case of string theory there is no particular reason to expect the coupling constant g_s to be small. So it is unlikely that a realistic vacuum could be analyzed accurately using only perturbation theory. For this reason, it is important to understand nonperturbative effects in string theory.

Superstrings

The *first superstring revolution* began in 1984 with the discovery that quantum mechanical consistency of a ten-dimensional theory with $\mathcal{N}=1$ supersymmetry requires a local Yang–Mills gauge symmetry based on one of two possible Lie algebras: $SO(32)$ or $E_8 \times E_8$. As is explained in Chapter 5, only for these two choices do certain quantum mechanical anomalies cancel. The fact that only these two groups are possible suggested that string theory has a very constrained structure, and therefore it might be very predictive. [2]

When one uses the superstring formalism for both left-moving modes and right-moving modes, the supersymmetries associated with the left-movers and the right-movers can have either opposite handedness or the same handedness. These two possibilities give different theories called the type IIA and type IIB superstring theories, respectively. A third possibility, called type I superstring theory, can be derived from the type IIB theory by modding out by its left–right symmetry, a procedure called orientifold projection. The strings that survive this projection are unoriented. The type I and type II superstring theories are described in Chapters 4 and 5 using formalisms with world-sheet and space-time supersymmetry, respectively.

A more surprising possibility is to use the formalism of the 26-dimensional bosonic string for the left-movers and the formalism of the 10-dimensional superstring for the right-movers. The string theories constructed in this way are called "heterotic." Heterotic string theory is discussed in Chapter 7. The mismatch in space-time dimensions may sound strange, but it is actually exactly what is needed. The extra 16 left-moving dimensions must describe a torus with very special properties to give a consistent theory. There are precisely two distinct tori that have the required properties, and they correspond to the Lie algebras $SO(32)$ and $E_8 \times E_8$.

Altogether, there are five distinct superstring theories, each in ten dimensions. Three of them, the type I theory and the two heterotic theories, have $\mathcal{N} = 1$ supersymmetry in the ten-dimensional sense. The minimal spinor in ten dimensions has 16 real components, so these theories have 16 conserved supercharges. The type I superstring theory has the gauge group $SO(32)$, whereas the heterotic theories realize both $SO(32)$ and $E_8 \times E_8$. The other two theories, type IIA and type IIB, have $\mathcal{N} = 2$ supersymmetry or equivalently 32 supercharges.

1.4 Modern developments in superstring theory

The realization that there are five different superstring theories was somewhat puzzling. Certainly, there is only one Universe, so it would be most satisfying if there were only one possible theory. In the late 1980s it was

---

[2] Anomaly analysis alone also allows $U(1)^{496}$ and $E_8 \times U(1)^{248}$. However, there are no string theories with these gauge groups.

realized that there is a property known as T-duality that relates the two type II theories and the two heterotic theories, so that they shouldn't really be regarded as distinct theories.

Progress in understanding nonperturbative phenomena was achieved in the 1990s. Nonperturbative S-dualities and the opening up of an eleventh dimension at strong coupling in certain cases led to new identifications. Once all of these correspondences are taken into account, one ends up with the best possible conclusion: there is a unique underlying theory. Some of these developments are summarized below and are discussed in detail in the later chapters.

T-duality

String theory exhibits many surprising properties. One of them, called T-duality, is discussed in Chapter 6. T-duality implies that in many cases two different geometries for the extra dimensions are physically equivalent! In the simplest example, a circle of radius R is equivalent to a circle of radius ℓ_s^2/R, where (as before) ℓ_s is the fundamental string length scale.

T-duality typically relates two different theories. For example, it relates the two type II and the two heterotic theories. Therefore, the type IIA and type IIB theories (also the two heterotic theories) should be regarded as a single theory. More precisely, they represent opposite ends of a continuum of geometries as one varies the radius of a circular dimension. This radius is not a parameter of the underlying theory. Rather, it arises as the vacuum expectation value of a scalar field, and it is determined dynamically.

There are also fancier examples of duality equivalences. For example, there is an equivalence of type IIA superstring theory compactified on a Calabi–Yau manifold and type IIB compactified on the "mirror" Calabi–Yau manifold. This mirror pairing of topologically distinct Calabi–Yau manifolds is discussed in Chapter 9. A surprising connection to T-duality will emerge.

S-duality

Another kind of duality – called S-duality – was discovered as part of the *second superstring revolution* in the mid-1990s. It is discussed in Chapter 8. S-duality relates the string coupling constant g_s to $1/g_s$ in the same way that T-duality relates R to ℓ_s^2/R. The two basic examples relate the type I superstring theory to the $SO(32)$ heterotic string theory and the type IIB superstring theory to itself. Thus, given our knowledge of the small g_s behavior of these theories, given by perturbation theory, we learn how these three theories behave when $g_s \gg 1$. For example, strongly coupled type I theory is equivalent to weakly coupled $SO(32)$ heterotic theory. In the type IIB case the theory is related to itself, so one is actually dealing with a symmetry. The string coupling constant g_s is given by the vacuum expectation value of $\exp \phi$, where ϕ is the dilaton field. S-duality, like T-duality, is actually a field transformation, $\phi \to -\phi$, and not just a statement about vacuum expectation values.

D-branes

When studied nonperturbatively, one discovers that superstring theory contains various p-branes, objects with p spatial dimensions, in addition to the fundamental strings. All of the p-branes, with the single exception of the fundamental string (which is a 1-brane), become infinitely heavy as $g_s \to 0$, and therefore they do not appear in perturbation theory. On the other hand, when the coupling g_s is not small, this distinction is no longer significant. When that is the case, all of the p-branes are just as important as the fundamental strings, so there is p-brane democracy.

The type I and II superstring theories contain a class of p-branes called D-branes, whose tension is proportional $1/g_s$. As was mentioned earlier, their defining property is that they are objects on which fundamental strings can end. The fact that fundamental strings can end on D-branes implies that quantum field theories of the Yang–Mills type, like the standard model, reside on the world volumes of D-branes. The Yang–Mills fields arise as the massless modes of open strings attached to the D-branes. The fact that theories resembling the standard model reside on D-branes has many interesting implications. For example, it has led to the speculation that the reason we experience four space-time dimensions is because we are confined to live on three-dimensional D-branes (D3-branes), which are embedded in a higher-dimensional space-time. Model-building along these lines, sometimes called the *brane-world* approach or scenario, is discussed in Chapter 10.

What is M-theory?

S-duality explains how three of the five original superstring theories behave at strong coupling. This raises the question: What happens to the other two superstring theories – type IIA and $E_8 \times E_8$ heterotic – when g_s is large? The answer, which came as quite a surprise, is that they grow an eleventh dimension of size $g_s \ell_s$. This new dimension is a circle in the type IIA case and a line interval in the heterotic case. When the eleventh dimension is large, one is outside the regime of perturbative string theory, and new techniques are required. Most importantly, a new type of quantum theory in 11 dimensions, called M-theory, emerges. At low energies it is approximated by a classical field theory called 11-dimensional supergravity, but M-theory is much more than that. The relation between M-theory and the two superstring theories previously mentioned, together with the T and S dualities discussed above, imply that the five superstring theories are connected by a web of dualities, as depicted in Fig. 1.3. They can be viewed as different corners of a single theory.

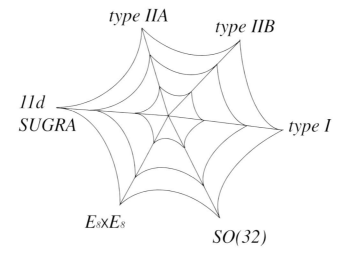

Fig. 1.3. Different string theories are connected through a web of dualities.

There are techniques for identifying large classes of superstring and M-theory vacua, and describing them exactly, but there is not yet a succinct and compelling formulation of the underlying theory that gives rise to these vacua. Such a formulation should be completely unique, with no adjustable dimensionless parameters or other arbitrariness. Many things that we usually take for granted, such as the existence of a space-time manifold, are likely to be understood as emergent properties of specific vacua rather than identifiable features of the underlying theory. If this is correct, then the missing formulation of the theory must be quite unlike any previous theory. Usual approaches based on quantum fields depend on the existence of an ambient space-time manifold. It is not clear what the basic degrees of freedom should be in a theory that does not assume a space-time manifold at the outset.

There is an interesting proposal for an exact quantum mechanical descrip-

1.4 Modern developments in superstring theory

tion of M-theory, applicable to certain space-time backgrounds, that goes by the name of *Matrix theory*. Matrix theory gives a dual description of M-theory in flat 11-dimensional space-time in terms of the quantum mechanics of $N \times N$ matrices in the large N limit. When n of the spatial dimensions are compactified on a torus, the dual Matrix theory becomes a quantum field theory in n spatial dimensions (plus time). There is evidence that this conjecture is correct when n is not too large. However, it is unclear how to generalize it to other compactification geometries, so Matrix theory provides only pieces of a more complete description of M-theory.

F-theory

As previously discussed, the type IIA and heterotic $E_8 \times E_8$ theories can be viewed as arising from a more fundamental eleven-dimensional theory, M-theory. One may wonder if the other superstring theories can be derived in a similar fashion. An approach, called F-theory, is described in Chapter 9. It utilizes the fact that ten-dimensional type IIB superstring theory has a nonperturbative $SL(2,\mathbb{Z})$ symmetry. Moreover, this is the modular group of a torus and the type IIB theory contains a complex scalar field τ that transforms under $SL(2,\mathbb{Z})$ as the complex structure of a torus. Therefore, this symmetry can be given a geometric interpretation if the type IIB theory is viewed as having an auxiliary two-torus T^2 with complex structure τ. The $SL(2,\mathbb{Z})$ symmetry then has a natural interpretation as the symmetry of the torus.

Flux compactifications

One question that already bothered Kaluza and Klein is why should the fifth dimension curl up? Another puzzle in those early days was the size of the circle, and what stabilizes it at a particular value. These questions have analogs in string theory, where they are part of what is called *the moduli-space problem*. In string theory the shape and size of the internal manifold is dynamically determined by the vacuum expectation values of scalar fields. String theorists have recently been able to provide answers to these questions in the context of flux compactifications , which is a rapidly developing area of modern string theory research. This is discussed in Chapter 10.

Even though the underlying theory (M-theory) is unique, it admits an enormous number of different solutions (or quantum vacua). One of these solutions should consist of four-dimensional Minkowski space-time times a compact manifold and accurately describes the world of particle physics.

One of the major challenges of modern string theory research is to find this solution.

It would be marvelous to identify the correct vacuum, and at the same time to understand *why* it is the right one. Is it picked out by some special mathematical property, or is it just an environmental accident of our particular corner of the Universe? The way this question plays out will be important in determining the extent to which the observed world of particle physics can be deduced from first principles.

Black-hole entropy

It follows from general relativity that macroscopic *black holes* behave like thermodynamic objects with a well-defined *temperature* and *entropy*. The entropy is given (in gravitational units) by 1/4 the area of the event horizon, which is the *Bekenstein–Hawking* entropy formula. In quantum theory, an entropy S ordinarily implies that there are a large number of quantum states (namely, $\exp S$ of them) that contribute to the corresponding microscopic description. So a natural question is whether this rule also applies to black holes and their higher-dimensional generalizations, which are called *black p-branes*. D-branes provide a set-up in which this question can be investigated.

In the early work on this subject, reliable techniques for counting microstates only existed for very special types of black holes having a large amount of supersymmetry. In those cases one found agreement with the entropy formula. More recently, one has learned how to analyze a much larger class of black holes and black p-branes, and even how to compute corrections to the area formula. This subject is described in Chapter 11. Many examples have been studied and no discrepancies have been found, aside from corrections that are expected. It is fair to say that these studies have led to a much deeper understanding of the thermodynamic properties of black holes in terms of string-theory microphysics, a fact that is one of the most striking successes of string theory so far.

AdS/CFT duality

A remarkable discovery made in the late 1990s is the exact equivalence (or duality) of conformally invariant quantum field theories and superstring theory or M-theory in special space-time geometries. A collection of coincident p-branes produces a space-time geometry with a horizon, like that of a black hole. In the vicinity of the horizon, this geometry can be approximated by a product of an *anti-de Sitter space* and a sphere. In the example that arises from considering N coincident D3-branes in the type IIB superstring theory, one obtains a duality between $SU(N)$ Yang–Mills theory with $\mathcal{N} = 4$ supersymmetry in four dimensions and type IIB superstring theory in a ten-dimensional geometry given by a product of a five-dimensional anti-de Sitter space (AdS_5) and a five-dimensional sphere (S^5). There are N units of five-form flux threading the five sphere. There are also analogous M-theory dualities.

These dualities are sometimes referred to as AdS/CFT dualities. AdS stands for *anti-de Sitter space*, a maximally symmetric space-time geometry with negative scalar curvature. CFT stands for *conformal field theory*, a quantum field theory that is invariant under the group of conformal transformations. This type of equivalence is an example of a *holographic duality*, since it is analogous to representing three-dimensional space on a two-dimensional emulsion. The study of these dualities is teaching us a great deal about string theory and M-theory as well as the dual quantum field theories. Chapter 12 gives an introduction to this vast subject.

String and M-theory cosmology

The field of *superstring cosmology* is emerging as a new and exciting discipline. String theorists and string-theory considerations are injecting new ideas into the study of cosmology. This might be the arena in which predictions that are specific to string theory first confront data.

In a quantum theory that contains gravity, such as string theory, the *cosmological constant*, Λ, which characterizes the energy density of the vacuum, is (at least in principle) a computable quantity. This energy (sometimes called *dark energy*) has recently been measured to fairly good accuracy, and found to account for about 70% of the total mass/energy in the present-day Universe. This fraction is an increasing function of time. The observed value of the cosmological constant/dark energy is important for cosmology, but it is extremely tiny when expressed in Planck units (about 10^{-120}). The first attempts to account for $\Lambda > 0$ within string theory and M-theory, based on compactifying 11-dimensional supergravity on time-independent compact manifolds, were ruled out by "no-go" theorems. However, certain nonperturbative effects allow these no-go theorems to be circumvented.

A viewpoint that has gained in popularity recently is that string theory can accommodate almost any value of Λ, but only solutions for which Λ is sufficiently small describe a Universe that can support life. So, if it were much larger, we wouldn't be here to ask the question. This type of reasoning is called *anthropic*. While this may be correct, it would be satisfying to have another explanation of why Λ is so small that does not require this type of reasoning.

Another important issue in cosmology concerns the accelerated expansion of the very early Universe, which is referred to as *inflation*. The observational case for inflation is quite strong, and it is an important question to understand how it arises from a fundamental theory. Before the period of inflation was the *Big Bang*, the origin of the observable Universe, and much effort is going into understanding that. Two radically different proposals are quantum tunneling from nothing and a collision of branes.

2
The bosonic string

This chapter introduces the simplest string theory, called the bosonic string. Even though this theory is unrealistic and not suitable for phenomenology, it is the natural place to start. The reason is that the same structures and techniques, together with a number of additional ones, are required for the analysis of more realistic superstring theories. This chapter describes the free (noninteracting) theory both at the classical and quantum levels. The next chapter discusses various techniques for introducing and analyzing interactions.

A string can be regarded as a special case of a p-brane, a p-dimensional extended object moving through space-time. In this notation a point particle corresponds to the $p = 0$ case, in other words to a zero-brane. Strings (whether fundamental or solitonic) correspond to the $p = 1$ case, so that they can also be called one-branes. Two-dimensional extended objects or two-branes are often called membranes. In fact, the name p-brane was chosen to suggest a generalization of a membrane. Even though strings share some properties with higher-dimensional extended objects at the classical level, they are very special in the sense that their two-dimensional world-volume quantum theories are renormalizable, something that is not the case for branes of higher dimension. This is a crucial property that makes it possible to base quantum theories on them. In this chapter we describe the string as a special case of p-branes and describe the properties that hold only for the special case $p = 1$.

2.1 p-brane actions

This section describes the free motion of p-branes in space-time using the principle of minimal action. Let us begin with a point particle or zero-brane.

Relativistic point particle

The motion of a relativistic particle of mass m in a curved D-dimensional space-time can be formulated as a variational problem, that is, an action principle. Since the classical motion of a point particle is along geodesics, the action should be proportional to the invariant length of the particle's trajectory

$$S_0 = -\alpha \int ds, \tag{2.1}$$

where α is a constant and $\hbar = c = 1$. This length is extremized in the classical theory, as is illustrated in Fig. 2.1.

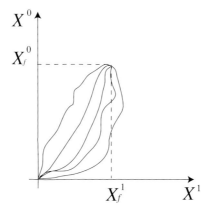

Fig. 2.1. The classical trajectory of a point particle minimizes the length of the world line.

Requiring the action to be dimensionless, one learns that α has the dimensions of inverse length, which is equivalent to mass in our units, and hence it must be proportional to m. As is demonstrated in Exercise 2.1, the action has the correct nonrelativistic limit if $\alpha = m$, so the action becomes

$$S_0 = -m \int ds. \tag{2.2}$$

In this formula the line element is given by

$$ds^2 = -g_{\mu\nu}(X)dX^\mu dX^\nu. \tag{2.3}$$

Here $g_{\mu\nu}(X)$, with $\mu, \nu = 0, \ldots, D-1$, describes the background geometry, which is chosen to have Minkowski signature $(-+\cdots+)$. The minus sign has been introduced here so that ds is real for a time-like trajectory. The particle's trajectory $X^\mu(\tau)$, also called the world line of the particle, is parametrized by a real parameter τ, but the action is independent of the choice of parametrization (see Exercise 2.2). The action (2.2) therefore takes the form

$$S_0 = -m \int \sqrt{-g_{\mu\nu}(X)\dot{X}^\mu \dot{X}^\nu} d\tau, \qquad (2.4)$$

where the dot represents the derivative with respect to τ.

The action S_0 has the disadvantage that it contains a square root, so that it is difficult to quantize. Furthermore, this action obviously cannot be used to describe a massless particle. These problems can be circumvented by introducing an action equivalent to the previous one at the classical level, which is formulated in terms of an auxiliary field $e(\tau)$

$$\widetilde{S}_0 = \frac{1}{2} \int d\tau \left(e^{-1} \dot{X}^2 - m^2 e \right), \qquad (2.5)$$

where $\dot{X}^2 = g_{\mu\nu}(X)\dot{X}^\mu \dot{X}^\nu$. Reparametrization invariance of $\widetilde{S}_0$ requires that $e(\tau)$ transforms in an appropriate fashion (see Exercise 2.3). The equation of motion of $e(\tau)$, given by setting the variational derivative of this action with respect to $e(\tau)$ equal to zero, is $m^2 e^2 + \dot{X}^2 = 0$. Solving for $e(\tau)$ and substituting back into $\widetilde{S}_0$ gives S_0.

Generalization to the p-brane action

The action (2.4) can be generalized to the case of a string sweeping out a two-dimensional world sheet in space-time and, in general, to a p-brane sweeping out a $(p+1)$-dimensional world volume in D-dimensional space-time. It is necessary, of course, that $p < D$. For example, a membrane or two-brane sweeps out a three-dimensional world volume as it moves through a higher-dimensional space-time. This is illustrated for a string in Fig. 2.2.

The generalization of the action (2.4) to a p-brane naturally takes the form

$$S_p = -T_p \int d\mu_p. \qquad (2.6)$$

Here T_p is called the p-brane tension and $d\mu_p$ is the $(p+1)$-dimensional volume element given by

$$d\mu_p = \sqrt{-\det G_{\alpha\beta}}\, d^{p+1}\sigma, \qquad (2.7)$$

where the induced metric is given by

$$G_{\alpha\beta} = g_{\mu\nu}(X)\partial_\alpha X^\mu \partial_\beta X^\nu \quad \alpha, \beta = 0, \ldots, p. \qquad (2.8)$$

To write down this form of the action, one has taken into account that p-brane world volumes can be parametrized by the coordinates $\sigma^0 = \tau$, which is time-like, and σ^i, which are p space-like coordinates. Since $d\mu_p$ has units of $(\text{length})^{p+1}$ the dimension of the p-brane tension is

$$[T_p] = (\text{length})^{-p-1} = \frac{\text{mass}}{(\text{length})^p}, \qquad (2.9)$$

or energy per unit p-volume.

EXERCISES

EXERCISE 2.1
Show that the nonrelativistic limit of the action (2.1) in flat Minkowski space-time determines the value of the constant α to be the mass of the point particle.

SOLUTION

In the nonrelativistic limit the action (2.1) becomes

$$S_0 = -\alpha \int \sqrt{dt^2 - d\vec{x}^2} = -\alpha \int dt \sqrt{1 - \vec{v}^2} \approx -\alpha \int dt \left(1 - \frac{1}{2}\vec{v}^2 + \ldots\right).$$

Comparing the above expansion with the action of a nonrelativistic point

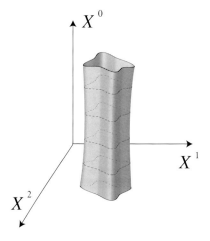

Fig. 2.2. The classical trajectory of a string minimizes the area of the world sheet.

particle, namely
$$S_{\rm nr} = \int dt \frac{1}{2}m\vec{v}^2,$$
gives $\alpha = m$. In the nonrelativistic limit an additional constant (the famous $E = mc^2$ term) appears in the above expansion of S_0. This constant does not contribute to the classical equations of motion. $\square$

EXERCISE 2.2
One important requirement for the point-particle world-line action is that it should be invariant under reparametrizations of the parameter τ. Show that the action S_0 is invariant under reparametrizations of the world line by substituting $\tau' = f(\tau)$.

SOLUTION
The action
$$S_0 = -m \int \sqrt{-\frac{dX^\mu}{d\tau}\frac{dX_\mu}{d\tau}} d\tau$$
can be written in terms of primed quantities by taking into account
$$d\tau' = \frac{df(\tau)}{d\tau}d\tau = \dot{f}(\tau)d\tau \quad \text{and} \quad \frac{dX^\mu}{d\tau} = \frac{dX^\mu}{d\tau'}\frac{d\tau'}{d\tau} = \frac{dX^\mu}{d\tau'} \cdot \dot{f}(\tau).$$
This gives,
$$S_0' = -m \int \sqrt{-\frac{dX^\mu}{d\tau'}\frac{dX_\mu}{d\tau'}} \dot{f}(\tau) \cdot \frac{d\tau'}{\dot{f}(\tau)} = -m \int \sqrt{-\frac{dX^\mu}{d\tau'}\frac{dX_\mu}{d\tau'}} \cdot d\tau',$$
which shows that the action S_0 is invariant under reparametrizations. $\square$

EXERCISE 2.3
The action $\widetilde{S}_0$ in Eq. (2.5) is also invariant under reparametrizations of the particle world line. Even though it is not hard to consider finite transformations, let us consider an infinitesimal change of parametrization
$$\tau \to \tau' = f(\tau) = \tau - \xi(\tau).$$
Verify the invariance of $\widetilde{S}_0$ under an infinitesimal reparametrization.

SOLUTION
The field X^μ transforms as a world-line scalar, $X^{\mu\prime}(\tau') = X^\mu(\tau)$. Therefore,

the first-order shift in X^μ is

$$\delta X^\mu = X^{\mu\prime}(\tau) - X^\mu(\tau) = \xi(\tau)\dot{X}^\mu.$$

Notice that the fact that X^μ has a space-time vector index is irrelevant to this argument. The auxiliary field $e(\tau)$ transforms at the same time according to

$$e'(\tau')d\tau' = e(\tau)d\tau.$$

Infinitesimally, this leads to

$$\delta e = e'(\tau) - e(\tau) = \frac{d}{d\tau}(\xi e).$$

Let us analyze the special case of a flat space-time metric $g_{\mu\nu}(X) = \eta_{\mu\nu}$, even though the result is true without this restriction. In this case the vector index on X^μ can be raised and lowered inside derivatives. The expression $\widetilde{S}_0$ has the variation

$$\delta \widetilde{S}_0 = \frac{1}{2}\int d\tau \left(\frac{2\dot{X}^\mu \delta \dot{X}_\mu}{e} - \frac{\dot{X}^\mu \dot{X}_\mu}{e^2}\delta e - m^2 \delta e \right).$$

Here $\delta \dot{X}_\mu$ is given by

$$\delta \dot{X}_\mu = \frac{d}{d\tau}\delta X_\mu = \dot{\xi}\dot{X}_\mu + \xi \ddot{X}_\mu.$$

Together with the expression for δe, this yields

$$\delta \widetilde{S}_0 = \frac{1}{2}\int d\tau \left[\frac{2\dot{X}^\mu}{e}\left(\dot{\xi}\dot{X}_\mu + \xi \ddot{X}_\mu\right) - \frac{\dot{X}^\mu \dot{X}_\mu}{e^2}\left(\dot{\xi}e + \xi \dot{e}\right) - m^2 \frac{d(\xi e)}{d\tau} \right].$$

The last term can be dropped because it is a total derivative. The remaining terms can be written as

$$\delta \widetilde{S}_0 = \frac{1}{2}\int d\tau \cdot \frac{d}{d\tau}\left(\frac{\xi}{e}\dot{X}^\mu \dot{X}_\mu \right).$$

This is a total derivative, so it too can be dropped (for suitable boundary conditions). Therefore, $\widetilde{S}_0$ is invariant under reparametrizations. □

EXERCISE 2.4

The reparametrization invariance that was checked in the previous exercise allows one to choose a gauge in which $e = 1$. As usual, when doing this one should be careful to retain the e equation of motion (evaluated for $e = 1$). What is the form and interpretation of the equations of motion for e and X^μ resulting from $\widetilde{S}_0$?

Solution

The equation of motion for e derived from the action principle for $\tilde{S}_0$ is given by the vanishing of the variational derivative

$$\frac{\delta \tilde{S}_0}{\delta e} = -\frac{1}{2}\left(e^{-2}\dot{X}^\mu \dot{X}_\mu + m^2\right) = 0.$$

Choosing the gauge $e(\tau) = 1$, we obtain the equation

$$\dot{X}^\mu \dot{X}_\mu + m^2 = 0.$$

Since $p^\mu = \dot{X}^\mu$ is the momentum conjugate to X^μ, this equation is simply the mass-shell condition $p^2 + m^2 = 0$, so that m is the mass of the particle, as was shown in Exercise 2.1. The variation with respect to X^μ gives the second equation of motion

$$-\frac{d}{d\tau}(g_{\mu\nu}\dot{X}^\nu) + \frac{1}{2}\partial_\mu g_{\rho\lambda}\dot{X}^\rho \dot{X}^\lambda$$

$$= -(\partial_\rho g_{\mu\nu})\dot{X}^\rho \dot{X}^\nu - g_{\mu\nu}\ddot{X}^\nu + \frac{1}{2}\partial_\mu g_{\rho\lambda}\dot{X}^\rho \dot{X}^\lambda = 0.$$

This can be brought to the form

$$\ddot{X}^\mu + \Gamma^\mu_{\rho\lambda}\dot{X}^\rho \dot{X}^\lambda = 0, \qquad (2.10)$$

where

$$\Gamma^\mu_{\rho\lambda} = \frac{1}{2}g^{\mu\nu}(\partial_\rho g_{\lambda\nu} + \partial_\lambda g_{\rho\nu} - \partial_\nu g_{\rho\lambda})$$

is the Christoffel connection (or Levi–Civita connection). Equation (2.10) is the geodesic equation. Note that, for a flat space-time, $\Gamma^\mu_{\rho\lambda}$ vanishes in Cartesian coordinates, and one recovers the familiar equation of motion for a point particle in flat space. Note also that the more conventional normalization $(\dot{X}^\mu \dot{X}_\mu + 1 = 0)$ would have been obtained by choosing the gauge $e = 1/m$. $\square$

Exercise 2.5

The action of a p-brane is invariant under reparametrizations of the $p+1$ world-volume coordinates. Show this explicitly by checking that the action (2.6) is invariant under a change of variables $\sigma^\alpha \to \sigma^\alpha(\tilde{\sigma})$.

Solution

Under this change of variables the induced metric in Eq. (2.8) transforms in the following way:

$$G_{\alpha\beta} = \frac{\partial X^\mu}{\partial \sigma^\alpha}\frac{\partial X^\nu}{\partial \sigma^\beta} g_{\mu\nu} = (f^{-1})_\alpha^\gamma \frac{\partial X^\mu}{\partial \widetilde\sigma^\gamma}(f^{-1})_\beta^\delta \frac{\partial X^\nu}{\partial \widetilde\sigma^\delta} g_{\mu\nu},$$

where

$$f_\beta^\alpha(\widetilde\sigma) = \frac{\partial \sigma^\alpha}{\partial \widetilde\sigma^\beta}.$$

Defining J to be the Jacobian of the world-volume coordinate transformation, that is, $J = \det f_\beta^\alpha$, the determinant appearing in the action becomes

$$\det\left(g_{\mu\nu}\frac{\partial X^\mu}{\partial \sigma^\alpha}\frac{\partial X^\nu}{\partial \sigma^\beta}\right) = J^{-2} \det\left(g_{\mu\nu}\frac{\partial X^\mu}{\partial \widetilde\sigma^\gamma}\frac{\partial X^\nu}{\partial \widetilde\sigma^\delta}\right).$$

The measure of the integral transforms according to

$$d^{p+1}\sigma = J d^{p+1}\widetilde\sigma,$$

so that the Jacobian factors cancel, and the action becomes

$$\widetilde S_p = -T_p \int d^{p+1}\widetilde\sigma \sqrt{-\det\left(g_{\mu\nu}\frac{\partial X^\mu}{\partial \widetilde\sigma^\gamma}\frac{\partial X^\nu}{\partial \widetilde\sigma^\delta}\right)}.$$

Therefore, the action is invariant under reparametrizations of the world-volume coordinates. □

2.2 The string action

This section specializes the discussion to the case of a string (or one-brane) propagating in D-dimensional flat Minkowski space-time. The string sweeps out a two-dimensional surface as it moves through space-time, which is called the *world sheet*. The points on the world sheet are parametrized by the two coordinates $\sigma^0 = \tau$, which is time-like, and $\sigma^1 = \sigma$, which is space-like. If the variable σ is periodic, it describes a closed string. If it covers a finite interval, the string is open. This is illustrated in Fig. 2.3.

The Nambu–Goto action

The space-time embedding of the string world sheet is described by functions $X^\mu(\sigma, \tau)$, as shown in Fig. 2.4. The action describing a string propagating in a flat background geometry can be obtained as a special case of the more general p-brane action of the previous section. This action, called the *Nambu–Goto action*, takes the form

$$S_{\text{NG}} = -T \int d\sigma d\tau \sqrt{(\dot X \cdot X')^2 - \dot X^2 X'^2}, \qquad (2.11)$$

where

$$\dot{X}^\mu = \frac{\partial X^\mu}{\partial \tau} \quad \text{and} \quad X^{\mu\prime} = \frac{\partial X^\mu}{\partial \sigma}, \qquad (2.12)$$

and the scalar products are defined in the case of a flat space-time by $A \cdot B = \eta_{\mu\nu} A^\mu B^\nu$. The integral appearing in this action describes the area of the world sheet. As a result, the classical string motion minimizes (or at least extremizes) the world-sheet area, just as classical particle motion makes the length of the world line extremal by moving along a geodesic.

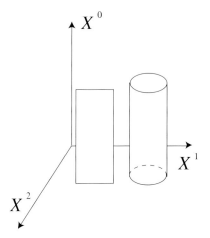

Fig. 2.3. The world sheet for the free propagation of an open string is a rectangular surface, while the free propagation of a closed string sweeps out a cylinder.

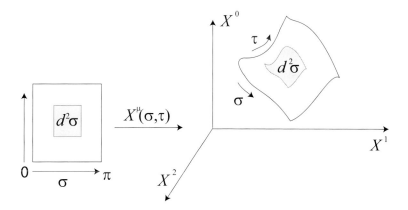

Fig. 2.4. The functions $X^\mu(\sigma, \tau)$ describe the embedding of the string world sheet in space-time.

The string sigma model action

Even though the Nambu–Goto action has a nice physical interpretation as the area of the string world sheet, its quantization is again awkward due to the presence of the square root. An action that is equivalent to the Nambu–Goto action at the classical level, because it gives rise to the same equations of motion, is the *string sigma model action*.[1]

The string sigma-model action is expressed in terms of an auxiliary world-sheet metric $h_{\alpha\beta}(\sigma,\tau)$, which plays a role analogous to the auxiliary field $e(\tau)$ introduced for the point particle. We shall use the notation $h_{\alpha\beta}$ for the world-sheet metric, whereas $g_{\mu\nu}$ denotes a space-time metric. Also,

$$h = \det h_{\alpha\beta} \quad \text{and} \quad h^{\alpha\beta} = (h^{-1})_{\alpha\beta}, \qquad (2.13)$$

as is customary in relativity. In this notation the string sigma-model action is

$$S_\sigma = -\frac{1}{2}T \int d^2\sigma \sqrt{-h}\, h^{\alpha\beta} \partial_\alpha X \cdot \partial_\beta X. \qquad (2.14)$$

At the classical level the string sigma-model action is equivalent to the Nambu–Goto action. However, it is more convenient for quantization.

EXERCISES

EXERCISE 2.6

Derive the equations of motion for the auxiliary metric $h_{\alpha\beta}$ and the bosonic field X^μ in the string sigma-model action. Show that classically the string sigma-model action (2.14) is equivalent to the Nambu–Goto action (2.11).

SOLUTION

As for the point-particle case discussed earlier, the auxiliary metric $h_{\alpha\beta}$ appearing in the string sigma-model action can be eliminated using its equations of motion. Indeed, since there is no kinetic term for $h_{\alpha\beta}$, its equation of motion implies the vanishing of the world-sheet energy–momentum tensor

---

[1] This action, traditionally called the *Polyakov action*, was discovered by Brink, Di Vecchia and Howe and by Deser and Zumino several years before Polyakov skillfully used it for path-integral quantization of the string.

$T_{\alpha\beta}$, that is,

$$T_{\alpha\beta} = -\frac{2}{T}\frac{1}{\sqrt{-h}}\frac{\delta S_\sigma}{\delta h^{\alpha\beta}} = 0.$$

To evaluate the variation of the action, the following formula is useful:

$$\delta h = -h h_{\alpha\beta} \delta h^{\alpha\beta},$$

which implies that

$$\delta\sqrt{-h} = -\frac{1}{2}\sqrt{-h}\, h_{\alpha\beta}\delta h^{\alpha\beta}. \tag{2.15}$$

After taking the variation of the action, the result for the energy–momentum tensor takes the form

$$T_{\alpha\beta} = \partial_\alpha X \cdot \partial_\beta X - \frac{1}{2} h_{\alpha\beta} h^{\gamma\delta} \partial_\gamma X \cdot \partial_\delta X = 0.$$

This is the equation of motion for $h_{\alpha\beta}$, which can be used to eliminate $h_{\alpha\beta}$ from the string sigma-model action. The result is the Nambu–Goto action. The easiest way to see this is to take the square root of minus the determinant of both sides of the equation

$$\partial_\alpha X \cdot \partial_\beta X = \frac{1}{2} h_{\alpha\beta} h^{\gamma\delta} \partial_\gamma X \cdot \partial_\delta X.$$

This gives

$$\sqrt{-\det(\partial_\alpha X \cdot \partial_\beta X)} = \frac{1}{2}\sqrt{-h}\, h^{\gamma\delta} \partial_\gamma X \cdot \partial_\delta X.$$

Finally, the equation of motion for X^μ, obtained from the Euler–Lagrange condition, is

$$\Delta X^\mu = -\frac{1}{\sqrt{-h}}\partial_\alpha\left(\sqrt{-h}\, h^{\alpha\beta}\partial_\beta X^\mu\right) = 0.$$

□

EXERCISE 2.7

Calculate the nonrelativistic limit of the Nambu–Goto action

$$S_{\mathrm{NG}} = -T\int d\tau d\sigma \sqrt{-\det G_{\alpha\beta}}, \qquad G_{\alpha\beta} = \partial_\alpha X^\mu \partial_\beta X_\mu$$

for a string in Minkowski space-time. Use the static gauge, which fixes the longitudinal directions $X^0 = \tau$, $X^1 = \sigma$, while leaving the transverse directions X^i free. Show that the kinetic energy contains only the transverse velocity. Determine the mass per unit length of the string.

Solution

In the static gauge

$$\det G_{\alpha\beta} = \det \begin{pmatrix} \partial_\tau X^\mu \partial_\tau X_\mu & \partial_\tau X^\mu \partial_\sigma X_\mu \\ \partial_\sigma X^\mu \partial_\tau X_\mu & \partial_\sigma X^\mu \partial_\sigma X_\mu \end{pmatrix}$$

$$= \det \begin{pmatrix} -1 + \partial_\tau X^i \partial_\tau X_i & \partial_\tau X^i \partial_\sigma X_i \\ \partial_\sigma X^i \partial_\tau X_i & 1 + \partial_\sigma X^i \partial_\sigma X_i \end{pmatrix}.$$

Then,

$$\det G_{\alpha\beta} \approx -1 + \partial_\tau X^i \partial_\tau X_i - \partial_\sigma X^i \partial_\sigma X_i + \ldots$$

Here the dots indicate higher-order terms that can be dropped in the nonrelativistic limit for which the velocities are small. In this limit the action becomes (after a Taylor expansion)

$$S_{\text{NG}} = -T \int d\tau d\sigma \sqrt{|-1 + \partial_\tau X^i \partial_\tau X_i - \partial_\sigma X^i \partial_\sigma X_i|}$$

$$\approx T \int d\tau d\sigma \left(-1 + \frac{1}{2} \partial_\tau X^i \partial_\tau X_i - \frac{1}{2} \partial_\sigma X^i \partial_\sigma X_i \right).$$

The first term in the parentheses gives $-m \int d\tau$, if L is the length of the σ interval and $m = LT$. This is the rest-mass contribution to the potential energy. Note that L is a distance in space, because of the choice of static gauge. Thus the tension T can be interpreted as the mass per unit length, or mass density, of the string. The last two terms of the above formula are the kinetic energy and the negative of the potential energy of a nonrelativistic string of tension T. □

Exercise 2.8

Show that if a *cosmological constant term* is added to the string sigma-model action, so that

$$S_\sigma = -\frac{T}{2} \int d^2\sigma \sqrt{-h} h^{\alpha\beta} \partial_\alpha X^\mu \partial_\beta X_\mu + \Lambda \int d^2\sigma \sqrt{-h},$$

it leads to inconsistent classical equations of motion.

Solution

The equation of motion for the world-sheet metric is

$$\frac{2}{\sqrt{-h}} \frac{\delta S_\sigma}{\delta h^{\gamma\delta}} = -T[\partial_\gamma X^\mu \partial_\delta X_\mu - \frac{1}{2} h_{\gamma\delta}(h^{\alpha\beta} \partial_\alpha X^\mu \partial_\beta X_\mu)] - \Lambda h_{\gamma\delta} = 0,$$

where we have used Eq. (2.15). Contracting with $h^{\gamma\delta}$ gives

$$h_{\gamma\delta}h^{\gamma\delta}\Lambda = T(\frac{1}{2}h_{\gamma\delta}h^{\gamma\delta} - 1)h^{\alpha\beta}\partial_\alpha X^\mu \partial_\beta X_\mu.$$

Since $h_{\gamma\delta}h^{\gamma\delta} = 2$, the right-hand side vanishes. Thus, assuming $h \ne 0$, consistency requires $\Lambda = 0$. In other words, adding a cosmological constant term gives inconsistent classical equations of motion. $\square$

EXERCISE 2.9

Show that the sigma-model form of the action of a p-brane, for $p \ne 1$, requires a cosmological constant term.

SOLUTION

Consider a p-brane action of the form

$$S_\sigma = -\frac{T_p}{2}\int d^{p+1}\sigma \sqrt{-h}\, h^{\alpha\beta}\partial_\alpha X \cdot \partial_\beta X + \Lambda_p \int d^{p+1}\sigma \sqrt{-h}. \quad (2.16)$$

The equation of motion for the world-volume metric is obtained exactly as in the previous exercise, with the result

$$T_p[\partial_\gamma X \cdot \partial_\delta X - \frac{1}{2}h_{\gamma\delta}(h^{\alpha\beta}\partial_\alpha X \cdot \partial_\beta X)] + \Lambda_p h_{\gamma\delta} = 0.$$

This equation is not so easy to solve directly, so let us instead investigate whether it is solved by equating the world-volume metric to the induced metric

$$h_{\alpha\beta} = \partial_\alpha X \cdot \partial_\beta X. \quad (2.17)$$

Substituting this ansatz in the previous equation and dropping common factors gives

$$T_p(1 - \frac{1}{2}h^{\alpha\beta}h_{\alpha\beta}) + \Lambda_p = 0.$$

Substituting $h^{\alpha\beta}h_{\alpha\beta} = p+1$, one learns that

$$\Lambda_p = \frac{1}{2}(p-1)T_p. \quad (2.18)$$

Thus, consistency requires this choice of Λ_p.[2] This confirms the previous result that $\Lambda_1 = 0$ and shows that $\Lambda_p \ne 0$ for $p \ne 1$. Substituting the value of the metric in Eq. (2.17) and the value of Λ_p in Eq. (2.18), one finds that Eq. (2.16) is equivalent classically to Eq. (2.6). For the special case of

---

[2] A different value is actually equivalent, if one makes a corresponding rescaling of $h_{\alpha\beta}$. However, this results in a multiplicative factor in the relation (2.17).

2.3 String sigma-model action: the classical theory

In this section we discuss the symmetries of the string sigma-model action in Eq. (2.14). This is helpful for writing the string action in a gauge in which quantization is particularly simple.

Symmetries

The string sigma-model action for the bosonic string in Minkowski space-time has a number of symmetries:

- *Poincaré transformations.* These are global symmetries under which the world-sheet fields transform as

$$\delta X^\mu = a^\mu{}_\nu X^\nu + b^\mu \quad \text{and} \quad \delta h^{\alpha\beta} = 0. \tag{2.19}$$

  Here the constants $a^\mu{}_\nu$ (with $a_{\mu\nu} = -a_{\nu\mu}$) describe infinitesimal Lorentz transformations and b^μ describe space-time translations.

- *Reparametrizations.* The string world sheet is parametrized by two coordinates τ and σ, but a change in the parametrization does not change the action. Indeed, the transformations

$$\sigma^\alpha \to f^\alpha(\sigma) = \sigma'^\alpha \quad \text{and} \quad h_{\alpha\beta}(\sigma) = \frac{\partial f^\gamma}{\partial \sigma^\alpha} \frac{\partial f^\delta}{\partial \sigma^\beta} h_{\gamma\delta}(\sigma') \tag{2.20}$$

  leave the action invariant. These local symmetries are also called *diffeomorphisms*. Strictly speaking, this implies that the transformations and their inverses are infinitely differentiable.

- *Weyl transformations.* The action is invariant under the rescaling

$$h_{\alpha\beta} \to e^{\phi(\sigma,\tau)} h_{\alpha\beta} \quad \text{and} \quad \delta X^\mu = 0, \tag{2.21}$$

  since $\sqrt{-h} \to e^\phi \sqrt{-h}$ and $h^{\alpha\beta} \to e^{-\phi} h^{\alpha\beta}$ give cancelling factors. This local symmetry is the reason that the energy–momentum tensor is traceless.

Poincaré transformations are global symmetries, whereas reparametrizations and Weyl transformations are local symmetries. The local symmetries can be used to choose a gauge, such as the static gauge discussed earlier, or else one in which some of the components of the world-sheet metric $h_{\alpha\beta}$ are of a particular form.

Gauge fixing

The gauge-fixing procedure described earlier for the point particle can be generalized to the case of the string. In this case the auxiliary field has three independent components, namely

$$h_{\alpha\beta} = \begin{pmatrix} h_{00} & h_{01} \\ h_{10} & h_{11} \end{pmatrix}, \qquad (2.22)$$

where $h_{10} = h_{01}$. Reparametrization invariance allows us to choose two of the components of h, so that only one independent component remains. But this remaining component can be gauged away by using the invariance of the action under Weyl rescalings. So in the case of the string there is sufficient symmetry to gauge fix $h_{\alpha\beta}$ completely. As a result, the auxiliary field $h_{\alpha\beta}$ can be chosen as

$$h_{\alpha\beta} = \eta_{\alpha\beta} = \begin{pmatrix} -1 & 0 \\ 0 & 1 \end{pmatrix}. \qquad (2.23)$$

Actually such a flat world-sheet metric is only possible if there is no topological obstruction. This is the case when the world sheet has vanishing Euler characteristic. Examples include a cylinder and a torus. When a flat world-sheet metric is an allowed gauge choice, the string action takes the simple form

$$S = \frac{T}{2} \int d^2\sigma (\dot{X}^2 - X'^2). \qquad (2.24)$$

The string actions discussed so far describe propagation in flat Minkowski space-time. Keeping this requirement, one could consider the following two additional terms, both of which are renormalizable (or super-renormalizable) and compatible with Poincaré invariance,

$$S_1 = \lambda_1 \int d^2\sigma \sqrt{-h} \quad \text{and} \quad S_2 = \lambda_2 \int d^2\sigma \sqrt{-h} R^{(2)}(h). \qquad (2.25)$$

S_1 is a cosmological constant term on the world sheet. This term is not allowed by the equations of motion (see Exercise 2.8). The term S_2 involves $R^{(2)}(h)$, the scalar curvature of the two-dimensional world-sheet geometry. Such a contribution raises interesting issues, which are explored in the next chapter. For now, let us assume that it can be ignored.

Equations of motion and boundary conditions
Equations of motion

Let us now suppose that the world-sheet topology allows a flat world-sheet metric to be chosen. For a freely propagating closed string a natural choice is an infinite cylinder. Similarly, the natural choice for an open string is an infinite strip. In both cases, the motion of the string in Minkowski space is governed by the action in Eq. (2.24). This implies that the X^μ equation of motion is the wave equation

$$\partial_\alpha \partial^\alpha X^\mu = 0 \quad \text{or} \quad \left(\frac{\partial^2}{\partial \sigma^2} - \frac{\partial^2}{\partial \tau^2} \right) X^\mu = 0. \tag{2.26}$$

Since the metric on the world sheet has been gauge fixed, the vanishing of the energy–momentum tensor, that is, $T_{\alpha\beta} = 0$ originating from the equation of motion of the world-sheet metric, must now be imposed as an additional constraint condition. In the gauge $h_{\alpha\beta} = \eta_{\alpha\beta}$ the components of this tensor are

$$T_{01} = T_{10} = \dot{X} \cdot X' \quad \text{and} \quad T_{00} = T_{11} = \frac{1}{2}(\dot{X}^2 + X'^2). \tag{2.27}$$

Using $T_{00} = T_{11}$, we see the vanishing of the trace of the energy–momentum tensor $\text{Tr} T = \eta^{\alpha\beta} T_{\alpha\beta} = T_{11} - T_{00}$. This is a consequence of Weyl invariance, as was mentioned before.

Boundary conditions

In order to give a fully defined variational problem, boundary conditions need to be specified. A string can be either closed or open. For convenience, let us choose the coordinate σ to have the range $0 \leq \sigma \leq \pi$. The stationary points of the action are determined by demanding invariance of the action under the shifts

$$X^\mu \to X^\mu + \delta X^\mu. \tag{2.28}$$

In addition to the equations of motion, there is the boundary term

$$-T \int d\tau \left[X'_\mu \delta X^\mu |_{\sigma=\pi} - X'_\mu \delta X^\mu |_{\sigma=0} \right], \tag{2.29}$$

which must vanish. There are several different ways in which this can be achieved. For an open string these possibilities are illustrated in Fig. 2.5.

- *Closed string.* In this case the embedding functions are periodic,

$$X^\mu(\sigma, \tau) = X^\mu(\sigma + \pi, \tau). \tag{2.30}$$

- *Open string with Neumann boundary conditions.* In this case the component of the momentum normal to the boundary of the world sheet vanishes, that is,

$$X'_\mu = 0 \quad \text{at} \quad \sigma = 0, \pi. \tag{2.31}$$

If this choice is made for all μ, these boundary conditions respect D-dimensional Poincaré invariance. Physically, they mean that no momentum is flowing through the ends of the string.

- *Open string with Dirichlet boundary conditions*. In this case the positions of the two string ends are fixed so that $\delta X^\mu = 0$, and

$$X^\mu|_{\sigma=0} = X_0^\mu \quad \text{and} \quad X^\mu|_{\sigma=\pi} = X_\pi^\mu, \tag{2.32}$$

where X_0^μ and X_π^μ are constants and $\mu = 1, \ldots, D - p - 1$. Neumann boundary conditions are imposed for the other $p+1$ coordinates. Dirichlet boundary conditions break Poincaré invariance, and for this reason they were not considered for many years. But, as is discussed in Chapter 6, there are circumstances in which Dirichlet boundary conditions are unavoidable. The modern interpretation is that X_0^μ and X_π^μ represent the positions of Dp-branes. A Dp-brane is a special type of p-brane on which a fundamental string can end. The presence of a Dp-brane breaks Poincaré invariance unless it is space-time filling ($p = D - 1$).

Solution to the equations of motion

To find the solution to the equations of motion and constraint equations it is convenient to introduce world-sheet *light-cone coordinates*, defined as

$$\sigma^\pm = \tau \pm \sigma. \tag{2.33}$$

In these coordinates the derivatives and the two-dimensional Lorentz metric take the form

$$\partial_\pm = \frac{1}{2}(\partial_\tau \pm \partial_\sigma) \quad \text{and} \quad \begin{pmatrix} \eta_{++} & \eta_{+-} \\ \eta_{-+} & \eta_{--} \end{pmatrix} = -\frac{1}{2}\begin{pmatrix} 0 & 1 \\ 1 & 0 \end{pmatrix}. \tag{2.34}$$

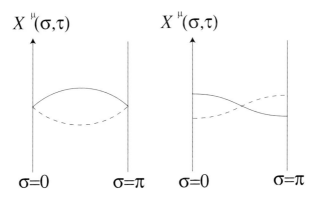

Fig. 2.5. Illustration of Dirichlet (left) and Neumann (right) boundary conditions. The solid and dashed lines represent string positions at two different times.

In light-cone coordinates the wave equation for X^μ is

$$\partial_+\partial_- X^\mu = 0. \tag{2.35}$$

The vanishing of the energy–momentum tensor becomes

$$T_{++} = \partial_+ X^\mu \partial_+ X_\mu = 0, \tag{2.36}$$

$$T_{--} = \partial_- X^\mu \partial_- X_\mu = 0, \tag{2.37}$$

while $T_{+-} = T_{-+} = 0$ expresses the vanishing of the trace, which is automatic. The general solution of the wave equation (2.35) is given by

$$X^\mu(\sigma,\tau) = X^\mu_{\text{R}}(\tau - \sigma) + X^\mu_{\text{L}}(\tau + \sigma), \tag{2.38}$$

which is a sum of *right-movers* and *left-movers*. To find the explicit form of X_{R} and X_{L} one should require $X^\mu(\sigma,\tau)$ to be real and impose the constraints

$$(\partial_- X_{\text{R}})^2 = (\partial_+ X_{\text{L}})^2 = 0. \tag{2.39}$$

The quantum version of these constraints will be discussed in the next section.

Closed-string mode expansion

The most general solution of the wave equation satisfying the closed-string boundary condition is given by

$$X^\mu_{\text{R}} = \frac{1}{2}x^\mu + \frac{1}{2}l_{\text{s}}^2 p^\mu(\tau - \sigma) + \frac{i}{2}l_{\text{s}}\sum_{n\neq 0}\frac{1}{n}\alpha^\mu_n e^{-2in(\tau-\sigma)}, \tag{2.40}$$

$$X^\mu_{\text{L}} = \frac{1}{2}x^\mu + \frac{1}{2}l_{\text{s}}^2 p^\mu(\tau + \sigma) + \frac{i}{2}l_{\text{s}}\sum_{n\neq 0}\frac{1}{n}\tilde{\alpha}^\mu_n e^{-2in(\tau+\sigma)}, \tag{2.41}$$

where x^μ is a *center-of-mass* position and p^μ is the total string momentum, describing the free motion of the string center of mass. The exponential terms represent the string excitation modes. Here we have introduced a new parameter, the *string length scale* l_{s}, which is related to the string tension T and the open-string Regge slope parameter α' by

$$T = \frac{1}{2\pi\alpha'} \quad \text{and} \quad \frac{1}{2}l_{\text{s}}^2 = \alpha'. \tag{2.42}$$

The requirement that X^μ_{R} and X^μ_{L} are real functions implies that x^μ and p^μ are real, while positive and negative modes are conjugate to each other

$$\alpha^\mu_{-n} = (\alpha^\mu_n)^\star \quad \text{and} \quad \tilde{\alpha}^\mu_{-n} = (\tilde{\alpha}^\mu_n)^\star. \tag{2.43}$$

The terms linear in σ cancel from the sum $X_R^\mu + X_L^\mu$, so that closed-string boundary conditions are indeed satisfied. Note that the derivatives of the expansions take the form

$$\partial_- X_R^\mu = l_s \sum_{m=-\infty}^{+\infty} \alpha_m^\mu e^{-2im(\tau-\sigma)} \tag{2.44}$$

$$\partial_+ X_L^\mu = l_s \sum_{m=-\infty}^{+\infty} \tilde{\alpha}_m^\mu e^{-2im(\tau+\sigma)}, \tag{2.45}$$

where

$$\alpha_0^\mu = \tilde{\alpha}_0^\mu = \frac{1}{2} l_s p^\mu. \tag{2.46}$$

These expressions are useful later. In order to quantize the theory, let us first introduce the canonical momentum conjugate to X^μ. It is given by

$$P^\mu(\sigma, \tau) = \frac{\delta S}{\delta \dot{X}_\mu} = T \dot{X}^\mu. \tag{2.47}$$

With this definition of the canonical momentum, the classical Poisson brackets are

$$\left[P^\mu(\sigma, \tau), P^\nu(\sigma', \tau) \right]_{\text{P.B.}} = \left[X^\mu(\sigma, \tau), X^\nu(\sigma', \tau) \right]_{\text{P.B.}} = 0, \tag{2.48}$$

$$\left[P^\mu(\sigma, \tau), X^\nu(\sigma', \tau) \right]_{\text{P.B.}} = \eta^{\mu\nu} \delta(\sigma - \sigma'). \tag{2.49}$$

In terms of $\dot{X}^\mu$

$$\left[\dot{X}^\mu(\sigma, \tau), X^\nu(\sigma', \tau) \right]_{\text{P.B.}} = T^{-1} \eta^{\mu\nu} \delta(\sigma - \sigma'). \tag{2.50}$$

Inserting the mode expansion for X^μ and $\dot{X}^\mu$ into these equations gives the Poisson brackets satisfied by the modes[3]

$$\left[\alpha_m^\mu, \alpha_n^\nu \right]_{\text{P.B.}} = \left[\tilde{\alpha}_m^\mu, \tilde{\alpha}_n^\nu \right]_{\text{P.B.}} = im \eta^{\mu\nu} \delta_{m+n,0} \tag{2.51}$$

and

$$\left[\alpha_m^\mu, \tilde{\alpha}_n^\nu \right]_{\text{P.B.}} = 0. \tag{2.52}$$

---

[3] The derivation of the commutation relations for the modes uses the Fourier expansion of the Dirac delta function

$$\delta(\sigma - \sigma') = \frac{1}{\pi} \sum_{n=-\infty}^{+\infty} e^{2in(\sigma-\sigma')}.$$

2.4 Canonical quantization

The world-sheet theory can now be quantized by replacing Poisson brackets by commutators

$$[\ldots]_{\text{P.B.}} \to i[\ldots]. \tag{2.53}$$

This gives

$$[\alpha_m^\mu, \alpha_n^\nu] = [\widetilde{\alpha}_m^\mu, \widetilde{\alpha}_n^\nu] = m\eta^{\mu\nu}\delta_{m+n,0}, \qquad [\alpha_m^\mu, \widetilde{\alpha}_n^\nu] = 0. \tag{2.54}$$

Defining

$$a_m^\mu = \frac{1}{\sqrt{m}}\alpha_m^\mu \quad \text{and} \quad a_m^{\mu\dagger} = \frac{1}{\sqrt{m}}\alpha_{-m}^\mu \quad \text{for} \quad m > 0, \tag{2.55}$$

the algebra satisfied by the modes is essentially the algebra of raising and lowering operators for quantum-mechanical harmonic oscillators

$$[a_m^\mu, a_n^{\nu\dagger}] = [\widetilde{a}_m^\mu, \widetilde{a}_n^{\nu\dagger}] = \eta^{\mu\nu}\delta_{m,n} \quad \text{for} \quad m, n > 0. \tag{2.56}$$

There is just one unusual feature: the commutators of time components have a negative sign, that is,

$$\left[a_m^0, a_m^{0\dagger}\right] = -1. \tag{2.57}$$

This results in negative norm states, which will be discussed in a moment. The spectrum is constructed by applying raising operators on the ground state, which is denoted $|0\rangle$. By definition, the ground state is annihilated by the lowering operators:

$$a_m^\mu |0\rangle = 0 \quad \text{for} \quad m > 0. \tag{2.58}$$

One can also specify the momentum k^μ carried by a state $|\phi\rangle$,

$$|\phi\rangle = a_{m_1}^{\mu_1\dagger} a_{m_2}^{\mu_2\dagger} \cdots a_{m_n}^{\mu_n\dagger} |0; k\rangle, \tag{2.59}$$

which is the eigenvalue of the momentum operator p^μ,

$$p^\mu |\phi\rangle = k^\mu |\phi\rangle. \tag{2.60}$$

It should be emphasized that this is *first quantization*, and all of these states (including the ground state) are one-particle states. Second quantization requires *string field theory*, which is discussed briefly at the end of Chapter 3.

The states with an even number of time-component operators have positive norm, while those that are constructed with an odd number of time-component operators have negative norm.[4] A simple example of a negative-norm state is given by

$$a_m^{0\dagger}|0\rangle \quad \text{with} \quad \text{norm} \quad \langle 0|a_m^0 a_m^{0\dagger}|0\rangle = -1, \qquad (2.61)$$

where the ground state is normalized as $\langle 0|0\rangle = 1$. In order for the theory to be physically sensible, it is essential that all physical states have positive norm. Negative-norm states in the physical spectrum of an interacting theory would lead to violations of causality and unitarity. The way in which the negative-norm states are eliminated from the physical spectrum is explained later in this chapter.

Open-string mode expansion

The general solution of the string equations of motion for an open string with Neumann boundary conditions is given by

$$X^\mu(\tau,\sigma) = x^\mu + l_s^2 p^\mu \tau + i l_s \sum_{m \neq 0} \frac{1}{m} \alpha_m^\mu e^{-im\tau} \cos(m\sigma). \qquad (2.62)$$

Mode expansions for other type of boundary conditions are given as homework problems. Note that, for the open string, only one set of modes α_m^μ appears, whereas for the closed string there are two independent sets of modes α_m^μ and $\tilde{\alpha}_m^\mu$. The open-string boundary conditions force the left- and right-moving modes to combine into standing waves. For the open string

$$2\partial_\pm X^\mu = \dot{X}^\mu \pm X'^\mu = l_s \sum_{m=-\infty}^{\infty} \alpha_m^\mu e^{-im(\tau \pm \sigma)}, \qquad (2.63)$$

where, $\alpha_0^\mu = l_s p^\mu$.

Hamiltonian and energy–momentum tensor

As discussed above, the string sigma-model action is invariant under various symmetries.

Noether currents

Recall that there is a standard method, due to Noether, for constructing a conserved current $\mathcal{J}_\alpha$ associated with a global symmetry transformation

$$\phi \to \phi + \delta_\varepsilon \phi, \qquad (2.64)$$

---

[4] States that have negative norm are sometimes called *ghosts*, but we reserve that word for the ghost fields that are arise from covariant BRST quantization in the next chapter.

where ϕ is any field of the theory and ε is an infinitesimal parameter. Such a transformation is a symmetry of the theory if it leaves the equations of motion invariant. This is the case if the action changes at most by a surface term, which means that the Lagrangian density changes at most by a total derivative. The Noether current is then determined from the change in the action under the above transformation

$$\mathcal{L} \to \mathcal{L} + \varepsilon \partial_\alpha \mathcal{J}^\alpha. \tag{2.65}$$

When ε is a constant, this change is a total derivative, which reflects the fact that there is a global symmetry. Then the equations of motion imply that the current is conserved, $\partial_\alpha \mathcal{J}^\alpha = 0$. The Poincaré transformations

$$\delta X^\mu = a^\mu{}_\nu X^\nu + b^\mu, \tag{2.66}$$

are global symmetries of the string world-sheet theory. Therefore, they give rise to conserved Noether currents. Applying the Noether method to derive the conserved currents associated with the Poincaré transformation of X^μ, one obtains

$$P^\mu_\alpha = T \partial_\alpha X^\mu, \tag{2.67}$$

$$J^{\mu\nu}_\alpha = T \left(X^\mu \partial_\alpha X^\nu - X^\nu \partial_\alpha X^\mu \right), \tag{2.68}$$

where the first current is associated with the translation symmetry, and the second one originates from the invariance under Lorentz transformations.

Hamiltonian

World-sheet time evolution is generated by the Hamiltonian

$$H = \int_0^\pi \left(\dot{X}_\mu P^\mu_0 - \mathcal{L} \right) d\sigma = \frac{T}{2} \int_0^\pi \left(\dot{X}^2 + X'^2 \right) d\sigma, \tag{2.69}$$

where

$$P^\mu_0 = \frac{\delta S}{\delta \dot{X}_\mu} = T \dot{X}^\mu, \tag{2.70}$$

was previously called $P^\mu(\sigma, \tau)$. Inserting the mode expansions, the result for the closed-string Hamiltonian is

$$H = \sum_{n=-\infty}^{+\infty} \left(\alpha_{-n} \cdot \alpha_n + \widetilde{\alpha}_{-n} \cdot \widetilde{\alpha}_n \right), \tag{2.71}$$

while for the open string the corresponding expression is

$$H = \frac{1}{2} \sum_{n=-\infty}^{+\infty} \alpha_{-n} \cdot \alpha_n. \tag{2.72}$$

2.4 Canonical quantization

These results hold for the classical theory. In the quantum theory there are ordering ambiguities that need to be resolved.

Energy momentum tensor

Let us now consider the mode expansions of the energy–momentum tensor. Inserting the closed-string mode expansions for X_L and X_R into the energy–momentum tensor Eqs (2.36), (2.37), one obtains

$$T_{--} = 2\, l_s^2 \sum_{m=-\infty}^{+\infty} L_m e^{-2im(\tau-\sigma)} \quad \text{and} \quad T_{++} = 2\, l_s^2 \sum_{m=-\infty}^{+\infty} \tilde{L}_m e^{-2im(\tau+\sigma)}, \tag{2.73}$$

where the Fourier coefficients are the *Virasoro generators*

$$L_m = \frac{1}{2} \sum_{n=-\infty}^{+\infty} \alpha_{m-n} \cdot \alpha_n \quad \text{and} \quad \tilde{L}_m = \frac{1}{2} \sum_{n=-\infty}^{+\infty} \tilde{\alpha}_{m-n} \cdot \tilde{\alpha}_n. \tag{2.74}$$

In the same way, one can get the result for the modes of the energy–momentum tensor of the open string. Comparing with the Hamiltonian, results in the expression

$$\frac{1}{2} H = L_0 + \tilde{L}_0 = \frac{1}{2} \sum_{n=-\infty}^{+\infty} \left(\alpha_{-n} \cdot \alpha_n + \tilde{\alpha}_{-n} \cdot \tilde{\alpha}_n \right), \tag{2.75}$$

for a closed string, while for an open string

$$H = L_0 = \frac{1}{2} \sum_{n=-\infty}^{+\infty} \alpha_{-n} \cdot \alpha_n. \tag{2.76}$$

The above results hold for the classical theory. Again, in the quantum theory one needs to resolve ordering ambiguities.

Mass formula for the string

Classically the vanishing of the energy–momentum tensor translates into the vanishing of all the Fourier modes

$$L_m = 0 \quad \text{for} \quad m = 0, \pm 1, \pm 2, \ldots \tag{2.77}$$

The classical constraint

$$L_0 = \tilde{L}_0 = 0, \tag{2.78}$$

can be used to derive an expression for the mass of a string. The relativistic mass-shell condition is

$$M^2 = -p_\mu p^\mu, \tag{2.79}$$

where p_μ is the total momentum of the string. This total momentum is given by

$$p^\mu = T \int_0^\pi d\sigma \dot{X}^\mu(\sigma), \qquad (2.80)$$

so that only the zero mode in the mode expansion of $\dot{X}^\mu(\sigma,\tau)$ contributes.

For the open string, the vanishing of L_0 then becomes

$$L_0 = \sum_{n=1}^{\infty} \alpha_{-n} \cdot \alpha_n + \frac{1}{2}\alpha_0^2 = \sum_{n=1}^{\infty} \alpha_{-n} \cdot \alpha_n + \alpha' p^2 = 0, \qquad (2.81)$$

which gives a relation between the mass of the string and the oscillator modes. For the open string one gets the relation

$$M^2 = \frac{1}{\alpha'} \sum_{n=1}^{\infty} \alpha_{-n} \cdot \alpha_n. \qquad (2.82)$$

For the closed string one has to take the left-moving and right-moving modes into account, and then one obtains

$$M^2 = \frac{2}{\alpha'} \sum_{n=1}^{\infty} \left(\alpha_{-n} \cdot \alpha_n + \widetilde{\alpha}_{-n} \cdot \widetilde{\alpha}_n \right). \qquad (2.83)$$

These are the mass-shell conditions for the string, which determine the mass of a given string state. In the quantum theory these relations get slightly modified.

The Virasoro algebra

Classical theory

In the classical theory the Virasoro generators satisfy the algebra

$$[L_m, L_n]_{\text{P.B.}} = i(m-n)L_{m+n}. \qquad (2.84)$$

The appearance of the Virasoro algebra is due to the fact that the gauge choice Eq. (2.23) has not fully gauge fixed the reparametrization symmetry. Let ξ^α be an infinitesimal parameter for a reparametrization and let Λ be an infinitesimal parameter for a Weyl rescaling. Then residual reparametrization symmetries satisfying

$$\partial^\alpha \xi^\beta + \partial^\beta \xi^\alpha = \Lambda \eta^{\alpha\beta}, \qquad (2.85)$$

still remain. These are the reparametrizations that are also Weyl rescalings. If one defines the combinations $\xi^\pm = \xi^0 \pm \xi^1$ and $\sigma^\pm = \sigma^0 \pm \sigma^1$, then one

2.4 Canonical quantization

finds that Eq. (2.85) is solved by

$$\xi^+ = \xi^+(\sigma^+) \quad \text{and} \quad \xi^- = \xi^-(\sigma^-). \tag{2.86}$$

The infinitesimal generators for the transformations $\delta\sigma^\pm = \xi^\pm$ are given by

$$V^\pm = \frac{1}{2}\xi^\pm(\sigma^\pm)\frac{\partial}{\partial \sigma^\pm}, \tag{2.87}$$

and a complete basis for these transformations is given by

$$\xi_n^\pm(\sigma^\pm) = e^{2in\sigma^\pm} \quad n \in \mathbb{Z}. \tag{2.88}$$

The corresponding generators $V_n^\pm$ give two copies of the Virasoro algebra. In the case of open strings there is just one Virasoro algebra, and the infinitesimal generators are

$$V_n = e^{in\sigma^+}\frac{\partial}{\partial \sigma^+} + e^{in\sigma^-}\frac{\partial}{\partial \sigma^-} \quad n \in \mathbb{Z}. \tag{2.89}$$

In the classical theory the equation of motion for the metric implies the vanishing of the energy–momentum tensor, that is, $T_{++} = T_{--} = 0$, which in terms of the Fourier components of Eq. (2.73) is

$$L_m = \frac{1}{2}\sum_{n=-\infty}^{+\infty} \alpha_{m-n} \cdot \alpha_n = 0 \quad \text{for} \quad m \in \mathbb{Z}. \tag{2.90}$$

In the case of closed strings, there are also corresponding $\tilde{L}_m$ conditions.

Quantum theory

In the quantum theory these operators are defined to be normal-ordered, that is,

$$L_m = \frac{1}{2}\sum_{n=-\infty}^{\infty} :\alpha_{m-n} \cdot \alpha_n:. \tag{2.91}$$

According to the normal-ordering prescription the lowering operators always appear to the right of the raising operators. In particular, L_0 becomes

$$L_0 = \frac{1}{2}\alpha_0^2 + \sum_{n=1}^{\infty} \alpha_{-n} \cdot \alpha_n. \tag{2.92}$$

Actually, this is the only Virasoro operator for which normal-ordering matters. Since an arbitrary constant could have appeared in this expression, one must expect a constant to be added to L_0 in all formulas, in particular the Virasoro algebra.

Using the commutators for the modes α_m^μ, one can show that in the quantum theory the Virasoro generators satisfy the relation

$$[L_m, L_n] = (m-n)L_{m+n} + \frac{c}{12}m(m^2-1)\delta_{m+n,0}, \qquad (2.93)$$

where $c = D$ is the space-time dimension. The term proportional to c is a quantum effect. This means that it appears after quantization and is absent in the classical theory. This term is called a *central extension*, and c is called a *central charge*, since it can be regarded as multiplying the unit operator, which when adjoined to the algebra is in the center of the extended algebra.

$SL(2,\mathbb{R})$ *subalgebra*

The Virasoro algebra contains an $SL(2,\mathbb{R})$ subalgebra that is generated by L_0, L_1 and L_{-1}. This is a noncompact form of the familiar $SU(2)$ algebra. Just as $SU(2)$ and $SO(3)$ have the same Lie algebra, so do $SL(2,\mathbb{R})$ and $SO(2,1)$. Thus, in the case of closed strings, the complete Virasoro algebra of both left-movers and right-movers contains the subalgebra $SL(2,\mathbb{R}) \times SL(2,\mathbb{R}) = SO(2,2)$. This is a noncompact version of the Lie algebra identity $SU(2) \times SU(2) = SO(4)$. The significance of this subalgebra will become clear in the next chapter.

Physical states

As was mentioned above, in the quantum theory a constant may need to be added to L_0 to parametrize the arbitrariness in the ordering prescription. Therefore, when imposing the constraint that the zero mode of the energy–momentum tensor should vanish, the only requirement in the case of the open string is that there exists some constant a such that

$$(L_0 - a)|\phi\rangle = 0. \qquad (2.94)$$

Here $|\phi\rangle$ is any physical on-shell state in the theory, and the constant a will be determined later. Similarly, for the closed string

$$(L_0 - a)|\phi\rangle = (\tilde{L}_0 - a)|\phi\rangle = 0. \qquad (2.95)$$

Mass operator

The constant a contributes to the mass operator. Indeed, in the quantum theory Eq. (2.94) corresponds to the mass-shell condition for the open string

$$\alpha' M^2 = \sum_{n=1}^\infty \alpha_{-n} \cdot \alpha_n - a = N - a, \qquad (2.96)$$

where

$$N = \sum_{n=1}^{\infty} \alpha_{-n} \cdot \alpha_n = \sum_{n=1}^{\infty} n a_n^{\dagger} \cdot a_n, \quad (2.97)$$

is called the *number operator*, since it has integer eigenvalues. For the ground state, which has $N = 0$, this gives $\alpha' M^2 = -a$, while for the excited states $\alpha' M^2 = 1 - a, 2 - a, \ldots$

For the closed string

$$\frac{1}{4}\alpha' M^2 = \sum_{n=1}^{\infty} \alpha_{-n} \cdot \alpha_n - a = \sum_{n=1}^{\infty} \tilde{\alpha}_{-n} \cdot \tilde{\alpha}_n - a = N - a = \tilde{N} - a. \quad (2.98)$$

Level matching

The normal-ordering constant a cancels out of the difference

$$(L_0 - \tilde{L}_0)|\phi\rangle = 0, \quad (2.99)$$

which implies $N = \tilde{N}$. This is the so-called *level-matching condition* of the bosonic string. It is the only constraint that relates the left- and right-moving modes.

Virasoro generators and physical states

In the quantum theory one cannot demand that the operator L_m annihilates all the physical states, for all $m \neq 0$, since this is incompatible with the Virasoro algebra. Rather, a physical state can only be annihilated by half of the Virasoro generators, specifically

$$L_m|\phi\rangle = 0 \quad m > 0. \quad (2.100)$$

Together with the mass-shell condition

$$(L_0 - a)|\phi\rangle = 0, \quad (2.101)$$

this characterizes a physical state $|\phi\rangle$. This is sufficient to give vanishing matrix elements of $L_n - a\delta_{n,0}$, between physical states, for all n. Since

$$L_{-m} = L_m^{\dagger}, \quad (2.102)$$

the hermitian conjugate of Eq. (2.100) ensures that the negative-mode Virasoro operators annihilate physical states on their left

$$\langle\phi|L_m = 0 \quad m < 0. \quad (2.103)$$

There are no normal-ordering ambiguities in the Lorentz generators[5]

$$J^{\mu\nu} = x^\mu p^\nu - x^\nu p^\mu - i \sum_{n=1}^{\infty} \frac{1}{n} \left(\alpha^\mu_{-n} \alpha^\nu_n - \alpha^\nu_{-n} \alpha^\mu_n \right), \qquad (2.104)$$

and therefore they can be interpreted as quantum operators without any quantum corrections. Using this expression, it is possible to check that

$$[L_m, J^{\mu\nu}] = 0, \qquad (2.105)$$

which implies that the physical-state condition is invariant under Lorentz transformations. Therefore, physical states must appear in complete Lorentz multiplets. This follows from the fact that, the formalism being discussed here is manifestly Lorentz covariant.

Absence of negative-norm states

The goal of this section is to show that a spectrum free of negative-norm states is only possible for certain values of a and the space-time dimension D. In order to carry out the analysis in a covariant manner, a crucial ingredient is the Virasoro algebra in Eq. (2.93).

In the quantum theory the values of a and D are not arbitrary. For some values negative-norm states appear and for other values the physical Hilbert space is positive definite. At the boundary where positive-norm states turn into negative-norm states, an increased number of zero-norm states appear. Therefore, in order to determine the allowed values for a and D, an effective strategy is to search for zero-norm states that satisfy the physical-state conditions.

Spurious states

A state $|\psi\rangle$ is called *spurious* if it satisfies the mass-shell condition and is orthogonal to all physical states

$$(L_0 - a)|\psi\rangle = 0 \quad \text{and} \quad \langle \phi | \psi \rangle = 0, \qquad (2.106)$$

where $|\phi\rangle$ represents any physical state in the theory. An example of a spurious state is

$$|\psi\rangle = \sum_{n=1}^{\infty} L_{-n} |\chi_n\rangle \quad \text{with} \quad (L_0 - a + n)|\chi_n\rangle = 0. \qquad (2.107)$$

---

5 J^{ij} generates rotations and J^{i0} generates boosts.

2.4 Canonical quantization

In fact, any such state can be recast in the form

$$|\psi\rangle = L_{-1}|\chi_1\rangle + L_{-2}|\chi_2\rangle \quad (2.108)$$

as a consequence of the Virasoro algebra (e.g. $L_{-3} = [L_{-1}, L_{-2}]$). Moreover, any spurious state can be put in this form. Spurious states $|\psi\rangle$ defined this way are orthogonal to every physical state, since

$$\langle\phi|\psi\rangle = \sum_{n=1}^{\infty}\langle\phi|L_{-n}|\chi_n\rangle = \sum_{n=1}^{\infty}\langle\chi_n|L_n|\phi\rangle^{\star} = 0. \quad (2.109)$$

If a state $|\psi\rangle$ is spurious *and* physical, then it is orthogonal to all physical states including itself

$$\langle\psi|\psi\rangle = \sum_{n=1}^{\infty}\langle\chi_n|L_n|\psi\rangle = 0. \quad (2.110)$$

As a result, such a state has zero norm.

Determination of a

When the constant a is suitably chosen, a class of zero-norm spurious states has the form

$$|\psi\rangle = L_{-1}|\chi_1\rangle \quad (2.111)$$

with

$$(L_0 - a + 1)|\chi_1\rangle = 0 \quad \text{and} \quad L_m|\chi_1\rangle = 0 \quad m > 0. \quad (2.112)$$

Demanding that $|\psi\rangle$ is physical implies

$$L_m|\psi\rangle = (L_0 - a)|\psi\rangle = 0 \quad \text{for} \quad m = 1, 2, \ldots \quad (2.113)$$

The Virasoro algebra implies the identity

$$L_1 L_{-1} = 2L_0 + L_{-1}L_1, \quad (2.114)$$

which leads to

$$L_1|\psi\rangle = L_1 L_{-1}|\chi_1\rangle = (2L_0 + L_{-1}L_1)|\chi_1\rangle = 2(a-1)|\chi_1\rangle = 0, \quad (2.115)$$

and hence $a = 1$. Thus $a = 1$ is part of the specification of the boundary between positive-norm and negative-norm physical states.

Determination of the space-time dimension

The number of zero-norm spurious states increases dramatically if, in addition to $a = 1$, the space-time dimension is chosen appropriately. To see this, let us construct zero-norm spurious states of the form

$$|\psi\rangle = \left(L_{-2} + \gamma L_{-1}^2\right)|\widetilde{\chi}\rangle. \tag{2.116}$$

This has zero norm for a certain γ, which is determined below. Here $|\psi\rangle$ is spurious if $|\widetilde{\chi}\rangle$ is a state that satisfies

$$(L_0 + 1)|\widetilde{\chi}\rangle = L_m|\widetilde{\chi}\rangle = 0 \quad \text{for} \quad m = 1, 2, \ldots \tag{2.117}$$

Now impose the condition that $|\psi\rangle$ is a physical state, that is, $L_1|\psi\rangle = 0$ and $L_2|\psi\rangle = 0$, since the rest of the constraints $L_m|\psi\rangle = 0$ for $m \geq 3$ are then also satisfied as a consequence of the Virasoro algebra. Let us first evaluate the condition $L_1|\psi\rangle = 0$ using the relation

$$\left[L_1, L_{-2} + \gamma L_{-1}^2\right] = 3L_{-1} + 2\gamma L_0 L_{-1} + 2\gamma L_{-1} L_0$$

$$= (3 - 2\gamma)L_{-1} + 4\gamma L_0 L_{-1}. \tag{2.118}$$

This leads to

$$L_1|\psi\rangle = L_1\left(L_{-2} + \gamma L_{-1}^2\right)|\widetilde{\chi}\rangle = \left[(3 - 2\gamma)L_{-1} + 4\gamma L_0 L_{-1}\right]|\widetilde{\chi}\rangle. \tag{2.119}$$

The first term vanishes for $\gamma = 3/2$ while the second one vanishes in general, because

$$L_0 L_{-1}|\widetilde{\chi}\rangle = L_{-1}(L_0 + 1)|\widetilde{\chi}\rangle = 0. \tag{2.120}$$

Therefore, the result of evaluating the $L_1|\psi\rangle = 0$ constraint is $\gamma = 3/2$. Let us next consider the $L_2|\psi\rangle = 0$ condition. Using

$$\left[L_2, L_{-2} + \frac{3}{2}L_{-1}^2\right] = 13L_0 + 9L_{-1}L_1 + \frac{D}{2} \tag{2.121}$$

gives

$$L_2|\psi\rangle = L_2\left(L_{-2} + \frac{3}{2}L_{-1}^2\right)|\widetilde{\chi}\rangle = \left(-13 + \frac{D}{2}\right)|\widetilde{\chi}\rangle. \tag{2.122}$$

Thus the space-time dimension $D = 26$ gives additional zero-norm spurious states.

Critical bosonic theory

The zero-norm spurious states are unphysical. The fact that they are spurious ensures that they decouple from all physical processes. In fact, all negative-norm states decouple, and all physical states have positive norm. Thus, the complete physical spectrum is free of negative-norm states when the two conditions $a = 1$ and $D = 26$ are satisfied, as is proved in the next section. The $a = 1$, $D = 26$ bosonic string theory is called *critical*, and one says that the *critical dimension* is 26. The spectrum is also free of negative-norm states for $a \leq 1$ and $D \leq 25$. In these cases the theory is called *noncritical*. Noncritical string theory is discussed briefly in the next chapter.

EXERCISES

EXERCISE 2.10
Find the mode expansion for angular-momentum generators $J^{\mu\nu}$ of an open bosonic string.

SOLUTION

Using the current in Eq. (2.68),

$$J^{\mu\nu} = \int_0^\pi J_0^{\mu\nu} d\sigma = T \int_0^\pi (X^\mu \dot{X}^\nu - X^\nu \dot{X}^\mu) d\sigma.$$

Now

$$X^\mu(\tau, \sigma) = x^\mu + l_s^2 p^\mu \tau + i l_s \sum_{m \neq 0} \frac{1}{m} \alpha_m^\mu e^{-im\tau} \cos(m\sigma),$$

$$\dot{X}^\mu(\tau, \sigma) = l_s^2 p^\mu + l_s \sum_{m \neq 0} \alpha_m^\mu e^{-im\tau} \cos(m\sigma),$$

and $T = 1/(\pi l_s^2)$. A short calculation gives

$$J^{\mu\nu} = x^\mu p^\nu - x^\nu p^\mu - i \sum_{m=1}^\infty \frac{1}{m} \left(\alpha_{-m}^\mu \alpha_m^\nu - \alpha_{-m}^\nu \alpha_m^\mu \right).$$

□

2.5 Light-cone gauge quantization

As discussed earlier, the bosonic string has residual diffeomorphism symmetries, even after choosing the gauge $h_{\alpha\beta} = \eta_{\alpha\beta}$, which consist of all the conformal transformations. Therefore, there is still the possibility of making an additional gauge choice. By making a particular noncovariant gauge choice, it is possible to describe a Fock space that is manifestly free of negative-norm states and to solve explicitly all the Virasoro conditions instead of imposing them as constraints.

Let us introduce light-cone coordinates for space-time[6]

$$X^{\pm} = \frac{1}{\sqrt{2}}(X^0 \pm X^{D-1}). \quad (2.123)$$

Then the D space-time coordinates X^{μ} consist of the null coordinates $X^{\pm}$ and the $D-2$ transverse coordinates X^i. In this notation, the inner product of two arbitrary vectors takes the form

$$v \cdot w = v_{\mu} w^{\mu} = -v^+ w^- - v^- w^+ + \sum_i v^i w^i. \quad (2.124)$$

Indices are raised and lowered by the rules

$$v^- = -v_+, \qquad v^+ = -v_-, \qquad \text{and} \qquad v^i = v_i. \quad (2.125)$$

Since two coordinates are treated differently from the others, Lorentz invariance is no longer manifest when light-cone coordinates are used.

What simplification can be achieved by using the residual gauge symmetry? In terms of $\sigma^{\pm}$ the residual symmetry corresponds to the reparametrizations in Eq. (2.86) of each of the null world-sheet coordinates

$$\sigma^{\pm} \to \xi^{\pm}(\sigma^{\pm}). \quad (2.126)$$

These transformations correspond to

$$\tilde{\tau} = \frac{1}{2}\left[\xi^+(\sigma^+) + \xi^-(\sigma^-)\right], \quad (2.127)$$

$$\tilde{\sigma} = \frac{1}{2}\left[\xi^+(\sigma^+) - \xi^-(\sigma^-)\right]. \quad (2.128)$$

This means that $\tilde{\tau}$ can be an arbitrary solution to the free massless wave equation

$$\left(\frac{\partial^2}{\partial\sigma^2} - \frac{\partial^2}{\partial\tau^2}\right)\tilde{\tau} = 0. \quad (2.129)$$

---

[6] It is convenient to include the $\sqrt{2}$ factor in the definition of space-time light-cone coordinates while omitting it in the definition of world-sheet light-cone coordinates.

2.5 Light-cone gauge quantization

Once $\tilde{\tau}$ is determined, $\tilde{\sigma}$ is specified up to a constant.

In the gauge $h_{\alpha\beta} = \eta_{\alpha\beta}$, the space-time coordinates $X^\mu(\sigma, \tau)$ also satisfy the two-dimensional wave equation. The light-cone gauge uses the residual freedom described above to make the choice

$$X^+(\tilde{\sigma}, \tilde{\tau}) = x^+ + l_s^2 p^+ \tilde{\tau}. \tag{2.130}$$

This corresponds to setting

$$\alpha_n^+ = 0 \quad \text{for} \quad n \neq 0. \tag{2.131}$$

In the following the tildes are omitted from the parameters $\tilde{\tau}$ and $\tilde{\sigma}$.

When this noncovariant gauge choice is made, there is a risk that a quantum-mechanical anomaly could lead to a breakdown of Lorentz invariance. So this needs to be checked. In fact, conformal invariance is essential for making this gauge choice, so it should not be surprising that a *Lorentz anomaly* in the light-cone gauge approach corresponds to a *conformal anomaly* in a covariant gauge that preserves manifest Lorentz invariance.

The light-cone gauge has eliminated the oscillator modes of X^+. It is possible to determine the oscillator modes of X^-, as well, by solving the Virasoro constraints $(\dot{X} \pm X')^2 = 0$. In the light-cone gauge these constraints become

$$\dot{X}^- \pm X^{-\prime} = \frac{1}{2p^+ l_s^2}(\dot{X}^i \pm X^{i\prime})^2. \tag{2.132}$$

This pair of equations can be used to solve for X^- in terms of X^i. In terms of the mode expansion for X^-, which for an open string is

$$X^- = x^- + l_s^2 p^- \tau + i l_s \sum_{n \neq 0} \frac{1}{n} \alpha_n^- e^{-in\tau} \cos n\sigma, \tag{2.133}$$

the solution is

$$\alpha_n^- = \frac{1}{p^+ l_s} \left(\frac{1}{2} \sum_{i=1}^{D-2} \sum_{m=-\infty}^{+\infty} : \alpha_{n-m}^i \alpha_m^i : -a\delta_{n,0} \right). \tag{2.134}$$

Therefore, in the light-cone gauge it is possible to eliminate both X^+ and X^- (except for their zero modes) and express the theory in terms of the transverse oscillators. Thus a critical string only has transverse excitations, just as a massless particle only has transverse polarization states. The convenient feature of the light-cone gauge in Eq. (2.130) is that it turns the Virasoro constraints into linear equations for the modes of X^-.

Mass-shell condition

In the light-cone gauge the open-string mass-shell condition is

$$M^2 = -p_\mu p^\mu = 2p^+ p^- - \sum_{i=1}^{D-2} p_i^2 = 2(N - a)/l_s^2, \qquad (2.135)$$

where

$$N = \sum_{i=1}^{D-2} \sum_{n=1}^{\infty} \alpha_{-n}^i \alpha_n^i. \qquad (2.136)$$

Let us now construct the physical spectrum of the bosonic string in the light-cone gauge.

In the light-cone gauge all the excitations are generated by acting with the transverse modes α_n^i. The first excited state, given by $\alpha_{-1}^i |0; p\rangle$, belongs to a $(D-2)$-component vector representation of the rotation group $SO(D-2)$ in the transverse space. As a general rule, Lorentz invariance implies that physical states form representations of $SO(D-1)$ for massive states and $SO(D-2)$ for massless states. Therefore, the bosonic string theory in the light-cone gauge can only be Lorentz invariant if the vector state $\alpha_{-1}^i |0; p\rangle$ is massless. This immediately implies that $a = 1$.

Having fixed the value of a, the next goal is to determine the space-time dimension D. A heuristic approach is to compute the normal-ordering constant appearing in the definition of L_0 directly. This constant can be determined from

$$\frac{1}{2} \sum_{i=1}^{D-2} \sum_{n=-\infty}^{+\infty} \alpha_{-n}^i \alpha_n^i = \frac{1}{2} \sum_{i=1}^{D-2} \sum_{n=-\infty}^{+\infty} : \alpha_{-n}^i \alpha_n^i : + \frac{1}{2}(D-2) \sum_{n=1}^{\infty} n. \qquad (2.137)$$

The second sum on the right-hand side is divergent and needs to be regularized. This can be achieved using ζ-function regularization. First, one considers the general sum

$$\zeta(s) = \sum_{n=1}^{\infty} n^{-s}, \qquad (2.138)$$

which is defined for any complex number s. For $\mathrm{Re}(s) > 1$, this sum converges to the Riemann zeta function $\zeta(s)$. This zeta function has a unique analytic continuation to $s = -1$, where it takes the value $\zeta(-1) = -1/12$. Therefore, after inserting the value of $\zeta(-1)$ in Eq. (2.137), the result for the additional term is

$$\frac{1}{2}(D-2) \sum_{n=1}^{\infty} n = -\frac{D-2}{24}. \qquad (2.139)$$

2.5 Light-cone gauge quantization

Using the earlier result that the normal-ordering constant a should be equal to 1, one gets the condition

$$\frac{D-2}{24} = 1, \tag{2.140}$$

which implies $D = 26$. Though it is not very rigorous, this is the quickest way to determined the values of a and D. The earlier analysis of the no-negative-norm states theorem also singled out $D = 26$. Another approach is to verify that the Lorentz generators satisfy the Lorentz algebra, which is not manifest in the light-cone gauge. The nontrivial requirement is

$$[J^{i-}, J^{j-}] = 0. \tag{2.141}$$

Once the α_n^- oscillators are eliminated, J^{i-} becomes cubic in transverse oscillators. The algebra is rather complicated, but the bottom line is that the commutator only vanishes for $a = 1$ and $D = 26$. Other derivations of the critical dimension are presented in the next chapter.

Analysis of the spectrum

Having determined the preferred values $a = 1$ and $D = 26$, one can now determine the spectrum of the bosonic string.

The open string

At the first few mass levels the physical states of the open string are as follows:

- For $N = 0$ there is a tachyon $|0; k\rangle$, whose mass is given by $\alpha' M^2 = -1$.
- For $N = 1$ there is a vector boson $\alpha^i_{-1}|0; k\rangle$. As was explained in the previous section, Lorentz invariance requires that it is massless. This state gives a vector representation of $SO(24)$.
- $N = 2$ gives the first states with positive (mass)^2. They are

$$\alpha^i_{-2}|0; k\rangle \quad \text{and} \quad \alpha^i_{-1}\alpha^j_{-1}|0; k\rangle, \tag{2.142}$$

with $\alpha' M^2 = 1$. These have 24 and $24 \cdot 25/2$ states, respectively. The total number of states is 324, which is the dimensionality of the symmetric traceless second-rank tensor representation of $SO(25)$, since $25 \cdot 26/2 - 1 = 324$. So, in this sense, the spectrum consists of a single massive spin-two state at this mass level.

All of these states have a positive norm, since they are built entirely from the transverse modes, which describe a positive-definite Hilbert space. In the light-cone gauge the fact that the negative-norm states have decoupled is made manifest. All of the massive representations can be rearranged in complete $SO(25)$ multiplets, as was just demonstrated for the first massive level. Lorentz invariance of the spectrum is guaranteed, because the Lorentz algebra is realized on the Hilbert space of transverse oscillators.

The number of states

The total number of physical states of a given mass is easily computed. For example, in the case of open strings, it follows from Eqs (2.135) and (2.136) with $a = 1$ that the number of physical states d_n whose mass is given by $\alpha' M^2 = n - 1$ is the coefficient of w^n in the power-series expansion of

$$\operatorname{tr} w^N = \prod_{n=1}^{\infty} \prod_{i=1}^{24} \operatorname{tr} w^{\alpha^i_{-n}\alpha^i_n} = \prod_{n=1}^{\infty}(1-w^n)^{-24}. \tag{2.143}$$

This number can be written in the form

$$d_n = \frac{1}{2\pi i} \oint \frac{\operatorname{tr} w^N}{w^{n+1}} dw. \tag{2.144}$$

The number of physical states d_n can be estimated for large n by a saddle-point evaluation. Since the saddle point occurs close to $w = 1$, one can use the approximation

$$\operatorname{tr} w^N = \prod_{n=1}^{\infty}(1-w^n)^{-24} \sim \exp\left(\frac{4\pi^2}{1-w}\right). \tag{2.145}$$

This is an approximation to the modular transformation formula

$$\eta(-1/\tau) = (-i\tau)^{1/2}\eta(\tau) \tag{2.146}$$

for the Dedekind eta function

$$\eta(\tau) = e^{i\pi\tau/12} \prod_{n=1}^{\infty}\left(1 - e^{2\pi i n\tau}\right), \tag{2.147}$$

as one sees by setting $w = e^{2\pi i\tau}$. Then one finds that, for large n,

$$d_n \sim \operatorname{const.} n^{-27/4} \exp(4\pi\sqrt{n}). \tag{2.148}$$

The exponential factor can be rewritten in the form $\exp(M/M_0)$ with

$$M_0 = (4\pi\sqrt{\alpha'})^{-1}. \tag{2.149}$$

The quantity M_0 is called the *Hagedorn temperature*. Depending on details that go beyond present considerations, it is either a maximum possible temperature or else the temperature of a phase transition.

The closed string

For the case of the closed string, there are two sets of modes (left-movers and right-movers), and the level-matching condition must be taken into account. The spectrum is easily deduced from that of the open string, since closed-string states are tensor products of left-movers and right-movers, each of which has the same structure as open-string states. The mass of states in the closed-string spectrum is given by

$$\alpha' M^2 = 4(N-1) = 4(\tilde{N}-1). \tag{2.150}$$

The physical states of the closed string at the first two mass levels are as follows:

- The ground state $|0;k\rangle$ is again a tachyon, this time with

$$\alpha' M^2 = -4. \tag{2.151}$$

- For the $N=1$ level there is a set of $24^2 = 576$ states of the form

$$|\Omega^{ij}\rangle = \alpha^i_{-1} \tilde{\alpha}^j_{-1} |0;k\rangle, \tag{2.152}$$

corresponding to the tensor product of two massless vectors, one left-moving and one right-moving. The part of $|\Omega^{ij}\rangle$ that is symmetric and traceless in i and j transforms under $SO(24)$ as a massless spin-two particle, the *graviton*. The trace term $\delta_{ij}|\Omega^{ij}\rangle$ is a massless scalar, which is called the *dilaton*. The antisymmetric part $|\Omega^{ij}\rangle - |\Omega^{ji}\rangle$ transforms under $SO(24)$ as an antisymmetric second-rank tensor. Each of these three massless states has a counterpart in superstring theories, where they play fundamental roles that are discussed in later chapters.

HOMEWORK PROBLEMS

PROBLEM 2.1
Consider the following classical trajectory of an open string

$$\begin{aligned} X^0 &= B\tau, \\ X^1 &= B\,\cos(\tau)\cos(\sigma), \\ X^2 &= B\,\sin(\tau)\cos(\sigma), \\ X^i &= 0, \quad i > 2, \end{aligned}$$

and assume the conformal gauge condition.

(i) Show that this configuration describes a solution to the equations of motion for the field X^μ corresponding to an open string with Neumann boundary conditions. Show that the ends of this string are moving with the speed of light.

(ii) Compute the energy $E = P^0$ and angular momentum J of the string. Use your result to show that

$$\frac{E^2}{|J|} = 2\pi T = \frac{1}{\alpha'}.$$

(iii) Show that the constraint equation $T_{\alpha\beta} = 0$ can be written as

$$(\partial_\tau X)^2 + (\partial_\sigma X)^2 = 0, \qquad \partial_\tau X^\mu \partial_\sigma X_\mu = 0,$$

and that this constraint is satisfied by the above solution.

Problem 2.2

Consider the following classical trajectory of an open string

$$\begin{aligned} X^0 &= 3A\tau, \\ X^1 &= A\cos(3\tau)\cos(3\sigma), \\ X^2 &= A\sin(a\tau)\cos(b\sigma), \end{aligned}$$

and assume the conformal gauge.

(i) Determine the values of a and b so that the above equations describe an open string that solves the constraint $T_{\alpha\beta} = 0$. Express the solution in the form

$$X^\mu = X_L^\mu(\sigma^-) + X_R^\mu(\sigma^+).$$

Determine the boundary conditions satisfied by this field configuration.

(ii) Plot the solution in the (X^1, X^2)-plane as a function of τ in steps of $\pi/12$.

(iii) Compute the center-of-mass momentum and angular momentum and show that they are conserved. What do you obtain for the relation between the energy and angular momentum of this string? Comment on your result.

Problem 2.3

Compute the mode expansion of an open string with Neumann boundary conditions for the coordinates $X^0, \ldots, X^{24}$, while the remaining coordinate X^{25} satisfies the following boundary conditions:

(i) Dirichlet boundary conditions at both ends

$$X^{25}(0, \tau) = X_0^{25} \quad \text{and} \quad X^{25}(\pi, \tau) = X_\pi^{25}.$$

What is the interpretation of such a solution? Compute the conjugate momentum P^{25}. Is this momentum conserved?

(ii) Dirichlet boundary conditions on one end and Neumann boundary conditions at the other end

$$X^{25}(0, \tau) = X_0^{25} \quad \text{and} \quad \partial_\sigma X^{25}(\pi, \tau) = 0.$$

What is the interpretation of this solution?

PROBLEM 2.4
Consider the bosonic string in light-cone gauge.

(i) Find the mass squared of the following on-shell open-string states:

$$|\phi_1\rangle = \alpha^i_{-1}|0; k\rangle, \qquad |\phi_2\rangle = \alpha^i_{-1}\alpha^j_{-1}|0; k\rangle,$$
$$|\phi_3\rangle = \alpha^i_{-3}|0; k\rangle, \qquad |\phi_4\rangle = \alpha^i_{-1}\alpha^j_{-1}\alpha^k_{-2}|0; k\rangle.$$

(ii) Find the mass squared of the following on-shell closed-string states:

$$|\phi_1\rangle = \alpha^i_{-1}\tilde{\alpha}^j_{-1}|0; k\rangle, \qquad |\phi_2\rangle = \alpha^i_{-1}\alpha^j_{-1}\tilde{\alpha}^k_{-2}|0; k\rangle.$$

(iii) What can you say about the following closed-string state?

$$|\phi_3\rangle = \alpha^i_{-1}\tilde{\alpha}^j_{-2}|0; k\rangle$$

PROBLEM 2.5
Use the mode expansion of an open string with Neumann boundary conditions in Eq. (2.62) and the commutation relations for the modes in Eq. (2.54) to check explicitly the equal-time commutators

$$[X^\mu(\sigma, \tau), X^\nu(\sigma', \tau)] = [P^\mu(\sigma, \tau), P^\nu(\sigma', \tau)] = 0,$$

while

$$[X^\mu(\sigma, \tau), P^\nu(\sigma', \tau)] = i\eta^{\mu\nu}\delta(\sigma - \sigma').$$

Hint: The representation $\delta(\sigma - \sigma') = \frac{1}{\pi}\sum_{n\in\mathbb{Z}} \cos(n\sigma)\cos(n\sigma')$ might be useful.

Problem 2.6

Exercise 2.10 showed that the Lorentz generators of the open-string world sheet are given by

$$J^{\mu\nu} = x^\mu p^\nu - x^\nu p^\mu - i \sum_{n=1}^{\infty} \frac{1}{n} \left(\alpha^\mu_{-n} \alpha^\nu_n - \alpha^\nu_{-n} \alpha^\mu_n \right).$$

Use the canonical commutation relations to verify the Poincaré algebra

$$[p^\mu, p^\nu] = 0,$$

$$[p^\mu, J^{\nu\sigma}] = -i\eta^{\mu\nu} p^\sigma + i\eta^{\mu\sigma} p^\nu,$$

$$[J^{\mu\nu}, J^{\sigma\lambda}] = -i\eta^{\nu\sigma} J^{\mu\lambda} + i\eta^{\mu\sigma} J^{\nu\lambda} + i\eta^{\nu\lambda} J^{\mu\sigma} - i\eta^{\mu\lambda} J^{\nu\sigma}.$$

Problem 2.7

Exercise 2.10 derived the angular-momentum generators $J^{\mu\nu}$ for an open bosonic string. Derive them for a closed bosonic string.

Problem 2.8

The open-string angular momentum generators in Exercise 2.10 are appropriate for covariant quantization. What are the formulas in the case of light-cone gauge quantization.

Problem 2.9

Show that the Lorentz generators commute with all Virasoro generators,

$$[L_m, J^{\mu\nu}] = 0.$$

Explain why this implies that the physical-state condition is invariant under Lorentz transformations, and states of the string spectrum appear in complete Lorentz multiplets.

Problem 2.10

Consider an on-shell open-string state of the form

$$|\phi\rangle = \left(A \alpha_{-1} \cdot \alpha_{-1} + B \alpha_0 \cdot \alpha_{-2} + C(\alpha_0 \cdot \alpha_{-1})^2 \right) |0; k\rangle,$$

where A, B and C are constants. Determine the conditions on the coefficients A, B and C so that $|\phi\rangle$ satisfies the physical-state conditions for $a = 1$ and arbitrary D. Compute the norm of $|\phi\rangle$. What conclusions can you draw from the result?

Problem 2.11
The open-string states at the $N = 2$ level were shown in Section 2.5 to form a certain representation of $SO(25)$. What does this result imply for the spectrum of the closed bosonic string at the $N_{\rm L} = N_{\rm R} = 2$ level?

Problem 2.12
Construct the spectrum of open and closed strings in light-cone gauge for level $N = 3$. How many states are there in each case? Without actually doing it (unless you want to), describe a strategy for assembling these states into irreducible $SO(25)$ multiplets.

Problem 2.13
We expect the central extension of the Virasoro algebra to be of the form
$$[L_m, L_n] = (m - n)L_{m+n} + A(m)\delta_{m+n,0},$$
because normal-ordering ambiguities only arise for $m + n = 0$.

(i) Show that if $A(1) \neq 0$ it is possible to change the definition of L_0, by adding a constant, so that $A(1) = 0$.

(ii) For $A(1) = 0$ show that the generators L_0 and $L_{\pm 1}$ form a closed subalgebra.

Problem 2.14
Derive an equation for the coefficients $A(m)$ defined in the previous problem that follows from the Jacobi identity
$$[[L_m, L_n], L_p] + [[L_p, L_m], L_n] + [[L_n, L_p], L_m] = 0.$$
Assuming $A(1) = 0$, prove that $A(m) = (m^3 - m)A(2)/6$ is the unique solution, and hence that the central charge is $c = 2A(2)$.

Problem 2.15
Verify that the Virasoro generators in Eq. (2.91) satisfy the Virasoro algebra. It is difficult to verify the central-charge term directly from the commutator. Therefore, a good strategy is to verify that $A(1)$ and $A(2)$ have the correct values. These can be determined by computing the ground-state matrix element of Eq. (2.93) for the cases $m = -n = 1$ and $m = -n = 2$.

3
Conformal field theory and string interactions

The previous chapter described the free bosonic string in Minkowski space-time. It was argued that consistency requires the dimension of space-time to be $D = 26$ (25 space and one time). Even then, there is a tachyon problem. When interactions are included, this theory might not have a stable vacuum. The justification for studying the bosonic string theory, despite its deficiencies, is that it is a good warm-up exercise before tackling more interesting theories that do have stable vacua. This chapter continues the study of the bosonic string theory, covering a lot of ground rather concisely.

One important issue concerns the possibilities for introducing more general backgrounds than flat 26-dimensional Minkowski space-time. Another concerns the development of techniques for describing interactions and computing scattering amplitudes in perturbation theory. We also discuss a quantum field theory of strings. In this approach field operators create and destroy entire strings. All of these topics exploit the conformal symmetry of the world-sheet theory, using the techniques of conformal field theory (CFT). Therefore, this chapter begins with an overview of that subject.

3.1 Conformal field theory

Until now it has been assumed that the string world sheet has a Lorentzian signature metric, since this choice is appropriate for a physically evolving string. However, it is extremely convenient to make a Wick rotation $\tau \to -i\tau$, so as to obtain a world sheet with Euclidean signature, and thereby make the world-sheet metric $h_{\alpha\beta}$ positive definite. Having done this, one can introduce complex coordinates (in local patches)

$$z = e^{2(\tau - i\sigma)} \quad \text{and} \quad \bar{z} = e^{2(\tau + i\sigma)} \qquad (3.1)$$

and regard the world sheet as a Riemann surface. The factors of two in the exponents reflect the earlier convention of choosing the periodicity of the closed-string parametrization to be $\sigma \to \sigma + \pi$. Replacing σ by $-\sigma$ in these formulas would interchange the identifications of left-movers and right-movers. Note that if the world sheet is the complex plane, Euclidean time corresponds to radial distance, with the origin representing the infinite past and the circle at infinity the infinite future. The residual symmetries in the conformal gauge, $\tau \pm \sigma \to f_\pm(\tau \pm \sigma)$, described in Chapter 2, now become conformal mappings $z \to f(z)$ and $\bar{z} \to \bar{f}(\bar{z})$. For example, the complex plane (minus the origin) is equivalent to an infinitely long cylinder, as shown in Fig. 3.1. Thus, we are led to consider conformally invariant two-dimensional field theories.

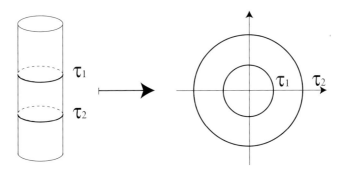

Fig. 3.1. Conformal mapping of an infinitely long cylinder onto a plane.

The conformal group in D dimensions

The main topic of this section is the conformal symmetry of two-dimensional world-sheet theories. However, conformal symmetry in other dimensions also plays an important role in recent string theory research (discussed in Chapter 12). Therefore, before specializing to two dimensions, let us consider the conformal group in D dimensions.

A D-dimensional manifold is called *conformally flat* if the invariant line element can be written in the form

$$ds^2 = e^{\omega(x)} dx \cdot dx. \qquad (3.2)$$

The dot product represents contraction with the Lorentz metric $\eta_{\mu\nu}$ in the case of a Lorentzian-signature pseudo-Riemannian manifold or with the Kronecker metric $\delta_{\mu\nu}$ in the case of a Euclidean-signature Riemannian manifold. The function $\omega(x)$ in the conformal factor is allowed to be x-dependent.

The *conformal group* is the subgroup of the group of general coordinate transformations (or diffeomorphisms) that preserves the conformal flatness of the metric. The important geometric property of conformal transformations is that they preserve angles while distorting lengths.

Part of the conformal group is obvious. Namely, it contains translations and rotations. By "rotations" we include Lorentz transformations (in the case of Lorentzian signature). Another conformal group transformation is a scale transformation $x^\mu \to \lambda x^\mu$, where λ is a constant. One can either regard this as changing ω, or else it can be viewed as a symmetry, if one also transforms ω appropriately at the same time.

Another class of conformal group transformations, called *special conformal transformations*, is less obvious. However, there is a simple way of deriving them. This hinges on noting that the conformal group includes an inversion element

$$x^\mu \to \frac{x^\mu}{x^2}. \tag{3.3}$$

This maps

$$dx \cdot dx \to \frac{dx \cdot dx}{(x^2)^2}, \tag{3.4}$$

so that the metric remains conformally flat.[1] The trick then is to consider a sequence of transformations: inversion – translation – inversion. In other words, one conjugates a translation ($x^\mu \to x^\mu + b^\mu$) by an inversion. This gives

$$x^\mu \to \frac{x^\mu + b^\mu x^2}{1 + 2b \cdot x + b^2 x^2}. \tag{3.5}$$

Taking b^μ to be infinitesimal, we get

$$\delta x^\mu = b^\mu x^2 - 2x^\mu b \cdot x. \tag{3.6}$$

Summarizing the results given above, the following infinitesimal transformations are conformal:

$$\delta x^\mu = a^\mu + \omega^\mu{}_\nu x^\nu + \lambda x^\mu + b^\mu x^2 - 2x^\mu (b \cdot x). \tag{3.7}$$

The parameters a^μ, $\omega^\mu{}_\nu$, λ and b^μ are infinitesimal constants. After lowering an index with $\eta_{\mu\nu}$ or $\delta_{\mu\nu}$, as appropriate, the parameters of infinitesimal

---

[1] Strictly speaking, in the case of Euclidean signature this requires regarding the point at infinity to be part of the manifold, a procedure known as *conformal compactification*. In the case of Lorentzian signature, a Wick rotation to Euclidean signature should be made first for the inversion to make sense.

rotations are required to satisfy $\omega_{\mu\nu} = -\omega_{\nu\mu}$. Altogether there are

$$D + \frac{1}{2}D(D-1) + 1 + D = \frac{1}{2}(D+2)(D+1) \qquad (3.8)$$

linearly independent infinitesimal conformal transformations, so this is the number of generators of the conformal group.

The number of conformal-group generators in D dimensions is the same as for the group of rotations in $D+2$ dimensions. In fact, by commuting the infinitesimal conformal transformations one can derive the Lie algebra, and it turns out to be a noncompact form of $SO(D+2)$. In the case of Lorentzian signature, the Lie algebra is $SO(D,2)$, while if the manifold is Euclidean it is $SO(D+1,1)$.

When $D > 2$ the algebras discussed above generate the entire conformal group, except that an inversion is not infinitesimally generated. Because of the inversion element, the groups have two disconnected components. When $D = 2$, the $SO(2,2)$ or $SO(3,1)$ algebra is a subalgebra of a much larger algebra.

The conformal group in two dimensions

As has already been remarked, conformal transformations in two dimensions consist of analytic coordinate transformations

$$z \to f(z) \quad \text{and} \quad \bar{z} \to \bar{f}(\bar{z}). \qquad (3.9)$$

These are angle-preserving transformations wherever f and its inverse function are holomorphic, that is, f is *biholomorphic*.

To exhibit the generators, consider infinitesimal conformal transformations of the form

$$z \to z' = z - \varepsilon_n z^{n+1} \quad \text{and} \quad \bar{z} \to \bar{z}' = \bar{z} - \bar{\varepsilon}_n \bar{z}^{n+1}, \quad n \in \mathbb{Z}. \qquad (3.10)$$

The corresponding infinitesimal generators are[2]

$$\ell_n = -z^{n+1}\partial \quad \text{and} \quad \bar{\ell}_n = -\bar{z}^{n+1}\bar{\partial}, \qquad (3.11)$$

where $\partial = \partial/\partial z$ and $\bar{\partial} = \partial/\partial \bar{z}$. These generators satisfy the classical Virasoro algebras

$$[\ell_m, \ell_n] = (m-n)\ell_{m+n} \quad \text{and} \quad [\bar{\ell}_m, \bar{\ell}_n] = (m-n)\bar{\ell}_{m+n}, \qquad (3.12)$$

while $[\ell_m, \bar{\ell}_n] = 0$. In the quantum case the Virasoro algebra can acquire

---

[2] For $n < -1$ these are defined on the punctured plane, which has the origin removed. Similarly, for $n > 1$, the point at infinity is removed. Note that ℓ_{-1}, ℓ_0 and ℓ_1 are special in that they are defined globally on the Riemann sphere.

a *central extension*, or *conformal anomaly*, with central charge c, in which case it takes the form

$$[L_m, L_n] = (m-n)L_{m+n} + \frac{c}{12}m(m^2-1)\delta_{m+n,0}. \qquad (3.13)$$

In a two-dimensional conformal field theory the Virasoro operators are the modes of the energy–momentum tensor, which therefore is the operator that generates conformal transformations. The term "central extension" means that the constant term can be understood to multiply the unit operator, which is adjoined to the Lie algebra. The expression "conformal anomaly" refers to the fact that in certain settings the central charge can be interpreted as signalling a quantum mechanical breaking of the classical conformal symmetry.

The conformal group is infinite-dimensional in two dimensions. However, as was pointed out in Chapter 2, it contains a finite-dimensional subgroup generated by $\ell_{0,\pm 1}$ and $\bar{\ell}_{0,\pm 1}$. This remains true in the quantum case. Infinitesimally, the transformations are

$$\begin{aligned} \ell_{-1}: & \quad z \to z - \varepsilon, \\ \ell_0: & \quad z \to z - \varepsilon z, \\ \ell_1: & \quad z \to z - \varepsilon z^2. \end{aligned} \qquad (3.14)$$

The interpretation of the corresponding transformations is that ℓ_{-1} and $\bar{\ell}_{-1}$ generate translations, $(\ell_0 + \bar{\ell}_0)$ generates scalings, $i(\ell_0 - \bar{\ell}_0)$ generates rotations and ℓ_1 and $\bar{\ell}_1$ generate special conformal transformations.

The finite form of the group transformations is

$$z \to \frac{az+b}{cz+d} \quad \text{with} \quad a,b,c,d \in \mathbb{C}, \quad ad-bc=1. \qquad (3.15)$$

This is the group $SL(2,\mathbb{C})/\mathbb{Z}_2 = SO(3,1)$.[3] The division by $\mathbb{Z}_2$ accounts for the freedom to replace the parameters a,b,c,d by their negatives, leaving the transformations unchanged. This is the two-dimensional case of $SO(D+1,1)$, which is the conformal group for $D > 2$ Euclidean dimensions. In the Lorentzian case it is replaced by $SO(2,2) = SL(2,\mathbb{R}) \times SL(2,\mathbb{R})$, where one factor pertains to left-movers and the other to right-movers. This finite-dimensional subgroup of the two-dimensional conformal group is called the *restricted conformal group*.

The previous chapter described the construction of the world-sheet energy–momentum tensor $T_{\alpha\beta}$. It was shown to satisfy $T_{+-} = T_{-+} = 0$ as a consequence of Weyl symmetry. Since the world-sheet theory has translation

---

3 By $SO(3,1)$ we really mean the connected component of the group. There is a similar qualification, as well as implicit division by $\mathbb{Z}_2$ factors, in the Lorentzian case that follows.

3.1 Conformal field theory

symmetry, this tensor is also conserved

$$\partial^\alpha T_{\alpha\beta} = 0. \tag{3.16}$$

After Wick rotation the light-cone indices $\pm$ are replaced by $(z, \bar{z})$. So the nonvanishing components are T_{zz} and $T_{\bar{z}\bar{z}}$, and the conservation conditions are

$$\bar{\partial} T_{zz} = 0 \quad \text{and} \quad \partial T_{\bar{z}\bar{z}} = 0. \tag{3.17}$$

Thus one is holomorphic and the other is antiholomorphic

$$T_{zz} = T(z) \quad \text{and} \quad T_{\bar{z}\bar{z}} = \widetilde{T}(\bar{z}). \tag{3.18}$$

The Virasoro generators are the modes of the energy–momentum tensor.

In the current notation, for $l_s = \sqrt{2\alpha'} = 1$, the right-moving part of the coordinate X^μ given in Chapter 2 becomes

$$X_R^\mu(\sigma, \tau) \to X_R^\mu(z) = \frac{1}{2} x^\mu - \frac{i}{4} p^\mu \ln z + \frac{i}{2} \sum_{n \neq 0} \frac{1}{n} \alpha_n^\mu z^{-n} \tag{3.19}$$

and similarly

$$X_L^\mu(\sigma, \tau) \to X_L^\mu(\bar{z}) = \frac{1}{2} x^\mu - \frac{i}{4} p^\mu \ln \bar{z} + \frac{i}{2} \sum_{n \neq 0} \frac{1}{n} \widetilde{\alpha}_n^\mu \bar{z}^{-n}. \tag{3.20}$$

The holomorphic derivatives take the simple form

$$\partial X^\mu(z, \bar{z}) = -\frac{i}{2} \sum_{n=-\infty}^{\infty} \alpha_n^\mu z^{-n-1} \tag{3.21}$$

and

$$\bar{\partial} X^\mu(z, \bar{z}) = -\frac{i}{2} \sum_{n=-\infty}^{\infty} \widetilde{\alpha}_n^\mu \bar{z}^{-n-1}. \tag{3.22}$$

Out of this one can compute the holomorphic component of the energy–momentum tensor

$$T_X(z) = -2 : \partial X \cdot \partial X : = \sum_{n=-\infty}^{+\infty} \frac{L_n}{z^{n+2}}. \tag{3.23}$$

Similarly,

$$\widetilde{T}_X(\bar{z}) = -2 : \bar{\partial} X \cdot \bar{\partial} X : = \sum_{n=-\infty}^{+\infty} \frac{\widetilde{L}_n}{\bar{z}^{n+2}}. \tag{3.24}$$

The subscript X has been introduced here to emphasize that these energy–momentum tensors are constructed out of the X^μ coordinates.

Since the two-dimensional conformal algebra is infinite-dimensional, there is an infinite number of conserved charges, which are essentially the Virasoro generators. For the infinitesimal conformal transformation

$$\delta z = \varepsilon(z) \quad \text{and} \quad \delta \bar{z} = \tilde{\varepsilon}(\bar{z}), \tag{3.25}$$

the associated conserved charge that generates this transformation is

$$Q = Q_\varepsilon + Q_{\tilde{\varepsilon}} = \frac{1}{2\pi i} \oint \left[T(z)\varepsilon(z)dz + \tilde{T}(\bar{z})\tilde{\varepsilon}(\bar{z})d\bar{z} \right]. \tag{3.26}$$

The integral is performed over a circle of fixed radius. The variation of a field $\Phi(z,\bar{z})$ under a conformal transformation is then given by

$$\delta_\varepsilon \Phi(z,\bar{z}) = [Q_\varepsilon, \Phi(z,\bar{z})] \quad \text{and} \quad \delta_{\tilde{\varepsilon}} \Phi(z,\bar{z}) = [Q_{\tilde{\varepsilon}}, \Phi(z,\bar{z})]. \tag{3.27}$$

Conformal fields and operator product expansions

The fields of a conformal field theory are characterized by their conformal dimensions, which specify how they transform under scale transformations. Φ is called a *conformal field* (also sometimes called a *primary field*) of conformal dimension $(h, \tilde{h})$ if

$$\Phi(z,\bar{z}) \to \left(\frac{\partial w}{\partial z}\right)^h \left(\frac{\partial \bar{w}}{\partial \bar{z}}\right)^{\tilde{h}} \Phi(w,\bar{w}) \tag{3.28}$$

under finite conformal transformations $z \to w(z)$. In other words, the $(h, \tilde{h})$ differential

$$\Phi(z,\bar{z})(dz)^h (d\bar{z})^{\tilde{h}} \tag{3.29}$$

is invariant under conformal transformations.

Equations (3.26) and (3.27) give

$$\delta_\varepsilon \Phi(w,\bar{w}) = \frac{1}{2\pi i} \oint dz\, \varepsilon(z) [T(z), \Phi(w,\bar{w})]. \tag{3.30}$$

This expression is somewhat formal, since we still have to specify the integration contour. The operator products $T(z)\Phi(w,\bar{w})$ and $\Phi(w,\bar{w})T(z)$ only have convergent series expansions for radially ordered operators. This means that the integral $\oint dz\, \varepsilon(z) T(z) \Phi(w,\bar{w})$ should be evaluated along a contour with $|z| > |w|$. This is the first contour displayed in Fig. 3.2. Similarly, $\oint dz\, \varepsilon(z) \Phi(w,\bar{w}) T(z)$ should be evaluated along a contour with $|z| < |w|$.[4]

---

[4] The point is that matrix elements of these products have convergent mode expansions when these inequalities are satisfied. The results can then be analytically continued to other regions.

3.1 Conformal field theory

This is the second contour in Fig. 3.2. The difference of these two expressions, which gives the commutator, corresponds to a z contour that encircles the point w.

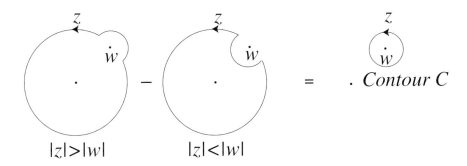

Fig. 3.2. Integration contour for the z integral in Eq. (3.30). Since the integrand is radially ordered, the z integral is performed on a small path encircling w.

Evaluation of this contour integral only requires knowing the singular terms in the operator product expansion (OPE) for $z \to w$. If the singularities are poles, all is well. The general idea is that a product of local operators in a quantum field theory defined at nearby locations (compared to any other operators) can be expanded in a series of local operators at one of their positions (or any other nearby position). In doing this, it is customary to write the terms that are most singular first, the next most singular second, and so forth. For our purposes, the terms that diverge as $z \to w$ are all that are required, and the rest of the terms are represented by dots. Sometimes the term that is finite in the limit is also of interest.

Equation (3.28) describes the transformation behavior of $\Phi(w, \bar{w})$ under conformal transformations. Infinitesimally, it becomes

$$\delta_\varepsilon \Phi(w, \bar{w}) = h\, \partial \varepsilon(w) \Phi(w, \bar{w}) + \varepsilon(w) \partial \Phi(w, \bar{w}), \tag{3.31}$$

$$\delta_{\tilde{\varepsilon}} \Phi(w, \bar{w}) = \tilde{h}\, \bar{\partial} \tilde{\varepsilon}(\bar{w}) \Phi(w, \bar{w}) + \tilde{\varepsilon}(\bar{w}) \bar{\partial} \Phi(w, \bar{w}). \tag{3.32}$$

Requiring that the charge Q induce these transformations determines the short-distance singularities in the OPE of T and $\tilde{T}$ with Φ

$$T(z) \Phi(w, \bar{w}) = \frac{h}{(z-w)^2} \Phi(w, \bar{w}) + \frac{1}{z-w} \partial \Phi(w, \bar{w}) + \ldots, \tag{3.33}$$

$$\tilde{T}(\bar{z}) \Phi(w, \bar{w}) = \frac{\tilde{h}}{(\bar{z}-\bar{w})^2} \Phi(w, \bar{w}) + \frac{1}{\bar{z}-\bar{w}} \bar{\partial} \Phi(w, \bar{w}) + \ldots \tag{3.34}$$

The dots represent nonsingular terms. These short-distance expansions determine the quantum energy–momentum tensor.

A free scalar field, such as $X^\mu(z)$, is a conformal field with $h = 0$. However, its OPE with itself is not meromorphic

$$X^\mu(z)X^\nu(w) = -\frac{1}{4}\eta^{\mu\nu} \ln(z-w) + \ldots \tag{3.35}$$

The field $\partial X^\mu(z)$, which is a conformal field of dimension $(1,0)$, has meromorphic OPEs with itself as well as with $X^\nu(w)$.

Recall that ∂X is the conformal field that enters in the energy–momentum tensor, where it gives a contribution $-2 : \partial X \cdot \partial X :$. The dots were defined in Chapter 2 to mean normal-ordering of the oscillators. An equivalent, but more elegant, viewpoint is that the dots represent removing the singular part as follows:

$$: \partial X^\mu(z) \partial X^\nu(z) := \lim_{w \to z} \left(\partial_z X^\mu(z) \partial_w X^\nu(w) + \frac{\eta^{\mu\nu}}{4(z-w)^2} \right). \tag{3.36}$$

These dots are sometimes omitted when the meaning is clear. Each such scalar field gives a contribution of 1 to the conformal anomaly c. So in D dimensions the X^μ coordinates give $c = \bar{c} = D$.

The OPE of the energy–momentum tensor with itself is

$$T(z)T(w) = \frac{c/2}{(z-w)^4} + \frac{2}{(z-w)^2}T(w) + \frac{1}{z-w}\partial T(w) + \ldots \tag{3.37}$$

Note that the energy–momentum tensor is not a conformal field unless $c = 0$. In that case $T(z)$ has dimension $(2,0)$ and $\widetilde{T}(\bar{z})$ has dimension $(0,2)$. Using the OPE in Eq. (3.37), it is possible to derive how the energy–momentum tensor transforms under a finite conformal transformation $z \to w(z)$. The result is

$$(\partial w)^2 T'(w) = T(z) - \frac{c}{12}S(w,z), \tag{3.38}$$

where

$$S(w,z) = \frac{2(\partial w)(\partial^3 w) - 3(\partial^2 w)^2}{2(\partial w)^2} \tag{3.39}$$

is called the *Schwarzian derivative*. $T'(w)$ denotes the transformed energy–momentum tensor.

Another important example of a conformal field is a free fermi field $\psi(z)$, which has $h = (1/2, 0)$ and the OPE

$$\psi(z)\psi(w) = \frac{1}{z-w}. \tag{3.40}$$

Such fields play an important role in the next chapter. A free fermi field has

$$T(z) = -\frac{1}{2} : \psi(z)\partial\psi(z) : \qquad (3.41)$$

which leads to $c = 1/2$.

The fact that a pair of fermi fields gives $c = 1$ is significant. When a free scalar field takes values on a circle of suitable radius, there is an equivalent theory in which the scalar field is replaced by a pair of fermi fields. The replacement of a boson by a pair of fermions is called *fermionization*, and its (more common) inverse is called *bosonization*. It is not our purpose to explore this in detail here, just to point out that the central charges match up. In fact, in the simplest case the formulas take the form[5]

$$\psi_\pm = : \exp(\pm i\phi) : . \qquad (3.42)$$

Here ϕ is a boson normalized in the usual way, so that the normal-ordered operator has dimension $1/2$. Clearly, for this expression to be single-valued, ϕ should have period 2π.

Given a holomorphic primary field $\Phi(z)$ of dimension h, one can associate a state $|\Phi\rangle$ that satisfies

$$L_0|\Phi\rangle = h|\Phi\rangle \quad \text{and} \quad L_n|\Phi\rangle = 0, \ n > 0. \qquad (3.43)$$

Such a state is called a *highest-weight state*. This *state–operator correspondence* is another very useful concept in conformal field theory. The relevant definition is

$$|\Phi\rangle = \lim_{z \to 0} \Phi(z)|0\rangle, \qquad (3.44)$$

where $|0\rangle$ denotes the conformal vacuum. Recall that $z = 0$ corresponds to the infinite past in Euclidean time. Writing a mode expansion

$$\Phi(z) = \sum_{n=-\infty}^{\infty} \frac{\Phi_n}{z^{n+h}}, \qquad (3.45)$$

the way this works is that

$$\Phi_n|0\rangle = 0 \quad \text{for} \quad n > -h \quad \text{and} \quad \Phi_{-h}|0\rangle = |\Phi\rangle. \qquad (3.46)$$

A highest-weight state $|\Phi\rangle$, taken together with the infinite collection of states of the form

$$L_{-n_1} L_{-n_2} \ldots L_{-n_k}|\Phi\rangle, \qquad (3.47)$$

---

[5] Strictly speaking, the right-hand side of this equation should contain another factor called a cocycle. However, this can often be ignored.

which are known as the *descendant states*, gives a representation of the (holomorphic) Virasoro algebra known as a *Verma module*.

Highest-weight states appeared in Chapter 2, where we learned that the physical open-string states of the bosonic string theory satisfy

$$(L_0 - 1)|\phi\rangle = 0 \tag{3.48}$$

and

$$L_n|\phi\rangle = 0 \quad \text{with} \quad n > 0. \tag{3.49}$$

Therefore, physical open-string states of the bosonic string theory correspond to highest-weight states with $h = 1$. This construction has a straightforward generalization to primary fields $\Phi(z, \bar{z})$ of dimension $(h, \tilde{h})$. In this case one has

$$(L_0 - h)|\Phi\rangle = (\tilde{L}_0 - \tilde{h})|\Phi\rangle = 0 \tag{3.50}$$

and

$$L_n|\Phi\rangle = \tilde{L}_n|\Phi\rangle = 0 \quad \text{with} \quad n > 0. \tag{3.51}$$

Therefore, physical closed-string states of the bosonic string theory correspond to highest-weight states with $h = \tilde{h} = 1$.

Kac–Moody algebras

Particularly interesting examples of conformal fields are the two-dimensional currents $J_\alpha^A(z, \bar{z})$, $A = 1, 2, \ldots, \dim G$, associated with a compact Lie group symmetry G in a conformal field theory. Current conservation implies that there is a holomorphic component $J^A(z)$ and an antiholomorphic component $\tilde{J}^A(\bar{z})$, just as was shown for T earlier. Let us consider the holomorphic current $J^A(z)$ only. The zero modes J_0^A are the generators of the Lie algebra of G with

$$[J_0^A, J_0^B] = if^{AB}{}_C J_0^C. \tag{3.52}$$

The algebra of the currents $J^A(z)$ is an infinite-dimensional extension of this, known as an *affine Lie algebra* or a *Kac–Moody algebra* $\widehat{G}$. These currents have conformal dimension $h = 1$, and therefore the mode expansion is

$$J^A(z) = \sum_{n=-\infty}^{\infty} \frac{J_n^A}{z^{n+1}} \quad A = 1, 2, \ldots, \dim G. \tag{3.53}$$

The Kac–Moody algebra is given by the OPE

$$J^A(z)J^B(w) \sim \frac{k\delta^{AB}}{2(z-w)^2} + \frac{if^{AB}{}_C J^C(w)}{z-w} + \ldots \tag{3.54}$$

or the equivalent commutation relations

$$[J_m^A, J_n^B] = \frac{1}{2} km \delta^{AB} \delta_{m+n,0} + i f^{AB}{}_C J_{m+n}^C. \qquad (3.55)$$

The parameter k in the Kac–Moody algebra, called the *level*, is analogous to the parameter c in the Virasoro algebra. For a $U(1)$ Kac–Moody algebra, $\widehat{U}(1)$, it can be absorbed in the normalization of the current. However, for a nonabelian group G, it has an absolute meaning once the normalization of the structure constants is specified. The energy–momentum tensor associated with an arbitrary Kac–Moody algebra is

$$T(z) = \frac{1}{k + \tilde{h}_G} \sum_{A=1}^{\dim G} : J^A(z) J^A(z) :, \qquad (3.56)$$

where the *dual Coxeter number* $\tilde{h}_G$ takes the value $n+1$ for $A_n = SU(n+1)$, $2n-1$ for $B_n = SO(2n+1)$ – except that it is 2 for $SO(3)$, $n+1$ for $C_n = Sp(2n)$, $2n-2$ for $D_n = SO(2n)$, 4 for G_2, 9 for F_4, 12 for E_6, 18 for E_7, and 30 for E_8. In the case of *simply-laced Lie groups*[6] the dual Coxeter number $\tilde{h}_G$ is equal to c_A, the *quadratic Casimir number* of the adjoint representation, which is defined (with our normalization conventions) by

$$f^{BC}{}_D f^{B'D}{}_C = c_A \delta^{BB'}. \qquad (3.57)$$

The central charge associated with this energy–momentum tensor is

$$c = \frac{k \dim G}{k + \tilde{h}_G}. \qquad (3.58)$$

For example, in the case of $\widehat{SU}(2)_k$, $\tilde{h}_G = 2$ and $c = 3k/(k+2)$.

Kac–Moody algebra representations of conformal symmetry are unitary if G is compact and the level k is a positive integer. These symmetry structures can be realized in *Wess–Zumino–Witten models*, which are σ models having the group manifold as target space.

Coset-space theories

Suppose that the Kac–Moody algebra $\widehat{G}_k$ has a subalgebra $\widehat{H}_l$. The level l is determined by the embedding of H in G. For example, if the simple roots of H are a subset of the simple roots of G, then $l = k$. If the Kac–Moody algebra is a direct product of the form $\widehat{G}_{k_1} \times \widehat{G}_{k_2}$ and $\widehat{H}_l$ is the diagonal subgroup, then $l = k_1 + k_2$. Let us denote the corresponding

---

[6] By definition, these are the Lie groups all of whose nonzero roots have the same length. They are the groups that are labeled by A, D, E in the Cartan classification.

energy–momentum tensors by $T_G(z)$ and $T_H(z)$. Now consider the difference of the two energy–momentum tensors

$$T(z) = T_G(z) - T_H(z). \qquad (3.59)$$

The modes of $T(z)$ are $L_m = L_m^G - L_m^H$. The nontrivial claim is that this difference defines an energy–momentum tensor, and therefore it gives a representation of the conformal group. If this is true, it is obviously unitary, since it is realized on a subspace of the positive-definite representation space of $\widehat{G}_k$.

The key to proving that $T(z)$ satisfies the Virasoro algebra is to show that it commutes with the currents that generate $\widehat{H}_l$. These currents $J^a(z)$, $a = 1, 2, \ldots, \dim H$, are a subset of the currents of $\widehat{G}_k$ and therefore have conformal dimension $h = 1$ with respect to T_G. In other words,

$$T_G(z) J^a(w) \sim \frac{J^a(w)}{(z-w)^2} + \frac{\partial J^a(w)}{z-w} + \ldots \qquad (3.60)$$

However, since they are also currents of $\widehat{H}_l$,

$$T_H(z) J^a(w) \sim \frac{J^a(w)}{(z-w)^2} + \frac{\partial J^a(w)}{z-w} + \ldots \qquad (3.61)$$

Taking the difference of these equations,

$$T(z) J^a(w) \sim O(1), \qquad (3.62)$$

or, in terms of modes, $[L_m, J_n^a] = 0$. Since $T_H(z)$ is constructed entirely out of the $\dim H$ currents $J^a(z)$, it follows that

$$T(z) T_H(w) \sim O(1), \qquad (3.63)$$

or, in terms of modes,

$$[L_m, L_n^H] = [L_m^G - L_m^H, L_n^H] = 0. \qquad (3.64)$$

Using this, together with the identity

$$[L_m, L_n] = [L_m^G, L_n^G] - [L_m^H, L_n^H] - [L_m^H, L_n] - [L_m, L_n^H], \qquad (3.65)$$

one finds that

$$[L_m, L_n] = (m-n) L_{m+n} + \frac{c}{12}(m^3 - m)\delta_{m+n,0}, \qquad (3.66)$$

where the central charge of $T(z)$ is

$$c = c_G - c_H. \qquad (3.67)$$

3.1 Conformal field theory

An immediate generalization of the construction above is for $\widehat{G}$ to be semisimple, that is, of the form

$$(\widehat{G}_1)_{k_1} \times (\widehat{G}_2)_{k_2} \times \ldots \times (\widehat{G}_n)_{k_n}. \tag{3.68}$$

As a specific example, consider the coset model given by

$$\frac{\widehat{SU}(2)_k \times \widehat{SU}(2)_l}{\widehat{SU}(2)_{k+l}}, \tag{3.69}$$

where the diagonal embedding is understood. This defines a chiral algebra with central charge

$$c = \frac{3k}{k+2} + \frac{3l}{l+2} - \frac{3(k+l)}{k+l+2}. \tag{3.70}$$

Minimal models

An interesting problem is the classification of all unitary representations of the conformal group. Since the group is infinite-dimensional this is rather nontrivial, and the complete answer is not known. A necessary requirement for a unitary representation is that $c > 0$. There is an infinite family of representations with $c < 1$, called minimal models, which have a central charge

$$c = 1 - \frac{6(p'-p)^2}{pp'}, \tag{3.71}$$

where p and p' are coprime positive integers (with $p' > p$) that characterize the minimal model. The minimal models are only unitary if $p' = p + 1 = m + 3$, so that

$$c = 1 - \frac{6}{(m+2)(m+3)} \quad m = 1, 2, \ldots \tag{3.72}$$

The explicit construction of unitary representations with these central charges (due to Goddard, Kent and Olive) uses the coset-space method of the preceding section.

Consider the coset model

$$\frac{\widehat{SU}(2)_1 \otimes \widehat{SU}(2)_m}{\widehat{SU}(2)_{m+1}}, \tag{3.73}$$

corresponding to Eq. (3.69) with $l = 1$. The central charge of the associated energy–momentum tensor $T(z)$ is

$$c = 1 + \frac{3m}{m+2} - \frac{3(m+1)}{m+3} = 1 - \frac{6}{(m+2)(m+3)}, \tag{3.74}$$

which is the desired result. The first nontrivial case is $m = 1$, which has $c = 1/2$. It has been proved that these are all of the unitary representations of the Virasoro algebra with $c < 1$.

To understand the structure of these unitary minimal models, one should also determine all of their highest-weight states. Equivalently, one can identify the primary fields that give rise to the highest-weight states by acting on the conformal vacuum $|0\rangle$. Since $|0\rangle$, itself, is a highest-weight state, the identity operator I is a primary field (with $h = 0$). Using the known $\widehat{SU}(2)_k$ representations, one can work out all of the primary fields of these minimal models. The result is a collection of conformal fields ϕ_{pq} with conformal dimensions h_{pq} given by

$$h_{pq} = \frac{[(m+3)p - (m+2)q]^2 - 1}{4(m+2)(m+3)}, \quad 1 \le p \le m+1 \quad \text{and} \quad 1 \le q \le p. \tag{3.75}$$

Because of the symmetry $(p, q) \to (m + 2 - p, m + 3 - q)$, an equivalent labeling is to allow $1 \le p \le m+1$, $1 \le q \le m+2$ and to restrict $p - q$ to even values. For example, the $m = 1$ theory, with $c = 1/2$, describes the two-dimensional Ising model at the critical point. It has primary fields with dimensions $h_{11} = 0$ (the identity operator), $h_{21} = 1/2$ (a free fermion), and $h_{22} = 1/16$ (a spin field).

Note that the minimal models have $c < 1$ and accumulate at $c = 1$. This limiting value $c = 1$ can be realized by a free boson X. There are actually a continuously infinite number of possibilities for $c = 1$ unitary representations, since the coordinate X can describe a circle of any radius.[7]

EXERCISES

EXERCISE 3.1

Use the oscillator expansion in Eq. (3.21) to derive the OPE:

$$\partial X^\mu(z) \partial X^\nu(w) = -\frac{1}{4} \frac{\eta^{\mu\nu}}{(z-w)^2} + \dots$$

SOLUTION

Since the singular part of the OPE of the two fields $\partial X^\mu(z)$ and $\partial X^\nu(w)$

[7] Chapter 6 shows that radius R and radius α'/R are equivalent.

is proportional to the identity operator, it can be determined by computing the correlation function

$$\langle \partial X^\mu(z) \partial X^\nu(w) \rangle = -\frac{1}{4} \sum_{m=-\infty}^{+\infty} \sum_{n=-\infty}^{+\infty} \langle 0 | \alpha_m^\mu \alpha_n^\nu | 0 \rangle z^{-m-1} w^{-n-1}.$$

Since the positive modes and the zero mode annihilate the vacuum on the right and the negative modes and the zero mode annihilate the vacuum on the left, this yields

$$-\frac{1}{4} \sum_{m,n=1}^{+\infty} \langle 0 | \alpha_m^\mu \alpha_{-n}^\nu | 0 \rangle z^{-m-1} w^{n-1} = -\frac{\eta^{\mu\nu}}{4} \sum_{m,n=1}^{+\infty} m \delta_{m,n} z^{-m-1} w^{n-1}$$

$$= -\frac{1}{4} \frac{\eta^{\mu\nu}}{(z-w)^2}.$$

Note that convergence requires $|w| < |z|$. □

EXERCISE 3.2
Derive the Virasoro algebra from Eq. (3.37), that is, from the OPE of the energy–momentum tensor with itself.

SOLUTION

The modes of the energy–momentum tensor are defined by

$$T(z) = \sum_{n=-\infty}^{+\infty} \frac{L_n}{z^{n+2}} \quad \text{or} \quad L_n = \oint \frac{dz}{2\pi i} z^{n+1} T(z),$$

where one uses Cauchy's theorem to invert the definition of the modes. The modes then satisfy

$$[L_m, L_n] = \left[\oint \frac{dz}{2\pi i} z^{m+1} T(z), \oint \frac{dw}{2\pi i} w^{n+1} T(w) \right].$$

One has to be a bit careful when defining the commutator of the above contour integrals. Let us do the z integral first while holding w fixed. When doing the z integral we assume that the integrand is radially ordered. As a result, the commutator is computed by considering the z integral along a small path encircling w (contour C in Fig. 3.2). Using Eq. (3.37), this gives

$$\oint \frac{dw}{2\pi i} w^{n+1} \oint_C \frac{dz}{2\pi i} z^{m+1} \left[\frac{c/2}{(z-w)^4} + \frac{2}{(z-w)^2} T(w) + \frac{1}{z-w} \partial T(w) + \ldots \right]$$

$$= \oint \frac{dw}{2\pi i} \left[\frac{c}{12}(m^3 - m)w^{m+n-1} + 2(m+1)w^{n+m+1}T(w) + w^{m+n+2}\partial T(w) \right].$$

Performing the integral over w on a path encircling the origin, yields the Virasoro algebra

$$[L_m, L_n] = (m-n)L_{m+n} + \frac{c}{12}(m^3 - m)\delta_{m+n,0}.$$

□

EXERCISE 3.3
Verify that the expressions (3.38) and (3.39) for the transformation of the energy–momentum tensor under conformal transformations are consistent with Eq. (3.37) for an infinitesimal transformation $w(z) = z + \varepsilon(z)$.

SOLUTION

Under the infinitesimal transformation $f(z) = z + \varepsilon(z)$, Eqs (3.38) and (3.39) reduce to $T(z) \to T(z) + \delta_\varepsilon T(z)$ with

$$\delta_\varepsilon T(z) = -2\partial \varepsilon(z) T(z) - \varepsilon(z)\partial T(z) - \frac{c}{12}\partial^3 \varepsilon(z).$$

On the other hand, using Eq. (3.30), the change of $T(w)$ under an infinitesimal conformal transformation is given by

$$\delta_\varepsilon T(w) = \oint \frac{dz}{2\pi i} \varepsilon(z)[T(z), T(w)] = \oint_C \frac{dz}{2\pi i} \varepsilon(z) T(z) T(w),$$

where the integration contour C is the one displayed in Fig. 3.2. Using Eq. (3.37), this becomes

$$\oint_C \frac{dz}{2\pi i} \varepsilon(z) \left[\frac{c/2}{(z-w)^4} + \frac{2T(w)}{(z-w)^2} + \frac{\partial T(w)}{z-w} \right]$$

$$= 2\partial \varepsilon(w) T(w) + \varepsilon(w) \partial T(w) + \frac{c}{12}\partial^3 \varepsilon(w).$$

But $\delta_\varepsilon T(w) = -\delta_\varepsilon T(z)$, since $z \sim w - \varepsilon(w)$. This shows that both methods yield the same result for $\partial_\varepsilon T(z)$ to first order in ε. □

EXERCISE 3.4
Show that Eqs (3.38) and (3.39) satisfy the group property by considering two successive conformal transformations.

SOLUTION

After two successive conformal transformations $w(u(z))$, one finds

$$(\partial w)^2 T(w) = T(z) - \frac{c}{12}S(u,z) - \frac{c}{12}(\partial u)^2 S(w,u),$$

where $\partial = \partial/\partial z$. In order to prove the group property, we need to verify that

$$S(w,z) = S(u,z) + (\partial u)^2 S(w,u).$$

This can be shown by substituting

$$\frac{dw}{du} = \left(\frac{du}{dz}\right)^{-1}\frac{dw}{dz} = \frac{w'}{u'}$$

and the corresponding expressions for the higher-order derivatives

$$\frac{d^2 w}{du^2} = \frac{w''u' - w'u''}{(u')^3}$$

$$\frac{d^3 w}{du^3} = \frac{w'''(u')^2 - 3w''u''u' - w'u'''u' + 3w'(u'')^2}{(u')^5}$$

into $S(w,u)$. □

3.2 BRST quantization

An interesting type of conformal field theory appears in the BRST analysis of the path integral.

In the Faddeev–Popov analysis of the path integral the choice of conformal gauge results in a Jacobian factor that can be represented by the introduction of a pair of fermionic ghost fields, called b and c, with conformal dimensions 2 and -1, respectively.[8] For these choices the b ghost transforms the same way as the energy–momentum tensor, and the c ghost transforms the same way as the gauge parameter.

These ghosts are a special case of the following set-up. A pair of holomorphic ghost fields $b(z)$ and $c(z)$, with conformal dimensions λ and $1-\lambda$, respectively, have an OPE

$$c(z)b(w) = \frac{1}{z-w} + \ldots \quad \text{and} \quad b(z)c(w) = \frac{\varepsilon}{z-w} + \ldots, \quad (3.76)$$

while $c(z)c(w)$ and $b(z)b(w)$ are nonsingular. The choice $\varepsilon = +1$ is made

---
[8] For details about the Faddeev–Popov gauge-fixing procedure we refer the reader to volume 1 of GSW or Polchinski.

when b and c satisfy fermi statistics, and the choice $\varepsilon = -1$ is made when they satisfy bose statistics. The conformal dimensions λ and $1-\lambda$ correspond to a contribution to the energy–momentum tensor of the form

$$T_{bc}(z) = -\lambda : b(z)\partial c(z) + \varepsilon(\lambda - 1) : c(z)\partial b(z) : . \tag{3.77}$$

This in turn implies a conformal anomaly

$$c(\varepsilon, \lambda) = -2\varepsilon(6\lambda^2 - 6\lambda + 1). \tag{3.78}$$

For the bosonic string theory, there is a single pair of ghosts (associated with reparametrization invariance) satisfying $\varepsilon = 1$ and $\lambda = 2$. Thus $c^{\text{gh}} = -26$ in this case, and the conformal anomaly from all other sources must total $+26$ in order to cancel the conformal anomaly. For example, 26 space-time coordinates X^μ, the choice made in the previous chapter, is a possibility.

One may saturate the central-charge condition in other ways. In critical string theories one chooses $D \leq 26$ space-time dimensions, and then adjoins a unitary CFT with $c = 26 - D$ to make up the rest of the required central charge. This CFT need not have a geometric interpretation. Nevertheless, it gives a consistent string theory (ignoring the usual problem of the tachyon). An alternative way of phrasing this is to say that such a construction gives another consistent quantum vacuum of the (unique) bosonic string theory. Without knowing the final definitive formulation of string theory, which is still lacking, it is not always clear when one has a new theory as opposed to a new vacuum of an old theory.

Chapter 4 considers theories with $\mathcal{N} = 1$ superconformal symmetry. For such theories the choice of superconformal gauge gives an additional pair of bosonic ghost fields with $\varepsilon = -1$ and $\lambda = 3/2$. Since $c(-1, 3/2) = 11$, the total ghost contribution to the conformal anomaly in this case is $c^{\text{gh}} = -26 + 11 = -15$. This must again be balanced by other contributions. For example, ten-dimensional space-time with a fermionic partner ψ^μ for each space-time coordinate X^μ gives $c = 10 \cdot (1 + 1/2) = 15$.

Let us now specialize to the bosonic string in 26 dimensions including the fermionic ghosts. The quantum world-sheet action of the gauge-fixed theory is

$$S_{\text{q}} = \frac{1}{2\pi} \int \left(2\partial X^\mu \bar{\partial} X_\mu + b\bar{\partial}c + \tilde{b}\partial\tilde{c} \right) d^2 z, \tag{3.79}$$

and the associated energy–momentum tensor is

$$T(z) = T_X(z) + T_{bc}(z), \tag{3.80}$$

where T_X is given in Eq. (3.23) and

$$T_{bc}(z) = -2 : b(z)\partial c(z) : + : c(z)\partial b(z) : . \qquad (3.81)$$

The quantum action has no conformal anomaly, because the OPE of T with itself has no central-charge term. The contribution of the ghosts cancels the contribution of the X coordinates.

The quantum action in Eq. (3.79) has a *BRST symmetry*, which is a global fermionic symmetry, given by

$$\begin{aligned} \delta X^\mu &= \eta c \partial X^\mu, \\ \delta c &= \eta c \partial c, \\ \delta b &= \eta T. \end{aligned} \qquad (3.82)$$

Most authors do not display the constant infinitesimal Grassmann parameter η. One reason for doing so is to keep track of minus signs that arise when anticommuting fermionic expressions past one another. There is also a complex-conjugate set of transformations that is not displayed.

The BRST charge that generates the transformations (3.82) is

$$Q_B = \frac{1}{2\pi i} \oint (cT_X + : bc\partial c :) \, dz. \qquad (3.83)$$

The integrand is only determined up to a total derivative, so a term proportional to $\partial^2 c$, which appears in the BRST current, can be omitted. In particular, this operator solves the equation

$$\{Q_B, b(z)\} = T(z), \qquad (3.84)$$

which is the quantum version of $\delta b = \eta T(z)$. There is also a conjugate BRST charge $\tilde{Q}_B$ given by complex conjugation. In terms of modes, the BRST charge has the expansion

$$Q_B = \sum_{m=-\infty}^{\infty} (L_{-m}^{(X)} - \delta_{m,0})c_m - \frac{1}{2} \sum_{m,n=-\infty}^{\infty} (m-n) : c_{-m}c_{-n}b_{m+n} : . \qquad (3.85)$$

Note the appearance of the combination $L_0 - 1$, the same combination that gives the mass-shell condition, in the coefficient of c_0.

Another useful quantity is *ghost number*. One assigns ghost number $+1$ to c, ghost number -1 to b and ghost number 0 to X^μ. This is an additive global symmetry of the quantum action, so there is a corresponding conserved ghost-number current and ghost-number charge. Thus, if one starts with a Fock-space state of a certain ghost number and acts on it with various oscillators, the ghost number of the resulting state is the initial ghost number plus the number of c-oscillator excitations minus the number of b-oscillator excitations.

The BRST charge has an absolutely crucial property. It is *nilpotent*, which means that

$$Q_B^2 = 0. \tag{3.86}$$

Some evidence in support of this result is the vanishing of iterated field variations (3.82). However, this test, while necessary, is not sufficiently refined to pick up terms that are beyond leading order in the α' expansion. Thus, it cannot distinguish between L_0 and $L_0 - 1$ or establish the necessity of 26 dimensions. This can be verified directly using the oscillator expansion, though the calculation is very tedious. A somewhat quicker method is to anticommute two of the integral representations using the various OPEs and using Cauchy's theorem to evaluate the contributions of the poles, though even this is a certain amount of work.

A complete proof of nilpotency that avoids difficult algebra goes as follows. Consider the identity

$$\{[Q_B, L_m], b_n\} = \{[L_m, b_n], Q_B\} + [\{b_n, Q_B\}, L_m]. \tag{3.87}$$

Using $[L_m, b_n] = (m-n)b_{m+n}$, $\{b_n, Q_B\} = L_n - \delta_{n,0}$ and the Virasoro algebra, one finds that the right-hand side vanishes for central charge $c = 0$. Thus $[Q_B, L_m]$ cannot contain any c-ghost modes. However, it has ghost number (the number of c modes minus the number of b modes) equal to 1, so this implies that it must vanish. Thus $c = 0$ implies that Q_B is conformally invariant. Next, consider the identity

$$[Q_B^2, b_n] = [Q_B, \{Q_B, b_n\}] = [Q_B, L_n]. \tag{3.88}$$

If Q_B is conformally invariant, the right-hand side vanishes. This implies that Q_B^2 has no c-ghost modes. Since it has ghost number equal to 2, this implies that it must vanish. Putting these facts together leads to the conclusion that Q_B is nilpotent if and only if $c = 0$.

Recall that the oscillators that arise in the mode expansions of the X^μ coordinates give a Fock space that includes many unphysical states including ones of negative norm, and it is necessary to impose the Virasoro constraints to define the subspace of physical states. Given this fact, the reader may wonder why it represents progress to add even more oscillators, the modes of the b and c ghost fields. This puzzle has a very beautiful answer.

The key is to focus on the nilpotency equation $Q_B^2 = 0$. It has the same mathematical structure as the equation satisfied by the exterior derivative in differential geometry $d^2 = 0$.[9] In that case one considers various types of differential forms ω. Ones that satisfy $d\omega = 0$ are called closed, and ones that can be written in the form $\omega = d\rho$ are called exact. Nilpotency of d implies that every exact form is closed. If there are closed forms that are not exact, this encodes topological information about the manifold $\mathcal{M}$ on which the differential forms are defined. One defines equivalence classes of closed forms by declaring two closed forms to be equivalent if and only if their difference is exact. These equivalence classes then define elements of the cohomology of $\mathcal{M}$. More specifically, an equivalence class of closed n-forms is an element of the nth cohomology group $H^n(\mathcal{M})$.

The idea is now clear. Physical string states are identified as BRST cohomology classes. Thus, in the enlarged Fock space that includes the b and c oscillators in addition to the α oscillators, one requires that a physical on-shell string state is annihilated by the operator Q_B, that is, it is BRST closed. Furthermore, if the difference of two BRST-closed states is BRST exact, so that it is given as Q_B applied to some state, then the two BRST-closed states represent the same physical state. In the case of closed strings, this applies to the holomorphic and antiholomorphic sectors separately.

Because of the ghost zero modes, b_0 and c_0, the ground state is doubly degenerate. Denoting the two states by $|\uparrow\rangle$ and $|\downarrow\rangle$, $c_0|\downarrow\rangle = |\uparrow\rangle$ and $b_0|\uparrow\rangle = |\downarrow\rangle$. Also, $c_0|\uparrow\rangle = b_0|\downarrow\rangle = 0$. The ghost number assigned to one of these two states is a matter of convention. The other is then determined. The most symmetrical choice is to assign the values $\pm 1/2$, which is what we do. This resolves the ambiguity of a constant in the ghost-number operator

$$U = \frac{1}{2\pi i} \oint : c(z) b(z) : dz = \frac{1}{2}(c_0 b_0 - b_0 c_0) + \sum_{n=1}^{\infty} (c_{-n} b_n - b_{-n} c_n). \quad (3.89)$$

Which one of the two degenerate ground states corresponds to the physical ground state (the tachyon)? The fields b and c are not on a symmetrical footing, so there is a definite answer, namely $|\downarrow\rangle$, as will become clear shortly. The definition of physical states can now be made precise: they correspond to BRST cohomology classes with ghost number equal to $-1/2$. In the case of open strings, this is the whole story. In the case of closed strings, this construction has to be carried out for the holomorphic (right-moving) and antiholomorphic (left-moving) sectors separately. The two sectors are then tensored with one another in the usual manner.

To make contact with the results of Chapter 2, let us construct a unique

---

[9] This is the proper analogy for open strings. In the case of closed strings, the better analogy relates Q_B and $\widetilde{Q}_B$ to the holomorphic and antiholomorphic differential operators ∂ and $\bar{\partial}$ of complex differential geometry.

representative of each cohomology class. A simple choice is given by the α oscillators and Virasoro constraints applied to the ground state $|\downarrow\rangle$. The way to achieve this is to select states $|\phi\rangle$ that satisfy $b_n|\phi\rangle = 0$ for $n = 0, 1, \ldots$ Note that this implies, in particular, that $|\downarrow\rangle$ is physical and $|\uparrow\rangle$ is not. Then the Virasoro constraints and the mass-shell condition follow from $Q_B|\phi\rangle = 0$ combined with $\{Q_B, b_n\} = L_n - \delta_{n,0}$. Note that $b_n|\phi\rangle = 0$ implies that $|\phi\rangle$ can contain no c-oscillator excitations. Then the ghost-number requirement excludes b-oscillator excitations as well. So these representatives precisely correspond to the physical states constructed in Chapter 2.

It was mentioned earlier that a pair of fermion fields can be equivalent to a boson field on a circle of suitable radius. Let us examine this bosonization for the ghosts. The claim is that it is possible to introduce a scalar field $\phi(z)$ such that the energy–momentum tensors T_{bc} and T_ϕ can be equated:

$$-\frac{1}{2}(\partial\phi)^2 + \frac{3i}{2}\partial^2\phi = c(z)\partial b(z) - 2b(z)\partial c(z), \quad (3.90)$$

and similarly for the antiholomorphic fields. The coefficient of the term proportional to $\partial^2\phi$ is chosen so that the central charge is -26. In particular, for the zero mode Eq. (3.90) gives

$$\frac{1}{2}\phi_0^2 + \sum_{n=1}^{\infty}\phi_{-n}\phi_n - 1/8 = \sum_{n=1}^{\infty}n(b_{-n}c_n + c_{-n}b_n). \quad (3.91)$$

The $-1/8$ is the difference of the normal-ordering constants of the boson and the fermions. The ϕ oscillators satisfy $[\phi_m, \phi_n] = m\delta_{m+n,0}$, as usual. Also, ϕ_0 is identified with the ghost-number operator U, which is the zero mode of the relation $-i\partial\phi = cb$. Note that $\frac{1}{2}\phi_0^2 - 1/8 = 0$ for ghost number $\pm 1/2$. More generally, $U = \phi_0$ takes values in $\mathbb{Z} + 1/2$. The integer spacing determines the periodicity of ϕ to be 2π, and the half-integer offset means that string wave functions must be antiperiodic in their ϕ dependence

$$\Psi(\phi(\sigma) + 2\pi) = -\Psi(\phi(\sigma)). \quad (3.92)$$

EXERCISES

EXERCISE 3.5
Show that the integrand in Eq. (3.79) changes by a total derivative under the transformations (3.82).

SOLUTION

Under the global fermionic symmetry the integrand $\mathcal{L}$ changes by

$$\delta \mathcal{L} = 2 \partial \delta X \cdot \bar{\partial} X + 2 \partial X \cdot \bar{\partial} \delta X + \delta b \bar{\partial} c + b \bar{\partial} \delta c = \delta \mathcal{L}_1 + \delta \mathcal{L}_3,$$

where the index on $\delta \mathcal{L}$ counts the number of fermionic fields. Using Eqs (3.82) we obtain

$$\delta \mathcal{L}_1 = 2\eta \partial(c \partial X) \cdot \bar{\partial} X + 2\eta \partial X \cdot \bar{\partial}(c \partial X) + \eta T_X \bar{\partial} c = 2\eta \partial \left(c \partial X^\mu \bar{\partial} X_\mu \right)$$

and

$$\delta \mathcal{L}_3 = \eta T_{bc} \bar{\partial} c - \eta b \bar{\partial}(c \partial c) = -\eta \partial \left(bc \bar{\partial} c \right),$$

which are total derivatives since η is constant. The result for the complex-conjugate fields can be derived similarly. □

3.3 Background fields

Among the background fields, three that are especially significant are associated with massless bosonic fields in the spectrum. They are the metric $g_{\mu\nu}(X)$, the antisymmetric two-form gauge field $B_{\mu\nu}(X)$ and the dilaton field $\Phi(X)$. The metric appears as a background field in the term

$$S_g = \frac{1}{4\pi\alpha'} \int_M \sqrt{h}\, h^{\alpha\beta} g_{\mu\nu}(X) \partial_\alpha X^\mu \partial_\beta X^\nu d^2 z. \tag{3.93}$$

In Chapter 2 only flat Minkowski space-time with $(g_{\mu\nu} = \eta_{\mu\nu})$ was considered, but other geometries are also of interest.

The antisymmetric two-form gauge field $B_{\mu\nu}$ appears as a background field in the term[10]

$$S_B = \frac{1}{4\pi\alpha'} \int_M \varepsilon^{\alpha\beta} B_{\mu\nu}(X) \partial_\alpha X^\mu \partial_\beta X^\nu d^2 z. \tag{3.94}$$

This term is only present in theories of oriented bosonic strings. The projection onto strings that are invariant under reversal of orientation (a procedure called orientifold projection) eliminates the B field from the string spectrum. In cases when this term is present, it can be regarded as a two-form analog of the coupling S_A of a one-form Maxwell field to the world line of a charged particle,

$$S_A = q \int A_\mu \dot{x}^\mu d\tau. \tag{3.95}$$

---

[10] The antisymmetric tensor density $\varepsilon^{\alpha\beta}$ has components $\varepsilon^{01} = -\varepsilon^{10} = 1$ and $\varepsilon^{00} = \varepsilon^{11} = 0$. The combination $\varepsilon^{\alpha\beta}/\sqrt{h}$ transforms as a tensor.

So the strings of such theories are charged in this sense. This is explored further in later chapters.

The dilaton appears in a term of the form

$$S_\Phi = \frac{1}{4\pi} \int_M \sqrt{h}\, \Phi(X) R^{(2)}(h)\, d^2z, \qquad (3.96)$$

where $R^{(2)}(h)$ is the scalar curvature of the two-dimensional string world sheet computed from the world-sheet metric $h_{\alpha\beta}$. The dilaton term S_Φ is one order higher than S_g and S_B in the α' expansion, since it is lacking the two explicit factors of X that appear in S_g and S_B.

The role of the dilaton

The dilaton plays a crucial role in defining the string perturbation expansion. The special role of the dilaton term is most easily understood by considering the particular case in which Φ is a constant. More generally, if it approaches a constant at infinity, it is possible to separate this constant mode from the rest of Φ and focus on its contribution.

The key observation is that, when Φ is a constant, the integrand in Eq. (3.96) is a total derivative. This means that the value of the integral is determined by the global topology of the world sheet, and this term does not contribute to the classical field equations. The topological invariant that arises here is an especially famous one. Namely,

$$\chi(M) = \frac{1}{4\pi} \int_M \sqrt{h}\, R^{(2)}(h)\, d^2z \qquad (3.97)$$

is the *Euler characteristic* of M. It is related to the number of handles n_h, the number of boundaries n_b and the number of cross-caps n_c of the Euclidean world sheet M by

$$\chi(M) = 2 - 2n_h - n_b - n_c. \qquad (3.98)$$

The simplest example is the sphere, which has $\chi = 2$, since it has no handles, boundaries or cross-caps. $\chi = 1$ is achieved for a disk, which has one boundary and for a projective plane, which has one cross-cap. One can derive a projective plane from a disk by decreeing that opposite points on the boundary of the disk are identified as equivalent. There are four distinct topologies that can give $\chi = 0$. They are a torus (one handle), an annulus or cylinder (two boundaries), a Moebius strip (one boundary and one cross-cap), and a Klein bottle (two cross-caps).

There are several distinct classes of string-theory perturbation expansions, which are distinguished by whether the fundamental strings are oriented or unoriented and whether or not the theory contains open strings in addition to closed strings. All of these options can be considered as different versions of the bosonic string theory. In a string theory that contains only closed strings there can be no world-sheet boundaries, since these are created by the ends of open strings. Also, in a theory of oriented strings the world sheet is necessarily orientable, and this implies that there can be no cross-caps.

In the simplest and most basic class of string theories, the fundamental strings are closed and oriented, and there are no open strings. This possibility is especially important as it is the case for type II superstring theories and heterotic string theories in ten-dimensional Minkowski spacetime, which are discussed in subsequent chapters. For such theories the only possible string world-sheet topologies are closed and oriented Riemann surfaces, whose topologies are uniquely characterized by the genus n_h (the number of handles). The genus corresponds precisely to the number of string loops. One can visualize this by imagining a slice through the world sheet, which exposes a collection of closed strings that are propagating inside the diagram.

A nice feature of theories of closed oriented strings is that there is just one string theory Feynman diagram at each order of the perturbation expansion, since the Euler characteristic is uniquely determined by the genus. The enormous number of Feynman diagrams in the field theories that approximate these string theories at low energy corresponds to various possible degenerations (or singular limits) of these Riemann surfaces. Another marvelous fact is that at each order of the perturbation theory (that is, for each genus) these amplitudes have no ultraviolet (UV) divergences. Thus these string theories are UV finite theories of quantum gravity. As yet, no other approach to quantum gravity has been found that can achieve this.

Another important possibility is that the fundamental strings are unoriented and they can be open as well as closed. This is the situation for type I superstring theory. The fact that the strings are unoriented is ultimately attributable to the presence of an object called an orientifold plane. In a similar spirit, the fact that open strings are allowed can be traced to the presence of objects called *D-branes*. D-branes are physical objects on which strings can end, and the presence of D-branes implies that strings are breakable. Thus, for example, in the type I superstring theory one has to include all possible world sheets that have boundaries and cross-caps as well as handles. Clearly this is a more complicated story than in the cases without boundaries and cross-caps. Moreover, the cancellation of ultraviolet divergences for such theories is only achieved when all diagrams of the same Euler characteristic are (carefully) combined. The remainder of this section applies to theories that contain only oriented closed strings, so that the relevant Riemann surface topologies are characterized entirely by the genus n_h.

Effective potential and moduli fields

The dependence of a string theory on the background values of scalar fields can be characterized, at least at energies that are well below the string scale $1/l_\mathrm{s}$, by an effective potential $V_\mathrm{eff}(\phi)$, where ϕ now refers to all low-mass or zero-mass scalar fields, and one imagines that high-mass fields have been integrated out. String vacua correspond to local minima of this function. Such minima may be only metastable if tunneling to lower minima is possible.

In a nongravitational theory, an additive constant in the definition of V_eff would not matter. However, in a gravitational theory the value of V_eff at each of the minima determines the energy density in the corresponding vacuum. This energy density acts as a source of gravity and influences the geometry of the space-time. The value of V_eff at a minimum determines the cosmological constant for that vacuum. The measured value in our Universe is exceedingly small, $\Lambda \sim 10^{-120}$ in Planck units. As such, it is completely irrelevant to particle physics. However, it plays an important role in cosmology. Explaining the observed vacuum energy, or *dark energy*, is a major challenge that has been a research focus in recent years.

If the effective potential has an isolated minimum then the matrix of second derivatives determines the masses of all the scalar fields to be positive. If, on the other hand, there are flat directions, one or more eigenvalues of the matrix of second derivatives vanishes and some of the scalar fields are massless. The vacuum expectation values (or *vevs*) of those fields can be varied continuously while remaining at a minimum. In this case one has a continuous *moduli space of vacua* and one speaks of a *flat potential*. If there are no massless scalars in the real world, the true vacuum should be an isolated point rather than part of a continuum. This seems likely to be the case for a realistic vacuum, because scalars in string theory typically couple with (roughly) gravitational strength. The classical tests of general relativity establish that the long-range gravitational force is pure tensor, without a scalar component, to better than 1% precision. It is difficult to accommodate a massless scalar in string theory without violating this constraint. So one of the major challenges in string phenomenology is to construct isolated vacua without any moduli. This is often referred to as the problem of *moduli stabilization*, which is discussed in Chapter 10.

3.4 Vertex operators

Vertex operators V_ϕ are world-sheet operators that represent the emission or absorption of a physical on-shell string mode $|\phi\rangle$ from a specific point on the string world sheet. There is a one-to-one mapping between physical states and vertex operators. Since physical states are highest-weight states, the corresponding vertex operators are primary fields, and the problem of constructing them is the inverse of the problem discussed earlier in connection with the state–operator correspondence. In the case of an open string, the vertex operator must act on a boundary of the world sheet, whereas for a closed string it acts on the interior. Thus, summing over all possible insertion points gives an expression of the form $g_o \oint V_\phi(s)ds$ in the open-string case. The idea here is that the integral is over a boundary that is parametrized by a real parameter s. In the closed-string case one has $g_s \int V_\phi(z, \bar{z})d^2z$, which is integrated over the entire world sheet. In each case, the index ϕ is meant to label the specific state that is being emitted or absorbed (including its 26-momentum). There is a string coupling constant g_s that accompanies each closed-string vertex operator. The open-string coupling constant g_o is related to it by $g_o^2 = g_s$. To compensate for the integration measure, and give a coordinate-independent result, a vertex operator must have conformal dimension 1 in the open-string case and $(1, 1)$ in the closed-string case.

If the emitted particle has momentum k^μ, the corresponding vertex operator should contain a factor of $\exp(ik \cdot x)$. To give a conformal field, this should be extended to $\exp(ik \cdot X)$. However, this expression needs to be normal-ordered. Once this is done, there is a nonzero conformal dimension, which (in the usual units $l_s = \sqrt{2\alpha'} = 1$) is equal to $k^2/2$ in the open-string case and $(k^2/8, k^2/8)$ in the closed-string case. The relation between these two results can be understood by recalling that the left-movers and the right-movers each carry half of the momentum in the closed-string case. These results are exactly what is expected for the vertex operators of the respective tachyons. For other physical states, the vertex operator contains an additional factor of dimension n or (n, n), where n is a positive integer. Let us now explain the rule for constructing these factors.

A Fock-space state has the form

$$|\phi\rangle = \prod_i \alpha^{\mu_i}_{-m_i} \prod_j \tilde{\alpha}^{\nu_j}_{-n_j} |0; k\rangle, \tag{3.99}$$

or (more generally) a superposition of such terms. The vertex operator of the tachyon ground state is $\exp(ik \cdot X)$ (with normal-ordering implicit). In the following we describe how to modify the ground-state vertex operator to account for the α^μ_{-m} factors. To do this notice that the contour integral identity

$$\alpha^\mu_{-m} = \frac{1}{\pi} \oint z^{-m} \partial X^\mu dz \qquad (3.100)$$

suggests that we simply replace

$$\alpha^\mu_{-m} \to \frac{2i}{(m-1)!} \partial^m X^\mu, \qquad m > 0. \qquad (3.101)$$

This is not an identity, of course. The right-hand side contains α^μ_{-m} plus an infinite series of z-dependent terms with positive and negative powers. So, according to this proposal, a general closed-string vertex operator is given by an expression of the form

$$V_\phi(z, \bar z) = \, : \prod_i \partial^{m_i} X^{\mu_i}(z) \prod_j \bar\partial^{n_j} X^{\nu_j}(\bar z) e^{ik \cdot X(z,\bar z)} : , \qquad (3.102)$$

or a superposition of such terms, where

$$\frac{k^2}{8} = 1 - \sum_i m_i = 1 - \sum_j n_j. \qquad (3.103)$$

It is not at all obvious that this ensures that V_ϕ has conformal dimension $(1,1)$. In fact, this is only the case if the original Fock-space state satisfies the Virasoro constraints.

Vertex operators can also be introduced in the formalism with Faddeev–Popov ghosts. In this case the physical state condition is $Q_B|\phi\rangle = \widetilde Q_B|\phi\rangle = 0$. Physical states are BRST closed, but not exact. The corresponding statement for vertex operators is that if ϕ is BRST closed, then $[Q_B, V_\phi] = [\widetilde Q_B, V_\phi] = 0$. Similarly, if ϕ is BRST exact, then V_ϕ can be written as the anticommutator of Q_B or $\widetilde Q_B$ with some operator.

The operator correspondences for the ghosts are

$$b_{-m} \to \frac{1}{(m-2)!} \partial^{m-1} b, \qquad m \geq 2 \qquad (3.104)$$

and

$$c_{-m} \to \frac{1}{(m+1)!} \partial^{m+1} c, \qquad m \geq -1. \qquad (3.105)$$

These rules reflect the fact that b is dimension 2 and c is dimension -1. In particular, the unit operator is associated with a state that is annihilated by b_m with $m \geq -1$ and by c_m with $m \geq 2$. Such a state is uniquely (up to normalization) given by $b_{-1}|\!\downarrow\rangle$, which has ghost number $-3/2$. Let us illustrate the implications of this by considering the tachyon. Since one

3.4 Vertex operators

must act on $b_{-1}|\downarrow\rangle$ with c_1 to obtain the tachyon state, it follows that in the BRST formalism the closed-string tachyon vertex operator takes the form

$$V_\text{t}(z,\bar z) = \, :c(z)\tilde c(\bar z)e^{ik\cdot X(z,\bar z)}: \, . \tag{3.106}$$

Let V_ϕ denote the dimension $(1,1)$ vertex operator for a physical state $|\phi\rangle$ described earlier. Then $c\tilde c V_\phi$ is the vertex operator corresponding to $|\phi\rangle$ in the formalism with ghosts, provided that one chooses the BRST cohomology class representative satisfying $b_m|\phi\rangle = 0$ for $m \geq 0$ discussed earlier. Since the c ghost has dimension -1 this operator has dimension $(0,0)$. As was explained, dimension $(1,1)$ ensures that the integrated expression $\int V_\phi\, d^2 z$ is invariant under conformal transformations. Similarly, the dimension $(0,0)$ unintegrated expression $c\tilde c V_\phi$ is also conformally invariant. For reasons that are explained in the next section, both kinds of vertex operators, integrated and unintegrated, are required.

Exercises

Exercise 3.6
By computing the OPE with the energy–momentum tensor determine the dimension of the vertex operator $V = \, :e^{ik\cdot X(z,\bar z)}:$.

Solution

In order to determine the dimension of the vertex operator V we only need the leading singularity of the OPE

$$T(z) :e^{ik\cdot X(w,\bar w)}: \, = -2 :\partial X^\mu(z)\partial X_\mu(z):: e^{ik\cdot X(w,\bar w)}: \, .$$

This can be computed using Eq. (3.35), which gives

$$\langle \partial X^\mu(z) X^\nu(w)\rangle = -\frac{1}{4}\frac{\eta^{\mu\nu}}{z-w}.$$

Here, $X^\nu(w)$ should be identified with the holomorphic part of $X^\nu(w,\bar w)$. From this it follows that

$$\partial X^\mu(z) :e^{ik\cdot X(w,\bar w)}: \, \sim \, \langle\partial X^\mu(z)\, ik\cdot X(w)\rangle :e^{ik\cdot X(w,\bar w)}:$$

$$\sim -\frac{i}{4}\frac{k^\mu}{z-w} :e^{ik\cdot X(w,\bar w)}: \, .$$

Therefore,
$$T(z) : e^{ik \cdot X(w,\bar{w})} : \sim \frac{k^2/8}{(z-w)^2} : e^{ik \cdot X(w,\bar{w})} : + \ldots$$

This shows that $h = k^2/8$. Similarly one can compute the OPE with $\tilde{T}(\bar{z})$ showing $(h, \bar{h}) = (k^2/8, k^2/8)$ for the closed string. In particular, this is the tachyon emission operator, which has dimension $(1,1)$, for $M^2 = -k^2 = -8$.
□

EXERCISE 3.7
Determine the conformal dimensions of the operator
$$V = f_{\mu\nu} : \partial X^\mu(w) \bar{\partial} X^\nu(\bar{w}) e^{ik \cdot X(w,\bar{w})} : .$$

What condition has to be imposed on $f_{\mu\nu}$ so that this vertex operator is a conformal field?

SOLUTION

The OPE of the energy–momentum tensor with the vertex operator is
$$-2 f_{\mu\nu} : \partial X^\rho(z) \partial X_\rho(z) :: \partial X^\mu(w) \bar{\partial} X^\nu(\bar{w}) e^{ik \cdot X(w,\bar{w})} : .$$

There are several contributions in the above OPE, which we denote by $\mathcal{K}_N$ where the index N denotes the contribution of order $(z-w)^{-N}$. First of all there is a cubic contribution
$$\mathcal{K}_3 = -\frac{i}{4} k^\mu f_{\mu\nu} \frac{\bar{\partial} X^\nu(\bar{w})}{(z-w)^3},$$

which is required to vanish if V is supposed to be a conformal field. As a result
$$k^\mu f_{\mu\nu} = 0.$$

The conformal dimension of V is then obtained from the $\mathcal{K}_2$ term, which takes the form
$$\mathcal{K}_2 = \frac{1 + k^2/8}{(z-w)^2} V.$$

The 1 term comes from contracting T with the prefactor and the $k^2/8$ term comes from contracting T with the exponential (as in the previous problem). This shows that V has conformal dimension $(h, \bar{h}) = (1 + k^2/8, 1 + k^2/8)$. □

3.5 The structure of string perturbation theory

The starting point for studying string perturbation theory is the world-sheet action with Euclidean signature. Before gauge fixing, it has the general form

$$S_{\text{WS}} = \int_M \mathcal{L}(h_{\alpha\beta}; X^\mu; \text{background fields}) \, d^2z \, . \tag{3.107}$$

As usual, $h_{\alpha\beta}$ is the two-dimensional world-sheet metric, and $X^\mu(z, \bar{z})$ describes the embedding of the world sheet M into the space-time manifold $\mathcal{M}$. Thus z is a local coordinate on the world sheet and X^μ are local coordinates of space-time. Working with a Euclidean signature world-sheet metric ensures that the functional integrals (to be defined) are converted to convergent Gaussian integrals. The background fields should satisfy the field equations to be consistent. When this is the case, the world-sheet theory has conformal invariance.

Partition functions and scattering amplitudes

Partition functions and on-shell scattering amplitudes can be formulated as path integrals of the form proposed by Polyakov

$$Z \sim \int Dh_{\alpha\beta} \int DX^\mu \cdots e^{-S[h, X, \ldots]}. \tag{3.108}$$

Here $\int Dh$ means the sum over all Riemann surfaces (M, h). However, this is a gauge theory, since S is invariant under diffeomorphisms and Weyl transformations. So one should really sum over Riemann surfaces modulo diffeomorphisms and Weyl transformations.[11]

World-sheet diffeomorphism symmetry allows one to choose a conformally flat world-sheet metric

$$h_{\alpha\beta} = e^\psi \delta_{\alpha\beta}. \tag{3.109}$$

When this is done, one must add the Faddeev–Popov ghost fields $b(z)$ and $c(z)$ to the world-sheet theory to represent the relevant Jacobian factors in the path integral. Then the local Weyl symmetry ($h_{\alpha\beta} \to \Lambda h_{\alpha\beta}$) allows one to fix ψ (locally) – say to zero. However, this is not possible globally, due to a topological obstruction:

$$\psi = 0 \Rightarrow R(h) = 0 \Rightarrow \chi(M) = 0. \tag{3.110}$$

So, such a choice is only possible for world sheets that admit a flat metric.

---

[11] In the case of superstrings in the RNS formalism, discussed in the next chapter, the action also has local world-sheet supersymmetry and super-Weyl symmetry, so these equivalences also need to be taken into account.

Among orientable Riemann surfaces without boundary, the only such case is $n_h = 1$ (the torus). For each genus n_h there are particular ψs compatible with $\chi(M) = 2 - 2n_h$ that are allowed. A specific choice of such a ψ corresponds to choosing a *complex structure* for M. Let us now consider the moduli space of inequivalent choices.

Riemann surfaces of different topology are certainly not diffeomorphic, so each value of the genus can be considered separately, giving a perturbative expansion of the form

$$Z = \sum_{n_h=0}^{\infty} Z_{n_h}. \tag{3.111}$$

This series is only an asymptotic expansion, as in ordinary quantum field theory. Moreover, there are additional nonperturbative contributions that it does not display. Sometimes some of these can be identified by finding suitable saddle points of the functional integral, as in the study of instantons.

A constant dilaton $\Phi(x) = \Phi_0$ contributes

$$S_{\text{dil}} = \Phi_0 \chi(M) = \Phi_0 (2 - 2n_h). \tag{3.112}$$

Thus Z_{n_h} contains a factor

$$\exp(-S_{\text{dil}}) = \exp(\Phi_0(2n_h - 2)) = g_s^{2n_h - 2}, \tag{3.113}$$

where the closed-string coupling constant is

$$g_s = e^{\Phi_0}. \tag{3.114}$$

Thus each handle contributes a factor of g_s^2.

This role of the dilaton is very important. It illustrates a very general lesson: all dimensionless parameters in string theory – including the value of the string coupling constant – can ultimately be traced back to the vacuum values of scalar fields. The underlying theory does not contain any dimensionless parameters. Rather, all dimensionless numbers that characterize specific string vacua are determined as the vevs of scalar fields.

The moduli space of Riemann surfaces

The gauge-fixed world-sheet theory, with a conformally flat metric, has two-dimensional conformal symmetry, which is generated by the Virasoro operators. In carrying out the Polyakov path integral, it is necessary to integrate over all conformally inequivalent Riemann surfaces of each topology. The choice of a complex structure for the Riemann surface precisely corresponds to the choice of a *conformal equivalence class*, so one needs to integrate over the moduli space of complex structures, which parametrizes these classes. In the case of superstrings the story is more complicated, because there are also fermionic moduli and various possible choices of spin structures. We will not explore these issues.

In order to compute an N-particle scattering amplitude, not just the partition function, it is necessary to specify N points on the Riemann surface. At each of them one inserts a vertex operator $V_\phi(z, \bar z)$ representing the emission or absorption of an asymptotic physical string state of type ϕ. Mathematicians like to regard such marked points as removed from the surface, and therefore they refer to them as *punctures*.

To compute the n_h-loop contribution to the amplitude requires integrating over the moduli space $\mathcal{M}_{n_\mathrm{h},N}$ of genus n_h Riemann surfaces with N punctures. According to a standard result in complex analysis, the Riemann–Roch theorem, the number of complex dimensions of this space is

$$\dim_{\mathbb{C}} \mathcal{M}_{n_\mathrm{h},N} = 3n_\mathrm{h} - 3 + N, \tag{3.115}$$

and the real dimension is twice this. Therefore, this is the dimension of the integral that represents the string amplitude. For $n_\mathrm{h} > 1$ it is very difficult to specify the integration region $\mathcal{M}_{n_\mathrm{h},N}$ explicitly and to define the integral precisely. However, this is just a technical problem, and not an issue of principle. The cases $n_\mathrm{h} = 0, 1$ are much easier, and they can be made very explicit.

In the case of genus 0 (or tree approximation), one can conformally map the Riemann sphere to the complex plane (plus a point at infinity). The $SL(2, \mathbb{C})$ group of conformal isometries is just sufficient to allow three of the punctures to be mapped to arbitrarily specified distinct positions. Then all that remains is to integrate over the coordinates of the other $N-3$ puncture positions. This counting of moduli agrees with Eq. (3.115) for the choice $n_\mathrm{h} = 0$. To achieve this in a way consistent with conformal invariance, one should use three unintegrated vertex operators and $N-3$ integrated vertex operators in the Polyakov path integral. These two types of vertex operators were described in the previous section. In the tree approximation, using the fact that the correlator of two X fields on the complex plane is a logarithm, one obtains the N-tachyon amplitude (or Shapiro–Virasoro amplitude)

$$A_N(k_1, k_2, \ldots, k_N) = g_\mathrm{s}^{N-2} \int d\mu_N(z) \prod_{i<j} |z_i - z_j|^{k_i \cdot k_j / 2}, \tag{3.116}$$

where

$$d\mu_N(z) = |(z_A - z_B)(z_B - z_C)(z_C - z_A)|^2$$

$$\times \delta^2(z_A - z_A^0)\delta^2(z_B - z_B^0)\delta^2(z_C - z_C^0)\prod_{i=1}^{N} d^2 z_i. \tag{3.117}$$

The formula is independent of z_A^0, z_B^0, z_C^0 due to the $SL(2,\mathbb{C})$ symmetry, which allows them to be mapped to arbitrary values.

In the case of a torus (genus one), the complex structure (or conformal equivalence class) is characterized by one complex number τ. The conformal isometry group in this case corresponds to translations, so the position of one puncture can be fixed. Thus, in the genus-one case the path integral should contain one unintegrated vertex operator and $N-1$ integrated vertex operators. This leaves an integral over τ and the coordinates of $N-1$ of the punctures for a total of N complex integrations in agreement with Eq. (3.115) for $n_h = 1$. For genus $n_h > 1$, there are no conformal isometries, and so all N vertex operators should be integrated. In all cases, the number of unintegrated vertex operators, and hence the number of c-ghost insertions is equal to the dimension of the space of conformal isometries. This also matches the number of c-ghost zero modes on the corresponding Riemann surface, so these insertions are just what is required to give nonvanishing integrals for the c-ghost zero modes.[12]

There also needs to be the right number of b-ghost insertions to match the number of b-ghost zero modes. This number is just the dimension of the moduli space. By combining these b-ghost factors with expressions called Beltrami differentials in the appropriate way, one obtains a moduli-space measure that is invariant under reparametrizations of the moduli space. The reader is referred to the literature (e.g., volume 1 of Polchinski) for further details.

Let us now turn to the definition of τ, the modular parameter of the torus, and the determination of its integration region (the genus-one moduli space). A torus can be characterized by specifying two periods in the complex plane,

$$z \sim z + w_1, \quad z \sim z + w_2. \tag{3.118}$$

The only restriction is that the two periods should be finite and nonzero, and their ratio should not be real. The torus is then identified with the complex plane $\mathbb{C}$ modulo a two-dimensional lattice $\Lambda_{(w_1,w_2)}$, where $\Lambda_{(w_1,w_2)} = \{mw_1 + nw_2, m, n \in \mathbb{Z}\}$,

$$T^2 = \mathbb{C}/\Lambda_{(w_1,w_2)}. \tag{3.119}$$

Rescaling by the conformal transformation $z \to z/w_2$, this torus is conformally equivalent to one whose periods are 1 and $\tau = w_1/w_2$, as shown in

---

[12] Recall that, for a Grassmann coordinate c_0, $\int dc_0 = 0$ and $\int c_0 dc_0 = 1$.

3.5 The structure of string perturbation theory

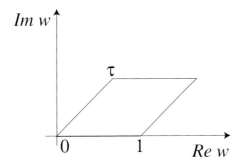

Fig. 3.3. When opposite edges of the parallelogram are identified, this becomes a torus.

Fig. 3.3. Without loss of generality (interchanging w_1 and w_2, if necessary), one can restrict τ to the upper half-plane $\mathcal{H}$ (Im $\tau > 0$). Now note that the alternative fundamental periods

$$w'_1 = aw_1 + bw_2 \quad \text{and} \quad w'_2 = cw_1 + dw_2 \qquad (3.120)$$

define the same lattice, if $a, b, c, d \in \mathbb{Z}$ and $ad - bc = 1$. In other words,

$$\begin{pmatrix} a & b \\ c & d \end{pmatrix} \in SL(2, \mathbb{Z}). \qquad (3.121)$$

This implies that a torus with modular parameter τ is conformally equivalent to one with modular parameter

$$\tau' = \frac{w'_1}{w'_2} = \frac{a\tau + b}{c\tau + d}. \qquad (3.122)$$

Accordingly, the moduli space of conformally inequivalent Riemann surfaces of genus one is

$$\mathcal{M}_{n_{\rm h}=1} = \mathcal{H}/PSL(2, \mathbb{Z}). \qquad (3.123)$$

The infinite discrete group $PSL(2, \mathbb{Z}) = SL(2, \mathbb{Z})/\mathbb{Z}_2$ is generated by the transformations $\tau \to \tau + 1$ and $\tau \to -1/\tau$. The division by $\mathbb{Z}_2$ takes account of the equivalence of an $SL(2, \mathbb{Z})$ matrix and its negative. The $PSL(2, \mathbb{Z})$ identifications give a tessellation of the upper half-plane $\mathcal{H}$.

A natural choice for the fundamental region $\mathcal{F}$ is

$$|\text{Re}\,\tau| \le 1/2, \quad \text{Im}\,\tau > 0, \quad |\tau| \ge 1, \qquad (3.124)$$

as shown in Fig. 3.4. The moduli space has three cusps or singularities,

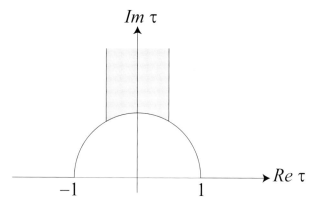

Fig. 3.4. The shaded region is the fundamental region of the modular group.

where there is a deficit angle, which are located at the τ values i, ∞, and $\omega = \exp(i\pi/3)$.[13] Therefore, it is not a smooth manifold.

If one uses the translation symmetry freedom to set $z_1 = 0$, then a one-loop amplitude takes the form

$$\int_{\mathcal{F}} \frac{d^2\tau}{(\operatorname{Im}\tau)^2} \int_{T^2} \mu(\tau, z) \langle V_1(0) V_2(z_2) \ldots V_N(z_N) \rangle d^2 z_2 \ldots d^2 z_N. \quad (3.125)$$

The angular brackets around the product of vertex operators denote a functional integration over the world-sheet fields. An essential consistency requirement is *modular invariance*. This means that the integrand should be invariant under the $SL(2, \mathbb{Z})$ transformations (also called *modular transformations*)

$$\tau \to \frac{a\tau + b}{c\tau + d}, \quad z_i \to \frac{z_i}{c\tau + d}, \quad (3.126)$$

so that the result is the same whether one integrates over the fundamental region $\mathcal{F}$ or any of its $SL(2, \mathbb{Z})$ images. It is a highly nontrivial fact that this works for all consistent string theories. In fact, it is one method of understanding why the only possible gauge groups for the heterotic string theory (with $\mathcal{N} = 1$ supersymmetry in ten-dimensional Minkowski space-time) are $SO(32)$ and $E_8 \times E_8$, as is discussed in Chapter 7.

There are higher-genus analogs of modular invariance, which must also be satisfied. This has not been explored in full detail, but enough is known about the various string theories to make a convincing case that they must be consistent. For now, let us make some general remarks about multiloop

---

13 The point $\omega^2 = \exp(2i\pi/3)$ may appear to be another cusp, but it differs from ω by 1, and therefore it represents the same point in the moduli space.

string amplitudes that are less detailed than the particular issue of modular invariance.

It is difficult to describe explicitly the moduli of higher-genus Riemann surfaces, and it is even harder to specify a fundamental region analogous to the one described above for genus one. However, the dimension of moduli space, which is the number of integrations, is not hard to figure out. It is as shown in Table 3.1. Note that in all cases the sum is $3n_h - 3 + N$, as stated in Eq. (3.115).

	moduli of $\mathcal{M}$	moduli of punctures
$n_h = 0$	0	$N - 3$
$n_h = 1$	1	$N - 1$
$n_h \geq 2$	$3n_h - 3$	N

Table 3.1. *The number of complex moduli for an n_h-loop N-particle closed-string amplitude.*

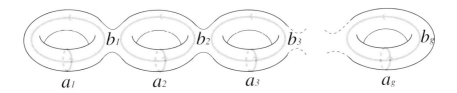

Fig. 3.5. Canonical basis of one-cycles for a genus-g Riemann surface.

The first homology group of a genus-n_h Riemann surface has $2n_h$ generators. It is convenient to introduce a canonical basis consisting of n_h a-cycles and n_h b-cycles, as shown in Fig. 3.5. There are also $2n_h$ one-forms that generate the first cohomology group. The complex structure of the Riemann surface can be used to divide these into n_h holomorphic and n_h antiholomorphic one-forms. Thus one obtains the fundamental result that a genus-n_h Riemann surface admits n_h linearly independent holomorphic one-forms. One can choose a basis ω_i, $i = 1, 2, \ldots, n_h$, of holomorphic one-forms by the requirement that

$$\oint_{a_i} \omega_j = \delta_{ij}. \tag{3.127}$$

The integrals around the b-cycles then give a matrix

$$\oint_{b_i} \omega_j = \Omega_{ij} \tag{3.128}$$

called the *period matrix*. For example, in the simple case of the torus $\omega = dz$ and $\Omega = \tau$. Two fundamental facts are that Ω is a symmetric matrix and that its imaginary part is positive definite. Symmetric matrices with a positive-definite imaginary part define a region called the *Siegel upper half plane*.

There is a group of equivalences for the period matrices that generalizes the $SL(2,\mathbb{Z})$ group of equivalences in the genus-one case. It acts in a particularly simple way on the period matrices. Specifically, one has

$$\Omega \to \Omega' = (A\Omega + B)(C\Omega + D)^{-1}, \qquad (3.129)$$

where A, B, C, D are $n_\mathrm{h} \times n_\mathrm{h}$ matrices and

$$\begin{pmatrix} A & B \\ C & D \end{pmatrix} \in Sp(n_\mathrm{h}, \mathbb{Z}). \qquad (3.130)$$

This group is called the *symplectic modular group*. The notation $Sp(n,\mathbb{Z})$ refers to $2n$-dimensional symplectic matrices with integer entries. Recall that symplectic transformations preserve an antisymmetric "metric"

$$\begin{pmatrix} A & B \\ C & D \end{pmatrix} \begin{pmatrix} 0 & 1 \\ -1 & 0 \end{pmatrix} \begin{pmatrix} A^T & C^T \\ B^T & D^T \end{pmatrix} = \begin{pmatrix} 0 & 1 \\ -1 & 0 \end{pmatrix}. \qquad (3.131)$$

In the one-loop case the modular parameter τ and the period matrix are the same thing. So integration over the moduli space of conformally inequivalent Riemann surfaces is the same as integration over a fundamental region defined by modular transformations. At higher genus the story is more complicated. The period matrix has complex dimension $\frac{1}{2}n_\mathrm{h}(n_\mathrm{h}+1)$ (since it is a complex symmetric matrix), whereas the moduli space has $3n_\mathrm{h} - 3$ complex dimensions. At genus 2 and 3 these dimensions are the same, and the relation between a fundamental region in the Siegel upper half plane and the moduli space can be worked out. For $n_\mathrm{h} > 3$, the moduli space is a subspace of finite codimension. Thus, even though the integrand can be written quite explicitly, it is a very nontrivial problem (known as the Riemann–Schottky problem) to determine which period matrices correspond to Riemann surfaces.

EXERCISES

EXERCISE 3.8
Explain why the point $\tau = i$ is a cusp of the moduli space of the torus.

3.5 The structure of string perturbation theory

SOLUTION

This can be understood by examining the identifications made in the moduli space. This is displayed in Fig. 3.6. Specifically, the identification $\tau \sim -1/\tau$ glues the left half of the unit circle to the right half, and it has $\tau = i$ as a fixed point.

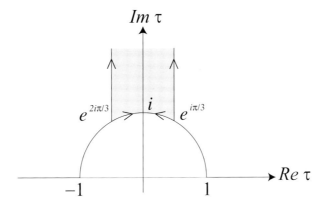

Fig. 3.6. Image of the fundamental domain of the torus. Opposite edges are glued together as indicated by the arrows. This explains why there are cusps in the moduli space.

□

EXERCISE 3.9
Show that $d^2\tau/(\mathrm{Im}\,\tau)^2$ is an $SL(2,\mathbb{Z})$-invariant measure on $\mathcal{M}$. Using this measure, compute the volume of $\mathcal{M}$.

SOLUTION

Under the $SL(2,\mathbb{Z})$ transformation in Eq. (3.122)

$$d^2\tau \to |c\tau + d|^{-4} d^2\tau \quad \text{and} \quad \mathrm{Im}\,\tau \to |c\tau + d|^{-2}\mathrm{Im}\,\tau,$$

which implies the invariance of the measure. Equivalently, one can check that the measure is invariant under the two transformations $\tau \to \tau + 1$ and $\tau \to -1/\tau$ which generate $SL(2,\mathbb{Z})$.

The volume of the moduli space is obtained from the integral

$$\mathcal{I} = \int_\mathcal{F} \frac{d^2\tau}{(\mathrm{Im}\,\tau)^2},$$

over the fundamental region. Letting $\tau = x + iy$ and defining $d^2\tau = dxdy$, this takes the form

$$\mathcal{I} = \int_{-1/2}^{+1/2} dx \int_{\sqrt{1-x^2}}^{\infty} \frac{dy}{y^2} = \int_{-1/2}^{+1/2} \frac{dx}{\sqrt{1-x^2}} = \frac{\pi}{3},$$

where we have set $\tau = x + iy$. □

3.6 The linear-dilaton vacuum and noncritical strings

An interesting example of a nontrivial background that preserves conformal symmetry is one in which the dilaton field depends linearly on the spatial coordinates. Letting y denote the direction along which it varies and x^μ the other $D-1$ space-time coordinates, the linear dilaton background is

$$\Phi(X^\mu, Y) = kY(z, \bar{z}), \qquad (3.132)$$

where k is a constant. After fixing the conformal gauge, the dilaton term no longer contributes to the world-sheet action, which remains independent of k, but it does contribute to the energy–momentum tensor.

The energy–momentum tensor for the linear-dilaton background is derived by varying the action with respect to the world-sheet metric before fixing the conformal gauge. The result is

$$T(z) = -2(\partial X^\mu \partial X_\mu + \partial Y \partial Y) + k\partial^2 Y. \qquad (3.133)$$

This expression gives a TT OPE that still has the correct structure to define a CFT. One peculiarity is that the OPE of T with Y has an extra term (proportional to k), which implies that ∂Y does not satisfy the definition of a conformal field.

Calling D the total space-time dimension (including Y), the central charge determined by the TT OPE turns out to be

$$c = \tilde{c} = D + 3k^2. \qquad (3.134)$$

Thus, the required value $c = 26$ can be achieved for $D < 26$ by choosing

$$k = \sqrt{\frac{26-D}{3}}. \qquad (3.135)$$

Of course, there is Lorentz invariance in only $D-1$ dimensions, since the Y direction is special. Theories with $k \neq 0$ are called *noncritical string theories*.

The extra term in T contributes to L_0, and hence to the equation of motion for the free tachyon field $t(x^\mu, y)$. For simplicity, let us consider solutions

3.6 The linear-dilaton vacuum and noncritical strings

that are independent of x^μ. Then the equation of motion $(L_0 - 1)|t\rangle = 0$ becomes

$$t''(y) - 2kt'(y) + 8t(y) = 0. \qquad (3.136)$$

Since this is a stationary (zero-energy) equation, the existence of oscillatory solutions is a manifestation of tachyonic behavior. This equation has solutions of the form $\exp(qy)$ for

$$q = q_\pm = k \pm \sqrt{(2-D)/3}. \qquad (3.137)$$

Thus, there is no oscillatory behavior for $D \leq 2$, and one expects to have a stable vacuum in this case. Since the Y field is present in any case, $D \geq 1$. Fractional values between 1 and 2 are possible if a unitary minimal model is used in place of X^μ.

These results motivate one to further modify the world-sheet theory in the case of $D \leq 2$ by adding a tachyon background term of the form $T_0 \exp(q_- Y)$. The resulting world-sheet theory is called a *Liouville field theory*. Despite its nonlinearity, it is classically integrable, and even the quantum theory is quite well understood (after many years of hard work).

Recall that the exponential of the dilaton field gives the strength of the string coupling. So the linear dilaton background describes a world in which strings are weakly coupled for large negative y and strongly coupled for large positive y. One could worry about the reliability of the formalism in such a set-up. However, the tachyon background or Liouville exponential e^{qy} suppresses the contribution of the strongly coupled region, and this keeps things under control. Toy models of this sort with $D = 1$ or $D = 2$ are simple enough that their study has proved valuable in developing an understanding of some of the intricacies of string theory such as the asymptotic properties of the perturbation expansion at high genus and some nonperturbative features.

A completely different methodology that leads to exactly the same world-sheet theory makes no reference to dilatons or tachyons at all. Rather, one simply adds a cosmological constant term to the world-sheet theory. This is a rather drastic thing to do, because it destroys the classical Weyl invariance of the theory. The consequence of this is that, when one uses diffeomorphism invariance to choose a conformally flat world-sheet metric $h_{\alpha\beta} = e^\omega \eta_{\alpha\beta}$, the field ω no longer decouples. Rather, it becomes dynamical and plays the same role as the field Y in the earlier discussion. This is an alternative characterization of noncritical string theories.

Exercises

EXERCISE 3.10

By computing the TT OPE in the linear-dilaton vacuum verify the value of the central charge given in Eq. (3.134).

SOLUTION

In order to compute the OPE, it is convenient to rewrite the energy–momentum tensor in Eq.(3.133) in the form

$$T(z) = T_0(z) + a_\mu \partial^2 X^\mu(z),$$

where $a_\mu = k\delta^i_\mu$, and i is the direction along which the dilaton varies. Since we are interested in the central charge, we only need the leading singularity in this OPE, which is given by

$$T(z)T(w) = T_0(z)T_0(w) + a_\mu a_\nu \partial^2 X^\mu(z) \partial^2 X^\nu(w) + \ldots$$

Now we use the results for the leading-order singularities

$$T_0(z)T_0(w) = \frac{D/2}{(z-w)^4} \quad \text{and} \quad \partial^2 X^\mu(z) \partial^2 X^\nu(w) = \frac{3}{2} \frac{\eta^{\mu\nu}}{(z-w)^4},$$

to get

$$T(z)T(w) = \frac{(D+3a^2)/2}{(z-w)^4} + \ldots$$

This shows that in the original notation the central charge is

$$c = D + 3k^2.$$

The same computation can be repeated to obtain the result $\tilde{c} = c$. □

3.7 Witten's open-string field theory

Witten's description of the field theory of the open bosonic string has many analogies with Yang–Mills theory. This is not really surprising inasmuch as open strings can be regarded as an infinite-component generalization of Yang–Mills fields. It is pedagogically useful to emphasize these analogies in describing the theory. The basic object in Yang–Mills theory is the vector potential $A^a_\mu(x^\rho)$, where μ is a Lorentz index and a runs over the generators

3.7 Witten's open-string field theory

of the symmetry algebra. By contracting with matrices $(\lambda^a)_{ij}$ that represent the algebra and differentials dx^μ one can define

$$A_{ij}(x^\rho) = \sum_{a,\mu} (\lambda^a)_{ij} A_\mu^a(x^\rho) dx^\mu, \qquad (3.138)$$

which is a matrix of one-forms. This is a natural quantity from a geometric point of view. The analogous object in open-string field theory is the string field

$$A[x^\rho(\sigma), c(\sigma)]. \qquad (3.139)$$

This is a functional field that creates or destroys an entire string with coordinates $x^\rho(\sigma), c(\sigma)$, where the parameter σ is taken to have the range $0 \leq \sigma \leq \pi$. The coordinate $c(\sigma)$ is the anticommuting ghost field described earlier in this chapter. In this formulation the conjugate antighost $b(\sigma)$ is represented by a functional derivative with respect to $c(\sigma)$.

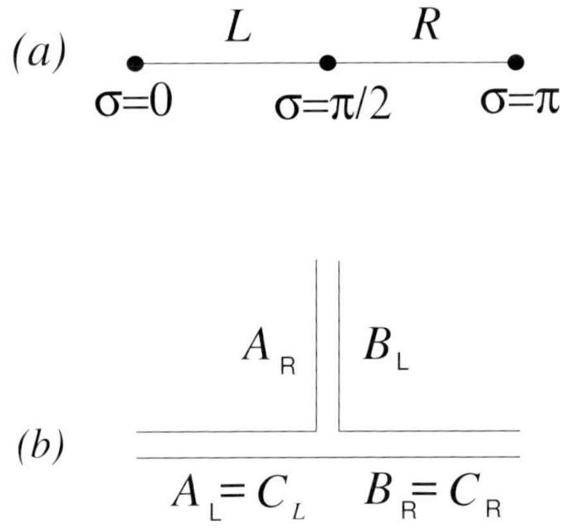

Fig. 3.7. An open string has a left side ($\sigma < \pi/2$) and a right side ($\sigma > \pi/2$) depicted in (a), which can be treated as matrix indices. The multiplication $A * B = C$ is depicted in (b).

The string field A can be regarded as a matrix (in analogy to A_{ij}) by regarding the coordinates with $0 \leq \sigma \leq \pi/2$ as providing the left matrix index and those with $\pi/2 \leq \sigma \leq \pi$ as providing the right matrix index as shown in part (a) of Fig. 3.7. One could also associate Chan–Paton quark-like charges with the ends of the strings,[14] which would then be included in the matrix labels as well, but such labels are not displayed. By not including such charges one is describing the $U(1)$ open-string theory. $U(1)$

---

[14] This is explained in Chapter 6.

gauge theory (without matter fields) is a free theory, but the string extension has nontrivial interactions.

In the case of Yang–Mills theory, two fields can be multiplied by the rule

$$\sum_k A_{ik} \wedge B_{kj} = C_{ij}. \qquad (3.140)$$

This is a combination of matrix multiplication and antisymmetrization of the tensor indices (the wedge product of differential geometry). This multiplication is associative but noncommutative. A corresponding rule for string fields is given by a $*$ product,

$$A * B = C. \qquad (3.141)$$

This infinite-dimensional matrix multiplication is depicted in part (b) of Fig. 3.7. One identifies the coordinates of the right half of string A with those of the left half of string B and functionally integrates over the coordinates of these identified half strings. This leaves string C consisting of the left half of string A and the right half of string B. It is also necessary to include a suitable factor involving the ghost coordinates at the midpoint $\sigma = \pi/2$.

A fundamental operation in gauge theory is exterior differentiation $A \to dA$. In terms of components

$$dA = \frac{1}{2}(\partial_\mu A_\nu - \partial_\nu A_\mu)dx^\mu \wedge dx^\nu, \qquad (3.142)$$

which contains the abelian field strengths as coefficients. Exterior differentiation is a nilpotent operation, $d^2 = 0$, since partial derivatives commute and vanish under antisymmetrization. The nonabelian Yang–Mills field strength is given by the matrix-valued two-form

$$F = dA + A \wedge A, \qquad (3.143)$$

or in terms of tensor indices,

$$F_{\mu\nu} = \partial_\mu A_\nu - \partial_\nu A_\mu + [A_\mu, A_\nu]. \qquad (3.144)$$

Let us now construct analogs of d and F for the open-string field. The operator that plays the roles of d is the nilpotent BRST operator Q_B, which can be written explicitly as a differential operator involving the coordinates $X(\sigma)$, $c(\sigma)$. Given the operator Q_B, there is an obvious formula for the string-theory field strength, analogous to the Yang–Mills formula, namely

$$F = Q_B A + A * A. \qquad (3.145)$$

The string field A describes physical string states, and therefore it has ghost number $-1/2$. Since Q_B has ghost number $+1$, it follows that F has ghost number $+1/2$. For $A*A$ to have the same ghost number, the $*$ operation must contribute $+3/2$ to the ghost number.

An essential feature of Yang–Mills theory is gauge invariance. Infinitesimal gauge transformation can be described by a matrix of infinitesimal parameters $\Lambda(x^\rho)$. The transformation rules for the potential and the field strength are then

$$\delta A = d\Lambda + [A, \Lambda] \tag{3.146}$$

and

$$\delta F = [F, \Lambda]. \tag{3.147}$$

There are completely analogous formulas for the string theory, namely

$$\delta A = Q_B \Lambda + [A, \Lambda] \tag{3.148}$$

and

$$\delta F = [F, \Lambda]. \tag{3.149}$$

In this case $[A, \Lambda]$ means $A*\Lambda - \Lambda*A$, of course. Since the infinitesimal parameter $\Lambda[x^\rho(\sigma), c(\sigma)]$ is a functional, it can be expanded in terms of an infinite number of ordinary functions. Thus the gauge symmetry of string theory is infinitely richer than that of Yang–Mills theory, as required for the consistency of the infinite spectrum of high-spin fields contained in the theory.

The next step is to formulate a gauge-invariant action. The key ingredient in doing this is to introduce a suitably defined integral. In the case of Yang–Mills theory one integrates over space-time and takes a trace over the matrix indices. Thus it is convenient to define $\int Y$ as $\int d^4x \text{Tr}(Y(x))$. In this notation the usual Yang–Mills action is

$$S \sim \int g^{\mu\rho} g^{\nu\lambda} F_{\mu\nu} F_{\rho\lambda}. \tag{3.150}$$

The definition of integration appropriate to string theory is a "trace" that identifies the left and right segments of the string field Lagrangian, specifically

$$\int Y = \int D^{26} X^\mu(\sigma) D\phi(\sigma) \exp\left(-\frac{3i}{2} \phi(\pi/2)\right) Y[X^\mu(\sigma), \phi(\sigma)]$$

$$\times \prod_{\sigma < \pi/2} \delta^{26}(X^\mu(\sigma) - X^\mu(\pi - \sigma)) \delta(\phi(\sigma) - \phi(\pi - \sigma)). \tag{3.151}$$

As indicated in part (a) of Fig. 3.8, this identifies the left and right segments of X. A ghost factor has been inserted at the midpoint. $\phi(\sigma)$ is the bosonized form of the ghosts described earlier. This ensures that $\int$ contributes $-3/2$ to the ghost number, as required. This definition of integration satisfies the important requirements

$$\int Q_B Y = 0 \quad \text{and} \quad \int [Y_1, Y_2] = 0. \tag{3.152}$$

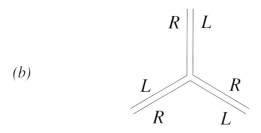

Fig. 3.8. Integration of a string functional requires identifying the left and right halves as depicted in (a). The three-string vertex, shown in (b), is based on two multiplications (star products) and one integration and treats the three strings symmetrically.

We now have the necessary ingredients to write a string action. Trying to emulate the Yang–Mills action runs into a problem, because no analog of the metric $g^{\mu\rho}$ has been defined. Rather than trying to find one, it proves more fruitful to look for a gauge-invariant action that does not require one. The simplest possibility is given by the Chern–Simons form

$$S \sim \int \left(A * Q_B A + \frac{2}{3} A * A * A \right). \tag{3.153}$$

In the context of ordinary Yang–Mills theory the integrand is a three-form, whose variation under a gauge transformation is closed, and therefore such a term can only be introduced in three dimensions, where it is interpreted as giving mass to the gauge field. In string theory the interpretation is different, though the mathematics is quite analogous, and the formula makes perfectly good sense. In fact, in both cases extremizing the action gives rise to the deceptively simple classical field equation $F = 0$.

The fact that the string equation of motion is $F = 0$ does not mean the theory is trivial. Dropping the interaction term, the equation of motion

3.7 Witten's open-string field theory

for the free theory is $Q_B A = 0$, which is invariant under the abelian gauge transformation $\delta A = Q_B \Lambda$, since $Q_B^2 = 0$. Once one requires that A be restricted to contain ghost number $-1/2$ fields, this precisely reproduces the known spectrum of the bosonic string. As was explained earlier, the physical states of the free theory are in one-to-one correspondence with BRST cohomology classes of ghost number $-1/2$.

The cubic string interaction is depicted in part (b) of Fig. 3.8. Two of the segment identifications are consequences of the $*$ products in $A * A * A$, and the third is a consequence of the integration. Altogether, this gives an expression that is symmetric in the three strings.

As was explained earlier, in string theory one is only interested in equivalence classes of metrics that are related by conformal mappings. It is always possible to find representatives of each equivalence class in which the metric is flat everywhere except at isolated points where the curvature is infinite. Such a metric describes a surface with conical singularities, which is not a manifold in the usual sense. In fact, it is an example of a class of surfaces called orbifolds. The string field theory construction of the amplitude automatically chooses a particular metric, which is of this type. The conical singularities occur at the string midpoints in the interaction. They have the property that a small circle of radius r about this point has circumference $3\pi r$. This is exactly what is required so that the Riemann surfaces constructed by gluing vertices and propagators have the correct integrated curvature, as required by Euler's theorem.

Witten's string field theory seems to be as simple and beautiful as one could hope for, though there are subtleties in defining it precisely that have been glossed over in the brief presentation given here. For the bosonic string theory, it does allow a computation of all processes with only open-string external lines to all orders in perturbation theory (at least in principle). The extension to open superstrings is much harder and has not been completed yet. It has been proved that the various Feynman diagrams generated by this field theory piece together so as to cover the relevant Riemann surface moduli spaces exactly once. In particular, this means that the contributions of closed strings in the interior of diagrams is properly taken into account. Moreover, the fact that this is a field-theoretic formulation means that it can be used to define amplitudes with off-shell open strings, which are otherwise difficult to define in string theory. This off-shell property has been successfully exploited in nonperturbative studies of tachyon condensation. However, since this approach is based on open-string fields, it is not applicable to theories that only have closed strings. Corresponding constructions for closed-string theories (mostly due to Zwiebach) are more complicated.

Homework Problems

Problem 3.1
Compute the commutator of an infinitesimal translation and an infinitesimal special conformal transformation in D dimensions. Identify the resulting transformations and their infinitesimal parameters.

Problem 3.2
Show that the transformations (3.14) give rise to the $D = 2$ case of the D-dimensional transformations in Eq. (3.7).

Problem 3.3
Show that the algebra of Lorentzian-signature conformal transformations in D dimensions is isomorphic to the Lie algebra $SO(D, 2)$.

Problem 3.4
Derive the OPE
$$T(z)X^\mu(w,\bar{w}) \sim \frac{1}{z-w}\partial X^\mu(w,\bar{w}) + \ldots$$
What does this imply for the conformal dimension of X^μ?

Problem 3.5
(i) Use the result of the previous problem to deduce the OPE of $T(z)$ with each of the following operators:
$$\partial X^\mu(w,\bar{w}) \qquad \bar{\partial}X^\mu(w,\bar{w}), \qquad \partial^2 X^\mu(w,\bar{w}).$$
(ii) What do these results imply for the conformal dimension $(h, \tilde{h})$ (if any) in each case?

Problem 3.6
Show that
$$[\alpha_m^\mu, \alpha_n^\nu] = [\tilde{\alpha}_m^\mu, \tilde{\alpha}_n^\nu] = m\eta^{\mu\nu}\delta_{m+n,0}, \qquad [\alpha_m^\mu, \tilde{\alpha}_n^\nu] = 0$$
by using the OPE of the field $\partial X^\mu(z,\bar{z})$ with itself and with $\bar{\partial}X^\mu(z,\bar{z})$.

Problem 3.7
Consider a conformal field $\Phi(z)$ of dimension h and a mode expansion of the

form
$$\Phi(z) = \sum_{n=-\infty}^{+\infty} \frac{\Phi_n}{z^{n+h}}.$$

Using contour-integral methods, like those of Exercise 3.2, evaluate the commutator $[L_m, \Phi_n]$.

Problem 3.8
Let $\Phi(z)$ be a holomorphic primary field of conformal dimension h in a conformal field theory with the mode expansion given in the previous problem. The conformal vacuum satisfies $\Phi_n|0\rangle = 0$ for $n > -h$. Use the results of the previous problem to prove that $|\Phi\rangle = \Phi_{-h}|0\rangle$ is a highest-weight state.

Problem 3.9

(i) Calculate the two-point functions $\langle 0|\phi_i(z_1, \bar{z}_1)\phi_j(z_2, \bar{z}_2)|0\rangle$ for an arbitrary pair of primary fields with conformal weights $(h_i, \tilde{h}_i)$ and $(h_j, \tilde{h}_j)$ taking into account that the Virasoro generators L_0 and $L_{\pm 1}$ annihilate the in and out vacua $|0\rangle$ and $\langle 0|$.

(ii) Show that the three-point function $\langle 0|\phi_i(z_1, \bar{z}_1)\phi_j(z_2, \bar{z}_2)\phi_k(z_3, \bar{z}_3)|0\rangle$ is completely determined in terms of the conformal weights of the fields up to an overall coefficient C_{ijk}.

Problem 3.10

(i) Show that in a unitary conformal field theory, that is, one with a positive-definite Hilbert space, the central charge satisfies $c > 0$, and the conformal dimensions of primary fields satisfy $h \geq 0$. Hint: evaluate $\langle \phi|[L_n, L_{-n}]|\phi\rangle$ for a highest-weight state $|\phi\rangle$.

(ii) Show that $h = \tilde{h} = 0$ if and only if $|\phi\rangle = |0\rangle$.

Problem 3.11
Verify the expression (3.78) for the central charge of a system of b, c ghosts by computing the OPE of the energy–momentum tensor T_{bc} with itself.

Problem 3.12
Verify the property $Q_B^2 = 0$ of the BRST charge by anticommuting two of the integral representations and using the various OPEs.

Problem 3.13

Consider a closed oriented bosonic string theory in flat 26-dimensional space-time. In this theory the integrated vertex operators are integrals of primary fields of conformal dimension $(1,1)$.

(i) What is the form of these vertex operators for physical states with $N_L = N_R = 1$?

(ii) Verify that these vertex operators lead to physical states $|\phi\rangle$ that satisfy the physical state conditions

$$(L_n - \delta_{n,0})|\phi\rangle = 0, \qquad (\tilde{L}_n - \delta_{n,0})|\phi\rangle = 0 \quad n \geq 0.$$

Problem 3.14

Carry out the BRST quantization for the first two levels ($N_L = N_R = 0$ and $N_L = N_R = 1$) of the closed bosonic string. In other words, identify the BRST cohomology classes that correspond to the physical states. Hint: analyze the left-movers and right-movers separately.

Problem 3.15

Identify the BRST cohomology classes that correspond to physical states for the third level ($N = 2$) of the open string.

Problem 3.16

The open-string field can be expanded as a Fock-space vector in the first-quantized Fock space given by the α and ghost oscillators. The first term in the expansion is $A = T(x)|\downarrow\rangle$, where $T(x)$ is the tachyon field. Expand the string field A in component fields displaying the next two levels remembering that the total ghost number should be $-1/2$. Expand the action of the free theory to level $N = 1$.

4
Strings with world-sheet supersymmetry

The bosonic string theory that was discussed in the previous chapters is unsatisfactory in two respects. First, the closed-string spectrum contains a tachyon. If one chooses to include open strings, then additional open-string tachyons appear. Tachyons are unphysical because they imply an instability of the vacuum. The elimination of open-string tachyons from the physical spectrum has been understood in terms of the decay of D-branes into closed-string radiation. However, the fate of the closed-string tachyon has not been determined yet.

The second unsatisfactory feature of the bosonic string theory is that the spectrum (of both open and closed strings) does not contain fermions. Fermions play a crucial role in nature, of course. They include the quarks and leptons in the standard model. As a result, if we would like to use string theory to describe nature, fermions have to be incorporated. In string theory the inclusion of fermions turns out to require *supersymmetry*, a symmetry that relates bosons and fermions, and the resulting string theories are called *superstring theories*. In order to incorporate supersymmetry into string theory two basic approaches have been developed[1]

- *The Ramond–Neveu–Schwarz* (RNS) *formalism* is supersymmetric on the string world sheet.
- *The Green–Schwarz* (GS) *formalism* is supersymmetric in ten-dimensional Minkowski space-time. It can be generalized to other background space-time geometries.

These two approaches are actually equivalent, at least for ten-dimensional Minkowski space-time. This chapter describes the RNS formulation of superstring theory, which is based on world-sheet supersymmetry.

[1] More recently, various alternative formalisms have been proposed by Berkovits.

4.1 Ramond–Neveu–Schwarz strings

In the RNS formalism the bosonic fields $X^\mu(\sigma,\tau)$ of the two-dimensional world-sheet theory discussed in the previous chapter are paired up with fermionic partners $\psi^\mu(\sigma,\tau)$. The new fields $\psi^\mu(\sigma,\tau)$ are two-component spinors on the world sheet and vectors under Lorentz transformations of the D-dimensional space-time. These fields are anticommuting which is consistent with spin and statistics, since they are spinors in the two-dimensional sense. Consistency with spin and statistics in $D = 10$ dimensions is also achieved, though that is less obvious at this point.

As was discussed in Chapter 2, the action for the bosonic string in conformal gauge is (for $\alpha' = 1/2$ or $T = 1/\pi$)

$$S = -\frac{1}{2\pi} \int d^2\sigma \partial_\alpha X_\mu \partial^\alpha X^\mu, \tag{4.1}$$

and this needs to be supplemented by Virasoro constraints. This is a free field theory in two dimensions. To generalize this action, let us introduce additional internal degrees of freedom describing fermions on the world sheet. Concretely, one can incorporate D Majorana fermions that belong to the vector representation of the Lorentz group $SO(D-1,1)$. In the representation of the two-dimensional Dirac algebra described below, a Majorana spinor is equivalent to a real spinor. The desired action is obtained by adding the standard Dirac action for D free massless fermions to the free theory of D massless bosons

$$S = -\frac{1}{2\pi} \int d^2\sigma \left(\partial_\alpha X_\mu \partial^\alpha X^\mu + \bar\psi^\mu \rho^\alpha \partial_\alpha \psi_\mu \right). \tag{4.2}$$

Here ρ^α, with $\alpha = 0, 1$, represent the two-dimensional Dirac matrices, which obey the Dirac algebra[2]

$$\{\rho^\alpha, \rho^\beta\} = 2\eta^{\alpha\beta}. \tag{4.3}$$

To be explicit, let us choose a basis in which these matrices take the form

$$\rho^0 = \begin{pmatrix} 0 & -1 \\ 1 & 0 \end{pmatrix} \quad \text{and} \quad \rho^1 = \begin{pmatrix} 0 & 1 \\ 1 & 0 \end{pmatrix}. \tag{4.4}$$

Classically, the fermionic world-sheet field ψ^μ is made of Grassmann numbers, which implies that it satisfies the anticommutation relations

$$\{\psi^\mu, \psi^\nu\} = 0. \tag{4.5}$$

---

[2] A Dirac algebra is known to mathematicians as a *Clifford algebra*. In GSW the definition of ρ^α differed by a factor of i and the anticommutator was $-2\eta^{\alpha\beta}$. As a result, some signs differ from those of GSW in subsequent formulas.

4.1 Ramond–Neveu–Schwarz strings

This changes after quantization, of course.

The spinor ψ^μ has two components ψ^μ_A, $A = \pm$,

$$\psi^\mu = \begin{pmatrix} \psi^\mu_- \\ \psi^\mu_+ \end{pmatrix}. \tag{4.6}$$

Here, and in the following, we define the Dirac conjugate of a spinor as

$$\bar\psi = \psi^\dagger \beta, \qquad \beta = i\rho^0, \tag{4.7}$$

which for a Majorana spinor is simply $\psi^T \beta$. Since the Dirac matrices are purely real, Eq. (4.4) is a Majorana representation, and the Majorana spinors ψ^μ are real (in the sense appropriate to Grassmann numbers)

$$\psi^\star_+ = \psi_+ \quad \text{and} \quad \psi^\star_- = \psi_-. \tag{4.8}$$

In this notation the fermionic part of the action is (suppressing the Lorentz index)

$$S_\mathrm{f} = \frac{i}{\pi} \int d^2\sigma \, (\psi_- \partial_+ \psi_- + \psi_+ \partial_- \psi_+), \tag{4.9}$$

where $\partial_\pm$ refer to the world-sheet light-cone coordinates $\sigma^\pm$ introduced in Chapter 2. The equation of motion for the two spinor components is the Dirac equation, which now takes the form

$$\partial_+ \psi_- = 0 \quad \text{and} \quad \partial_- \psi_+ = 0. \tag{4.10}$$

These equations describe left-moving and right-moving waves. For spinors in two dimensions, these are the Weyl conditions. Thus the fields $\psi_\pm$ are Majorana–Weyl spinors.[3]

EXERCISES

EXERCISE 4.1
Show that one can rewrite the fermionic part of the action in Eq. (4.2) in the form in Eq. (4.9).

SOLUTION

Taking $\partial_\pm = \frac{1}{2}(\partial_0 \pm \partial_1)$ and the explicit form of the two-dimensional Dirac

[3] Group theoretically, they are two inequivalent real one-dimensional spinor representations of the two-dimensional Lorentz group $Spin(1,1)$.

matrices (4.4) into account, one obtains

$$\rho^\alpha \partial_\alpha = \begin{pmatrix} 0 & \partial_1 - \partial_0 \\ \partial_1 + \partial_0 & 0 \end{pmatrix} = 2 \begin{pmatrix} 0 & -\partial_- \\ \partial_+ & 0 \end{pmatrix}.$$

From the definition $\bar\psi = \psi^\dagger i\rho^0$, it follows that $\bar\psi = i(\psi_+, -\psi_-)$. The action in Eq. (4.9) is then obtained after carrying out the matrix multiplication. □

4.2 Global world-sheet supersymmetry

The action in Eq. (4.2) is invariant under the infinitesimal transformations

$$\delta X^\mu = \bar\varepsilon \psi^\mu, \tag{4.11}$$

$$\delta \psi^\mu = \rho^\alpha \partial_\alpha X^\mu \varepsilon, \tag{4.12}$$

where ε is a constant infinitesimal Majorana spinor that consists of anticommuting Grassmann numbers. Writing the spinors in components

$$\varepsilon = \begin{pmatrix} \varepsilon_- \\ \varepsilon_+ \end{pmatrix}, \tag{4.13}$$

the *supersymmetry transformations* take the form

$$\delta X^\mu = i(\varepsilon_+ \psi^\mu_- - \varepsilon_- \psi^\mu_+), \tag{4.14}$$

$$\delta \psi^\mu_- = -2\partial_- X^\mu \varepsilon_+, \tag{4.15}$$

$$\delta \psi^\mu_+ = 2\partial_+ X^\mu \varepsilon_-. \tag{4.16}$$

The symmetry holds up to a total derivative that can be dropped for suitable boundary conditions. Since ε is not dependent on σ and τ, this is a global symmetry of the world-sheet theory.[4] The supersymmetry transformations (4.11) mix the bosonic and fermionic world-sheet fields. This fermionic symmetry of the two-dimensional RNS world-sheet action was noted by Gervais and Sakita in 1971 at about the same time that the four-dimensional super-Poincaré algebra was introduced by Golfand and Likhtman in the Soviet Union. Prior to these works, it was believed to be impossible to have a symmetry that relates particles of different spin in a relativistic field theory.

---

4 This is the world-sheet theory in conformal gauge. There is a more fundamental formulation in which the world-sheet supersymmetry is a local symmetry. In conformal gauge it gives rise to the theory considered here.

Superspace

Exercise 4.2 shows that the action (4.2) is invariant under the supersymmetry transformations. The supersymmetry of component actions, such as this one, is not manifest. The easiest way to make this symmetry manifest is by rewriting the action using a superspace formalism. *Superspace* is an extension of ordinary space-time that includes additional anticommuting (Grassmann) coordinates, and *superfields* are fields defined on superspace. The superfield formulation entails adding an off-shell degree of freedom to the world-sheet theory, without changing the physical content. This has the advantage of ensuring that the algebra of supersymmetry transformations closes off-shell, that is, without use of the equations of motion.

The superfield formulation is very convenient for making supersymmetry manifest (and simplifying calculations) in theories that have a relatively small number of conserved supercharges. The number of supercharges is two in the present case. When the number is larger than four, as is necessarily the case for supersymmetric theories when the space-time dimension is greater than four, a superfield formulation can become very unwieldy or even impossible.

The super-world-sheet coordinates are given by $(\sigma^\alpha, \theta_A)$, where

$$\theta_A = \begin{pmatrix} \theta_- \\ \theta_+ \end{pmatrix} \tag{4.17}$$

are anticommuting Grassmann coordinates

$$\{\theta_A, \theta_B\} = 0, \tag{4.18}$$

which form a Majorana spinor. Upper and lower spinor indices need not be distinguished here, so $\theta^A = \theta_A$. Frequently these indices are not displayed. For the usual bosonic world-sheet coordinates let us define $\sigma^0 = \tau$ and $\sigma^1 = \sigma$. One can then introduce a superfield $Y^\mu(\sigma^\alpha, \theta)$. The most general such function has a series expansion in θ of the form

$$Y^\mu(\sigma^\alpha, \theta) = X^\mu(\sigma^\alpha) + \bar{\theta}\psi^\mu(\sigma^\alpha) + \frac{1}{2}\bar{\theta}\theta B^\mu(\sigma^\alpha), \tag{4.19}$$

where $B^\mu(\sigma^\alpha)$ is an auxiliary field whose inclusion does not change the physical content of the theory. This field is needed to make supersymmetry manifest. A term with more powers of θ would automatically vanish as a consequence of the anticommutation properties of the Grassmann numbers θ_A. Since $\bar{\psi}\theta = \bar{\theta}\psi$ for Majorana spinors, a term linear in θ would be equivalent to the linear term in $\bar{\theta}$ appearing above.

The generators of supersymmetry transformations of the super-world-sheet coordinates, called *supercharges*, are

$$Q_A = \frac{\partial}{\partial \bar{\theta}^A} - (\rho^\alpha \theta)_A \partial_\alpha. \tag{4.20}$$

The world-sheet supersymmetry transformations given above can be expressed in terms of Q_A. Acting on superspace, $\bar\varepsilon Q$ generates the transformations

$$\delta \theta^A = \left[\bar\varepsilon Q, \theta^A\right] = \varepsilon^A, \tag{4.21}$$

$$\delta \sigma^\alpha = \left[\bar\varepsilon Q, \sigma^\alpha\right] = -\bar\varepsilon \rho^\alpha \theta = \bar\theta \rho^\alpha \varepsilon, \tag{4.22}$$

of the superspace coordinates. In this way a supersymmetry transformation is interpreted as a geometrical transformation of superspace (see Exercise 4.3). The supercharge Q acts on the superfield according to

$$\delta Y^\mu = \left[\bar\varepsilon Q, Y^\mu\right] = \bar\varepsilon Q Y^\mu. \tag{4.23}$$

Expanding this equation in components and using the two-dimensional Fierz transformation

$$\theta_A \bar\theta_B = -\frac{1}{2} \delta_{AB} \bar\theta_C \theta_C, \tag{4.24}$$

one gets the supersymmetry transformations

$$\delta X^\mu = \bar\varepsilon \psi^\mu, \tag{4.25}$$

$$\delta \psi^\mu = \rho^\alpha \partial_\alpha X^\mu \varepsilon + B^\mu \varepsilon, \tag{4.26}$$

$$\delta B^\mu = \bar\varepsilon \rho^\alpha \partial_\alpha \psi^\mu. \tag{4.27}$$

The first two formulas reduce to the supersymmetry transformations in Eqs (4.11) and (4.12), which do not contain the auxiliary field B^μ, if one uses the field equation $B^\mu = 0$.

The action can be written in superfield language using the *supercovariant derivative*

$$D_A = \frac{\partial}{\partial \bar\theta^A} + (\rho^\alpha \theta)_A \partial_\alpha. \tag{4.28}$$

Note that $\{D_A, Q_B\} = 0$, and therefore the supercovariant derivative $D_A \Phi$ of an arbitrary superfield Φ transforms under supersymmetry in the same way as Φ itself. The desired action, written in terms of superfields, is

$$S = \frac{i}{4\pi} \int d^2\sigma d^2\theta \bar{D} Y^\mu D Y_\mu. \tag{4.29}$$

4.2 Global world-sheet supersymmetry

The definition of integration over Grassmann coordinates is described below. It has the property that the θ integral of a θ derivative is zero. This superspace action has manifest supersymmetry, since the variation gives

$$\delta S = \frac{i}{4\pi} \int d^2\sigma d^2\theta \, \bar{\varepsilon} Q (\bar{D} Y^\mu D Y_\mu). \tag{4.30}$$

Both terms in the definition of Q give total derivatives: one term is a total σ^α derivative and the other term is a total θ^A derivative. Depending on the σ boundary conditions the world-sheet supersymmetry can be broken or unbroken. Both cases are of interest. There are no boundary terms associated with the Grassmann integrations.

The superspace formula for the action can be written in components by substituting the component expansion of Y and carrying out the Grassmann integrations. The basic rule for Grassmann integration in the case of a single coordinate is

$$\int d\theta (a + \theta b) = b. \tag{4.31}$$

In the present case there are two Grassmann coordinates, and the only nonzero integral is

$$\int d^2\theta \, \bar{\theta}\theta = -2i. \tag{4.32}$$

The component form of the action can be derived by using this rule as well as the expansions

$$DY^\mu = \psi^\mu + \theta B^\mu + \rho^\alpha \theta \partial_\alpha X^\mu - \frac{1}{2}\bar{\theta}\theta \rho^\alpha \partial_\alpha \psi^\mu, \tag{4.33}$$

$$\bar{D} Y^\mu = \bar{\psi}^\mu + B^\mu \bar{\theta} - \bar{\theta} \partial_\alpha X^\mu \rho^\alpha + \frac{1}{2}\bar{\theta}\theta \partial_\alpha \bar{\psi}^\mu \rho^\alpha. \tag{4.34}$$

One finds

$$S = -\frac{1}{2\pi} \int d^2\sigma \left(\partial_\alpha X_\mu \partial^\alpha X^\mu + \bar{\psi}^\mu \rho^\alpha \partial_\alpha \psi_\mu - B_\mu B^\mu \right). \tag{4.35}$$

This action implies that the equation of motion for B^μ is $B^\mu = 0$, as was asserted earlier. As a result, the auxiliary field B^μ can be eliminated from the theory leaving Eq. (4.2). The price of doing this is the loss of manifest supersymmetry as well as off-shell closure of the supersymmetry algebra.

Exercises

Exercise 4.2
Verify that the action (4.2) is invariant under the supersymmetry transformations (4.11) up to a total derivative.

Solution

Suppressing Lorentz indices, it is straightforward to vary the action

$$S = \frac{1}{\pi} \int d^2\sigma \left(2\partial_+ X \partial_- X + i\psi_- \partial_+ \psi_- + i\psi_+ \partial_- \psi_+\right).$$

The terms proportional to ε_+ are

$$\delta_+ S = \frac{2i}{\pi} \varepsilon_+ \int d^2\sigma \left(\partial_+ \psi_- \partial_- X + \partial_+ X \partial_- \psi_- - \partial_- X \partial_+ \psi_- + \psi_- \partial_+ \partial_- X\right) \approx 0.$$

The equivalence to 0 is a consequence of the fact that the integrand is a total derivative. The terms proportional to ε_- work in a similar manner. $\square$

Exercise 4.3
Show that the commutator of two supersymmetry transformations (4.11) amounts to a translation along the string world sheet by evaluating the commutators $[\delta_1, \delta_2] X^\mu$ and $[\delta_1, \delta_2] \psi^\mu$.

Solution

Using the supersymmetry transformations

$$\delta X^\mu = \bar{\varepsilon} \psi^\mu, \qquad \delta \psi^\mu = \rho^\alpha \partial_\alpha X^\mu \varepsilon,$$

we first compute the commutator acting on the fermionic field

$$[\delta_{\varepsilon_1}, \delta_{\varepsilon_2}] \psi^\mu = \delta_{\varepsilon_1}(\delta_{\varepsilon_2} \psi^\mu) - \delta_{\varepsilon_2}(\delta_{\varepsilon_1} \psi^\mu) = \delta_{\varepsilon_1}(\rho^\alpha \partial_\alpha X^\mu \varepsilon_2) - \delta_{\varepsilon_2}(\rho^\alpha \partial_\alpha X^\mu \varepsilon_1)$$

$$= \rho^\alpha \varepsilon_2 \partial_\alpha \delta_{\varepsilon_1} X^\mu - \rho^\alpha \varepsilon_1 \partial_\alpha \delta_{\varepsilon_2} X^\mu = \rho^\alpha (\varepsilon_2 \bar{\varepsilon}_1 - \varepsilon_1 \bar{\varepsilon}_2) \partial_\alpha \psi^\mu.$$

Using the spinor identity $\varepsilon_2 \bar{\varepsilon}_1 - \varepsilon_1 \bar{\varepsilon}_2 = -\bar{\varepsilon}_1 \rho_\beta \varepsilon_2 \rho^\beta$ and the anticommutation relations of the Dirac matrices, this becomes

$$-\bar{\varepsilon}_1 \rho_\beta \varepsilon_2 \rho^\alpha \rho^\beta \partial_\alpha \psi^\mu = -2\bar{\varepsilon}_1 \rho^\alpha \varepsilon_2 \partial_\alpha \psi^\mu + \bar{\varepsilon}_1 \rho_\beta \varepsilon_2 \rho^\beta \rho^\alpha \partial_\alpha \psi^\mu.$$

4.2 Global world-sheet supersymmetry

The first term is interpreted as a translation by the amount

$$a^\alpha = -2\bar\varepsilon_1 \rho^\alpha \varepsilon_2.$$

Note that this is an even element of the Grassmann algebra, but not an ordinary number. So the notion of translation has to be generalized in this way. The second term vanishes using the equation of motion $\rho^\alpha \partial_\alpha \psi^\mu = 0$. This is what we are referring to when we say that the algebra only closes on-shell. When the auxiliary field is included, one achieves off-shell closure of the algebra.

The commutator acting on the bosonic field can be computed in a similar way

$$[\delta_{\varepsilon_1}, \delta_{\varepsilon_2}] X^\mu = \bar\varepsilon_2 \delta_{\varepsilon_1} \psi^\mu - \bar\varepsilon_1 \delta_{\varepsilon_2} \psi^\mu = -2\bar\varepsilon_1 \rho^\alpha \varepsilon_2 \partial_\alpha X^\mu,$$

where we have used the identity $\bar\varepsilon_1 \rho^\alpha \varepsilon_2 = -\bar\varepsilon_2 \rho^\alpha \varepsilon_1$. This is a translation by the same a^α as before. □

EXERCISE 4.4

Use the supersymmetry transformation for the superfield (4.23) to derive the supersymmetry transformation for the component fields (4.25)–(4.27).

SOLUTION

The supersymmetry variation of the superfield is

$$\delta Y^\mu = [\bar\varepsilon Q, Y^\mu(\sigma, \theta)] = \bar\varepsilon Q Y^\mu(\sigma, \theta),$$

where

$$Q_A = \frac{\partial}{\partial \bar\theta^A} - (\rho^\alpha \theta)_A \partial_\alpha.$$

So we obtain

$$\delta Y^\mu(\sigma, \theta) = \bar\varepsilon^A Q_A \left(X^\mu(\sigma) + \bar\theta \psi^\mu(\sigma) + \frac{1}{2} \bar\theta \theta B^\mu(\sigma) \right)$$

$$= \bar\varepsilon^A \psi^\mu_A(\sigma) - \bar\varepsilon^A (\rho^\alpha \theta)_A \partial_\alpha X^\mu(\sigma) + \bar\varepsilon^A \theta_A B^\mu(\sigma) - \bar\varepsilon^A (\rho^\alpha \theta)_A \bar\theta^B \partial_\alpha \psi^\mu_B(\sigma)$$

$$= \bar\varepsilon \psi^\mu(\sigma) + \bar\theta \rho^\alpha \varepsilon \partial_\alpha X^\mu(\sigma) + \bar\theta \varepsilon B^\mu(\sigma) + \frac{1}{2} \bar\theta \theta \bar\varepsilon \rho^\alpha \partial_\alpha \psi^\mu(\sigma).$$

From here we can read off the supersymmetry transformations for the component fields by matching the different terms in the θ expansion. □

Exercise 4.5

Derive the component form of the action in Eq. (4.35) from the superspace action in Eq. (4.29).

Solution

The supercovariant derivatives acting on superfields DY^μ and $\bar{D}Y^\mu$ are given in Eqs (4.33) and (4.34). We now multiply these expressions and substitute into Eq. (4.29). Since only terms quadratic in θ survive integration, the nonzero terms in Eq. (4.29) are

$$S = \frac{i}{4\pi} \int d^2\sigma d^2\theta \left(-\bar{\psi}^\mu \rho^\alpha \partial_\alpha \psi_\mu \bar{\theta}\theta + B^\mu B_\mu \bar{\theta}\theta - \bar{\theta}\rho^\alpha \partial_\alpha X^\mu \rho^\beta \theta \partial_\beta X_\mu \right).$$

The last term simplifies according to

$$\bar{\theta}\rho^\alpha \partial_\alpha X^\mu \rho^\beta \theta \partial_\beta X_\mu = \partial^\alpha X^\mu \partial_\alpha X_\mu \bar{\theta}\theta.$$

Therefore, by using Eq. (4.32) one obtains Eq. (4.35) for the component action. □

4.3 Constraint equations and conformal invariance

Let us now proceed as in Chapter 2. From the equations of motion we can derive the mode expansion of the fields and use canonical quantization to construct the spectrum of the theory. The problem of negative-norm states appears also in the supersymmetric theory. Recall that in the case of the bosonic string theory the spectrum seemed to contain negative-norm states, but these were shown to be unphysical. Specifically, in Chapter 2 it was shown that the negative-norm states decouple and Lorentz invariance is maintained for $D = 26$. The RNS string has a *superconformal symmetry* that allows us to proceed in a similar manner. The negative-norm states are eliminated by using the super-Virasoro constraints that follow from the superconformal symmetry in the critical dimension $D = 10$. Alternatively, one can use it to fix a light-cone gauge and maintain Lorentz invariance for $D = 10$.

In order to discuss the appropriate generalization of conformal invariance for the RNS string, let us start by constructing the conserved currents associated with the global symmetries of the action. These are the energy–momentum tensor (associated with translation symmetry) and the supercurrent (associated with supersymmetry). In particular, the energy–

4.3 Constraint equations and conformal invariance

momentum tensor of the RNS string is

$$T_{\alpha\beta} = \partial_\alpha X^\mu \partial_\beta X_\mu + \frac{1}{4}\bar\psi^\mu \rho_\alpha \partial_\beta \psi_\mu + \frac{1}{4}\bar\psi^\mu \rho_\beta \partial_\alpha \psi_\mu - (\text{trace}). \tag{4.36}$$

The conserved current associated with the global world-sheet supersymmetry of the RNS string is the world-sheet supercurrent. It can be constructed using the Noether method. Specifically, taking the supersymmetry parameter ε to be nonconstant, one finds that up to a total derivative the variation of the action (4.2) takes the form

$$\delta S \sim \int d^2\sigma (\partial_\alpha \bar\varepsilon) J^\alpha, \tag{4.37}$$

where

$$J^\alpha_A = -\frac{1}{2}(\rho^\beta \rho^\alpha \psi_\mu)_A \partial_\beta X^\mu. \tag{4.38}$$

This current satisfies

$$(\rho_\alpha)_{AB} J^\alpha_B = 0 \tag{4.39}$$

as a consequence of the identity $\rho_\alpha \rho^\beta \rho^\alpha = 0$. This is the analog of the tracelessness of the $T_{\alpha\beta}$. In fact, it can be traced back to local super-Weyl invariance in the formalism with local world-sheet supersymmetry. As a result, J^α_A has only two independent components, which can be denoted J_+ and J_-.

Written in terms of world-sheet light-cone coordinates, the nonzero components of the energy–momentum tensor in Eq. (4.36) are

$$T_{++} = \partial_+ X_\mu \partial_+ X^\mu + \frac{i}{2}\psi^\mu_+ \partial_+ \psi_{+\mu}, \tag{4.40}$$

$$T_{--} = \partial_- X_\mu \partial_- X^\mu + \frac{i}{2}\psi^\mu_- \partial_- \psi_{-\mu}. \tag{4.41}$$

Similarly, the nonzero components of the supercurrent in Eq. (4.38) are

$$J_+ = \psi^\mu_+ \partial_+ X_\mu \quad \text{and} \quad J_- = \psi^\mu_- \partial_- X_\mu. \tag{4.42}$$

The supercurrent (4.38) is conserved, $\partial_\alpha J^\alpha_A = 0$, as a consequence of the equations of motion, which leads to

$$\partial_- J_+ = \partial_+ J_- = 0. \tag{4.43}$$

The energy–momentum tensor satisfies analogous relations

$$\partial_- T_{++} = \partial_+ T_{--} = 0. \tag{4.44}$$

These relations follow immediately from the equations of motion $\partial_+ \partial_- X^\mu =$

0 and $\partial_+\psi^\mu_- = \partial_-\psi^\mu_+ = 0$. However, the requirements of superconformal symmetry actually lead to stronger conditions than these, namely the vanishing of the supercurrent and the energy–momentum tensor.

In order to quantize the theory, one can introduce canonical anticommutation relations for the fermionic world-sheet fields

$$\{\psi^\mu_A(\sigma,\tau), \psi^\nu_B(\sigma',\tau)\} = \pi\eta^{\mu\nu}\delta_{AB}\delta(\sigma-\sigma') \qquad (4.45)$$

in addition to the commutation relations for the bosonic world-sheet fields $X^\mu(\sigma,\tau)$ given in Chapter 2. Because $\eta^{00} = -1$, there are negative-norm states that originate from the time-like fermion ψ^0 in the same way as for the time-like boson X^0. These must not appear in the physical spectrum, if one wants a sensible causal theory.

Once again there is sufficient symmetry to eliminate the unwanted negative-norm states. In the case of the bosonic theory the conditions $T_{+-} = T_{-+} = 0$ followed from Weyl invariance, while $T_{++} = T_{--} = 0$ followed from the equations of motion for the world-sheet metric. The latter conditions were shown to imply conformal invariance. This symmetry could be used to choose the light-cone gauge, which gives a manifestly positive-norm spectrum in the quantum theory. Let us try to follow the same steps in the RNS case.

The first step is to formulate the constraint equations that can be used to eliminate the time-like components of ψ^μ and X^μ. In the bosonic case the time-like component was eliminated in 26 dimensions by using the Virasoro constraints $T_{++} = T_{--} = 0$. In the supersymmetric case it is natural to try the same procedure again and to eliminate the time-like components by using suitably generalized Virasoro conditions. In the RNS theory the corresponding conditions are

$$J_+ = J_- = T_{++} = T_{--} = 0. \qquad (4.46)$$

One way of understanding this is in terms of the consistency with the algebra of the currents. However, a deeper understanding can be achieved by starting from a world-sheet action that has local supersymmetry. This can be constructed by gauging the world-sheet supersymmetry by introducing a world-sheet Rarita–Schwinger gauge field, in addition to a world-sheet zweibein, which replaces the world-sheet metric for theories with spinors. The formulas are given in Section 4.3.4 of GSW. Just as the equations of motion of the metric in conformal gauge give the vanishing of the energy–momentum tensor, so the equations of motion of the Rarita–Schwinger field give the vanishing of the supercurrent.

Exercises

Exercise 4.6
Verify the form of the energy–momentum tensor in Eqs (4.40) and (4.41).

Solution

These conserved currents should be a consequence of the world-sheet translation symmetry of the action

$$S = \frac{1}{\pi} \int d^2\sigma \left(2\partial_+ X \cdot \partial_- X + i\psi_- \cdot \partial_+ \psi_- + i\psi_+ \cdot \partial_- \psi_+\right)$$

derivable by the Noether method. An infinitesimal translation is given by $\delta X = a^\alpha \partial_\alpha X$ and $\delta \psi_A = a^\alpha \partial_\alpha \psi_A$. We focus here on $\delta_+ X = a^+ \partial_+ X$ and $\delta_+ \psi_A = a^+ \partial_+ \psi_A$, since the a^- transformations work in exactly the same way.

$$\delta_+ \left(2\partial_+ X \cdot \partial_- X + i\psi_- \cdot \partial_+ \psi_- + i\psi_+ \cdot \partial_- \psi_+\right)$$
$$= a^+ \left(-2\partial_-(\partial_+ X \cdot \partial_+ X) + i\partial_+(\psi_+ \cdot \partial_- \psi_+) - i\partial_-(\psi_+ \cdot \partial_+ \psi_+)\right)$$

up to a total derivative. Identifying this with

$$-2a^+(\partial_- T_{++} + \partial_+ T_{-+})$$

gives the desired result

$$T_{++} = \partial_+ X \cdot \partial_+ X + \frac{i}{2}\psi_+ \cdot \partial_+ \psi_+.$$

It also appears to give $T_{-+} = -\frac{i}{2}\psi_+ \cdot \partial_- \psi_+$. However, this vanishes by an equation of motion. Similarly, the a^- variation leads to

$$T_{--} = \partial_- X \cdot \partial_- X + \frac{i}{2}\psi_- \cdot \partial_- \psi_-.$$

$\square$

Exercise 4.7
Verify the form of the supercurrent in Eq. (4.42).

Solution

The method is the same as in the previous exercise. This time we want to find the currents associated with the supersymmetry transformations in Eqs (4.14)–(4.16). It is sufficient to consider the ε_- transformations, since the ε_+ ones work in an identical way. Therefore, we consider

$$\delta_- X^\mu = i\varepsilon_- \psi_+^\mu,$$

$$\delta_- \psi_+^\mu = -2\partial_+ X^\mu \varepsilon_- \quad \text{and} \quad \delta_- \psi_-^\mu = 0.$$

Using these rules,

$$\delta_- \left(2\partial_+ X \cdot \partial_- X + i\psi_- \cdot \partial_+ \psi_- + i\psi_+ \cdot \partial_- \psi_+\right) = -4i\varepsilon_- \partial_- (\psi_+ \cdot \partial_+ X)$$

up to a total derivative. Thus, choosing the normalization appropriately, this shows that $J_+ = \psi_+ \cdot \partial_+ X$. Similarly, the expression $J_- = \psi_- \cdot \partial_- X$ is obtained by considering an ε_+ transformation. □

4.4 Boundary conditions and mode expansions

The possible boundary conditions and mode expansions for the bosonic fields X^μ are exactly the same as for the case of the bosonic string theory, so that discussion is not repeated here.

Suppressing the Lorentz index μ, the action for the fermionic fields ψ^μ in light-cone world-sheet coordinates is

$$S_{\text{f}} \sim \int d^2\sigma \left(\psi_- \partial_+ \psi_- + \psi_+ \partial_- \psi_+\right). \tag{4.47}$$

By considering variations of the fields $\psi_\pm$ one finds that the action is stationary if the equations of motion (4.10) are satisfied. The boundary terms in the variation of the action,

$$\delta S \sim \int d\tau \left(\psi_+ \delta\psi_+ - \psi_- \delta\psi_-\right)|_{\sigma=\pi} - \left(\psi_+ \delta\psi_+ - \psi_- \delta\psi_-\right)|_{\sigma=0}, \tag{4.48}$$

must also vanish. There are several ways to achieve this, which are discussed in the next two subsections.

Open strings

In the case of open strings the two terms in (4.48), corresponding to the two ends of the string, must vanish separately. This requirement is satisfied if at each end of the string

$$\psi_+^\mu = \pm \psi_-^\mu. \tag{4.49}$$

4.4 Boundary conditions and mode expansions

The overall relative sign between ψ_+^μ and ψ_-^μ is a matter of convention. Therefore, without loss of generality, one can choose to set

$$\psi_+^\mu|_{\sigma=0} = \psi_-^\mu|_{\sigma=0}. \tag{4.50}$$

The relative sign at the other end then becomes meaningful, and there are two possible cases:

- *Ramond boundary condition*: In this case one chooses at the second end of the string

$$\psi_+^\mu|_{\sigma=\pi} = \psi_-^\mu|_{\sigma=\pi}. \tag{4.51}$$

As is shown later, Ramond (or R) boundary conditions give rise to space-time fermions. The mode expansion of the fermionic field in the R sector takes the form

$$\psi_-^\mu(\sigma,\tau) = \frac{1}{\sqrt{2}} \sum_{n\in\mathbb{Z}} d_n^\mu e^{-in(\tau-\sigma)}, \tag{4.52}$$

$$\psi_+^\mu(\sigma,\tau) = \frac{1}{\sqrt{2}} \sum_{n\in\mathbb{Z}} d_n^\mu e^{-in(\tau+\sigma)}. \tag{4.53}$$

The Majorana condition requires these expansions to be real, and hence $d_{-n}^\mu = d_n^{\mu\dagger}$. The normalization factor is chosen for later convenience.

- *Neveu–Schwarz boundary condition*: This boundary condition corresponds to choosing a relative minus sign at the second end of the string, namely

$$\psi_+^\mu|_{\sigma=\pi} = -\psi_-^\mu|_{\sigma=\pi}. \tag{4.54}$$

As is shown later, Neveu–Schwarz (or NS) boundary conditions give rise to space-time bosons. The mode expansion in the NS sector is

$$\psi_-^\mu(\sigma,\tau) = \frac{1}{\sqrt{2}} \sum_{r\in\mathbb{Z}+1/2} b_r^\mu e^{-ir(\tau-\sigma)}, \tag{4.55}$$

$$\psi_+^\mu(\sigma,\tau) = \frac{1}{\sqrt{2}} \sum_{r\in\mathbb{Z}+1/2} b_r^\mu e^{-ir(\tau+\sigma)}. \tag{4.56}$$

In the following, the letters m and n are used for integers while r and s are used for half-integers, that is,

$$m,n \in \mathbb{Z} \quad\text{while}\quad r,s \in \mathbb{Z} + \frac{1}{2}. \tag{4.57}$$

Closed strings

Closed-string boundary conditions give two sets of fermionic modes, corresponding to the left- and right-moving sectors. There are two possible periodicity conditions

$$\psi_\pm(\sigma) = \pm\psi_\pm(\sigma + \pi), \tag{4.58}$$

each of which makes the boundary term vanish. The positive sign in the above relation describes periodic boundary conditions while the negative sign describes antiperiodic boundary conditions. It is possible to impose the periodicity (R) or antiperiodicity (NS) of the right- and left-movers separately. This means that, for the right-movers, one can choose

$$\psi_-^\mu(\sigma,\tau) = \sum_{n\in\mathbb{Z}} d_n^\mu e^{-2in(\tau-\sigma)} \quad \text{or} \quad \psi_-^\mu(\sigma,\tau) = \sum_{r\in\mathbb{Z}+1/2} b_r^\mu e^{-2ir(\tau-\sigma)}, \tag{4.59}$$

while for the left-movers one can choose

$$\psi_+^\mu(\sigma,\tau) = \sum_{n\in\mathbb{Z}} \tilde{d}_n^\mu e^{-2in(\tau+\sigma)} \quad \text{or} \quad \psi_+^\mu(\sigma,\tau) = \sum_{r\in\mathbb{Z}+1/2} \tilde{b}_r^\mu e^{-2ir(\tau+\sigma)}. \tag{4.60}$$

Corresponding to the different pairings of the left- and right-movers there are four distinct closed-string sectors. States in the NS–NS and R–R sectors are space-time bosons, while states in the NS–R and R–NS sectors are space-time fermions.

4.5 Canonical quantization of the RNS string

The modes in the Fourier expansion of the space-time coordinates satisfy the same commutation relations as in the case of the bosonic string, namely

$$[\alpha_m^\mu, \alpha_n^\nu] = m\delta_{m+n,0}\eta^{\mu\nu}. \tag{4.61}$$

For the closed string there is again a second set of modes $\tilde{\alpha}_m^\mu$.

The fermionic coordinates obey the free Dirac equation on the world sheet. As a result, the canonical anticommutation relations are those given in Eq. (4.45), which imply that the Fourier coefficients satisfy

$$\{b_r^\mu, b_s^\nu\} = \eta^{\mu\nu}\delta_{r+s,0} \quad \text{and} \quad \{d_m^\mu, d_n^\nu\} = \eta^{\mu\nu}\delta_{m+n,0}. \tag{4.62}$$

Since the space-time metric appears on the right-hand side in the above commutation relations, the time components of the fermionic modes give rise to negative-norm states, just like the time components of the bosonic modes.

These negative-norm states are decoupled as a consequence of the appropriate generalization of conformal invariance. Specifically, the conformal symmetry of the bosonic string generalizes to a superconformal symmetry of the RNS string, which is just what is required.

The oscillator ground state in the two sectors is defined by

$$\alpha_m^\mu |0\rangle_R = d_m^\mu |0\rangle_R = 0 \quad \text{for} \quad m > 0 \qquad (4.63)$$

and

$$\alpha_m^\mu |0\rangle_{NS} = b_r^\mu |0\rangle_{NS} = 0 \quad \text{for} \quad m, r > 0. \qquad (4.64)$$

Excited states are constructed by acting with the negative modes (or raising modes) of the oscillators. Acting with the negative modes increases the mass of the states. In the NS sector there is a unique ground state, which corresponds to a state of spin 0 in space-time. Since all the oscillators transform as space-time vectors, the excited states that are obtained by acting with raising operators are also space-time bosons.

By contrast, in the R sector the ground state is degenerate. The operators d_0^μ can act without changing the mass of a state, because they commute with the number operator N, defined below, whose eigenvalue determines the mass squared. Equation (4.62) tells us that these zero modes satisfy the algebra

$$\{d_0^\mu, d_0^\nu\} = \eta^{\mu\nu}. \qquad (4.65)$$

Aside from a factor of two, this is identical to the Dirac algebra

$$\{\Gamma^\mu, \Gamma^\nu\} = 2\eta^{\mu\nu}. \qquad (4.66)$$

As a result, the set of ground states in the R sector must furnish a representation of this algebra. This means that there is a set of degenerate ground states, which can be written in the form $|a\rangle$, where a is a spinor index, such that

$$d_0^\mu |a\rangle = \frac{1}{\sqrt{2}} \Gamma_{ba}^\mu |b\rangle. \qquad (4.67)$$

Hence the R-sector ground state is a space-time fermion. Since all of the oscillators (α_n^μ and d_n^μ) are space-time vectors, and every state in the R sector can be obtained by acting with raising operators on the R-sector ground state, all R-sector states are space-time fermions.

Super-Virasoro generators and physical states

The super-Virasoro generators are the modes of the energy–momentum tensor $T_{\alpha\beta}$ and the supercurrent J^α_A. For the open string they are given by

$$L_m = \frac{1}{\pi}\int_{-\pi}^{\pi} d\sigma\, e^{im\sigma} T_{++} = L_m^{(b)} + L_m^{(f)}. \tag{4.68}$$

- The contribution coming from the bosonic modes is

$$L_m^{(b)} = \frac{1}{2}\sum_{n\in\mathbb{Z}} :\alpha_{-n}\cdot\alpha_{m+n}: \qquad m\in\mathbb{Z}. \tag{4.69}$$

- The contribution of the fermionic modes in the NS sector is

$$L_m^{(f)} = \frac{1}{2}\sum_{r\in\mathbb{Z}+1/2}\left(r+\frac{m}{2}\right):b_{-r}\cdot b_{m+r}: \qquad m\in\mathbb{Z}. \tag{4.70}$$

The modes of the supercurrent in the NS sector are

$$G_r = \frac{\sqrt{2}}{\pi}\int_{-\pi}^{\pi} d\sigma\, e^{ir\sigma} J_+ = \sum_{n\in\mathbb{Z}}\alpha_{-n}\cdot b_{r+n} \qquad r\in\mathbb{Z}+\frac{1}{2}. \tag{4.71}$$

The operator L_0 can be written in the form

$$L_0 = \frac{1}{2}\alpha_0^2 + N, \tag{4.72}$$

where the number operator N is given by

$$N = \sum_{n=1}^{\infty} \alpha_{-n}\cdot\alpha_n + \sum_{r=1/2}^{\infty} r\, b_{-r}\cdot b_r. \tag{4.73}$$

As in the bosonic theory of Chapter 2, the eigenvalue of N determines the mass squared of an excited string state.

- In the R sector

$$L_m^{(f)} = \frac{1}{2}\sum_{n\in\mathbb{Z}}\left(n+\frac{m}{2}\right):d_{-n}\cdot d_{m+n}: \qquad m\in\mathbb{Z}, \tag{4.74}$$

while the modes of the supercurrent are

$$F_m = \frac{\sqrt{2}}{\pi}\int_{-\pi}^{\pi} d\sigma\, e^{im\sigma} J_+ = \sum_{n\in\mathbb{Z}}\alpha_{-n}\cdot d_{m+n} \qquad m\in\mathbb{Z}. \tag{4.75}$$

Note that there is no normal-ordering ambiguity in the definition of F_0.

The algebra satisfied by the modes of the energy–momentum tensor and supercurrent can now be determined. For the modes of the supercurrent in the R sector one obtains the super-Virasoro algebra

$$[L_m, L_n] = (m-n)L_{m+n} + \frac{D}{8}m^3 \delta_{m+n,0}, \qquad (4.76)$$

$$[L_m, F_n] = \left(\frac{m}{2} - n\right) F_{m+n}, \qquad (4.77)$$

$$\{F_m, F_n\} = 2L_{m+n} + \frac{D}{2}m^2 \delta_{m+n,0}, \qquad (4.78)$$

while in the NS sector one gets the super-Virasoro algebra

$$[L_m, L_n] = (m-n)L_{m+n} + \frac{D}{8}m(m^2-1)\delta_{m+n,0}, \qquad (4.79)$$

$$[L_m, G_r] = \left(\frac{m}{2} - r\right) G_{m+r}, \qquad (4.80)$$

$$\{G_r, G_s\} = 2L_{r+s} + \frac{D}{2}\left(r^2 - \frac{1}{4}\right) \delta_{r+s,0}. \qquad (4.81)$$

When quantizing the RNS string one can only require that the positive modes of the Virasoro generators annihilate the physical state. So in the NS sector the physical-state conditions are

$$G_r|\phi\rangle = 0 \qquad r > 0, \qquad (4.82)$$

$$L_m|\phi\rangle = 0 \qquad m > 0, \qquad (4.83)$$

$$(L_0 - a_{\rm NS})|\phi\rangle = 0. \qquad (4.84)$$

The last of these conditions implies that $\alpha' M^2 = N - a_{\rm NS}$, where M is the mass of a state $|\phi\rangle$ and N is replaced by its eigenvalue for this state. Similarly, in the R sector the physical-state conditions are

$$F_n|\phi\rangle = 0 \qquad n \geq 0, \qquad (4.85)$$

$$L_m|\phi\rangle = 0 \qquad m > 0, \qquad (4.86)$$

$$(L_0 - a_{\rm R})|\phi\rangle = 0. \qquad (4.87)$$

In the above formulas $a_{\rm NS}$ and $a_{\rm R}$ are constants introduced to allow for a normal-ordering ambiguity, which must be determined. In fact, the value $a_{\rm R} = 0$ in the R sector is immediately deduced from the identity $L_0 = F_0^2$ and the F_0 equation. The F_0 equation can be written in the form

$$\left(p\cdot\Gamma + \frac{2\sqrt{2}}{l_s}\sum_{n=1}^{\infty}(\alpha_{-n}\cdot d_n + d_{-n}\cdot \alpha_n)\right)|\phi\rangle = 0. \tag{4.88}$$

This is a stringy generalization of the Dirac equation, known as the *Dirac–Ramond equation*.

EXERCISES

EXERCISE 4.8
Verify that the F_0 constraint can be rewritten as the Dirac–Ramond equation (4.88), as stated above.

SOLUTION
Since
$$\alpha_0^\mu = \frac{1}{2}l_s p^\mu, \qquad d_0^\mu = \frac{1}{\sqrt{2}}\Gamma^\mu$$

and
$$F_0 = \sum_{n=-\infty}^{\infty}\alpha_{-n}\cdot d_n = \alpha_0\cdot d_0 + \sum_{n=1}^{\infty}(\alpha_{-n}\cdot d_n + d_{-n}\cdot \alpha_n),$$

the equation $F_0|\phi\rangle = 0$ takes the form given in Eq. (4.88). $\square$

EXERCISE 4.9
Verify that the NS sector super-Virasoro generators L_1, L_0, L_{-1}, $G_{-1/2}$ and $G_{1/2}$ form a closed superalgebra.

SOLUTION
It is easy to see from inspection of the NS sector super-Virasoro algebra given in Eqs (4.79)–(4.81) that the commutation and anticommutation relations of these operators give a closed superalgebra. In particular, they imply that $G_{1/2}^2 = \frac{1}{2}\{G_{1/2}, G_{1/2}\} = L_1$ and $G_{-1/2}^2 = \frac{1}{2}\{G_{-1/2}, G_{-1/2}\} = L_{-1}$. The name of this superalgebra with three even generators and two odd generators is $SU(1,1|1)$ or $OSp(1|2)$. $\square$

Absence of negative-norm states

As in the discussion of the bosonic string in Chapter 2, there are specific values of a and D for which additional zero-norm states appear in the spectrum. The critical dimension turns out to be $D = 10$, while the result for a depends on the sector:

$$a_{\mathrm{NS}} = \frac{1}{2} \quad \text{and} \quad a_{\mathrm{R}} = 0. \tag{4.89}$$

As before, the theory is only Lorentz invariant in the light-cone gauge if a_{NS}, a_{R} and D take these values.

Let us consider a few simple examples of zero-norm spurious states. Recall that these are states that are orthogonal to physical states and decouple from the theory even though they satisfy the physical state conditions.

- *Example 1*: Consider NS-sector states of the form

$$|\psi\rangle = G_{-1/2}|\chi\rangle, \tag{4.90}$$

with $|\chi\rangle$ satisfying the conditions

$$G_{1/2}|\chi\rangle = G_{3/2}|\chi\rangle = \left(L_0 - a_{\mathrm{NS}} + \frac{1}{2}\right)|\chi\rangle = 0. \tag{4.91}$$

The last of these conditions is equivalent to $(L_0 - a_{\mathrm{NS}})|\psi\rangle = 0$. To ensure that $|\psi\rangle$ is physical, it is therefore sufficient to require that $G_{1/2}|\psi\rangle = G_{3/2}|\psi\rangle = 0$. The $G_{3/2}$ condition is an immediate consequence of the corresponding conditions for $|\chi\rangle$. So only the $G_{1/2}$ condition needs to be checked:

$$G_{1/2}|\psi\rangle = G_{1/2}G_{-1/2}|\chi\rangle = (2L_0 - G_{-1/2}G_{1/2})|\chi\rangle = (2a_{\mathrm{NS}} - 1)|\chi\rangle. \tag{4.92}$$

Requiring this to vanish gives $a_{\mathrm{NS}} = 1/2$. This choice gives a family of zero-norm spurious states $|\psi\rangle$. Such a state satisfies the conditions for a physical state with $a_{\mathrm{NS}} = 1/2$. Moreover, $|\psi\rangle$ is orthogonal to all physical states, including itself, since

$$\langle \alpha|\psi\rangle = \langle \alpha|G_{-1/2}|\chi\rangle = \langle \chi|G_{1/2}|\alpha\rangle^\star = 0, \tag{4.93}$$

for any physical state $|\alpha\rangle$. Therefore, for $a_{\mathrm{NS}} = 1/2$ these are zero-norm spurious states.

- *Example 2*: Now let us construct a second class of NS-sector zero-norm spurious states. Consider states of the form

$$|\psi\rangle = \left(G_{-3/2} + \lambda G_{-1/2}L_{-1}\right)|\chi\rangle. \tag{4.94}$$

Suppose further that the state $|\chi\rangle$ satisfies

$$G_{1/2}|\chi\rangle = G_{3/2}|\chi\rangle = (L_0 + 1)|\chi\rangle = 0. \tag{4.95}$$

The L_0 condition incorporates the previous result, $a = 1/2$. Using the super-Virasoro algebra one can compute the following relations:

$$G_{1/2}|\psi\rangle = (2 - \lambda)L_{-1}|\chi\rangle, \tag{4.96}$$

$$G_{3/2}|\psi\rangle = (D - 2 - 4\lambda)|\chi\rangle, \tag{4.97}$$

which have to vanish if $|\psi\rangle$ is a physical state. Therefore, by the same reasoning as in the previous example, one concludes that $|\psi\rangle$ is a zero-norm spurious state if $\lambda = 2$ and $D = 10$.

- *Example 3*: It was already explained that $a_R = 0$ in the R sector as a consequence of $F_0^2 = L_0$. It is possible to construct a family of zero-norm spurious states to confirm the choice $D = 10$ in this sector. Such a set of zero-norm states can be built from R-sector states of the form

$$|\psi\rangle = F_0 F_{-1}|\chi\rangle, \tag{4.98}$$

where

$$F_1|\chi\rangle = (L_0 + 1)|\chi\rangle = 0. \tag{4.99}$$

This state satisfies $F_0|\psi\rangle = 0$. If it is also annihilated by L_1, then it is a physical state with zero-norm. It is easy to check that

$$L_1|\psi\rangle = (\frac{1}{2}F_1 + F_0 L_1)F_{-1}|\chi\rangle = \frac{1}{4}(D - 10)|\chi\rangle. \tag{4.100}$$

This vanishes for $D = 10$ giving us another family of zero-norm spurious states for this space-time dimension.

4.6 Light-cone gauge quantization of the RNS string

As in the case of the bosonic string, after gauge fixing there is a residual symmetry that can be used to impose the light-cone gauge condition

$$X^+(\sigma, \tau) = x^+ + p^+ \tau. \tag{4.101}$$

This is true for the RNS string as well. Moreover, there is also a residual fermionic symmetry that can be used to set[5]

$$\psi^+(\sigma, \tau) = 0, \tag{4.102}$$

---

[5] This formula is correct in the NS sector. In the R sector one should keep the zero mode, which is a Dirac matrix.

4.6 Light-cone gauge quantization of the RNS string

at the same time. Because of the Virasoro constraint, the coordinate X^- is not an independent degree of freedom in the light-cone gauge (except for its zero mode). The same is true for ψ^- when the RNS theory is analyzed in light-cone gauge. Therefore, all the independent physical excitations are obtained in light-cone gauge by acting on the ground states with the transverse raising modes of the bosonic and fermionic oscillators.

Analysis of the spectrum

This subsection describes the first few states of the open string in the light-cone gauge. Remember that the fermionic fields have two possible boundary conditions, giving rise to the NS and R sectors.

The Neveu–Schwarz sector

Recalling that $a_{\mathrm{NS}} = 1/2$, the mass formula in the NS sector is

$$\alpha' M^2 = \sum_{n=1}^{\infty} \alpha^i_{-n} \alpha^i_n + \sum_{r=1/2}^{\infty} r b^i_{-r} b^i_r - \frac{1}{2}. \quad (4.103)$$

The first two states in this sector are as follows:

- The ground state is annihilated by the positive lowering modes, that is, it satisfies

$$\alpha^i_n |0; k\rangle_{\mathrm{NS}} = b^i_r |0; k\rangle_{\mathrm{NS}} = 0 \quad \text{for} \quad n, r > 0 \quad (4.104)$$

and

$$\alpha^\mu_0 |0; k\rangle_{\mathrm{NS}} = \sqrt{2\alpha'} k^\mu |0; k\rangle_{\mathrm{NS}}. \quad (4.105)$$

The ground state in the NS sector is a scalar in space-time. From the mass formula it becomes clear that the mass m of the NS-sector ground state is given by

$$\alpha' M^2 = -\frac{1}{2}. \quad (4.106)$$

As a result, the ground state of the RNS string in the NS sector is once again a tachyon. The next subsection describes how this state is eliminated from the spectrum.

- In order to construct the first excited state in the NS sector, one acts with the raising operators having the smallest associated frequency, namely $b^i_{-1/2}$, on the ground state

$$b^i_{-1/2} |0; k\rangle_{\mathrm{NS}}. \quad (4.107)$$

Since this is in light-cone gauge, the index i labels the $D-2 = 8$ transverse directions. The operator b^i_{-r} raises the value of $\alpha' M^2$ by r units, whereas α^i_{-m} would raise it by m (a positive integer) units. This is the reason why the first excited state is built by acting with a $b^i_{-1/2}$ operator. This operator is a transverse vector in space-time. Since it is acting on a bosonic ground state that is a space-time scalar, the resulting state is a space-time vector. Note that there are eight polarization states, as required for a massless vector in ten dimensions. Using the same reasoning as for the bosonic string, one can use this state in order to independently determine the value of $a_{\rm NS}$. Indeed, since the above state is a space-time vector of $SO(8)$ it must be massless. In general, its mass is given by

$$\alpha' M^2 = \frac{1}{2} - a_{\rm NS}. \tag{4.108}$$

So requiring that this state is massless, as required by Lorentz invariance, once again gives $a_{\rm NS} = 1/2$.

The Ramond sector

In the light-cone gauge description of the R sector the mass-shell condition is

$$\alpha' M^2 = \sum_{n=1}^{\infty} \alpha^i_{-n} \alpha^i_n + \sum_{n=1}^{\infty} n d^i_{-n} d^i_n. \tag{4.109}$$

In this sector the states are as follows:
- The ground state is the solution of

$$\alpha^i_n |0; k\rangle_{\rm R} = d^i_n |0; k\rangle_{\rm R} = 0 \quad \text{for} \quad n > 0, \tag{4.110}$$

as well as the massless Dirac equation. The states have a spinor index that is not displayed. As was discussed above, the solution of these equations is not unique, since the zero modes satisfy the ten-dimensional Dirac algebra. Thus the solution to these constraints gives a $Spin(9, 1)$ spinor. The operation of multiplying with d^μ_0 is then nothing else than multiplying with a ten-dimensional Dirac matrix, which is a 32×32 matrix. Therefore, the ground state in the R sector is described by a 32-component spinor.

In ten dimensions spinors can be restricted by Majorana and Weyl conditions. The Majorana condition is already implicit, but the possibility of Weyl projection goes beyond what has been explained so far. Taking this into account, there are two alternative ground states corresponding to the two possible ten-dimensional chiralities. One could also imagine that both chiralities are allowed, though that turns out not to be the case. This is not the whole story, since the Dirac–Ramond equation (4.88) must also be solved. For the ground state, the excited oscillators do not contribute, and so this reduces to the massless Dirac equation. Solving this eliminates half of the components of the $Spin(9,1)$ spinor leaving a $Spin(8)$ spinor. Thus in the end, the minimal possibility for a Ramond ground state has eight physical degrees of freedom corresponding to an irreducible spinor of $Spin(8)$. This choice, rather than an R-sector ground state consisting of more degrees of freedom, turns out to be necessary.

- The excited states in the R sector are obtained by acting with α^i_{-n} or d^i_{-n} on the R-sector ground state. Since these operators are space-time vectors, the resulting states are also space-time spinors. The possibilities are restricted further by the GSO condition described below.

Zero-point energies

In Chapter 2 we learned that the parameter a in the mass-shell condition for the bosonic string, $(L_0 - a)|\phi\rangle = 0$, is $a = 1$. The reason for this was traced to the fact that there are 24 transverse periodic bosonic degrees of freedom on the world sheet, each of which contributes a zero-point energy $\frac{1}{2}\zeta(-1) = -1/24$.

The NS sector of the RNS string has $a_{NS} = 1/2$, which means that the total zero-point energy is $-1/2$. Of this, $-8/24 = -1/3$ is attributable to eight transverse periodic bosons. The remaining $-1/6$ is due to the eight transverse antiperiodic world-sheet fermions, each of which contributes $-1/48$.

The R sector of the RNS string has $a_R = 0$, which means that the total zero-point energy is 0. The contribution of each transverse periodic world-sheet boson is $-1/24$, and the contribution of each transverse periodic world-sheet fermion is $+1/24$. The reason that these cancel is world-sheet supersymmetry, which remains unbroken for R boundary conditions.

The fermionic zero-point energies deduced here can also be obtained by (less rigorous) zeta-function methods like that described in Section 2.5. One can also show that an antiperiodic boson, which was not needed here, but can arise in other contexts, would give $+1/48$.

The GSO projection

The previous section described the spectrum of states of the RNS string that survives the super-Virasoro constraints. But it is important to realize that this spectrum has several problems. For one thing, in the NS sector the ground state is a tachyon, that is, a particle with imaginary mass.

Also, the spectrum is not space-time supersymmetric. For example, there is no fermion in the spectrum with the same mass as the tachyon. Unbroken supersymmetry is required for a consistent interacting theory, since the spectrum contains a massless gravitino, which is the quantum of the gauge field for local supersymmetry. This inconsistency manifests itself in a variety of ways. It is analogous to coupling massless Yang–Mills fields to incomplete gauge multiplets, which leads to a breakdown of gauge invariance and causality. This subsection explains how to turn the RNS string theory into a consistent theory, by truncating (or projecting) the spectrum in a very specific way that eliminates the tachyon and leads to a supersymmetric theory in ten-dimensional space-time. This projection is called the *GSO projection*, since it was introduced by Gliozzi, Scherk and Olive.

In order to describe the truncation of the spectrum, let us first define an operator called *G-parity*.[6] In the NS sector the definition is given by

$$G = (-1)^{F+1} = (-1)^{\sum_{r=1/2}^{\infty} b^i_{-r} b^i_r + 1} \quad \text{(NS)}. \tag{4.111}$$

Note that F is the number of b-oscillator excitations, which is the world-sheet fermion number. So this operator determines whether a state has an even or an odd number of world-sheet fermion excitations. In the R sector the corresponding definition is

$$G = \Gamma_{11} (-1)^{\sum_{n=1}^{\infty} d^i_{-n} d^i_n} \quad \text{(R)}, \tag{4.112}$$

where

$$\Gamma_{11} = \Gamma_0 \Gamma_1 \ldots \Gamma_9 \tag{4.113}$$

is the ten-dimensional analog of the Dirac matrix γ_5 in four dimensions.

The matrix Γ_{11} satisfies

$$(\Gamma_{11})^2 = 1 \quad \text{and} \quad \{\Gamma_{11}, \Gamma^\mu\} = 0. \tag{4.114}$$

Spinors that satisfy

$$\Gamma_{11} \psi = \pm \psi \tag{4.115}$$

are said to have positive or negative chirality. The chirality projection operators are

$$P_\pm = \frac{1}{2} (1 \pm \Gamma_{11}). \tag{4.116}$$

A spinor with a definite chirality is called a Weyl spinor.

---

6 This name was introduced in the original NS paper which hoped to use this theory to describe hadrons. This operator was identified there with the G-parity operator for hadrons. Here its role is entirely different.

The GSO projection consists of keeping only the states with a positive G-parity in the NS sector, that is, those states with

$$(-1)^{F_{\text{NS}}} = -1, \tag{4.117}$$

while the states with a negative G-parity should be eliminated. In other words, all NS-sector states should have an odd number of b-oscillator excitations. In the R sector one can project on states with positive or negative G-parity depending on the chirality of the spinor ground state. The choice is purely a matter of convention.

The GSO projection eliminates the open-string tachyon from the spectrum, since it has negative G-parity

$$G|0\rangle_{\text{NS}} = -|0\rangle_{\text{NS}}. \tag{4.118}$$

The first excited state, $b^i_{-1/2}|0\rangle_{\text{NS}}$, on the other hand, has positive G-parity and survives the projection. After the GSO projection, this massless vector boson becomes the ground state of the NS sector. This matches nicely with the fact that the ground state in the fermionic sector is a massless spinor. This is a first indication that the spectrum could be space-time supersymmetric after performing the GSO projection. At this point the GSO projection may appear to be an *ad hoc* condition, but actually it is essential for consistency. It is possible to derive this by demanding one-loop and two-loop modular invariance. A much simpler argument is to note that it leaves a supersymmetric spectrum. As has already been emphasized, the closed-string spectrum contains a massless gravitino (or two) and therefore the interacting theory wouldn't be consistent without supersymmetry. In particular, this requires an equal number of physical bosonic and fermionic modes at each mass level. In order to check whether this is plausible, let us examine the lowest-lying states in the spectrum.

The ground state in the R sector is a massless spinor while the ground state in the NS sector is a massless vector. Let us compare the number of physical degrees of freedom. The ground state in the NS sector after the GSO projection is $b^\mu_{-1/2}|0, k\rangle$, which has only eight propagating degrees of freedom. This is most easily seen in the light-cone gauge, where one just has the eight transverse excitations $b^i_{-1/2}|0, k\rangle$, as was discussed earlier. This must match the number of fermionic degrees of freedom.

A fermion in ten dimensions has 32 complex components, since in general a spinor in D dimensions would have $2^{D/2}$ complex components (when D is even). However, the spinors can be further restricted by Majorana and Weyl conditions, each of which gives a reduction by a factor of two. Moreover, in ten dimensions the two conditions are compatible, so there exist Majorana–Weyl spinors with 16 real components.[7] In a Majorana representation the Majorana condition is just the statement that the spinor is real. Therefore, this restriction leaves 32 real components in ten dimensions. The Weyl condition implies that the spinor has a definite chirality. In other words, it is an eigenstate of the chirality operator Γ_{11}. As we have said, in ten dimensions the Majorana and the Weyl conditions can be satisfied at the same time, and Majorana–Weyl spinors have 16 real components. Imposing the Dirac equation eliminates half of these components leaving eight real components. This agrees with the number of degrees of freedom in the ground state of the NS sector. Therefore, the ground state, the massless sector, has an equal number of physical on-shell bosonic and fermionic degrees of freedom. They form two inequivalent real eight-dimensional representations of $Spin(8)$. The equality of number of bosons and fermions is a necessary, but not sufficient, condition for these states to form a supersymmetry multiplet. The proof of supersymmetry is described in the next chapter.

It is far from obvious, but nonetheless true, that the GSO projection leaves an equal number of bosons and fermions at each mass level, as required by space-time supersymmetry. This constitutes strong evidence, but not a proof, of space-time supersymmetry. This is presented in the next chapter, which describes the Green–Schwarz (GS) formalism. That formalism has the advantage of making the space-time supersymmetry manifest.

The massless closed-string spectrum

To analyze the closed-string spectrum, it is necessary to consider left-movers and right-movers. As a result, there are four possible sectors: R–R, R–NS, NS–R and NS–NS. By projecting onto states with a positive G-parity in the NS sector, the tachyon is eliminated. For the R sector we can project onto states with positive or negative G-parity depending on the chirality of the ground state on which the states are built. Thus two different theories can be obtained depending on whether the G-parity of the left- and right-moving R sectors is the same or opposite.

In the type IIB theory the left- and right-moving R-sector ground states have the same chirality, chosen to be positive for definiteness. Therefore, the two R sectors have the same G-parity. Let us denote each of them by $|+\rangle_R$. In this case the massless states in the type IIB closed-string spectrum

---

[7] The rules for the possible types of spinors depend on the space-time dimension modulo 8. This is known to mathematicians as Bott periodicity. Thus the situation in ten dimensions is quite similar to the two-dimensional case discussed earlier.

4.6 Light-cone gauge quantization of the RNS string

are given by

$$|+\rangle_R \otimes |+\rangle_R, \tag{4.119}$$

$$\tilde{b}^i_{-1/2}|0\rangle_{NS} \otimes b^j_{-1/2}|0\rangle_{NS}, \tag{4.120}$$

$$\tilde{b}^i_{-1/2}|0\rangle_{NS} \otimes |+\rangle_R, \tag{4.121}$$

$$|+\rangle_R \otimes b^i_{-1/2}|0\rangle_{NS}. \tag{4.122}$$

Since $|+\rangle_R$ represents an eight-component spinor, each of the four sectors contains $8 \times 8 = 64$ physical states.

For the type IIA theory the left- and right-moving R-sector ground states are chosen to have the opposite chirality. The massless states in the spectrum are given by

$$|-\rangle_R \otimes |+\rangle_R, \tag{4.123}$$

$$\tilde{b}^i_{-1/2}|0\rangle_{NS} \otimes b^j_{-1/2}|0\rangle_{NS}, \tag{4.124}$$

$$\tilde{b}^i_{-1/2}|0\rangle_{NS} \otimes |+\rangle_R, \tag{4.125}$$

$$|-\rangle_R \otimes b^i_{-1/2}|0\rangle_{NS}. \tag{4.126}$$

The states are very similar to the ones of the type IIB string except that now the fermionic states come with two different chiralities.

The massless spectrum of each of the type II closed-string theories contains two Majorana–Weyl gravitinos, and therefore they form $\mathcal{N} = 2$ supergravity multiplets. Each of the states in these multiplets plays an important role in the theory. There are 64 states in each of the four massless sectors, that we summarize below.

- NS–NS sector: This sector is the same for the type IIA and type IIB cases. The spectrum contains a scalar called the dilaton (one state), an antisymmetric two-form gauge field (28 states) and a symmetric traceless rank-two tensor, the graviton (35 states).
- NS–R and R–NS sectors: Each of these sectors contains a spin 3/2 gravitino (56 states) and a spin 1/2 fermion called the dilatino (eight states). In the IIB case the two gravitinos have the same chirality, whereas in the type IIA case they have opposite chirality.
- R–R sector: These states are bosons obtained by tensoring a pair of Majorana–Weyl spinors. In the IIA case, the two Majorana–Weyl spinors have opposite chirality, and one obtains a one-form (vector) gauge field

(eight states) and a three-form gauge field (56 states). In the IIB case the two Majorana–Weyl spinors have the same chirality, and one obtains a zero-form (that is, scalar) gauge field (one state), a two-form gauge field (28 states) and a four-form gauge field with a self-dual field strength (35 states).

EXERCISES

EXERCISE 4.10

Show that there are the same number of physical degrees of freedom in the NS and R sectors at the first massive level after GSO projection.

SOLUTION

At this level, the NS states have $N = 3/2$ and the R states have $N = 1$. The G-parity constraint in the NS sector requires there to be an odd number of b-oscillator excitations. In the R sector, the constraint correlates the number of d-oscillator excitations with the chirality of the spinor.

Now let us count the number of physical bosonic and fermionic states that survive the GSO projection. On the bosonic side (the NS sector) the states at this level (in light-cone gauge) are

$$\alpha^i_{-1} b^j_{-1/2} |0\rangle, \qquad b^i_{-1/2} b^j_{-1/2} b^k_{-1/2} |0\rangle, \qquad b^i_{-3/2} |0\rangle,$$

which gives a total of $64 + 56 + 8 = 128$ states. Since these are massive states they must combine into $SO(9)$ representations. In fact, it turns out that they give two $SO(9)$ representations, $\mathbf{128} = \mathbf{44} \oplus \mathbf{84}$. On the fermionic side (the R sector) the states are

$$\alpha^i_{-1} |\psi_0\rangle, \qquad d^i_{-1} |\psi'_0\rangle,$$

which again makes $64 + 64 = 128$ states, so that there is agreement with the number of degrees of freedom on the bosonic side. Note that $|\psi_0\rangle$ and $|\psi'_0\rangle$ denote a pair of Majorana–Weyl spinors of opposite chirality, each of which has 16 real components. However, there are only eight physical degrees of freedom, because the Dirac–Ramond equation $F_0 |\psi\rangle = 0$ gives a factor of two reduction. These 128 fermionic states form an irreducible spinor representation of $Spin(9)$. This massive supermultiplet in ten dimensions, consisting of 128 bosons and 128 fermions, is identical to the massless supergravity multiplet in 11 dimensions. □

EXERCISE 4.11

Construct generating functions that encode the number of physical degrees of freedom in the NS and R sectors at all levels after GSO projection.

SOLUTION

Let us denote the number of degrees of freedom with $\alpha' M^2 = n$ in the NS and R sectors of an open superstring by $d_{\text{NS}}(n)$ and $d_{\text{R}}(n)$, respectively. Then the generating functions are

$$f_{\text{NS}}(w) = \sum_{n=0}^{\infty} d_{\text{NS}}(n) w^n \quad \text{and} \quad f_{\text{R}}(w) = \sum_{n=0}^{\infty} d_{\text{R}}(n) w^n.$$

Before GSO projection, the degeneracies in the NS sector are given by $\operatorname{tr} w^{N-1/2}$, where N is given in Eq. (4.73), except that in light-cone gauge there are only transverse oscillators. The basic key to evaluating the traces is to use the fact that for a bosonic oscillator

$$\operatorname{tr} w^{a^\dagger a} = 1 + w + w^2 + \ldots = \frac{1}{1-w}$$

and for a fermionic oscillator

$$\operatorname{tr} w^{b^\dagger b} = 1 + w.$$

Since there are eight transverse dimensions for $D = 10$, it therefore follows that

$$\operatorname{tr} w^{N-1/2} = \frac{1}{\sqrt{w}} \prod_{m=1}^{\infty} \left(\frac{1 + w^{m-1/2}}{1 - w^m} \right)^8.$$

To take account of the GSO projection we need to eliminate the contributions due to an even number of b-oscillator excitations. This is achieved by taking

$$f_{\text{NS}}(w) = \frac{1}{2\sqrt{w}} \left[\prod_{m=1}^{\infty} \left(\frac{1 + w^{m-1/2}}{1 - w^m} \right)^8 - \prod_{m=1}^{\infty} \left(\frac{1 - w^{m-1/2}}{1 - w^m} \right)^8 \right].$$

The analysis in the R sector works in a similar manner. In this case the effect of the GSO projection is to reduce the degeneracy associated with zero modes from 16 to 8. Thus one obtains

$$f_R(w) = 8 \prod_{m=1}^{\infty} \left(\frac{1+w^m}{1-w^m}\right)^8.$$

In 1829, Jacobi proved that $f_{NS}(w) = f_R(w)$. □

4.7 SCFT and BRST

In the study of the bosonic string theory in Chapter 3, it proved useful to focus on the interpretation of the world-sheet action in the conformal gauge as a conformal field theory. This reasoning extends nicely to the RNS string, where the symmetry gets enlarged to a superconformal symmetry. The Euclideanized conformal-gauge bosonic string action was written (in units $l_s = \sqrt{2\alpha'} = 1$) in the form

$$S = \frac{1}{\pi} \int \partial X^\mu \bar{\partial} X_\mu d^2 z. \tag{4.127}$$

Then the holomorphic energy–momentum tensor took the form

$$T = -2 : \partial X^\mu \partial X_\mu := \sum_{n=-\infty}^{\infty} \frac{L_n}{z^{n+2}}. \tag{4.128}$$

The Virasoro algebra, characterizing the conformal symmetry, is encoded in the OPE

$$T(z)T(w) = \frac{c/2}{(z-w)^4} + \frac{2}{(z-w)^2} T(w) + \frac{1}{z-w} \partial T(w), \tag{4.129}$$

where the central charge c equals D, the dimension of the space-time.

Superconformal field theory

The generalization of these formulas to the RNS superstring is quite straightforward. The gauge-fixed world-sheet action becomes

$$S_{\text{matter}} = \frac{1}{2\pi} \int \left(2\partial X^\mu \bar{\partial} X_\mu + \frac{1}{2}\psi^\mu \bar{\partial}\psi_\mu + \frac{1}{2}\tilde{\psi}^\mu \partial\tilde{\psi}_\mu\right) d^2 z, \tag{4.130}$$

where ψ and $\tilde{\psi}$ correspond to ψ_+ and ψ_- in the Lorentzian description. The holomorphic energy–momentum tensor takes the form (B stands for bosonic)

$$T_B(z) = -2\partial X^\mu(z)\partial X_\mu(z) - \frac{1}{2}\psi^\mu(z)\partial\psi_\mu(z) = \sum_{n=-\infty}^{\infty} \frac{L_n}{z^{n+2}}, \tag{4.131}$$

4.7 SCFT and BRST

which now has central charge $c = 3D/2$. The conformal field $\psi^\mu(z)$ is a free fermion. As explained in Chapter 3, it has conformal dimension $h = 1/2$ and the OPE

$$\psi^\mu(z)\psi^\nu(w) \sim \frac{\eta^{\mu\nu}}{z-w}. \tag{4.132}$$

In the superconformal gauge, this theory also has a conserved $h = 3/2$ supercurrent, whose holomorphic part is denoted $T_F(z)$ (F stands for fermionic)

$$T_F(z) = 2i\psi^\mu(z)\partial X_\mu(z) = \sum_{r=-\infty}^{\infty} \frac{G_r}{z^{r+3/2}}. \tag{4.133}$$

This mode expansion is appropriate to the NS sector. In the R sector G_r, which has half-integer modes, would be replaced by F_n, which has integer modes. Together with the energy–momentum tensor, which is now denoted $T_B(z)$, it forms a superconformal algebra with OPE

$$T_F(z)T_F(w) \sim \frac{\hat{c}}{4(z-w)^3} + \frac{T_B(w)}{2(z-w)} + \ldots \tag{4.134}$$

where $c = \frac{3}{2}\hat{c}$, so that $\hat{c} = D$. One has $c = 3D/2 = 15$ because each bosonic field contributes one unit and each fermionic field contributes half a unit of central charge. It is convenient to use a superspace formulation involving a single Grassmann parameter θ. It can be regarded as a holomorphic Grassmann coordinate that corresponds to θ_+ in the Lorentzian description. One can then combine T_F and T_B into a single expression

$$T(z,\theta) = T_F(z) + \theta T_B(z) \tag{4.135}$$

whose OPE is

$$T(z_1,\theta_1)T(z_2,\theta_2) \sim \frac{\hat{c}}{4z_{12}^3} + \frac{3\theta_{12}}{2z_{12}^2}T(z_2,\theta_2) + \frac{D_2 T(z_2,\theta_2)}{2z_{12}} + \frac{\theta_{12}}{z_{12}}\partial_2 T(z_2,\theta_2) + \ldots, \tag{4.136}$$

where $z_{12} = z_1 - z_2 - \theta_1\theta_2$ and $\theta_{12} = \theta_1 - \theta_2$. Also,

$$D = \frac{\partial}{\partial \theta} + \theta \frac{\partial}{\partial z}. \tag{4.137}$$

This describes the entire superconformal algebra. Note that θ_{12} and z_{12} are invariant under the supersymmetry transformations $\delta\theta_i = \varepsilon$, $\delta z_i = \theta_i \varepsilon$. A superfield $\Phi(z,\theta)$ with components of conformal dimension h and $h + \frac{1}{2}$ satisfies

$$T(z_1,\theta_1)\Phi(z_2,\theta_2) \sim h\frac{\theta_{12}}{z_{12}^2}\Phi(z_2,\theta_2) + \frac{1}{2z_{12}}D_2\Phi + \frac{\theta_{12}}{z_{12}}\partial_2\Phi + \ldots \tag{4.138}$$

BRST symmetry

Superconformal field theory appears naturally when discussing the path-integral quantization of supersymmetric strings. In the quantum theory, it is convenient to add Faddeev–Popov ghosts to represent the Jacobian factors in the path integral associated with gauge fixing. Rather than discuss the path-integral quantization in detail, let us focus on the resulting superconformal field theory.

As discussed in Chapter 3, in the bosonic string theory the Faddeev–Popov ghosts consist of a pair of fermionic fields b and c with conformal dimensions 2 and -1, respectively. These arose from gauge fixing the world-sheet diffeomorphism symmetry. In the case of the RNS string there is also a local supersymmetry on the world sheet that has been gauge-fixed, and as a result an additional pair of Faddeev–Popov ghosts is required. They are bosonic ghost fields, called β and γ, with conformal dimensions 3/2 and $-1/2$, respectively. They have the OPE

$$\gamma(z)\beta(w) \sim \frac{1}{z-w}. \tag{4.139}$$

Since these are bosonic fields, this is equivalent to

$$\beta(z)\gamma(w) \sim -\frac{1}{z-w}. \tag{4.140}$$

The gauge-fixed quantum action includes all of these fields. It is $S = S_{\text{matter}} + S_{\text{ghost}}$, where S_{matter} is the expression in Eq. (4.130) and

$$S_{\text{ghost}} = \frac{1}{2\pi}\int (b\bar\partial c + \bar b \partial \bar c + \beta\bar\partial\gamma + \bar\beta\partial\bar\gamma) d^2 z. \tag{4.141}$$

The fields c and γ have ghost number $+1$, while the fields b and β have ghost number -1. The bosonic ghosts β and γ are required to have the same moding as the fermi field ψ^μ – integer modes in the R sector and half-integer modes in the NS sector. When the factors of z^{-h} are taken into account, this implies that $\psi^\mu(z)$, $\beta(z)$ and $\gamma(z)$ involve integer powers of z and are single-valued in the NS sector. whereas in the R sector they involve half-integer powers and are double-valued.

The superconformal symmetry operators of this system are also given as the sum of matter and ghost contributions. The ghost fields give the following contributions:

$$T_{\text{B}}^{\text{ghost}} = -2b\partial c + c\partial b - \frac{3}{2}\beta\partial\gamma - \frac{1}{2}\gamma\partial\beta, \tag{4.142}$$

$$T_{\text{F}}^{\text{ghost}} = -2b\gamma + c\partial\beta + \frac{3}{2}\beta\partial c. \tag{4.143}$$

These contribute $\hat{c} = -10$, and so the superconformal anomaly cancels for $D = 10$.

As in the case of the bosonic string theory, the quantum action has a global fermionic symmetry, namely BRST symmetry. In this case the transformations that leave the Lagrangian invariant up to a total derivative are

$$\delta X^\mu = \eta(c\partial X^\mu - \frac{i}{2}\gamma\psi^\mu), \tag{4.144}$$

$$\delta\psi^\mu = \eta(c\partial\psi^\mu - \frac{1}{2}\psi^\mu\partial c + 2i\gamma\partial X^\mu), \tag{4.145}$$

$$\delta c = \eta(c\partial c - \gamma^2), \tag{4.146}$$

$$\delta b = \eta T_{\rm B}, \tag{4.147}$$

$$\delta\gamma = \eta(c\partial\gamma - \frac{1}{2}\gamma\partial c), \tag{4.148}$$

$$\delta\beta = \eta T_{\rm F}. \tag{4.149}$$

These transformations are generated by the BRST charge

$$Q_{\rm B} = \frac{1}{2\pi i}\oint (cT_{\rm B}^{\rm matter} + \gamma T_{\rm F}^{\rm matter} + bc\partial c - \frac{1}{2}c\gamma\partial\beta - \frac{3}{2}c\beta\partial\gamma - b\gamma^2)dz. \tag{4.150}$$

The transformations of b and β, in particular, correspond to the basic equations

$$\{Q_{\rm B}, b(z)\} = T_{\rm B}(z) \tag{4.151}$$

and

$$[Q_{\rm B}, \beta(z)] = T_{\rm F}(z). \tag{4.152}$$

As in the case of the bosonic string, the BRST charge is nilpotent, $Q_{\rm B}^2 = 0$, in the critical dimension $D = 10$. The proof is a straightforward analog of the one given for the bosonic string theory and is left as a homework problem. One first uses Jacobi identities to prove that $[\{Q_{\rm B}, G_r\}, \beta_s]$ and $[\{Q_{\rm B}, G_r\}, b_m]$ vanish if $\hat{c} = 0$. This implies that $\{Q_{\rm B}, G_r\}$ cannot depend on the γ or c ghosts. Since it has positive ghost number, this implies that it vanishes. It follows (using the superconformal algebra and Jacobi identities) that $[Q_{\rm B}, L_n]$ must also vanish. Hence $Q_{\rm B}$ is superconformally invariant for $\hat{c} = 0$. In this case $[Q_{\rm B}^2, b_n] = [Q_{\rm B}, L_n] = 0$ and $[Q_{\rm B}^2, \beta_r] = \{Q_{\rm B}, G_r\} = 0$, which implies that $Q_{\rm B}^2$ cannot depend on the c or γ ghosts. Since it also has positive ghost number, it vanishes. Thus nilpotency follows from $\hat{c} = 0$.

As a result of nilpotency, it is again possible to describe the physical states in terms of BRST cohomology classes. In the NS sector, the β, γ system has half-integer moding, and so there is a two-fold vacuum degeneracy due to the zero modes b_0 and c_0, just as in the case of the bosonic string. As in that case, physical states are required to have ghost number $-1/2$. The case of the R sector is more subtle, because in that sector there are additional zero modes β_0 and γ_0, which give rise to an infinite degeneracy. Without going into details, let us just give a hint about how this is handled. The degeneracy due to the β_0–γ_0 Fock space is interpreted as giving infinitely many equivalent descriptions of each physical state in different *pictures*. There is an integer label that characterizes the picture, and there are *picture-changing operators* that enable one to map back and forth between adjacent pictures. In formulating path integrals for amplitudes, there are some restrictions on which pictures can be used for the vertex operators that enter into the calculation.

Homework Problems

Problem 4.1
Consider a massless supersymmetric particle (or superparticle) propagating in D-dimensional Minkowski space-time. It is described by D bosonic fields $X^\mu(\tau)$ and D Majorana fermions $\psi^\mu(\tau)$. The action is

$$S_0 = \int d\tau \left(\frac{1}{2} \dot{X}^\mu \dot{X}_\mu - i\psi^\mu \dot{\psi}_\mu \right).$$

(i) Derive the field equations for X^μ, ψ^μ.

(ii) Show that the action is invariant under the global supersymmetry transformations

$$\delta X^\mu = i\varepsilon \psi^\mu, \qquad \delta \psi^\mu = \frac{1}{2} \varepsilon \dot{X}^\mu,$$

where ε is an infinitesimal real constant Grassmann parameter.

(iii) Suppose that δ_1 and δ_2 are two infinitesimal supersymmetry transformations with parameters ε_1 and ε_2, respectively. Show that the commutator $[\delta_1, \delta_2]$ gives a τ translation by an amount $\delta\tau$. Determine $\delta\tau$ and explain why $\delta\tau$ is real.

Problem 4.2
In Problem 4.1, supersymmetry was only a global symmetry, as ε did not depend on τ. To construct an action in which this symmetry is local, one

needs to include the auxiliary field e and its fermionic partner, which we denote by χ. The action takes the form

$$\tilde{S}_0 = \int d\tau \left(\frac{\dot{X}^\mu \dot{X}_\mu}{2e} + \frac{i\dot{X}^\mu \psi_\mu \chi}{e} - i\psi^\mu \dot{\psi}_\mu \right).$$

(i) Show that this action is reparametrization invariant, that is, it is invariant under the following infinitesimal transformations with parameter $\xi(\tau)$:

$$\delta X^\mu = \xi \dot{X}^\mu, \qquad \delta \psi^\mu = \xi \dot{\psi}^\mu,$$

$$\delta e = \frac{d}{d\tau}(\xi e), \qquad \delta \chi = \frac{d}{d\tau}(\xi \chi).$$

(ii) Show explicitly that the action is invariant under the local supersymmetry transformations

$$\delta X^\mu = i\varepsilon \psi^\mu, \qquad \delta \psi^\mu = \frac{1}{2e}(\dot{X}^\mu - i\chi\psi^\mu)\varepsilon,$$

$$\delta \chi = \dot{\varepsilon}, \qquad \delta e = -i\chi\varepsilon.$$

(iii) Show that in the gauge $e = 1$ and $\chi = 0$, one recovers the action in Problem 4.1 and the constraint equations $\dot{X}^2 = 0$, $\dot{X} \cdot \psi = 0$.

PROBLEM 4.3

Consider quantization of the superparticle action in Problem 4.1.

(i) Show that canonical quantization gives the equal-τ commutation and anticommutation relations

$$[X^\mu, \dot{X}^\nu] = i\eta^{\mu\nu} \quad \text{and} \quad \{\psi^\mu, \psi^\nu\} = \eta^{\mu\nu}.$$

(ii) Explain why this describes a space-time fermion.

(iii) What is the significance of the constraints $\dot{X}^2 = 0$ and $\dot{X} \cdot \psi = 0$ obtained in Problem 4.2?

PROBLEM 4.4

Show the invariance of the action (4.35) under the supersymmetry transformations (4.25)–(4.27).

PROBLEM 4.5

Derive the mass formulas for states in the R and NS sector of the RNS open superstring.

Problem 4.6

Verify the constants $a_R = 0$, $a_{NS} = 1/2$ for the critical RNS superstring by using zeta-function regularization to compute the world-sheet fermion zero-point energies, as suggested in Section 4.6.

Problem 4.7

Consider the RNS string in ten-dimensional Minkowski space-time. Show that after the GSO projection the NS and R sectors have the same number of physical degrees of freedom at the second massive level. Determine the explicit form of the states in the light-cone gauge. In other words, repeat the analysis of Exercise 4.10 for the next level.

Problem 4.8

Given a pair of two-dimensional Majorana spinors ψ and χ, prove that

$$\psi_A \bar{\chi}_B = -\frac{1}{2}\left(\bar{\chi}\psi \delta_{AB} + \bar{\chi}\rho_\alpha \psi \rho^\alpha_{AB} + \bar{\chi}\rho_3 \psi (\rho_3)_{AB}\right),$$

where $\rho_3 = \rho_0 \rho_1$.

Problem 4.9

Derive the NS-sector Lorentz transformation generators in the light-cone gauge.

Problem 4.10

Show that Eqs (4.131) and Eq. (4.133) lead to the mode expansions of L_n and G_r given earlier.

Problem 4.11

Using the energy–momentum tensor T_B in Eq. (4.131) and the supercurrent T_F in Eq. (4.133), verify that Eq. (4.134) holds with $\hat{c} = 10$.

Problem 4.12

Work out the OPEs that correspond to the coefficients of the various powers of θ in Eqs (4.136) and (4.138).

Problem 4.13

Show that the total action $S = S_{\text{matter}} + S_{\text{ghost}}$, given in Eqs (4.130) and (4.141), is invariant under the BRST transformations of Eqs (4.144)–(4.149).

Problem 4.14

Consider the ghost contributions to the super-Virasoro generators in the NS sector.

(i) Work out the mode expansions for ghost contributions to L_n and G_r implied by Eqs (4.142) and (4.143).

(ii) Prove that these generate a super-Virasoro algebra with $\hat{c} = -10$.

Problem 4.15

Using the method sketched in the text, show that the BRST charge in Eq. (4.150) is nilpotent for the critical dimension $D = 10$.

5
Strings with space-time supersymmetry

After the GSO projection the spectrum of the ten-dimensional RNS superstring has an equal number of bosons and fermions at each mass level. This is strong circumstantial evidence that the theory has space-time supersymmetry, even though this symmetry is extremely obscure in the RNS formalism. This suggests that there should exist a different formulation of the theory in which space-time supersymmetry becomes manifest. This chapter begins by describing the Green–Schwarz (GS) formulation of superstring theory, which achieves this.

Since the bosonic string theory is defined in terms of maps of the string world sheet into space-time, a natural supersymmetric generalization to consider is based on maps of the string world sheet into superspace, so that the basic world-sheet fields are

$$X^\mu(\sigma, \tau) \quad \text{and} \quad \Theta^a(\sigma, \tau). \tag{5.1}$$

This is the approach implemented in the GS formalism.

The GS formalism has advantages and disadvantages compared to the RNS formalism. The basic disadvantage of the GS formalism stems from the fact that it is very difficult to quantize the world-sheet action in a way that maintains space-time Lorentz invariance as a manifest symmetry. However, it can be quantized in the light-cone gauge. This is sufficient for analyzing the physical spectrum. It is also sufficient for studying tree and one-loop amplitudes. An advantage of the GS formalism is that the GSO projection is automatically built in without having to make any truncations, and space-time supersymmetry is manifest. Moreover, in contrast to the RNS formalism, the bosonic and fermionic strings are unified in a single Fock space.

5.1 The D0-brane action

Let us begin with a warm-up exercise that shares some features with the GS superstring but is quite a bit simpler, specifically a space-time supersymmetric world-line action for a point particle of mass m. The example of particular interest, called the D0-brane, is a massive point particle that appears as a nonperturbative excitation in the type IIA theory. The D0-brane is a special case of more general Dp-branes, which are the subject of the next chapter.

Recall that the action for a massive point particle in flat Minkowski space-time has the form

$$S = -m \int \sqrt{-\dot{X}_\mu \dot{X}^\mu} d\tau. \qquad (5.2)$$

Our goal here is to find a generalization of this action describing a massive point particle that is supersymmetric in space-time. Any number, $\mathcal{N}$, of supersymmetries can be described by introducing $\mathcal{N}$ anticommuting spinor coordinates $\Theta^{Aa}(\tau)$ with $A = 1, \ldots, \mathcal{N}$. The index a labels the components of the space-time spinor in D dimensions. For a general Dirac spinor $a = 1, \ldots, 2^{D/2}$ if D is even. In the following it is assumed that the spinors are Majorana. This is the case of most interest, and it simplifies the formulas, because one can use identities such as $\bar{\psi}_1 \Gamma^\mu \psi_2 = -\bar{\psi}_2 \Gamma^\mu \psi_1$. In the important case of ten dimensions, there exist Majorana–Weyl spinors, so a Weyl constraint can be imposed at the same time.

Supersymmetry can be represented in terms of infinitesimal supersymmetry transformations of superspace

$$\delta \Theta^{Aa} = \varepsilon^{Aa}, \qquad (5.3)$$

$$\delta X^\mu = \bar{\varepsilon}^A \Gamma^\mu \Theta^A. \qquad (5.4)$$

Here, summation on the repeated index A is understood. Supersymmetry is a nontrivial extension of the usual symmetries of space-time. In particular, a simple computation shows that the commutator of two infinitesimal supersymmetry transformations gives

$$[\delta_1, \delta_2]\Theta^A = 0 \quad \text{and} \quad [\delta_1, \delta_2]X^\mu = -2\bar{\varepsilon}_1^A \Gamma^\mu \varepsilon_2^A = a^\mu. \qquad (5.5)$$

This shows that the commutator of two infinitesimal supersymmetry transformations is an infinitesimal space-time translation of X^μ by a^μ. The supergroup obtained by adjoining supersymmetry transformations to the Poincaré group is called the *super-Poincaré group*, and the generators define the super-Poincaré algebra. It is made manifest in the formulas that follow. These symmetries are global symmetries of the world-line action, so ε^{Aa} is independent of τ.

In order to construct the supersymmetric action, let us define the supersymmetric combination

$$\Pi_0^\mu = \dot{X}^\mu - \bar{\Theta}^A \Gamma^\mu \dot{\Theta}^A. \tag{5.6}$$

The subscript 0 refers to the fact that both terms involve time derivatives. The corresponding formula for a Dp-brane is

$$\Pi_\alpha^\mu = \partial_\alpha X^\mu - \bar{\Theta}^A \Gamma^\mu \partial_\alpha \Theta^A, \quad \alpha = 0, 1, \ldots, p. \tag{5.7}$$

In the case of the D0-brane, $p = 0$, and so the index α can only take the value 0.

Since Π_0^μ is invariant under supersymmetry transformations, a space-time supersymmetric action can be constructed by making the replacement

$$\dot{X}^\mu \to \Pi_0^\mu \tag{5.8}$$

in the action (5.2). As a result, one obtains the action

$$S_1 = -m \int \sqrt{-\Pi_0 \cdot \Pi_0} \, d\tau. \tag{5.9}$$

This action is invariant under global super-Poincaré transformations and local diffeomorphisms of the world line.

The D0-branes are massive supersymmetric point particles that appears in the type IIA theory. Therefore, since this is a ten-dimensional theory, in the following we assume that $D = 10$. Since the type IIA theory has $\mathcal{N} = 2$ space-time supersymmetry, there are two spinor coordinates, Θ^{1a} and Θ^{2a}, which are both Majorana–Weyl and have opposite chirality. One can define a Majorana (but not Weyl) spinor

$$\Theta = \Theta^1 + \Theta^2, \tag{5.10}$$

and obtain Θ^1 and Θ^2 by projecting onto each chirality

$$\Theta^1 = \frac{1}{2}(1 + \Gamma_{11})\Theta \quad \text{and} \quad \Theta^2 = \frac{1}{2}(1 - \Gamma_{11})\Theta, \tag{5.11}$$

where, as in Chapter 4,

$$\Gamma_{11} = \Gamma_0 \Gamma_1 \ldots \Gamma_9 \tag{5.12}$$

satisfies $\Gamma_{11}^2 = 1$ and $\{\Gamma_{11}, \Gamma^\mu\} = 0$. In this case one can write

$$\Pi_0^\mu = \dot{X}^\mu - \bar{\Theta} \Gamma^\mu \dot{\Theta}, \tag{5.13}$$

since the cross terms between opposite-chirality spinors vanish.

5.1 The D0-brane action

It turns out that the action S_1, by itself, does not give the desired theory. This can be seen by deriving the equations of motion associated with X^μ and Θ^A. The canonical conjugate momentum to X^μ is

$$P_\mu = \frac{\delta S_1}{\delta \dot X^\mu} = \frac{m}{\sqrt{-\Pi_0 \cdot \Pi_0}} \left(\dot X_\mu - \bar\Theta \Gamma_\mu \dot\Theta \right). \tag{5.14}$$

The X^μ equations of motion imply

$$\dot P_\mu = 0. \tag{5.15}$$

Not all the components of the momentum are independent. Squaring both sides of Eq. (5.14) gives the mass-shell condition

$$P^2 = -m^2. \tag{5.16}$$

On the other hand, the equation of motion for Θ is

$$P \cdot \Gamma \dot\Theta = 0. \tag{5.17}$$

Multiplying this with $P \cdot \Gamma$ gives $m^2 \dot\Theta = 0$, so for $m \neq 0$ one obtains $\dot\Theta = 0$. There is nothing obviously wrong with this. However, the factor $P \cdot \Gamma$ is singular in the massless case. This corresponds to saturation of a BPS bound, a circumstance that reflects enhanced supersymmetry. This suggests that another contribution to the action may be missing whose inclusion would ensure saturation of a BPS bound and enhanced supersymmetry in the massive case as well.

Suppose that there is a second contribution to the action that changes Eq. (5.17) to

$$(P \cdot \Gamma + m\Gamma_{11})\dot\Theta = 0. \tag{5.18}$$

This equation only forces half the components of Θ to be constant without constraining the other half at all. The reason is that half of the eigenvalues of $P \cdot \Gamma + m\Gamma_{11}$ are zero. As evidence of this consider its square

$$(P \cdot \Gamma + m\Gamma_{11})^2 = (P \cdot \Gamma)^2 + m\{P \cdot \Gamma, \Gamma_{11}\} + (m\Gamma_{11})^2 = P^2 + m^2 = 0. \tag{5.19}$$

Thus the number of independent equations is only half the number of components of Θ. This suggests that there are local fermionic symmetries such that half the components of Θ are actually gauge degrees of freedom.

The missing contribution to the action that gives this additional term in the Θ equation of motion is

$$S_2 = -m \int \bar\Theta \Gamma_{11} \dot\Theta \, d\tau. \tag{5.20}$$

The choice of the sign of this term is arbitrary. If this choice describes a

D0-brane, then the opposite sign would describe an anti-D0-brane. To summarize, the complete space-time supersymmetric action for a point particle of mass m is

$$S = S_1 + S_2 = -m \int \sqrt{-\Pi_0 \cdot \Pi_0} \, d\tau - m \int \bar{\Theta}\Gamma_{11}\dot{\Theta} \, d\tau. \tag{5.21}$$

Kappa symmetry

The action S is invariant under super-Poincaré transformations and diffeomorphisms of the world line. By adding the contribution S_2, the point-particle action gains a new symmetry, called κ symmetry, which is a local fermionic symmetry. κ symmetry involves a variation $\delta\Theta$, whose form is determined later, combined with a transformation of the bosonic variables given by

$$\delta X^\mu = \bar{\Theta}\Gamma^\mu \delta\Theta = -\delta\bar{\Theta}\Gamma^\mu\Theta. \tag{5.22}$$

This determines the transformation of Π_0^μ to be

$$\delta\Pi_0^\mu = -2\delta\bar{\Theta}\Gamma^\mu\dot{\Theta}. \tag{5.23}$$

The variation of the action S_1 in Eq. (5.9) under a κ transformation is

$$\delta S_1 = m \int \frac{\Pi_0 \cdot \delta\Pi_0}{\sqrt{-\Pi_0^2}} \, d\tau. \tag{5.24}$$

Using (5.23) and the fact that Γ_{11} squares to 1, we obtain

$$\delta S_1 = -2m \int \frac{\Pi_0^\mu \delta\bar{\Theta}\Gamma_\mu \dot{\Theta}}{\sqrt{-\Pi_0^2}} \, d\tau = -2m \int \delta\bar{\Theta}\gamma\Gamma_{11}\dot{\Theta} \, d\tau, \tag{5.25}$$

where

$$\gamma = \frac{\Gamma \cdot \Pi_0}{\sqrt{-\Pi_0^2}} \Gamma_{11}. \tag{5.26}$$

Since

$$\gamma^2 = \frac{(\Gamma \cdot \Pi_0)^2}{\Pi_0^2} = 1, \tag{5.27}$$

γ can be used to construct projection operators

$$P_\pm = \frac{1}{2}(1 \pm \gamma). \tag{5.28}$$

The second contribution to the action, S_2 in Eq. (5.20), has the variation

$$\delta S_2 = -2m \int \delta\bar{\Theta}\Gamma_{11}\dot{\Theta} \, d\tau. \tag{5.29}$$

Thus

$$\delta(S_1 + S_2) = -2m \int \delta\bar{\Theta}(1+\gamma)\Gamma_{11}\dot{\Theta}\, d\tau = -4m \int \delta\bar{\Theta}P_+\Gamma_{11}\dot{\Theta}\, d\tau. \quad (5.30)$$

For a transformation $\delta\Theta$ that takes the form

$$\delta\bar{\Theta} = \bar{\kappa}P_-, \quad (5.31)$$

with $\kappa(\tau)$ an arbitrary Majorana spinor, the action is invariant. So this describes a local symmetry of the action. To summarize, the D0-brane action S is invariant under the transformations

$$\delta\bar{\Theta} = \bar{\kappa}P_- \quad \text{and} \quad \delta X^\mu = -\bar{\kappa}P_-\Gamma^\mu\Theta. \quad (5.32)$$

The local fermionic κ symmetry implies that half of the components of Θ are decoupled and can be gauged away. The key point to realize is that without this symmetry there would be the wrong number of propagating fermionic degrees of freedom. What is required is a local fermionic symmetry that effectively eliminates half of the components of Θ.

EXERCISES

EXERCISE 5.1
Given two Majorana spinors Θ_1 and Θ_2 prove that

$$\bar{\Theta}_1\Gamma_\mu\Theta_2 = -\bar{\Theta}_2\Gamma_\mu\Theta_1.$$

SOLUTION

In a Majorana representation the Dirac matrices are real. Since Γ_0 is antihermitian and the spatial components Γ_i are hermitian, this implies that Γ_0 is antisymmetric and Γ_i is symmetric. Using these facts and the Dirac algebra, it follows that the charge-conjugation matrix $\mathcal{C} = \Gamma_0$ satisfies

$$\mathcal{C}\Gamma_\mu\mathcal{C}^{-1} = -\Gamma_\mu^T.$$

For Majorana spinors

$$\bar{\Theta}_1\Gamma_\mu\Theta_2 = \Theta_1^\dagger\Gamma_0\Gamma_\mu\Theta_2 = \Theta_1^T\mathcal{C}\Gamma_\mu\Theta_2.$$

This can be written in the form

$$-\Theta_2^T\Gamma_\mu^T\mathcal{C}^T\Theta_1 = -\Theta_2^T\mathcal{C}\Gamma_\mu\Theta_1 = -\bar{\Theta}_2\Gamma_\mu\Theta_1,$$

which proves the desired result.

More generally, the same reasoning gives

$$\bar{\Theta}_1 \Gamma_{\mu_1 \cdots \mu_n} \Theta_2 = (-1)^{n(n+1)/2} \bar{\Theta}_2 \Gamma_{\mu_1 \cdots \mu_n} \Theta_1,$$

where we define

$$\Gamma_{\mu_1 \mu_2 \cdots \mu_n} = \Gamma_{[\mu_1} \Gamma_{\mu_2} \cdots \Gamma_{\mu_n]}$$

and square brackets denote antisymmetrization of the enclosed indices. □

EXERCISE 5.2
Check explicitly that the commutator of two supersymmetry transformations gives the result claimed in Eq. (5.5).

SOLUTION

Under the supersymmetry transformations in Eqs (5.3) and (5.4) the fermionic coordinate transformation is $\delta \Theta^A = \varepsilon^A$. Therefore, $\delta_1 \delta_2 \Theta^A = \delta_1 \varepsilon_2^A = 0$, which implies that $[\delta_1, \delta_2] \Theta^A = 0$. Similarly,

$$\delta_1 \delta_2 X^\mu = \delta_1 \left(\bar{\varepsilon}_2^A \Gamma^\mu \Theta^A \right) = \bar{\varepsilon}_2^A \Gamma^\mu \varepsilon_1^A.$$

As a result,

$$[\delta_1, \delta_2] X^\mu = \bar{\varepsilon}_2^A \Gamma^\mu \varepsilon_1^A - \bar{\varepsilon}_1^A \Gamma^\mu \varepsilon_2^A = -2 \bar{\varepsilon}_1^A \Gamma^\mu \varepsilon_2^A,$$

where we have used the result of the previous exercise. □

EXERCISE 5.3
Show that Π_0^μ, as defined in Eq. (5.6), is invariant under the supersymmetry transformations in Eqs (5.3) and (5.4).

SOLUTION

From the definition of Π_0^μ it follows that

$$\delta(\dot{X}^\mu - \bar{\Theta}^A \Gamma^\mu \dot{\Theta}^A) = \frac{d}{d\tau}(\bar{\varepsilon}^A \Gamma^\mu \Theta^A) - \bar{\varepsilon}^A \Gamma^\mu \dot{\Theta}^A - \bar{\Theta}^A \Gamma^\mu \dot{\varepsilon}^A$$

$$= \bar{\varepsilon}^A \Gamma^\mu \dot{\Theta}^A - \bar{\varepsilon}^A \Gamma^\mu \dot{\Theta}^A = 0.$$

□

EXERCISE 5.4
Derive the equations of motion for X^μ and Θ^A obtained from the action S_1.

Solution

The momentum corresponding to the X_μ coordinate is

$$P^\mu = \frac{\delta \mathcal{L}}{\delta \dot{X}_\mu} = m \frac{\Pi^\mu}{\sqrt{-\Pi^2}}.$$

As a result, the equation of motion for X_μ is

$$\dot{P}^\mu = 0.$$

The equation of motion for the fermionic field is

$$\frac{d}{d\tau} \frac{\delta \mathcal{L}}{\delta \dot{\bar{\Theta}}^A} - \frac{\delta \mathcal{L}}{\delta \bar{\Theta}^A} = 0.$$

This gives

$$\frac{d}{d\tau}\left(P_\mu \Gamma^\mu \Theta^A\right) + P_\mu \Gamma^\mu \dot{\Theta}^A = 0,$$

or, using $\dot{P}^\mu = 0$,

$$P \cdot \Gamma \dot{\Theta}^A = 0.$$

□

5.2 The supersymmetric string action

As was discussed in Chapter 4, there are two string theories with $\mathcal{N} = 2$ supersymmetry in ten dimensions, called the type IIA and type IIB superstring theories. Since in each case the supersymmetry is $\mathcal{N} = 2$, there are two fermionic coordinates Θ^1 and Θ^2. For the type IIA theory these spinors have opposite chirality while for the type IIB theory they have the same chirality, that is,

$$\Gamma_{11} \Theta^A = (-1)^{A+1} \Theta^A \quad \text{type IIA} \tag{5.33}$$

$$\Gamma_{11} \Theta^A = \Theta^A \quad \text{type IIB}. \tag{5.34}$$

The two spinors Θ^{Aa}, $A = 1, 2$, are Majorana–Weyl spinors.

In order to construct the GS world-sheet action for the type II superstrings, let us start with the bosonic Nambu–Goto action (for $\alpha' = 1/2$ or $T = 1/\pi$)

$$S_{\mathrm{NG}} = -\frac{1}{\pi} \int d^2\sigma \sqrt{-\det\left(\partial_\alpha X^\mu \partial_\beta X_\mu\right)}. \tag{5.35}$$

The obvious guess is that the supersymmetric string action takes the form

$$S_1 = -\frac{1}{\pi} \int d^2\sigma \sqrt{-G}, \tag{5.36}$$

with $G = \det G_{\alpha\beta}$, $G_{\alpha\beta} = \Pi_\alpha \cdot \Pi_\beta$ and

$$\Pi_\alpha^\mu = \partial_\alpha X^\mu - \bar{\Theta}^A \Gamma^\mu \partial_\alpha \Theta^A. \tag{5.37}$$

This expression is supersymmetric even if the number of supersymmetries is different from $\mathcal{N} = 2$. In the general case the index A takes the values $A = 1, \ldots, \mathcal{N}$. However, the case of interest to us has $D = 10$, $\mathcal{N} = 2$, and the spinors Θ^A are Majorana–Weyl spinors with 16 independent real components (though we use a 32-component notation).

As in the case of the D-particle, the action S_1 is not the complete answer, because it is not invariant under κ transformations. As before, a second term S_2 has to be added in order to produce local κ symmetry and thereby decouple half of the components of the fermionic variables. The action S_1 is invariant under global super-Poincaré transformations as well as local reparametrizations (diffeomorphisms) of the world sheet. These properties must be preserved by the new term S_2.

Kappa symmetry

In analogy to the discussion of the D0-brane, the bosonic variables transform under κ transformations according to

$$\delta X^\mu = \bar{\Theta}^A \Gamma^\mu \delta \Theta^A = -\delta \bar{\Theta}^A \Gamma^\mu \Theta^A, \tag{5.38}$$

which implies

$$\delta \Pi_\alpha^\mu = -2\delta \bar{\Theta}^A \Gamma^\mu \partial_\alpha \Theta^A. \tag{5.39}$$

Using (5.39) one obtains

$$\delta S_1 = \frac{2}{\pi} \int d^2\sigma \sqrt{-G} G^{\alpha\beta} \Pi_\alpha^\mu \delta \bar{\Theta}^A \Gamma_\mu \partial_\beta \Theta^A. \tag{5.40}$$

The next step is to construct a second contribution to the action S_2 that also has global super-Poincaré symmetry and local diffeomorphism symmetry. Moreover, its kappa variation δS_2 should combine nicely with δS_1 so as to ensure kappa symmetry of the sum. The analysis can be rather messy if one does it by brute force. It makes a lot more sense, however, if one focuses on the crucial geometrical aspects of the problem in the manner that follows. This methodology is generally applicable to problems of this type.

There is a large class of world-volume theories for which the action takes

5.2 The supersymmetric string action

the form $S_1 + S_2$, where S_1 is of the Nambu–Goto type and S_2 is of the Chern–Simons or Wess–Zumino type. These characterizations concern the way in which the diffeomorphism symmetry is implemented. S_1 has the structure of a supersymmetrized volume. The term S_2, on the other hand, is naturally described as the integral of a two-form

$$S_2 = \int \Omega_2 = \frac{1}{2} \int d^2\sigma \epsilon^{\alpha\beta} \Omega_{\alpha\beta}, \tag{5.41}$$

where Ω_2 does not depend on the world-sheet metric. More generally, for a p-brane it would be an integral of a $(p+1)$-form. Such a geometric structure has manifest diffeomorphism symmetry.

The way to make the symmetries of the problem manifest is to formally introduce an additional dimension and consider the three-form $\Omega_3 = d\Omega_2$. As a mathematical device, one may imagine that there is a three-dimensional region D whose boundary is the string world sheet M. The region D has no physical significance. In mathematical notation, $M = \partial D$. Then by Stokes' theorem

$$\int_D \Omega_3 = \int_M \Omega_2. \tag{5.42}$$

The advantage of this is that the symmetries of the problem are manifest in Ω_3. The differential form Ω_3 is like a characteristic class in that it is closed and invariant under the symmetries in question. The differential form Ω_2 is the corresponding Chern–Simons form. In general it is not invariant under the corresponding symmetry transformations. However, its variation is a total derivative, which is sufficient for our purposes.

A key formula in this subject is an identity satisfied by a Majorana–Weyl spinor Θ in ten dimensions

$$\Gamma^\mu d\Theta \, d\bar{\Theta} \Gamma_\mu d\Theta = 0. \tag{5.43}$$

In our notation wedge products are implicit, so the left-hand side of this equation is a three-form. This formula is crucial to the existence of supersymmetric Yang–Mills theory in ten dimensions, and it is also required in the analysis that follows. It is proved by considering Fierz transformations of the spinors, which are given in the appendix of Chapter 10.

Let us focus on the implementation of global space-time supersymmetry. There are three one-forms that are supersymmetric, namely $d\Theta^1$, $d\Theta^2$, and $\Pi^\mu = dX^\mu - \bar{\Theta}^A \Gamma^\mu d\Theta^A$. So Ω_3 should be a Lorentz-invariant three-form constructed out of these. Up to a normalization constant, c, to be determined later, the appropriate choice is

$$\Omega_3 = c(d\bar{\Theta}^1 \Gamma_\mu d\Theta^1 - d\bar{\Theta}^2 \Gamma_\mu d\Theta^2)\Pi^\mu. \tag{5.44}$$

The crucial minus sign in this formula is determined from the requirement that Ω_3 should be closed, that is, $d\Omega_3 = 0$. To see this substitute the explicit formula $d\Pi^\mu = -(d\bar\Theta^1 \Gamma^\mu d\Theta^1 + d\bar\Theta^2 \Gamma^\mu d\Theta^2)$ into

$$d\Omega_3 = c(d\bar\Theta^1 \Gamma_\mu d\Theta^1 - d\bar\Theta^2 \Gamma_\mu d\Theta^2) d\Pi^\mu. \tag{5.45}$$

The minus sign ensures the cancellation of the cross terms that have two powers of $d\Theta^1$ and two powers of $d\Theta^2$. The terms that are quartic in $d\Theta^1$ or $d\Theta^2$, on the other hand, vanish due to Eq. (5.43).

Let us now compute the kappa symmetry variation of Ω_3,

$$\delta\Omega_3 = 2c(d\delta\bar\Theta^1 \Gamma_\mu d\Theta^1 - d\delta\bar\Theta^2 \Gamma_\mu d\Theta^2)\Pi^\mu$$
$$-2c(d\bar\Theta^1 \Gamma_\mu d\Theta^1 - d\bar\Theta^2 \Gamma_\mu d\Theta^2)\delta\bar\Theta^A \Gamma^\mu d\Theta^A. \tag{5.46}$$

Using Eq. (5.43) again, the second line of this expression can be recast in the form

$$-2c(\delta\bar\Theta^1 \Gamma_\mu d\Theta^1 - \delta\bar\Theta^2 \Gamma_\mu d\Theta^2) d\Pi^\mu. \tag{5.47}$$

Therefore,

$$\delta\Omega_3 = d\left[2c(\delta\bar\Theta^1 \Gamma_\mu d\Theta^1 - \delta\bar\Theta^2 \Gamma_\mu d\Theta^2)\Pi^\mu\right], \tag{5.48}$$

and thus

$$\delta\Omega_2 = 2c(\delta\bar\Theta^1 \Gamma_\mu d\Theta^1 - \delta\bar\Theta^2 \Gamma_\mu d\Theta^2)\Pi^\mu. \tag{5.49}$$

To be explicit, setting $c = 1/\pi$ gives

$$\delta S_2 = \frac{2}{\pi} \int d^2\sigma \varepsilon^{\alpha\beta}(\delta\bar\Theta^1 \Gamma_\mu \partial_\alpha \Theta^1 - \delta\bar\Theta^2 \Gamma_\mu \partial_\alpha \Theta^2)\Pi^\mu_\beta. \tag{5.50}$$

The term S_2 is required to have this variation, since then the variation of the entire action under κ transformations takes the form

$$\delta S = \frac{4}{\pi} \int d^2\sigma \varepsilon^{\alpha\beta}(\delta\bar\Theta^1 P_+ \Gamma_\mu \partial_\alpha \Theta^1 - \delta\bar\Theta^2 P_- \Gamma_\mu \partial_\alpha \Theta^2)\Pi^\mu_\beta. \tag{5.51}$$

The orthogonal projection operators $P_\pm$ are defined by

$$P_\pm = \frac{1}{2}(1 \pm \gamma) \tag{5.52}$$

with

$$\gamma = -\frac{\varepsilon^{\alpha\beta} \Pi^\mu_\alpha \Pi^\nu_\beta \Gamma_{\mu\nu}}{2\sqrt{-G}}. \tag{5.53}$$

It now follows that the action is invariant under the transformations

$$\delta\bar\Theta^1 = \bar\kappa^1 P_- \quad \text{and} \quad \delta\bar\Theta^2 = \bar\kappa^2 P_+ \tag{5.54}$$

for arbitrary MW spinors κ^1 and κ^2 of appropriate chirality.

Let us now construct the term S_2. Using Eq. (5.44), $\Omega_3 = d\Omega_2$ can be solved for Ω_2. The solution, unique up to an irrelevant exact expression, is

$$\Omega_2 = c(\bar{\Theta}^1 \Gamma_\mu d\Theta^1 - \bar{\Theta}^2 \Gamma_\mu d\Theta^2) dX^\mu - c\bar{\Theta}^1 \Gamma_\mu d\Theta^1 \bar{\Theta}^2 \Gamma^\mu d\Theta^2. \qquad (5.55)$$

Note that changing the sign of c corresponds to interchanging Θ^1 and Θ^2, and therefore the choice is a matter of convention. The term S_2 can be reconstructed from this formula in the manner indicated in Eq. (5.41). Altogether, the κ-invariant action for the string is then

$$S = S_1 + S_2. \qquad (5.56)$$

Other p-branes, some of which are discussed in Chapter 6, also have world-volume actions with local κ symmetry. One example is the supermembrane in $D = 11$ supergravity (or M theory). Other examples contain additional world-volume fields besides X^μ and Θ. For example, the Dp-brane world-volume actions also contain $U(1)$ gauge fields. This gauge field could be ignored in the special case $p = 0$ discussed earlier.

EXERCISES

EXERCISE 5.5

Show that γ, defined in Eq. (5.53), satisfies $\gamma^2 = 1$, as required for $P_\pm = (1 \pm \gamma)/2$ to be orthogonal projection operators.

SOLUTION

The square of γ is

$$\gamma^2 = -\frac{1}{4G} \left(\varepsilon^{\alpha\beta} \Pi_\alpha^\mu \Pi_\beta^\nu \Gamma_{\mu\nu} \right)^2 = -\frac{1}{8G} \varepsilon^{\alpha_1\beta_1} \varepsilon^{\alpha_2\beta_2} \Pi_{\alpha_1}^{\mu_1} \Pi_{\beta_1}^{\nu_1} \Pi_{\alpha_2}^{\mu_2} \Pi_{\beta_2}^{\nu_2} \{\Gamma_{\mu_1\nu_1}, \Gamma_{\mu_2\nu_2}\}.$$

Using the identity

$$\{\Gamma_{\mu_1\nu_1}, \Gamma_{\mu_2\nu_2}\} = -2\eta_{\mu_1\mu_2}\eta_{\nu_1\nu_2} + 2\eta_{\mu_1\nu_2}\eta_{\nu_1\mu_2} + 2\Gamma_{\mu_1\nu_1\mu_2\nu_2},$$

and noting that the $\Gamma_{\mu_1\nu_1\mu_2\nu_2}$ term does not contribute, one obtains

$$\gamma^2 = \frac{1}{4G} \varepsilon^{\alpha_1\beta_1} \varepsilon^{\alpha_2\beta_2} \left(G_{\alpha_1\alpha_2} G_{\beta_1\beta_2} - G_{\alpha_1\beta_2} G_{\beta_1\alpha_2} \right) = 1.$$

□

5.3 Quantization of the GS action

The GS action is difficult to quantize covariantly, since the equations of motion are nonlinear in the coordinates X^μ and Θ^A. Also, the canonical variables satisfy constraints as a consequence of the local κ symmetry. These constraints are a mixture of first and second class (in Dirac's classification). Standard methods require disentangling the two types of constraints and treating them differently. However, this separation cannot be achieved covariantly. Many proposals for overcoming these difficulties have been made over the years, most of which were unsuccessful. More recently, Berkovits has found a scheme based on *pure spinors*, that does seem to work, but it is not yet understood well enough to include here.

The following analysis uses the light-cone gauge in which the equations of motion become linear, and the quantization of the theory becomes tractable. This gauge choice is very convenient for analyzing the physical spectrum of the theory. It can also be used to compute tree and one-loop amplitudes. However, to be perfectly honest, it is very awkward for most other purposes.

The light-cone gauge

As in the case of the bosonic string, the diffeomorphism symmetry can be used to choose the conformally flat gauge in which the world-sheet metric takes the form

$$h_{\alpha\beta} = e^\phi \eta_{\alpha\beta}. \tag{5.57}$$

After choosing this gauge the action is still invariant under superconformal transformations. As explained in Section 2.5, this residual symmetry allows one to choose the light-cone gauge in which the oscillators α_n^+, with $n \neq 0$, vanish, and therefore

$$X^+ = x^+ + p^+ \tau. \tag{5.58}$$

As before, this leaves only the transverse coordinates X^i with $i = 1, \ldots, 8$ as independent degrees of freedom. As a result, the theory contains eight bosonic degrees of freedom corresponding to the eight transverse directions in ten dimensions. In the world-sheet theory these appear as eight left-movers and eight right-movers.

Let's consider the fermionic degrees of freedom. As was discussed earlier, a generic spinor in ten dimensions has 32 complex components. Imposing Majorana and Weyl conditions reduces this to 16 real components, which is the content of a Majorana–Weyl spinor. In the present set-up there are two Majorana–Weyl spinors Θ^A, which therefore have a total of 32 real components. A factor of two reduction is provided by the local κ symmetry, which can be used to gauge away half of the 32 fermionic degrees of freedom. The final factor of two that leaves eight real degrees of freedom, for both left-movers and right-movers, is provided by the equations of motion.

A natural and convenient gauge choice is

$$\Gamma^+ \Theta^A = 0, \quad \text{where} \quad \Gamma^\pm = \frac{1}{\sqrt{2}}(\Gamma^0 \pm \Gamma^9). \tag{5.59}$$

This reduces the number of fermionic degrees of freedom for each of the two Θs to eight. Note that $\eta_{+-} = -1$, so that $\Gamma^+ = -\Gamma_-$ and $\Gamma^- = -\Gamma_+$. This gauge choice meshes nicely with gauge-fixing X^+, since $\delta X^+ = \bar{\varepsilon}^A \Gamma^+ \Theta^A$ vanishes. It could be justified by constructing a local κ transformation that implements this choice.

The GS action, in the version with a world-sheet metric, is

$$S = S_1 + S_2, \tag{5.60}$$

with

$$S_1 = -\frac{1}{2\pi} \int d^2\sigma \sqrt{-h} h^{\alpha\beta} \Pi_\alpha \cdot \Pi_\beta \tag{5.61}$$

and

$$S_2 = \frac{1}{\pi} \int d^2\sigma \varepsilon^{\alpha\beta} \left[-\partial_\alpha X^\mu (\bar{\Theta}^1 \Gamma_\mu \partial_\beta \Theta^1 - \bar{\Theta}^2 \Gamma_\mu \partial_\beta \Theta^2) - \bar{\Theta}^1 \Gamma^\mu \partial_\alpha \Theta^1 \bar{\Theta}^2 \Gamma_\mu \partial_\beta \Theta^2 \right]. \tag{5.62}$$

The equations of motion for the superstring in the GS formalism are highly nonlinear and given by

$$\Pi_\alpha \cdot \Pi_\beta = \frac{1}{2} h_{\alpha\beta} h^{\gamma\delta} \Pi_\gamma \cdot \Pi_\delta, \tag{5.63}$$

$$\Gamma \cdot \Pi_\alpha P_-^{\alpha\beta} \partial_\beta \Theta^1 = \Gamma \cdot \Pi_\alpha P_+^{\alpha\beta} \partial_\beta \Theta^2 = 0, \tag{5.64}$$

$$\partial_\alpha \left[\sqrt{-h} \left(h^{\alpha\beta} \partial_\beta X^\mu - 2 P_-^{\alpha\beta} \bar{\Theta}^1 \Gamma^\mu \partial_\beta \Theta^1 - 2 P_+^{\alpha\beta} \bar{\Theta}^2 \Gamma^\mu \partial_\beta \Theta^2 \right) \right] = 0, \tag{5.65}$$

where

$$P_\pm^{\alpha\beta} = \frac{1}{2} \left(h^{\alpha\beta} \pm \frac{\varepsilon^{\alpha\beta}}{\sqrt{-h}} \right). \tag{5.66}$$

Once the gauge choices (5.58) and (5.59) are imposed, the equations of motion for the string become linear. The basic reason for this simplification is that the term

$$\bar{\Theta}^A \Gamma^\mu \partial_\alpha \Theta^A \tag{5.67}$$

vanishes for $\mu = i, +$ and is nonvanishing only for $\mu = -$. Using the fermion gauge choice (5.59), the first equation in (5.64) takes the form

$$(\Gamma_\mu \Pi^\mu_\alpha) P^{\alpha\beta}_- \partial_\beta \Theta^1 = (\Gamma_+ \Pi^+_\alpha + \Gamma_i \Pi^i_\alpha) P^{\alpha\beta}_- \partial_\beta \Theta^1 = 0. \tag{5.68}$$

Multiplying this result by Γ^+ gives

$$\Gamma^+ (\Gamma_+ \Pi^+_\alpha + \Gamma_i \Pi^i_\alpha) P^{\alpha\beta}_- \partial_\beta \Theta^1 = 2\Pi^+_\alpha P^{\alpha\beta}_- \partial_\beta \Theta^1 = 0. \tag{5.69}$$

Using $\Pi^+_\alpha = p^+ \delta_{\alpha,0}$ this gives

$$P^{0\beta}_- \partial_\beta \Theta^1 = 0. \tag{5.70}$$

Using the definition of $P^{\alpha\beta}_-$ and the gauge choice $h_{\alpha\beta} = e^\phi \eta_{\alpha\beta}$, this takes the form

$$\left(\frac{\partial}{\partial \tau} + \frac{\partial}{\partial \sigma} \right) \Theta^1 = 0. \tag{5.71}$$

This is the equation of motion for Θ^1 in the light-cone gauge. It is considerably simpler than the covariant equation of motion. Since this equation is linear, it can be solved explicitly. In a similar way, the equations of motion for X^i and Θ^2 also become linear. One learns, in particular, that Θ^1 and Θ^2 describe waves that propagate in opposite directions along the string. This fact can be traced back to the relative minus sign between the Θ^1 and Θ^2 dependence in S_2.

The light-cone gauge action

The superstring theories considered here have ten-dimensional Lorentz invariance, but in the light-cone gauge only an $SO(8)$ transverse rotational symmetry is manifest. The eight surviving components of each Θ form an eight-dimensional spinor representation of this transverse $SO(8)$ group (or more precisely its $Spin(8)$ covering group). There are two inequivalent spinor representations of $Spin(8)$, which are denoted by $\mathbf{8_s}$ and $\mathbf{8_c}$. These two representations describe spinors of opposite eight-dimensional chirality. The ten-dimensional chirality of the spinors $\Theta^{1,2}$ determines whether an $\mathbf{8_s}$ or $\mathbf{8_c}$ representation survives in the light-cone gauge. Using the symbol S for the surviving components of Θ, multiplied by a factor proportional to $\sqrt{p^+}$, the choices are

$$\text{IIA}: \sqrt{p^+} \Theta^A \to \mathbf{8_s} + \mathbf{8_c} = (S^a_1, S^{\dot{a}}_2), \tag{5.72}$$

$$\text{IIB}: \sqrt{p^+} \Theta^A \to \mathbf{8_s} + \mathbf{8_s} = (S^a_1, S^a_2). \tag{5.73}$$

In the above formulas the letters $a, b, \ldots$ label the indices of a spinor in the $\mathbf{8_s}$ representation and dotted indices $\dot{a}, \dot{b}, \ldots$ label spinors in the $\mathbf{8_c}$ representation. This should be contrasted with the result obtained for the RNS formalism in light-cone gauge, where the fermionic fields on the world sheet are vectors of $SO(8)$ rather than spinors. In the case of the group $Spin(8)$ there is a triality symmetry that relates the vector representation to the two spinor representations. This symmetry is manifested as an S_3 symmetry of the $Spin(8)$ Dynkin diagram shown in Fig. 5.1.[1]

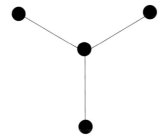

Fig. 5.1. Dynkin diagram for $SO(8) = D_4$. Triality refers to the symmetries of this diagram.

In the notation where the components of Θ that survive after gauge fixing are denoted by S, as described above, the equations of motion take the very simple form

$$\partial_+\partial_- X^i = 0, \qquad \partial_+ S_1^a = 0 \quad \text{and} \quad \partial_- S_2^{a \text{ or } \dot{a}} = 0. \qquad (5.74)$$

These equations are identical to those for the fields X^i, ψ_+^i and ψ_-^i in the RNS formalism. The only difference is that now the fermions are in the spinor representation of $Spin(8)$, whereas in the RNS formalism they were in the vector representation. By using the triality symmetry discussed above one can transform space-time spinors into vectors. As a result, these equations of motion are very similar to those for superstrings in the RNS formalism, though there are important differences in their usage.

The light-cone gauge action that gives rise to the above equations of motion is

$$S = -\frac{1}{2\pi} \int d^2\sigma \partial_\alpha X_i \partial^\alpha X^i + \frac{i}{\pi} \int d^2\sigma (S_1^a \partial_+ S_1^a + S_2^a \partial_- S_2^a), \qquad (5.75)$$

for the type IIB string. For the type IIA string one replaces S_2^a by $S_2^{\dot{a}}$. In the IIB case one can combine S_1 and S_2 into a two-component Majorana

---

[1] The representation theory of the groups $Spin(2n)$ is described in Appendix **5.A** of GSW.

world-sheet spinor giving the action

$$S = -\frac{1}{2\pi} \int d^2\sigma (\partial_\alpha X_i \partial^\alpha X^i + \bar{S}^a \rho^\alpha \partial_\alpha S^a), \quad (5.76)$$

where ρ^α are the two-dimensional Dirac matrices described in Chapter 4. As a result, the light-cone gauge superstring in the GS formalism looks almost the same as in the RNS formalism. An important difference is the fact that, whereas the RNS formalism required two sectors (R and NS), the entire spectrum is obtained from a single sector in the GS approach. It is an interesting fact that before gauge fixing the GS fermions transform as world-sheet scalars, but after gauge fixing they transform as world-sheet spinors.

Canonical quantization

Canonical quantization of the coordinates X^i is the same as in the case of the bosonic string or of the RNS string. Therefore, the equations of motion and the boundary conditions are solved by the same oscillator expansions. The fermionic coordinates satisfy anticommutation relations

$$\left\{S^{Aa}(\sigma,\tau), S^{Bb}(\sigma',\tau)\right\} = \pi \delta^{ab} \delta^{AB} \delta(\sigma - \sigma'), \quad (5.77)$$

where $A, B = 1, 2$ and $a, b = 1, \ldots, 8$. To determine the quantization conditions for the coefficients in the mode expansions of the fermionic fields, one must first choose boundary conditions for the fermionic coordinates. These determine the structure of the mode expansions, just as for the bosonic coordinates. There are several different possibilities:

Open type I superstring

The bosonic fields of the open or type I superstring satisfy Neumann boundary conditions at $\sigma = 0, \pi$. When they are required to end on lower-dimensional hypersurfaces (D-branes), Dirichlet boundary conditions, which are another possibility, are discussed in Chapter 6. The corresponding boundary conditions for the fermionic fields S^1 and S^2 require that they are related at the ends of the strings.

In order to keep the fermionic zero mode, which is necessary for unbroken space-time supersymmetry, there is no arbitrariness for the choice of sign in the boundary conditions. This is in contrast to the situation in the RNS approach. Space-time supersymmetry is only possible for the same relative sign choice at both ends. Thus the appropriate boundary conditions are

$$S^{1a}|_{\sigma=0} = S^{2a}|_{\sigma=0} \quad \text{and} \quad S^{1a}|_{\sigma=\pi} = S^{2a}|_{\sigma=\pi}. \quad (5.78)$$

5.3 Quantization of the GS action

Since the space-time supersymmetry transformation is $\delta\Theta^A = \varepsilon^A$ (where the ε^A are constants) the above boundary conditions are only compatible with supersymmetry if $\varepsilon^1 = \varepsilon^2$. As a result, open strings only have an $\mathcal{N} = 1$ supersymmetry. Such open strings occur in the type I superstring theory, which therefore is a theory with $\mathcal{N} = 1$ supersymmetry. The mode expansions for the fermionic fields of an open string satisfying Eqs (5.74) and (5.78) are

$$S^{1a} = \frac{1}{\sqrt{2}} \sum_{n=-\infty}^{\infty} S_n^a e^{-in(\tau-\sigma)}, \tag{5.79}$$

$$S^{2a} = \frac{1}{\sqrt{2}} \sum_{n=-\infty}^{\infty} S_n^a e^{-in(\tau+\sigma)}. \tag{5.80}$$

After quantization, the coefficients in the above mode expansions satisfy

$$\{S_m^a, S_n^b\} = \delta_{m+n,0} \delta^{ab}. \tag{5.81}$$

The reality condition implies $S_{-m}^a = (S_m^a)^\dagger$.

Closed strings

Closed strings require the periodicity

$$S^{Aa}(\sigma, \tau) = S^{Aa}(\sigma + \pi, \tau), \tag{5.82}$$

since this is the only boundary condition that is compatible with supersymmetry. As a result, the mode expansions become

$$S^{1a} = \sum_{-\infty}^{\infty} S_n^a e^{-2in(\tau-\sigma)}, \tag{5.83}$$

$$S^{2a} = \sum_{-\infty}^{\infty} \tilde{S}_n^a e^{-2in(\tau+\sigma)}. \tag{5.84}$$

Each set of modes satisfies the same canonical anticommutation relations as in Eq. (5.81). S^1 and S^2 belong to different spinor representations, $\mathbf{8_s}$ and $\mathbf{8_c}$, for the type IIA theory and to the same spinor representation, $\mathbf{8_s}$ or $\mathbf{8_c}$, for the type IIB theory. A left–right symmetrization (or orientifold projection) of the closed type IIB superstring gives a truncated spectrum that describes the closed type I superstring with $\mathcal{N} = 1$ supersymmetry.

The free string spectrum

Let us now examine the spectrum of free GS strings with space-time supersymmetry in flat ten-dimensional Minkowski space-time starting with the type I open-string states. This is useful for closed strings, as well, since closed-string left-movers and right-movers have essentially the same structure as open strings. This implies that closed-string states can be constructed as tensor products of open-string states, just as for the bosonic string.

Open type I superstrings

Open type I superstrings satisfy the mass-shell condition

$$\alpha' M^2 = \sum_{n=1}^{\infty} \left(\alpha^i_{-n} \alpha^i_n + n S^a_{-n} S^a_n \right). \tag{5.85}$$

Note that there is no extra constant (previously called a), since the normal-ordering constants for the bosonic and fermionic modes cancel exactly. As a result, there is no tachyon in the spectrum, and so no analog of the GSO projection is required to eliminate a tachyon. Moreover, the ground state is degenerate since the operator S^a_0 commutes with the mass operator. The ground-state spectrum must provide a representation of the zero-mode Clifford algebra

$$\{S^a_0, S^b_0\} = \delta^{ab} \qquad a, b = 1, \dots, 8. \tag{5.86}$$

The representation consists of a massless vector $\mathbf{8_v}$, which we denote by $|i\rangle$, $i = 1, \dots, 8$, and a massless spinor partner $|\dot{a}\rangle$, $\dot{a} = 1, \dots, 8$, which belongs to the $\mathbf{8_c}$. These are related according to[2]

$$|\dot{a}\rangle = \Gamma^i_{\dot{a}b} S^b_0 |i\rangle \quad \text{and} \quad |i\rangle = \Gamma^i_{\dot{a}b} S^b_0 |\dot{a}\rangle. \tag{5.87}$$

This construction is identical to the one used for the zero modes of the Ramond sector in the RNS formalism in Chapter 4. The difference is that the role of a vector and spinor representation has been interchanged. However, because of the triality symmetry of $Spin(8)$, the mathematics is the same.

This is exactly the massless spectrum required by supersymmetry that was found earlier. This time it has been achieved in a single sector, without any GSO-like projection. The excited levels at positive mass are obtained by acting on the massless states with the negative modes (S^a_{-n} and α^i_{-n}) in the usual way. The methodology of this construction ensures that the

---

[2] The eight 8×8 Dirac matrices $\Gamma^i_{\dot{a}b}$ are the Clebsch–Gordon coefficients for combining the three inequivalent $\mathbf{8}$s of $Spin(8)$ into a singlet.

5.3 Quantization of the GS action

supersymmetry generators can be expressed in terms of these oscillators, and therefore the physical spectrum is guaranteed to be supersymmetric.

Type II superstring theories

Type II superstrings, on the other hand, have the following spectrum. The ground state for the closed string is also massless and is given by the tensor product of left- and right-movers. Since the ground state for the open string is the 16-dimensional multiplet given by $\mathbf{8_v} + \mathbf{8_c}$, there are $256 = 16 \times 16$ states in the closed-string ground state. The resulting supermultiplets are different for the type IIA and type IIB theories.

In the case of the type IIA theory one should form the tensor product of two supermultiplets in which the spinors have opposite chirality

$$(\mathbf{8_v} + \mathbf{8_c}) \otimes (\mathbf{8_v} + \mathbf{8_s}). \tag{5.88}$$

This tensor product gives rise to the following bosonic fields:

$$\mathbf{8_v} \otimes \mathbf{8_v} = \mathbf{1} + \mathbf{28} + \mathbf{35} \quad \text{and} \quad \mathbf{8_s} \otimes \mathbf{8_c} = \mathbf{8_v} + \mathbf{56_t}, \tag{5.89}$$

while the tensor products of $\mathbf{8_v} \otimes \mathbf{8_s}$ and $\mathbf{8_v} \otimes \mathbf{8_c}$ give rise to the corresponding fermionic superpartners. The product of the two vectors $\mathbf{8_v} \otimes \mathbf{8_v}$ decomposes into a scalar, an antisymmetric rank-two tensor and a symmetric traceless tensor. The corresponding fields are the dilaton, antisymmetric tensor and the graviton. The product of the two spinors of opposite chirality, denoted ζ and χ, is evaluated by constructing the independent tensors

$$\bar{\zeta}\Gamma_i\chi \quad \text{and} \quad \bar{\zeta}\Gamma_{ijk}\chi. \tag{5.90}$$

These describe $8 + 56 = 64$ fermionic states, which is the expected number. Equation (5.89) describes the massless bosons of the ten-dimensional type IIA theory. This is the same bosonic content that is obtained when 11-dimensional supergravity is dimensionally reduced to ten dimensions. Furthermore, the fermions also match. This relationship has rather deep significance, as it suggests a connection between the two theories. This is explored in Chapter 8.

The spectrum of massless particles of the type IIB theory is given by the tensor product of two supermultiplets in which the spinors have the same chirality. The massless ground states are then given by

$$(\mathbf{8_v} + \mathbf{8_c}) \otimes (\mathbf{8_v} + \mathbf{8_c}). \tag{5.91}$$

This gives rise to the following bosonic fields:

$$\mathbf{8_v} \otimes \mathbf{8_v} = \mathbf{1} + \mathbf{28} + \mathbf{35} \quad \text{and} \quad \mathbf{8_c} \otimes \mathbf{8_c} = \mathbf{1} + \mathbf{28} + \mathbf{35_+}. \tag{5.92}$$

Here $\mathbf{35}_+$ describes a fourth-rank self-dual antisymmetric tensor. This spectrum does not arise from dimensional reduction of a higher-dimensional theory.

The above results show that both type II theories have the same field content in the NS–NS sector,[3] namely (in a mixed notation)

$$\mathbf{8_v} \otimes \mathbf{8_v} = \phi \oplus B_{\mu\nu} \oplus G_{\mu\nu}, \tag{5.93}$$

which are the dilaton, antisymmetric tensor and the graviton. In the R–R sector the type IIA and type IIB theories are different. The type IIA theory contains odd rank potentials, namely $\mathbf{8_v}$ and $\mathbf{56_t}$, which are one-form and three-form potentials. The type IIB theory, on the other hand, contains even rank potentials, namely $\mathbf{1}$, $\mathbf{28}$ and $\mathbf{35}_+$, which are a zero-form potential corresponding to an R–R scalar, a two-form potential and a four-form potential with a self-dual field strength in ten dimensions.

EXERCISES

EXERCISE 5.6
Show that $\bar{\Theta}^A \Gamma^\mu \partial_\alpha \Theta^A$ vanishes for $\mu = +, i$.

SOLUTION

The vanishing is obvious for $\mu = +$, because $\Gamma^+ \Theta^A = 0$. To see it for the case $\mu = i$ insert $1 = -(\Gamma^+ \Gamma^- + \Gamma^- \Gamma^+)/2$. Then Γ^+ multiplies Θ^A either from the left or the right to give zero, and so each of the two terms vanishes. □

EXERCISE 5.7
Show that Θ^1 and Θ^2 propagate in opposite directions along the string.

SOLUTION

Θ^1 and Θ^2 satisfy the equations of motion

$$(\partial_\tau + \partial_\sigma) \Theta^1 = (\partial_\tau - \partial_\sigma) \Theta^2 = 0.$$

---

[3] The GS formalism doesn't have distinct sectors, but these are the states of the NS–NS sector in the RNS formalism.

EXERCISE 5.8
Work out the decomposition of the tensor products $\mathbf{8_s} \otimes \mathbf{8_s}$ and $\mathbf{8_c} \otimes \mathbf{8_s}$.

SOLUTION

This problem involves evaluating tensor products of representations of the Lie group $Spin(8) = D_4$. Recall that it has three eight-dimensional representations, denoted $\mathbf{8_v}$, $\mathbf{8_s}$ and $\mathbf{8_c}$. These are related to one another by the triality automorphism group. The tensor product

$$\mathbf{8_v} \otimes \mathbf{8_v} = \mathbf{1} + \mathbf{28} + \mathbf{35}$$

is ordinary $SO(8)$ group theory: the decomposition of a second rank tensor t_{ij} into a trace, antisymmetric and symmetric-traceless parts. The product $\mathbf{8_s} \otimes \mathbf{8_s}$ works the same way for a tensor t_{ab}. One obtains

$$\mathbf{8_s} \otimes \mathbf{8_s} = \mathbf{1} + \mathbf{28} + \mathbf{35_-}.$$

The $\mathbf{1}$ and $\mathbf{28}$ are triality-invariant representations. However, there are three 35-dimensional representations related by triality. The $\mathbf{35_+}$ and $\mathbf{35_-}$ can be described alternatively as the self-dual and anti-self-dual parts of a fourth-rank antisymmetric tensor

$$t_{ijkl} = \pm \frac{1}{4!} \varepsilon^{ijkli'j'k'l'} t_{i'j'k'l'}.$$

Each of these has $\frac{1}{2}\binom{8}{4} = 35$ independent components.

The tensor product $\mathbf{8_c} \otimes \mathbf{8_s}$ contains an $\mathbf{8_v}$ given by $\Gamma^i_{ab} t_{ab}$, where Γ^i_{ab} is the invariant tensor described in the text. The remaining 56 components of the product form an irreducible representation $\mathbf{56_t}$. It has an alternative description as a third-rank antisymmetric tensor t_{ijk}, which has $\binom{8}{3} = 56$ independent components.

□

5.4 Gauge anomalies and their cancellation

In the early 1980s it appeared that superstrings could not describe parity-violating theories, because of quantum inconsistencies called anomalies. The

1984 discovery that the anomalies could cancel in certain cases was important for convincing many theorists that string theory is a promising approach to unification. In the years that have passed since then, string theory has been studied intensively, and many issues are understood much better now. In particular, it is possible to present the anomaly cancellation mechanism in a more elegant way than in the original papers. The improvements that are incorporated in the following discussion include an improved understanding of the association of specific terms with specific string world sheets as well as some mathematical tricks.

When a symmetry of a classical theory is broken by radiative corrections, so that there is no choice of local counterterms that can be added to the low-energy effective action to restore the symmetry, the symmetry is called *anomalous*. Anomalies arise from divergent Feynman diagrams, with a classically conserved current attached, that do not admit a regulator compatible with conservation of the current. Anomalies only arise at one-loop order (Adler–Bardeen theorem) in diagrams with a chiral fermion or boson going around the loop. Their origin can be traced to the behavior of Jacobian factors in the path-integral measure.

There are two categories of anomalies. The first category consists of anomalies that break a global symmetry. An example is the axial part of the flavor $SU(2) \times SU(2)$ symmetry of QCD. These anomalies are *good* in that they do not imply any inconsistency. Rather, they make it possible to carry out certain calculations to high precision. The classic example is the rate for the decay $\pi^0 \to \gamma\gamma$. The second category of anomalies consists of ones that break a local gauge symmetry. These are *bad*, in that they imply that the quantum theory is inconsistent. They are our concern here.

Parity-violating theories with chiral fields only exist in space-times with an even dimension. If the dimension is $D = 2n$, then anomalies can occur in Feynman diagrams with one current and n gauge fields attached to a chiral field circulating around the loop. In four dimensions these are triangle diagrams, and in ten dimensions these are hexagon diagrams, as shown in Fig. 5.2. The resulting nonconservation of the current J^μ takes the form

$$\partial_\mu J^\mu = a\epsilon^{\mu_1\mu_2\cdots\mu_{2n}} F_{\mu_1\mu_2} \cdots F_{\mu_{2n-1}\mu_{2n}}, \tag{5.94}$$

where a is some constant.

In string theory there are various world-sheet topologies that correspond to one-loop diagrams, as was discussed in Chapter 3. In the case of type II or heterotic theories the only possibility is a torus. For the type I superstring theory it can be a torus, a Klein bottle, a cylinder or a Moebius strip. However, the anomaly analysis can be carried out entirely in terms of a low-

5.4 Gauge anomalies and their cancellation

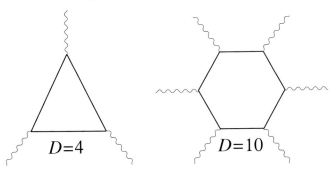

Fig. 5.2. Diagrams contributing to the gauge anomaly in four and ten dimensions. Each of these diagrams contains one current, while the remaining insertions are gauge fields.

energy effective action, which is what we do here. Even so, it is possible to interpret the type I result in terms of string world sheets. The torus turns out not to contribute to the anomaly. For the other world-sheet topologies, it is convenient to imagine them as made by piecing together boundary states $|B\rangle$ and cross-cap states $|C\rangle$. Cross-caps can be regarded as boundaries that have opposite points identified. In this way $\langle B|B\rangle$ represents a surface with two boundaries, which is a cylinder, $\langle B|C\rangle$ and $\langle C|B\rangle$ represent surfaces with one boundary and one cross-cap, which is a Moebius strip, and $\langle C|C\rangle$ represents a surface with two cross-caps, which is a Klein bottle. The correct relative weights of the Feynman diagrams are encoded in the combinations

$$(\langle B| + \langle C|) \times (|B\rangle + |C\rangle). \tag{5.95}$$

The consistency of the $SO(32)$ type I theory arises from a cancellation between the boundary and cross-cap contributions. It should also be pointed out that the modern interpretation of the boundary state is in terms of a world sheet that ends on a *D-brane*, whereas the cross-cap state corresponds to a world sheet that ends on an object called an *orientifold plane*. These are discussed in Chapter 6.

Chiral fields

As we learned earlier, in ten dimensions (in contrast to four dimensions) there exist spinors that are simultaneously Majorana and Weyl. Another difference between four and ten dimensions is that in ten dimensions it is also possible to have chiral bosons! To be specific, consider a fourth rank antisymmetric tensor field $A_{\mu\nu\rho\lambda}$, which is conveniently represented as a four-form A_4. Then the five-form field strength $F_5 = dA_4$ has a gauge invariance analogous to that of the Maxwell field, namely $\delta A_4 = d\Lambda_3$, where Λ_3 is a three-form. Moreover, one can covariantly eliminate half of the degrees of freedom associated with this field by requiring that F is self-dual (or anti-self-dual). Because the self-duality condition involves the ε symbol, the resulting degrees of freedom are not reflection invariant, and they therefore describe a chiral boson. When interactions are taken into account, the self-duality condition of the free theory is deformed by interaction terms. This construction in ten dimensions is consistent with Lorentzian signature, whereas in four dimensions a two-form field strength can be self-dual for Euclidean signature (a fact that is crucial for constructing instantons).

Differential forms and characteristic classes

To analyze anomalies it is extremely useful to use differential forms and characteristic classes. For example, Yang–Mills gauge fields are Lie-algebra-valued one-forms:

$$A = \sum_{\mu,a} A_\mu^a(x) \lambda^a dx^\mu. \tag{5.96}$$

Here the λ^a are matrices in a convenient representation (call it ρ) of the Lie algebra $\mathcal{G}$. The field strengths are Lie-algebra-valued two-forms:

$$F = \frac{1}{2} \sum_{\mu\nu} F_{\mu\nu} dx^\mu \wedge dx^\nu = dA + A \wedge A. \tag{5.97}$$

Note that this definition constrains F and A to be antihermitian in the case that the representation is complex and antisymmetric for real representations.[4] Under an infinitesimal Yang–Mills gauge transformation

$$\delta_\Lambda A = d\Lambda + [A, \Lambda] \quad \text{and} \quad \delta_\Lambda F = [F, \Lambda], \tag{5.98}$$

where Λ is an infinitesimal Lie-algebra-valued zero-form.

Gravity (in the vielbein formalism) is described in an almost identical manner. The *spin connection* one-form

$$\omega = \sum_{\mu,a} \omega_\mu^a(x) \lambda^a dx^\mu \tag{5.99}$$

is a gauge field for local Lorentz symmetry. The λ^a are chosen to be in the

---

[4] To make contact with the hermitian fields that appear in the low-energy effective actions in a later chapter the fields have to be rescaled by a factor of i.

fundamental representation of the Lorentz algebra ($D \times D$ matrices). The curvature two-form is

$$R = d\omega + \omega \wedge \omega. \tag{5.100}$$

Under an infinitesimal local Lorentz transformation (with infinitesimal parameter Θ)

$$\delta_\Theta \omega = d\Theta + [\omega, \Theta] \quad \text{and} \quad \delta_\Theta R = [R, \Theta]. \tag{5.101}$$

Characteristic classes are differential forms, constructed out of F and R, that are closed and gauge invariant. Thus $X(R, F)$ is a characteristic class provided that

$$dX(R, F) = 0 \quad \text{and} \quad \delta_\Lambda X(R, F) = \delta_\Theta X(R, F) = 0. \tag{5.102}$$

Some examples are

$$\mathrm{tr}(F \wedge \ldots \wedge F) \equiv \mathrm{tr}(F^k), \tag{5.103}$$

$$\mathrm{tr}(R \wedge \ldots \wedge R) \equiv \mathrm{tr}(R^k), \tag{5.104}$$

as well as polynomials constructed out of these building blocks using wedge products.

Characterization of anomalies

Yang–Mills anomalies and local Lorentz symmetry anomalies (also called *gravitational anomalies*) in $D = 2n$ dimensions are encoded in a characteristic class that is a $(2n + 2)$-form, denoted I_{2n+2}. You can't really antisymmetrize $2n + 2$ indices in $2n$ dimensions, so these expressions are a bit formal, though they can be given a precise mathematical justification. In any case, the physical anomaly is characterized by a $2n$-form G_{2n}, which certainly does exist. The precise formula is

$$\delta S_\mathrm{eff} = \int G_{2n}. \tag{5.105}$$

Here, S_eff represents the quantum effective action and the variation δ is an infinitesimal gauge transformation. The formulas for G_{2n} are rather ugly and subject to the ambiguity of local counterterms and total derivatives. On the other hand, by pretending that there are two extra dimensions, one uniquely encodes the anomalies in beautiful expressions I_{2n+2}. Moreover, any G_{2n} that is deduced from an I_{2n+2} by the formulas that follow is guaranteed to satisfy the Wess–Zumino consistency conditions.

The anomaly G_{2n} is obtained from I_{2n+2} (in a coordinate patch) by the descent equations

$$I_{2n+2} = d\omega_{2n+1} \tag{5.106}$$

and

$$\delta\omega_{2n+1} = dG_{2n}. \tag{5.107}$$

Here δ represents a combined gauge transformation (that is, $\delta = \delta_\Lambda + \delta_\Theta$). The ambiguities in the determination of the Chern–Simons form ω_{2n+1} and the anomaly form G_{2n} from these equations are just as they should be and do not pose a problem. The total anomaly is a sum of contributions from each of the chiral fields in the theory, and it can be encoded in a characteristic class

$$I_{2n+2} = \sum_\alpha I_{2n+2}^{(\alpha)}. \tag{5.108}$$

The formulas for every possible anomaly contribution $I_{2n+2}^{(\alpha)}$ were worked out by Alvarez-Gaumé and Witten. Dropping an overall normalization factor, because the goal is to achieve cancellation, their results are as follows:

- A left-handed Weyl fermion belonging to the ρ representation of the Yang–Mills gauge group contributes

$$I_{1/2}(R, F) = \left[\hat{A}(R)\, \mathrm{tr}_\rho e^{iF}\right]_{2n+2}. \tag{5.109}$$

The notation $[\cdots]_{2n+2}$ means that one should extract the $(2n+2)$-form part of the enclosed expression, which is a sum of differential forms of various orders. The factor $\mathrm{tr}_\rho e^{iF}$ is called a *Chern character*. The *Dirac roof genus* $\hat{A}(R)$ is given by

$$\hat{A}(R) = \prod_{i=1}^n \frac{\lambda_i/2}{\sinh\lambda_i/2}, \tag{5.110}$$

where the λ_i are the *eigenvalue two-forms* of the curvature:

$$R \sim \begin{pmatrix} 0 & \lambda_1 & & & & & \\ -\lambda_1 & 0 & & & & & \\ & & 0 & \lambda_2 & & & \\ & & -\lambda_2 & 0 & & & \\ & & & & \cdot & & \\ & & & & & 0 & \lambda_n \\ & & & & & -\lambda_n & 0 \end{pmatrix}. \tag{5.111}$$

5.4 Gauge anomalies and their cancellation

The first few terms in the expansion of $\hat{A}(R)$ are

$$\hat{A}(R) = 1 + \frac{1}{48}\operatorname{tr} R^2 + \frac{1}{16}\left[\frac{1}{288}(\operatorname{tr} R^2)^2 + \frac{1}{360}\operatorname{tr} R^4\right] + \ldots \qquad (5.112)$$

- A left-handed Weyl gravitino, which is always a singlet of any Yang–Mills groups, contributes $I_{3/2}(R)$, where

$$I_{3/2}(R) = \left(\sum_j 2\cosh\lambda_j - 1\right)\prod_i \frac{\lambda_i/2}{\sinh\lambda_i/2}. \qquad (5.113)$$

- A self-dual tensor gives a contribution denoted $I_A(R)$, where

$$I_A(R) = -\frac{1}{8}L(R), \qquad (5.114)$$

where the Hirzebruch L-function is defined by

$$L(R) = \prod_{i=1}^{n} \frac{\lambda_i}{\tanh\lambda_i}. \qquad (5.115)$$

In each case a chiral field of the opposite chirality (right-handed instead of left-handed) gives an anomaly contribution of the opposite sign. An identity that will be used later is

$$\hat{A}(R/2) = \sqrt{L(R/4)\hat{A}(R)}, \qquad (5.116)$$

which is an immediate consequence of Eqs (5.110) and (5.115).

Type IIB superstring theory

Type IIB superstring theory is a ten-dimensional parity-violating theory, whose massless chiral fields consist of two left-handed Majorana–Weyl gravitinos (or, equivalently, one Weyl gravitino), two right-handed Majorana–Weyl spinors (or dilatinos) and a self-dual boson. Thus the total anomaly is given by the 12-form part of

$$I(R) = I_{3/2}(R) - I_{1/2}(R) + I_A(R). \qquad (5.117)$$

An important result of the Alvarez-Gaumé and Witten paper is that this 12-form vanishes, so that this theory is anomaly-free. The proof requires showing that the expression

$$\left(2\sum_{j=1}^{5}\cosh\lambda_j - 2\right)\prod_{i=1}^{5}\frac{\lambda_i/2}{\sinh\lambda_i/2} - \frac{1}{8}\prod_{i=1}^{5}\frac{\lambda_i}{\tanh\lambda_i} \qquad (5.118)$$

contains no terms of sixth order in the λ_i. This involves three nontrivial cancellations. The relevance of this fact to the type I theory is that it allows us to represent $I_{3/2}(R)$ by $I_{1/2}(R) - I_A(R)$. This is only correct for the 12-form part, but that is all that is needed.

Type I superstring theory

Type I superstring theory has 16 conserved supercharges, which form a Majorana–Weyl spinor in ten dimensions. The massless fields of type I superstring theory consist of a supergravity multiplet in the closed-string sector and a super Yang–Mills multiplet in the open-string sector.

The supergravity multiplet

The supergravity multiplet contains three bosonic fields: the metric (35), a two-form (28), and a scalar dilaton (1). The parenthetical numbers are the number of physical polarization states represented by these fields. None of these is chiral. It also contains two fermionic fields: a left-handed Majorana–Weyl gravitino (56) and a right-handed Majorana–Weyl dilatino (8). These are chiral and contribute an anomaly given by

$$I_{\text{sugra}} = \frac{1}{2}\left[I_{3/2}(R) - I_{1/2}(R)\right]_{12} = -\frac{1}{2}\left[I_A(R)\right]_{12} = \frac{1}{16}\left[L(R)\right]_{12}. \quad (5.119)$$

The super Yang–Mills multiplet

The super Yang–Mills multiplet contains the gauge fields and left-handed Majorana–Weyl fermions (gauginos), each of which belongs to the adjoint representation of the gauge group. Classically, the gauge group of a type I superstring theory can be any orthogonal or symplectic group. In the following we only consider the case of $SO(N)$, since it is the one for which the desired anomaly cancellation can be achieved. In this case the adjoint representation corresponds to antisymmetric $N \times N$ matrices, and has dimension $N(N-1)/2$.

Adding the anomaly contribution of the gauginos to the supergravity contribution given above yields

$$I = \left[\frac{1}{2}\hat{A}(R)\,\text{Tr}\,e^{iF} + \frac{1}{16}L(R)\right]_{12}. \quad (5.120)$$

The symbol Tr is used to refer to the adjoint representation, whereas the symbol tr is used (later) to refer to the N-dimensional fundamental representation.

5.4 Gauge anomalies and their cancellation

The Chern-character factorization property

$$\text{tr}_{\rho_1 \times \rho_2} e^{iF} = \left(\text{tr}_{\rho_1} e^{iF}\right)\left(\text{tr}_{\rho_2} e^{iF}\right) \tag{5.121}$$

allows us to deduce that, for $SO(N)$,

$$\text{Tr}\, e^{iF} = \frac{1}{2}\left(\text{tr}\, e^{iF}\right)^2 - \frac{1}{2}\text{tr}\, e^{2iF} = \frac{1}{2}\left(\text{tr}\cos F\right)^2 - \frac{1}{2}\text{tr}\cos 2F. \tag{5.122}$$

The last step used the fact that the trace of an odd power of F vanishes, since the matrix is antisymmetric.

Substituting Eq. (5.122) into Eq. (5.120) gives the anomaly as the 12-form part of

$$I = \frac{1}{4}\hat{A}(R)\left(\text{tr}\cos F\right)^2 - \frac{1}{4}\hat{A}(R)\text{tr}\cos 2F + \frac{1}{16}L(R). \tag{5.123}$$

Since this is of sixth order in Rs and Fs, the following expression has the same 12-form part:

$$I' = \frac{1}{4}\hat{A}(R)\left(\text{tr}\cos F\right)^2 - 16\hat{A}(R/2)\text{tr}\cos F + 256 L(R/4). \tag{5.124}$$

Moreover, using Eq. (5.116), this can be recast as a perfect square

$$I' = Y^2 \quad \text{where} \quad Y = \frac{1}{2}\sqrt{\hat{A}(R)}\,\text{tr}\cos F - 16\sqrt{L(R/4)}. \tag{5.125}$$

There is no choice of N for which $[I']_{12} = [I]_{12}$ vanishes. However, as is explained later, it is possible to introduce a local counterterm that cancels the anomaly if $[I]_{12}$ factorizes into a product of a four-form and an eight-form.

Indeed, a priori, Y is a sum of forms $Y_0 + Y_4 + Y_8 + \ldots$. However, if the constant term vanishes ($Y_0 = 0$), then

$$[I]_{12} = [(Y_4 + Y_8 + \ldots)^2]_{12} = 2Y_4 Y_8, \tag{5.126}$$

as required. To evaluate the constant term Y_0, note that L and $\hat{A}$ are each equal to 1 plus higher-order forms and that $\text{tr}\cos F = N + \ldots$ Thus

$$Y_0 = \frac{N - 32}{2}, \tag{5.127}$$

and the desired factorization only works for the choice $N = 32$ in which case the gauge algebra is $SO(32)$.

Let us express Y as a sum of two terms $Y_B + Y_C$, where

$$Y_B = \frac{1}{2}\sqrt{\hat{A}(R)}\,\text{tr}\cos F \tag{5.128}$$

and
$$Y_C = -16\sqrt{L(R/4)}. \tag{5.129}$$

This decomposition has a simple interpretation in terms of string world sheets. Y_B is the boundary – or D-brane – contribution. It carries all the dependence on the gauge fields. Y_C is the cross-cap – or orientifold plane – contribution. Note that

$$I' = Y^2 = Y_B^2 + 2Y_BY_C + Y_C^2 \tag{5.130}$$

displays the anomaly contributions arising from distinct world-sheet topologies: the cylinder, the Moebius strip, and the Klein bottle, as shown in Fig. 5.3.

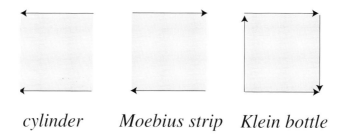

cylinder   *Moebius strip*   *Klein bottle*

Fig. 5.3. World-sheet topologies contributing to the anomaly in type I superstring theory. Opposite edges with arrows are identified with the arrow aligned.

Cancellation of the anomaly requires a local counterterm, S_{ct}, with the property that

$$\delta S_{\text{ct}} = -\int G_{10}, \tag{5.131}$$

where G_{10} is the anomaly ten-form that follows, via the descent equations, from $[I]_{12} = 2Y_4Y_8$. As was mentioned earlier, there are inconsequential ambiguities in the determination of G_{10} from $[I]_{12}$. A convenient choice in the present case is

$$G_{10} = 2G_2Y_8, \tag{5.132}$$

where G_2 is a two-form that is related to Y_4 by the descent equations $Y_4 = d\omega_3$ and $\delta\omega_3 = dG_2$. This works because Y_8 is closed and gauge invariant. Specifically, for the normalizations given here,

$$Y_4 = \frac{1}{4}(\text{tr} R^2 - \text{tr} F^2) \tag{5.133}$$

5.4 Gauge anomalies and their cancellation

and $\omega_3 = (\omega_{3\mathrm{L}} - \omega_{3\mathrm{Y}})/4$, where

$$d\omega_{3\mathrm{L}} = \mathrm{tr}\, R^2 \quad \text{and} \quad d\omega_{3\mathrm{Y}} = \mathrm{tr}\, F^2. \tag{5.134}$$

The type I supergravity multiplet contains a two-form gauge field denoted C_2. It is the only R–R sector field of the type IIB supergravity multiplet that survives the orientifold projection. In terms of its index structure, it would seem that the field C_2 should be invariant under Yang–Mills gauge transformations and local Lorentz transformations. However, it does transform nontrivially under each of them in just such a way as to cancel the anomaly. Specifically, writing the counterterm as

$$S_{\mathrm{ct}} = \mu \int C_2 Y_8, \tag{5.135}$$

Eq. (5.131) is satisfied provided that

$$\mu \delta C_2 = -2 G_2. \tag{5.136}$$

The coefficient μ is a parameter whose value depends on normalization conventions that are not specified here.

One consequence of the nontrivial gauge transformation properties of the field C_2 is that the naive kinetic term $\int |dC_2|^2$ must be modified to give gauge invariance. The correct choice is $\int |\widetilde{F}_3|^2$, where

$$\widetilde{F}_3 = dC_2 + 2\mu^{-1}\omega_3. \tag{5.137}$$

Note that ω_3 contains both Yang–Mills and Lorentz Chern–Simons forms. Only the former is present in the classical supergravity theory.

The case of $E_8 \times E_8$

The preceding discussion presented the anomaly analysis for the type I theory in a way where the physical meaning of the various terms could be understood. In order to describe the situation for the $E_8 \times E_8$ theory, it is useful to begin by backing up and presenting the same result from a more "brute force" viewpoint.

Writing down the various contributions to the anomaly 12-form characteristic class, one finds that the required factorization into a product of a four-form and an eight-form ($I_{12} \sim Y_4 Y_8$) requires that two conditions be satisfied: (1) the dimension of the gauge group must be 496 to ensure cancellation of $\mathrm{tr}\, R^6$ terms, a condition that is satisfied by both $SO(32)$ and

$E_8 \times E_8$; (2) $\mathrm{Tr} F^6$ must be re-expressible as follows:

$$\mathrm{Tr} F^6 = \frac{1}{48} \mathrm{Tr} F^2 \mathrm{Tr} F^4 - \frac{1}{14,400} (\mathrm{Tr} F^2)^3. \tag{5.138}$$

This identity is satisfied in the case of $SO(32)$ because of the following identities relating adjoint representation traces to fundamental representation traces:

$$\mathrm{Tr} F^2 = 30 \, \mathrm{tr} F^2, \tag{5.139}$$

$$\mathrm{Tr} F^4 = 24 \, \mathrm{tr} F^4 + 3(\mathrm{tr} F^2)^2, \tag{5.140}$$

$$\mathrm{Tr} F^6 = 15 \, \mathrm{tr} F^2 \, \mathrm{tr} F^4. \tag{5.141}$$

These identities follow from Eq. (5.122). Given these formulas, the factorized anomaly can be written in the form $I_{12} \sim X_4 X_8$,[5] where

$$X_4 = \mathrm{tr} R^2 - \frac{1}{30} \mathrm{Tr} F^2 \tag{5.142}$$

and

$$X_8 = \frac{1}{8} \mathrm{tr} R^4 + \frac{1}{32} (\mathrm{tr} R^2)^2 - \frac{1}{240} \mathrm{tr} R^2 \mathrm{Tr} F^2 + \frac{1}{24} \mathrm{Tr} F^4 - \frac{1}{7200} (\mathrm{Tr} F^2)^2. \tag{5.143}$$

In the case of $E_8 \times E_8$, Eq. (5.138) is also satisfied, and X_4 and X_8 are again given by Eqs (5.142) and (5.143). To see this one needs to understand first that

$$\mathrm{Tr} F^{2n} = \mathrm{Tr} F_1^{2n} + \mathrm{Tr} F_2^{2n}, \tag{5.144}$$

where the subscripts 1 and 2 refer to the two individual E_8 factors. In other words,

$$F = \begin{pmatrix} F_1 & 0 \\ 0 & F_2 \end{pmatrix}. \tag{5.145}$$

Thus this formula re-expresses the trace of a 496-dimensional matrix as the sum of the traces of two 248-dimensional matrices.

The following identities hold for each of the two E_8 groups:

$$\mathrm{Tr} F_i^4 = \frac{1}{100} (\mathrm{Tr} F_i^2)^2 \quad \text{and} \quad \mathrm{Tr} F_i^6 = \frac{1}{7200} (\mathrm{Tr} F_i^2)^3 \quad i = 1, 2. \tag{5.146}$$

Using these relations it is straightforward to verify Eq. (5.138). These formulas have a certain black-magic quality. It would be more satisfying to obtain a deeper understanding of where they come from, as was done in the

---
[5] We have introduced $X_4 = 4Y_4$ and $X_8 = 48Y_8$.

5.4 Gauge anomalies and their cancellation

type I case. Such an understanding was achieved by Hořava and Witten in 1995, and it is very different from that of the type I theory.

The key observation of Hořava and Witten was that at strong coupling the $E_8 \times E_8$ heterotic string theory grows an eleventh dimension that is a line interval of length $g_s l_s$. In the detailed construction, which is described in Chapter 8, it is convenient to represent the line interval as an $S^1/\mathbb{Z}_2$ orbifold. Since the size of this dimension is proportional to the string coupling, it is invisible in perturbation theory, where the space-time appears to be ten-dimensional. However, a deeper understanding of the anomaly cancellation can be achieved by reconsidering it from an 11-dimensional viewpoint.

Theories in an odd number of space-time dimensions ordinarily are not subject to anomalies. However, in the case of the M-theory set-up appropriate to the $E_8 \times E_8$ theory, the space-time has two ten-dimensional boundaries, and there can be anomalies that are localized on these boundaries. The picture one gets is that each of the E_8 factors is associated with one of the boundaries. Thus, one set of E_8 gauge fields is localized on one boundary and the other set of E_8 gauge fields is localized on the other boundary. This gives a very nice intuitive understanding of why the gauge group is the direct product of two identical groups. Indeed, Hořava and Witten carried out the anomaly analysis in detail and showed that, when M-theory has a ten-dimensional boundary, there must be an E_8 vector supermultiplet confined to that boundary. No other choice of gauge group is consistent with the anomaly analysis. This is one of many deep connections between M-theory and the Lie group E_8.

Since the two E_8 groups are spatially separated, the anomaly analysis should work for each of them separately. This requires that the factorized anomaly 12-form should be re-expressible as the sum of two factorized anomaly 12-forms

$$X_4 X_8 = X_4^{(1)} X_8^{(1)} + X_4^{(2)} X_8^{(2)}, \tag{5.147}$$

where the first term on the right-hand side only involves the gauge fields of the first E_8, and the second term only involves the gauge fields of the second E_8. It is a matter of some straightforward algebra to verify that this identity is satisfied for the choices

$$X_4^{(i)} = \frac{1}{2} \text{tr} R^2 - \frac{1}{30} \text{Tr} F_i^2 \quad i = 1, 2 \tag{5.148}$$

and

$$X_8^{(i)} = \frac{1}{8} \text{tr} R^4 + \frac{1}{32} (\text{tr} R^2)^2 - \frac{1}{120} \text{tr} R^2 \text{Tr} F_i^2 + \frac{1}{3600} (\text{Tr} F_i^2)^2 \quad i = 1, 2. \tag{5.149}$$

Finally, the local counterterms that complete the anomaly analysis have the structure $\sum_i \int B^{(i)} \wedge X_8^{(i)}$, where the integral is over the ith boundary. The field $B^{(i)}$ is obtained from the M-theory three-form field $A_{\mu\nu\rho}$ by setting one index equal to 11 (the compact direction) and restricting to the ith boundary.

EXERCISES

EXERCISE 5.9

Let us consider supergravity theories in six dimensions with $\mathcal{N} = 1$ supersymmetry. Let us further assume that the minimal supergravity multiplet is coupled to a tensor multiplet as well as n_H hypermultiplets and n_V vector multiplets. Show that a necessary condition for anomaly cancellation is

$$n_H - n_V = 244.$$

SOLUTION

The fields of the gravity and tensor multiplets combine to give a graviton $g_{\mu\nu}$, a two-form $B_{\mu\nu}$, a scalar, a left-handed gravitino and a right-handed dilatino. The reason for combining these two multiplets is that one of them gives the self-dual part of $H = dB$ and the other gives the anti-self-dual part. A vector multiplet contains a vector gauge field and a left-handed gaugino. A hypermultiplet contains four scalars and a right-handed hyperino. Therefore, the total purely gravitational anomaly is given by the eight-form part of

$$I_{3/2}(R) + (n_V - n_H - 1)I_{1/2}(R).$$

Using the formulas in the text, the eight-form parts of $I_{1/2}(R)$ and $I_{3/2}(R)$ are

$$I_{1/2}^{(8)}(R) = \frac{1}{128 \cdot 180}(4\,\mathrm{tr}R^4 + 5(\mathrm{tr}R^2)^2),$$

$$I_{3/2}^{(8)}(R) = \frac{1}{128 \cdot 180}(980\,\mathrm{tr}R^4 - 215(\mathrm{tr}R^2)^2).$$

By the same reasoning as in the text, a necessary requirement for anomaly cancellation is that the total anomaly factorizes into a product of two four-forms. A necessary requirement for this to be possible is the cancellation of the $\mathrm{tr}R^4$ terms, since $\mathrm{tr}R^4$ cannot be factorized. This requirement gives the condition $980 + 4(n_V - n_H - 1) = 0$, which simplifies to $n_H - n_V = 244$. $\square$

Exercise 5.10

The type IIA NS5-brane, which is introduced in Chapter 8, has a six-dimensional world-volume theory with $(0,2)$ supersymmetry. This means that both supercharges have the same chirality. As a result, the theory is chiral, and there is an anomaly associated with it. The resulting anomaly cannot be canceled by the methods described in this chapter. Instead, the brane has interactions with fields of the ten-dimensional bulk that lead to an *anomaly-inflow mechanism* that cancels the anomaly. Determine the form of this interaction required for the cancellation.

Solution

The NS5-brane world volume has $\mathcal{N} = 2$ supersymmetry, and the field content is given by two matter multiplets. The first multiplet contains four scalars and a right-handed fermion. The second multiplet contains one anti-self-dual tensor, a single scalar and another right-handed fermion. Of these fields only the fermions and the anti-self-dual tensor are chiral and contribute to the anomaly. Since the theory is six-dimensional, the anomalies are characterized by eight-forms. For this problem, it is desirable to keep track of the overall normalization, which was not relevant in the previous discussions. For this purpose it is convenient to express the anomalies in terms of Pontryagin classes. These are defined by the formula

$$p(R) = \det\left(1 + \frac{R}{2\pi}\right) = \prod_{i=1}^{n}\left(1 + (\lambda_i/2\pi)^2\right).$$

Thus

$$p_1 = -\frac{1}{2}\frac{1}{(2\pi)^2}\mathrm{tr}R^2 \quad \text{and} \quad p_2 = \frac{1}{8}\frac{1}{(2\pi)^4}\left((\mathrm{tr}R^2)^2 - 2\mathrm{tr}R^4\right)$$

and so forth.

Expressed in terms of Pontryagin classes, including the overall normalization factor, the anomalies are

$$I^{(8)}_{1/2} = \frac{1}{5760}\left(7p_1^2 - 4p_2\right) \quad \text{and} \quad I^{(8)}_A = \frac{1}{5760}\left(16p_1^2 - 112p_2\right).$$

So the total anomaly on the NS5-brane world volume is

$$I_8 = 2I^{(8)}_{1/2} + I^{(8)}_A = \frac{1}{192}\left(p_1^2 - 4p_2\right) = \frac{1}{192}\frac{1}{(2\pi)^4}\left(\mathrm{tr}R^4 - \frac{1}{4}(\mathrm{tr}R^2)^2\right).$$

The descent equations in this case can be written in the form $I_8 = d\omega_7$ and $\delta\omega_7 = dG_6$. The anomaly G_6 is a certain six-form that depends on the infinitesimal parameter of a local Lorentz transformation.

The type IIA theory contains a massless antisymmetric tensor B_2 in the NS–NS sector with a field strength $H_3 = dB_2$. Now suppose that the low-energy effective action of the type IIA theory contains the term

$$\int H_3 \wedge \omega_7.$$

Under infinitesimal local Lorentz transformations this expression has the variation

$$\delta \int H_3 \wedge \omega_7 = \int H_3 \wedge dG_6 = -\int dH_3 \wedge G_6.$$

The NS5-brane is a source for the gauge field B_2, a fact that can be expressed in the form

$$dH_3 = \delta_W,$$

where δ_W is a four-dimensional delta function with support on the 5-brane world volume. Therefore, in the presence of a 5-brane the variation of this term under an infinitesimal local Lorentz transformation is $-\int G_6$. This term exactly cancels the anomaly contribution due to the chiral fields on the 5-brane world volume. Therefore, quantum consistency requires the ten-dimensional interaction $\int H_3 \wedge \omega_7$.

Let us jump ahead in the story and mention that the strong-coupling limit of the type IIA theory is an 11-dimensional theory called M-theory. In the strong-coupling limit the type IIA NS 5-brane goes over to the M5-brane in 11 dimensions. Also, the two-form B_2 becomes part of a three-form potential A_3 with a four-form field strength $F_4 = dA_3$. The corresponding interaction in M-theory that cancels the world-volume anomaly of the M5-brane has the form

$$\int F_4 \wedge \omega_7.$$

□

HOMEWORK PROBLEMS

PROBLEM 5.1
Show that the action in Eq. (5.21) is invariant under a reparametrization of the world line.

PROBLEM 5.2

In order to obtain a nontrivial massless limit of Eq. (5.21), it is useful to first restore the auxiliary field $e(\tau)$ described in Chapter 2.

(i) Re-express the massive D0-brane action with the auxiliary field $e(\tau)$.
(ii) Find the massless limit of the D0-brane action.[6]
(iii) Verify the κ symmetry of the massless D0-brane action.

PROBLEM 5.3

Prove that, for a pair of Majorana spinors, Θ_1 and Θ_2, the flip symmetry is given by

$$\bar{\Theta}_1 \Gamma_{\mu_1 \cdots \mu_n} \Theta_2 = (-1)^{n(n+1)/2} \bar{\Theta}_2 \Gamma_{\mu_1 \cdots \mu_n} \Theta_1,$$

as asserted at the end of Exercise 5.2.

PROBLEM 5.4

Derive the relevant Fierz transformation identities for Majorana–Weyl spinors in ten dimensions and use them to prove that

$$\Gamma^\mu d\Theta \, d\bar{\Theta} \Gamma_\mu d\Theta = 0.$$

PROBLEM 5.5

Verify that the action (5.41) with Ω_2 given by Eq. (5.55) is invariant under supersymmetry transformations.

PROBLEM 5.6

Prove the identity

$$\{\Gamma_{\mu_1 \nu_1}, \Gamma_{\mu_2 \nu_2}\} = -2\eta_{\mu_1 \mu_2} \eta_{\nu_1 \nu_2} + 2\eta_{\mu_1 \nu_2} \eta_{\nu_1 \mu_2} + 2\Gamma_{\mu_1 \nu_1 \mu_2 \nu_2},$$

invoked in Exercise 5.5.

PROBLEM 5.7

Verify that the action (5.62) is supersymmetric.

PROBLEM 5.8

Construct the conserved supersymmetry charges for open strings in the light-cone gauge formalism of Section 5.3 and verify that they satisfy the supersymmetry algebra. Hint: the 16 supercharges are given by two eight-component spinors, Q^+ and Q^-. The Q^+s anticommute to P^+, the Q^-s

---

[6] This is sometimes called the Brink–Schwarz superparticle.

Problem 5.9

(i) Show that $\text{tr} F \wedge F$ is closed and gauge invariant.

(ii) This quantity is a characteristic class proportional to c_2, the second Chern class. Since it is closed, in a local coordinate patch one can write $\text{tr} F \wedge F = d\omega_3$, where ω_3 is a Chern–Simons three-form. Show that

$$\omega_3 = \text{tr}\left(A \wedge dA + \frac{2}{3} A \wedge A \wedge A\right).$$

(iii) Similarly, one can write $\text{tr} F^4 = d\omega_7$. Find ω_7.

Problem 5.10

Check the identity in Eq. (5.122) for $SO(N)$.

Problem 5.11

(i) Using the definition of Y in Eq. (5.125), obtain an expression for Y_4.

(ii) Apply the descent formalism to obtain a formula for G_2 in Eq. (5.132).

Problem 5.12

Prove the relations given in Eqs (5.139)–(5.141).

Problem 5.13

Verify that the identity (5.138) is satisfied for the gauge group $E_8 \times E_8$.

Problem 5.14

There is no string theory known with the gauge groups $E_8 \times U(1)^{248}$ or $U(1)^{496}$. Nevertheless, the anomalies cancel in these cases as well. Prove that this is the case. Hint: infer the result from the fact that the anomalies cancel for $E_8 \times E_8$.

Problem 5.15

Prove that $\text{Tr} F^4 = \frac{1}{100}(\text{Tr} F^2)^2$ for the adjoint representation of E_8. Hint: use the $Spin(16)$ decomposition $\mathbf{248} = \mathbf{120} + \mathbf{128}$.

6
T-duality and D-branes

String theory is not only a theory of fundamental one-dimensional strings. There are also a variety of other objects, called branes, of various dimensionalities. The list of possible branes, and their stability properties, depends on the specific theory and vacuum configuration under consideration. One clue for deciphering the possibilities is provided by the spectrum of massless particles. Chapters 4 and 5 described the spectra of massless states that appear in the type I and type II superstring theories in ten-dimensional Minkowski space-time. In particular, it was shown that several antisymmetric tensor (or differential form) gauge fields appear in the R–R sector of each of the type II theories. These tensor fields couple naturally to higher-dimensional extended objects, called D-branes. However, this is not the defining property of D-branes. Rather, the defining property is that D-branes are objects on which open strings can end. A string that does not touch a D-brane must be a closed loop. Those D-branes that have charge couplings to antisymmetric tensor gauge fields are stable, whereas those that do not usually are unstable.

One way of motivating the necessity of D-branes is based on T-duality, so this chapter starts with a discussion of T-duality of the bosonic string theory. Under T-duality transformations, closed bosonic strings transform into closed strings of the same type in the T-dual geometry. The situation is different for open strings, however. The key is to focus on the type of boundary conditions imposed at the ends of the open strings. Even though the only open-string boundary conditions that are compatible with Poincaré invariance (in all directions) are of Neumann type, Dirichlet boundary conditions inevitably appear in the equivalent T-dual reformulation. Open strings with Dirichlet boundary conditions in certain directions have ends with specified positions in those directions, which means that they have to end on specified hypersurfaces. Even though this violates Lorentz invariance, there is a good physical reason for them to end in this manner. The reason this is sensible is that they are ending on other physical objects that are also part of the theory, which are called Dp-branes. The letter D stands for Dirichlet, and p denotes the number of spatial dimensions of the D-brane. For example, as discussed in the previous chapter, a D0-brane is a point particle. When the time direction is also taken into account, the world volume of a Dp-brane has $p+1$ dimensions.

Much of the importance of Dp-branes stems from the fact that they provide a remarkable way of introducing nonabelian gauge symmetries in string theory: nonabelian gauge fields naturally appear confined to the world volume of multiple coincident Dp-branes. Moreover, Dp-branes are useful for discovering dualities that relate apparently different string theories. T-duality is introduced in this chapter, because it can be understood in perturbative string theory. Most other string dualities are nonperturbative. The general subject of string dualities is discussed in more detail in Chapter 8.

6.1 The bosonic string and Dp-branes

T-duality and closed strings

In order to introduce the notion of T-duality, let us first consider the simplest example, namely the bosonic string with one of the 25 spatial directions forming a circle of radius R. Altogether, the space-time geometry is chosen to be 25-dimensional Minkowski space-time times a circle ($\mathbb{R}^{24,1} \times S^1$). Sometimes one describes this as *compactification on a circle of radius R*. In this case a T-duality transformation inverts the radius of the circle, that is, it maps $R \to \tilde{R} = \alpha'/R$, and it leaves the mass formula for the string invariant provided that the string winding number is exchanged with the Kaluza–Klein excitation number. Let us now explore how this works.

To describe a closed bosonic string in a theory compactified on a circle of radius R, one takes periodic boundary conditions for one of the coordinates

$$X^{25}(\sigma + \pi, \tau) = X^{25}(\sigma, \tau) + 2\pi R W, \qquad W \in \mathbb{Z}, \qquad (6.1)$$

where W is the *winding number*. The winding number W indicates the number of times the string winds around the circle and its sign encodes the direction, as shown in Fig. 6.1. Let us now consider the mode expansion for a closed string with winding number W. The expansion of the coordinates X^μ, for $\mu = 0, \ldots, 24$, does not change compared to the expansion in flat 26-dimensional Minkowski space given in Chapter 2. However, the expansion of $X^{25}(\sigma, \tau)$ has to be changed, by adding a term linear in σ, in order to

6.1 The bosonic string and Dp-branes

incorporate the boundary condition (6.1). The expansion is

$$X^{25}(\sigma, \tau) = x^{25} + 2\alpha' p^{25}\tau + 2RW\sigma + \ldots, \qquad (6.2)$$

where the coefficient of σ is chosen to satisfy (6.1). The dots refer to the oscillator terms, which are not modified by the compactification.

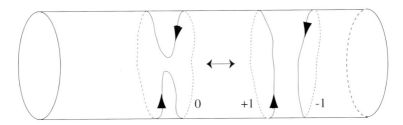

Fig. 6.1. Strings winding around a compact direction.

Since one dimension is compact, the momentum eigenvalue along that direction, p^{25}, is quantized. Remember that the quantum mechanical wave function contains the factor $\exp(ip^{25}x^{25})$. As a result, if x^{25} is increased by $2\pi R$, corresponding to going once around the circle, the wave function should return to its original value. In other words, it should be single-valued on the circle. This implies that the momentum in the 25 direction is of the form

$$p^{25} = \frac{K}{R}, \qquad K \in \mathbb{Z}. \qquad (6.3)$$

The integer K is called the *Kaluza–Klein excitation number*. Splitting the expansion into left- and right-movers,

$$X^{25}(\sigma, \tau) = X_L^{25}(\tau + \sigma) + X_R^{25}(\tau - \sigma), \qquad (6.4)$$

gives

$$X_R^{25}(\tau - \sigma) = \frac{1}{2}(x^{25} - \tilde{x}^{25}) + (\alpha' \frac{K}{R} - WR)(\tau - \sigma) + \ldots, \qquad (6.5)$$

$$X_L^{25}(\tau + \sigma) = \frac{1}{2}(x^{25} + \tilde{x}^{25}) + (\alpha' \frac{K}{R} + WR)(\tau + \sigma) + \ldots, \qquad (6.6)$$

where $\tilde{x}^{25}$ is a constant that cancels in the sum. In terms of the zero modes α_0^{25} and $\tilde{\alpha}_0^{25}$, defined in Chapter 2, the mode expansion is

$$X_R^{25}(\tau - \sigma) = \frac{1}{2}(x^{25} - \tilde{x}^{25}) + \sqrt{2\alpha'}\alpha_0^{25}(\tau - \sigma) + \ldots, \qquad (6.7)$$

$$X_L^{25}(\tau + \sigma) = \frac{1}{2}(x^{25} + \tilde{x}^{25}) + \sqrt{2\alpha'}\tilde{\alpha}_0^{25}(\tau + \sigma) + \ldots, \qquad (6.8)$$

where

$$\sqrt{2\alpha'}\alpha_0^{25} = \alpha'\frac{K}{R} - WR, \tag{6.9}$$

$$\sqrt{2\alpha'}\tilde{\alpha}_0^{25} = \alpha'\frac{K}{R} + WR. \tag{6.10}$$

The mass formula for the string with one dimension compactified on a circle can be interpreted from a 25-dimensional viewpoint in which one regards each of the Kaluza–Klein excitations (labelled by K) as distinct particles. The 25-dimensional mass squared is given by

$$M^2 = -\sum_{\mu=0}^{24} p_\mu p^\mu. \tag{6.11}$$

On the other hand, the requirement that the operators $L_0 - 1$ and $\tilde{L}_0 - 1$ annihilate on-shell physical states still holds. The expressions for L_0 and $\tilde{L}_0$ include contributions from all 26 dimensions, including the 25th. As a result, the equations $L_0 = 1$ and $\tilde{L}_0 = 1$ become

$$\frac{1}{2}\alpha' M^2 = (\tilde{\alpha}_0^{25})^2 + 2N_L - 2 = (\alpha_0^{25})^2 + 2N_R - 2. \tag{6.12}$$

Taking the sum and difference of these formulas, and using Eqs (6.9) and (6.10), gives

$$N_R - N_L = WK \tag{6.13}$$

and

$$\alpha' M^2 = \alpha'\left[\left(\frac{K}{R}\right)^2 + \left(\frac{WR}{\alpha'}\right)^2\right] + 2N_L + 2N_R - 4. \tag{6.14}$$

Note that Eq. (6.13) shows how the usual level-matching condition $N_L = N_R$ is modified for closed strings with both nonzero winding number W and nonzero Kaluza–Klein momentum K.

Equations (6.13) and (6.14) are invariant under interchange of W and K, provided that one simultaneously sends $R \to \tilde{R} = \alpha'/R$. This symmetry of the bosonic string is called *T-duality*. It suggests that compactification on a circle of radius R is physically equivalent to compactification on a circle of radius $\tilde{R}$. In fact, this turns out to be exactly true for the full interacting string theory, at least perturbatively.[1]

In the example considered here, T-duality maps two theories of the same

---

[1] It is unclear whether the bosonic string theory actually exists nonperturbatively (due to the closed-string tachyon), so that it is only sensible to discuss this theory at the perturbative level. However, the corresponding statements for superstrings are true nonperturbatively, as well.

type (one with a circle of radius R and one with a circle of radius $\widetilde{R} = \alpha'/R$) into one another. The physical equivalence of a circle of radius R and a circle of radius $\widetilde{R}$ is a clear indication that ordinary geometric concepts and intuitions can break down in string theory at the string scale. This is not so surprising once one realizes that this is the characteristic size of the objects that are probing the geometry. Note that the $W \leftrightarrow K$ interchange means that momentum excitations in one description correspond to winding-mode excitations in the dual description and *vice versa*.

Omitting the superscript 25, the transformation can be expressed as

$$\alpha_0 \to -\alpha_0 \quad \text{and} \quad \tilde{\alpha}_0 \to \tilde{\alpha}_0, \tag{6.15}$$

as becomes clear from Eqs (6.9) and (6.10). In fact, it is not just the zero mode, but the entire right-moving part of the compact coordinate that flips sign under the T-duality transformation

$$X_\mathrm{R} \to -X_\mathrm{R} \quad \text{and} \quad X_\mathrm{L} \to X_\mathrm{L}. \tag{6.16}$$

It is evident that this is a symmetry of the theory as physical quantities such as the energy–momentum tensor and correlation functions are invariant under this transformation. Equivalently, X is mapped into

$$\widetilde{X}(\sigma, \tau) = X_\mathrm{L}(\tau + \sigma) - X_\mathrm{R}(\tau - \sigma), \tag{6.17}$$

which has an expansion

$$\widetilde{X}(\sigma, \tau) = \tilde{x} + 2\alpha' \frac{K}{R}\sigma + 2RW\tau + \ldots \tag{6.18}$$

Note that the coordinate x, which parametrizes the original circle with periodicity $2\pi R$, has been replaced by a coordinate $\tilde{x}$. It is clear that this parametrizes the dual circle with periodicity $2\pi \widetilde{R}$, because its conjugate momentum is $\tilde{p}^{25} = RW/\alpha' = W/\widetilde{R}$.

T-duality and the sigma model

The conclusion that T-duality interchanges $X(\tau, \sigma)$ and $\widetilde{X}(\tau, \sigma)$ can also be understood from a world-sheet viewpoint. Consider the following world-sheet action:

$$\int (\frac{1}{2} V^\alpha V_\alpha - \epsilon^{\alpha\beta} X \partial_\beta V_\alpha) \, d^2\sigma, \tag{6.19}$$

where an overall constant coefficient is omitted, because the considerations that follow are classical. Varying X, which acts as a Lagrange multiplier, gives the equation of motion $\epsilon^{\alpha\beta} \partial_\beta V_\alpha = 0$, which can be solved by setting

$V_\alpha = \partial_\alpha \widetilde{X}$, for an arbitrary function $\widetilde{X}$. Substituting this into the action gives

$$\frac{1}{2} \int \partial^\alpha \widetilde{X} \partial_\alpha \widetilde{X} \, d^2\sigma. \tag{6.20}$$

Alternatively, varying V_α in the original action gives the equation of motion $V_\alpha = -\epsilon_\alpha{}^\beta \partial_\beta X$. Substituting this into the original action and using

$$\epsilon^{\alpha\beta} \epsilon_\alpha{}^\gamma = -\eta^{\beta\gamma}, \tag{6.21}$$

where the minus sign is due to the Lorentzian signature, gives

$$\frac{1}{2} \int \partial^\alpha X \partial_\alpha X \, d^2\sigma. \tag{6.22}$$

If we compare the two formulas for V_α we get

$$\partial_\alpha \widetilde{X} = -\epsilon_\alpha{}^\beta \partial_\beta X, \tag{6.23}$$

which is equivalent to the rule in Eq. (6.16). This type of world-sheet analysis of T-duality is repeated in a more general setting including background fields later in this chapter. Toroidal generalizations are discussed in the next chapter.

T-duality and open strings

Boundary conditions

The dynamics of a bosonic string in 26-dimensional Minkowski space-time is described in conformal gauge by the action

$$S = -\frac{1}{4\pi\alpha'} \int d\tau d\sigma \, \eta^{\alpha\beta} \partial_\alpha X^\mu \partial_\beta X_\mu. \tag{6.24}$$

For a small variation δX^μ, the variation of the action consists of a bulk term, whose vanishing gives the equations of motion, plus a boundary contribution

$$\delta S = -\frac{1}{2\pi\alpha'} \int d\tau \, \partial_\sigma X_\mu \delta X^\mu \Big|_{\sigma=0}^{\sigma=\pi}. \tag{6.25}$$

As was discussed in Chapter 2, making this boundary variation vanish requires imposing suitable boundary conditions at the ends of open strings. The only choice of boundary conditions that is compatible with invariance under Poincaré transformations in all 26 dimensions is Neumann boundary conditions for all components of X^μ

$$\frac{\partial}{\partial \sigma} X^\mu(\sigma, \tau) = 0, \quad \text{for} \quad \sigma = 0, \pi. \tag{6.26}$$

A natural question to ask at this point is what happens when a T-duality transformation is applied to a theory containing open strings. The first thing to note about open strings in a theory that is compactified on a circle is that they have no winding modes. Topologically, an open string can always be contracted to a point, so winding number is not a meaningful concept. Since the winding modes were crucial to relate the closed-string spectra of two bosonic theories using T-duality, one should not expect open strings to transform in the same way. Let us look at this in more detail.

In order to find the T-dual of an open string with Neumann boundary conditions, recall that in Chapter 2 we saw that the mode expansion for a space-time coordinate with Neumann boundary conditions is

$$X(\tau, \sigma) = x + p\tau + i \sum_{n \neq 0} \frac{1}{n} \alpha_n e^{-in\tau} \cos(n\sigma), \quad (6.27)$$

where we have set $l_s = 1$ or equivalently $\alpha' = 1/2$. It is convenient to split the mode expansion into left- and right-movers, just as was done for closed strings. The expansions for these fields are

$$X_R(\tau - \sigma) = \frac{x - \tilde{x}}{2} + \frac{1}{2} p(\tau - \sigma) + \frac{i}{2} \sum_{n \neq 0} \frac{1}{n} \alpha_n e^{-in(\tau - \sigma)}, \quad (6.28)$$

$$X_L(\tau + \sigma) = \frac{x + \tilde{x}}{2} + \frac{1}{2} p(\tau + \sigma) + \frac{i}{2} \sum_{n \neq 0} \frac{1}{n} \alpha_n e^{-in(\tau + \sigma)}. \quad (6.29)$$

Compactifying, once again, on a circle of radius R and carrying out a T-duality transformation gives

$$X_R \to -X_R \quad \text{and} \quad X_L \to X_L. \quad (6.30)$$

For the dual coordinate in the 25 direction this implies

$$\tilde{X}(\tau, \sigma) = X_L - X_R = \tilde{x} + p\sigma + \sum_{n \neq 0} \frac{1}{n} \alpha_n e^{-in\tau} \sin(n\sigma). \quad (6.31)$$

Now let us read off the properties of the T-dual theory. First, the dual open string has no momentum in the 25 direction, since Eq. (6.31) contains no term linear in τ. Therefore, the coordinate of the T-dual open string only undergoes oscillatory motion. Next, Eq. (6.31) can be used to read off the boundary conditions satisfied by the T-dual open string in the circular $\tilde{X}$ direction. At $\sigma = 0, \pi$ the position of the string is fixed, since the oscillator terms vanish. This means that T-duality maps Neumann boundary conditions into Dirichlet boundary conditions (and *vice versa*) in the relevant directions, as can be seen by comparing the original field (6.27) with the T-dualized field (6.31). Explicitly, the boundary conditions are

$$\widetilde{X}(\tau,0) = \tilde{x} \quad \text{and} \quad \widetilde{X}(\tau,\pi) = \tilde{x} + \frac{\pi K}{R} = \tilde{x} + 2\pi K \widetilde{R}, \qquad (6.32)$$

where we have used $p = K/R$ and $\widetilde{R} = \alpha'/R = 1/(2R)$ for the dual radius. Observe that this string wraps the dual circle K times. This winding mode is topologically stable, since the end points of the string are fixed by the Dirichlet boundary conditions. Therefore, this string cannot unwind without breaking.

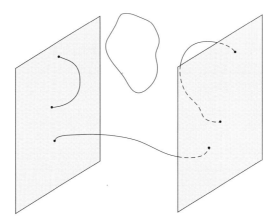

Fig. 6.2. Dp-branes and open strings ending on them.

D-branes

T-duality has transformed a bosonic open string with Neumann boundary conditions on a circle of radius R to a bosonic open string with Dirichlet boundary conditions on a circle of radius $\widetilde{R}$. We started with a string that has momentum and no winding in the circular direction and ended up with a string that has winding but no momentum in the dual circular direction. The ends of the dual open string are attached to the hyperplane $\widetilde{X} = \tilde{x}$, and they can wrap around the circle an integer number of times. The hyperplane $\widetilde{X} = \tilde{x}$ is an example of a *Dirichlet-brane* or a *D-brane* for short. In general, a D-brane is defined as a hypersurface on which an open string can end, as illustrated in Fig. 6.2. The important point to appreciate, though, is that this is not just an arbitrary location in empty space. Rather, it is a physical object. Usually one specifies the dimension of the brane and calls it a Dp-brane, where p denotes the number of spatial dimensions. In the example given here $p = 24$. By applying a T-duality transformation to open bosonic strings with Neumann boundary conditions in all directions, we learned that in the dual theory the corresponding open strings have Dirichlet boundary conditions along the dual circle and therefore end on a D24-brane.

This reasoning can be iterated by taking other directions to be circular and performing T-duality transformations in those directions, as well. Starting with n such circles (or an n-torus) one ends up with a T-dual description in which the open strings have Dirichlet boundary conditions in n directions. This implies that the string ends on a D$(25-n)$-brane. What does this mean for the open strings in the original description, which had Neumann boundary conditions for all directions? Clearly this is just the $n = 0$ case, so those open strings should be regarded as ending on a space-time-filling D25-brane. In general, one can consider a set-up in which there are a number of D-branes of various dimensions. They are replaced by D-branes of other dimensions in T-dual formulations.

To summarize: the general rule that we learn from the previous discussion is that if a D-brane wraps a circle that is T-dualized, then it doesn't wrap the T-dual circle and *vice versa*.

Open-string tachyons

An important feature of the bosonic string theory is the existence of tachyons in the spectrum. As we saw in Chapter 2, this is true both for the closed-string spectrum and the open-string spectrum. It is also true for open strings that satisfy Dirichlet boundary conditions in some directions, as is shown in Exercise 6.1.

Tachyons imply a quantum instability. The negative value of M^2 means that one is studying the theory at a point in field space where the effective potential is either at a maximum or a saddle point. This raises the following question: Where is the true vacuum? In the case of the open-string tachyons, it has been argued that the corresponding Dp-branes decay into closed-string radiation. Thus, once the string coupling is turned on, the bosonic string theory doesn't really contain any D-branes (and hence any open strings) as stable objects. Unless the coupling is very small, these D-branes decay rapidly. This picture has been borne out by detailed computations in Witten's open-string field theory. The basic idea is to find a string field configuration that minimizes the energy density and to show that its depth relative to the unstable tachyonic vacuum equals the energy density (or tension) of the space-time-filling D-brane. Using an approximation technique, called the level-truncation method, agreement to better than 1% accuracy has been achieved.

Chan–Paton charges, Wilson lines and multiple branes

In the preceding construction a single Dp-brane appeared naturally after applying T-duality to an open string with Neumann boundary conditions. This section shows that, when several Dp-branes are present instead of a single one, something rather interesting happens, namely nonabelian gauge symmetries emerge in the theory.

An open string can carry additional degrees of freedom at its end points, called *Chan–Paton charges*. These are degrees of freedom that were originally introduced, when string theory was being developed as a model for strong interactions, to describe flavor quantum numbers of quarks and antiquarks attached to the ends of an open string. The original idea was to describe the global $SU(2)$ isotopic spin symmetry acting on a quark–antiquark pair located at the ends of the string, but it was eventually realized that the construction actually gives a gauge symmetry.

Fig. 6.3. Chan–Paton charges at the ends of an open string.

The Chan–Paton factors associate N degrees of freedom with each of the end points of the string. For the case of oriented open strings, which is the case we have discussed so far, the two ends of the string are distinguished, and so it makes sense to associate the fundamental representation $\mathbf{N}$ with the $\sigma = 0$ end and the antifundamental representation $\overline{\mathbf{N}}$ with the $\sigma = \pi$ end, as indicated in Fig. 6.3. In this way one describes the gauge group $U(N)$.

For strings that are unoriented, such as type I superstrings, the representations associated with the two ends have to be the same, and this forces the symmetry group to be one with a real fundamental representation, specifically an orthogonal or symplectic group. Each state is either symmetric or antisymmetric under orientation reversal, an operation that interchanges the two ends. If the massless vectors correspond to antisymmetric states, then there are $N(N-1)/2$ of them and the group is $SO(N)$. On the other hand, if they are symmetric, there are $N(N+1)/2$ of them and the group is

$USp(N)$. Since symplectic matrices are even-dimensional, the latter groups only exist for N even.

Let us consider the case of oriented bosonic open strings. In this case, every state in the open-string spectrum now has an additional N^2 multiplicity. In particular, the N^2 massless vector states describe the $U(N)$ gauge fields. Since the charges that are associated with the ends of a string are associated with an unbroken gauge symmetry, they are conserved. Also, the energy–momentum tensor does not depend on the new degrees of freedom, so the conformal invariance of theory is unaffected. In general, the Chan–Paton charges are nondynamical in the world-sheet theory, so that a unique index is associated with each world-sheet boundary in a scattering process such as the one depicted in Fig. 6.4. This three-point scattering amplitude (and similarly for other open-string amplitudes) contains an extra factor

$$\delta^{ii'}\delta^{jj'}\delta^{kk'}\lambda^1_{ij}\lambda^2_{j'k}\lambda^3_{k'i'} = \text{Tr}\lambda^1\lambda^2\lambda^3, \tag{6.33}$$

coming from the Chan–Paton matrices. The λ matrices encode the charge states of the strings as described below. For a boundary on the interior of the string world sheet, one should sum over the associated Chan–Paton index, which gives a factor of N. This guarantees that the scattering amplitudes are invariant under the $U(N)$ symmetry.

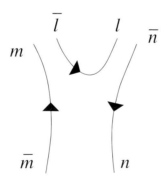

Fig. 6.4. An interaction involving three open strings.

A basis of open-string states in $\mathbb{R}^{25,1}$ can be labeled by Fock-space states ϕ (as usual), momentum k, and a pair of integers $i, j = 1, 2, \ldots, N$ labeling the Chan–Paton charges at the left and right ends of the string

$$|\phi, k, ij\rangle. \tag{6.34}$$

This state transforms with charge $+1$ under $U(1)_i$ and charge -1 under $U(1)_j$. To describe an arbitrary string state, we need to introduce N^2 hermitian matrices, the Chan–Paton matrices λ_{ij}, which are representation matrices of the $U(N)$ algebra. An arbitrary state can then be expressed as a linear combination

$$|\phi, k, \lambda\rangle = \sum_{i,j=1}^{N} |\phi, k, ij\rangle \lambda_{ij}. \tag{6.35}$$

String states become matrices transforming in the adjoint representation of $U(N)$. There are now N^2 tachyons, N^2 massless vector bosons and so on.

In a theory compactified on a circle, a flat potential[2] can have nontrivial physical effects analogous to the Aharanov–Bohm effect. If the component of the gauge potential along the circle takes nonzero constant values, it gives a holonomy matrix, or Wilson line,

$$U = \exp i \int_0^{2\pi R} A dx. \tag{6.36}$$

Diagonalizing the hermitian matrix A by a constant gauge transformation allows it to be written in the form

$$A = -\frac{1}{2\pi R} \mathrm{diag}(\theta_1, \theta_2, \ldots, \theta_N). \tag{6.37}$$

The presence of nonzero gauge fields, characterized by the Wilson line, breaks the $U(N)$ gauge symmetry to the subgroup commuting with U. For example, if the eigenvalues of U are all distinct, the symmetry is broken from $U(N)$ to $U(1)^N$.

In the presence of Wilson lines the momentum assigned to a string state $|\phi, k, ij\rangle$ gets shifted so that the wave function becomes

$$e^{ip2\pi R} = e^{-i(\theta_i - \theta_j)}. \tag{6.38}$$

This is derived in Exercise 6.2 and explored further in a homework problem. Therefore, the momentum in the circular direction becomes fractional

$$p = \frac{K}{R} - \frac{\theta_i - \theta_j}{2\pi R}, \quad K \in \mathbb{Z}. \tag{6.39}$$

Applying the T-duality rules, one obtains the result that the θ_is describe the angular positions along the dual circle of N D24-branes. Indeed, since the momentum number gets mapped to the winding number, the fractional Kaluza–Klein excitation number introduced by the Wilson line is mapped to a fractional winding number. A fractional winding number means that the open string winds over a fraction of the circle, which is appropriate for

---

2 A flat potential is one that gives a vanishing field strength, that is, $F = dA + iA \wedge A = 0$. The factor of i appears when A is chosen to be hermitian (rather than antihermitian).

an open string connecting two separated D-branes. Only when $\theta_i = \theta_j$ do we have an integer number of windings. This is illustrated in Fig. 6.5.

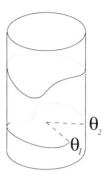

Fig. 6.5. Strings with fractional and integer winding number.

The mode expansion of the dual ij open string becomes

$$\tilde{X}^{25}_{ij} = \tilde{x}_0 + \theta_i \tilde{R} + 2\tilde{R}\sigma \left(K + \frac{\theta_j - \theta_i}{2\pi} \right) + \ldots, \qquad (6.40)$$

so that one end is at $\tilde{x}_0 + \theta_i \tilde{R}$ and the other end is at $\tilde{x}_0 + \theta_j \tilde{R}$. This is interpreted as an open string whose $\sigma = 0$ end is attached to the ith D-brane and whose $\sigma = \pi$ end is attached to the jth D-brane. Note that diagonal strings wind an integer number of times around the circle while off-diagonal strings generally do not.

The spectrum

The masses of the particles in the ij open-string spectrum of the bosonic string theory compactified on a circle are[3]

$$M^2_{ij} = \left(\frac{K}{R} + \frac{\theta_j - \theta_i}{2\pi R} \right)^2 + \frac{1}{\alpha'}(N - 1). \qquad (6.41)$$

This formula follows from the mass-shell condition and the fact that the p^{25} component of the momentum is shifted according to Eq. (6.39).

Equation (6.41) shows that if all of the θ_is are different, the only massless vector states are ones that arise from strings starting and ending on the same D-brane without wrapping the circle. All other vector string states are massive. Therefore, when no D-branes coincide, there are N different massless $U(1)$ vectors given by the diagonal strings with $K = 0$. As a result, the unbroken gauge symmetry is $U(1)^N$.

[3] The number operator N should not be confused with the rank of the gauge group.

If two θ_is are equal, so that two of the D-branes coincide, two extra off-diagonal string states become massless. This enhances the gauge symmetry from $U(1) \times U(1)$ to $U(2)$. If the D-branes are moved apart, the gauge symmetry is broken to $U(1) \times U(1)$, with the off-diagonal noncommuting gauge bosons becoming massive through a stringy Higgs mechanism. More generally, if $N_0 \leq N$ D-branes coincide, then the unbroken gauge symmetry contains a $U(N_0)$ factor. Therefore, the possibility of having multiple coincident D-branes gives a way of realizing nonabelian gauge symmetries in string theory. This fact is of fundamental importance. A collection of five parallel D-branes, which gives $U(1)^5$ gauge symmetry, is shown in Fig. 6.6.

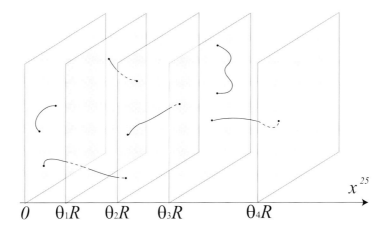

Fig. 6.6. A collection of D-branes with some attached strings.

Let us find the concrete form of some of the states in more detail. Massless states generically come from open strings that can shrink to a point. These strings start and end on the same brane (or collection of coincident branes), and they are naturally regarded as living on the world volume of the brane (or branes). Concretely, one type of massless state that appears in the spectrum is the scalar particle arising from an oscillator excitation in the circle direction

$$\alpha^{25}_{-1}|0, k\rangle, \tag{6.42}$$

which corresponds to a scalar field $A_{25}(\vec{x})$. The rest of the components are tangential to the D24-brane

$$\alpha^{\mu}_{-1}|0, k\rangle \quad \text{with} \quad \mu = 0, \ldots, 24, \tag{6.43}$$

and correspond to a vector field $A_\mu(\vec{x})$. Here $\vec{x} = (x^0, \ldots, x^{24})$ denotes the coordinates on the Dp-brane. These are all 25 coordinates other than the coordinate $\tilde{x}^{25}$, of the circle, which is fixed at the position of the D-brane. So these states describe a gauge field on the D24-brane.

When A_{25} is allowed to depend on the 25 noncompact space-time coordinates, the transverse displacement of the D24-brane in the $\tilde{x}^{25}$ direction can vary along its world volume. Therefore, an A_{25} background configuration can describe a curved D-brane world volume. More generally, starting with a flat rigid Dp-brane, transverse deformations are described by the values of the $25 - p$ world-volume fields that correspond to massless scalar open-string states. These scalar fields are the $25 - p$ transverse components of the higher-dimensional gauge field, and their values describe the transverse position of the D-brane. These scalar fields on the D-brane world volume can be interpreted as the Goldstone bosons associated with spontaneously broken translation symmetry in the transverse directions. The translation symmetry is broken by the presence of the D-branes.

This discussion illustrates the fact that condensates (or vacuum expectation values) of massless string modes can have a geometrical interpretation. There is a similar situation for gravity itself. String theory defined on a flat space-time background gives a massless graviton in the closed-string spectrum, and the corresponding field is the space-time metric. The metric can take values that differ from the Lorentz metric, thereby describing a curved space-time geometry. The significant difference in the case of D-branes is that their geometry is controlled by open-string scalar fields.

EXERCISES

EXERCISE 6.1
Compute the mass squared of the ground state of an open string attached to a flat Dp-brane in $\mathbb{R}^{25,1}$.

SOLUTION

Let us label the coordinates that satisfy Neumann boundary conditions by an index $i = 0, \ldots, p$ and the coordinates that satisfy Dirichlet boundary conditions at both ends by an index $I = p+1, \ldots, 25$. The mode expansions for left- and right-movers are, as usual,

$$X_L^\mu = \frac{x^\mu + \tilde{x}^\mu}{2} + \frac{1}{2}l_s^2 p^\mu(\tau + \sigma) + \frac{i}{2}l_s \sum_{m \neq 0} \frac{1}{m}\alpha_m^\mu e^{-im(\tau+\sigma)},$$

$$X_R^\mu = \frac{x^\mu - \tilde{x}^\mu}{2} + \frac{1}{2}l_s^2 p^\mu(\tau - \sigma) + \frac{i}{2}l_s \sum_{m\neq 0} \frac{1}{m}\alpha_m^\mu e^{-im(\tau-\sigma)}.$$

The mode expansions for the fields with Neumann and Dirichlet boundary conditions are

$$X^i = X_L^i + X_R^i \quad \text{and} \quad X^I = X_L^I - X_R^I,$$

respectively. The two ends of the string have

$$X^I(0,\tau) = X^I(\pi,\tau) = \tilde{x}^I,$$

which specifies the position of the D-brane. In uncompactified space-time there can be no winding modes, so $p^I = 0$.

The energy–momentum tensor

$$T_{++} = \partial_+ X^i \partial_+ X_i + \partial_+ X^I \partial_+ X_I = \partial_+ X_L^\mu \partial_+ X_{L\mu}$$

has the same mode expansion as in Chapter 2, independent of p, and thus the Virasoro generators, the zero-point energy, and the mass formula, are the same as before

$$M^2 = 2(N-1)/l_s^2.$$

The main difference is that this is now the mass of a state in the $(p+1)$-dimensional world volume of the Dp-brane, whereas in Chapter 2 only the space-time-filling $p = 25$ case was considered. The mass squared of the open-string ground state therefore is

$$M^2 = -2/l_s^2 = -1/\alpha'.$$

$\square$

EXERCISE 6.2
Consider a relativistic point particle with mass m and electric charge e moving in an electromagnetic potential $A_\mu(x)$. The action describing this particle is

$$S = \int \left(-m\sqrt{-\dot{X}^\mu \dot{X}_\mu} - e\dot{X}^\mu A_\mu\right)d\tau.$$

Suppose that one direction is compactified on a circle of radius R. Show that a constant vector potential along this direction, given by

$$A = -\frac{\theta}{2\pi R},$$

leads to a fractional momentum component along the compact direction

$$p = \frac{K}{R} + \frac{e\theta}{2\pi R},$$

where K is an integer.

SOLUTION

In the gauge $\tau = X^0 = t$ the action takes the form

$$S = \int \Big(-m\sqrt{1-v^2} - e(A_0 + \vec{A}\cdot\vec{v})\Big) dt,$$

where $v^i = \dot{X}^i$. The canonical momentum conjugate to the compact coordinate X, which is one of the X^is, is

$$P = \frac{\delta S}{\delta \dot{X}} = p - eA = p + \frac{e\theta}{2\pi R},$$

where

$$p = \frac{m\dot{X}}{\sqrt{1-v^2}}$$

is the physical momentum. The wave function of the charged particle includes a factor containing the canonical momentum

$$\Psi(x) \sim e^{iPX},$$

since $P \sim -i\partial/\partial X$. This must be single-valued, and thus $P = K/R$, where K is an integer. This gives

$$p = \frac{K}{R} - \frac{e\theta}{2\pi R}.$$

$\square$

6.2 D-branes in type II superstring theories

D-branes also exist in superstring theories. Indeed, just as in the bosonic theory, adding D-branes to the type IIA or type IIB vacuum configuration gives a theory that has closed strings in the bulk plus open strings that end on the D-branes. Certain D-branes in superstring theories exhibit an important feature that does not occur in the bosonic string theory. Namely, they carry a conserved charge that ensures their stability. In such a case, the spectrum of open strings that start and end on the D-brane is tachyon-free.

When D-branes are present, some of the symmetries of the superstring vacuum are broken. For example, consider starting with the Minkowski space vacuum of a type II superstring theory, which has ten-dimensional Poincaré invariance. Adding a flat Dp-brane, and neglecting its back reaction on the geometry, breaks the ten-dimensional $SO(1,9)$ Lorentz symmetry to $SO(1,p) \times SO(9-p)$. Moreover, some or all of the supersymmetry is also broken by the addition of the Dp-brane.

Recall that both of the type II superstring theories, in the ten-dimensional Minkowski vacuum, have $\mathcal{N} = 2$ supersymmetry. Since each supercharge corresponds to a Majorana–Weyl spinor, with 16 real components, there are a total of 32 conserved supercharges. However, the maximum number of unbroken supersymmetries that is possible for vacua containing D-branes is 16. There are several ways of seeing this. A simple one is to note that the massless open strings form a vector supermultiplet, and such supermultiplets only exist with 16 or fewer conserved supercharges. Thus, when D-branes are added to type II superstring vacua, not only is translational invariance in the transverse directions broken, but at least 16 of the original 32 supersymmetries must also be broken.

Form fields and p-brane charges

The five superstring theories and M-theory contain a variety of massless antisymmetric tensor gauge fields, which can be represented as differential forms. An n-form gauge field is given by

$$A_n = \frac{1}{n!} A_{\mu_1 \mu_2 \cdots \mu_n} dx^{\mu_1} \wedge dx^{\mu_2} \wedge \cdots \wedge dx^{\mu_n}. \tag{6.44}$$

These can be regarded as generalizations of an ordinary Maxwell field, which corresponds to the case $n = 1$. With this in mind, one defines the $(n+1)$-form field strength by $F_{n+1} = dA_n$, where

$$F_{n+1} = \frac{1}{(n+1)!} F_{\mu_1 \mu_2 \cdots \mu_{n+1}} dx^{\mu_1} \wedge dx^{\mu_2} \wedge \cdots \wedge dx^{\mu_{n+1}}. \tag{6.45}$$

Such a field strength is invariant under a gauge transformation of the form $\delta A_n = d\Lambda_{n-1}$, since the square of an exterior derivative vanishes ($d^2 = 0$).

Maxwell theory

Recall that classical electromagnetism is described by Maxwell's equations, which can be written in the form

$$dF = 0 \quad \text{and} \quad d \star F = 0, \tag{6.46}$$

in the absence of charges and currents. Here F is the two-form field strength describing the electric and magnetic fields. Notice that the above equations are symmetric under the interchange of F and $\star F$.

More generally, one should include electric and magnetic sources. Electrically charged particles (or electric monopoles) exist, but magnetic monopoles have not been observed yet. Most likely, magnetic monopoles exist with masses much higher than have been probed experimentally. When sources are included, Maxwell's equations become

$$dF = \star J_m \quad \text{and} \quad d \star F = \star J_e. \tag{6.47}$$

In each case $J = J_\mu dx^\mu$ is a one-form related to the current and charge density as

$$J_\mu = (\rho, \vec{j}), \tag{6.48}$$

with $\mu = 0, \ldots, 3$ in the case of four dimensions. For a point-like electric charge the charge density is described by a delta function $\rho = e\delta^{(3)}(\vec{r})$, where e denotes the electric charge. Similarly, a point-like magnetic source has an associated magnetic charge, which we denote by g. These charges can be defined in terms of the field strength

$$e = \int_{S^2} \star F \quad \text{and} \quad g = \int_{S^2} F, \tag{6.49}$$

where the integrations are carried out over a two-sphere surrounding the charges.

Electric and magnetic charges are not independent. Indeed, as Dirac pointed out in 1931, the wave function of an electrically charged particle moving in the field of a magnetic monopole is uniquely defined only if the electric charge e is related to the magnetic charge g by the Dirac quantization condition[4]

$$e \cdot g \in 2\pi \mathbb{Z}. \tag{6.50}$$

The derivation of this result is described in Exercise 6.3.

Generalization to p-branes

The preceding considerations can be generalized to p-branes that couple to $(p+1)$-form gauge fields in D dimensions. To determine the possibilities for stable p-branes, it is worthwhile to consider the types of conserved charges that they can carry. This entails generalizing the statement that a point particle (or 0-brane) can carry a charge such that it acts as a source for a

---

[4] For *dyons*, which carry both electric and magnetic charge, the Dirac quantization rule generalizes to Witten's rule: $e_1 g_2 - e_2 g_1 = 2\pi n$.

one-form gauge field, that is, a Maxwell field $A = A_\mu dx^\mu$. There are two aspects to this. On the one hand, a charged particle couples to the gauge field in a way that is described by the interaction

$$S_\text{int} = e\int A = e\int d\tau A_\mu \frac{dx^\mu}{d\tau}, \tag{6.51}$$

where e is the electric charge. On the other hand, the charge of the particle can be determined by Gauss's law. This entails surrounding the particle with a two-sphere and integrating the electric field over the sphere. Defining the field strength by $F = dA$, as usual, the relevant integral is $\int_{S^2} \star F$. Note that F is a two-form and in D dimensions its Hodge dual $\star F$ is a $(D-2)$-form. In terms of components

$$(\star F)^{\mu_1 \mu_2 \cdots \mu_{D-2}} = \frac{\varepsilon^{\mu_1 \mu_2 \cdots \mu_D}}{2\sqrt{-g}} F_{\mu_{D-1}\mu_D}. \tag{6.52}$$

In general, a $(D-2)$-sphere can surround a point in D-dimensional Lorentzian space-time. For example, an electrically charged D0-brane in the type IIA theory can be surrounded by an eight-sphere S^8. The magnetic dual of an electrically charged point particle carries a magnetic charge that is measured by integrating the magnetic flux over a sphere that surrounds it. This is simply $\int F$, which in the case of a Maxwell field is a two-dimensional integral. In D dimensions a two-sphere S^2 can surround a $(D-4)$-brane. In four dimensions this is a point particle, but in the ten-dimensional type IIA theory the magnetic dual of the D0-brane is a D6-brane.

The preceding can be generalized to an n-form gauge field A_n with an $(n+1)$-form field strength $F_{n+1} = dA_n$. An n-form gauge field can couple electrically to the world volume of a brane whose world volume has $n = p+1$ dimensions

$$S_\text{int} = \mu_p \int A_{p+1}, \tag{6.53}$$

where μ_p is the p-brane charge and the pullback from the bulk to the brane is understood. In other words,

$$\int A_{p+1} = \frac{1}{(p+1)!}\int A_{\mu_1 \cdots \mu_{p+1}} \frac{\partial x^{\mu_1}}{\partial \sigma^0}\cdots \frac{\partial x^{\mu_{p+1}}}{\partial \sigma^p} d^{p+1}\sigma. \tag{6.54}$$

This generalizes Eq. (6.51), which has $p = 0$ and $e = \mu_0$. This brane is electrically charged as can be seen by evaluating the electric charge using Gauss's law $\mu_p = \int \star F_{p+2}$. In D dimensions this is an integral over a sphere S^{D-p-2}, which is the dimension required to surround a p-brane. The charge of the magnetic dual branes can be measured by computing the flux $\int F_{p+2}$ through a surrounding S^{p+2}. In D dimensions an S^{p+2} can surround a

$(D-p-4)$-brane. Thus, in the case of ten dimensions, the magnetic dual of a p-brane is a $(6-p)$-brane.

The Dirac quantization condition for point-like charges in $D = 4$, $eg = 2\pi n$, has a straightforward generalization to the charges carried by a dual pair of p-branes. For our normalization conventions, in ten dimensions one has

$$\mu_p \mu_{6-p} \in 2\pi \mathbb{Z}. \tag{6.55}$$

This is derived by a generalization of the usual proof that is described in Exercise 6.4. The basic idea is to require that the wave function of an electric brane is well defined in the field of the magnetic brane. In all superstring theory and M-theory examples it turns out that a single p-brane carries the minimum allowed quantum of charge. In other words, the product of the charges of a single p-brane and a single dual $(6-p)$-brane is exactly 2π.

Stable D-branes in type II superstring theories

Specializing to the case of ten dimensions, the preceding considerations tell us that an n-form gauge field can couple electrically to a p-brane with $p = n - 1$ and magnetically to a p-brane with $p = 7 - n$. Since the R–R sector of the type IIA theory contains gauge fields with $n = 1$ and $n = 3$, this theory should contain stable branes that carry the corresponding charges. These are Dp-branes with $p = 0, 2, 4, 6$. Since this is giving even integers, it is natural to consider $p = 8$, as well. Larger even values are not possible, since the dimension of the brane cannot exceed the dimension of the space-time. The existence of a D8-brane would seem to require a nine-form gauge field with a ten-form field strength. Such a field is nondynamical, and therefore it did not arise when we analyzed the physical degrees of freedom of type IIA supergravity. In fact, stable D8-branes do occur in special circumstances, which are discussed later in this chapter.

In the case of the type IIB theory the R–R sector contains n-form gauge fields with $n = 0, 2, 4$. Applying the rules given above the zero-form should couple electrically to a (-1)-brane. This is an object that is localized in time as well as in space. It is interpreted as a D-instanton, which makes sense in the Euclideanized theory. Its magnetic dual is a D7-brane which is well defined in the Lorentzian signature theory. However, since a D7-brane has codimension 2 it gives rise to a deficit angle in the geometry, just as occurs for a point mass in three-dimensional general relativity. The two-form couples electrically to a D1-brane (also called a D-string) and magnetically to a D5-brane. The four-form couples both electrically and magnetically to a D3-brane. However, these are not distinct D-branes. Since the field strength is self-dual, $F_5 = \star F_5$, the D3-brane carries a self-dual charge. In addition, one can also introduce space-time-filling D9-branes in the IIB theory, though there are consistency conditions that restrict when they can occur. Altogether, the conclusion is that type IIB superstring theory admits stable Dp-branes, carrying conserved charges, for odd values of p.

The stable D-branes (with p even in the IIA theory or odd in the IIB theory) preserve half of the supersymmetry (16 supersymmetries). Therefore, they are sometimes called *half-BPS D-branes*. This fact implies that the associated open-string spectrum has this much supersymmetry, and therefore it must be tachyon-free. To be explicit, let Q_1 and Q_2 be the two supersymmetry charges of the string theory. These are Majorana–Weyl spinors, which have opposite chirality in the IIA case and the same chirality in the IIB case. Now suppose a Dp-brane extends along the directions $0, 1, \ldots, p$. Then the supersymmetry that is conserved is the linear combination

$$Q = Q_1 + \Gamma^{01\cdots p} Q_2, \qquad (6.56)$$

where the sign of the second term depends on conventions. Note that in all cases the two terms have the same chirality, since the Dirac matrix flips the chirality of the Q_2 term when p is even (the IIA case) but not when p is odd (the IIB case).

To recapitulate, conserved R–R charges, supersymmetry, stability, and absence of tachyons are all features of these type II Dp-branes.

Non-BPS D-branes

The type II superstring theories also admit Dp-branes with "wrong" values of p, meaning that p is odd in the IIA theory or even in the IIB theory. These Dp-branes do not carry conserved charges and are unstable. They break all of the supersymmetry and give an open-string spectrum that includes a tachyon. The features of these branes are the same as those of Dp-branes with any value of p in the bosonic string theory. In the context of superstring theories, D-branes of this type are sometimes referred to as non-BPS D-branes.

Type II superstrings and T-duality

T-duality for the closed bosonic string theory, compactified on a circle of radius R, maps the theory to an identical theory on a dual circle of radius $\tilde{R} = \alpha'/R$. In this sense the theory is self-dual under T-duality, and there is a $\mathbb{Z}_2$ symmetry at the self-dual radius $R_{\text{sd}} = \sqrt{\alpha'}$. Let us now examine the same T-duality transformation for type II superstring theories. It will turn out that the type IIA theory is mapped to the type IIB theory and *vice versa*. Of course, if several directions are compactified on circles it is possible to carry out several T-dualities. In this case an even number of transformations gives back the same type II theory that one started with (on the dual torus). This is a symmetry if the torus is self-dual.

Returning to the case of a single circle, imagine that the X^9 coordinate of a type II theory is compactified on a circle of radius R and that a T-duality transformation is carried out for this coordinate. The transformation of the bosonic coordinates is the same as for the bosonic string, namely

$$X_{\text{L}}^9 \to X_{\text{L}}^9 \quad \text{and} \quad X_{\text{R}}^9 \to -X_{\text{R}}^9, \tag{6.57}$$

which interchanges momentum and winding numbers. In the RNS formalism, world-sheet supersymmetry requires the world-sheet fermion ψ^9 to transform in the same way as its bosonic partner X^9, that is,

$$\psi_{\text{L}}^9 \to \psi_{\text{L}}^9 \quad \text{and} \quad \psi_{\text{R}}^9 \to -\psi_{\text{R}}^9. \tag{6.58}$$

This implies that after T-duality the chirality of the right-moving Ramond-sector ground state is reversed (see Exercise 6.5). The relative chirality of the left-moving and right-moving ground states is what distinguishes the type IIA and type IIB theories. Since only one of these is reversed, it follows that if the type IIA theory is compactified on a circle of radius R, a T-duality transformation gives the type IIB theory on a circle of radius $\tilde{R}$.

In the light-cone gauge formulation, only X^i and ψ^i, $i = 1, \ldots, 8$, are independent dynamical degrees of freedom. In this case a T-duality transformation along any of those directions works as described above, but one along the x^9 direction is more awkward to formulate.

Now let us examine what happens to type II Dp-branes when the theory is T-dualized. Since the half-BPS Dp-branes of the type IIA theory have p even, while the half-BPS Dp-branes of the type IIB theory have p odd, these D-branes are mapped into one another by T-duality transformations. A similar statement can also be made for the non-BPS Dp-branes. The relevant analysis is the same as for the bosonic string. Let us review the analysis for a pair of flat parallel Dp-branes that fill the dimensions x^μ, with $\mu = 0, \ldots, p$, and have definite values of the other transverse coordinates. An open string connecting these two Dp-branes satisfies Neumann boundary conditions in $p+1$ dimensions

$$\partial_\sigma X^\mu|_{\sigma=0} = \partial_\sigma X^\mu|_{\sigma=\pi} = 0, \quad \mu = 0, \ldots, p, \tag{6.59}$$

and Dirichlet boundary conditions for the transverse coordinates

$$X^i|_{\sigma=0} = d_1^i \quad \text{and} \quad X^i|_{\sigma=\pi} = d_2^i, \quad i = p+1, \ldots, 9, \quad (6.60)$$

where d_1^i and d_2^i are constants. These boundary conditions imply that the mode expansions are

$$X^\mu(\tau, \sigma) = x^\mu + p^\mu \tau + i \sum_{n \neq 0} \frac{1}{n} \alpha_n^\mu \cos n\sigma e^{-in\tau}, \quad (6.61)$$

$$X^i(\tau, \sigma) = d_1^i + (d_2^i - d_1^i)\frac{\sigma}{\pi} + \sum_{n \neq 0} \frac{1}{n} \alpha_n^i \sin n\sigma e^{-in\tau}. \quad (6.62)$$

Now consider a T-duality transformation along the circular X^9 direction. The transformation $X_R^9 \to -X_R^9$ interchanges Dirichlet and Neumann boundary conditions. Running the previous analysis in the reverse direction, one learns that in the dual description there is a pair of D-branes that wrap the dual circle and that the $U(2)$ gauge symmetry is broken to $U(1) \times U(1)$ by a pair of Wilson lines. As in the bosonic theory, Dp-branes that were localized on the original circle of radius R are wrapped on the dual circle of radius $\tilde{R}$.

Thus the general rule is that under T-duality the branes that are wrapped and those that are unwrapped are interchanged. If T-duality is performed in one of the directions of the original theory on which a p-brane is wrapped, then T-duality transforms this p-brane into a $(p-1)$-brane, which is localized on the dual circle. This is consistent with the requirement that the half-BPS Dp-branes of the type IIA theory, which have p even, are mapped into the half-BPS Dp-branes of the type IIB theory, which have p odd. Starting with any one of these half-BPS D-branes, all of the others can be accessed by repeated T-duality transformations.

Mapping of coupling constants

T-duality of the type IIA and type IIB superstring theories is a perturbative duality, which holds order by order in the string perturbation expansion. When the type IIA theory is compactified on a circle of radius R and the type IIB theory is compactified on a circle of radius $\tilde{R}$, the two theories are related by the T-duality identification $R\tilde{R} = \alpha'$. This amounts to inverting the dimensionless parameter $\sqrt{\alpha'}/R$. Let us now examine the mapping of the string coupling constants implied by T-duality. To do this it is sufficient to consider the coupling constant dependence of the NS–NS part of the low-energy effective action of the type IIA theory, which has the form

$$\frac{1}{g_s^2} \int d^{10}x \mathcal{L}_{NS}. \tag{6.63}$$

For the NS–NS part of the type IIB theory, one has the same formula, with the IIA string coupling g_s replaced by the IIB string coupling $\tilde{g}_s$. The explicit formula for the Lagrangian $\mathcal{L}_{NS}$ is given in Chapter 8. Compactifying each of these theories on a circle, and keeping only the zero-mode contributions on the circle gives

$$\frac{2\pi R}{g_s^2} \int d^9 x \mathcal{L}_{NS} \tag{6.64}$$

in the type IIA case, and

$$\frac{2\pi \tilde{R}}{\tilde{g}_s^2} \int d^9 x \mathcal{L}_{NS} \tag{6.65}$$

in the type IIB case. T-duality implies that these two expressions should be the same. Using the T-duality relation $R\tilde{R} = \alpha'$, one obtains the relation between the coupling constants

$$\tilde{g}_s = \frac{\sqrt{\alpha'}}{R} g_s. \tag{6.66}$$

Although derived here by examining certain terms in the low-energy expansion, the relation in Eq. (6.66) is completely general. Since the two string coupling constants are proportional, a perturbative expansion in g_s in type IIA corresponds to a perturbative expansion in $\tilde{g}_s$ in type IIB.

K-theory

Since D-branes carry conserved R–R charges that are sources for R–R gauge fields, which are differential forms, one might suppose that the charges could be identified with cohomology classes of gauge field configurations. This is roughly, but not precisely, correct. The appropriate mathematical generalization uses K-theory, and classifies D-brane charges by K-theory classes.

Type II D-branes

Consider a collection of coincident type II D-branes – N Dp-branes and N' $\overline{Dp}$-branes. $\overline{Dp}$ denotes an *antibrane*, which is the charge-conjugate of the Dp-brane. The important world-volume fields can be combined in a superconnection

$$\mathcal{A} = \begin{pmatrix} A & T \\ T & A' \end{pmatrix}, \tag{6.67}$$

where A is a connection on a $U(N)$ vector bundle E, A' is a connection on a $U(N')$ vector bundle E', and T is a section of $E^* \otimes E'$ that describes an $N \times N'$ matrix of tachyon fields. The $(p+1)$-dimensional world volume of the branes, X, is the base of E and E'. The three types of fields arise as modes of the three types of open strings: those connecting branes to branes, those connecting antibranes to antibranes, and those connecting branes to antibranes.

If the gauge field bundles E and E' are topologically equivalent ($E \sim E'$) complete annihilation should be possible. This requires $N = N'$ so that the total charge is zero. Moreover, the tachyon field matrix should take a value $T = T_0$ that gives the true minimum of the tachyon potential. If there is complete annihilation, the minimum of the tachyon potential energy $V(T)$ should be negative and exactly cancel the energy density of the branes so that the total energy is zero

$$V(T_0) + 2NT_{\mathrm{D}p} = 0. \tag{6.68}$$

As a specific example, consider the case $p = 9$ in the type IIB theory. Consistency of the quantum theory (tadpole cancellation) requires that the total R–R 9-brane charge should vanish, and thus $N = N'$. So we must have an equal number of D9-branes and $\overline{\mathrm{D}9}$-branes filling the ten-dimensional space-time X. Associated with this there are a pair of vector bundles (E, E'), where E and E' are rank-N complex vector bundles.

We now want to define equivalence of pairs (E, E') and (F, F') whenever the associated 9-brane systems can be related by brane–antibrane annihilation and creation. In particular, $E \sim E'$ corresponds to pure vacuum, and therefore

$$(E, E') \sim 0 \Leftrightarrow E \sim E'. \tag{6.69}$$

If we add more D9-branes and $\overline{\mathrm{D}9}$-branes with identical vector bundles H, this should not give anything new, since they are allowed to annihilate. This means that

$$(E \oplus H, E' \oplus H) \sim (E, E'). \tag{6.70}$$

In this way we form equivalence classes of pairs of bundles. These classes form an abelian group. For example, (E', E) belongs to the inverse class of the class containing (E, E'). If N and N' are unrestricted, the group is called $K(X)$. However, the group that we have constructed above is the subgroup of $K(X)$ defined by requiring $N = N'$. This subgroup is called $\widetilde{K}(X)$. Thus type IIB D-brane charges should be classified by elements of $\widetilde{K}(X)$. Let us examine whether this works.

6.2 D-branes in type II superstring theories

The formalism is quite general, but we only consider the relatively simple case of Dp-branes that are hyperplanes in flat $\mathbb{R}^{9,1}$. For this purpose it is natural to decompose the space into tangential and normal directions

$$\mathbb{R}^{9,1} = \mathbb{R}^{p,1} \times \mathbb{R}^{9-p}, \tag{6.71}$$

and consider bundles that are independent of the tangential $\mathbb{R}^{p,1}$ coordinates. If the fields fall sufficiently at infinity, so that the energy is normalizable, then we can add the point at infinity thereby compactifying the normal space so that it becomes topologically a sphere S^{9-p}. Then the relevant base space for the Dp-brane bundles is $X = S^{9-p}$. We can now invoke the mathematical results:

$$\widetilde{K}(S^{9-p}) = \begin{cases} \mathbb{Z} & p = \text{odd} \\ 0 & p = \text{even} \end{cases}. \tag{6.72}$$

This precisely accounts for the R–R charge of all the stable (BPS) Dp-branes of the type IIB theory on $\mathbb{R}^{9,1}$. It should be noted that the unstable non-BPS type IIB D-branes, discussed earlier, carry no conserved charges, and they do not show up in this classification.

Suppose now that some dimensions form a compact manifold Q of dimension q, so that the total space-time is $\mathbb{R}^{9-q,1} \times Q$. Then the construction of a Dp-brane requires compactifying the normal space $\mathbb{R}^{9-p-q} \times Q$ to give $S^{9-p-q} \times Q$. This involves adjoining a copy of Q at infinity. In this case the appropriate mathematical objects to classify D-brane charges are relative K-theory groups $K(S^{9-p-q} \times Q, Q)$. In particular, if $Q = S^1$, we have $K(S^{8-p} \times S^1, S^1)$. Mathematically, it is known that this relative K-theory group can be decomposed into two pieces

$$K(X \times S^1, S^1) = K^{-1}(X) \oplus \widetilde{K}(X). \tag{6.73}$$

The physical interpretation of this formula is very nice. $\widetilde{K}(S^{8-p})$ classifies the type IIB D-branes that are wrapped on the circle, whereas

$$K^{-1}(S^{8-p}) \cong \widetilde{K}(S^{9-p}) \tag{6.74}$$

classifies unwrapped D-branes. So, altogether, in nine dimensions there are additive D-brane charges for all $p < 8$.

The type IIA case is somewhat more subtle, since the space-time-filling D9-branes are unstable in this case. The right K-theory group in this case is $K^{-1}(X)$, the same group that appeared in the previous paragraph. The mathematical results

$$K^{-1}(S^{9-p}) = \begin{cases} \mathbb{Z} & \text{for } p = \text{even} \\ 0 & \text{for } p = \text{odd} \end{cases} \tag{6.75}$$

account for all the stable type IIA Dp-branes embedded in $\mathbb{R}^{9,1}$. Compactifying the type IIA theory on a circle gives the relative K-theory group

$$K^{-1}(X \times S^1, S^1) = \widetilde{K}(X) \oplus K^{-1}(X). \tag{6.76}$$

This time $K^{-1}(X)$ describes wrapped D-branes and $\widetilde{K}(X)$ describes unwrapped ones. This result matches the type IIB result in exactly the way required by T-duality (wrapped $\leftrightarrow$ unwrapped).

EXERCISES

EXERCISE 6.3
Derive the Dirac quantization condition (6.50) for point particles in four-dimensional space-time.

SOLUTION

In the case of Maxwell theory in $D = 4$ the vector potential is a one-form A_1 whose field strength is a two-form $F_2 = dA_1$. Let us denote the dual of this field strength $\star F_2$, which is also a two-form, by $\widetilde{F}_2$. Then Gauss's law is the statement that if a two-sphere S^2 surrounds an electric charge e, one has $\int_{S^2} \widetilde{F}_2 = e$. Similarly, if it surrounds a magnetic charge g, $\int_{S^2} F_2 = g$.

Now consider the wave function $\psi(x)$ of an electrically charged particle, with charge e, in the field of a magnetic monopole of charge g. Such a wave function has the form

$$\psi(x) = \exp\left(ie \int_{x_0}^x A_1\right) \psi_0(x),$$

where the integral is along some path to the end point x. The choice of base point x_0 (and the contour) gives an overall x-independent phase that doesn't matter. This formula can be understood as follows: the minimal coupling $J \cdot A$ ensures that the vector potential enters the Schrödinger equation only via the covariant derivative $D_\mu = \partial_\mu - ieA_\mu$. Then the phase factor isolates the non-gauge-invariant part of $\psi(x)$; the function $\psi_0(x)$ is gauge invariant.

Now consider the change in this wave function as x traces out a small circle γ. One obtains

$$\psi(x) \to U(\gamma)\psi(x), \quad U(\gamma) = e^{ie \oint_\gamma A_1},$$

where the contour integral is around the circle γ. Let D denote a disk whose boundary is γ. By Stokes' theorem,

$$\oint_\gamma A_1 = \int_D F_2.$$

However, the choice of D is not unique, and any choice must give the same answer for the wave function to be well defined. Let D' be another choice that passes on the other side of the magnetic charge. Then the difference $D - D'$ is topologically a two-sphere that surrounds the magnetic charge. In other words,

$$\int_D F_2 - \int_{D'} F_2 = \int_{D-D'} F_2 = g.$$

Thus the holonomy group element $U(\gamma)$ is well defined only if $\exp(ieg) = 1$. This gives the Dirac quantization condition

$$eg \in 2\pi \mathbb{Z}.$$

There is a mathematical issue that has been suppressed in the preceding discussion. Namely, the field of a monopole gives a topologically nontrivial $U(1)$ bundle. This means that the region exterior to the monopole can be covered by two open sets, O and O', on which the gauge field is A and A', respectively. On the overlap $O \cup O'$, the two gauge fields differ by a gauge transformation: $A - A' = d\Lambda$.[5] It also means that the "wave function" is not a function, but rather a *section of a line bundle*. In the use of Stokes' theorem the field A should be used for the extension to D, which is assumed to be interior to O, and the field A' should be used for the extension to D', which is assumed to be interior to O'. By explicitly integrating the difference of A and A' along γ and requiring that $U(\gamma)$ is unique, one can give an alternative proof of the quantization condition. □

EXERCISE 6.4

Generalize the reasoning of the preceding exercise to prove the Dirac quantization condition for p-branes in Eq. (6.55).

SOLUTION

Equation (6.55) applies to ten dimensions. Let us be a bit more general, and consider D dimensions instead. Given an electrically charged p-brane with charge μ_p, there is a $(p+1)$-form gauge field that has the minimal coupling

---

[5] If one only uses one field A it is singular along a line, called a *Dirac string*, which runs from the monopole to infinity. It should be emphasized that a Dirac string is a mathematical artefact and not a physical object.

$\mu_p \int A_{p+1}$ to the brane. The gauge-invariant field strength is $F_{p+2} = dA_{p+1}$ and its dual is

$$\widetilde{F}_{D-p-2} = \star F_{p+2}.$$

Gauss's law is the statement that if we loop the p-brane once with a sphere S^{D-p-2}, then the charge is given by

$$\mu_p = \int_{S^{D-p-2}} \widetilde{F}_{D-p-2}.$$

The magnetic dual of this brane is a $(D-p-4)$-brane that can be encircled by a sphere S^{p+2}. Gauss's law gives its magnetic charge

$$\mu_{D-p-4} = \int_{S^{p+2}} F_{p+2}.$$

Requiring that both branes have nonnegative dimension gives $0 \leq p \leq D-4$.

Now let's consider a probe electric p-brane in the field of a magnetic $(D-p-4)$-brane. For the argument that follows, the topology of the magnetic brane doesn't matter, but it is extremely convenient to choose the electric brane to be topologically a sphere S^p. Let us denote this p-cycle by β. Then, for the same reason as in the previous exercise, the wave function of the p-brane has the form

$$\psi(\beta) = \exp\left(i\mu_p \int_{\beta_0}^{\beta} A_{p+1}\right) \psi_0(\beta)$$

where ψ_0 is gauge invariant. The lower limit is a fixed p-cycle β_0 and the integral is over a region that is a "cylinder" whose topology is a line interval times S^p. As before, it does not matter how this is chosen.

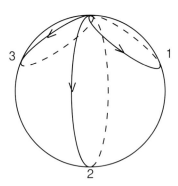

Fig. 6.7. This illustrates, for the case $p = 1$, how a loop of p-dimensional spheres can trace out a $(p+1)$-dimensional sphere γ.

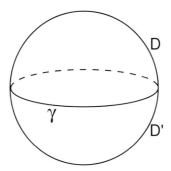

Fig. 6.8. This illustrates, for the case $p = 0$, that the difference of two $(p+2)$-dimensional balls D and D' with a common boundary γ (a $(p+1)$-dimensional sphere) that pass on opposite sides of a magnetic brane is a $(p+2)$-dimensional sphere that encircles the magnetic brane.

Now we need to generalize the step in the previous exercise in which the electric charge traced out a circle. What we want is for the p-brane to trace out a surface γ that is topologically a sphere S^{p+1}. The way to achieve this is shown in Fig. 6.7. For a vanishingly small cycle β this gives the result that

$$\psi(\beta) \to U(\gamma)\psi(\beta), \quad U(\gamma) = \exp\left(i\mu_p \int_\gamma A_{p+1}\right).$$

Now let D be a ball whose boundary is γ. Stokes' theorem gives

$$\oint_\gamma A_{p+1} = \int_D F_{p+2}.$$

Again D is not unique, and we can consider two different choices D and D' that pass on opposite sides of the magnetic brane. Their difference is topologically a sphere S^{p+2} that surrounds the magnetic brane, as indicated in Fig. 6.8. Thus,

$$\int_D F_{p+2} - \int_{D'} F_{p+2} = \int_{D-D'} F_{p+2} = \mu_{D-p-4}.$$

Now requiring that $U(\gamma)$ is well defined gives $\exp(i\mu_p \mu_{D-p-4}) = 1$, and hence

$$\mu_p \mu_{D-p-4} \in 2\pi \mathbb{Z}.$$

□

EXERCISE 6.5

Show that a T-duality transformation reverses the chirality of the right-moving Ramond-sector ground state.

SOLUTION

T-duality reverses the sign of the right-moving bosons

$$X_R^9 \to -X_R^9.$$

World-sheet supersymmetry requires the fermions to transform in the same way as the bosons, that is,

$$\psi_R^9 \to -\psi_R^9.$$

In particular, the zero mode of ψ_R^9 in the Ramond sector transforms is reversed

$$d_0^9 \to -d_0^9.$$

In Chapter 4 we learned that there is a relation between R-sector zero modes and ten-dimensional Dirac matrices

$$\Gamma^\mu = \sqrt{2} d_0^\mu.$$

Thus, under a T-duality transformation

$$\Gamma_\mu \to \Gamma_\mu \text{ (for } \mu \neq 9) \quad \text{and} \quad \Gamma_9 \to -\Gamma_9.$$

We conclude that the chirality operator behaves as

$$\Gamma_{11} = \Gamma_0 \Gamma_1 \cdots \Gamma_9 \to -\Gamma_{11},$$

so the chirality of the right-moving Ramond ground state is reversed. This may seem paradoxical until one realizes that both ten-dimensional chiralities correspond to nonchiral spinors in nine dimensions. □

EXERCISE 6.6

T-duality has been described for superstrings in the RNS formulation. How do the world-sheet fields transform under a T-duality transformation in the x^j direction in the light-cone GS formulation?

SOLUTION

The world-sheet fields consist of left-movers X_L^i and S_1^a and right-movers X_R^i and $S_2^{\dot{a}}$ (type IIA) or S_2^a (type IIB). As always, the left-movers are unchanged, and the only nontrivial bosonic transformation is $X_R^j \to -X_R^j$. So the issue boils down to finding the transformation rule for the S_2 s. There is really only one sensible possibility. In Chapter 5 we introduced the Dirac matrices $\Gamma^i_{a\dot b} = \Gamma^i_{\dot b a}$, which were also interpreted as Clebsch–Gordon coefficients for coupling the three inequivalent eight-dimensional representations of $Spin(8)$. Clearly, the rule

$$S_2^{\dot a} \to \Gamma^j_{\dot b a} S_2^b \text{ (for IIA)} \quad \text{and} \quad S_2^a \to \Gamma^j_{a \dot b} S_2^{\dot b} \text{ (for IIB)}$$

respects the symmetries of the problem and maps the type IIA theory to the type IIB theory and *vice versa*. Also, it squares to the trivial transformation because $\Gamma^j_{a\dot b} \Gamma^j_{\dot b c} = \delta_{ac}$ and $\Gamma^j_{\dot a b} \Gamma^j_{b \dot c} = \delta_{\dot a \dot c}$, where the index j is unsummed.

For multiple T-dualities, such as along x^1 and x^2, there is a sign ambiguity. Depending on the order, one could get $S_2 \to \Gamma_1 \Gamma_2 S_2$ or $S_2 \to \Gamma_2 \Gamma_1 S_2 = -\Gamma_1 \Gamma_2 S_2$. However, the sign reversal $S_2 \to -S_2$ is a trivial symmetry of both the type IIA and IIB theories, so this is inconsequential. □

EXERCISE 6.7

T-duality transforms a p-brane into a $(p-1)$-brane if a direction along the brane is T-dualized, while it transforms a p-brane into a $(p+1)$-brane if a direction orthogonal to the brane is T-dualized. Let us analyze this statement for a concrete brane configuration. Consider a system of one D0-brane, D2-brane, D4-brane and D6-brane. The last three branes are extended along the $(8,9)$, $(6,7,8,9)$ and $(4,5,6,7,8,9)$ directions, respectively. What brane configurations can be obtained after T-duality?

SOLUTION

The relative orientation of the different branes is illustrated in the table below.

	0	1	2	3	4	5	6	7	8	9
D6	×				×	×	×	×	×	×
D4	×				×	×			×	×
D2	×								×	×
D0	×									

Let us just consider transformations along a single circle. Then the original type IIA configuration gets mapped to a type IIB configuration. A T-duality transformation along the 1, 2 or 3 directions gives a D7, D5, D3, D1 configuration. A T-duality transformation along the 4 or 5 directions gives a D5, D3, D3, D1 configuration. A T-duality transformation along the 6 or 7 directions gives a D5, D5, D3, D1 configuration. A T-duality transformation along the 8 or 9 directions gives a D5, D3, D1, D1 configuration.

One might also consider a T-duality transformation along the time direction. However, this only makes sense in the context of finite temperature, where one has a periodic Euclidean time coordinate. That would lead one to an object that is localized in the time direction. Quite aside from the issue of T-duality, one could consider an object that fills some spatial directions and is localized in time and the other spatial directions. This is a higher-dimensional analog of an instanton, called an *S-brane*. Like instantons, it is not a physical object, but rather a possible stationary point of a path-integral that could play a role in the nonperturbative physics.

□

6.3 Type I superstring theory
Orientifold projection

Type I superstring theory can be understood as arising from a projection of type IIB superstring theory. Type IIB superstrings are oriented, and their world sheets are orientable. The world-sheet parity transformation

$$\Omega : \sigma \to -\sigma \quad (6.77)$$

reverses the orientation of the world sheet. World-sheet parity exchanges the left- and right-moving modes of the world-sheet fields X^μ and ψ^μ. This $\mathbb{Z}_2$ transformation is a symmetry of the type IIB theory and not of the type IIA theory, because only in the IIB case do the left- and right-moving fermions carry the same space-time chirality. When one gauges this $\mathbb{Z}_2$ symmetry, the type I theory results. The projection operator

$$P = \frac{1}{2}(1 + \Omega) \quad (6.78)$$

retains the left–right symmetric parts of physical states, which implies that the resulting type I closed strings are unoriented.

The type I closed-string spectrum is obtained by keeping the states that are even under the world-sheet parity transformation and eliminating the ones that are odd. The massless type IIB closed-string states in the NS–NS sector are given by the tensor product of two vectors. Only states that are symmetric in the two vectors survive the orientifold projection. These are the dilaton and the graviton, while the antisymmetric tensor B_2 is eliminated.

The two gravitino fields of type IIB superstring theory, Ψ_1^μ and Ψ_2^μ, are associated with the Fock-space states

$$b^\mu_{-1/2}|0;a\rangle \quad \text{and} \quad \tilde{b}^\mu_{-1/2}|a;0\rangle. \tag{6.79}$$

Here the label a denotes a spinor index for a Ramond-sector ground state. Under world-sheet parity $|0;a\rangle \leftrightarrow |a;0\rangle$, and left-moving and right-moving excitations are exchanged, which implies that only the sum $\Psi_1^\mu + \Psi_2^\mu$ survives the projection. Similarly, one of the two type IIB dilatinos survives, so that one is left with a total of $56 + 8 = 64$ massless fermionic degrees of freedom. The fact that only one gravitino survives implies that the type I theory has half as much supersymmetry as the type IIB theory (16 conserved supercharges instead of 32). This supersymmetry corresponds to the diagonal sum of the left-moving and right-moving supersymmetries of the type IIB theory.

Which massless R–R sector states survive the world-sheet parity projection can be determined by counting degrees of freedom. Since there is a massless gravitino field in the spectrum, the theory must be supersymmetric, and therefore the number of massless fermionic and bosonic degrees of freedom have to be equal. The only way to achieve this is to require that C_0 and C_4 are eliminated while the two-form C_2 survives. To summarize, after the projection the massless closed-string bosonic fields are the graviton and the dilaton in the NS–NS sector and the two-form C_2 in the R–R sector. This gives a total of $35 + 1 + 28 = 64$ bosonic degrees of freedom, which matches the number of fermionic degrees of freedom. Together, these give the $\mathcal{N} = 1$ supergravity multiplet.

In addition, it is necessary to add a twisted sector – the type I open strings. These are strings whose ends are associated with the fixed points of $\sigma \to -\sigma$, which are at $\sigma = 0$ and $\sigma = \pi$.[6] Since this applies for all X^μ, and open strings always end on D-branes, the existence of these open strings signals the presence of space-time-filling D9-branes. The open strings must also respect the Ω symmetry, so they are also unoriented.

The type IIB fundamental string (F-string) is a stable BPS object that carries a conserved charge that couples to B_2. Since the orientifold projection eliminates B_2, the type I fundamental string is not a stable BPS object. It can break. However, the amplitude for breaking is proportional to the string coupling constant. So at weak coupling, which is assumed in perturbation theory, type I superstrings are long-lived. At strong coupling, fundamental type I strings cease to be a useful concept, since they quickly disintegrate.

---

6 To obtain the usual open-string σ interval of length π, one should start with a closed-string coordinate σ of period 2π, which is double the choice that has been made previously.

Anomalies

As was explained in Chapter 5, type I supergravity in ten-dimensional Minkowski space-time by itself is inconsistent due to gravitational anomalies. Moreover, the only way to eliminate anomalies is to couple it to super Yang–Mills theory with an $SO(32)$ or $E_8 \times E_8$ gauge group. Only the group $SO(32)$ is possible for type I superstrings, and it can be realized by including open strings with Chan–Paton charges corresponding to this gauge group. Under world-sheet parity the open-string coordinates ψ^μ can transform with either sign. Taking into account the Chan–Paton degrees of freedom, represented by labels i, j, the transformation rule for open-string states becomes

$$\Omega b^\mu_{-1/2}|0, ij\rangle = \pm b^\mu_{-1/2}|0, j\, i\rangle, \tag{6.80}$$

because the world-sheet parity transformation interchanges the two ends of the string. If one chooses the plus sign in Eq. (6.80), then the projection picks out symmetric matrices, which corresponds to a symplectic gauge group. If, on the other hand, one chooses the minus sign the projection leaves antisymmetric matrices, which corresponds to an orthogonal group. So this is the choice that is needed to describe the anomaly-free supersymmetric $SO(32)$ theory.

Another way of interpreting the preceding conclusion is as follows. The orientifold projection results in the appearance of a space-time-filling orientifold plane. The plus sign in Eq. (6.80) results in the appearance of an $O9^+$ plane with $+16$ units of D9-brane charge, whereas the minus sign in Eq. (6.80) results in the appearance of an $O9^-$ plane with -16 units of D9-brane charge. Consistency requires the cancellation of this D9-brane charge. This corresponds to the cancellation of R–R tadpoles, which also ensures the cancellation of all gauge anomalies. This cancellation can be achieved in the first case (the plus sign) by the addition of 16 anti-D9-branes. This results in a theory with $USp(32)$ gauge symmetry. However, the presence of anti-D9-branes breaks all of the supersymmetry. In the second case (the minus sign) consistency is achieved by adding 16 D9-branes, which results in $SO(32)$ gauge symmetry. As discussed above, this preserves one of the two type IIB supersymmetries.

The tension of both kinds of O9-planes is $-16T_{D9}$. Therefore, in both cases the total energy density of the vacuum is zero. In the supersymmetric $SO(32)$ case this is ensured to all orders in the string coupling constant by supersymmetry. In the nonsupersymmetric $USp(32)$ case, perturbative corrections to the free theory are expected to generate a nonzero vacuum energy.

Other type I D-branes

The only massless R–R field in the type I spectrum is C_2. Therefore, aside from the D9-branes, the only stable type IIB D-branes that survive the orientifold projection are the ones that couple to this field. They are the D1-brane and its magnetic dual, the D5-brane.

The world-volume theories of these D-branes are more complicated than in the type IIB case. The basic reason is that there are additional massless modes that arise from open strings that connect the D1-brane or the D5-brane to the 16 D9-branes. Moreover, this is taking place in the presence of an O9$^-$ plane.

Let us consider first a system of N coincident D1-branes. In the type IIB theory the world-volume theory would be a maximally supersymmetric $U(N)$ gauge theory. However, due to the presence of the orientifold plane in the type I theory, the gauge symmetry is enhanced to $SO(2N)$, and there is half as much unbroken supersymmetry as in the type IIB case. Moreover, the world-volume theory contains massless matter supermultiplets that arise as modes of open strings connecting the D1-branes to the D9-branes. These transform as $(\mathbf{2N}, \mathbf{32})$ under $SO(2N) \times SO(32)$. The $SO(32)$ gauge symmetry of the ten-dimensional bulk is a global symmetry of the D1-brane world-volume theory.

The analysis of the world-volume theory of a system of N coincident D5-branes is carried out in a similar manner. The $U(N)$ gauge symmetry that is present in the type IIB case is enhanced to $USp(2N)$ due to the O9$^-$ plane, and the amount of unbroken supersymmetry is cut in half. Moreover, there are massless supermultiplets that arise as modes of open strings connecting the D5-branes to the D9-branes. They transform as $(\mathbf{2N}, \mathbf{32})$ under $USp(2N) \times SO(32)$.

The K-theory analysis of possible charges of type I D-branes, which is not presented here, accounts for all of the D-branes listed above. Moreover, it also predicts the existence of a stable point particle in $\mathbb{R}^{9,1}$ that carries a $\mathbb{Z}_2$ charge and is not supersymmetric. Thus this particle is a stable non-BPS D0-brane. This particle, like all D-branes, is a nonperturbative excitation of the theory. Moreover, it belongs to a spinor representation of the gauge group. Its existence implies that, nonperturbatively, the gauge group is actually $Spin(32)/\mathbb{Z}_2$ rather than $SO(32)$. The stability of this particle is ensured by the fact that it is the lightest state belonging to a spinor representation. The mod 2 conservation rule is also an obvious consequence of the group theory: two spinors can combine to give tensor representations. In Chapter 8 it is argued that type I superstring theory is dual to one of the heterotic string theories. The non-BPS D0-brane of the type I theory corresponds to a perturbative excitation of the dual heterotic theory.

The type I' theory

Let us now examine the T-dual description of the type I theory on a space-time of the form $\mathbb{R}^{8,1} \times S^1$, where the circle has radius R. Since the type IIB theory is T dual to the type IIA theory, and the type I theory is an orientifold projection of the type IIB theory, one should not be surprised to learn that the result is a certain orientifold projection of the type IIA theory compactified on the dual circle $\widetilde{S}^1$ of radius $\widetilde{R} = \alpha'/R$. The resulting T-dual version is called the type I' theory. The name type IA is also used.

Recall that T-duality for a type II theory compactified on a circle corresponds to the world-sheet transformation

$$X_R \to -X_R, \qquad \psi_R \to -\psi_R, \tag{6.81}$$

for the component of X and ψ along the circle. This implies that

$$X = X_L + X_R \to \widetilde{X} = X_L - X_R. \tag{6.82}$$

In the case of type II theories, $\widetilde{X}$ describes the dual circle $\widetilde{S}^1$. In the type I theory world-sheet parity Ω, which corresponds to $X_L \leftrightarrow X_R$, is gauged. Evidently, in the T-dual formulation this corresponds to

$$\widetilde{X} \to -\widetilde{X}. \tag{6.83}$$

Therefore, the gauging of Ω gives an orbifold projection of the dual circle, $\widetilde{S}^1/\mathbb{Z}_2$. More precisely, the $\mathbb{Z}_2$ action is an orientifold projection that combines $\widetilde{X} \to -\widetilde{X}$ with Ω. As noted earlier, Ω is not a symmetry of the IIA theory, since left-moving and right-moving fermions have opposite chirality. However, the simultaneous spatial reflection $\widetilde{X} \to -\widetilde{X}$ compensates for this mismatch.

The quotient $\widetilde{S}^1/\mathbb{Z}_2$ describes half of a circle. In other words, it is the interval $0 \leq \widetilde{X} \leq \pi \widetilde{R}$. The other half of the circle is present as a mirror image that is also Ω reflected. Altogether, the statement of T-duality is the equivalence of the compactified IIB orientifold

$$(\mathbb{R}^{8,1} \times S^1)/\Omega, \tag{6.84}$$

with the type IIA orientifold

$$(\mathbb{R}^{8,1} \times S^1)/\Omega \cdot \mathcal{I}, \tag{6.85}$$

where the symbol $\mathcal{I}$ represents the reflection $\widetilde{X} \to -\widetilde{X}$.

6.3 Type I superstring theory

The fixed-point set in the type I' construction consists of a pair of orientifold 8-planes located at $\widetilde{X} = 0$ and $\widetilde{X} = \pi\widetilde{R}$. Each of these carries -8 units of R–R charge. Consistency of the type I' theory requires adding 16 D8-branes, which are localized at points in the interval $0 \leq \widetilde{X} \leq \pi\widetilde{R}$ while filling the nine noncompact space-time dimensions. Clearly, these D8-branes are the T-duals of the D9-branes of the type I description.

The positions of the D8-branes along the interval are determined in the type I description by Wilson lines in the Cartan subalgebra of $SO(32)$. Since this group has rank 16, its Cartan subalgebra has 16 generators. The corresponding Wilson lines take values in compact $U(1)$ groups, so these values can be characterized by angles θ_I that are defined modulo 2π. These angles determine the dual positions of the D8-branes to be

$$\widetilde{X}_I = \theta_I \widetilde{R}, \quad I = 1, 2, \ldots, 16. \tag{6.86}$$

The $SO(32)$ gauge symmetry is broken by the Wilson lines. In terms of the type I' description the unbroken gauge symmetry is given by the following rules:

- When N D8-branes coincide in the interior of the interval, this corresponds to an unbroken $U(N)$ gauge group.
- When N D8-branes coincide with an O8$^-$ plane they give an unbroken $SO(2N)$ gauge group.

In both cases the gauge bosons arise as zero modes of D8–D8 open strings. In the second case the mirror-image D8-branes also contribute.

The case of trivial Wilson lines (all $\theta_I = 0$) corresponds to having all 16 D8-branes (and their mirror images) coincide with one of the O8$^-$ planes. This gives $SO(32)$ gauge symmetry, of course. In addition, there are two $U(1)$ factors. The corresponding gauge fields arise as components of the ten-dimensional metric and C_2 field: $g_{\mu 9}$ and $C_{\mu 9}$.

Somewhat more generally, consider the Wilson lines given by

$$\theta_I = 0 \text{ for } I = 1, \ldots, 8 + N \quad \text{and} \quad \theta_I = \pi \text{ for } I = 9 + N, \ldots, 16. \tag{6.87}$$

This corresponds to having $8 + N$ D8-branes coincide with the O8$^-$ plane at $\widetilde{X} = 0$ and $8 - N$ D8-branes coincide with the O8$^-$ plane at $\widetilde{X} = \pi\widetilde{R}$. Generically, according to the rules given above, this gives rise to the gauge symmetry

$$SO(16 + 2N) \times SO(16 - 2N) \times U(1)^2. \tag{6.88}$$

However, for the particular value of the radius $\widetilde{R} = \sqrt{g_s N \alpha'/8}$ one finds the gauge symmetry enhancement[7]

$$SO(16-2N) \times U(1) \to E_{9-N}. \qquad (6.89)$$

This is a nonperturbative symmetry enhancement. As such, it cannot be explained using the tools that have been described so far. It is best understood in terms of the S-dual heterotic string described in Chapter 8.

EXERCISES

EXERCISE 6.8

Show that under a T-duality transformation in the x^9 direction the worldsheet parity operator Ω of the type IIB theory transforms as follows:

$$\Omega \text{ in IIB} \quad \to \quad I_9 \Omega \text{ in IIA},$$

where I_9 inverts the sign of the ninth coordinates $X^9 \to -X^9$ and $\psi^9 \to -\psi^9$. How does ΩI_9 act on the type IIA space-time fermions?

SOLUTION

The orientifold projection Ω in type IIB corresponds to $T_9 \Omega T_9$ in type IIA, because the T_9 operations map back and forth between type IIA and type IIB. Therefore, the desired result is obtained if one can verify the identity

$$T_9 \Omega T_9 = I_9 \Omega.$$

This identity holds because

$$T_9 \Omega T_9 : (X_L^9, X_R^9) \to (X_L^9, -X_R^9) \to (-X_R^9, X_L^9) \to (-X_R^9, -X_L^9)$$

and

$$I_9 \Omega : (X_L^9, X_R^9) \to (X_R^9, X_L^9) \to (-X_R^9, -X_L^9).$$

The fermi coordinate ψ^9 transforms in exactly the same way.

The combined operation $I_9 \Omega$ maps R–NS type IIA space-time spinors to NS–R space-time spinors of the same chirality. The operation Ω interchanges the R–NS and NS–R fermions, and the operation I_9 reverses their chirality. This is what must happen in order to define a nontrivial projection operator. □

---

[7] E_6, E_7, and E_8 are exceptional Lie groups. The meaning of E_n with $n < 6$ can be inferred by extrapolating Dynkin diagrams. This gives $E_5 = SO(10)$, $E_4 = SU(5)$, $E_3 = SU(3) \times SU(2)$, $E_2 = SU(2) \times U(1)$ and $E_1 = SU(2)$.

6.4 T-duality in the presence of background fields

The previous sections have discussed T-duality for string theories compactified on a circle with the assumption that the remaining space-time dimensions are described by Minkowski space-time and that all other background fields vanish. In this section we shall discuss the generalization of the T-duality transformations along a circle in curved space-times with background fields. The first part considers NS–NS background fields: the graviton $g_{\mu\nu}$, two-form tensor $B_{\mu\nu}$ and dilaton Φ, while the second part considers the nontrivial R–R background fields.

NS–NS sector fields

The massless fields that appear in the closed bosonic-string spectrum or the NS–NS sector of either type II superstring consist of the space-time metric $g_{\mu\nu}$, the two-form $B_{\mu\nu}$ and the dilaton Φ. So far we have only considered a flat background with vanishing $B_{\mu\nu}$. The value of $\exp(\Phi)$ gives the string coupling constant g_s, which has been assumed to be constant and small. One can analyze more general possibilities by introducing the background fields into the world-sheet action. This cannot be done in an arbitrary way, since the action only has the required conformal symmetry for backgrounds that are consistent solutions of the theory. One possibility that works is for all of the background fields to be constants. There are more general possibilities, which are explored in this section.

The appropriate generalization of the world-sheet action in conformal gauge that includes NS–NS background fields is

$$S = S_g + S_B + S_\Phi, \tag{6.90}$$

with

$$S_g = -\frac{1}{4\pi\alpha'} \int d^2\sigma \sqrt{-h} h^{\alpha\beta} g_{\mu\nu} \partial_\alpha X^\mu \partial_\beta X^\nu, \tag{6.91}$$

$$S_B = \frac{1}{4\pi\alpha'} \int d^2\sigma \varepsilon^{\alpha\beta} B_{\mu\nu} \partial_\alpha X^\mu \partial_\beta X^\nu, \tag{6.92}$$

$$S_\Phi = \frac{1}{4\pi} \int d^2\sigma \sqrt{-h} \Phi R^{(2)}. \tag{6.93}$$

The first term replaces the Minkowski metric with the more general space-time metric in the obvious way. The second term expresses the fact that the fundamental string carries NS–NS two-form charge, just as the half-BPS

D-branes carry R–R charge. In differential form notation for the pullback field, it is proportional to $\int B_2$. The coefficient says that the two-form charge is equal to the string tension. For suitable normalization conventions, this is required by supersymmetry. The Φ term is higher-order in the α' expansion. Note also that both the B term and the Φ term are total derivatives for constant fields. Even so, they have an important influence on the physics. The S_B term contributes to the world-sheet canonical momenta and hence to the canonical commutation relations. The dilaton determines the string coupling constant precisely due to the term S_Φ, as was discussed in Chapter 3.

If the background fields are independent of the circular coordinate (for example, X^9 in the case of the superstring), the T-dual world-sheet theory can be derived by a duality transformation of the X^9 coordinate. The formulas can be derived by using the Lagrange multiplier method introduced in Section 6.1. Introducing a Lagrange multiplier $\tilde{X}^9$, consider the action

$$4\pi\alpha' S = \int d^2\sigma \left[\sqrt{-h} h^{\alpha\beta} \left(-g_{99} V_\alpha V_\beta - 2g_{9\mu} V_\alpha \partial_\beta X^\mu - g_{\mu\nu} \partial_\alpha X^\mu \partial_\beta X^\nu \right) + \right.$$

$$\left. \varepsilon^{\alpha\beta}(B_{9\mu} V_\alpha \partial_\beta X^\mu + B_{\mu\nu} \partial_\alpha X^\mu \partial_\beta X^\nu) + \tilde{X}^9 \varepsilon^{\alpha\beta} \partial_\alpha V_\beta + \alpha' \sqrt{-h} R^{(2)} \Phi(X) \right]. \tag{6.94}$$

In the above action $\mu, \nu = 0, \ldots, 8$ refer to all space-time coordinates except X^9. The $\tilde{X}^9$ equation of motion,

$$\varepsilon^{\alpha\beta} \partial_\alpha V_\beta = 0, \tag{6.95}$$

is solved by writing $V_\beta = \partial_\beta X^9$. Substituting this into the action returns us to the original action (6.90). On the other hand, using the V_α equations of motion to eliminate this field, gives the dual action

$$\tilde{S} = S_{\tilde{g}} + S_{\tilde{B}} + S_{\tilde{\Phi}}, \tag{6.96}$$

where the background fields of the dual theory are given by

$$\tilde{g}_{99} = \frac{1}{g_{99}}, \quad \tilde{g}_{9\mu} = \frac{B_{9\mu}}{g_{99}}, \quad \tilde{g}_{\mu\nu} = g_{\mu\nu} + \frac{B_{9\mu} B_{9\nu} - g_{9\mu} g_{9\nu}}{g_{99}}.$$

$$\tilde{B}_{9\mu} = -\tilde{B}_{\mu 9} = \frac{g_{9\mu}}{g_{99}}, \quad \tilde{B}_{\mu\nu} = B_{\mu\nu} + \frac{g_{9\mu} B_{9\nu} - B_{9\mu} g_{9\nu}}{g_{99}}. \tag{6.97}$$

The dilaton transformation rule requires a different analysis. We argued in Section 6.2 that the type IIA and type IIB coupling constants are related by $\tilde{g}_s = g_s \sqrt{\alpha'}/R$. For the identifications $g_{99} = R^2/\alpha'$ and $\tilde{g}_{99} = \tilde{R}^2/\alpha'$, this implies that

$$\tilde{\Phi} = \Phi - \frac{1}{2} \log g_{99}, \tag{6.98}$$

at least if we assume $g_{9\mu} = \tilde{g}_{9\mu} = 0$. Equation (6.66) can be understood as the vacuum expectation value of this relation.

R–R sector fields

The massless spectrum of each of the superstring theories also contains bosonic fields in the R–R sector. There is an obstruction to describing their coupling to the string world sheet in the RNS formulation, a fact that is a fundamental limitation of this approach. They can be coupled to the world sheet in the GS formulation, in which case they have couplings of the form $\bar{\Theta}\Gamma^{\mu_1\cdots\mu_n}\Theta F_{\mu_1\cdots\mu_n}$.

A possible approach to understanding the behavior of R–R background fields under T-duality is to go back to the construction of these fields as bilinears in fermionic fields in the GS formulation of the superstring and use the fact that under T-duality the right-moving fermions are multiplied by a Dirac matrix (see Exercise 6.6). Alternatively, since they couple to D-branes, one can use the T-duality properties of D-branes to deduce the transformation rules. Either method leads to the same conclusion. In total, the effect of T-duality on the R–R tensor fields of the type IIA theory is to give the following type IIB R–R fields:

$$\tilde{C}_9 = C, \qquad \tilde{C}_\mu = C_{\mu 9}, \qquad \tilde{C}_{\mu\nu 9} = C_{\mu\nu}, \qquad \tilde{C}_{\mu\nu\lambda} = C_{\mu\nu\lambda 9}. \qquad (6.99)$$

As a result, the odd-form potentials of the type IIA theory are mapped to the even-form potentials of the type IIB theory. These formulas can be read backwards to describe the transformations in the other direction, that is, from type IIB to type IIA. These formulas are only valid for trivial NS–NS backgrounds ($B_{\mu\nu} = 0$, $g_{\mu\nu} = \eta_{\mu\nu}$ and constant Φ). Otherwise, they need to be generalized.

6.5 World-volume actions for D-branes

Let us now turn to the construction of world-volume actions for D-branes. The basic idea is that modes of the open strings that start and end on a given D-brane can be described by fields that are restricted to the world volume of the D-brane. In order to describe the dynamics of the D-brane at energies that are low compared to the string scale, only the massless open-string modes need to be considered, and one can construct a low-energy effective action based entirely on them. Thus, associated with a Dp-brane, there is a $(p+1)$-dimensional effective field theory of massless fields (scalars, spinors, and vectors), that captures the low-energy dynamics of the D-brane in question.

Restricting our attention to the half-BPS D-branes, p is even for the type IIA theory and odd for the type IIB theory. As was explained, these are the stable D-branes that preserve half of the space-time supersymmetry. Associated with such a brane there is a world-volume theory that has 16 conserved supercharges. The way to construct this theory is to use the GS formalism with κ symmetry. This construction is carried out here for a flat space-time background.[8] In fact, this was done already in Chapter 5 for the case of a D0-brane in the type IIA theory.

There are a number of interesting generalizations. One is the extension to a curved background, as well as the coupling to background fields in both the NS–NS and R–R sectors. Such actions are described later, but only for the truncation to the bosonic sector, which has no κ symmetry. An extension that is especially interesting is the generalization to multiple coincident D-branes. In this case the world-volume theory has a nonabelian gauge symmetry, and there are interesting new phenomena that emerge.

Kappa symmetric D-brane actions

The D-brane world-volume theories that follow contain the same ingredients as in Chapter 5 as well as one new ingredient. The familiar ingredients are the functions

$$X^\mu(\sigma),$$

which describe the embedding of the D-brane in ten-dimensional Minkowski space-time. Here the coordinates σ^α, $\alpha = 0, 1, \ldots, p$, parametrize the Dp-brane world volume. The other familiar ingredient is a pair of Majorana–Weyl spinors,

$$\Theta^{1a}(\sigma) \quad \text{and} \quad \Theta^{2a}(\sigma),$$

which extends the mapping to $\mathcal{N} = 2$ superspace. The new ingredient is an abelian world-volume gauge field $A_\alpha(\sigma)$.

Counting of degrees of freedom

There are several ways of understanding the necessity of the gauge field. Perhaps the best one is to realize that it is part of the spectrum of the open string that starts and ends on the D-brane. As a check, one can verify

---

[8] It can be generalized to other backgrounds, provided that they satisfy the classical supergravity field equations.

6.5 World-volume actions for D-branes

that there are an equal number of physical bosonic and fermionic degrees of freedom, as required by supersymmetry. In fact, after all local symmetries are taken into account, the physical content should be the same as in maximally supersymmetric Maxwell theory, which also has 16 conserved supercharges. That theory has eight propagating fermionic states and eight propagating bosonic states. In ten dimensions the relevant massless supermultiplet in the open-string spectrum consists of a massless vector and a Majorana–Weyl spinor.

The fields Θ^{Aa} have 32 real components. Kappa symmetry gives a factor of two reduction and the Dirac equation implies that half of the remaining 16 components are independent propagating degrees of freedom. This is correct counting for all values of p. The bosonic degrees of freedom come partly from X^μ and partly from A_α. Taking account of the $p+1$ diffeomorphism symmetries that are built into the world-volume theory, only $10 - (p+1) = 9 - p$ components of the X^μ are propagating degrees of freedom. These are the components that describe transverse excitations of the Dp-brane. The gauge field A_α has $p + 1$ components, but for a gauge-invariant theory two of them are nondynamical, so A contributes $p - 1$ physical degrees of freedom. Altogether, the total number of physical bosonic degrees of freedom is $(9 - p) + (p - 1) = 8$, as required by supersymmetry.

Born–Infeld action

Before the advent of quantum mechanics, Born and Infeld proposed a nonlinear generalization of Maxwell theory in an attempt to eliminate the infinite classical self-energy of a charged point particle. They suggested replacing the Maxwell action by

$$S_{\text{BI}} \sim \int \sqrt{-\det(\eta_{\alpha\beta} + kF_{\alpha\beta})}\, d^4\sigma, \qquad (6.100)$$

where k is a constant. Expanding in powers of F gives a constant plus the Maxwell action plus higher powers of F. The Born–Infeld action was an inspired guess in that exactly this structure appears in low-energy effective D-brane actions. They were led to this structure by realizing that it would be generally covariant if the Lorentz metric were replaced by an arbitrary space-time metric. This reasoning does not give a unique result, however.

To see evidence that such a formula is required in string theory, consider specializing to the two-dimensional D1-brane case and supposing that the spatial dimension is a circle. Evaluating the determinant in this case gives

$$\int \sqrt{1 - k^2 F_{01}^2}\, d^2\sigma. \qquad (6.101)$$

By T-duality there should be a dual interpretation in terms of a D0-brane on a dual circle. In this case it was shown in Chapter 5 that

$$A_1 = -\frac{1}{2\pi\alpha'}\widetilde{X}^1, \qquad (6.102)$$

where $\widetilde{X}^1$ is the coordinate on the dual circle. This gives a field strength

$$F_{01} = -\frac{1}{2\pi\alpha'}v \quad \text{where} \quad v = \dot{\widetilde{X}}^1. \qquad (6.103)$$

Here v is the velocity of the D0-brane on the dual circle. The spatial integration gives a constant factor, and one is left with the action for a relativistic particle (compare with Chapter 2)

$$-m\int \sqrt{1-v^2}\, dt, \qquad (6.104)$$

for the choice

$$k = 2\pi\alpha'. \qquad (6.105)$$

Thus the Born–Infeld structure is required for Lorentz invariance of the T-dual description.

Generalizing to $p+1$ dimensions, the Born–Infeld structure combines nicely with the usual Nambu–Goto structure for a Dp-brane (discussed in Chapter 2) to give the action

$$S_1 = -T_{\mathrm{D}p}\int d^{p+1}\sigma \sqrt{-\det(G_{\alpha\beta} + k\mathcal{F}_{\alpha\beta})}, \qquad (6.106)$$

where $T_{\mathrm{D}p}$ is the tension (or energy density), and $k = 2\pi\alpha'$. For type II superstrings in Minkowski space-time supersymmetry is incorporated by defining

$$G_{\alpha\beta} = \eta_{\mu\nu}\Pi^\mu_\alpha \Pi^\nu_\beta, \qquad (6.107)$$

where

$$\Pi^\mu_\alpha = \partial_\alpha X^\mu - \bar{\Theta}^A \Gamma^\mu \partial_\alpha \Theta^A. \qquad (6.108)$$

This is the same supersymmetric combination introduced in Chapter 5. Also,

$$\mathcal{F}_{\alpha\beta} = F_{\alpha\beta} + b_{\alpha\beta}, \qquad (6.109)$$

where $F = dA$ is the usual Maxwell field strength and the two-form b is a Θ-dependent term that is required in order that $\mathcal{F}$ is supersymmetric. The concrete expression, whose verification is a homework problem, is

$$b = (\bar{\Theta}^1 \Gamma_\mu d\Theta^1 - \bar{\Theta}^2 \Gamma_\mu d\Theta^2)(dX^\mu - \frac{1}{2}\bar{\Theta}^A \Gamma^\mu d\Theta^A). \qquad (6.110)$$

6.5 World-volume actions for D-branes

An action with the general structure of S_1 is usually referred to as a DBI action, referring to Dirac, Born and Infeld, even though it would make sense to refer to Nambu and Goto, as well. As in the examples described in Chapter 5, a Chern–Simons action S_2 still needs to be added in order to implement κ symmetry. The form of S_2 is determined below.

D-brane tensions

As was already mentioned, the DBI Lagrangian density can be expanded in powers of the field strength. The first term is proportional to $\sqrt{-\det G_{\alpha\beta}}$. A convenient gauge choice is the static gauge in which the diffeomorphism symmetry is used to set the first $p+1$ components of X^μ equal to the world-volume coordinates σ^α, while the other $9-p$ components survive as scalar fields on the world volume that describe transverse excitations of the brane. In the static gauge, the Lagrangian density consists of the constant term $-T_{\mathrm{D}p}$ plus field-dependent terms. Thus the Hamiltonian density, which gives the energy density of the brane, is $+T_{\mathrm{D}p}$ plus positive field-dependent terms. The zero-point energies of the world-volume fields exactly cancel, thanks to supersymmetry, so this remains true in the quantum theory. The Maxwell term (the term quadratic in k in the expansion of S_1) can be written in the form (see Exercise 6.6)

$$S_{\mathrm{Maxwell}} = -\frac{1}{4g^2} \int F_{\alpha\beta} F^{\alpha\beta} d^{p+1}\sigma. \tag{6.111}$$

Here g is the gauge coupling in $p+1$ dimensions, which is proportional to the dimensionless open-string coupling constant g_{open}, since the gauge field is an open-string excitation. The open-string coupling is related in turn to the closed-string coupling g_{s} by $g_{\mathrm{s}} = g_{\mathrm{open}}^2$. These facts imply that the Dp-brane tension is given by

$$T_{\mathrm{D}p} = \frac{c_p}{g_{\mathrm{s}}}. \tag{6.112}$$

The numerical factor c_p is derived below.

The tension of a Dp-brane (in the string frame) is proportional to $1/g_{\mathrm{s}}$. This shows that D-branes are nonperturbative excitations of string theory, which become very heavy at weak coupling. This justifies treating them as rigid objects in the weak-coupling limit. The tension of a D-brane increases more slowly for $g_{\mathrm{s}} \to 0$ than more conventional solitons, such as the NS5-brane, the magnetic dual of the fundamental string, whose tension is proportional to $1/g_{\mathrm{s}}^2$. When the growth is this rapid, there is no longer a weak-coupling regime in which it is a valid approximation to neglect the gravitational back reaction on the geometry in the vicinity of the brane.

One reason D-branes are useful probes of string geometry is that a tension proportional to $1/g_s$ does allow for such a regime. Chapter 12 considers a situation in which the number of D-branes N is increased at the same time as $g_s \to 0$ with $N \sim 1/g_s$. The gravitational effects of the D-branes survive in this limit.

The same type of reasoning used earlier to relate the type IIA and IIB string coupling constants can be used to determine D-brane tensions. T-duality exchanges a wrapped Dp-brane in the type IIA theory and an unwrapped D$(p-1)$-brane in the type IIB theory (and *vice versa*). Using this fact, compactification of the D-brane action on a circle gives (for p even) the relation $2\pi R T_{\mathrm{D}p} = T_{\mathrm{D}(p-1)}$, or

$$\frac{2\pi R c_p}{g_s} = \frac{c_{p-1}}{\tilde{g}_s}. \tag{6.113}$$

Inserting the relation between the string coupling constants in Eq. (6.66) gives

$$c_p = \frac{1}{2\pi\sqrt{\alpha'}} c_{p-1}. \tag{6.114}$$

If one sets $T_{\mathrm{D}0} = (g_s\sqrt{\alpha'})^{-1}$, a result that is derived in Chapter 8, then one obtains the precise formula

$$T_{\mathrm{D}p} = \frac{1}{g_s(2\pi)^p(\alpha')^{(p+1)/2}}. \tag{6.115}$$

As before, it is understood that the type IIA string coupling constant is used if p is even, and the type IIB coupling constant is used if p is odd.

The construction of S_2

Supersymmetric D-brane actions require κ symmetry in order to have the right number of fermionic degrees of freedom. As in the examples of Chapter 5, this requires the addition of a Chern–Simons term, which can be written as the integral of a $(p+1)$-form

$$S_2 = \int \Omega_{p+1}. \tag{6.116}$$

However, as in the case of the superstring, it is easier to construct the $(p+2)$-form $d\Omega_{p+1}$. It is manifestly invariant under supersymmetry, whereas the supersymmetry variation of Ω_{p+1} is a total derivative.

The analysis is rather lengthy, but it involves the same techniques that were described for simpler examples in Chapter 5. Let us settle here for a description of the result. The answer takes the form

$$d\Omega_{p+1} = d\bar{\Theta}^A \mathcal{T}_p^{AB} d\Theta^B, \tag{6.117}$$

where $\mathcal{T}_p^{AB}$ is a 2×2 matrix of p-form valued Dirac matrices and $A, B = 1, 2$ is summed. Comparing to the result for the D0-brane given in Chapter 5, gives in that case

$$\Omega_1 = -m\bar{\Theta}\Gamma_{11}d\Theta = m(\bar{\Theta}^1 d\Theta^2 - \bar{\Theta}^2 d\Theta^1), \tag{6.118}$$

which implies that

$$\mathcal{T}_0 = m \begin{pmatrix} 0 & 1 \\ -1 & 0 \end{pmatrix}. \tag{6.119}$$

The formula for D-brane tensions gives the identification

$$m = T_{\text{D0}} = \frac{1}{g_s \sqrt{\alpha'}}. \tag{6.120}$$

Now let us present the general result for S_2. It turns out to be simpler to give all the results at once rather than to enumerate them one by one. In other words, the expression for

$$\mathcal{T}^{AB} = \sum_{p=0}^{\infty} \mathcal{T}_p^{AB} \tag{6.121}$$

can be written relatively compactly.[9] In the type IIA case the sum is over even values of p, and in the type IIB case the sum is over odd values of p. Given $\mathcal{T}$, which is a sum of differential forms of various orders, one simply extracts the p-form part to obtain $\mathcal{T}_p$ and construct the Chern–Simons term S_2 of the Dp-brane action. Forms of order higher than 9 are not relevant. The expression for $\mathcal{T}$ turns out to have the form

$$\mathcal{T}^{AB} = m\, e^{2\pi\alpha' \mathcal{F}} f^{AB}(\psi), \tag{6.122}$$

where $\mathcal{F}$ is given in Eq. (6.109), and ψ is a matrix-valued one-form given by

$$\psi = \frac{1}{\sqrt{2\pi\alpha'}} \Gamma_\mu \Pi_\alpha^\mu d\sigma^\alpha. \tag{6.123}$$

In the type IIA case

$$f(\psi) = \begin{pmatrix} 0 & \cos\psi \\ -\cosh\psi & 0 \end{pmatrix} \tag{6.124}$$

---

[9] Recall that sums of differential forms of various orders were encountered earlier in the anomaly discussion of Chapter 5.

and in the type IIB case

$$f(\psi) = \begin{pmatrix} 0 & \sin\psi \\ \sinh\psi & 0 \end{pmatrix}. \qquad (6.125)$$

The formulas for the functions f ensure that the matrix is symmetric or antisymmetric for the appropriate powers of ψ, as required when $\mathcal{T}$ is sandwiched between Majorana–Weyl spinors. It is not obvious that the formulas for $d\Omega_{p+1}$ presented here are closed. However, with a certain amount of effort, this can be proved and the formulas for Ω_{p+1} can be extracted.

The static gauge

As was briefly mentioned earlier, the static gauge consists of using the diffeomorphism symmetry of the Dp-brane action to identify $p+1$ of the space-time coordinates X^μ with the world-volume coordinates σ^α. Let us then relabel the remaining $9-p$ coordinates as $2\pi\alpha'\Phi^i$ to emphasize the fact that they are scalar fields of the world-volume theory with mass dimension equal to one. Doing this, the bosonic part of the DBI action collapses to the form

$$S_{\text{DBI}} = -T_{\text{D}p}\int d^{p+1}\sigma\sqrt{-\det(\eta_{\alpha\beta} + k^2\partial_\alpha\Phi^i\partial_\beta\Phi^i + kF_{\alpha\beta})}, \qquad (6.126)$$

where $k = 2\pi\alpha'$, as before.

Now let us generalize this result to include fermion degrees of freedom, by considering first the D9-brane case. This requires making a gauge choice for the κ symmetry. A particularly nice choice, which maintains manifest Lorentz invariance, is to use this freedom to set one of the two Θ^As equal to zero. This completely kills the Chern–Simons term, because the matrices $f^{(A)}$ and $f^{(B)}$ are entirely off-diagonal. Making this gauge choice in the special case $p = 9$ and renaming the remaining Majorana–Weyl Θ variable as $k\lambda$ gives the action S_{D9} equal to

$$T_{\text{D}9}\int d^{10}\sigma\sqrt{-\det\left(\eta_{\alpha\beta} + kF_{\alpha\beta} - 2k^2\bar\lambda\Gamma_\alpha\partial_\beta\lambda + k^3\bar\lambda\Gamma^\gamma\partial_\alpha\lambda\bar\lambda\Gamma_\gamma\partial_\beta\lambda\right)}. \qquad (6.127)$$

It is truly remarkable that this nonlinear extension of ten-dimensional super-Maxwell theory has exact unbroken supersymmetry. In addition to the usual 16 linearly realized supersymmetries of super-Maxwell theory, it also has 16 nonlinearly realized supersymmetries that represent the spontaneously broken supersymmetries that gave rise to λ as a Goldstone fermion. Put differently, this action combines features of the Born–Infeld theory with features of the Volkov–Akulov theory of the Goldstone fermion.

The static gauge Dp-brane actions with $p < 9$ can be obtained in a similar manner. However, a quicker method is to note that they can be obtained by dimensional reduction of the gauge-fixed D9-brane action in Eq. (6.127). Dimensional reduction simply means dropping the dependence of the world-volume fields on $9-p$ of the coordinates. This works for both even and odd values of p. For example, dimensional reduction of Eq. (6.127) to four dimensions gives an exactly supersymmetric nonlinear extension of $\mathcal{N}=4$ super Maxwell theory. The supersymmetry transformations are complicated, because the gauge-fixing procedure contributes induced κ transformations to the original ε transformations of the fields.

Bosonic D-brane actions with background fields

The D-brane actions obtained in the previous section are of interest as they describe D-branes in flat space. However, one frequently needs a generalization that describes the D-brane in a more general background in which the various bosonic massless supergravity fields are allowed to take arbitrary values. These actions exhibit interesting features, that we shall now address.

The abelian case

The background fields in the NS–NS sector are the space-time metric $g_{\mu\nu}$, the two-form $B_{\mu\nu}$ and the dilaton Φ. These can be pulled back to the world volume

$$P[g+B]_{\alpha\beta} = (g_{\mu\nu} + B_{\mu\nu})\partial_\alpha X^\mu \partial_\beta X^\nu. \tag{6.128}$$

Henceforth, for ease of writing, pullbacks are implicit, and this is denoted $g_{\alpha\beta} + B_{\alpha\beta}$. Note that this $g_{\alpha\beta}$ is the bosonic restriction of the quantity that was called $G_{\alpha\beta}$ previously. With this definition, the DBI term in static gauge takes the form

$$S_{\mathrm{D}p} = -T_{\mathrm{D}p} \int d^{p+1}\sigma e^{-\Phi_0} \sqrt{-\det(g_{\alpha\beta} + B_{\alpha\beta} + k^2 \partial_\alpha \Phi^i \partial_\beta \Phi^i + kF_{\alpha\beta})}. \tag{6.129}$$

Since the string coupling constant g_s is already included in the tension $T_{\mathrm{D}p}$, the dilaton field is shifted by a constant so that it has vanishing expectation value ($\Phi = \log g_s + \Phi_0$). This is the significance of the subscript. Note that invariance under a two-form gauge transformation

$$\delta B = d\Lambda \tag{6.130}$$

requires a compensating shift of the gauge field A.

The possibility of R–R background fields should also be considered. They do not contribute to the DBI action, but they play an important role in the Chern–Simons term. Let us denote an n-form R–R field by C_n and the corresponding field strength by $F_{n+1} = dC_n$. Previously, it was stated that the complete list of these fields in type II superstring theories involves only $n = 0, 1, 2, 3, 4$. However, it is convenient to introduce redundant fields C_n for $n = 5, 6, 7, 8$. This makes it possible to treat electric and magnetic couplings in a more symmetrical manner and leads to more elegant formulas. The idea is to generalize the self-duality of the five-form field strength by requiring that

$$\star F_{n+1} = F_{9-n}. \quad (6.131)$$

This requires that the R–R gauge fields are harmonic. This can be generalized to allow for interactions by including additional terms in the definitions of the field strengths $F_{n+1} = dC_n + \ldots$

The C_n fields are differential forms in ten-dimensional space-time. However, they can also be pulled back to the D-brane world volume, after which they are represented by the same symbols. Then the Chern–Simons term must contain a contribution

$$\mu_p \int C_{p+1}, \quad (6.132)$$

where μ_p denotes the Dp-brane charge, since a Dp-brane couples electrically to the R–R field C_{p+1}. However, this is not the entire Chern–Simons term. In the presence of a background B field or world-volume gauge fields, the D-brane also couples to R–R potentials of lower rank. This can be described most elegantly in terms of the total R–R potential

$$C = \sum_{n=0}^{8} C_n. \quad (6.133)$$

The result then turns out to be

$$S_{\mathrm{CS}} = \mu_p \int \left(C \, e^{B+kF} \right)_{p+1}. \quad (6.134)$$

The subscript means that one should extract the $(p+1)$-form piece of the integrand. Since B and F are two-forms, only odd-rank R–R fields contribute for even p (the IIA case) and only even-rank R–R fields contribute for odd p (the IIB case). The B and F fields appear in the same combination as in the DBI term, and so the two-form gauge invariance still works in the same way. The structure of the Chern–Simons term implies that a Dp-brane in the presence of suitable backgrounds can also carry induced charge of the type that is associated with a D$(p-2n)$-brane for $n = 0, 1, \ldots$ Generically, this charge is smeared over the $(p+1)$-dimensional world volume, though in special cases it may be concentrated on a lower-dimensional hypersurface, for example a brane within a brane.

In the presence of space-time curvature the Chern–Simons term contains an additional factor involving differential forms constructed from the curvature tensor. We won't describe this factor, since it would require a rather long digression. It reduces to 1 in a flat space-time, which is the case considered here.

The nonabelian case

When N Dp-branes coincide, the world-volume theory is a $U(N)$ gauge theory. Almost all studies of nonabelian D-brane actions use the static gauge from the outset, since otherwise it is unclear how to implement diffeomorphism invariance and κ symmetry. In the static gauge the world-volume fields are just those of a maximally supersymmetric vector supermultiplet: gauge fields, scalars and spinors, all in the adjoint representation of $U(N)$. If one only wants to describe the leading nontrivial terms in a weak-field expansion, the result is exactly super Yang–Mills theory. This approximation is sufficient for many purposes including the important examples of Matrix theory, based on D0-branes, and AdS/CFT duality, based on D3-branes, which are discussed in Chapter 12.

When one tries to include higher powers of fields to give formulas that correctly describe nonabelian D-brane physics for strong fields, the subject can become mathematically challenging and physically confusing. The reason it can be confusing concerns the domain of validity of DBI-type actions. They are meant to capture the physics in the regime of approximation in which the background fields and the world-volume gauge fields are allowed to be arbitrarily large, but whose variation is small over distances of order the string scale. The requirement of slow variation is meant to justify dropping terms involving derivatives of the world-volume fields. The tricky issue in the nonabelian case is that one should use covariant derivatives to maintain gauge invariance, but there are relations of the form

$$[D_\alpha, D_\beta] \sim F_{\alpha\beta}. \qquad (6.135)$$

This makes it somewhat ambiguous whether a term is derivative or not, and so it is not obvious how to suppress rapid variation while allowing strong fields. Nonetheless, some success has been achieved, which will now be described.

Henceforth all fermion fields are set to zero and only bosonic actions are considered. In addition to the background fields g, B, Φ and C, the desired actions contain adjoint gauge fields A and $9 - p$ adjoint scalars Φ^i, both of which are represented as hermitian $N \times N$ matrices. The notation that is used is

$$A_\alpha = \sum_n A_\alpha^{(n)} T_n \quad \text{and} \quad \Phi^i = \sum_n \Phi^{i(n)} T_n, \tag{6.136}$$

where T_n are N^2 hermitian $N \times N$ matrices satisfying $\text{Tr}(T_m T_n) = N\delta_{mn}$. We also define[10]

$$F_{\alpha\beta} = \partial_\alpha A_\beta - \partial_\beta A_\alpha + i[A_\alpha, A_\beta], \tag{6.137}$$

$$D_\alpha \Phi^i = \partial_\alpha \Phi^i + i[A_\alpha, \Phi^i]. \tag{6.138}$$

Let us start with the nonabelian D9-brane action, which is relatively simple, because there are no scalar fields. In this case the proposed DBI term is

$$S_1 = -T_{D9} \int d^{10}\sigma e^{-\Phi_0} \text{Tr}\left(\sqrt{-\det(g_{\alpha\beta} + B_{\alpha\beta} + kF_{\alpha\beta})}\right). \tag{6.139}$$

This innocent-looking formula requires explanation. The determinant refers to the 10×10 matrix labelled by the Lorentz indices. However, the expression inside the determinant is also an $N \times N$ matrix, assuming that g and B are multiplied by unit matrices. The understanding is that the square root of the determinant is computed for each of the N^2 matrix elements, though only the diagonal entries are required, since the trace of the resulting $N \times N$ matrix needs to be taken. This is the simplest prescription that makes sense, and it has survived a number of checks. For example, if one chooses the positive branch of the square root in each case, then the trace is N plus field-dependent terms. This gives an energy density of N times the tension of a single brane, as one expects for N coincident branes.

In similar fashion, the proposed nonabelian D9-brane Chern–Simons term is

$$S_2 = \mu_9 \int \text{Tr}\left(C e^{B+kF}\right)_{10}. \tag{6.140}$$

Starting from this ansatz for the $p = 9$ case, Myers was able to deduce a unique formula for all the $p < 9$ cases by implementing consistency with T-duality. This required allowing the background fields to be functionals of the nonabelian coordinates and the introduction of nonabelian pullbacks. The formula that was obtained in this way has a complicated Φ dependence. Rather than describing it in detail, we settle here for pointing out an interesting feature of the result: in the abelian case a Dp-brane can couple to

---

[10] When the T_ns are chosen to be antihermitian, the factors of i do not appear.

6.5 World-volume actions for D-branes 241

the R–R potentials $C_{p-1}, C_{p-3}, \ldots$ in addition to the usual C_{p+1}. The surprising result in the nonabelian case is that the Dp-brane can also couple to the higher-rank R–R potentials $C_{p+3}, C_{p+5}, \ldots$

The Myers effect

The coupling of nonabelian D-branes to higher-rank R–R potentials has some interesting physical consequences. The simplest example, due to Myers, concerns N coincident D0-branes in the presence of constant four-form flux $F_4 = dC_3$. The flux is chosen to be electric, meaning that the only nonzero components have a time index and three spatial indices F_{0ijk}. It is sufficient to restrict the nonvanishing components to three spatial directions and write $F_{0ijk} = f\epsilon_{ijk}$, where f is a constant. All other background fields are set to zero, and the background geometry is assumed to be ten-dimensional Minkowski space-time. The result to be described concerns the point-like D0-brane system becoming polarized into a fuzzy two-sphere by the electric field.

The relevant terms that need to be considered are a kinetic energy term proportional to $\text{Tr}(\dot{\Phi}^i \dot{\Phi}^i)$, which comes from the DBI term, and a potential energy term

$$V(\Phi) \sim -\frac{1}{4}\text{Tr}([\Phi^i, \Phi^j][\Phi^i, \Phi^j]) - \frac{i}{3}f\epsilon_{ijk}\text{Tr}(\Phi^i \Phi^j \Phi^k). \tag{6.141}$$

The first term in the potential comes from the DBI action, and the second term in the potential, which is the coupling to the R–R four-form electric field, comes from the nonabelian CS action. Now let us look for a static solution for which the potential is extremal, which requires

$$[[\Phi^i, \Phi^j], \Phi^j] + if\epsilon_{ijk}[\Phi^j, \Phi^k] = 0. \tag{6.142}$$

A class of solutions of this equation is obtained by letting $\Phi^i = f\alpha^i/2$, where α^i is an N-dimensional representation of $SU(2)$ satisfying

$$[\alpha^i, \alpha^j] = 2i\epsilon_{ijk}\alpha^k. \tag{6.143}$$

This gives many possible solutions (besides zero) if N is large – one for each partition of N. However, the one of lowest energy is given by the N-dimensional irreducible representation of $SU(2)$, which satisfies

$$\text{Tr}(\alpha^i \alpha^j) = \frac{1}{3}N(N^2 - 1)\delta_{ij}. \tag{6.144}$$

Recall that in the abelian theory $2\pi\alpha'\Phi^i$ is interpreted as a transverse coordinate of the D-brane. In the nonabelian theory this becomes an $N \times N$

matrix, so this identification is not so straightforward anymore. In the absence of the four-form electric field, the preferred configurations that minimize the potential have $[\Phi^i, \Phi^j] = 0$. This allows one to define a moduli space on which these matrices are simultaneously diagonal. One can interpret the diagonal entries as characterizing the positions of the N D-branes. The pattern of $U(N)$ symmetry breaking is encoded in the degeneracies of these positions.

In the presence of the four-form flux, the Φ^i no longer commute at the extrema of the potential, and so the classical interpretation of the D-brane positions breaks down. There is an irreducible fuzziness in the description of their positions. One can say that the mean-square value of the ith coordinate (averaged over all N D-branes) is given by

$$\langle (X^i)^2 \rangle = \frac{1}{N}(2\pi\alpha')^2 \text{Tr}[(\Phi^i)^2]. \qquad (6.145)$$

Summing over the three coordinates gives a "fuzzy sphere" whose radius R squared is the sum of three such terms. Substituting the ground-state solution gives

$$R^2 = (\pi\alpha' f)^2 (N^2 - 1). \qquad (6.146)$$

For large N the sphere becomes less fuzzy, and the radius is approximately $R = \pi\alpha' f N$. Specifically, the uncertainty δR is proportional to $1/N$. So the radius is proportional to the strength of the electric field and the number of D0-branes. If one used a reducible representation of $SU(2)$ instead, one would find a set of concentric fuzzy spheres, one for each irreducible component. However, such solutions are energetically disfavored.

The fuzzy sphere has an alternative interpretation as a spherical D2-brane with N dissolved D0-branes. For large N this can be analyzed using the abelian D2-brane theory. The total D2-brane charge is zero, though there is a nonzero D2-charge electric dipole moment, which couples to the four-form electric field. The previous results can be reproduced, at least for large N, in this picture.

EXERCISES

EXERCISE 6.9
Expand (6.106) to quartic order in k and show that the quadratic term gives the Maxwell action (6.111).

Solution

Because

$$\det(G_{\alpha\beta} + kF_{\alpha\beta}) = \det(G_{\alpha\beta} + kF_{\alpha\beta})^T = \det(G_{\alpha\beta} - kF_{\alpha\beta}),$$

this is an even function of k. Using a matrix notation, let us define

$$M = kG^{-1}F.$$

Then

$$\sqrt{-\det(G + kF)} = \sqrt{-\det G}\sqrt{\det(1 + M)}$$

$$= \sqrt{-\det G}\Big[\det(1 - M^2)\Big]^{1/4}.$$

Next, we use the identity

$$\log\det(1 - M^2) = \operatorname{tr}\log(1 - M^2) = -\operatorname{tr}\Big(M^2 + \frac{1}{2}M^4 + \dots\Big).$$

Thus

$$\Big[\det(1 - M^2)\Big]^{1/4} = \exp\Big(-\frac{1}{4}\operatorname{tr}M^2 - \frac{1}{8}\operatorname{tr}M^4 + \dots\Big)$$

$$= 1 - \frac{1}{4}\operatorname{tr}M^2 - \frac{1}{8}\operatorname{tr}M^4 + \frac{1}{32}(\operatorname{tr}M^2)^2 + \dots$$

The final form of the action has a constant energy-density term, a quadratic Maxwell-type term, plus higher-order corrections

$$S_1 = -T_{\mathrm{D}p}\int d^{p+1}\sigma\sqrt{-\det(G_{\alpha\beta} + kF_{\alpha\beta})}$$

$$= -T_{\mathrm{D}p}\int d^{p+1}\sigma\sqrt{-\det G}\Big(1 + \frac{k^2}{4}F_{\alpha\beta}F^{\alpha\beta}$$

$$-\frac{k^4}{8}(F_{\alpha\beta}F^{\alpha\beta})^2 + \frac{k^4}{32}F_{\alpha\beta}F^{\beta\gamma}F_{\gamma\delta}F^{\delta\alpha} + \dots\Big).$$

Indices are raised in this formula using the inverse of the induced metric $G_{\alpha\beta}$. The Maxwell term has the normalization $-\frac{1}{4g^2}\int F_{\alpha\beta}F^{\alpha\beta}d^{p+1}\sigma$ for the identification $g^2 = (2\pi)^{p-2}\ell_s^{p-3}g_s$. □

Exercise 6.10

Consider the static-gauge DBI action for a Dp-brane given in Eq. (6.126)

$$S_{\mathrm{DBI}} = -T_{\mathrm{D}p}\int d^{p+1}\sigma\sqrt{-\det(\eta_{\alpha\beta} + k^2\partial_\alpha\Phi^i\partial_\beta\Phi^i + kF_{\alpha\beta})}.$$

What types of charged soliton solutions is this theory expected to have? What are their physical interpretations?

SOLUTION

This is a $(p+1)$-dimensional theory containing a $U(1)$ gauge field. A one-form gauge field can couple electrically to a point-like charge in any dimension. Furthermore, as we have learned, it can couple magnetically to a $(p-3)$-brane for $D = p+1$. Therefore, a solitonic 0-brane solution could be an electric source of the gauge field and a solitonic $(p-3)$-brane solution could be a magnetic source of the gauge field.

If these solitons do actually exist (finding them is homework), then they should have an interpretation from the point of view of the ten-dimensional superstring theory that contains the Dp-brane. The defining property of a D-brane is that a fundamental string can end on it. Moreover, the fundamental string carries a unit of Chan–Paton electric charge at its end. Thus the electrically charged 0-brane soliton should be interpreted as the end of a fundamental string.

Recall that the scalars Φ^i can be interpreted as transverse displacements of the D-brane. Using this fact, the solution that one finds actually exhibits a spike sticking out from the D-brane that asymptotically approaches zero thickness. So the solution allows one to see the entire string, not just its end point. In fact, the solution describes a smooth transition from a p-dimensional D-brane to a one-dimensional string.

The magnetic solution is somewhat similar. In this case a $(p-3)$-brane soliton is the end of a D$(p-2)$-brane. In other words, a D$(p-2)$-brane can end on a Dp-brane. When it does so, its end, which has $p-2$ spatial dimensions, is interpreted as a magnetic source of the $U(1)$ gauge field in the Dp-brane world-volume theory. Again, the explicit soliton solution allows one to see the entire D$(p-2)$-brane protruding from the Dp-brane. □

HOMEWORK PROBLEMS

PROBLEM 6.1

Consider the type IIA and type IIB superstring theories compactified on a circle so that the space-time is $M_{10} = \mathbb{R}^{8,1} \times S^1$, where $\mathbb{R}^{8,1}$ denotes nine-dimensional Minkowski space-time. Show that the spectrum of the type IIA theory for radius R agrees with the spectrum of the type IIB theory for radius $\tilde{R} = \alpha'/R$.

PROBLEM 6.2
Equations (6.38) and (6.39) describe a generalization of the result of Exercise 6.2 from a $U(1)$ gauge field to a $U(N)$ gauge field

$$A = -\frac{1}{2\pi R}\mathrm{diag}(\theta_1, \theta_2, \ldots, \theta_N),$$

where the θs are again constants. Derive these equations.

PROBLEM 6.3
The T-duality rules for R–R sector tensor fields can be derived by taking into account that the field strengths are constructed as bilinears in Majorana–Weyl spinors in the covariant RNS approach. Explicitly,

$$F_{\mu_1\ldots\mu_n} = \bar{\psi}_\mathrm{L} \Gamma_{\mu_1\ldots\mu_n} \psi_\mathrm{R}.$$

(i) Explain why n is even for the type IIA theory and odd for the type IIB theory.
(ii) Explain why (in differential form notation) $F_n = \star F_{10-n}$.
(iii) Show that, for both the type IIA and type IIB theories, the number of independent components of the tensor fields agrees with the number of degrees of freedom of a tensor product of two Weyl–Majorana spinors in ten dimensions.

PROBLEM 6.4
Show that the Dirac equations for ψ_L and ψ_R in the previous problem imply that the field equations and Bianchi identities for the field strengths are satisfied, that is,

$$\partial_{[\mu} F_{\mu_1\ldots\mu_n]} = 0, \qquad \partial^\mu F_{\mu\mu_2\ldots\mu_n} = 0.$$

Also, show that, when these equations for F_n are re-expressed as equations for F_{10-n}, the field equation and Bianchi identity are interchanged.

PROBLEM 6.5
Derive the T-duality transformation formulas for NS–NS background fields in (6.97). You may ignore the dilaton term and set $h_{\alpha\beta} = \eta_{\alpha\beta}$. Verify that if the transformation is repeated a second time, one recovers the original field configuration.

Problem 6.6

Show that the Born–Infeld action (6.100) gives a finite classical self-energy for a charged point particle. Hint: show that the solution to the equations of motion with a point particle of charge e at the origin is given by

$$E_r = F_{rt} = \frac{e}{\sqrt{(r^4 + r_0^4)}}, \qquad r_0^2 = 2\pi\alpha' e.$$

Problem 6.7

Consider the DBI action in Eq. (6.106).

(i) Derive the equation of motion for the gauge field.
(ii) Expand this equation in powers of k to obtain the leading correction to the usual Maxwell field equation of electrodynamics in the absence of sources. You may use the result of Exercise 6.9.

Problem 6.8

Consider a D0–D8 system in the type I' theory, where the D0-brane is coincident with the D8-brane. There are also other D8-branes and O8-planes parallel to the D8-brane, as described in Section 6.3.

(i) Determine the zero-point energy of a D0–D8 open string in the NS sector.
(ii) Describe the supersymmetries that are preserved by this configuration. How many of them are there? Hint: Eq. (6.56) shows which supersymmetries are preserved by a single D-brane. The problem here is to determine which ones are preserved by both of the D-branes.

Problem 6.9

Consider a type I' configuration in which N_1 D8-branes are coincident at $X'_L = \theta_L R'$ and the remaining $N_2 = 16 - N_1$ D8-branes are coincident at $X'_R = \theta_R R'$.

(i) What is the gauge symmetry for generic positions X'_L and X'_R?
(ii) What is the maximum enhanced gauge symmetry that can be achieved for $N_1 = N_2 = 8$? How are the D8-branes positioned in this case?

Problem 6.10

Show that the right-hand side of Eq. (6.117) is closed.

Problem 6.11

(i) Determine how the two-form b, defined in Eq. (6.110), transforms under a supersymmetry transformation.

(ii) Determine the supersymmetry transformation of the gauge field A for which the field strength $\mathcal{F}$, defined in Eq. (6.109), is supersymmetric, that is, invariant under supersymmetry transformations.

Problem 6.12

By taking account of the pullback on the Dp-brane world volume show that the action (6.129) is invariant under (6.130) if a compensating shift of the gauge field A is made.

Problem 6.13

Consider the static-gauge DBI action for a Dp-brane given in Eq. (6.126) that was discussed in Exercise 6.10.

(i) Find the action for a D3-brane in spherical coordinates (t, r, θ, ϕ) for the special case in which the only nonzero fields are $A_t(r)$ and one scalar $\Phi(r)$.

(ii) Obtain the equations of motion for $A_t(r)$ and $\Phi(r)$.

(iii) Find a solution of the equations of motion that corresponds to an electric charge at the origin, and deduce the profile of the string that is attached to the D3-brane. For what range of r are the DBI approximations justified?

Problem 6.14

As in the preceding problem, consider the static-gauge DBI action for a Dp-brane given in Eq. (6.126) that was discussed in Exercise 6.10.

(i) Find the action for a D3-brane in spherical coordinates (t, r, θ, ϕ) for the special case in which the only nonzero fields are $A_\phi(\theta)$ and one scalar $\Phi(r)$.

(ii) Obtain the equations of motion for $A_\phi(\theta)$ and $\Phi(r)$.

(iii) Find a solution of the equations of motion that corresponds to a magnetic charge at the origin and deduce the profile of the D-string that is attached to the D3-brane. For what range of r are the DBI approximations justified?

PROBLEM **6.15**

Compute the minimum of the potential function in Eq. (6.141) when the N-dimensional representation of $SU(2)$ is irreducible. What is the minimum of the potential if the N-dimensional representation of $SU(2)$ is the sum of two irreducible representations? How does it compare to the previous result? Describe the fuzzy sphere configuration in this case.

7
The heterotic string

The preceding chapters have described bosonic strings as well as type I and type II superstrings. In the case of the bosonic string, one was led to 26-dimensional Minkowski space-time by the requirement of cancellation of the conformal anomaly of the world-sheet theory. Similar reasoning led to the conclusion that the type I and type II superstring theories should have $D = 10$.

In all of these theories the world-sheet degrees of freedom can be divided into left-movers and right-movers, though in the case of open strings these are required to combine so as to give standing waves. In the case of the type II superstring theories, the left-moving and right-moving modes introduce independent conserved supersymmetry charges, each of which is a Majorana–Weyl spinor with 16 real components. Thus, the type II superstring theories have two such conserved charges, or $\mathcal{N} = 2$ supersymmetry, which means that they have 32 conserved supercharges. The type IIA and type IIB theories are distinguished by whether the two Majorana–Weyl spinors have the same (IIB) or opposite (IIA) chirality. In the case of the type I theory, as well as related theories whose construction involves an orientifold projection, the only conserved supercharge that survives the projection is the sum of the left-moving and right-moving supercharges of the type IIB theory. Thus these theories have $\mathcal{N} = 1$ supersymmetry in ten dimensions.

There is an alternative method of constructing supersymmetrical string theories in ten dimensions with $\mathcal{N} = 1$ supersymmetry, which is the topic of this chapter. These theories, known as *heterotic string theories*, implement this supersymmetry by combining the left-moving degrees of freedom of the 26-dimensional bosonic string theory with the right-moving degrees of freedom of the ten-dimensional superstring theory. It is surprising at first sight that this is a sensible thing to do, but it leads to interesting new superstring theories. Since heterotic string theories have $\mathcal{N} = 1$ supersymmetry in ten dimensions, they are subject to the consistency conditions required by anomaly cancellation that were described in Chapter 5. This means that their spectrum must contain massless super Yang–Mills multiplets based on either an $SO(32)$ or $E_8 \times E_8$ gauge group.

The heterotic construction is the only construction of a ten-dimensional superstring with $E_8 \times E_8$ gauge symmetry, though there is an interesting connection to M-theory, which is explored in Chapter 8. On the other hand, the heterotic string provides an alternative realization of $SO(32)$ gauge symmetry, which is the gauge group for the type I superstring derived in Chapter 5. Chapter 8 shows that these two $SO(32)$ theories are actually dual descriptions of the same theory.

7.1 Nonabelian gauge symmetry in string theory

String theory naturally gives rise to the most interesting types of local gauge symmetries. These symmetries are general coordinate invariance, associated with a spin 2 quantum (the graviton), local supersymmetries associated with spin 3/2 quanta (gravitinos), and Yang–Mills gauge invariances associated with spin 1 quanta (gauge particles). Experimentally, the only one of these that is unconfirmed is supersymmetry, though there is some indirect evidence for it. It would be astonishing if such a wonderful opportunity were not utilized by Nature. In fact, if string theory is correct, supersymmetry must play a role at least at the Planck scale, if not at lower energies. What is certainly observed, and therefore should be incorporated in string theory, is local gauge symmetry. Indeed, the standard model of elementary particles, which describes the strong, weak and electromagnetic interactions, is based on $SU(3) \times SU(2) \times U(1)$ local gauge symmetry.

D-branes and orientifold planes

In the description of string theory presented so far, only one mechanism for realizing nonabelian gauge symmetries was described. It involved open strings ending on D-branes whose ends carry Chan–Paton charges. The $SO(32)$ gauge symmetry of the type I superstring theory is achieved in this way. In this theory the open strings end on a collection of 16 space-time-filling D9-branes, and there is also a space-time-filling orientifold plane. However, even though $SO(32)$ is a very large group, it is not a very good starting point for embedding the standard model. The possibilities for achieving nonabelian gauge symmetry utilizing D-branes and orientifold planes become much more elaborate in the context of compactification of extra dimensions.

In the case of the type II superstring theories, after compactification of the extra dimensions, various D-branes may fill the four noncompact dimensions and wrap various cycles[1] in the compact dimensions. As was explained in Chapter 6, N coincident D-branes have a $U(N)$ gauge symmetry on their world volume. If, in addition, there are orientifold planes or singularities in the compactification, other types of gauge groups can also arise. Thus by incorporating various collections of D-branes, and perhaps orientifold planes, a rich variety of gauge theories can be achieved. This is one of the main approaches that is being studied for constructing a realistic string model of elementary particles. Such constructions are explored in later chapters.

Isometries of the internal space

Another possibility for generating gauge symmetry is for the compactification space to have isometries. Then the zero modes of the ten-dimensional graviton on the compact manifold give rise to gauge fields in the noncompact dimensions that realize the symmetry of the manifold as a gauge symmetry. This is a basic feature of Kaluza–Klein compactification. For example, if the compact space is an N-torus T^N, one obtains a $U(1)^N$ gauge symmetry. Similarly an N-sphere S^N gives rise to an $SO(N+1)$ gauge symmetry and a projective space with N complex dimensions CP^N gives $SU(N+1)$ gauge symmetry. The case of S^5 plays an important role in the AdS/CFT correspondence in Chapter 12.

Heterotic strings

The heterotic string theories described in this chapter utilize yet another mechanism, special to theories of strings, for implementing local gauge symmetry. The heterotic theories are oriented closed strings, and the properties of the left-moving and right-moving modes are different. As was mentioned above, the supersymmetry charges are carried by the right-moving currents of the string. The heterotic theories realize Yang–Mills gauge symmetries in a similar way. Namely, the conserved charges of Yang–Mills gauge symmetries are carried by the left-moving currents of the string. Thus the charges are distributed democratically along closed strings. This is to be contrasted with the case of the type I superstring theory, where gauge-symmetry charges are localized at the end points of open strings.

---

1 Supersymmetric cycles are discussed in Chapter 9.

7.2 Fermionic construction of the heterotic string

In this section we would like to construct the action for the heterotic string. The conformal gauge action describing the bosonic string is

$$S = -\frac{1}{2\pi} \int d^2\sigma \partial_\alpha X_\mu \partial^\alpha X^\mu, \tag{7.1}$$

where the dimension of space-time is $D = 26$. This is supplemented by Virasoro constraints for both the left-moving and right-moving modes. The corresponding conformal gauge action for superstrings in the RNS formalism is

$$S = -\frac{1}{2\pi} \int d^2\sigma (\partial_\alpha X_\mu \partial^\alpha X^\mu + \bar{\psi}^\mu \rho^\alpha \partial_\alpha \psi_\mu). \tag{7.2}$$

In this case $D = 10$, and there are super-Virasoro constraints for both the left-moving and right-moving modes. The world-sheet fields ψ^μ are ten two-component Majorana spinors. This superstring action has world-sheet supersymmetry. Space-time supersymmetry arises by including both the R and NS sectors and imposing the GSO projection, as explained in Chapter 5.

In order to incorporate gauge degrees of freedom, let us consider a slightly different extension of the bosonic string theory. Specifically, let us add world-sheet fermions that are singlets under Lorentz transformation in space-time but which carry some internal quantum numbers. Introducing n Majorana fermions λ^A with $A = 1, \ldots, n$, consider the action

$$S = -\frac{1}{2\pi} \int d^2\sigma \left(\partial_\alpha X_\mu \partial^\alpha X^\mu + \bar{\lambda}^A \rho^\alpha \partial_\alpha \lambda^A \right). \tag{7.3}$$

This theory has an obvious global $SO(n)$ symmetry under which the λ^A transform in the fundamental representation and the coordinates X^μ are invariant. Since a fermion contributes half a unit to the central charge, the requirement that the total central charge should be 26 is satisfied provided that $D + n/2 = 26$. This is one way of describing a compactification of the bosonic string theory to $D < 26$.

Examining this theory more carefully, one sees that the symmetry is actually larger than $SO(n)$. Indeed, writing the terms out explicitly in world-sheet light-cone coordinates gives

$$S = \frac{1}{\pi} \int d^2\sigma \left(2\partial_+ X_\mu \partial_- X^\mu + i\lambda_-^A \partial_+ \lambda_-^A + i\lambda_+^A \partial_- \lambda_+^A \right). \tag{7.4}$$

Written this way, it is evident that the theory actually has an (unwanted) $SO(n)_\mathrm{L} \times SO(n)_\mathrm{R}$ global symmetry under which the left-movers and right-movers transform independently. One could try to discard the right-movers,

7.2 Fermionic construction of the heterotic string

for example, and work only with left-moving fermions, which would leave only the $SO(n)_L$ global symmetry. The problem with this, of course, is that then it would not be possible to satisfy the central-charge conditions for both the left-movers and right-movers at the same time.

Until now we have discussed bosonic strings, for which the critical dimension is 26, and superstrings, for which the critical dimension is 10. In both cases the world-sheet left-movers and right-movers are completely decoupled. This independence of the left-movers and right-movers was utilized by Gross, Harvey, Martinec and Rohm to propose a type of string theory in which the bosonic string structure is used for the left-movers and the superstring structure is used for the right-movers. They named this hybrid theory the *heterotic string*.

Space-time supersymmetry is implemented in the right-moving sector that corresponds to the superstring. Associated with this sector there are right-moving super-Virasoro constraints and a GSO projection of the usual sort. This ensures the absence of tachyons, which are removed by space-time supersymmetry.

Since the left-moving modes correspond to the bosonic string theory, the left-moving central charge should be 26, and there are constraints given by a left-moving Virasoro algebra. One possibility, known as the *fermionic construction of the heterotic string* is to have ten bosonic left-movers and 32 fermionic left-movers λ^A, since this gives a central charge $10 + 32/2 = 26$. This description makes it clear that this is a ten-dimensional theory, since the ten coordinates X^μ have both left-moving and right-moving degrees of freedom. The rest of the degrees of freedom are described by left-moving and right-moving fermions. This formulation of the heterotic string theory is pursued in the remainder of this section.

There is an equivalent *bosonic construction of the heterotic string*, which uses 26 left-moving bosonic coordinates. It is surprising in that it appears that the number of space-time dimensions is different for the left-moving and right-moving sectors. In this description one could wonder how many space-time dimensions there really are. However, as we have already asserted, this description is equivalent to the fermionic description in which it is clear that this is a ten-dimensional theory. The bosonic description of the heterotic string is given in Section 7.4.

The action for the heterotic string in the fermionic formulation is

$$S = \frac{1}{\pi} \int d^2\sigma (2\partial_+ X_\mu \partial_- X^\mu + i\psi^\mu \partial_+ \psi_\mu + i\sum_{A=1}^{32} \lambda^A \partial_- \lambda^A), \quad (7.5)$$

where $\mu = 0, \ldots, 9$ labels the vector representation of the ten-dimensional Lorentz group $SO(9,1)$, while λ^A are Lorentz singlets. Both sets of fermions are one-component Majorana–Weyl spinors from the point of view of the two-dimensional world-sheet Lorentz group.

Once one solves the equations of motion, there are ten right-moving bosons $X_R^\mu(\tau - \sigma)$ and ten left-moving bosons $X_L^\mu(\tau + \sigma)$. In addition, there are ten right-moving fermions $\psi^\mu(\tau - \sigma)$ and 32 left-moving fermions $\lambda^A(\tau + \sigma)$. These fields give a right-moving central charge $\hat{c} = 3c/2 = 10$ and a left-moving central charge $c = 26$, since each Majorana fermion contributes $c = 1/2$. Once the b and c ghosts are introduced for left-movers and right-movers and the β and γ ghosts are introduced for right-movers only the central charges cancel. Thus, one has a right-moving superconformal symmetry and a left-moving conformal symmetry.

As we remarked earlier, this action has a manifest $SO(32)$ symmetry under which the λ^A transform in the fundamental representation. This global symmetry of the world-sheet theory gives rise to a corresponding local gauge symmetry of the space-time theory. At this point it may appear rather mysterious how one could ever hope to achieve an $E_8 \times E_8$ gauge symmetry. The key to discriminating the different possibilities is the choice of GSO projections for the λ^A, as is explained in Section 7.2.

For the right-moving modes there is a world-sheet supersymmetry, whose transformations are given by

$$\delta X^\mu = i\varepsilon\psi^\mu \qquad \text{and} \qquad \delta\psi^\mu = -2\varepsilon\partial_- X^\mu. \tag{7.6}$$

This is what survives in conformal gauge of the local supersymmetry that is present before gauge fixing. This original local supersymmetry is the reason that the right-moving constraints are given by a super-Virasoro algebra. There is no supersymmetry for the left-movers.

The $SO(32)$ heterotic string

Let us start with an analysis of the $SO(32)$ heterotic string using methods that are similar to those used for the superstring in Chapter 4.

Right-movers

The right-moving modes of the hetcrotic string satisfy supcr-Virasoro constraints like those of right-moving modes of type II superstrings. As in that case, there is an NS and an R sector. In both sectors one should impose the GSO projections that were described in Chapter 5.

7.2 Fermionic construction of the heterotic string

- An on-shell physical state $|\phi\rangle$ in the NS sector must satisfy the conditions

$$G_r|\phi\rangle = L_m|\phi\rangle = \left(L_0 - \frac{1}{2}\right)|\phi\rangle = 0, \qquad r, m > 0, \qquad (7.7)$$

where the various super-Virasoro generators are given by the same formulas as in Chapter 4. The mass-shell condition is given by the L_0 equation

$$\left(L_0 - \frac{1}{2}\right)|\phi\rangle = \left(\frac{p^2}{8} + N_\mathrm{R} - \frac{1}{2}\right)|\phi\rangle = 0, \qquad (7.8)$$

where

$$N_\mathrm{R} = \sum_{n=1}^{\infty} \alpha_{-n} \cdot \alpha_n + \sum_{r=1/2}^{\infty} r b_{-r} \cdot b_r. \qquad (7.9)$$

- In the R sector the physical-state conditions are

$$F_m|\phi\rangle = L_m|\phi\rangle = 0, \qquad m \geq 0, \qquad (7.10)$$

which includes the mass-shell condition

$$L_0|\phi\rangle = \left(\frac{p^2}{8} + N_\mathrm{R}\right)|\phi\rangle = 0, \qquad (7.11)$$

where

$$N_\mathrm{R} = \sum_{n=1}^{\infty} (\alpha_{-n} \cdot \alpha_n + n d_{-n} \cdot d_n). \qquad (7.12)$$

Alternatively, if one uses the light-cone GS formalism of Chapter 5, then there is a very simple description of the right-moving modes that does not involve combining separate sectors or imposing GSO projections. Rather one simply has

$$L_0|\phi\rangle = \left(\frac{p^2}{8} + N_\mathrm{R}\right)|\phi\rangle = 0, \qquad (7.13)$$

where

$$N_\mathrm{R} = \sum_{n=1}^{\infty} (\alpha_{-n}^i \alpha_n^i + n S_{-n}^a S_n^a). \qquad (7.14)$$

As explained in Chapter 5, the transverse index i and the spinor index a each take eight values. The mass-shell condition in this formalism is

$$M^2 = 8 N_\mathrm{R}, \qquad (7.15)$$

which already shows that there are no tachyons, as expected from supersymmetry.

Left-movers

The left-moving fermionic fields λ^A can have periodic or antiperiodic boundary conditions, just like the fermionic coordinates ψ^μ in the RNS formalism. Periodic boundary conditions define the P sector, which is the analog of the R sector of the superstring. The mode expansion in the P sector is

$$\lambda^A(\tau + \sigma) = \sum_{n \in \mathbb{Z}} \lambda_n^A e^{-2in(\tau+\sigma)}. \tag{7.16}$$

These modes satisfy the anticommutation relations

$$\{\lambda_m^A, \lambda_n^B\} = \delta^{AB}\delta_{m+n,0}. \tag{7.17}$$

Antiperiodic boundary conditions define the A sector, which is the analog of the NS sector of the superstring. The mode expansion in the A sector is

$$\lambda^A(\tau + \sigma) = \sum_{r \in \mathbb{Z}+1/2} \lambda_r^A e^{-2ir(\tau+\sigma)}. \tag{7.18}$$

These modes satisfy the anticommutation relations

$$\{\lambda_r^A, \lambda_s^B\} = \delta^{AB}\delta_{r+s,0}. \tag{7.19}$$

The left-moving modes of the heterotic string satisfy Virasoro constraints

$$\tilde{L}_m|\phi\rangle = (\tilde{L}_0 - \tilde{a})|\phi\rangle = 0, \quad m > 0. \tag{7.20}$$

If one goes to light-cone gauge and solves the Virasoro constraints, then only the eight transverse components $\tilde{\alpha}_n^i$ are relevant. For the left-movers the A and P sectors need to be treated separately.

- For the P sector

$$\left(\tilde{L}_0 - \tilde{a}_P\right)|\phi\rangle = \left(\frac{p^2}{8} + N_L - \tilde{a}_P\right)|\phi\rangle = 0, \tag{7.21}$$

where

$$N_L = \sum_{n=1}^{\infty}(\tilde{\alpha}_{-n} \cdot \tilde{\alpha}_n + n\lambda_{-n}^A \lambda_n^A). \tag{7.22}$$

- In the A sector we have

$$\left(\tilde{L}_0 - \tilde{a}_A\right)|\phi\rangle = \left(\frac{p^2}{8} + N_L - \tilde{a}_A\right)|\phi\rangle = 0, \tag{7.23}$$

where

$$N_L = \sum_{n=1}^{\infty} \tilde{\alpha}_{-n} \cdot \tilde{\alpha}_n + \sum_{r=1/2}^{\infty} r\lambda_{-r}^A \lambda_r^A. \tag{7.24}$$

Now let us compute the left-moving normal-ordering constants $\tilde{a}_A$ and $\tilde{a}_P$. The general rule is most easily understood in the light-cone gauge, where only physical degrees of freedom contribute. The normal-ordering constant due to the zero-point energy of a periodic boson is $1/24$, for an antiperiodic fermion is $1/48$ and for a periodic fermion is $-1/24$.[2] Using these rules, we obtain the following value for the normal-ordering constants:

$$\tilde{a}_A = \frac{8}{24} + \frac{32}{48} = 1, \tag{7.25}$$

$$\tilde{a}_P = \frac{8}{24} - \frac{32}{24} = -1. \tag{7.26}$$

Thus, the mass formula for the states in the A sector is

$$\frac{1}{8}M^2 = N_R = N_L - 1, \tag{7.27}$$

and in the P sector it is

$$\frac{1}{8}M^2 = N_R = N_L + 1. \tag{7.28}$$

These equations show that massless states must have $N_R = 0$. Therefore, in the A sector there are massless states, which have to satisfy $N_L = 1$. On the other hand, there are no massless states in the P sector, since N_L cannot be negative.

Massless spectrum

Massless states are constructed by taking the tensor product of right-moving modes with $N_R = 0$ and left-moving modes with $N_L = 1$ in the A sector, as there are no massless states in the P sector.

- For the right-moving sector the states with $N_R = 0$ are those of the $D = 10$ vector supermultiplet, as in the superstring theories. Explicitly, in light-cone gauge notation, the massless modes in the $N_R = 0$ sector are

$$|i\rangle_R \quad \text{and} \quad |\dot{a}\rangle_R, \tag{7.29}$$

which are the ground states in the bosonic and fermionic sectors corresponding to the vector $\mathbf{8_v}$ and the spinor $\mathbf{8_c}$ representations of the transverse rotation group $Spin(8)$.
- The left-moving modes with $N_L = 1$ consist of

$$\tilde{\alpha}^i_{-1}|0\rangle_L, \tag{7.30}$$

[2] The derivation of these normal-ordering constants is given in Chapter 4.

which is an $SO(32)$ singlet and an $SO(8)$ vector, and

$$\lambda^A_{-1/2}\lambda^B_{-1/2}|0\rangle_L, \qquad (7.31)$$

which is an antisymmetric rank-two tensor of dimension $32 \times 31/2 = 496$. The latter states are Lorentz singlets and transform in the adjoint representation of the gauge group $SO(32)$.

Since the heterotic string theory is a closed-string theory, the physical states are given by the tensor product of right-movers and left-movers. Let us consider the contributions of $\tilde{\alpha}^i_{-1}|0\rangle_L$ first. In the *bosonic sector* this gives the massless states

$$|i\rangle_R \otimes \tilde{\alpha}^j_{-1}|0\rangle_L. \qquad (7.32)$$

These 64 states can be decomposed into a symmetric traceless part (graviton), an antisymmetric tensor and a scalar (dilaton). In the *fermionic sector* the massless states are

$$|\dot{a}\rangle_R \otimes \tilde{\alpha}^j_{-1}|0\rangle_L, \qquad (7.33)$$

which decomposes into a gravitino with 56 components and a dilatino with eight components. Altogether, these 64 bosons and 64 fermions form the $\mathcal{N} = 1$ supergravity multiplet.

The tensor product of the other 496 $N_L = 1$ left-moving states of the form in Eq. (7.31) with the 16 right-moving states with $N_R = 0$ gives the $D = 10$ vector supermultiplet for the gauge group $SO(32)$. It is important that all the massless vector states have appeared in the adjoint representation, as is required by Yang–Mills theory. The massless fermionic gauginos, which are their supersymmetry partners, are also in the adjoint representation.

A GSO-type projection

The modes in the A sector are constrained to satisfy $N_R = N_L - 1$. This condition has interesting implications for the left-movers. Recall that N_R as defined in Eq. (7.14) has only integer eigenvalues, whereas N_L can have half-integer eigenvalues, which arise whenever an odd number of λ^A oscillators act on the Fock-space vacuum. The relation $N_R = N_L - 1$ implies that these half-integer eigenvalues do not contribute to the physical spectrum. This projecting out of states with an odd number of λ^A oscillators is reminiscent of the GSO projection. In fact, it is a projection of exactly the same type, since it says that there must be an even number of λ^A-oscillator excitations. A similar projection condition is required for the P sector by one-loop unitarity. It cannot be discovered just from level-matching, since P-sector modes are integral in the first place. The projection condition in this case is $(-1)^F = 1$, where

$$(-1)^F = \bar{\lambda}_0 (-1)^{\sum_1^\infty \lambda_{-n}^A \lambda_n^A} \tag{7.34}$$

and

$$\bar{\lambda}_0 = \lambda_0^1 \lambda_0^2 \ldots \lambda_0^{32} \tag{7.35}$$

is the product of the fermionic zero modes. The $N_\text{L} = 0$ level contributes only half of the 2^{16} modes that one might otherwise expect. This corresponds to an irreducible spinor representation of $Spin(32)$, which is the universal covering group of $SO(32)$. In fact, because of this projection condition only one of the two possible conjugacy classes of spinors occurs in the physical spectrum. As a result, the gauge group of the theory is most precisely described as $Spin(32)/\mathbb{Z}_2$. This means that two of the four conjugacy classes of $Spin(32)$ survive: the adjoint conjugacy class (corresponding to the root lattice) and one for the two spinor conjugacy classes. The conjugacy class containing the vector **32** representation and the other spinor conjugacy class do not occur.

The $E_8 \times E_8$ heterotic string

The anomaly analysis in Chapter 5 showed that in addition to $SO(32)$ there is one other compact Lie group,[3] namely $E_8 \times E_8$, for which there could be a consistent supersymmetric gauge theory in ten dimensions. This group shows much more promise for phenomenological applications, since the gauge group of the standard model, $SU(3) \times SU(2) \times U(1)$, fits inside E_8 through a nice chain of embeddings:

$$SU(3) \times SU(2) \times U(1) \subset SU(5) \subset SO(10) \subset E_6 \subset E_7 \subset E_8. \tag{7.36}$$

The various groups that appear in this sequence are precisely the ones that have been most studied as candidates for grand unification symmetry groups. This gives additional motivation for trying to realize an $E_8 \times E_8$ gauge symmetry in the fermionic description of the heterotic string theory.

The construction of the $SO(32)$ heterotic string retained the manifest $SO(32)$ symmetry of the world-sheet action at all stages of the analysis by assigning the same boundary conditions (A or P) to all of the 32 left-moving fermions λ^A in each of the sectors. So there was just one P sector and one A sector. If maintaining the $SO(32)$ symmetry is no longer an objective, then

---

[3] For a brief introduction to Lie groups, and E_8 in particular, see Polchinski Section 11.4 and GSW, Appendix 6.A, respectively.

it is natural to consider introducing sectors in which there are A boundary conditions for some of the fermions and P boundary conditions for the rest of them. Of course, as long as the goal is to achieve a supersymmetric theory, there should be no change in the treatment of the right-moving fermions ψ^μ or S^a. So let us now explore the possibilities for introducing different λ^A sectors.

Boundary conditions for fermions

Suppose that n of the fermions λ^A satisfy the same boundary conditions, either A or P, and the other $(32-n)$ fermions independently satisfy A or P boundary conditions. If this results in a consistent theory, this would be expected to break the $SO(32)$ symmetry group to the subgroup $SO(n) \times SO(32-n)$.

There are four different sectors, denoted AA, AP, PA and PP, where the first label refers to the boundary condition of the first n components of λ^A and the second label refers to the boundary condition of the remaining $(32-n)$ components. As a result, there are four different choices for the normal-ordering constant $\tilde{a}$. Recall again that the normal-ordering constant for a boson is $+1/24$, while a periodic fermion has $-1/24$ and an antiperiodic fermion has $+1/48$. Taking this into account, the values for the normal-ordering constants are

$$\tilde{a}_{AA} = \frac{8}{24} + \frac{n}{48} + \frac{32-n}{48} = 1, \tag{7.37}$$

$$\tilde{a}_{AP} = \frac{8}{24} + \frac{n}{48} - \frac{32-n}{24} = \frac{n}{16} - 1, \tag{7.38}$$

$$\tilde{a}_{PA} = \frac{8}{24} - \frac{n}{24} + \frac{32-n}{48} = 1 - \frac{n}{16}, \tag{7.39}$$

$$\tilde{a}_{PP} = \frac{8}{24} - \frac{n}{24} - \frac{32-n}{24} = -1. \tag{7.40}$$

The sectors labeled AA and PP are the same as the ones labeled A and P in the $SO(32)$ theory discussed in the previous section, but the ones labeled AP and PA are new.

In each sector there is a level-matching condition of the form $N_R = N_L - \tilde{a}$. The eigenvalues of N_R are always integers, and the eigenvalues of N_L can be integers or half-integers. Therefore, there are no solutions unless $\tilde{a}$ is an integer or half-integer, which implies that n must be a multiple of 8. In this notation the $n = 32$ or 0 case corresponds to the theory constructed in the previous section, as stated above. The cases $n = 8$ and $n = 24$ would lead

7.2 Fermionic construction of the heterotic string

to a spectrum that is inconsistent due to gauge anomalies. Therefore, only the $n = 16$ case remains to be considered.

The $n = 16$ case

This is the case of most interest. It would naively appear to have an $SO(16) \times SO(16)$ gauge symmetry, but it turns out that each $SO(16)$ factor is enhanced to an E_8. The AP and PA sectors have $\tilde{a} = 0$ for $n = 16$. This value makes it possible to contribute states to the massless spectrum, a fact that proves to be very important in understanding the symmetry enhancement.

Let us now examine the massless spectrum in the $n = 16$ case. *The right-movers* have $N_R = 0$ (in the light-cone GS description) and contribute a vector supermultiplet, which should be tensored with the massless states of the left-moving sectors. *The left-movers* can have the boundary conditions:

- The PP sector does not contribute to the massless spectrum, as before.
- The AA sector, on the other hand, does contribute states with $N_L = 1$. These include states of the form

$$\tilde{\alpha}^i_{-1}|0\rangle_L \tag{7.41}$$

and

$$\lambda^A_{-1/2}\lambda^B_{-1/2}|0\rangle_L. \tag{7.42}$$

The eight states (7.41), when tensored with the right-moving vector multiplet give the $\mathcal{N} = 1$ gravity supermultiplet, just as in the case of the $SO(32)$ theory. The 496 states in (7.42) are exactly those that gave the $SO(32)$ gauge supermultiplets previously. They will do so again unless some of them are projected out. To see what is required, let us examine how they transform under $SO(16) \times SO(16)$:

$$\begin{array}{ll} (\mathbf{120}, \mathbf{1}) & \text{if} \quad A, B = 1, \ldots, 16, \\ (\mathbf{1}, \mathbf{120}) & \text{if} \quad A, B = 17, \ldots, 32, \\ (\mathbf{16}, \mathbf{16}) & \text{if} \quad A = 1, \ldots, 16, \ B = 17, \ldots, 32. \end{array} \tag{7.43}$$

Here $\mathbf{120} = 16 \times 15/2$ denotes the antisymmetric rank-two tensor in the adjoint representation of $SO(16)$ and $\mathbf{16}$ denotes the vector representation. Clearly, if we want to keep only the $SO(16) \times SO(16)$ gauge fields, then we need a rule that says that the $(\mathbf{120},\mathbf{1})$ and the $(\mathbf{1},\mathbf{120})$ multiplets are physical, while the $(\mathbf{16}, \mathbf{16})$ multiplet is unphysical. The way to do this is to require that the number of λ^A excitations involving the first set of 16 components and the second set of 16 components should each be even. This is more restrictive than just requiring that their sum is even, and it eliminates the (**16**, **16**) multiplet while retaining the other two multiplets. This rule, which is required to obtain the desired gauge symmetry, corresponds to using one and the same GSO projection for all sectors.

- Now consider the massless states in the PA and AP sectors. Since $\tilde{a} = 0$, states in the massless sector should have $N_L = 0$. The 16 components of λ^A with periodic boundary conditions have zero modes. Therefore, as usual, the Fock-space ground states should furnish a spinor representation of the corresponding $SO(16)$ group (more precisely, its $Spin(16)$ covering group). If we denote the two inequivalent spinor representations of $SO(16)$ by **128** and **128′**, then the possible additional massless states transform as

$$\text{PA} : (\mathbf{128}, \mathbf{1}) \oplus (\mathbf{128'}, \mathbf{1}),$$
$$\text{AP} : (\mathbf{1}, \mathbf{128}) \oplus (\mathbf{1}, \mathbf{128'}). \tag{7.44}$$

However, as in previous cases, not all of these states survive in the physical spectrum. There is a GSO-like projection that eliminates some of them.

The 32 fermions λ^A are divided into two sets of 16. As we already learned from studying the AA sector, separate projection conditions should be imposed for each of these two sets. Indeed, given the previous results, the rule is pretty clear. The analysis of the AA sector showed that for a set of 16 λ^A with A boundary conditions, there should be an even number of λ^A excitations. Now this needs to be supplemented with the corresponding rule for P boundary conditions.

The rules for the A and P sectors are the same as in the $SO(32)$ theory, except that they are applied to each set of 16 components separately. Thus, for example, if the first 16 λ^A have P boundary conditions, then a physical state is required to be an eigenstate, with eigenvalue equal to one, of the operator

$$(-1)^{F_1} = \bar{\lambda}_0^{(1)} (-1)^{\sum_{n=1}^{\infty} \sum_{A=1}^{16} \lambda_{-n}^A \lambda_n^A}, \tag{7.45}$$

where

$$\bar{\lambda}_0^{(1)} = \lambda_0^1 \lambda_0^2 \ldots \lambda_0^{16}. \tag{7.46}$$

The rule is the same for the second set of 16, of course. If they have P boundary conditions, then a physical state is required to be an eigenstate, with eigenvalue equal to one, of the operator

$$(-1)^{F_2} = \bar{\lambda}_0^{(2)} (-1)^{\sum_{n=1}^{\infty} \sum_{A=17}^{32} \lambda_{-n}^A \lambda_n^A}, \tag{7.47}$$

where
$$\bar{\lambda}_0^{(2)} = \lambda_0^{17}\lambda_0^{18}\ldots\lambda_0^{32}. \tag{7.48}$$

This rule eliminates one of the two spinors from each of the AP and PA sectors. Therefore, their surviving contribution to the massless spectrum is

$$(\mathbf{128},\mathbf{1}) \oplus (\mathbf{1},\mathbf{128}). \tag{7.49}$$

Each of the left-moving multiplets (7.49) is tensored with the right-moving vector multiplet and therefore contributes additional massless vectors. To understand what this means let us focus on the massless vector fields. The massless spectrum contains vector fields that transform as $(\mathbf{120},\mathbf{1}) + (\mathbf{128},\mathbf{1})$ as well as ones that transform as $(\mathbf{1},\mathbf{120}) + (\mathbf{1},\mathbf{128})$. The only way this can make sense is if these 248 states form the adjoint representation of a Lie group.

Here is where E_8 enters the picture. This Lie group is the largest of the five exceptional compact simple Lie groups in the Cartan classification. It has rank eight and dimension 248. Moreover, it contains an $SO(16)$ subgroup with respect to which the adjoint decomposes as $\mathbf{248} = \mathbf{120} + \mathbf{128}$. This is exactly the content that we found, so it is extremely plausible that the heterotic theory with the projections described here gives a consistent supersymmetric string theory in ten dimensions with $E_8 \times E_8$ gauge symmetry.

This suggests that there exists a consistent heterotic string theory with $E_8 \times E_8$ gauge symmetry. First indications appeared already from the anomaly analysis in Chapter 5, where this gauge group is one of the two possibilities that was singled out. The GSO-like projections introduced here are a straightforward generalization of those that gave the $SO(32)$ heterotic theory (as well as those of the RNS string), and they give precisely the necessary massless spectrum.

EXERCISES

EXERCISE 7.1
Consider left-moving currents

$$J^a(z) = \frac{1}{2}T^a_{AB}\lambda^A(z)\lambda^B(z),$$

where $\lambda^A(z)$ are free fermi fields that transform in a real representation R of a Lie group G. The representation matrices T^a satisfy the Lie algebra

$$[T^a, T^b] = i f^{abc} T^c.$$

Verify that the currents have the OPE

$$J^a(z) J^b(w) = \frac{k \delta^{ab}}{2(z-w)^2} + i \frac{f^{abc}}{z-w} J^c(w) + \ldots,$$

where the level k is given by

$$\mathrm{tr}(T^a T^b) = k \delta^{ab}.$$

This defines a level k Kac–Moody algebra of the type discussed in Section 3.1. Show that $k = 1$ in the special case of the vector representation of $SO(n)$.

SOLUTION

The free fermion fields satisfy the OPE

$$\lambda^A(z) \lambda^B(w) = \frac{\delta^{AB}}{z-w},$$

and as a result the leading term in the OPE is given by

$$\langle \frac{1}{2} T^a_{AB} \lambda^A(z) \lambda^B(z) \frac{1}{2} T^b_{CD} \lambda^C(w) \lambda^D(w) \rangle = \frac{1}{2} \frac{\mathrm{tr}(T^a T^b)}{(z-w)^2} = \frac{k \delta^{ab}}{2(z-w)^2}.$$

Note that the two contractions $(AC)(BD)$ and $(AD)(BC)$ have contributed equally due to the antisymmetry of the representation matrices and anticommutation of the fermi fields. The second term in the OPE works in a similar manner with four possible single contractions contributing. These combine in pairs to give commutators of the representation matrices, which are evaluated using the Lie algebra.

In the special case of the vector representation of $SO(n)$ there are n free fermi fields each of which contributes $1/2$ to the central charge giving a total of $n/2$. This should be compared with the general formula for the central charge at level k given in Section 3.1

$$c = \frac{k \dim G}{k + \tilde{h}_G}.$$

In the present case $\dim G = n(n-1)/2$ and $\tilde{h}_G = n - 2$. Thus, for $n > 2$, $c = n/2$ corresponds to $k = 1$. $\square$

EXERCISE **7.2**

Derive the mass formulas for states in the A and P sector, Eqs (7.27) and (7.28), respectively.

SOLUTION

Consider first the A sector. Because the right-moving sector is a superstring, from Chapter 4 we know that the mass formula is

$$\frac{1}{8}M^2 = N_R.$$

For the left-moving sector we have $\tilde{a}_A = 1$, which leads to the mass-shell condition

$$\tilde{L}_0 - 1 = 0 \rightarrow \frac{1}{8}M^2 = N_L - 1.$$

Thus altogether

$$\frac{1}{8}M^2 = N_R = N_L - 1.$$

In the P sector the left-movers have $\tilde{a}_P = -1$, and so the same reasoning gives

$$\frac{1}{8}M^2 = N_R = N_L + 1.$$

$\Box$

7.3 Toroidal compactification

Chapter 6 examined compactification on a circle in considerable detail. It was shown that the consequences for string theory are much more than one might expect based on classical geometric reasoning. One important lesson was the existence of T-duality, which relates radius R to radius α'/R. Another lesson was the existence of D-branes associated with open strings, which emerge after T-duality.

What is demonstrated here is that the generalization to compactification on an n-dimensional torus T^n adds additional interesting structure. The T-duality group becomes enlarged to an infinite discrete group, and there are interesting new possibilities for realizing nonabelian gauge symmetry. The details depend on which string theory one considers.

The bosonic string

Let us consider closed bosonic strings on a toroidally compactified space-time. Specifically, the space-time manifold is described by the metric

$$ds^2 = \sum_{\mu,\nu=0}^{d-1} \eta_{\mu\nu} dX^\mu dX^\nu + \sum_{I,J=1}^{n} G_{IJ} dY^I dY^J, \qquad (7.50)$$

with $d + n = 26$. Here the first term describes flat Minkowski space-time parametrized by coordinates X^μ and the second term describes the "internal" torus T^n with coordinates Y^I, each of which has period 2π.

The physical sizes and angles that characterize the T^n can be encoded in the constant internal metric G_{IJ}. For example, in the special case of a rectangular torus the n internal circles are all perpendicular and the internal metric is diagonal resulting in

$$G_{IJ} = R_I^2 \delta_{IJ}, \qquad (7.51)$$

where R_I is the radius of the Y^I circle. A more general internal metric with off-diagonal elements would describe a torus with nonorthogonal circles.

A closed bosonic string is described by the embedding maps $X^\mu(\sigma, \tau)$ and $Y^I(\sigma, \tau)$, where $0 \leq \sigma \leq \pi$. The fact that the string is closed implies

$$\begin{aligned} X^\mu(\sigma + \pi, \tau) &= X^\mu(\sigma, \tau), \\ Y^I(\sigma + \pi, \tau) &= Y^I(\sigma, \tau) + 2\pi W^I \quad \text{with} \quad W^I \in \mathbb{Z}. \end{aligned} \qquad (7.52)$$

Here W^I are the winding numbers which give the number of times (and direction) that the string winds around each of the cycles of the torus.

Mode expansions

The mode expansion for the external and internal components of X is a slight generalization of the expansion for a string compactified on a circle in Chapter 6. The mode expansions (for $l_s = 1$) for the noncompact coordinates take the form

$$\begin{aligned} X^\mu(\sigma, \tau) &= X_L^\mu(\tau + \sigma) + X_R^\mu(\tau - \sigma), \\ X_L^\mu(\tau + \sigma) &= \tfrac{1}{2} x^\mu + p_L^\mu(\tau + \sigma) + \tfrac{i}{2} \sum_{n \neq 0} \tfrac{1}{n} \tilde{\alpha}_n^\mu e^{-2in(\tau+\sigma)}, \\ X_R^\mu(\tau - \sigma) &= \tfrac{1}{2} x^\mu + p_R^\mu(\tau - \sigma) + \tfrac{i}{2} \sum_{n \neq 0} \tfrac{1}{n} \alpha_n^\mu e^{-2in(\tau-\sigma)}, \end{aligned} \qquad (7.53)$$

where

$$p_L^\mu = p_R^\mu = \frac{1}{2} p^\mu. \qquad (7.54)$$

7.3 Toroidal compactification

The compact coordinates $Y^I(\sigma, \tau)$ have analogous expansions given by

$$Y^I(\sigma, \tau) = Y_L^I(\tau + \sigma) + Y_R^I(\tau - \sigma),$$

$$Y_L^I(\tau + \sigma) = \tfrac{1}{2} y^I + p_L^I(\tau + \sigma) + \tfrac{i}{2} \sum_{n \neq 0} \tfrac{1}{n} \tilde{\alpha}_n^I e^{-2in(\tau+\sigma)}, \qquad (7.55)$$

$$Y_R^I(\tau - \sigma) = \tfrac{1}{2} y^I + p_R^I(\tau - \sigma) + \tfrac{i}{2} \sum_{n \neq 0} \tfrac{1}{n} \alpha_n^I e^{-2in(\tau-\sigma)}.$$

Notice that in these formulas p_L^I and p_R^I do not have to be equal. Thus the first terms in the expansion of Y^I are

$$Y^I(\sigma, \tau) = Y_L^I(\tau+\sigma) + Y_R^I(\tau-\sigma) = y^I + (p_L^I + p_R^I)\tau + (p_L^I - p_R^I)\sigma + \ldots, \qquad (7.56)$$

and the second equation in (7.52) implies that the difference between p_L^I and p_R^I is an integer given by the winding number

$$p_L^I - p_R^I = 2W^I \qquad \text{with} \qquad W^I \in \mathbb{Z}. \qquad (7.57)$$

Moreover, the sum of p_L^I and p_R^I, the momenta along the circle directions, must be quantized, so that e^{ipy} is single-valued. In the simplest case of a rectangular torus and no background B fields, this implies that

$$p_L^I + p_R^I = K_I \qquad \text{with} \qquad K_I \in \mathbb{Z}. \qquad (7.58)$$

These quantized internal momenta correspond to the Kaluza–Klein excitations.

Mode expansions with constant background fields

The above results hold for the case of no background B fields and a diagonal internal metric. Now consider turning on constant background values for the antisymmetric two-form B_{IJ} and the internal metric G_{IJ}. To derive the expressions for the momenta in terms of the winding numbers as well as Kaluza–Klein quantum numbers, the relevant part of the world-sheet action for strings in this background needs to be taken into account

$$S = -\frac{1}{2\pi} \int d^2\sigma \left(G_{IJ} \eta^{\alpha\beta} - B_{IJ} \varepsilon^{\alpha\beta} \right) \partial_\alpha Y^I \partial_\beta Y^J. \qquad (7.59)$$

This action gives the canonical momentum density

$$p_I = \frac{\delta S}{\delta \dot{Y}^I} = \frac{1}{\pi} \left(G_{IJ} \dot{Y}^J + B_{IJ} Y'^J \right). \qquad (7.60)$$

This momentum density integrates to give the total momentum vector

K_I, which is an integer, because Y^I are periodic. Using the mode expansion for the internal coordinates appearing in Eq. (7.56), one obtains

$$K_I = \int_0^\pi p_I d\sigma = G_{IJ}(p_L^J + p_R^J) + B_{IJ}(p_L^J - p_R^J) \quad \text{with} \quad K_I \in \mathbb{Z}. \quad (7.61)$$

This is the generalization of Eq. (7.58). Equations (7.57) and (7.61) can be solved for the left-moving and right-moving momenta resulting in

$$\begin{aligned} p_L^I &= W^I + G^{IJ}\left(\tfrac{1}{2}K_J - B_{JK}W^K\right), \\ p_R^I &= -W^I + G^{IJ}\left(\tfrac{1}{2}K_J - B_{JK}W^K\right), \end{aligned} \quad (7.62)$$

where, as usual, G with superscript indices denotes the inverse matrix.

The mass spectrum and level-matching condition

The starting point for determining the mass spectrum and level-matching condition of the toroidally compactified bosonic string is again the physical-state conditions

$$(L_0 - 1)|\Phi\rangle = \left(\tilde{L}_0 - 1\right)|\Phi\rangle = 0, \quad (7.63)$$

which now take the form

$$\frac{1}{8}M^2 = \frac{1}{2}G_{IJ}p_L^I p_L^J + N_L - 1 = \frac{1}{2}G_{IJ}p_R^I p_R^J + N_R - 1. \quad (7.64)$$

Here the number operators are the usual expressions (independent of the background fields)

$$N_R = \sum_{m=1}^\infty \alpha_{-m} \cdot \alpha_m \quad \text{and} \quad N_L = \sum_{m=1}^\infty \tilde{\alpha}_{-m} \cdot \tilde{\alpha}_m. \quad (7.65)$$

The difference of the two equations in (7.64) gives the *level-matching condition*

$$N_R - N_L = \frac{1}{2}G_{IJ}(p_L^I p_L^J - p_R^I p_R^J) = W^I K_I. \quad (7.66)$$

Taking the sum of the same two equations, the mass operator becomes

$$M^2 = M_0^2 + 4(N_R + N_L - 2) \quad \text{with} \quad M_0^2 = 2G_{IJ}(p_L^I p_L^J + p_R^I p_R^J). \quad (7.67)$$

A convenient way of rewriting M_0^2, which is useful for exhibiting the symmetries of the spectrum, is obtained by substituting Eqs (7.62) into Eq. (7.67). Suppressing indices, this gives

$$\frac{1}{2}M_0^2 = (W \ K)\mathcal{G}^{-1}\begin{pmatrix} W \\ K \end{pmatrix}, \quad (7.68)$$

where[4]

$$\mathcal{G}^{-1} = \begin{pmatrix} 2(G - BG^{-1}B) & BG^{-1} \\ -G^{-1}B & \frac{1}{2}G^{-1} \end{pmatrix}, \qquad (7.69)$$

or the inverse

$$\mathcal{G} = \begin{pmatrix} \frac{1}{2}G^{-1} & -G^{-1}B \\ BG^{-1} & 2(G - BG^{-1}B) \end{pmatrix}. \qquad (7.70)$$

Note that these are $2n \times 2n$ matrices written in terms of $n \times n$ blocks.

The $O(n, n; \mathbb{Z})$ duality group

Compactification of the bosonic string on tori T^n has a beautiful symmetry, called $O(n, n; \mathbb{Z})$, which generalizes the T-duality symmetry of circle compactifications. This $O(n, n; \mathbb{Z})$ symmetry of the spectrum is best described in terms of the matrix $\mathcal{G}$. Indeed, for a nonorthogonal torus the $R \to 1/R$ duality of the circle compactification generalizes to the inversion symmetry

$$W^I \leftrightarrow K_I, \quad \mathcal{G} \leftrightarrow \mathcal{G}^{-1}. \qquad (7.71)$$

This symmetry becomes clear from the expressions (7.68) – (7.70). Additional discrete shift symmetries are given by

$$B_{IJ} \to B_{IJ} + \frac{1}{2} N_{IJ} \quad \text{with} \quad W^I \to W^I, \; K_I \to K_I + N_{IJ} W^J, \qquad (7.72)$$

where N_{IJ} is an antisymmetric matrix of integers. These transformations are symmetries, because they leave p_L^I and p_R^I in (7.62) unchanged. Since N is antisymmetric these symmetries only appear when $n > 1$.

Altogether, the inversion symmetry and the shift symmetries generate the infinite discrete group $O(n, n; \mathbb{Z})$. By definition, the group $O(n, n; \mathbb{R})$ consists of matrices A satisfying

$$A^T \begin{pmatrix} 0 & 1_n \\ 1_n & 0 \end{pmatrix} A = \begin{pmatrix} 0 & 1_n \\ 1_n & 0 \end{pmatrix}, \qquad (7.73)$$

where 1_n denotes an $n \times n$ unit matrix. The group $O(n, n; \mathbb{Z})$ is the subgroup of $O(n, n; \mathbb{R})$ consisting of those matrices all of whose matrix elements are integers. Note that if $\mathcal{G}$ is integral, then $\mathcal{G}^{-1}$ is automatically integral, as well. The group $O(n, n; \mathbb{Z})$ is an infinite group (for $n > 1$). This group is generated by the geometric duality subgroup $SL(n, \mathbb{Z})$, which just corresponds to a change of basis for the defining periods of the torus, and the nongeometric transformation $\mathcal{G} \leftrightarrow \mathcal{G}^{-1}$.

A convenient way of rewriting the above symmetry transformations is the

---

[4] The various factors of 2 and 1/2 in these formulas are a consequence of the choice $\alpha' = 1/2$. They could be eliminated by redefining G and B by a factor of 2.

following. Under the T-duality group $O(n,n;\mathbb{Z})$, the symmetry is realized as

$$\mathcal{G} \to A\mathcal{G}A^T \quad \text{and} \quad \begin{pmatrix} W \\ K \end{pmatrix} \to \begin{pmatrix} W' \\ K' \end{pmatrix} = A \begin{pmatrix} W \\ K \end{pmatrix}. \tag{7.74}$$

This preserves the result for the mass spectrum in Eqs (7.67) and (7.68) as well as the level-matching condition in Eq. (7.66). The best way to see this is to rewrite the level-matching condition in the form

$$W^I K_I = \frac{1}{2} (W \ K) \begin{pmatrix} 0 & 1_n \\ 1_n & 0 \end{pmatrix} \begin{pmatrix} W \\ K \end{pmatrix} \tag{7.75}$$

and use Eq. (7.73). In terms of $O(n,n;\mathbb{Z})$ transformations, the inversion symmetry corresponds to the matrix

$$\text{inversion}: \quad A = \begin{pmatrix} 0 & 1_n \\ 1_n & 0 \end{pmatrix}, \tag{7.76}$$

and the shift symmetry corresponds to the matrix

$$\text{shift}: \quad A = \begin{pmatrix} 1_n & 0 \\ N_{IJ} & 1_n \end{pmatrix}. \tag{7.77}$$

The claim is that an arbitrary $O(n,n;\mathbb{Z})$ transformation can be represented by a succession of these two types of transformations. This is the T-duality group for the toroidally compactified bosonic string theory.

The moduli space

The compactification moduli space is parametrized by the n^2 parameters G_{IJ}, B_{IJ}. The sum $G_{IJ} + B_{IJ}$ is an $n \times n$ real matrix. The only restriction on this matrix is that its symmetric part is positive definite. This space of matrices can be represented as a homogeneous space, in other words as a coset space G/H. The appropriate choice is (see Exercise 7.5 for more details)

$$\mathcal{M}^0_{n,n} = O(n,n;\mathbb{R})/[O(n;\mathbb{R}) \times O(n;\mathbb{R})]. \tag{7.78}$$

This is not the whole story, however. Points in this moduli space that are related by an $O(n,n;\mathbb{Z})$ T-duality transformation are identified as physically equivalent.[5] Thus the physical moduli space is

$$\mathcal{M}_{n,n} = \mathcal{M}^0_{n,n}/O(n,n;\mathbb{Z}). \tag{7.79}$$

Since $\mathcal{M}^0_{n,n}$ is a homogeneous space, $\mathcal{M}^0_{n,n}$ is a smooth manifold of dimension n^2. On the other hand, $\mathcal{M}_{n,n}$ has singularities (or cusps) corresponding

---
[5] This is an example of a discrete gauge symmetry.

to fixed points of $O(n,n;\mathbb{Z})$ transformations. At these special values of (G_{IJ}, B_{IJ}) the spectrum has additional massless gauge bosons, and there is unbroken nonabelian gauge symmetry.

Enhanced gauge symmetry

Nonabelian gauge symmetries can arise from toroidal compactifications. From a Kaluza–Klein viewpoint this is very surprising. In a point-particle theory the gauge symmetries one would expect are those that correspond to isometries of the compact dimensions. The isometry of T^n is simply $U(1)^n$ and this is abelian. So the feature in question is a purely stringy one involving winding modes in addition to Kaluza–Klein excitations.

This section considers the bosonic string theory compactified on a T^n as before. The extension to the heterotic string is given in the following section. The basic idea in both cases is that for generic values of the moduli the gauge symmetry is abelian. In the case of the bosonic string theory it is actually $U(1)^{2n}$, so there are $2n$ massless $U(1)$ gauge bosons in the spectrum. Half of them arise from reduction of the 26-dimensional metric (namely, components of the form $g_{\mu I}$) and half of them arise from reduction of the 26-dimensional two-form (namely, components of the form $B_{\mu I}$).

At specific loci in the moduli space there appear additional massless particles including massless gauge bosons. When this happens there is symmetry enhancement resulting in nonabelian gauge symmetry. For example, in the $n = 1$ case, the symmetry is enhanced from $U(1) \times U(1)$ to $SU(2) \times SU(2)$ at the self-dual radius. Let us explore how this happens.

The self-dual radius

In order to consider enhanced gauge symmetry of the bosonic string theory compactified on a circle of radius R, let us assume that the coordinate X^{25} is compact and the remaining coordinates are noncompact. The spectrum is described by the mass formula

$$M^2 = \frac{K^2}{R^2} + 4R^2 W^2 + 4(N_{\rm L} + N_{\rm R} - 2), \tag{7.80}$$

as well as the level-matching condition

$$N_{\rm R} - N_{\rm L} = KW. \tag{7.81}$$

As before, W describes the number of times the string winds around the circle. Let us now explore some of the low-mass states in the spectrum of this theory.

- The Fock-space ground state, with $K = W = 0$, gives the tachyon with $M^2 = -8$, as usual.
- At the massless level with $N_R = N_L = 1$ and $K = W = 0$ there are the 25-dimensional graviton, antisymmetric tensor and dilaton, represented by

$$\alpha^\mu_{-1} \tilde\alpha^\nu_{-1} |0\rangle, \tag{7.82}$$

with the oscillators in the 25 noncompact directions. Here and in the following $\mu, \nu = 0, \ldots, 24$. There are also two massless vector states given by

$$|V_1^\mu\rangle = \alpha^\mu_{-1} \tilde\alpha_{-1} |0\rangle, \tag{7.83}$$

$$|V_2^\mu\rangle = \alpha_{-1} \tilde\alpha^\mu_{-1} |0\rangle, \tag{7.84}$$

where $\tilde\alpha_{-1}$, without space-time index, denotes the oscillator in the direction 25. As usual for a vector particle, these states satisfy the L_1 and $\tilde L_1$ Virasoro constraints provided that their polarization vectors are orthogonal to their momenta. These are Kaluza–Klein states that arise from the 26-dimensional graviton and antisymmetric tensor. These two fields give a $U(1)_L \times U(1)_R$ gauge symmetry. The symmetric linear combination, which comes from the graviton, couples electrically to the Kaluza–Klein charge K. Similarly, the antisymmetric combination, which comes from the B field, couples electrically to winding number W. The state

$$|\phi\rangle = \alpha_{-1} \tilde\alpha_{-1} |0\rangle \tag{7.85}$$

describes a massless scalar field.
- Let us now consider states with $W = K = \pm 1$. The level-matching condition in this case is $N_R = N_L + 1$. Choosing the first instance, namely $N_L = 0$ and $N_R = 1$, there are two vector states

$$|V^\mu_{++}\rangle = \alpha^\mu_{-1}|+1, +1\rangle \quad \text{and} \quad |V^\mu_{--}\rangle = \alpha^\mu_{-1}|-1, -1\rangle, \tag{7.86}$$

where we have introduced the notation $|K, W\rangle$. In addition there are two scalars

$$|\phi_{++}\rangle = \alpha_{-1}|+1, +1\rangle \quad \text{and} \quad |\phi_{--}\rangle = \alpha_{-1}|-1, -1\rangle. \tag{7.87}$$

The mass of these states depends on the radius of the circle and is given by

$$M^2 = \frac{1}{R^2} + 4R^2 - 4 = \left(\frac{1}{R} - 2R\right)^2. \tag{7.88}$$

Note that this vanishes for $R^2 = 1/2 = \alpha'$, which is precisely the self-dual radius of the T-duality transformation $R \to \alpha'/R$.

In the same way we can consider the states which have $K = -W = \pm 1$. Then there are again two vectors

$$|V^\mu_{+-}\rangle = \tilde{\alpha}^\mu_{-1}|+1,-1\rangle \quad \text{and} \quad |V^\mu_{-+}\rangle = \tilde{\alpha}^\mu_{-1}|-1,+1\rangle, \tag{7.89}$$

and two scalars

$$|\phi_{+-}\rangle = \tilde{\alpha}_{-1}|+1,-1\rangle \quad \text{and} \quad |\phi_{-+}\rangle = \tilde{\alpha}_{-1}|-1,+1\rangle. \tag{7.90}$$

The mass of these states is also given by Eq. (7.88). Altogether, at the self-dual radius there are four additional massless vectors in the spectrum in addition to the two that are present for any radius. The interpretation is that there is enhanced gauge symmetry for this particular value of the radius. The gauge group $U(1) \times U(1)$, which is present in general, is a subgroup of the enhanced symmetry group, which in this case is $SU(2) \times SU(2)$. This is explored in Exercise 7.3. The three vectors that involve an α^μ_{-1} excitation are associated with a right-moving $SU(2)$ on the string world sheet. Similarly, the other three involve a $\tilde{\alpha}^\mu_{-1}$ excitation and are associated with a left-moving $SU(2)$ on the string world sheet. The case of $SU(3) \times SU(3)$ is studied in Exercise 7.8.

This enhancement of gauge symmetry at the self-dual radius is a "stringy" effect. For other values of the radius the gauge symmetry is broken to $U(1)_\text{L} \times U(1)_\text{R}$. The four gauge bosons $|V^\mu_{\pm\pm}\rangle$ eat the four scalars $|\phi_{\pm\pm}\rangle$ as part of a stringy Higgs effect. On the other hand, the $U(1)_\text{L} \times U(1)_\text{R}$ gauge bosons, as well as the associated scalar $|\phi\rangle$, remain massless for all values of the radius. This neutral scalar has a flat potential (meaning that the potential function does not depend on it), which corresponds to the freedom of choosing the radius of the circle to be any value with no cost in energy. Altogether, the spectrum of the bosonic string compactified on a circle is characterized by a single parameter R, called the *modulus* of the compactification. It is the radius of the circle, whose value is determined by the vacuum expectation value of the scalar field $|\phi\rangle$.

As was explained in the previous section, the T-duality symmetry of the bosonic string theory requires that the moduli space of the theory compactified on a circle be defined as the quotient space of the positive line $R > 0$ modulo the identification of R and $1/(2R)$. Therefore, the point of enhanced gauge symmetry, which is the fixed point of the T-duality transformation, is also the singular point of the moduli space.

In the case of type II superstrings, compactification on a T^n again gives rise to $2n$ abelian gauge fields. However, unlike the bosonic string theory, there is no possibility of symmetry enhancement. One way of understanding this is to note that all $2n$ of the gauge fields belong to the supergravity multiplet in $10-n$ dimensions, and this cannot be extended to include additional gauge fields. Another way of understanding this is to observe that symmetry enhancement in the bosonic string utilized winding and Kaluza–Klein excitations so that $N_L = N_R \pm 1$. The same relations in the case of type II superstrings imply that the mass is strictly positive.

Toroidal compactification of the heterotic string is studied in Section 7.4. It is shown that compactification to $10-n$ dimensions gives n right-moving $U(1)$ currents and $16+n$ left-moving $U(1)$ currents. Moreover, there can be no symmetry enhancement for the right-moving current algebra, but there can be symmetry enhancement for the left-moving current algebra. In fact, in the special case $n=0$, the $U(1)^{16}$ is necessarily enhanced, to either $SO(32)$ or $E_8 \times E_8$.

One-loop modular invariance

Chapter 3 showed that one-loop amplitudes are given by integrals over the moduli space of genus-one (toroidal) Riemann surfaces. This space is parametrized by a modular parameter τ whose imaginary part is positive. An important consistency requirement is that the integral should have modular invariance. In other words, it should be of the form

$$\int_{\mathcal{F}} \frac{d^2\tau}{(\mathrm{Im}\,\tau)^2} I(\tau, \ldots), \qquad (7.91)$$

where $\mathcal{F}$ denotes a fundamental region of the modular group. Modular invariance requires that I is invariant under the $PSL(2,\mathbb{Z})$ modular transformations

$$\tau \to \tau' = \frac{a\tau + b}{c\tau + d}, \qquad (7.92)$$

where $a, b, c, d \in \mathbb{Z}$ and $ad - bc = 1$, since the measure $d^2\tau/(\mathrm{Im}\,\tau)^2$ is invariant. Two examples of modular transformations are shown in Fig. 7.1. This ensures that it is equivalent to define the integral over the region $\mathcal{F}$ or any of its images under a modular transformation. In other words, the value of the integral is independent of the particular choice of a fundamental region. This property is satisfied by the bosonic string theory in 26-dimensional Minkowski space-time. Accepting that result, we propose to examine here whether it continues to hold when the theory is compactified on T^n with arbitrary background fields G_{IJ} and B_{IJ}.

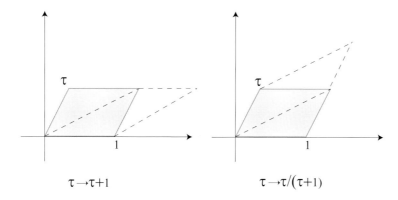

Fig. 7.1. Two examples of modular transformations of the torus. The right-hand dotted parallelogram has been rotated for clarity of presentation.

In the computation of the amplitude the key factor that needs to be considered is the partition function

$$\text{Tr}\left(q^{L_0}\bar{q}^{\tilde{L}_0}\right), \tag{7.93}$$

where

$$q = e^{2\pi i\tau}. \tag{7.94}$$

Toroidal compactification only changes the contribution of the zero modes to the partition function, which becomes

$$\text{Tr}\left(q^{\frac{1}{2}p_R^2}\bar{q}^{\frac{1}{2}p_L^2}\right) = \sum_{W^I, K_I} e^{\pi i\tau_1(p_R^2 - p_L^2)} e^{-\pi\tau_2(p_L^2 + p_R^2)}, \tag{7.95}$$

where $\tau = \tau_1 + i\tau_2$, and p_L and p_R are defined in Eqs (7.62). This factor replaces the momentum integration

$$\int \exp(-\pi\tau_2 p^2) d^n p = (\tau_2)^{-n/2}, \tag{7.96}$$

in the noncompact case. Therefore, to establish modular invariance of the toroidally compactified bosonic string, it is necessary to prove modular invariance of

$$F(\tau; G, B) = (\tau_2)^{n/2}\text{Tr}\left(q^{\frac{1}{2}p_R^2}\bar{q}^{\frac{1}{2}p_L^2}\right). \tag{7.97}$$

Modular invariance of $F(\tau; G, B)$

Modular invariance of $F(\tau; G, B)$ is verified by checking that it is invariant under the two transformations $\tau \to \tau + 1$ and $\tau \to -1/\tau$.

- Invariance under $\tau \to \tau + 1$ is verified using Eq. (7.95), since

$$p_R^2 - p_L^2 = 2(N_L - N_R) = -2W^I K_I \qquad (7.98)$$

is an even integer.

- Invariance under $\tau \to -1/\tau$ is the next step to check. The key to this is to make use of the Poisson resummation formula which states that if A is a positive definite $m \times m$ symmetric matrix and

$$f(A) = \sum_{\{M\}} \exp\left(-\pi M^T A M\right), \qquad (7.99)$$

where M represents a vector made of m integers $M_1, M_2, \ldots, M_m$, each of which is summed from $-\infty$ to $+\infty$, then

$$f(A) = \frac{1}{\sqrt{\det A}} f(A^{-1}). \qquad (7.100)$$

The derivation of the Poisson resummation formula is very beautiful and relatively easy to prove, so the proof is given in the appendix at the end of this chapter.

Now let us apply the Poisson resummation formula to our problem. The function $F(\tau, G, B)$ takes the form

$$F(\tau, G, B) = (\tau_2)^{n/2} f(A), \qquad (7.101)$$

where A is the $2n \times 2n$ matrix

$$A = \tau_2 \begin{pmatrix} 2(G - BG^{-1}B) & BG^{-1} \\ -G^{-1}B & \frac{1}{2}G^{-1} \end{pmatrix} + i\tau_1 \begin{pmatrix} 0 & 1_n \\ 1_n & 0 \end{pmatrix}. \qquad (7.102)$$

It is now a straightforward calculation, described in Exercise 7.4, to compute the determinant and the inverse of this matrix. The results are

$$\det A = |\tau|^{2n} \qquad (7.103)$$

and

$$A^{-1} = \tilde{\tau}_2 \begin{pmatrix} \frac{1}{2}G^{-1} & -G^{-1}B \\ BG^{-1} & 2(G - BG^{-1}B) \end{pmatrix} + i\tilde{\tau}_1 \begin{pmatrix} 0 & 1_n \\ 1_n & 0 \end{pmatrix}, \qquad (7.104)$$

where

$$\tilde{\tau} = -\frac{1}{\tau} = \frac{-\tau_1 + i\tau_2}{|\tau|^2}. \qquad (7.105)$$

7.3 Toroidal compactification

Interchanging the first n rows and columns with the second n rows and columns brings $A(\tau)^{-1}$ into agreement with $A(\tilde{\tau})$. One deduces that

$$F(\tau; G, B) = F\left(-\frac{1}{\tau}; G, B\right), \qquad (7.106)$$

which establishes one-loop modular invariance of the toroidally compactified theory.

Even self-dual lattices

The reason that the proof of modular invariance, given above, was successful can be traced to the fact that the moduli space $\mathcal{M}$ can be regarded as parametrizing the space of even self-dual lattices $\Gamma_{n,n}$ of signature (n,n). In order to explain what this means, a few basic facts about lattices are reviewed in the next section.

A brief introduction to lattices

In general, a lattice is defined as a set of points in a vector space V, which we take to be $\mathbb{R}^{(p,q)}$, (that is, $\mathbb{R}^{p+q}$ with Lorentzian inner product) of the form

$$\Lambda = \left\{\sum_{i=1}^{m} n_i e_i, n_i \in \mathbb{Z}\right\}, \qquad (7.107)$$

where $m = p + q$ and $\{e_i\}$ are the basis vectors of Λ. The metric on the lattice is defined by

$$g_{ij} = e_i \cdot e_j. \qquad (7.108)$$

This metric contains the information about the lengths of the basis vectors and their angles.

The *dual lattice* is defined by

$$\Lambda^\star = \{w \in V \text{ such that } w \cdot v \in \mathbb{Z}, \text{ for all } v \in \Lambda\}. \qquad (7.109)$$

This is illustrated in Fig. 7.2. If we call a set of basis vectors of the dual lattice $\{e_i^\star\}$, then the dual lattice is given by

$$\Lambda^\star = \{\sum_{i=1}^{m} n_i e_i^\star, n_i \in \mathbb{Z}\}. \qquad (7.110)$$

The basis vectors of the dual lattice can be chosen to satisfy

$$e_i^\star \cdot e_j = \delta_{ij}. \qquad (7.111)$$

278   The heterotic string

The metric on the dual lattice is therefore given by

$$g^\star_{ij} = e^\star_i \cdot e^\star_j, \qquad (7.112)$$

which is the inverse of g_{ij}. A lattice is called

- *unimodular* if $\mathrm{Vol}(\Lambda) = \sqrt{|\det g|} = 1$,
- *integral* if $v \cdot w \in \mathbb{Z}$ for all $v, w \in \Lambda$,
- *even* if Λ is integral and v^2 is even for all $v \in \Lambda$,
- *self-dual* if $\Lambda = \Lambda^\star$.

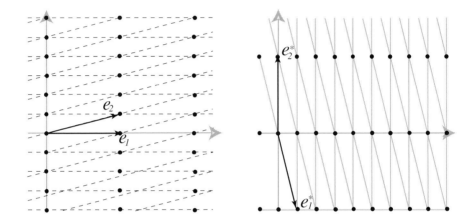

Fig. 7.2. A lattice and the dual lattice.

Lattices and toroidal compactifications

The momenta $p^I = (p^I_\mathrm{L}, p^I_\mathrm{R})$ in toroidal compactifications of the bosonic string live on a lattice $\Gamma_{n,n}$ with Lorentzian signature which turns out to be even and self-dual:

- The signature of the lattice is $((+1)^n, (-1)^n)$ since the length-squared of a $2n$-component vector of the form $p = (p^I_\mathrm{L}, p^I_\mathrm{R})$ is defined by

$$p^2 = p^2_\mathrm{L} - p^2_\mathrm{R}. \qquad (7.113)$$

Here the individual squares p^2_L and p^2_R are computed using the metric G_{IJ}.[6]

---

[6] This lattice can be represented by a lattice with metric $\eta^{ab} = ((+1)^n, (-1)^n)$, which was discussed in the previous section, by writing $G_{IJ} = e^a_I e^b_J \eta^{ab}$ and contracting the momenta with the e^a_I.

- Using Eq. (7.66), one obtains

$$p^2 = 2W^I K_I \in 2\mathbb{Z}. \tag{7.114}$$

This is an even integer. Thus, for fixed values of the moduli, the set of all possible vectors p forms a lattice in $2n$ dimensions all of whose sites have even length-squared. This is the condition for an *even lattice*. It ensures that the level-matching condition is satisfied.

- Moreover, the lattice is self-dual, since the lattice generated by $p = (p_L^I, p_R^I)$ is equivalent to the lattice generated by the winding and the Kaluza–Klein excitation numbers. Indeed the mass formula can be rewritten in the form

$$M^2 = 2(W^I \ K_I)\mathcal{G}^{-1}\begin{pmatrix} W^I \\ K_I \end{pmatrix} + 4(N_R + N_L - 2). \tag{7.115}$$

We saw that this formula has a duality symmetry that exchanges

$$W^I \leftrightarrow K_I, \quad \mathcal{G} \leftrightarrow \mathcal{G}^{-1}. \tag{7.116}$$

So this duality inverts the metric on the lattice, and as a result the lattice is *self-dual*. The self-duality condition ensures that the invariance under the modular transformation $\mathcal{G} \to \mathcal{G}^{-1}$, which was discussed in the previous section, is satisfied.

The lattice defined here is an even and self-dual lattice of signature (n, n). One can ask the following mathematical question: For what signatures (n_1, n_2) do even self-dual lattices exist? The answer is that n_1 and n_2 must differ by a multiple of 8. This result is relevant to the bosonic formulation of the heterotic string, where the left-moving dimension is 26 and the right-moving dimension is 10, so that their difference is 16.

Type II superstrings

There is a very similar construction for type II superstrings. In this case, the geometry is $\mathbb{R}^d \times T^n$ with $d + n = 10$. By the same reasoning as in the bosonic string, one finds an $O(n, n; \mathbb{Z})$ duality group. However, there is one new issue, which was already encountered in Chapter 6. This is that the inversion element of the duality group also reverses the relative chirality of the two fermionic coordinates, which are denoted θ in the GS formalism. In particular, recall that θ^1 and θ^2 (corresponding to left- and right-moving modes) have opposite chirality for the type IIA theory and the same chirality for the type IIB theory. In the case of circle compactification, we found that the moduli space is characterized entirely by the radius R of the circle. However, all radii $R > 0$ are allowed, in contrast to the bosonic string theory where R and $\widetilde{R} = 1/(2R)$ are equivalent.

As in the case of the bosonic string theory, one can form a moduli space $\mathcal{M}_{n,n}$ of inequivalent toroidal compactifications of type II superstrings as a quotient of $\mathcal{M}^0_{n,n}$ by a suitable duality group. The appropriate duality group is smaller in this case than in the bosonic theory. It is only a subgroup of the $O(n,n;\mathbb{Z})$ transformation group. Specifically, it is the subgroup that preserves the chirality of the spinors. This reduces the group to $SO(n,n;\mathbb{Z})$.

To summarize, one could say that the distinction between type IIA and type IIB dissolves after T^n compactification, and there is a single moduli space for the pair constructed in the way indicated here, but this moduli space is twice as large as in the case of the bosonic string theory. Chapter 8 shows that, when other dualities are taken into account, the duality group $SO(n,n;\mathbb{Z})$ is extended to $E_{n+1}(\mathbb{Z})$, which is a discrete subgroup of a noncompact exceptional group.

EXERCISES

EXERCISE 7.3
Consider the bosonic string theory compactified on a circle of radius $R = \sqrt{\alpha'}$. Verify that there is $SU(2) \times SU(2)$ gauge symmetry by constructing the conserved currents. Show that the modes of the currents satisfy a level-one Kac–Moody algebra.

SOLUTION

To do this let us focus on the holomorphic right-moving currents, since the antiholomorphic left-moving currents work in an identical fashion. Let us define
$$J^\pm(z) = e^{\pm 2iX^{25}(z)/\sqrt{\alpha'}}$$
and
$$J^3(z) = i\sqrt{2/\alpha'}\,\partial X^{25}(z).$$
The coefficients in the exponent have been chosen to ensure that $J^\pm(z)$ have conformal dimension $h = 1$. These currents are single valued at the self-dual radius $R = \sqrt{\alpha'}$, because $X^{25}(z)$ contains the zero mode $\frac{1}{2}x^{25}$. Note that in the text we have been setting $\alpha' = 1/2$.

Now one can compute the OPEs of these currents using the rules discussed in Chapter 3. Defining $J^\pm(z) = (J^1(z) \pm iJ^2(z))/\sqrt{2}$, one obtains

$$J^i(z)J^j(w) \sim \frac{\delta^{ij}}{(z-w)^2} + i\varepsilon^{ijk}\frac{J^k(w)}{z-w} + \ldots$$

Defining the modes by

$$J^i(z) = \sum_{n\in\mathbb{Z}} J^i_n z^{-n-1} \quad \text{or} \quad J^i_n = \oint \frac{dz}{2\pi i} z^n J^i(z),$$

as appropriate for $h = 1$ operators, it is possible to verify using the techniques described in Chapter 3 that

$$\left[J^i_m, J^j_n\right] = i\varepsilon^{ijk} J^k_{m+n} + m\delta^{ij}\delta_{m+n,0},$$

which is a level-one $SU(2)$ Kac–Moody algebra. □

EXERCISE 7.4
T-duality, which inverts $\mathcal{G}$, can be translated into transformations on the background fields G and B. Show that $\mathcal{G} \leftrightarrow \mathcal{G}^{-1}$ (a statement about $2n \times 2n$ matrices) is equivalent to $G + B \leftrightarrow \frac{1}{4}(G+B)^{-1}$ (a statement about $n \times n$ matrices).

SOLUTION

In order to check this, a new metric $\widetilde{G}$ and tensor field $\widetilde{B}$, which are related to the old fields by

$$\widetilde{G} + \widetilde{B} = \frac{1}{4}(G+B)^{-1},$$

are introduced. Taking the symmetric and antisymmetric parts leads to

$$\widetilde{G} = \frac{1}{8}\left[(G+B)^{-1} + (G-B)^{-1}\right]$$

and

$$\widetilde{B} = \frac{1}{8}\left[(G+B)^{-1} - (G-B)^{-1}\right].$$

By simple manipulations, these can be rewritten in the form

$$\widetilde{G} = \frac{1}{4}\left(G - BG^{-1}B\right)^{-1} \quad \text{and} \quad \widetilde{B} = -G^{-1}B\widetilde{G}.$$

Using these expressions for $\widetilde{G}$ and $\widetilde{B}$ and comparing Eqs (7.69) and (7.70) one concludes that

$$\widetilde{\mathcal{G}} = \mathcal{G}^{-1}$$

Exercise 7.5

Starting with Eq. (7.95) fill in the details of the derivation of Eq. (7.106). In particular, derive the expressions for the determinant (7.103) and the inverse matrix (7.104).

Solution

Starting with

$$\text{Tr}\left(q^{\frac{1}{2}p_R^2}\bar{q}^{\frac{1}{2}p_L^2}\right) = \sum_{\{W^I, K_I\}} e^{\pi i \tau_1(p_R^2 - p_L^2)} e^{-\pi \tau_2(p_L^2 + p_R^2)},$$

and using

$$p_R^2 - p_L^2 = -2W^I K_I \quad \text{and} \quad p_R^2 + p_L^2 = (W\ K)\, G^{-1} \begin{pmatrix} W \\ K \end{pmatrix},$$

one obtains that the formula for the trace is equivalent to

$$\sum_{\{M\}} \exp\left(-\pi M^T A M\right),$$

with

$$A = \begin{pmatrix} 2\tau_2(G - BG^{-1}B) & i\tau_1 1_n + \tau_2 BG^{-1} \\ i\tau_1 1_n - \tau_2 G^{-1}B & \frac{1}{2}\tau_2 G^{-1} \end{pmatrix} \quad \text{and} \quad M = \begin{pmatrix} W \\ K \end{pmatrix}.$$

The determinant of A can be obtained by using the fact that the determinant of a block matrix

$$\begin{pmatrix} M_1 & M_2 \\ M_3 & M_4 \end{pmatrix} = \begin{pmatrix} 1_n & M_2 M_4^{-1} \\ 0 & 1_n \end{pmatrix} \begin{pmatrix} M_1 - M_2 M_4^{-1} M_3 & 0 \\ M_3 & 1_n \end{pmatrix} \begin{pmatrix} 1_n & 0 \\ 0 & M_4 \end{pmatrix}$$

is given by

$$\det(M_1 - M_2 M_4^{-1} M_3) \det M_4.$$

This gives

$$\det A = |\tau|^{2n}.$$

The result for A^{-1} in Eq. (7.104) can be verified by checking that it gives the unit matrix when multiplied with A. The identities $\tau_1 \tilde{\tau}_2 + \tau_2 \tilde{\tau}_1 = 0$ and $\tau_2 \tilde{\tau}_2 - \tau_1 \tilde{\tau}_1 = 1$, which follow from $\tau \tilde{\tau} = -1$, are useful. □

Exercise 7.6

Show that $G_{IJ} + B_{IJ}$ has the right number of components to parametrize the coset space $\mathcal{M}_{n,n}^0$.

Solution

The moduli space $\mathcal{M}_{n,n}^0$ is given in terms of a lattice spanned by the left-moving and right-moving momenta (p_L, p_R) under the restriction that

$$p_L^2 - p_R^2 \in 2\mathbb{Z}.$$

This condition is left invariant by the group of $O(n, n, ; \mathbb{R})$ transformations, but the mass formula

$$M^2 = 2(p_L^2 + p_R^2) - 8 + \text{oscillators}$$

is not. The invariance of the mass formula is rather given by $O(n, \mathbb{R}) \times O(n, \mathbb{R})$. As a result, the moduli space is given by the quotient space

$$O(n, n; \mathbb{R})/O(n; \mathbb{R}) \times O(n, \mathbb{R}).$$

Taking into account that $O(n, \mathbb{R})$ has dimension $n(n-1)/2$ and $O(n, n, ; \mathbb{R})$ has dimension $n(2n-1)$, we see that the dimension of the moduli space is n^2.

On the other hand, the metric G is a symmetric tensor with $n(n+1)/2$ parameters while the antisymmetric B field has $n(n-1)/2$ independent components. In total, this gives n^2 components, as we wanted to show. □

Exercise 7.7

Compute the matrix $\mathcal{G}$ for the special case of compactification on a circle and compare with the results derived in Chapter 6.

Solution

In the special case of $n = 1$ one simply has a circle of radius R, and $G_{11} = R^2$. Then $\mathcal{G}$ reduces to the 2×2 matrix

$$\mathcal{G} = \begin{pmatrix} 1/(2R^2) & 0 \\ 0 & 2R^2 \end{pmatrix},$$

so that

$$M_0^2 = (2WR)^2 + (K/R)^2,$$

in agreement with the result obtained in Chapter 6 (for $\alpha' = 1/2$). The first term is the winding contribution and the second term is the Kaluza–Klein

Exercise 7.8

Consider the bosonic string compactified on a two-torus T^2. Where in the moduli space do enhanced gauge symmetries appear? What are the corresponding gauge groups?

Solution

A T^2 compactification is determined by the moduli

$$G = \begin{pmatrix} G_{11} & G_{12} \\ G_{12} & G_{22} \end{pmatrix} \quad \text{and} \quad B = B_{12} \begin{pmatrix} 0 & 1 \\ -1 & 0 \end{pmatrix}.$$

These four real parameters can be traded for two complex parameters τ and ρ by using the identifications

$$\tau = \tau_1 + i\tau_2 = \frac{G_{12}}{G_{22}} + i\frac{\sqrt{\det G}}{G_{22}}$$

and

$$\rho = \rho_1 + i\rho_2 = B_{12} + i\sqrt{\det G}.$$

Each of these transforms as an $SL(2,\mathbb{Z})$ modulus under T-duality transformations. The reason for this can be traced to the identity

$$SO(2,2;\mathbb{Z}) = SL(2,\mathbb{Z}) \times SL(2,\mathbb{Z}),$$

which is the discrete version of the identity $SO(2,2) = SL(2,\mathbb{R}) \times SL(2,\mathbb{R})$, which appeared in Section 2.2 in a different context. These relations can be inverted yielding

$$G + B = \frac{\rho_2}{\tau_2}\begin{pmatrix} \tau_1^2 + \tau_2^2 & \tau_1 \\ \tau_1 & 1 \end{pmatrix} + \rho_1 \begin{pmatrix} 0 & 1 \\ -1 & 0 \end{pmatrix}.$$

The moduli space of the torus is given by the fundamental domain displayed in Fig. 7.3.

For generic moduli the gauge group is $U(1)_L^2 \times U(1)_R^2$. The points of enhanced symmetries correspond to the singular points of the fundamental domain. We give several examples below, without attempting to give a systematic and exhaustive analysis. In particular, we focus on examples with $B = 0$, which have identical spectra for left-movers and right-movers.

7.3 Toroidal compactification

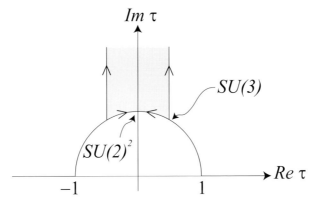

Fig. 7.3. Fundamental domain of the torus displaying the discrete identifications. Points in the τ plane where enhanced symmetries appear for $\rho = i$ are displayed.

Suppose that $\rho_1 = \tau_1 = 0$, so that $B = 0$ and

$$G = \begin{pmatrix} \rho_2 \tau_2 & 0 \\ 0 & \rho_2/\tau_2 \end{pmatrix}.$$

If $\rho_2 = \tau_2$ or $\rho_2 = 1/\tau_2$ then one of the two entries is one, which means that one of the two circles is at the self-dual radius, and there is an enhanced $SU(2)$ gauge symmetry for both left-movers and right-movers. These two relations are satisfied simultaneously if

$$(\tau, \rho) = (i, i).$$

In this case both circles are at the self-dual radius and the enhanced symmetry is $SU(2) \times SU(2)$ for both left-movers and right-movers giving $SU(2)^4$ altogether.

Another point of enhanced symmetry appears when

$$(\tau, \rho) = (-\frac{1}{2} + i\frac{\sqrt{3}}{2}, i).$$

In this case $B = 0$ and

$$G = \frac{1}{\sqrt{3}} \begin{pmatrix} 2 & -1 \\ -1 & 2 \end{pmatrix}.$$

Here G is proportional to the Cartan matrix of $SU(3)$. (For an introduction to the theory of roots and weights of Lie algebras see the review article by Goddard and Olive.) As a consequence both the left-movers and the right-movers contain the massless vectors required for $SU(3)$ enhanced gauge symmetry. Thus, altogether, the gauge symmetry in this case is $SU(3) \times SU(3)$.

$\square$

7.4 Bosonic construction of the heterotic string

Let us now consider the heterotic string in a formalism in which the current algebra is represented by bosons. The left-moving sector of the heterotic string corresponds to the bosonic string theory while the right-moving sector corresponds to the superstring theory. For the theory in ten-dimensional Minkowski space-time, the left-moving coordinates consist of ten bosonic fields $X_L^\mu(\tau + \sigma)$, $\mu = 0, \ldots, 9$, describing excitations in the noncompact dimensions and 16 bosonic fields $X_L^I(\tau + \sigma)$, $I = 1, \ldots, 16$, describing excitations on a 16-dimensional torus T^{16}. The torus is characterized by the momenta of the internal bosons p_L^I. They take discrete values that lie on a 16-dimensional lattice Γ_{16} spanned by 16 basis vectors $\{e_1^I, e_2^I, \ldots, e_{16}^I\}$

$$\mathbf{p}_L \in \Gamma_{16}, \qquad p_L^I = \sum_i n_i e_i^I, \qquad n_i \in \mathbb{Z}. \qquad (7.117)$$

One-loop modular invariance requires that Γ_{16} be a Euclidean even self-dual lattice. This ensures that the partition function

$$\Theta_\Gamma(\tau) = \sum_{p \in \Gamma} e^{i\pi\tau p^2} \qquad (7.118)$$

is a modular form of weight eight, which means that, for a modular transformation $\tau \to \tau'$ of the usual form, in Eq. (7.92)

$$\Theta_\Gamma(\tau') = (c\tau + d)^8 \Theta_\Gamma(\tau). \qquad (7.119)$$

Remarkably, in 16 dimensions there are only two Euclidean even self-dual lattices.

In eight dimensions there is a unique Euclidean even self-dual lattice, denoted Γ_8. The lattice Γ_8 is the root lattice of the Lie group E_8. This beautiful result is at the heart of the reason for the appearance of this Lie group in heterotic string theory. This implies that one way to make an even self-dual lattice in 16 dimensions is to form $\Gamma_8 \times \Gamma_8$, the product of two E_8 lattices.

The second even self-dual lattice in 16 dimensions, denoted Γ_{16}, is the weight lattice of $Spin(32)/\mathbb{Z}_2$. It contains the weights of two of the four conjugacy classes of $Spin(32)$. One conjugacy class is the root lattice of $SO(32)$. The second conjugacy class is one of the two spinor conjugacy classes of $Spin(32)$. Even though the $Spin(32)/\mathbb{Z}_2$ lattice Γ_{16} is different from the $E_8 \times E_8$ lattice, it gives exactly the same partition function $\Theta_\Gamma(\tau)$. This fact implies that the two heterotic string theories have the same number of physical states at every mass level.

Toroidal compactification of the heterotic string

Let us consider the heterotic string toroidally compactified to leave $10 - n$ noncompact dimensions. Noncompact dimensions must have both left- and right-movers. Therefore, it is necessary to compactify n of the right-moving dimensions and $16 + n$ of the left-moving dimensions. The compact dimensions in this set-up are characterized by $\Lambda_{16+n,n}$, which is the lattice that describes the discrete momenta and winding modes associated with $16 + n$ left-moving compact dimensions and n right-moving compact dimensions. Such a lattice is often called a *Narain lattice*. We are interested in classifying lattices of this signature that are even and self-dual. The reason is that this is exactly what is required to ensure one-loop modular invariance of scattering amplitudes.

For $n > 0$ there is a moduli space of dimension $(16 + n)n$ given by

$$\mathcal{M}_{16+n,n} = \mathcal{M}^0_{16+n,n}/O(16+n,n;\mathbb{Z}), \qquad (7.120)$$

where

$$\mathcal{M}^0_{16+n,n} = \frac{O(16+n,n;\mathbb{R})}{O(16+n,\mathbb{R}) \times O(n,\mathbb{R})}. \qquad (7.121)$$

The infinite discrete group $O(16+n,n;\mathbb{Z})$ is the T-duality group for the toroidally compactified heterotic string theory.

At a generic point in the moduli space the gauge symmetry consists of one $U(1)$ gauge field for each dimension of the lattice, giving $U(1)^{16+n} \times U(1)^n$. The left-moving gauge fields belong to vector supermultiplets, and the right-moving gauge fields belong to the supergravity multiplet. Once again, there is enhanced nonabelian gauge symmetry at the singularities of $\mathcal{M}$. However, in the case of the heterotic string only the left-moving gauge fields, which belong to vector supermultiplets, can become nonabelian in Minkowski space-time. When this happens, the rank remains $16 + n$. There is an enormously rich set of possibilities. One class of examples of nonabelian gauge groups that can be realized for special loci in the moduli space is $SO(32 + 2n)$. In particular, $SO(44)$ is possible for $d = 4$. The proof requires finding the locus in the moduli space where there are massless vectors with the appropriate $U(1)$ charges to give the nonzero roots of the adjoint representation of the group in question.

Duality and the heterotic string

Let us conclude this chapter by mentioning a beautiful and important relation between the two heterotic theories. The distinction between the $E_8 \times E_8$ and $SO(32)$ heterotic theories only exists in ten dimensions. After toroidal compactification, there is a single moduli space. In other words, the moduli space in ten dimensions consists of two points, whereas in $10-n$ dimensions it is a connected space of dimension $(16+n)n$.

This can be interpreted as implying that the $E_8 \times E_8$ and $SO(32)$ heterotic theories are related by T-duality. This is analogous to the relationship between the two type II superstring theories. To see what this means, let us consider compactification to nine dimensions. In this case the moduli space $\mathcal{M}_{17,1}$ has 17 dimensions. One scalar is the metric component g_{99} which encodes the radius of the circle. The other 16 moduli are the gauge field components A_9^I, which are the Wilson lines. For generic values of these moduli, the left-moving gauge symmetry is $U(1)^{17}$. However, on various loci in the moduli space enhanced gauge symmetry occurs. $E_8 \times E_8 \times U(1)$ and $SO(32) \times U(1)$ are just two of the many possibilities. An elementary method of exploring the possibilities is to construct Fock-space descriptions of the gauge fields with $N_R = N_L = 0$ and $p_L^2 = 2$ along the lines described for the bosonic string in 25 dimensions.

EXERCISES

EXERCISE 7.9
Use the Poisson resummation formula to prove that an even self-dual lattice Γ_{16} has a partition function $\Theta_{\Gamma_{16}}(\tau)$ that is a modular form of weight eight.

SOLUTION

The modular group is generated by the two transformations

$$\tau \to \tau + 1 \quad \text{and} \quad \tau \to -\frac{1}{\tau},$$

so it is sufficient to just consider them. Since the lattice Γ_{16} is even, the partition function

$$\Theta_{\Gamma_{16}}(\tau) = \sum_{p \in \Gamma_{16}} \exp\left(i\pi\tau p^2\right)$$

is invariant under $\tau \to \tau + 1$. In order to check how the partition function behaves under the second transformation, we rewrite it in terms of a vector N with components n_i and the matrix

$$A_{ij} = -i\tau G_{ij},$$

where

$$G_{ij} = e^I{}_i e_{Ij}$$

and $e^I{}_i$ are the basis vectors that appear in Eq. (7.117). This gives

$$\Theta_{\Gamma_{16}}(\tau) = \sum_{p \in \Gamma_{16}} \exp\left(-\pi N^T A N\right).$$

Applying the Poisson resummation formula yields

$$\frac{1}{\sqrt{\det A}} \sum_{p \in \Gamma_{16}} \exp\left(-\pi N^T A^{-1} N\right).$$

Since the lattice is self-dual $\det G = 1$. Also, replacing the matrix G by its inverse corresponds to replacing the basis vectors by the dual basis vectors, which span the same lattice. Therefore, the result simplifies to

$$\tau^{-8} \sum_{p \in \Gamma_{16}} \exp\left(-\pi N^T A^{-1} N\right) = \tau^{-8} \sum_{p \in \Gamma_{16}} \exp\left(-\frac{i\pi}{\tau} p^2\right),$$

which is exactly the transformation obtained from (7.119) for $\tau \to -1/\tau$. □

EXERCISE 7.10
Use the bosonic formulation of the heterotic string to construct the first massive level of the $E_8 \times E_8$ heterotic string.

SOLUTION

The mass formula for the heterotic string is

$$\frac{1}{8} M^2 = N_R = N_L - 1 + \frac{1}{2} \sum_{I=1}^{16} (p^I)^2.$$

For the first massive level, $M^2 = 8$, there are three possibilities:

(i)

$$N_R = 1, \quad N_L = 2, \quad \sum_{I=1}^{16} (p^I)^2 = 0$$

There are 324 possible left-moving states:

$$\tilde{\alpha}^I_{-1}\tilde{\alpha}^J_{-1}|0\rangle_L, \quad \tilde{\alpha}^I_{-2}|0\rangle_L, \quad \tilde{\alpha}^i_{-1}\tilde{\alpha}^j_{-1}|0\rangle_L, \quad \tilde{\alpha}^i_{-2}|0\rangle_L, \quad \tilde{\alpha}^i_{-1}\tilde{\alpha}^I_{-1}|0\rangle_L$$

(ii)
$$N_R = 1, \quad N_L = 1, \quad \sum_{I=1}^{16}(p^I)^2 = 2$$

In this case there are 24×480 possible left-moving states:

$$\tilde{\alpha}^I_{-1}|p^J, \sum_{J=1}^{16}(p^J)^2 = 2\rangle_L, \quad \tilde{\alpha}^i_{-1}|p^I, \sum_{I=1}^{16}(p^J)^2 = 2\rangle_L.$$

(iii)
$$N_R = 1, \quad N_L = 0, \quad \sum_{I=1}^{16}(p^I)^2 = 4$$

In this case there are 129×480 possible left-moving states:

$$|p^I, \sum_{I=1}^{16}(p^I)^2 = 4\rangle_L.$$

The total number of left-moving states is 73 764. In each case the right-movers have $N_R = 1$, so these are the 256 states

$$\alpha^i_{-1}|j\rangle_R, \quad \alpha^i_{-1}|a\rangle_R, \quad ; S^a_{-1}|i\rangle_R, \quad S^a_{-1}|b\rangle_R.$$

The spectrum of the heterotic string at this mass level is given by the tensor product of the left-movers and the right-movers, a total of almost 20 000 000 states. □

Appendix: The Poisson resummation formula

Let A be a positive definite $m \times m$ symmetric matrix and define

$$f(A) = \sum_{\{M\}} \exp\left(-\pi M^T A M\right). \tag{7.122}$$

Here M represents a vector made of m integers $M_1, M_2, \ldots, M_m$ each of which is summed from $-\infty$ to $+\infty$. The Poisson resummation formula is

$$f(A) = \frac{1}{\sqrt{\det A}} f(A^{-1}). \tag{7.123}$$

To derive this formula it is convenient to add dependence on m variables x^i and define

$$f(A, x) = \sum_{\{M\}} \exp\left(-\pi(M+x)^T A(M+x)\right). \tag{7.124}$$

This function is periodic, with period 1, in each of the x^i. Therefore, it must have a Fourier series expansion of the form

$$f(A, x) = \sum_{\{N\}} C_N(A) \exp(2\pi i N^T x). \tag{7.125}$$

The next step is to evaluate the Fourier coefficients:

$$C_N(A) = \int_0^1 f(A, x) e^{-2\pi i N^T x} d^m x. \tag{7.126}$$

Inserting the series expansion of $f(A, x)$ in Eq. (7.124) gives

$$C_N(A) = \int_{-\infty}^{\infty} \exp(-\pi x^T A x - 2\pi i N^T x) d^m x = \frac{\exp(-\pi N^T A^{-1} N)}{\sqrt{\det A}}. \tag{7.127}$$

Note that the summations in Eq. (7.125) have been taken into account by extending the range of the integrations. It therefore follows that

$$f(A) = \sum_{\{N\}} C_N(A) = \frac{1}{\sqrt{\det A}} f(A^{-1}) \tag{7.128}$$

as desired.

Homework Problems

Problem 7.1
Section 7.1 discussed several possibilities for generating nonabelian gauge symmetries in string theory. Show that in the context of toroidally compactified type II superstring theories, the only massless gauge fields are abelian.

Problem 7.2
It is possible to compactify the 26-dimensional bosonic string to ten dimensions by replacing 16 dimensions with 32 Majorana fermions. The 32 left-moving fermions and the 32 right-moving fermions each give a level-one

$SO(32)$ current algebra. Making the same GSO projection as in the left-moving sector of the heterotic string, find the ground state and the massless states of this theory.

PROBLEM 7.3
Exercise 7.1 introduced free-fermion representations of current algebras and showed that fermions in the fundamental representation of $SO(n)$ give a level-one current algebra.

(i) Find the level of the current algebra for fermions in the adjoint representation of $SO(n)$.
(ii) Find the level of the current algebra for fermions in a spinor representation of $SO(16)$.

PROBLEM 7.4
Generalize the analysis of Exercise 7.6 to the heterotic string. In particular, verify that the Wilson lines, together with the B and G fields, have the right number of parameters to describe the moduli space $\mathcal{M}^0_{16+n,n}$ in Eq. (7.121).

PROBLEM 7.5
In addition to the $SO(32)$ and $E_8 \times E_8$ heterotic string theories, there is a third tachyon-free ten-dimensional heterotic string theory that has an $SO(16) \times SO(16)$ gauge group. This theory is not supersymmetric. Invent a plausible set of GSO projection rules for the fermionic formulation of this theory that gives an $SO(16) \times SO(16)$ gauge group and does not give any gravitinos. Find the complete massless spectrum.

PROBLEM 7.6
The $SO(16) \times SO(16)$ heterotic string theory, constructed in the previous problem, is a chiral theory. Using the rules described in Chapter 5, construct the anomaly 12-form. Show that anomaly cancellation is possible by showing that this 12-form factorizes into the product of a four-form and an eight-form.

PROBLEM 7.7
The ten-dimensional $SO(32)$ and $E_8 \times E_8$ string theories have the same number of states at the massless level. Construct the spectrum at the first excited level explicitly in each case using the formulation with 32 left-moving fermions. What is the number of left-moving states at the first excited level

in each case? Show that the numbers are the same and that they agree with the result obtained in Exercise 7.10.

PROBLEM 7.8

(i) Consider a two-dimensional lattice generated by the basis vectors

$$e_1 = (1,1) \quad \text{and} \quad e_2 = (1,-1)$$

with a standard Euclidean scalar product. Construct the dual lattice $\Lambda^\star$. Is Λ: unimodular, integral, even or self-dual? How about $\Lambda^\star$?

(ii) Find a pair of basis vectors that generate a two-dimensional even self-dual *Lorentzian* lattice.

PROBLEM 7.9

Consider the Euclideanized world-sheet theory for a string coordinate X

$$S[X] = \frac{1}{\pi} \int_M \partial X \bar\partial X d^2 z.$$

Suppose that X is circular, so that $X \sim X + 2\pi R$ and that the world sheet M is a torus so that $z \sim z+1 \sim z+\tau$. Define winding numbers W_1 and W_2 by

$$X(z+1, \bar z+1) = X(z,\bar z) + 2\pi R W_1,$$

$$X(z+\tau, \bar z+\bar\tau) = X(z,\bar z) + 2\pi R W_2.$$

(i) Find the classical solution X_{cl} with these winding numbers.
(ii) Evaluate the action $S_{\text{cl}}(W_1, W_2) = S[X_{\text{cl}}]$.
(iii) Recast the classical partition function

$$Z_{\text{cl}} = \sum_{W_1, W_2} e^{-S_{\text{cl}}(W_1, W_2)}$$

by performing a Poisson resummation. Is the result consistent with T-duality?

PROBLEM 7.10

Consider a Euclidean lattice generated by basis vectors e_i, $i = 1,\ldots, 8$, whose inner products $e_i \cdot e_j$ are described by the following metric:

$$\begin{pmatrix} 2 & -1 & 0 & 0 & 0 & 0 & 0 & 0 \\ -1 & 2 & -1 & 0 & 0 & 0 & 0 & 0 \\ 0 & -1 & 2 & -1 & 0 & 0 & 0 & 0 \\ 0 & 0 & -1 & 2 & -1 & 0 & 0 & 0 \\ 0 & 0 & 0 & -1 & 2 & -1 & 0 & -1 \\ 0 & 0 & 0 & 0 & -1 & 2 & -1 & 0 \\ 0 & 0 & 0 & 0 & 0 & -1 & 2 & 0 \\ 0 & 0 & 0 & 0 & -1 & 0 & 0 & 2 \end{pmatrix}.$$

This is the Cartan matrix for the Lie group E_8.

(i) Find a set of basis vectors that gives this metric.
(ii) Prove that the lattice is even and self-dual. It is the E_8 lattice.

PROBLEM 7.11

As stated in Section 7.4, in 16 dimensions there are only two Euclidean even self-dual lattices. One of them, the $E_8 \times E_8$ lattice, is given by combining two of the E_8 lattices in the previous problem. Construct the other even self-dual lattice in 16 dimensions and show that it is the $Spin(32)/\mathbb{Z}_2$ weight lattice.

PROBLEM 7.12

Show that the spectrum of the bosonic string compactified on a two-torus parametrized using the two complex coordinates τ and ρ defined in Exercise 7.8 is invariant under the set of duality transformations $SL(2,\mathbb{Z})_\tau \times SL(2,\mathbb{Z})_\rho$ generated by

$$\begin{array}{ll} \tau \to \tau + 1 & \rho \to \rho + 1 \\ \tau \to -1/\tau & \rho \to -1/\rho \end{array}.$$

Moreover, show that the spectrum is invariant under the following interchanges of coordinates:

$$U : (\tau, \rho) \to (\rho, \tau) \quad \text{and} \quad V : (\sigma, \tau) \to (-\bar{\sigma}, -\bar{\tau}).$$

These results imply that the moduli space is given by two copies of the moduli space of a single torus dividing out by the symmetries U and V.

PROBLEM 7.13

Consider the bosonic string compactified on a square T^3

$$ds^2 = R^2(dx^2 + dy^2 + dz^2),$$

where the coordinates x, y, z each have period 2π. Suppose there is also a nonvanishing three-form $H_{xyz} = N$, where N is an integer. For example, $B_{xy} = Nz$.

(i) Using the T-duality rules for background fields derived in Chapter 6, carry out a T-duality transformation in the x direction followed by another one in the y direction. What is the form of the resulting metric and B fields?

(ii) One can regard the T^3 as a T^2, parametrized by x and y, fibered over the z-circle. Going once around the z-circle is trivial in the original background. What happens when we go once around the z-circle after the two T-dualities are performed?

(iii) The background after the T-dualities has been called *nongeometrical*. Explain why. Hint: use the results of the preceding problem.

PROBLEM 7.14

Consider the compactification of each of the two supersymmetric heterotic string theories on a circle of radius R. As discussed in Section 7.4, the moduli space is 17-dimensional and at generic points the left-moving gauge symmetry is $U(1)^{17}$. However, at special points there are enhanced symmetries. Assume that the gauge fields in the compact dimensions, that is, the Wilson lines, are chosen in each case to give $SO(16) \times SO(16) \times U(1)$ left-moving gauge symmetry. Show that the two resulting nine-dimensional theories are related by a T-duality transformation that inverts the radius of the circle. This is very similar to the T-duality relating the type IIA and IIB superstring theories compactified on a circle.

PROBLEM 7.15

(i) Compactifying the $E_8 \times E_8$ heterotic string on a six-torus to four dimensions leads to a theory with $\mathcal{N} = 4$ supersymmetry in four dimensions. Verify this statement and assemble the resulting massless spectrum into four-dimensional supermultiplets.

(ii) Repeat the analysis for the type IIA or type IIB superstring. What is the amount of supersymmetry in four dimensions in this case? What is the massless supermultiplet structure in this case?

8
M-theory and string duality

During the "Second Superstring Revolution," which took place in the mid-1990s, it became evident that the five different ten-dimensional superstring theories are related through an intricate web of dualities. In addition to the T-dualities that were discussed in Chapter 6, there are also S-dualities that relate various string theories at strong coupling to a corresponding dual description at weak coupling. Moreover, two of the superstring theories (the type IIA superstring and the $E_8 \times E_8$ heterotic string) exhibit an eleventh dimension at strong coupling and thus approach a common 11-dimensional limit, a theory called *M-theory*. In the decompactification limit, this 11-dimensional theory does not contain any strings, so it is not a string theory.

Low-energy effective actions

This chapter presents several aspects of M-theory, including its low-energy limit, which is 11-dimensional supergravity, as well as various nonperturbative string dualities. Some of these dualities can be illustrated using low-energy effective actions. These are supergravity theories that describe interactions of the massless fields in the string-theory spectrum. It is not obvious, *a priori*, that this should be a useful approach for analyzing nonperturbative features of string theory, since extrapolations from weak coupling to strong coupling are ordinarily beyond control. However, if one restricts such extrapolations to quantities that are protected by supersymmetry, one can learn a surprising amount in this way.

BPS branes

A second method of testing proposed duality relations is to exploit the various supersymmetric or Bogomolny–Prasad–Sommerfield (BPS) p-branes that these theories possess and the matching of the corresponding spectra of states. As we shall illustrate below, *saturation of a BPS bound* can lead to shortened supersymmetry multiplets, and then reliable extrapolations from weak coupling to strong coupling become possible. This makes it possible to carry out detailed matching of p-branes and their tensions in dual theories.

The concept of a BPS bound and its saturation can be illustrated by massive particles in four dimensions. The $\mathcal{N}$-extended supersymmetry algebra, restricted to the space of particles of mass $M > 0$ at rest in $D = 4$, takes the form

$$\{Q_\alpha^I, Q_\beta^{\dagger J}\} = 2M\delta^{IJ}\delta_{\alpha\beta} + 2iZ^{IJ}\Gamma^0_{\alpha\beta}, \qquad (8.1)$$

where Z^{IJ} is the *central-charge* matrix. $I, J = 1, \ldots, \mathcal{N}$ labels the supersymmetries and $\alpha, \beta = 1, 2, 3, 4$ labels the four components of each Majorana spinor supercharge. The central charges are conserved quantities that commute with all the other generators of the algebra. They can appear only in theories with *extended* supersymmetry, that is, theories that have more supersymmetry than the *minimal* $\mathcal{N} = 1$ case, because the central-charge matrix is antisymmetric $Z^{IJ} = -Z^{JI}$. The central charges are electric and magnetic charges that couple to the gauge fields belonging to the supergravity multiplet.

By a transformation of the form $Z \to U^T Z U$, where U is a unitary matrix, the antisymmetric matrix Z^{IJ} can be brought to the canonical form

$$Z^{IJ} = \begin{pmatrix} 0 & Z_1 & 0 & 0 & \\ -Z_1 & 0 & 0 & 0 & \cdots \\ 0 & 0 & 0 & Z_2 & \\ 0 & 0 & -Z_2 & 0 & \\ & \vdots & & & \ddots \end{pmatrix} \qquad (8.2)$$

with $|Z_1| \geq |Z_2| \geq \ldots \geq 0$. The structure of Eq. (8.1) implies that the $2\mathcal{N} \times 2\mathcal{N}$ matrix

$$\begin{pmatrix} M & Z \\ Z^\dagger & M \end{pmatrix} \qquad (8.3)$$

should be positive semidefinite. This in turn implies that the eigenvalues $M \pm |Z_i|$ have to be nonnegative. Therefore, the mass is bounded from below by the central charges, which gives the *BPS bound*

$$M \geq |Z_1|. \qquad (8.4)$$

States that have $M = |Z_1|$ are said to saturate the BPS bound. They belong to a *short supermultiplet* or BPS representation. States with $M > |Z_1|$

belong to a *long supermultiplet*. The zeroes that appear in the supersymmetry algebra when $M = |Z_1|$ are responsible for the multiplet shortening. A further refinement in the description of BPS states keeps track of the number of central charges that equal the mass. Thus, for example, in the $\mathcal{N} = 4$ case, states with $M = |Z_1| = |Z_2|$ are called *half-BPS* and ones with $M = |Z_1| > |Z_2|$ are called *quarter-BPS*. These fractions refer to the number of supersymmetries that are unbroken when these particles are present.

The preceding discussion is specific to point particles in four dimensions, but it generalizes to p-branes in D dimensions. The important point to remember from Chapter 6 is that a charged p-brane has a $(p+1)$-form conserved current, and hence a p-form charge. To analyze such cases the supersymmetry algebra needs to be generalized to cases appropriate to D dimensions and p-form central charges. Calling them central is a bit of a misnomer in this case, because for $p > 0$ they carry Lorentz indices and therefore do not commute with Lorentz transformations.

One very important conclusion from the BPS bound given above is that BPS states, which have $M = |Z_1|$ and belong to a short multiplet, are stable. The mass is tied to a central charge, and this relation does not change as parameters are varied if the supersymmetry is unbroken. The only way in which this could fail is if another representation becomes degenerate with the BPS multiplet, so that they can pair up to give a long representation. The idea is actually more general than supersymmetry. This is what happens in the Higgs mechanism, where a massless vector (a short representation of the Lorentz group) joins up with a scalar to give a massive vector (a long representation) as a parameter in the Higgs potential is varied. The thing that is different about supersymmetric examples is that short multiplets can be massive. In any case, the conclusion is that so long as such a joining of multiplets does not happen, it is possible to follow BPS states from weak coupling to strong coupling with precise control. This is very important for testing conjectures about the behavior of string theories at strong coupling, as we shall see in this chapter.

EXERCISES

EXERCISE 8.1

The $\mathcal{N} = 1$ supersymmetry algebra in four dimensions does not have a central extension. The explicit form of this algebra, with the supercharges

M-theory and string duality

expressed as two-component Weyl spinors Q_α and $\overline{Q}_{\dot\beta} = Q^\dagger_\beta$, is

$$\{Q_\alpha, \overline{Q}_{\dot\beta}\} = 2\sigma^\mu_{\alpha\dot\beta} P_\mu, \quad \text{and} \quad \{Q_\alpha, Q_\beta\} = \{\overline{Q}_{\dot\alpha}, \overline{Q}_{\dot\beta}\} = 0.$$

Determine the irreducible massive representations of this algebra.

SOLUTION

As in the text, for massive states we can work in the rest frame, where the momentum vector is $P_\mu = (-M, 0, 0, 0)$. Then the algebra becomes

$$\{Q_\alpha, \overline{Q}_{\dot\beta}\} = 2M\delta_{\alpha\dot\beta} = 2M \begin{pmatrix} 1 & 0 \\ 0 & 1 \end{pmatrix}.$$

This algebra is a Clifford algebra, so it is convenient to rescale the operators to obtain a standard form for the algebra

$$b_\alpha = \frac{1}{\sqrt{2M}} Q_\alpha \quad \text{and} \quad b^\dagger_\alpha = \frac{1}{\sqrt{2M}} \overline{Q}_{\dot\alpha}.$$

The supersymmetry algebra then becomes

$$\{b_\alpha, b^\dagger_\beta\} = \delta_{\alpha\beta}, \quad \{b_\alpha, b_\beta\} = \{b^\dagger_\alpha, b^\dagger_\beta\} = 0.$$

As a result, b_α and $b^\dagger_\alpha$ act as fermionic lowering and raising operators, and we obtain all the states in the supermultiplet by acting with raising operators $b^\dagger_\alpha$ on the Fock-space ground state $|\Omega\rangle$, which satisfies the condition $b_\alpha|\Omega\rangle = 0$. Then, if $|\Omega\rangle$ represents a state of spin j, a state of spin $j \pm \frac{1}{2}$ is created by acting with the fermionic operators $b^\dagger_\alpha|\Omega\rangle$. If the ground state $|\Omega\rangle$ has spin 0 (a boson), then $b^\dagger_\alpha|\Omega\rangle$ represent the two states of a spin 1/2 fermion. Moreover $b^\dagger_1 b^\dagger_2|\Omega\rangle$ gives a second spin 0 state. In general, for a ground state of spin $j > 0$ and multiplicity $2j+1$, this construction gives the $4(2j+1)$ states of a massive representation of $\mathcal{N} = 1$ supersymmetry in $D = 4$ with spins $j - 1/2, j, j, j + 1/2$. □

EXERCISE 8.2

Determine the multiplet structure for massive states of $\mathcal{N} = 2$ supersymmetry in four dimensions in the presence of the central charge. In particular derive the form of the short and long multiplets.

SOLUTION

For $\mathcal{N} = 2$ supersymmetry the central charge is $Z^{IJ} = Z\varepsilon^{IJ}$. For simplicity, let us assume that Z is real and nonnegative. Using this form of the central charge, the supersymmetry algebra in the rest frame can be written in the form

$$\{Q_\alpha^I, \overline{Q}_{\dot\beta}^J\} = 2M\delta_{\alpha\dot\beta}\delta^{IJ},$$

$$\{Q_\alpha^I, Q_\beta^J\} = 2Z\varepsilon_{\alpha\beta}\varepsilon^{IJ},$$

$$\{\overline{Q}_{\dot\alpha}^I, \overline{Q}_{\dot\beta}^J\} = 2Z\varepsilon_{\dot\alpha\dot\beta}\varepsilon^{IJ},$$

where $I, J = 1, 2$. We rearrange these generators and define

$$b_\alpha^\pm = Q_\alpha^1 \pm \varepsilon_{\alpha\beta}\overline{Q}_\beta^2 \quad \text{and} \quad (b_\alpha^\pm)^\dagger = \overline{Q}_\alpha^1 \pm \varepsilon_{\alpha\beta}Q_\beta^2.$$

Note that this construction identifies dotted and undotted indices. This is sensible because a massive particle at rest breaks the $SL(2,\mathbb{C})$ Lorentz group to the $SU(2)$ rotational subgroup, so that the **2** and **$\bar{2}$** representations become equivalent. It is then easy to verify that the only nonzero anticommutators of these generators are

$$\{b_\alpha^+, (b_\beta^+)^\dagger\} = 4\delta_{\alpha\beta}(M+Z) \quad \text{and} \quad \{b_\alpha^-, (b_\beta^-)^\dagger\} = 4\delta_{\alpha\beta}(M-Z).$$

These anticommutation relations give the BPS bound for $\mathcal{N} = 2$ theories, which takes the form

$$M \geq Z.$$

If this bound is not saturated, we can act with $(b_\alpha^\pm)^\dagger$ on a spin j ground state $|\Omega\rangle$ to create the $16(2j+1)$ states of a long supermultiplet. However, if the BPS bound is saturated, that is, if $M = Z$, then the physical states in the supermultiplet are created by acting only with $(b_\alpha^+)^\dagger$. This reduces the number of states to $4(2j+1)$ and creates a short supermultiplet. The case $j = 0$ gives a *half hypermultiplet*. Such a multiplet is always paired with its TCP conjugate to give a hypermultiplet with four scalars and two spinors. The case $j = 1/2$ gives a vector multiplet. □

8.1 Low-energy effective actions

Previous chapters have described how the spectrum of states of the various superstring theories behaves in the weak-coupling limit. The masses of all states other than the massless ones become very large for $\alpha' \to 0$, which corresponds to large string tension. Equivalently, at least in a Minkowski space background where there is no other scale, this corresponds to the low-energy limit, since the only dimensionless parameter is $\alpha' E^2$. In the low-energy limit, it is a good approximation to replace string theory by a *supergravity theory* describing the interactions of the massless modes only, as the massive modes are too heavy to be observed. This section describes the supergravity theories arising in the low-energy limit of string theory. These theories are not fundamental, but they do capture some of the important features of the more fundamental string theories.

Renormalizability

By conventional power counting, effective supergravity theories are nonrenormalizable. A good guide to assessing this is to examine the dimensions of various terms in the action. The Einstein–Hilbert action, for example, in D dimensions takes the form

$$S = \frac{1}{16\pi G_D} \int \sqrt{-g} R d^D x. \qquad (8.5)$$

The curvature has dimensions (length)$^{-2}$, and therefore the D-dimensional Newton constant G_D must have dimension (length)$^{D-2}$. This is proportional to the square of the gravitational coupling constant, which therefore has negative mass dimension for $D > 2$. Ordinarily, barring some miracle, this is an indication of nonrenormalizability.[1] It has been shown by explicit calculation that no such miracle occurs in the case of pure gravity in $D = 4$. There is no good reason to expect miraculous cancellations in other cases with $D > 3$, either, though it would be nice to prove that they don't occur.

Nonrenormalizability is okay for theories whose only intended use is as effective actions for describing the low-energy physics of a more fundamental theory (string theory or M-theory). The infinite number of higher-order quantum corrections to these actions can be ignored for most purposes at low energies. Some of these quantum corrections are important, however. In fact, some of them already arose in the anomaly analysis of Chapter 5.

M-theory certainly requires an infinite number of higher-dimension corrections to 11-dimensional supergravity. Such an expansion is unambiguously determined by M-theory (up to field redefinitions) if one assumes a simple space-time topology, such as $\mathbb{R}^{10,1}$. In Chapter 9 it is shown that in $\mathbb{R}^{10,1}$ there are R^4 terms, in particular. The present chapter describes dualities relating M-theory to type IIA and type IIB superstring theory. These have been used to determine the precise form of the R^4 corrections to $D = 11$ supergravity required by M-theory.

---

[1] Actually, pure gravity for $D = 3$ appears to be a consistent quantum theory. However, a graviton in three dimensions has no physical polarization states, so that theory is essentially topological.

Eleven-dimensional supergravity

The low-energy effective action of M-theory, called *11-dimensional supergravity*, is our starting point. This theory was constructed in 1978 and studied extensively in subsequent years, but it was only in the mid-1990s that this theory found its place on the string theory map.

In its heyday (around 1980) there were two major reasons for being skeptical about $D = 11$ supergravity. The first was its evident lack of renormalizability, which led to the belief that it does not approximate a well-defined quantum theory. The second was its lack of chirality, that is, its left–right symmetry, which suggested that it could not have a vacuum with the chiral structure required for a realistic model. Within the conventional Kaluza–Klein framework being explored at that time, both of these objections were justified. However, we now view $D = 11$ supergravity as a low-energy effective description of M-theory. As such, there are good reasons to believe that there is a well-defined quantum interpretation. The situation with regard to chirality is also changed. Among the new ingredients are the branes, the M2-brane and the M5-brane, as well as end-of-the-world 9-branes. As was mentioned in Chapter 5, and is discussed further in this chapter, the latter appear in the strong-coupling description of the $E_8 \times E_8$ heterotic string theory and introduce left–right asymmetry consistent with anomaly cancellation requirements. There are also nonperturbative dualities, which is discussed in this chapter, that relate M-theory to chiral superstring theories. Moreover, it is now understood that compactification on manifolds with suitable singularities, which would not be well defined in a pure Kaluza–Klein supergravity context, can result in chirality in four dimensions.

Field content

Compared to the massless spectrum of the ten-dimensional superstring theories, the field content of 11-dimensional supergravity is relatively simple. First, since it contains gravity, there is a graviton, which is a symmetric traceless tensor of $SO(D-2)$, the little group for a massless particle. It has

$$\frac{1}{2}(D-1)(D-2) - 1 = \frac{1}{2}D(D-3) = 44 \tag{8.6}$$

physical degrees of freedom (or polarization states). The first term counts the number of independent components of a symmetric $(D-2) \times (D-2)$ matrix and 1 is subtracted due to the constraint of tracelessness. Since this theory contains fermions, it is necessary to use the vielbein formalism and represent the graviton by a vielbein field E_M^A. This can also be called an *elfbein field* in the case of 11 dimensions, since *viel* is German for many, and *elf* is German for 11. The indices $M, N, \ldots$ are used for base-space (curved) vectors in 11 dimensions, and the indices $A, B, \ldots$ are used for tangent-space (flat) vectors. The former transform nontrivially under general coordinate transformations, and the latter transform nontrivially under local Lorentz transformations.[2]

The gauge field for local supersymmetry is the gravitino field Ψ_M, which has an implicit spinor index in addition to its explicit vector index. For each value of M, it is a 32-component Majorana spinor. When spinors are included, the little group becomes the covering group of $SO(9)$, which is $Spin(9)$. It has a real spinor representation of dimension 16. Group theoretically, the $Spin(9)$ Kronecker product of a vector and a spinor is $\mathbf{9 \times 16 = 128 + 16}$. The analogous construction in four dimensions gives spin 3/2 plus spin 1/2. As Rarita and Schwinger showed in the case of a free vector-spinor field in four dimensions, there is a local gauge invariance of the form $\delta \Psi_M = \partial_M \varepsilon$, which ensures that the physical degrees of freedom are pure spin 3/2. The kinetic term for a free gravitino field Ψ_M in any dimension has the structure

$$S_\Psi \sim \int \overline{\Psi}_M \Gamma^{MNP} \partial_N \Psi_P \, d^D x.$$

Due to the antisymmetry of Γ^{MNP}, for $\delta \Psi_M = \partial_M \varepsilon$ this is invariant up to a total derivative.

In the case of 11 dimensions this local symmetry implies that the physical degrees of freedom correspond only to the **128**. Therefore, this is the number of physical polarization states of the gravitino in 11 dimensions. In the interacting theory this local symmetry is identified as local supersymmetry. This amount of supersymmetry gives 32 conserved supercharges, which form a 32-component Majorana spinor. This is the dimension of the minimal spinor in 11 dimensions, so there couldn't be less supersymmetry than that in a Lorentz-invariant vacuum. Also, if there were more supersymmetry, the representation theory of the algebra would require the existence of massless states with spin greater than two. It is believed to be impossible to construct consistent interacting theories with such higher spins in Minkowski space-time. For this reason, one does not expect to find nontrivial supersymmetric theories for $D > 11$.

In order for the $D = 11$ supergravity theory to be supersymmetric, there must be an equal number of physical bosonic and fermionic degrees of freedom. The missing bosonic degrees of freedom required for supersymmetry

---

[2] The reader not familiar with these concepts can consult the appendix of Chapter 9 for some basics. These also appeared in the anomaly analysis of Chapter 5.

are obtained from a rank-3 antisymmetric tensor, A_{MNP}, which can be represented as a three-form A_3. As usual for such form fields, the theory has to be invariant under the gauge transformations

$$A_3 \to A_3 + d\Lambda_2, \tag{8.7}$$

where Λ_2 is a two-form. As is always the case for antisymmetric tensor gauge fields, including the Maxwell field, the gauge invariance ensures that the indices for the independent physical polarizations are transverse. In the case of a three-form in 11 dimensions this means that there are $9 \cdot 8 \cdot 7/3! = 84$ physical degrees of freedom. Together with the graviton, this gives $44 + 84 = 128$ propagating bosonic degrees of freedom, which matches the number of propagating fermionic degrees of freedom of the gravitino, which is the only fermi field in the theory.

Action

The requirement of invariance under A_3 gauge transformations, together with general coordinate invariance and local Lorentz invariance, puts strong constraints on the form of the action. As in all supergravity theories, dimensional analysis determines that the number of derivatives plus half the number of fermi fields is equal to two for each term in the action. This requirement reduces the arbitrariness to a few numerical coefficients. Finally, the requirement of local supersymmetry leads to a unique supergravity theory in $D = 11$ (up to normalization conventions). In fact, it is so strongly constrained that its existence appears quite miraculous.

The bosonic part of the 11-dimensional supergravity action is

$$2\kappa_{11}^2 S = \int d^{11}x \sqrt{-G} \left(R - \frac{1}{2}|F_4|^2 \right) - \frac{1}{6} \int A_3 \wedge F_4 \wedge F_4, \tag{8.8}$$

where R is the scalar curvature, $F_4 = dA_3$ is the field strength associated with the potential A_3, and κ_{11} denotes the 11-dimensional gravitational coupling constant. The relation between the 11-dimensional Newton's constant G_{11}, the gravitational constant κ_{11} and the 11-dimensional Planck length ℓ_p is[3]

$$16\pi G_{11} = 2\kappa_{11}^2 = \frac{1}{2\pi}(2\pi \ell_p)^9. \tag{8.9}$$

The last term in Eq. (8.8), which has a Chern–Simons structure, is independent of the elfbein (or the metric). The first term does depend on the elfbein, but only in the metric combination

$$G_{MN} = \eta_{AB} E_M^A E_N^B. \tag{8.10}$$

[3] The coefficients in these relations are the most commonly used conventions.

8.1 Low-energy effective actions

The quantity $|F_4|^2$ is defined by the general rule

$$|F_n|^2 = \frac{1}{n!} G^{M_1 N_1} G^{M_2 N_2} \cdots G^{M_n N_n} F_{M_1 M_2 \cdots M_n} F_{N_1 N_2 \cdots N_n}. \tag{8.11}$$

Supersymmetry transformations

The complete action of 11-dimensional supergravity is invariant under local supersymmetry transformations under which the fields transform according to

$$\begin{aligned}
\delta E_M^A &= \bar{\varepsilon} \Gamma^A \Psi_M, \\
\delta A_{MNP} &= -3 \bar{\varepsilon} \Gamma_{[MN} \Psi_{P]}, \\
\delta \Psi_M &= \nabla_M \varepsilon + \tfrac{1}{12} \left(\Gamma_M \mathbf{F}^{(4)} - 3 \mathbf{F}_M^{(4)} \right) \varepsilon.
\end{aligned} \tag{8.12}$$

Here we have introduced the definitions

$$\mathbf{F}^{(4)} = \frac{1}{4!} F_{MNPQ} \Gamma^{MNPQ} \tag{8.13}$$

and

$$\mathbf{F}_M^{(4)} = \frac{1}{2} [\Gamma_M, \mathbf{F}^{(4)}] = \frac{1}{3!} F_{MNPQ} \Gamma^{NPQ}. \tag{8.14}$$

Straightforward generalizations of this notation are used in the following. The formula for $\delta \Psi_M$ displays the terms that are of leading order in fermi fields. Additional terms of the form $(\text{fermi})^2 \varepsilon$ have been dropped. The Dirac matrices satisfy

$$\Gamma_M = E_M^A \Gamma_A, \tag{8.15}$$

where Γ_A are the numerical (coordinate-independent) matrices that obey the flat-space Dirac algebra. Also, the square brackets represent antisymmetrization of the indices with unit weight. For example,

$$\Gamma_{[MN} \Psi_{P]} = \frac{1}{3} (\Gamma_{MN} \Psi_P + \Gamma_{NP} \Psi_M + \Gamma_{PM} \Psi_N). \tag{8.16}$$

Another convenient notation that has been used here is

$$\Gamma^{M_1 M_2 \cdots M_n} = \Gamma^{[M_1} \Gamma^{M_2} \cdots \Gamma^{M_n]}. \tag{8.17}$$

The covariant derivative that appears in Eq. (8.12) involves the spin connection ω and is given by

$$\nabla_M \varepsilon = \partial_M \varepsilon + \frac{1}{4} \omega_{MAB} \Gamma^{AB} \varepsilon. \tag{8.18}$$

The spin connection can be expressed in terms of the elfbein by

$$\omega_{MAB} = \frac{1}{2}(-\Omega_{MAB} + \Omega_{ABM} - \Omega_{BMA}), \qquad (8.19)$$

where

$$\Omega_{MN}{}^A = 2\partial_{[N} E_{M]}^A. \qquad (8.20)$$

In fact, these relations are valid in any dimension. Depending on conventions, the spin connection may also contain terms that are quadratic in fermi fields. Such terms are neglected here, since they are not relevant to the issues that we discuss.

Supersymmetric solutions

One might wonder why the supersymmetry transformations have been presented without also presenting the fermionic terms in the action. After all, it is the complete action including the fermionic terms that is supersymmetric. The justification is that one of the main uses of this action, and others like it, is to construct classical solutions. For this purpose, only the bosonic terms in the action are required, since a classical solution always has vanishing fermionic fields.

One is also interested in knowing how many of the supersymmetries survive as vacuum symmetries of the solution. Given a supersymmetric solution, there exist spinors, called *Killing spinors*, that characterize the supersymmetries of the solution. The concept is similar to that of Killing vectors, which characterize bosonic symmetries. Killing vectors are vectors that appear as parameters of infinitesimal general coordinate transformations under which the fields are invariant for a specific solution. In analogous fashion, Killing spinors are spinors that parametrize infinitesimal supersymmetry transformations under which the fields are invariant for a specific field configuration. Since the supersymmetry variations of the bosonic fields always contain one or more fermionic fields, which vanish classically, these variations are guaranteed to vanish. Thus, in exploring supersymmetry of solutions, the terms of interest are the variations of the fermionic fields that do not contain any fermionic fields. In the case at hand this means that Killing spinors ε are given by solutions of the equation

$$\delta\Psi_M = \nabla_M \varepsilon + \frac{1}{12}\left(\Gamma_M \mathbf{F}^{(4)} - 3\mathbf{F}_M^{(4)}\right)\varepsilon = 0, \qquad (8.21)$$

and the bosonic terms that have been included in Eq. (8.12) determine the possible supersymmetric solutions.

M-branes

An important feature of M-theory (and 11-dimensional supergravity) is the presence of the three-form gauge field A_3. As has been explained in Chapter 6, such fields couple to branes, which in turn are sources for the gauge field. In this case ($n = 3$ and $D = 11$) the three-form can couple electrically to a two-brane, called the M2-brane, and magnetically to a five-brane, called the M5-brane. If the tensions saturate a BPS bound (as they do), these are stable supersymmetric branes whose tensions can be computed exactly. By focusing attention on BPS M-branes, it is possible to learn various facts about M-theory that go beyond the low-energy effective-action expansion. In fact, we will even discover an M-theory version of T-duality that shows the limitations of a geometrical description.

The only scale in M-theory is the 11-dimensional Planck length ℓ_p. Therefore, the M-brane tensions can be determined, up to numerical factors, by dimensional analysis. The exact results, which are confirmed by duality arguments relating M-branes to branes in type II superstring theories, turn out to be

$$T_{M2} = 2\pi(2\pi\ell_p)^{-3} \quad \text{and} \quad T_{M5} = 2\pi(2\pi\ell_p)^{-6}. \quad (8.22)$$

As is the case with all BPS branes, an M-brane can be excited so that it is no longer BPS, but then it would be unstable and radiate until reaching the minimal BPS energy density in (8.22).

Type IIA supergravity

The action of 11-dimensional supergravity is related to the actions of the various ten-dimensional supergravity theories, which are the low-energy effective descriptions of superstring theories. The most direct connection is between 11-dimensional supergravity and type IIA supergravity. The deep reason is that M-theory compactified on a circle of radius R corresponds to type IIA superstring theory in ten dimensions with coupling constant $g_s = R/\sqrt{\alpha'}$. This duality is discussed later in this chapter.[4] For now, the important consequence is that it implies that type IIA supergravity can be obtained from 11-dimensional supergravity by *dimensional reduction*. Dimensional reduction is achieved by taking one dimension to be a circle and only keeping the zero modes in the Fourier expansions of the various fields. This is to be contrasted with *compactification*, where all the modes are kept

---

4 In particular, it turns out that the type IIA superstring can be obtained from the M2-brane by wrapping one dimension of the membrane on the circle to give a string in the other ten dimensions.

in the lower-dimensional theory. In fact, the type IIA supergravity action was originally constructed by dimensional reduction. This is the easiest method, so it is utilized in the following.

Fermionic fields

As we already discussed in Chapter 5, the massless fermions of type IIA supergravity consist of two Majorana–Weyl gravitinos of opposite chirality and two Majorana–Weyl dilatinos of opposite chirality. These fermionic fields can be obtained by taking an 11-dimensional Majorana gravitino and dimensionally reducing it to ten dimensions. The 32-component Majorana spinors Ψ_M give a pair of 16-component Majorana–Weyl spinors of opposite chirality. Then the first ten components give the two ten-dimensional gravitinos and Ψ_{11} gives the two ten-dimensional dilatinos. Each type IIA dilatino has eight physical polarizations, because the Dirac equation implies that half of the 16 components describe independent propagating modes. For the counting to add up, it is clear that each of the gravitinos must have 56 physical degrees of freedom. These are the dimensions of irreducible representations of $Spin(8)$, so the discussion given here can be understood group theoretically as the decomposition of the **128** representation of $Spin(9)$ into irreducible representations of the subgroup $Spin(8)$. Altogether, there are 128 fermionic degrees of freedom, just as in 11 dimensions. This preservation of degrees of freedom is a general feature of dimensional reduction on circles or tori.

Bosonic fields

Let us now consider the dimensional reduction of the bosonic fields of 11-dimensional supergravity, the metric and the three-form. Greek letters $\mu, \nu, \ldots$ refer to the first ten components of the 11-dimensional indices M, N, which are chosen to take the values $0, 1, \ldots, 9, 11$. Note that we skip the index value 10. The metric is decomposed according to

$$G_{MN} = e^{-2\Phi/3} \begin{pmatrix} g_{\mu\nu} + e^{2\Phi} A_\mu A_\nu & e^{2\Phi} A_\mu \\ e^{2\Phi} A_\nu & e^{2\Phi} \end{pmatrix}, \quad (8.23)$$

where all of the fields depend on the ten-dimensional space-time coordinates x^μ only. The exponential factors of the scalar field Φ, which turns out to be the dilaton, are introduced for later convenience. From the decomposition of the 11-dimensional metric (8.23) one gets a ten-dimensional metric $g_{\mu\nu}$, a $U(1)$ gauge field A_μ and a scalar dilaton field Φ. Equation (8.23) can be recast in the form

$$ds^2 = G_{MN} dx^M dx^N = e^{-2\Phi/3} g_{\mu\nu} dx^\mu dx^\nu + e^{4\Phi/3} (dx^{11} + A_\mu dx^\mu)^2. \quad (8.24)$$

In terms of the elfbein E^A_M this reduction takes the form

$$E^A_M = \begin{pmatrix} e^{-\Phi/3} e^a_\mu & 0 \\ e^{2\Phi/3} A_\mu & e^{2\Phi/3} \end{pmatrix}, \tag{8.25}$$

where e^a_μ is the ten-dimensional zehnbein. The corresponding inverse elfbein, which is useful in the following, is given by

$$E^M_A = \begin{pmatrix} e^{\Phi/3} e^\mu_a & 0 \\ -e^{\Phi/3} A_a & e^{-2\Phi/3} \end{pmatrix}. \tag{8.26}$$

The three-form in $D = 11$ gives rise to a three-form and a two-form in $D = 10$

$$A^{(11)}_{\mu\nu\rho} = A_{\mu\nu\rho} \quad \text{and} \quad A^{(11)}_{\mu\nu 11} = B_{\mu\nu}, \tag{8.27}$$

with the corresponding field strengths given by

$$F^{(11)}_{\mu\nu\rho\lambda} = F_{\mu\nu\rho\lambda} \quad \text{and} \quad F^{(11)}_{\mu\nu\rho 11} = H_{\mu\nu\rho}. \tag{8.28}$$

The dimensional reduction can lead to somewhat lengthy formulas due to the nondiagonal form of the metric. A useful trick for dealing with this is to convert first to tangent-space indices, since the reduction of the tangent-space metric is trivial. With this motivation, let us expand

$$F^{(11)}_{ABCD} = E^M_A E^N_B E^P_C E^Q_D F^{(11)}_{MNPQ}. \tag{8.29}$$

There are two cases depending on whether the indices (A, B, C, D) are purely ten-dimensional or one of them is 11-dimensional

$$F^{(11)}_{abcd} = e^{4\Phi/3}(F_{abcd} + 4A_{[a} H_{bcd]}) = e^{4\Phi/3} \widetilde{F}_{abcd}, \tag{8.30}$$

$$F^{(11)}_{abc 11} = e^{\Phi/3} H_{abc}.$$

It follows that upon dimensional reduction the 11-dimensional field strength is a combination of a four-form and a three-form field strength

$$\mathbf{F}^{(4)} = e^{4\Phi/3} \widetilde{\mathbf{F}}^{(4)} + e^{\Phi/3} \mathbf{H}^{(3)} \Gamma_{11}, \tag{8.31}$$

where Γ_{11} is the ten-dimensional chirality operator. The quantities $\widetilde{\mathbf{F}}^{(4)}$ and $\mathbf{H}^{(3)}$ are defined in the same way as $\mathbf{F}^{(4)}$ in Eq. (8.13). Using differential-form notation, the rescaled tensor field can be written as

$$\widetilde{F}_4 = dA_3 + A_1 \wedge H_3. \tag{8.32}$$

Notice that for the four-form $\widetilde{F}_4$ to be invariant under the $U(1)$ gauge transformation $\delta A_1 = d\Lambda$, the three-form potential should transform as $\delta A_3 = d\Lambda \wedge B$. Then

$$\delta \widetilde{F}_4 = d(d\Lambda \wedge B) + d\Lambda \wedge H_3 = 0. \tag{8.33}$$

In addition, the four-form $\widetilde{F}_4$ is invariant under the more obvious gauge transformation $\delta A_3 = d\Lambda_2$.

Coupling constants

The vacuum expectation value of $\exp \Phi$ is the type IIA superstring coupling constant g_s. From Eq. (8.24) we see that if a distance in string units is 1, say, then the same distance measured in 11d Planck units is $g_s^{-1/3}$. For small g_s, this is large. It follows that the Planck length is smaller than the string length if g_s is small. As a result,[5]

$$\ell_p = g_s^{1/3} \ell_s \quad \text{with} \quad \ell_s = \sqrt{\alpha'}. \tag{8.34}$$

In ten dimensions the relation between Newton's constant, the gravitational coupling constant and the string length and coupling constant is

$$16\pi G_{10} = 2\kappa_{10}^2 = \frac{1}{2\pi} (2\pi \ell_s)^8 g_s^2. \tag{8.35}$$

Dimensional reduction on a circle of radius R_{11} gives a relation between Newton's constant in ten and 11 dimensions

$$G_{11} = 2\pi R_{11} G_{10}. \tag{8.36}$$

Using Eqs (8.9) and (8.34), one deduces that the radius of the circle is

$$R_{11} = g_s^{2/3} \ell_p = g_s \ell_s. \tag{8.37}$$

These formulas are confirmed again later in this chapter when the type IIA D0-brane is identified with the first Kaluza–Klein excitation on the circle. Let us also define

$$2\kappa^2 = \frac{1}{2\pi} (2\pi \ell_s)^8, \tag{8.38}$$

which agrees with $2\kappa_{10}^2$ up to a factor of g_s^2, that is, $\kappa_{10}^2 = \kappa^2 g_s^2$.

---

[5] Chapter 2 introduced a string length scale $l_s = \sqrt{2\alpha'}$, which has been used until now. Here it is convenient to introduce a string length scale $\ell_s = \sqrt{\alpha'}$, which is used throughout this chapter. Note the change of font. Both conventions are used in the literature, and there is little to be gained from eliminating one of them.

Action

The bosonic action in the *string frame* for the $D = 10$ type IIA supergravity theory is obtained from the bosonic $D = 11$ action once the integration over the compact coordinate is carried out. The result contains three distinct types of terms

$$S = S_{\text{NS}} + S_{\text{R}} + S_{\text{CS}}. \tag{8.39}$$

The first term is

$$S_{\text{NS}} = \frac{1}{2\kappa^2} \int d^{10}x \sqrt{-g}\, e^{-2\Phi} \left(R + 4\partial_\mu \Phi \partial^\mu \Phi - \frac{1}{2}|H_3|^2 \right). \tag{8.40}$$

Note that the coefficient is $1/2\kappa^2$, which does not contain any powers of the string coupling constant g_s. This string-frame action is characterized by the exponential dilaton dependence in front of the curvature scalar. By a Weyl rescaling of the metric, this action can be transformed to the *Einstein frame* in which the Einstein term has the conventional form. This is a homework problem.

The remaining two terms in the action S involve the R–R fields and are given by

$$S_{\text{R}} = -\frac{1}{4\kappa^2} \int d^{10}x \sqrt{-g} \left(|F_2|^2 + |\widetilde{F}_4|^2 \right), \tag{8.41}$$

$$S_{\text{CS}} = -\frac{1}{4\kappa^2} \int B_2 \wedge F_4 \wedge F_4. \tag{8.42}$$

As a side remark, let us point out the following: a general rule, discussed in Chapter 3, is that a world sheet of Euler characteristic χ gives a contribution with a dilaton dependence $\exp(\chi \Phi)$, which leads to the correct dependence on the string coupling constant. All terms in the classical action Eq. (8.39) correspond to a spherical world sheet with $\chi = -2$, because they describe the leading order of the expansion in g_s. Notice, however, that the terms S_{R} and S_{CS}, which involve R–R fields, do not contain the expected factor of $e^{-2\Phi}$. This is only a consequence of the way the R–R fields have been defined. One could rescale C_1 and F_2 by $C_1 = e^{-\Phi} \widetilde{C}_1$ and $F_2 = e^{-\Phi} \widetilde{F}_2$, where $\widetilde{F}_2 = d\widetilde{C}_1 - d\Phi \wedge \widetilde{C}_1$ and make analogous redefinitions for C_3 and F_4. Then the factor of $e^{-2\Phi}$ would appear in all terms. However, this field redefinition is not usually made, so the action that is displayed is in the form that is most commonly found in the literature.

Supersymmetry transformations

Let us now examine the supersymmetry transformations of the fermi fields to leading order in these fields. We first rewrite the gravitino variation in

Eq. (8.12) in the form

$$\delta \Psi_A = E_A^\mu \partial_\mu \varepsilon + \frac{1}{4}\omega_{ABC}\Gamma^{BC}\varepsilon + \frac{1}{24}\left(3\mathbf{F}^{(4)}\Gamma_A - \Gamma_A \mathbf{F}^{(4)}\right)\varepsilon, \qquad (8.43)$$

where we are using 11-dimensional tangent-space indices. To interpret the previous expression in terms of ten-dimensional quantities, we need to work out the various pieces of the spin connection, which (to avoid confusion) is now denoted $\omega_{ABC}^{(11)}$. Using Eq. (8.19), one finds that

$$\omega_{aBC}^{(11)}\Gamma^{BC} = e^{\Phi/3}(\omega_{abc}\Gamma^{bc} - \frac{2}{3}\Gamma_a{}^\mu \partial_\mu \Phi) + e^{4\Phi/3}F_{ab}\Gamma^b\Gamma_{11} \qquad (8.44)$$

and

$$\omega_{11BC}^{(11)}\Gamma^{BC} = -\frac{1}{2}e^{4\Phi/3}F_{bc}\Gamma^{bc} - \frac{4}{3}e^{\Phi/3}\Gamma^\mu \Gamma_{11}\partial_\mu \Phi. \qquad (8.45)$$

Using these equations

$$e^{-\Phi/3}\delta\Psi_{11} = -\frac{1}{4}e^\Phi \mathbf{F}^{(2)}\varepsilon - \frac{1}{3}\partial_\mu \Phi \Gamma^\mu \Gamma_{11}\varepsilon + \frac{1}{12}e^\Phi \widetilde{\mathbf{F}}^{(4)}\Gamma_{11}\varepsilon + \frac{1}{6}\mathbf{H}^{(3)}\varepsilon \quad (8.46)$$

and

$$e^{-\Phi/3}\delta\Psi_a = e_a^\mu \nabla_\mu \varepsilon - \frac{1}{6}\Gamma_a{}^\mu \partial_\mu \Phi \varepsilon + \frac{1}{4}e^\Phi F_{ab}\Gamma^b\Gamma_{11}\varepsilon$$

$$+\frac{1}{24}e^\Phi(3\widetilde{\mathbf{F}}^{(4)}\Gamma_a - \Gamma_a \widetilde{\mathbf{F}}^{(4)})\varepsilon - \frac{1}{24}(3\mathbf{H}^{(3)}\Gamma_a + \Gamma_a \mathbf{H}^{(3)})\Gamma_{11}\varepsilon. \qquad (8.47)$$

To obtain the supersymmetry transformations in the desired form, we define new spinors as follows:

$$\tilde\lambda = e^{-\Phi/6}\Psi_{11}, \qquad (8.48)$$

$$\widetilde\Psi_\mu = e^{-\Phi/6}(\Psi_\mu + \frac{1}{2}\Gamma_\mu \Gamma_{11}\Psi_{11}) \qquad (8.49)$$

and $\tilde\varepsilon = \exp(\Phi/6)\varepsilon$. The final expressions for the supersymmetry transformations then become[6]

$$\delta\lambda = \left(-\frac{1}{3}\Gamma^\mu \partial_\mu \Phi \Gamma_{11} + \frac{1}{6}\mathbf{H}^{(3)} - \frac{1}{4}e^\Phi \mathbf{F}^{(2)} + \frac{1}{12}e^\Phi \widetilde{\mathbf{F}}^{(4)}\Gamma_{11}\right)\varepsilon \qquad (8.50)$$

and

$$\delta\Psi_\mu = \left(\nabla_\mu - \frac{1}{4}\mathbf{H}_\mu^{(3)}\Gamma_{11} - \frac{1}{8}e^\Phi F_{\nu\rho}\Gamma_\mu{}^{\nu\rho}\Gamma_{11} + \frac{1}{8}e^\Phi \mathbf{F}^{(4)}\Gamma_\mu\right)\varepsilon. \qquad (8.51)$$

The second term in $\delta\Psi_\mu$ has an interpretation as torsion.[7] Because of the Γ_{11}

---

6 In order to make the equations less cluttered, we have removed the tildes from the fermionic fields and ε.
7 Torsion is defined in the appendix of Chapter 9.

factor, the torsion has opposite sign for the opposite chiralities $\frac{1}{2}(1\pm\Gamma_{11})\Psi_\mu$. The spinors λ, Ψ_μ and ε are each Majorana spinors. As such they could be decomposed into a pair of Majorana–Weyl spinors of opposite chirality, though there is no advantage in doing so. Therefore, they describe two dilatinos, two gravitinos and $\mathcal{N}=2$ supersymmetry in ten dimensions.

Type IIB supergravity

Unlike type IIA supergravity, the type IIB theory cannot be obtained by reduction from 11-dimensional supergravity. The guiding principles to construct this theory come from supersymmetry as well as gauge invariance. One challenging feature of the type IIB theory is that the self-dual five-form field strength introduces an obstruction to formulating the action in a manifestly covariant form. One strategy for dealing with this is to focus on the field equations instead, since they can be written covariantly. Alternatively, one can write an action that needs to be supplemented by a self-duality constraint.

Field content

Chapter 5 derived the massless spectrum of the type IIB superstring, which gives the particle content of type IIB supergravity. The fermionic part of the spectrum consists of two left-handed Majorana–Weyl gravitinos (or, equivalently, one Weyl gravitino) and two right-handed Majorana–Weyl dilatinos (or, equivalently, one Weyl dilatino). The NS–NS bosons consist of the metric (or zehnbein), the two-form B_2 (with field strength $H_3 = dB_2$) and the dilaton Φ. The R–R sector consists of form fields C_0, C_2 and C_4. The latter has a self-dual field strength $\widetilde{F}_5$.

The self-dual five-form

The presence of the self-dual five-form introduces a significant complication for writing down a classical action for type IIB supergravity. The basic issue, which also exists for analogous self-dual tensors in two and six dimensions, is that an action of the form

$$\int |F_5|^2 \, d^{10}x \tag{8.52}$$

does not incorporate the self-duality constraint, and therefore it describes twice the desired number of propagating degrees of freedom. The introduction of a Lagrange multiplier field to implement the self-duality condition does not help, because the Lagrange multiplier field itself ends up reintroducing the components it was intended to eliminate.

There are several different ways of dealing with the problem of the self-dual field. The original approach is to not construct an action, but only the field equations and the supersymmetry transformations. This is entirely adequate for most purposes, since the supergravity theory is only an effective theory, and not a quantum theory that one inserts in a path integral. The basic idea is that the supersymmetric variation of an equation of motion should give another equation of motion (or combination of equations of motion). By pursuing this systematically, it turns out to be possible to determine the supersymmetry transformations and the field equations simultaneously. In fact, the equations are highly overconstrained, so one obtains many consistency checks.

It is possible to formulate a manifestly covariant action with the correct degrees of freedom if, following Pasti, Sorokin, and Tonin (PST), one introduces an auxiliary scalar field and a compensating gauge symmetry in a suitable manner. The extra gauge symmetry can be used to set the auxiliary scalar field equal to one of the space-time coordinates as a gauge choice, but then the resulting gauge-fixed theory does not have manifest general coordinate invariance in one of the directions. Nonetheless, it is a correct theory, at least for space-time topologies for which this gauge choice is globally well defined.

An action

We do not follow the PST approach here, but instead present an action that gives the correct equations of motion when one imposes the self-duality condition as an extra constraint. Such an action is not supersymmetric, however, because (without the constraint) it has more bosonic than fermionic degrees of freedom. Moreover, the constraint cannot be incorporated into the action for the reasons discussed above.

The way to discover this action is to first construct the supersymmetric equations of motion, and then to write down an action that reproduces those equations when the self-duality condition is imposed by hand. The bosonic part of the type IIB supergravity action obtained in this way takes the form

$$S = S_{\rm NS} + S_{\rm R} + S_{\rm CS}. \tag{8.53}$$

Here $S_{\rm NS}$ is the same expression as for the type IIA supergravity theory in Eq. (8.40), while the parts of the action describing the massless R–R sector fields are given by

$$S_{\rm R} = -\frac{1}{4\kappa^2}\int d^{10}x \sqrt{-g}\left(|F_1|^2 + |\widetilde{F}_3|^2 + \frac{1}{2}|\widetilde{F}_5|^2\right), \tag{8.54}$$

8.1 Low-energy effective actions

$$S_{\text{CS}} = -\frac{1}{4\kappa^2} \int C_4 \wedge H_3 \wedge F_3. \tag{8.55}$$

In these formulas $F_{n+1} = dC_n$, $H_3 = dB_2$ and

$$\widetilde{F}_3 = F_3 - C_0\, H_3, \tag{8.56}$$

$$\widetilde{F}_5 = F_5 - \frac{1}{2} C_2 \wedge H_3 + \frac{1}{2} B_2 \wedge F_3. \tag{8.57}$$

These are the gauge-invariant combinations analogous to $\widetilde{F}_4$ in the type IIA theory. In each case the R–R fields that appear here differ by field redefinitions from the ones that couple simply to the D-brane world volumes, as described in Chapter 6. The five-form satisfying the self-duality condition is $\widetilde{F}_5$, that is,

$$\widetilde{F}_5 = \star \widetilde{F}_5. \tag{8.58}$$

This condition has to be imposed as a constraint that supplements the equations of motion that follow from the action.

Supersymmetry transformations

Even though the action we presented is not the bosonic part of a supersymmetric action, the field equations, including the constraint, are. In other words, as explained earlier, the supersymmetry variations of these equations vanish if after the variation one imposes the equations themselves. The supersymmetry transformations of type IIB supergravity are required in later chapters, so we present them here.

Let us represent the dilatino and gravitino fields by Weyl spinors λ and Ψ_μ, respectively. Similarly, the infinitesimal supersymmetry parameter is represented by a Weyl spinor ε. The supersymmetry transformations of the fermi fields of type IIB supergravity (to leading order in fermi fields) are

$$\delta\lambda = \frac{1}{2}\left(\partial_\mu \Phi - i e^\Phi \partial_\mu C_0\right) \Gamma^\mu \varepsilon + \frac{1}{4}\left(i e^\Phi \widetilde{\mathbf{F}}^{(3)} - \mathbf{H}^{(3)}\right)\varepsilon^\star \tag{8.59}$$

and

$$\delta\Psi_\mu = \left(\nabla_\mu + \frac{i}{8} e^\Phi \mathbf{F}^{(1)} \Gamma_\mu + \frac{i}{16} e^\Phi \widetilde{\mathbf{F}}^{(5)} \Gamma_\mu\right)\varepsilon - \frac{1}{8}\left(2\mathbf{H}^{(3)}_\mu + i e^\Phi \widetilde{\mathbf{F}}^{(3)} \Gamma_\mu\right)\varepsilon^\star. \tag{8.60}$$

Global $SL(2,\mathbb{R})$ symmetry

Type IIB supergravity has a noncompact global symmetry $SL(2,\mathbb{R})$. This is not evident in the equations above, so let us sketch what is required to make it apparent. The theory has two two-form potentials, B_2 and C_2, which transform as a doublet under the $SL(2,\mathbb{R})$ symmetry group. Therefore, to rewrite the action in a way that the symmetry is manifest, let us rename the two-form potentials $B_2 = B_2^{(1)}$ and $C_2 = B_2^{(2)}$ and introduce a two-component vector notation

$$B_2 = \begin{pmatrix} B_2^{(1)} \\ B_2^{(2)} \end{pmatrix}. \tag{8.61}$$

Similarly, $H_3 = dB_2$ is also a two-component column vector. Under a transformation by

$$\Lambda = \begin{pmatrix} d & c \\ b & a \end{pmatrix} \in SL(2,\mathbb{R}), \tag{8.62}$$

the B fields transform linearly by the rule

$$B_2 \to \Lambda B_2. \tag{8.63}$$

Since the parameters in Λ are constants, H_3 transforms in the same way. The complex scalar field τ, defined by

$$\tau = C_0 + ie^{-\Phi}, \tag{8.64}$$

is useful because it transforms nonlinearly by the familiar rule

$$\tau \to \frac{a\tau + b}{c\tau + d}. \tag{8.65}$$

The field C_0 is sometimes referred to as an axion, because of the shift symmetry $C_0 \to C_0 +$ constant of the theory (in the supergravity approximation), and then the complex field τ is referred to as an *axion–dilaton field*.

The action can be conveniently written in terms of the symmetric $SL(2,\mathbb{R})$ matrix

$$\mathcal{M} = e^{\Phi} \begin{pmatrix} |\tau|^2 & -C_0 \\ -C_0 & 1 \end{pmatrix}, \tag{8.66}$$

which transforms by the simple rule

$$\mathcal{M} \to (\Lambda^{-1})^T \mathcal{M} \Lambda^{-1}. \tag{8.67}$$

The canonical *Einstein-frame metric* $g_{\mu\nu}^{\mathrm{E}}$ and the four-form C_4 are $SL(2,\mathbb{R})$ invariant. Note that since the dilaton transforms, the type IIB *string-frame metric* $g_{\mu\nu}$ in the action (8.53), which is related to the canonical Einstein metric by

$$g_{\mu\nu} = e^{\Phi/2} g_{\mu\nu}^{\mathrm{E}}, \tag{8.68}$$

is not $SL(2, \mathbb{R})$ invariant. The transformation of the scalar curvature term under this change of variables is given by

$$\frac{1}{2\kappa^2} \int d^{10}x \sqrt{-g}\, e^{-2\Phi} R \to \frac{1}{2\kappa^2} \int d^{10}x \sqrt{-g}(R - \frac{9}{2}\partial^\mu \Phi \partial_\mu \Phi), \tag{8.69}$$

where the string-frame metric is used in the first expression and the Einstein-frame metric is used in the second one.

Using the quantities defined above, the type IIB supergravity action can be recast in the form

$$S = \frac{1}{2\kappa^2} \int d^{10}x \sqrt{-g} \left(R - \frac{1}{12} H^T_{\mu\nu\rho} \mathcal{M} H^{\mu\nu\rho} + \frac{1}{4}\mathrm{tr}(\partial^\mu \mathcal{M} \partial_\mu \mathcal{M}^{-1}) \right)$$

$$- \frac{1}{8\kappa^2} \left(\int d^{10}x \sqrt{-g} |\tilde{F}_5|^2 + \int \varepsilon_{ij} C_4 \wedge H_3^{(i)} \wedge H_3^{(j)} \right), \tag{8.70}$$

where the metric g^E is used throughout. This action is manifestly invariant under global $SL(2, \mathbb{R})$ transformations.

The self-duality equation, $\tilde{F}_5 = \star \tilde{F}_5$, which is imposed as a constraint in this formalism, is also $SL(2, \mathbb{R})$ invariant. To see this, first note that the Hodge dual that defines $\star \tilde{F}_5$ is invariant under a Weyl rescaling, so that it doesn't matter whether it is defined using the string-frame metric or the Einstein-frame metric. The definition of $\tilde{F}_5$ in Eq. (8.57) can be recast in the manifestly $SL(2, \mathbb{R})$ invariant form

$$\tilde{F}_5 = F_5 + \frac{1}{2} \varepsilon_{ij} B_2^{(i)} \wedge H_3^{(j)}. \tag{8.71}$$

The invariance of the self-duality equation then follows.

Type I supergravity

Field content

As explained in Chapter 6, type I superstring theory arises as an orientifold projection of the type IIB superstring theory. This involves a truncation of the type IIB closed-string spectrum to the left–right symmetric states as well as the addition of a twisted sector consisting of open strings. The massless closed-string sector is $\mathcal{N} = 1$ supergravity in ten dimensions and the massless open-string sector is $\mathcal{N} = 1$ super Yang–Mills theory with gauge group $SO(32)$ in ten dimensions. Therefore, the low-energy effective action should describe the interactions of these two supermultiplets to leading order in the α' expansion.

Restricting to the bosonic sector of the theory, the massless fields of type I superstring theory in ten dimensions consist of

$$g_{\mu\nu}, \ \Phi, \ C_2 \ \text{and} \ A_\mu. \tag{8.72}$$

Here $g_{\mu\nu}$ is the graviton, Φ is the dilaton, C_2 is the R–R two-form and A_μ is the $SO(32)$ Yang–Mills gauge field coming from the twisted sector.

Action

In the string frame, the bosonic part of the supersymmetric Lagrangian describing the low-energy limit of the type I superstring is

$$S = \frac{1}{2\kappa^2} \int d^{10}x \sqrt{-g} \left[e^{-2\Phi}(R + 4\partial_\mu \Phi \partial^\mu \Phi) - \frac{1}{2}|\tilde{F}_3|^2 - \frac{\kappa^2}{g^2} e^{-\Phi} \mathrm{tr}(|F_2|^2) \right]. \tag{8.73}$$

Here $F_2 = dA + A \wedge A$ is the Yang–Mills field strength corresponding to the gauge field $A = A_\mu dx^\mu$. Moreover,

$$\tilde{F}_3 = dC_2 + \frac{\ell_s^2}{4} \omega_3, \tag{8.74}$$

as explained in the anomaly analysis of Chapter 5.[8] In the full string theory the Chern–Simons term is

$$\omega_3 = \omega_{\mathrm{L}} - \omega_{\mathrm{YM}}, \tag{8.75}$$

where

$$\omega_{\mathrm{L}} = \mathrm{tr}(\omega \wedge d\omega + \frac{2}{3} \omega \wedge \omega \wedge \omega) \tag{8.76}$$

and

$$\omega_{\mathrm{YM}} = \mathrm{tr}(A \wedge dA + \frac{2}{3} A \wedge A \wedge A). \tag{8.77}$$

Here ω_{L} is the Lorentz Chern–Simons term (ω is the spin connection) and ω_{YM} is the Yang–Mills Chern–Simons term. However, the Lorentz Chern–Simons term is higher-order in derivatives, so only the Yang–Mills Chern–Simons term is part of the low-energy effective supergravity theory.

The parameter g, introduced in Eq. (8.73), is related to the ten-dimensional Yang–Mills coupling constant g_{YM} by

$$\frac{g_{\mathrm{YM}}^2}{4\pi} = \frac{g^2}{4\pi} g_{\mathrm{s}} = (2\pi \ell_{\mathrm{s}})^6 g_{\mathrm{s}}. \tag{8.78}$$

In type I superstring theory, g_{YM} is an open-string coupling, and therefore

---

[8] The conventions here correspond to setting the parameter μ that was introduced in Section 5.4 equal to $8/\ell_s^2$. The gauge field A is antihermitian as in Chapter 5.

it is proportional to $\sqrt{g_s}$. As discussed in Chapter 3, this is a consequence of the fact that open strings couple to world-sheet boundaries, whereas closed strings couple to interior points of the string world sheet.[9] In the heterotic string theory, considered in the next section, the counting is a bit different. There g_{YM} is a closed-string coupling, and therefore it is proportional to g_s.

Note that the first two terms of Eq. (8.73) come from a spherical world sheet (with $\chi = -2$), whereas the last term comes from a disk world sheet (with $\chi = -1$). The third term involves an R–R field and therefore is independent of Φ, as discussed earlier.

The action (8.73) describes $\mathcal{N} = 1$ supergravity coupled to $SO(32)$ super Yang–Mills theory in ten dimensions. As such, it only contains the leading terms in the low-energy expansion of the effective action of the type I superstring theory. In this particular case, some of the higher-order corrections to this action are already known from the anomaly analysis. Specifically, as mentioned above, the Chern–Simons term in the definition of $\widetilde{F}_3$ contains both a Yang–Mills and a Lorentz contribution in the full theory, but the Lorentz Chern–Simons term is higher-order in derivatives, and therefore it is not included in the leading low-energy effective action. A local counterterm proportional to

$$\int C_2 \wedge Y_8, \tag{8.79}$$

also required by anomaly cancellation, consists entirely of terms of higher dimension than are included in the action given above.[10]

Supersymmetry transformations

Let us now consider the supersymmetry transformations that leave the type I effective action invariant. The terms involving the supergravity multiplet can be obtained by truncation of the type IIB supersymmetry transformations given earlier. The type IIB formulas used complex fermi fields such as $\lambda = \lambda_1 + i\lambda_2$, and similarly for Ψ_μ and the supersymmetry parameter ε. In the truncation to type I the combinations that survive are Majorana–Weyl fields given by sums such as $\lambda = \lambda_1 + \lambda_2$, and similarly for Ψ_μ and the supersymmetry parameter ε. Using this rule, the type IIB formulas imply that the transformations of the fermions in the supergravity multiplet are

---

9 This rule can be understood in terms of the genus of the relevant world-sheet diagrams.
10 The precise form of Y_8 can be found in Chapter 5.

given in the type I case by

$$\delta\Psi_\mu = \nabla_\mu \varepsilon - \tfrac{1}{8} e^\Phi \widetilde{\mathbf{F}}^{(3)} \Gamma_\mu \varepsilon,$$

$$\delta\lambda = \tfrac{1}{2} \partial\!\!\!/\Phi\varepsilon + \tfrac{1}{4} e^\Phi \widetilde{\mathbf{F}}^{(3)} \varepsilon, \qquad (8.80)$$

$$\delta\chi = -\tfrac{1}{2} \mathbf{F}^{(2)} \varepsilon.$$

The last equation represents the supersymmetry transformation of the adjoint fermions χ in the super Yang–Mills multiplet. As always, there are corrections to these formulas that are quadratic in fermi fields, but these are not needed to construct Killing spinor equations.

Heterotic supergravity

Chapter 7 derived the particle spectrum of the heterotic string theories in ten-dimensional Minkowski space-time. The massless field content of the $SO(32)$ heterotic string theory is exactly the same as that of the type I superstring theory. The massless fields of the $E_8 \times E_8$ heterotic string differ only by the replacement of the gauge group, though the differences are more substantial for the massive excitations.

Action

The bosonic part of the low-energy effective action of both of the heterotic theories in the ten-dimensional string frame is given by

$$S = \frac{1}{2\kappa^2} \int d^{10}x \sqrt{-g}\, e^{-2\Phi} \left[R + 4\partial_\mu \Phi \partial^\mu \Phi - \tfrac{1}{2} |\widetilde{H}_3|^2 - \frac{\kappa^2}{30g^2} \mathrm{Tr}(|F_2|^2) \right]. \qquad (8.81)$$

Note that the entire action comes from a spherical world sheet in this case, and heterotic theories have no R–R fields, which explains why every term contains a factor of $\exp(-2\Phi)$. F_2 is the field strength corresponding to the gauge groups $SO(32)$ or $E_8 \times E_8$ and

$$\widetilde{H}_3 = dB_2 + \frac{\ell_s^2}{4} \omega_3 \qquad (8.82)$$

satisfies the relation

$$d\widetilde{H}_3 = \frac{\ell_s^2}{4} \left(\mathrm{tr}\, R \wedge R - \frac{1}{30} \mathrm{Tr}\, F \wedge F \right). \qquad (8.83)$$

However, as noted in the type I context, the Lorentz term is not part of the leading low-energy effective theory. The gauge theory trace denoted Tr

8.1 Low-energy effective actions

is evaluated using the 496-dimensional adjoint representation. As was discussed in Chapter 5, this can be re-expressed in terms of the 32-dimensional fundamental representation of $SO(32)$, for which the trace is denoted tr, by using the identity

$$\text{tr} F \wedge F = \frac{1}{30} \text{Tr} F \wedge F. \tag{8.84}$$

Sometimes this notation is used in the $E_8 \times E_8$ theory, as well, even though this group doesn't have a 32-dimensional representation. In this notation, the cohomology classes of $\text{tr} R \wedge R$ and $\text{tr} F \wedge F$ must be equal, since $d\tilde{H}_3$ is exact.

Supersymmetry transformations

The heterotic string effective action has $\mathcal{N} = 1$ local supersymmetry in ten dimensions, which means that the gravitino field Ψ_μ is a Majorana–Weyl spinor. There is also a Majorana–Weyl dilatino field λ. The bosonic parts of the transformation formulas of the fermi fields, which is what is required to read off the Killing spinor equations, are

$$\begin{aligned}
\delta\Psi_\mu &= \nabla_\mu \varepsilon - \tfrac{1}{4}\tilde{\mathbf{H}}^{(3)}_\mu \varepsilon, \\
\delta\lambda &= -\tfrac{1}{2}\Gamma^\mu \partial_\mu \Phi \varepsilon + \tfrac{1}{4}\tilde{\mathbf{H}}^{(3)} \varepsilon, \\
\delta\chi &= -\tfrac{1}{2}\mathbf{F}^{(2)}\varepsilon.
\end{aligned} \tag{8.85}$$

The first two transformations can be deduced from the type IIB supersymmetry transformations by truncating to an $\mathcal{N} = 1$ subsector and keeping only the NS–NS fields. A nice feature of this formulation is that the $\tilde{H}_3$ contribution to $\delta\Psi_\mu$ can be interpreted as torsion.

EXERCISES

EXERCISE 8.3
The previous section described the global symmetry of the type IIB supergravity action using a matrix $\mathcal{M}$. Verify the identities

$$\frac{1}{4}\text{tr}(\partial^\mu \mathcal{M} \partial_\mu \mathcal{M}^{-1}) = -\frac{\partial^\mu \tau \partial_\mu \bar{\tau}}{2(\text{Im}\tau)^2} = -\frac{1}{2}\left(\partial^\mu \Phi \partial_\mu \Phi + e^{2\Phi}\partial^\mu C_0 \partial_\mu C_0\right).$$

Verify the $SL(2, \mathbb{R})$ invariance of this expression.

SOLUTION

By definition $\tau = C_0 + ie^{-\Phi}$ and

$$\mathcal{M} = e^{\Phi}\begin{pmatrix} |\tau|^2 & -C_0 \\ -C_0 & 1 \end{pmatrix}.$$

As a result,

$$\mathcal{M}^{-1} = e^{\Phi}\begin{pmatrix} 1 & C_0 \\ C_0 & |\tau|^2 \end{pmatrix}.$$

So

$$\frac{1}{4}\mathrm{tr}(\partial^{\mu}\mathcal{M}\partial_{\mu}\mathcal{M}^{-1}) = \frac{1}{2}\partial_{\mu}\left(e^{\Phi}|\tau|^2\right)\partial^{\mu}\left(e^{\Phi}\right) - \frac{1}{2}\partial_{\mu}\left(C_0 e^{\Phi}\right)\partial^{\mu}\left(C_0 e^{\Phi}\right)$$

$$= -\frac{1}{2}\left(\partial^{\mu}\Phi\partial_{\mu}\Phi + e^{2\Phi}\partial^{\mu}C_0\partial_{\mu}C_0\right).$$

Also,

$$-\frac{\partial^{\mu}\tau\partial_{\mu}\bar{\tau}}{2(\mathrm{Im}\tau)^2} = -\frac{1}{2}e^{2\Phi}\partial^{\mu}\left(C_0 + ie^{-\Phi}\right)\partial_{\mu}\left(C_0 - ie^{-\Phi}\right)$$

$$= -\frac{1}{2}\left(\partial^{\mu}\Phi\partial_{\mu}\Phi + e^{2\Phi}\partial^{\mu}C_0\partial_{\mu}C_0\right).$$

This establishes the required identities. The $SL(2,\mathbb{R})$ symmetry is manifest for $\mathrm{tr}(\partial^{\mu}\mathcal{M}\partial_{\mu}\mathcal{M}^{-1})$, because when one substitutes $\mathcal{M} \to (\Lambda^{-1})^T \mathcal{M} \Lambda^{-1}$ the constant Λ factors cancel using the cyclicity of the trace. □

EXERCISE 8.4
Verify that the action in Eq. (8.70) agrees with Eq. (8.53).

SOLUTION

First we need the action (8.53) in the Einstein frame. Using Eqs (8.68) and (8.69), it is given by $S = S_{\mathrm{NS}} + S_{\mathrm{R}} + S_{\mathrm{CS}}$, where

$$S_{\mathrm{NS}} = \frac{1}{2\kappa^2}\int d^{10}x\sqrt{-g}\left(R - \frac{1}{2}\partial_{\mu}\Phi\partial^{\mu}\Phi - \frac{1}{2}e^{-\Phi}|H_3|^2\right)$$

$$S_{\mathrm{R}} = -\frac{1}{4\kappa^2}\int d^{10}x\sqrt{-g}\left(e^{2\Phi}|F_1|^2 + e^{\Phi}|\tilde{F}_3|^2 + \frac{1}{2}|\tilde{F}_5|^2\right)$$

$$S_{\mathrm{CS}} = -\frac{1}{4\kappa^2}\int C_4 \wedge H_3 \wedge F_3.$$

We only need to rewrite the first two terms in Eq. (8.70) and compare them with the corresponding terms in the above actions, since the last two terms obviously agree. These terms are

$$-\tfrac{1}{12} H^T_{\mu\nu\rho} \mathcal{M} H^{\mu\nu\rho} + \tfrac{1}{4}\mathrm{tr}(\partial^\mu \mathcal{M} \partial_\mu \mathcal{M}^{-1})$$

$$= -\tfrac{1}{2} e^\Phi \left(|\tau|^2 |H_3|^2 + |F_3|^2 - 2 C_0 F \cdot H \right) - \tfrac{1}{2} \left(\partial^\mu \Phi \partial_\mu \Phi + e^{2\Phi} \partial^\mu C_0 \partial_\mu C_0 \right)$$

$$= -\tfrac{1}{2} \left(e^{-\Phi} |H_3|^2 + e^\Phi (F_3 - C_0 H_3)^2 \right) - \tfrac{1}{2} \left(\partial^\mu \Phi \partial_\mu \Phi + e^{2\Phi} \partial^\mu C_0 \partial_\mu C_0 \right).$$

Using $\widetilde{F}_3 = F_3 - C_0 H_3$, it becomes manifest that all terms match. □

8.2 S-duality

S-duality is a transformation that relates a string theory with coupling constant g_s to a (possibly) different theory with coupling constant $1/g_s$. This is analogous to the way that T-duality relates a circular dimension of radius R to one of radius ℓ_s^2/R. In each case the parameter is given by the vacuum expectation value of a scalar field. Thus the duality, at a more fundamental level, can be understood in terms of field transformations.

The symmetry of Maxwell's equation under the interchange of electric and magnetic quantities, combined with the Dirac quantization condition, already hints at the possibility of such a duality in field theory. This strong–weak (or electric–magnetic) duality symmetry generalizes to nonabelian gauge theories. The cleanest example is $\mathcal{N} = 4$ supersymmetric Yang–Mills (SYM) theory, which is a conformally invariant quantum theory, a fact that plays an important role in Chapter 12. In fact, when one includes a θ term

$$S_\theta = \frac{\theta}{16\pi^2} \int F^a \wedge F^a \tag{8.86}$$

in the definition of the $\mathcal{N} = 4$ SYM theory (as one should), this theory has an $SL(2,\mathbb{Z})$ duality under which the complex coupling constant

$$\tau = \frac{\theta}{2\pi} + i \frac{4\pi}{g_{\mathrm{YM}}^2} \tag{8.87}$$

transforms as a modular parameter. The fact that the theory is conformally invariant ensures that τ is a constant independent of any renormalization scale. The simple electric–magnetic duality $g_{\mathrm{YM}} \to 4\pi/g_{\mathrm{YM}}$ corresponds to the special case $\tau \to -1/\tau$ evaluated for $\theta = 0$. There has been extensive progress in recent times in understanding electric–magnetic dualities of other supersymmetric gauge theories, starting with the important work of Seiberg and Witten in 1994 for $\mathcal{N} = 2$ gauge theories.

A double expansion

In order to understand the various string dualities and their relationships it is useful to view string theory as a simultaneous expansion in two parameters:[11]

- One parameter is the Regge slope (or inverse string tension) α'. An expansion in α' is an expansion in "stringiness" about the point-particle limit. Mathematically, it is the perturbation expansion that corresponds to quantum-mechanical treatment of the string world-sheet theory, even though it concerns the classical physics of a string. (Recall that the world-sheet action has a coefficient $1/\alpha'$, so that α' plays a role analogous to Planck's constant.) Since α' has dimensions of (length)^2, the dimensionless expansion parameter can be $\alpha' p^2$, where p is a characteristic momentum or energy, or α'/L^2, where L is a characteristic length scale, such as the size of a compact dimension.

- The second expansion is the one in the string coupling constant g_s, which is the expectation value of the exponentiated dilaton field. This is the expansion in the number of string loops or, equivalently, the genus of the string world sheet.

S-duality and T-duality are quite analogous. However, S-duality seems deeper in that it is nonperturbative in the string loop expansion, whereas T-duality holds order by order in the loop expansion. In particular, it is valid in the leading (tree or classical) approximation.

Type I superstring – SO(32) heterotic string duality

The low-energy effective actions for the type I and $SO(32)$ heterotic theories are very similar. In particular, they are mapped into one another by the simple transformation

$$\Phi \to -\Phi \qquad (8.88)$$

combined with a Weyl rescaling of the metric

$$g_{\mu\nu} \to e^{-\Phi} g_{\mu\nu}. \qquad (8.89)$$

Thus the canonical Einstein metric $g^{\mathrm{E}}_{\mu\nu} = e^{-\Phi/2} g_{\mu\nu}$ is an invariant combination. All other bosonic fields remain unchanged ($A \leftrightarrow A$ and $B_2 \leftrightarrow C_2$).

---

[11] The discussion that follows applies to any of the superstring theories.

This leads to the conjecture that the two string theories (not just their low-energy limits) are actually dual to one another, which means that they are descriptions in two different regions of the parameter space of one and the same quantum theory. Since the string coupling constant is the vev of $\exp(\Phi)$ in each case, Eq. (8.88) implies that the type I superstring coupling constant is the reciprocal of the $SO(32)$ heterotic string coupling constant,

$$g_s^I g_s^H = 1. \tag{8.90}$$

Thus, when one of the two theories is weakly coupled, the other one is strongly coupled. This, of course, makes proving the type I–heterotic duality difficult. Some checks, beyond the analysis of the effective actions described above, can be made and no discrepancy has been found. More significantly, this is one link in an intricate overconstrained web of dualities. If any of them were wrong, the whole story would fall apart.

Nonperturbative test

As an example of a nonperturbative test of the duality, consider the D-string of the type I theory, whose tension is

$$T_{D1} = \frac{1}{g_s} \frac{1}{2\pi \ell_s^2}. \tag{8.91}$$

Let us test the conjecture that this string actually *is* the $SO(32)$ heterotic string, whose tension is

$$T_{F1} = \frac{1}{2\pi \ell_s^2}, \tag{8.92}$$

continued from weak coupling to strong coupling. The D-string is a supersymmetric object that saturates a BPS bound, and therefore the tension formula ought to be exact for all values of g_s. To compare these formulas one must realize that although the physical values of ℓ_s are the same in the two cases, they are being measured in different metrics, as a consequence of the Weyl rescaling in Eq. (8.89). Thus

$$\ell_s \to \ell_s \sqrt{g_s}. \tag{8.93}$$

Combined with the rule $g_s \to 1/g_s$, this indeed implies that the tensions T_{D1} and T_{F1} agree. Note that the transformation $g_s \to 1/g_s$, $\ell_s \to \ell_s \sqrt{g_s}$ squares to the identity, and so it is the same as its inverse.

The tensions of the magnetically-charged 5-branes that are dual to these strings can be compared in similar fashion. This is guaranteed to work by what has already been said, but let's check it anyway. In the type I theory

$$T_{\text{D5}} = \frac{1}{g_s(2\pi)^5 \ell_s^6},\tag{8.94}$$

and in the heterotic theory

$$T_{\text{NS5}} = \frac{1}{(g_s)^2(2\pi)^5 \ell_s^6}.\tag{8.95}$$

Once again, these map into one another in the required fashion.

The fundamental type I string

Having seen that the $SO(32)$ heterotic string can be identified with the type I D-string, one might wonder whether one can also identify a counterpart for the fundamental type I string in the $SO(32)$ heterotic theory. To answer this it is important to understand the essential difference between the two types of strings. The type I F-string does not carry a conserved charge, and it is not supersymmetric. The two-form B_2, which is the field that couples to a fundamental type IIB string, is removed from the spectrum by the orientifold projection. There are two ways of thinking about the reason that a type I F-string can break, both of which are correct. One is that there are space-time-filling D9-branes, and fundamental strings can break on D-branes. The other one is that since it does not carry a conserved charge, and it is not supersymmetric, there is no conservation law that prevents it from breaking. The amplitude for breaking a type I string is proportional to $\sqrt{g_s}$, so these strings can be long-lived for sufficiently small coupling constant. This is good enough for making them the fundamental objects on which to base a perturbation expansion. However, if the type I coupling constant is large, the type I F-strings are no longer a useful concept, since they disintegrate as shown in Fig. 8.1. Accordingly, there is no trace of them in the weakly-coupled heterotic description.

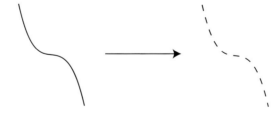

Fig. 8.1. The fundamental type I string disintegrates at strong coupling.

Type IIB S-duality

Type IIB supergravity has a global $SL(2,\mathbb{R})$ symmetry that was described earlier. However this symmetry of the low-energy effective action is not shared by the full type IIB superstring theory. Indeed, it is broken by a variety of stringy and quantum effects to the infinite discrete subgroup $SL(2,\mathbb{Z})$. One way of seeing this is to think about stable strings in this theory. Since there are two two-form gauge fields B_2 (NS–NS two-form) and C_2 (R–R two-form) there are two types of charge that a string can carry. The F-string (or fundamental string) has charge $(1,0)$, which means that it has one unit of the charge that couples to B_2 and none of the charge that couples to C_2. In similar fashion, the D-string couples to C_2 and has charge $(0,1)$. Since the two-forms form a doublet of $SL(2,\mathbb{R})$ it follows that these strings also transform as a doublet. In general, they transform into (p,q) strings, which carry both kinds of charge. The restriction to the $SL(2,\mathbb{Z})$ subgroup is essential to ensure that these charges are integers, as is required by the Dirac quantization conditions.

Symmetry under $g_s \to 1/g_s$

Recall that in type IIB supergravity the complex field

$$\tau = C_0 + ie^{-\Phi} \qquad (8.96)$$

transforms nonlinearly under $SL(2,\mathbb{R})$ transformations. This remains true in the full string theory, but only for the discrete subgroup $SL(2,\mathbb{Z})$. In particular, the transformation $\tau \to -1/\tau$, evaluated at $C_0 = 0$, changes the sign of the dilaton, which implies that the string coupling constant maps to its inverse. This is an S-duality transformation like the one that relates the type I superstring and $SO(32)$ heterotic string theories. In this case it relates the type IIB superstring theory to itself. Moreover, it is only one element of the infinite duality group $SL(2,\mathbb{Z})$. This duality group bears a striking resemblance to that of the $\mathcal{N} = 4$ SYM theory discussed at the beginning of this section. In Chapter 12 it is shown that this is not an accident.

(p,q) strings

The (p,q) strings are all on an equal footing, so they are all supersymmetric, in particular. This implies that each of their tensions saturates a BPS bound given by supersymmetry, and this uniquely determines what their tensions are. In the string frame, the result turns out to be

$$T_{(p,q)} = |p - q\tau_B| T_{F1} = T_{F1} \sqrt{\left(p - q\frac{\theta_0}{2\pi}\right)^2 + \frac{q^2}{g_s^2}}, \tag{8.97}$$

where we have defined the vev

$$\tau_B = \langle \tau \rangle = \frac{\theta_0}{2\pi} + \frac{i}{g_s} \tag{8.98}$$

and

$$T_{F1} = T_{(1,0)} = \frac{1}{2\pi \ell_s^2}. \tag{8.99}$$

This result can be derived by constructing the (p,q) strings as solitonic solutions of the type IIB supergravity field equations. The fact that these equations are only approximations to the superstring equations doesn't matter for getting the tension right, since it is a consequence of supersymmetry. Later, we confirm this tension formula by deriving it from a duality that relates the type IIB theory to M-theory.

Note that the F-string tension formula is valid for all values of θ_0, but the usual D-string tension formula

$$T_{D1} = T_{(0,1)} = \frac{T_{F1}}{g_s} \tag{8.100}$$

is only valid for $\theta_0 = 0$. Note also that a (p,q) string with $\theta_0 = 2\pi$ is equivalent to a $(p-q,q)$ string with $\theta_0 = 0$.

These (p,q) string tensions satisfy a triangle inequality

$$T_{(p_1+p_2,q_1+q_2)} \leq T_{(p_1,q_1)} + T_{(p_2,q_2)}, \tag{8.101}$$

and equality requires that the vectors (p_1, q_1) and (p_2, q_2) are parallel. One way of stating the conclusion is that a (p,q) string can be regarded as a bound state of p F-strings and q D-strings. It has lower tension than any other configuration with the same charges if and only if p and q are coprime. If there is a common divisor, there exists a multiple-string configuration with the same charges and tension.

Other BPS states

Let us briefly consider the $SL(2,\mathbb{Z})$ properties of the other BPS type IIB branes:

- The D3-brane carries a charge that couples to the $SL(2,\mathbb{Z})$ singlet field C_4. Therefore, it transforms as an $SL(2,\mathbb{Z})$ singlet, as well. This fact has the interesting consequence that any (p,q) string can end on a D3-brane,

since an $SL(2, \mathbb{Z})$ transformation that turns an F-string into a (p, q) string leaves the D3-brane invariant.
- There exist stable supersymmetric (p, q) 5-branes, which are the magnetic duals of (p, q) strings. Their $SL(2, \mathbb{Z})$ properties are quite similar to those of the (p, q) strings.
- The D7-brane couples magnetically to C_0. This field transforms in a rather complicated way under $SL(2, \mathbb{Z})$, so it is not immediately obvious how to classify 7-branes. Although this issue won't be pursued here, the classification is important, because certain nonperturbative vacua of type IIB superstring theory (described by F-theory) contain various 7-branes. This is addressed later.

The definition of a D-brane as a p-brane on which an F-string can end has to be interpreted carefully for $p = 1$. A naive interpretation of "a fundamental string ending on a D-string" would suggest a junction of three string segments, one of which is $(1, 0)$ and two of which are $(0, 1)$. This is not correct, however, because the charge on the end of the fundamental string results in flux that must go into one or the other of the attached string segments, changing the string charge in the process. In short, the three-string junction must satisfy charge conservation. This means that an allowed junction of three strings with charges $(p^{(i)}, q^{(i)})$ with $i = 1, 2, 3$ has to satisfy

$$\sum_i p^{(i)} = \sum_i q^{(i)} = 0. \tag{8.102}$$

Mathematically, this is just like momentum conservation at a vertex in a Feynman diagram (in two dimensions). The junction configuration is stable if the angles are chosen so that the three tensions, treated as vectors, add to zero. It is possible to build complex string webs using such junctions.

8.3 M-theory

The term M-theory was introduced by Witten to refer to the "mysterious" or "magical" quantum theory in 11 dimensions whose leading low-energy effective action is 11-dimensional supergravity. M-theory is not yet fully formulated, but the evidence for its existence is very compelling. It is as fundamental (but not more) as type IIB superstring theory, for example. In fact, the latter is somewhat better understood precisely because it is a string theory and therefore admits a well-defined perturbation expansion. This section describes a duality that relates M-theory compactified on a torus to type IIB superstring theory compactified on a circle. Since this duality requires a particular geometric set-up, it only allows solutions (or quantum vacua) of one theory to be recast in terms of the other theory for appropriate classes of geometries.

The description of M-theory in terms of an effective action is clearly not fundamental, so string theorists are searching for alternative formulations. One proposal for an exact nonperturbative formulation of M-theory, known as *Matrix theory*, is discussed in Chapter 12. It is not the whole story, however, since it is only applicable for a limited class of background geometries. A more general approach, called *AdS/CFT duality*, also is discussed in Chapter 12.

Type IIA superstring theory at strong coupling

The low-energy limit of type IIA superstring theory is type IIA supergravity, and this supergravity theory can be obtained by dimensional reduction of 11-dimensional supergravity, as has already been discussed. However, the correspondence between type IIA superstring theory and M-theory is much deeper than that. So let us take a closer look at the strong-coupling limit of the type IIA superstring theory.

D0-branes

Type IIA superstring theory has stable nonperturbative excitations, the D0-branes, whose mass in the string frame is given by $(\ell_s g_s)^{-1}$. The claim is that this can be interpreted from the viewpoint of M-theory compactified on a circle as the first Kaluza–Klein excitation of the massless supergravity multiplet. The entire 256-dimensional supermultiplet is sometimes referred to as the *supergraviton*. To examine this claim, let us consider 11-dimensional supergravity (or M-theory) compactified on a circle. The mass of the supergraviton in 11 dimensions is zero

$$M_{11}^2 = -p_M p^M = 0, \qquad M = 0, 1, \ldots, 9, 11. \tag{8.103}$$

In ten dimensions this takes the form

$$M_{10}^2 = -p_\mu p^\mu = p_{11}^2, \qquad \mu = 0, 1, \ldots, 9. \tag{8.104}$$

The momentum on the circle in the eleventh direction is quantized, $p_{11} = N/R_{11}$, and therefore the spectrum of ten-dimensional masses is

$$(M_N)^2 = (N/R_{11})^2 \qquad \text{with} \qquad N \in \mathbb{Z} \tag{8.105}$$

representing a tower of Kaluza–Klein excitations. These states also form short (256-dimensional) supersymmetry multiplets, so that they are all BPS states, and carry N units of a conserved $U(1)$ charge. For $N = 1$ the correspondence with the D0-brane requires that

$$R_{11} = \ell_s g_s, \qquad (8.106)$$

in agreement with the result presented in Section 8.1. The D0-branes are nonperturbative excitations of the type IIA theory, since their tensions diverge as $g_s \to 0$. Therefore, this correspondence provides a test of the duality between the type IIA theory and 11-dimensional M-theory that goes beyond the perturbative regime.

Since $R_{11} = \ell_s g_s$, the radius of the compactification is proportional to the string coupling constant. This means that the perturbative regime of the type IIA superstring theory in which $g_s \to 0$ corresponds to the limit $R_{11} \to 0$. Conversely, the strong-coupling limit, that is, the limit $g_s \to \infty$, corresponds to decompactification of the circular eleventh dimension giving a theory in which all ten spatial dimensions are on an equal footing. The 11-dimensional theory obtained in this way is M-theory, and the low-energy limit of M-theory is 11-dimensional supergravity.

Turning the argument around, this is powerful evidence in support of a nontrivial result concerning the existence of bound states of D0-branes. The Nth Kaluza–Klein excitation gives a multiplet of stable particles in ten dimensions that have N units of charge. Therefore, they can be regarded as bound states of N D0-branes. However, these are a very special type of bound state, one that has zero binding energy. There is no room for any binding energy, since these states saturate a BPS bound, which means they are as light as they are allowed to be for a state with N units of D0-brane charge. It also means that the mass formula in Eq. (8.105) is exact for all values of g_s. As discussed earlier, the only way in which the BPS mass formula could be violated would be for the short supermultiplet to turn into a long supermultiplet. However, the degrees of freedom that would be needed for this to happen are not present in this case.

Bound states with zero binding energy are called *threshold bound states*, and the question of whether or not they are stable is a very delicate matter. From the Kaluza–Klein viewpoint it is clear that they should be stable, but from the point of view of the dynamics of D0-branes in the type IIA theory, it is not at all obvious. In fact, the proof is highly technical involving an index theorem for a family of non-Fredholm operators. Moreover, the result is specific to this particular problem. There are other instances in which coincident BPS states do not form threshold bound states. An example that we already encountered concerns the type IIB (p, q) strings. These strings are only stable bound states if p and q are coprime.

M-branes

The BPS branes of M-theory are the M2-brane and the M5-brane. M-theory on $\mathbb{R}^{11}$ does not contain any strings. This raises the following question: What happens to the type IIA fundamental string for large coupling, when the theory turns into M-theory? The only plausible guess is that the type IIA F-string is actually an M2-brane with a circular dimension wrapping the circular eleventh dimension. Since tension is energy density, this identification requires that

$$T_{F1} = 2\pi R_{11} T_{M2}. \tag{8.107}$$

This relation is satisfied by the tensions

$$T_{F1} = \frac{1}{2\pi \ell_s^2} \quad \text{and} \quad T_{M2} = \frac{2\pi}{(2\pi \ell_p)^3}, \tag{8.108}$$

as can be verified using $R_{11} = \ell_s g_s$ and $\ell_p = g_s^{1/3} \ell_s$. All of these relations were presented earlier, and the proposal presented here confirms that they are correct. Various other branes can be matched in a similar manner. For example, the D4-brane is identified to be an M5-brane with one dimension wrapped on the spatial circle.

Another interesting fact can be deduced by considering the M-theory origin of a type IIA configuration in which an F-string ends on a D4-brane. In view of the above, this clearly corresponds to an M2-brane ending on an M5-brane, where each of the M-branes is wrapped around the circular dimension. One reason that a type IIA F-string can end on a D-brane is that the flux associated with the charge at the end of the string is carried away by the one-form gauge field of the D-brane world-volume theory. That being the case, one can ask what is the corresponding mechanism for M-branes. The end of the M2-brane is a string inside the M5-brane. So the world-volume theory of the M5-brane must contain a two-form gauge field A_2 to carry away the associated flux. That is indeed the case. In fact, the corresponding field strength F_3 is self-dual, just like the five-form field strength in type IIB supergravity.

The D6-brane

The preceding discussion explained the M-theory origin of the type IIA Dp-branes for $p = 0, 2, 4$ in terms of wrapped or unwrapped M-branes. This raises the question of how one should understand the D6-brane from an M-theory point of view. Clearly, unlike the other D-branes, it cannot be related to the M2-brane or the M5-brane in any simple way. The key to answering this question is to recall that the D6-brane is the magnetic dual of the D0-brane and that the D0-brane is interpreted as a Kaluza–Klein excitation along the x^{11} circle. The D0-brane carries electric charge with respect to the $U(1)$ gauge field $C_\mu = g_{\mu 11}$. Therefore, the D6-brane should couple magnetically to this same gauge field.

This problem was solved long ago for the case of pure gravity in five dimensions compactified on a circle. In this case, the challenge is to construct the five-dimensional metric that describes the *Kaluza–Klein monopole*, that is, a magnetically charged soliton in four dimensions. By tensoring this solution with $\mathbb{R}^6$, exactly the same construction applies to the 11-dimensional problem. The extra six flat dimensions constitute the spatial directions of the D6-brane world volume.

The relevant five-dimensional geometry that is Ricci-flat and nonsingular in five dimensions is given by

$$ds_5^2 = -dt^2 + ds_{\text{TN}}^2, \tag{8.109}$$

where the Taub–NUT metric is

$$ds_{\text{TN}}^2 = V(r)\left(dr^2 + r^2 d\Omega_2^2\right) + \frac{1}{V(r)}\left(dy + R\sin^2(\theta/2)\, d\phi\right)^2. \tag{8.110}$$

Here $d\Omega_2^2 = d\theta^2 + \sin^2\theta d\phi^2$ is the metric of a round unit two-sphere, and

$$V(r) = 1 + \frac{R}{2r}. \tag{8.111}$$

Also, the magnetic field is given by

$$\vec{B} = -\vec{\nabla}V = \vec{\nabla}\times\vec{A} \quad \text{with} \quad A_\phi = R\sin^2(\theta/2), \tag{8.112}$$

where we have displayed only the nonvanishing component of the vector potential. The Taub–NUT metric is nonsingular at $r=0$ if the coordinate y has period $2\pi R$. Thus the actual radius of the circle is

$$\tilde{R}(r) = V(r)^{-1/2} R, \tag{8.113}$$

which approaches R for $r \to \infty$ and zero as $r \to 0$.

The mass of the soliton described by the Taub–NUT metric can be computed by integrating the energy density T_{00}. For the purpose of understanding the tension of the D6-brane, we can add six more flat dimensions and obtain

$$T_{D6} = \frac{2\pi R}{16\pi G_{11}}\int d^3 x \nabla^2 V. \tag{8.114}$$

Since the integral gives $2\pi R$,

$$T_{D6} = \frac{(2\pi R)^2}{16\pi G_{11}} = \frac{2\pi R}{16\pi G_{10}} = \frac{2\pi}{(2\pi \ell_s)^7 g_s}, \quad (8.115)$$

where we have used $R = g_s \ell_s$. This agrees with the value obtained in Chapter 6.

There is a simple generalization of the above, the multi-center Taub–NUT metric, that describes a system of N parallel D6-branes. The metric in this case is

$$ds^2 = V(\vec{x}) d\vec{x} \cdot d\vec{x} + \frac{1}{V(\vec{x})} \left(dy + \vec{A} \cdot d\vec{x} \right)^2, \quad (8.116)$$

where

$$\vec{B} = -\vec{\nabla} V = \vec{\nabla} \times \vec{A} \quad \text{and} \quad V(\vec{x}) = 1 + \frac{R}{2} \sum_{\alpha=1}^{N} \frac{1}{|\vec{x} - \vec{x}_\alpha|}. \quad (8.117)$$

Since this system is BPS, the tension and magnetic charge are just N times the single D6-brane values.

A similar construction applies to other string theories compactified on circles. Indeed, the type IIB superstring theory compactified on a circle contains a Kaluza–Klein 5-brane, constructed in the same way as the D6-brane, which is the magnetic dual of the Kaluza–Klein 0-brane. A T-duality transformation along the circular dimension transforms the type IIB theory into the type IIA theory compactified on the dual circle. The Kaluza–Klein 0-brane is dual to a fundamental type IIA string wound on the dual circle. Therefore, the Kaluza–Klein 5-brane must map to the magnetic dual of the fundamental IIA string, which is the type IIA NS5-brane.

$E_8 \times E_8$ heterotic string theory at strong coupling

Let us briefly review the Hořava–Witten picture of the strongly coupled $E_8 \times E_8$ heterotic string theory. One starts with the strongly coupled type IIA superstring theory, or equivalently M-theory on $\mathbb{R}^{9,1} \times S^1$, and mods out by a certain $\mathbb{Z}_2$ symmetry, much like one does in deriving the type I superstring theory from the type IIB superstring theory. The appropriate $\mathbb{Z}_2$ symmetry in this case includes the following reversals:

$$x^{11} \to -x^{11} \quad \text{and} \quad A_3 \to -A_3. \quad (8.118)$$

In particular, modding out by this $\mathbb{Z}_2$ action implies that the zero mode of the Fourier expansion of $A_{\mu\nu\rho}$ in the x^{11} direction is eliminated from the spectrum, while the zero mode of

$$B_{\mu\nu} = A_{\mu\nu 11} \tag{8.119}$$

survives. This is required, of course, to account for the fact that $\mathcal{N} = 1$ supergravity in ten dimensions contains a massless two-form but no massless three-form. The heterotic string coupling constant g_s is given by

$$g_s = R_{11}/\ell_s, \tag{8.120}$$

just as in the case of the type IIA theory.

The space $S^1/\mathbb{Z}_2$ can be regarded as a line segment from $x^{11} = 0$ to $x^{11} = \pi R_{11}$. The two end points are the fixed points of the orbifold. Their presence leads to an interesting physical picture: the 11-dimensional space-time can be viewed as a slab of thickness πR_{11}. The two ten-dimensional boundaries are the orbifold singularities where the super Yang–Mills fields are localized. The two boundaries are sometimes called *end-of-the-world 9-branes*. Each of them carries the gauge fields for an E_8 group. This is a very intuitive way of understanding why this theory has a gauge group that is a product of two identical factors. The fact that the boundaries carry E_8 gauge supermultiplets is required for anomaly cancellation. There are no anomalies in odd dimensions, except at a boundary. In this case the boundary anomaly cancels only for the gauge group E_8. No other choice works, as was explained in Chapter 5.

There is an alternative route by which one can deduce that M-theory compactified on $S^1/\mathbb{Z}_2$ is dual to the $E_8 \times E_8$ heterotic string in ten dimensions. It uses the following sequence of dualities that have been introduced previously: (1) T-duality between the $E_8 \times E_8$ heterotic string and the $SO(32)$ heterotic string; (2) S-duality between the $SO(32)$ heterotic string and the type I superstring; (3) T-duality between the type I superstring and the type I$'$ superstring; (4) identification of the type I$'$ superstring as a type IIA orientifold; (5) duality between the type IIA superstring and M-theory on a circle. Quantitative details of this construction are described in Exercise 8.6.

M2-branes, with the topology of a cylinder, are allowed to terminate on a boundary of the space-time, so that the boundary of the M2-brane is a closed loop inside the end-of-the-world 9-brane. In this picture, an $E_8 \times E_8$ heterotic string is a cylindrical M2-brane suspended between the two space-time boundaries, with one E_8 associated with each boundary. This cylinder is well approximated by a string living in ten dimensions when the separation πR_{11} is small, as indicated in Fig. 8.2. Since perturbation theory in g_s is an expansion about $R_{11} = 0$, the fact that there really are 11 dimensions and that the string is actually a membrane is invisible in that approach. The tension of the heterotic string is therefore

$$T_{\rm H} = 2\pi R_{11} T_{\rm M2} = (2\pi \ell_s^2)^{-1}. \tag{8.121}$$

All of these statements are straightforward counterparts of statements concerning the strongly coupled type IIA superstring theory.

There are two possible strong coupling limits of the $E_8 \times E_8$ heterotic string theory. One possibility is a limit in which both boundaries go to infinity, so that one ends up with an $\mathbb{R}^{11}$ space-time geometry. This is the same limit as one obtains by starting with type IIA superstring theory and letting $R_{11} \to \infty$. The strongly coupled $E_8 \times E_8$ heterotic string and the type IIA superstring theory are identical in the 11-dimensional bulk. The only thing that distinguishes them is the existence of boundaries in the former case. The second possibility is to hold one boundary fixed as $R_{11} \to \infty$. This limit leads to a semi-infinite eleventh dimension. Since there is just one boundary in this limit, there is just one E_8 gauge group. This limit has received very little attention in the literature. It is also possible to consider 11-dimensional geometries with more than two boundaries, and therefore more than two E_8 groups.

In studies of possible phenomenological applications of the strongly coupled $E_8 \times E_8$ heterotic string, a subject sometimes called *heterotic M-theory*, one considers compactification of six more spatial dimensions (usually on a Calabi–Yau space). An interesting possibility that does not arise in

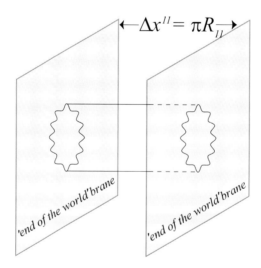

Fig. 8.2. A cylindrical M2-brane suspended between two end-of-the-world 9-branes is approximated by an $E_8 \times E_8$ heterotic string as $R_{11} \to 0$.

the weakly coupled ten-dimensional description is that moduli of this six-dimensional space, as well as other moduli (such as the vev of the dilaton), can vary along the length of the compact eleventh dimension. Thus, for example, one E_8 theory can be more strongly coupled than the other one. This is explored further in Chapter 10.

EXERCISES

EXERCISE 8.5
Use T-duality to deduce the tension of the type IIB Kaluza–Klein 5-brane.

SOLUTION

The type IIB KK5-brane is T-dual to the NS5-brane in the type IIA theory. In the type IIA theory one can form the dimensionless combination

$$\frac{T_{\text{NS5}}}{T_{\text{D2}}^2} = \frac{1}{2\pi}.$$

Since this is a dimensionless number, it is preserved under T-duality irrespective of the coordinate frame used to measure distances. Under the T-duality

$$T_{\text{NS5}} \to T_{\text{KK5}} \quad \text{and} \quad T_{\text{D2}} \to 2\pi R_9 T_{\text{D3}}.$$

Therefore, in the type IIB string frame

$$T_{\text{KK5}} = \frac{1}{2\pi}(2\pi R_9 T_{\text{D3}})^2 = \frac{R_9^2}{g_s^2 (2\pi)^5 \ell_s^8}. \qquad (8.122)$$

It is an interesting fact that this is proportional to the square of the radius.
□

EXERCISE 8.6
Show that the duality between M-theory on $S^1/\mathbb{Z}_2$ and the $E_8 \times E_8$ heterotic string is a consequence of previously discussed dualities.

SOLUTION

Consider the $E_8 \times E_8$ heterotic string with string coupling g_s and x^9 coordinate compactified on a circle of radius R_9. This is T-dual to the $SO(32)$

heterotic string on a circle of radius
$$R'_9 = \ell_s^2/R_9$$
and coupling
$$g'_s = \ell_s g_s/R_9.$$

As discussed in Chapter 7, Wilson lines need to be turned on to give the desired $E_8 \times E_8$ gauge symmetry. Applying an S-duality transformation then gives the type I string with
$$g''_s = \frac{1}{g'_s} = \frac{R_9}{\ell_s g_s}$$
and
$$R''_9 = R'_9 \sqrt{g''_s} = R'_9/\sqrt{g'_s} = (\ell_s)^{3/2}/\sqrt{R_9 g_s}.$$
Another T-duality then gives the type I$'$ theory with
$$R'''_9 = \ell_s^2/R''_9 = \sqrt{R_9 \ell_s g_s}$$
and
$$g'''_s = \ell_s g''_s/R''_9 = (R_9/\ell_s)^{3/2} g_s^{-1/2}.$$

In the bulk this is the type IIA theory, so we can use the type IIA/M-theory duality to introduce $R_{11} = g'''_s \ell_s$ and $\ell_p = (g'''_s)^{1/3}\ell_s$. A little algebra then gives the relations
$$R'''_9/\ell_p = (g_s)^{2/3}$$
and
$$R_{11} = \frac{R_9^2}{R'''_9}.$$

Now we can decompactify $R_9 \to \infty$ at fixed R'''_9 and ℓ_p. Note that $R_{11} \to \infty$ at the same time. On the one hand, this gives the ten-dimensional $E_8 \times E_8$ heterotic string, with coupling constant g_s, while on the other hand it gives a dual M-theory description with a compact eleventh dimension that is an interval of length $\pi R'''_9$ satisfying the expected relation $R'''_9 = (g_s)^{2/3}\ell_p$. □

8.4 M-theory dualities

The previous section showed that the strongly coupled type IIA superstring and the strongly coupled $E_8 \times E_8$ heterotic string have a simple M-theory interpretation. There are additional dualities involving M-theory that relate it to the other superstring theories as well as to itself.

An M-theory/type IIB superstring duality

M-theory compactified on a circle gives the type IIA superstring theory, while type IIA superstring theory on a circle corresponds to type IIB superstring theory on a dual circle. Putting these two facts together it follows that there should be a duality between M-theory on a two-torus T^2 and type IIB superstring theory on a circle S^1. The M-theory torus is characterized by an area A_M and a modulus τ_M, while the IIB circle has radius R_B. Let us explore this duality directly without reference to the type IIA theory. Specifically, the plan is to compare various BPS states and branes in nine dimensions.

Since all of the (p,q) strings in type IIB superstring theory are related by $SL(2,\mathbb{Z})$ transformations,[12] they are all equivalent, and any one of them can be weakly coupled. However, when one is weakly coupled, all of the others are necessarily strongly coupled. Let us consider an arbitrary (p,q) string and write down the spectrum of its nine-dimensional excitations in the limit of weak coupling using the standard string theory formulas given in Chapter 6:

$$M_B^2 = \left(\frac{K}{R_B}\right)^2 + (2\pi R_B W T_{(p,q)})^2 + 4\pi T_{(p,q)}(N_L + N_R). \qquad (8.123)$$

As before, K is the Kaluza–Klein excitation number and W is the string winding number. N_L and N_R are excitation numbers of left-moving and right-moving oscillator modes, and the level-matching condition is

$$N_R - N_L = KW. \qquad (8.124)$$

The plan is to use the formula above for all the (p,q) strings simultaneously. However, the formula is completely meaningless at strong coupling, and at most one of the strings is weakly coupled. The appropriate trick in this case is to consider only BPS states, that is, ones belonging to short supersymmetry multiplets, since their mass formulas can be reliably extrapolated to strong coupling. They are easy to identify, being given by either $N_L = 0$ or $N_R = 0$. (Ones with $N_L = N_R = 0$ are ultrashort.) In this way, one obtains exact mass formulas for a very large part of the spectrum – much more than appears in any perturbative limit. Of course, the appearance of this rich spectrum of BPS states depends crucially on the compactification.

There is a unique correspondence between the three integers W, p, q, where p and q are coprime, and an arbitrary pair of integers n_1, n_2 given by $(n_1, n_2) = (Wp, Wq)$. The integer W is the greatest common divisor of n_1

---
[12] It is assumed here that p and q are coprime.

and n_2. The only ambiguity is whether to choose W or $-W$, but since W is the (oriented) winding number and the $(-p, -q)$ string is the orientation-reversed (p, q) string, the two choices are actually equivalent. Thus BPS states are characterized by three integers K, n_1, n_2 and oscillator excitations corresponding to $N_L = |WK|$, tensored with a 16-dimensional short multiplet from the $N_R = 0$ sector (or *vice versa*). Note that the combination $|W|T_{(p,q)}$, which appears in Eq. (8.123), can be rewritten using Eq. (8.97) in the form

$$|W|T_{(p,q)} = |n_1 - n_2 \tau_B| T_{F1}. \tag{8.125}$$

Let us now consider M-theory compactified on a torus. If the two periods in the complex plane, which define the torus, are $2\pi R_{11}$ and $2\pi R_{11} \tau_M$, then

$$A_M = (2\pi R_{11})^2 \operatorname{Im} \tau_M \tag{8.126}$$

is the area of the torus. In terms of coordinates $z = x + iy$ on the torus, a single-valued wave function has the form

$$\psi_{n_1, n_2} \sim \exp\left\{ \frac{i}{R_{11}} \left[n_2 x - \frac{n_2 \operatorname{Re} \tau_M - n_1}{\operatorname{Im} \tau_M} y \right] \right\}. \tag{8.127}$$

These characterize Kaluza–Klein excitations. The contribution to the mass-squared is given by the eigenvalue of $-\partial_x^2 - \partial_y^2$,

$$M_{KK}^2 = \frac{1}{R_{11}^2} \left[n_2^2 + \frac{(n_2 \operatorname{Re} \tau_M - n_1)^2}{(\operatorname{Im} \tau_M)^2} \right] = \frac{|n_1 - n_2 \tau_M|^2}{(R_{11} \operatorname{Im} \tau_M)^2}. \tag{8.128}$$

Clearly, this term has the right structure to match the type IIB string winding-mode terms, described above, for the identification

$$\tau_M = \tau_B. \tag{8.129}$$

The normalization of M_{KK}^2 and the winding-mode contribution to M_B^2 is not the same, but that is because they are measured in different metrics. The matching tells us how to relate the two metrics, a formula to be presented soon.

The identification in Eq. (8.129) is a pleasant surprise, because it implies that the nonperturbative $SL(2, \mathbb{Z})$ symmetry of type IIB superstring theory, after compactification on a circle, has a dual M-theory interpretation as the modular group of a toroidal compactification! Modular transformations of the torus are certainly symmetries, since they correspond to the disconnected components of the diffeomorphism group. Once the symmetry is established for finite R_B, it should also persist in the decompactification limit $R_B \to \infty$.

To go further requires an M-theory counterpart of the term $(K/R_B)^2$ in the type IIB superstring mass formula (8.123). Here there is also a natural candidate: wrapping M-theory M2-branes so as to cover the torus K times. If the M2-brane tension is T_{M2}, this gives a contribution $(A_M T_{M2} K)^2$ to the mass-squared. Matching the normalization of this term and the Kaluza–Klein term gives two relations. One learns that the metrics in nine dimensions are related by

$$g^{(M)} = \beta^2 g^{(B)}, \qquad (8.130)$$

where

$$\beta^2 = \frac{2\pi R_{11} T_{M2}}{T_{F1}}, \qquad (8.131)$$

and that the compactification volumes are related by

$$\frac{g_s^2}{T_{F1} R_B^2} = T_{M2}(2\pi R_{11})^3 = T_{M2}\left(\frac{A_M}{\operatorname{Im} \tau_M}\right)^{3/2}. \qquad (8.132)$$

Since all the other factors are constants, this gives (for fixed $\tau_B = \tau_M$) the scaling law $R_B \sim A_M^{-3/4}$.

There still are the oscillator excitations of the type IIB superstring BPS mass formula to account for. Their M-theory counterparts must be excitations of the wrapped M2-brane. Unfortunately, the quantization of the M2-brane is not understood well enough to check this, though this must surely be possible.

Matching BPS brane tensions in nine dimensions

We can carry out additional tests of the proposed duality, and learn interesting new relations at the same time, by matching BPS p-branes with $p > 0$ in nine dimensions. Only some of the simpler cases are described here. Let us start with strings in nine dimensions. Trivial reduction of the type IIB strings, that is, not wrapped on the circular dimension gives strings with the same charges (p, q) and tensions $T_{(p,q)}$ in nine dimensions. The interesting question is how these should be interpreted in M-theory. The way to answer this is to start with an M2-brane of toroidal topology in M-theory and to wrap one of its cycles on a (p, q) homology cycle of the spatial torus. The minimal length of such a cycle is[13]

$$L_{(p,q)} = 2\pi R_{11} |p - q\tau_M|. \qquad (8.133)$$

---

[13] This is understood most easily by considering the covering space of the torus, which is the plane tiled by parallelograms. A closed geodesic curve on the torus is represented by a straight line between equivalent points in the covering space, as shown in Fig. 8.3.

Thus, this wrapping gives a string in nine dimensions, whose tension is

$$T^{(11)}_{(p,q)} = L_{(p,q)} T_{\text{M2}}. \tag{8.134}$$

The superscript 11 emphasizes that this is measured in the 11-dimensional metric. To compare with the type IIB string tensions, we use Eqs (8.130) and (8.131) to deduce that

$$T_{(p,q)} = \beta^{-2} T^{(11)}_{(p,q)}. \tag{8.135}$$

This agrees with the result given earlier, showing that this is a correct interpretation.

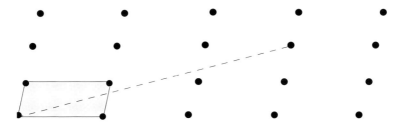

Fig. 8.3. In the covering space of the torus, which is the plane tiled by parallelograms, a closed geodesic curve on the torus is represented by a straight line between equivalent points.

To match 2-branes in nine dimensions requires wrapping the type IIB D3-brane on the circle and comparing to the unwrapped M2-brane. The wrapped D3-brane gives a 2-brane with tension $2\pi R_\text{B} T_{\text{D3}}$. Including the metric conversion factor, the matching gives

$$T_{\text{M2}} = 2\pi R_\text{B} \beta^3 T_{\text{D3}}. \tag{8.136}$$

Combining this with Eqs (8.131) and (8.132) gives the identity

$$T_{\text{D3}} = \frac{(T_{\text{F1}})^2}{2\pi g_\text{s}}, \tag{8.137}$$

in agreement with the tension formulas in Chapter 6. It is remarkable that the M-theory/type IIB superstring theory duality not only relates M-theory tensions to type IIB superstring theory tensions, but it even implies a relation involving only type IIB tensions.

Wrapping the M5-brane on the spatial torus gives a 3-brane in nine dimensions, which can be identified with the unwrapped type IIB D3-brane in nine dimensions. This gives

$$T_{\text{M5}} A_{\text{M}} = \beta^4 T_{\text{D3}}, \tag{8.138}$$

8.4 M-theory dualities

which combined with Eqs (8.131) and (8.137) implies that

$$T_{M5} = \frac{1}{2\pi}(T_{M2})^2. \tag{8.139}$$

This corresponds to satisfying the Dirac quantization condition with the minimum allowed product of charges. It also provides a check of the tensions in (8.22).

An M-theory/$SO(32)$ superstring duality

There is a duality that is closely related to the one just considered that relates M-theory compactified on $(S^1/\mathbb{Z}_2) \times S^1$ to the $SO(32)$ theory compactified on S^1. Because of the similarity of the two problems, fewer details are provided this time.

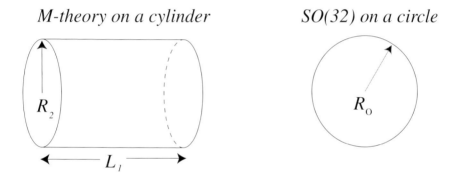

Fig. 8.4. Duality between M-theory on a cylinder and $SO(32)$ on a circle.

Since $S^1/\mathbb{Z}_2$ can be regarded as a line interval I, $(S^1/\mathbb{Z}_2) \times S^1$ can be regarded as a cylinder. Let us choose its height to be L_1 and its circumference to be $L_2 = 2\pi R_2$. The circumference of the circle on which the dual $SO(32)$ theory is compactified is $L_O = 2\pi R_O$ as measured in the ten-dimensional Einstein metric. This is illustrated in Fig. 8.4.

The $SO(32)$ theory in ten dimensions has both a type I and a heterotic description, which are S dual. As before, the duality can be explored by matching supersymmetry-preserving (BPS) branes in nine dimensions. Recall that in the $SO(32)$ theory, there is just one two-form field, and the p-branes that couple to it are the $SO(32)$ heterotic string and its magnetic dual, which is a solitonic 5-brane. (From the type I viewpoint, both of these are D-branes.) The heterotic string can give a 0-brane or a 1-brane in nine dimensions, and the dual 5-brane can give a 5-brane or a 4-brane in nine dimensions. In each case, the issue is simply whether or not one cycle of the brane wraps around the spatial circle.

Now let us find the corresponding nine-dimensional p-branes from the M-theory viewpoint and explore what can be learned from matching tensions. The $E_8 \times E_8$ string arises in ten dimensions from wrapping the M2-brane on I. Subsequent reduction on a circle can give a 0-brane or a 1-brane. The story for the M5-brane is just the reverse. Whereas the M2-brane must wrap the I dimension, the M5-brane must not do so. As a result, it gives a 5-brane or a 4-brane in nine dimensions according to whether or not it wraps around the S^1 dimension. So, altogether, both pictures give the electric–magnetic dual pairs $(0, 5)$ and $(1, 4)$ in nine dimensions.

From the p-brane matching one learns that the $SO(32)$ heterotic string coupling constant is

$$g_s^{(HO)} = \frac{L_1}{L_2}. \tag{8.140}$$

Thus, the $SO(32)$ heterotic string is weakly coupled when the spatial cylinder of the M-theory compactification is a thin ribbon ($L_1 \ll L_2$). This is consistent with the earlier conclusion that the $E_8 \times E_8$ heterotic string is weakly coupled when L_1 is small. Conversely, the type I superstring is weakly coupled for $L_2 \ll L_1$, in which case the spatial cylinder is long and thin. The $\mathbb{Z}_2$ transformation that inverts the modulus of the cylinder, L_1/L_2, corresponds to the type I/heterotic S duality of the $SO(32)$ theory. Since it is not a symmetry of the cylinder it implies that two different-looking string theories are S dual. This is to be contrasted with the $SL(2, \mathbb{Z})$ modular group symmetry of the torus, which accounts for the self-duality of the type IIB theory.

The p-brane matching in nine dimensions also gives the relation

$$L_1 L_2^2 T_{M2} = \left(\frac{T_1^{(HO)} L_O^2}{2\pi}\right)^{-1}, \tag{8.141}$$

which is the analog of Eq. (8.132). As in that case, it tells us that, for fixed modulus L_1/L_2, one has the scaling law $L_O \sim A_C^{-3/4}$, where $A_C = L_1 L_2$ is the area of the cylinder. Equation (8.139) relating T_{M2} and T_{M5} is reobtained, and one also learns that

$$T_5^{(HO)} = \frac{1}{(2\pi)^2} \left(\frac{L_2}{L_1}\right)^2 (T_1^{(HO)})^3. \tag{8.142}$$

In the heterotic string-frame metric, where $T_1^{(HO)}$ is a constant, this implies that

$$T_5^{(\mathrm{HO})} \sim (g_s^{(\mathrm{HO})})^{-2}, \qquad (8.143)$$

as is typical of a soliton. In the type I string-frame metric, on the other hand, it implies that

$$T_{\mathrm{D1}} \sim 1/g_s^{(\mathrm{I})} \quad \text{and} \quad T_{\mathrm{D5}} \sim 1/g_s^{(\mathrm{I})}, \qquad (8.144)$$

consistent with the fact that both are D-branes from the type I viewpoint.

U-duality

It is natural to seek type II counterparts of the $O(n,n;\mathbb{Z})$ and $O(16+n,n;\mathbb{Z})$ duality groups that were found in Chapter 7 for toroidal compactification of the bosonic and heterotic string theories, respectively. A clue is provided by the fact that the massless sector of type II superstring theories are maximal supergravity theories (ones with 32 conserved supercharges), with a noncompact global symmetry group.

In the case of type IIB supergravity in ten dimensions the noncompact global symmetry group is $SL(2,\mathbb{R})$, as was shown earlier in this chapter. Toroidal compactification leads to theories with maximal supersymmetry in lower dimensions.[14] So, for example, toroidal compactification of the type IIB theory to four dimensions and truncation to zero modes (dimensional reduction) leads to $\mathcal{N} = 8$ supergravity. $\mathcal{N} = 8$ supergravity has a noncompact E_7 symmetry. More generally, for $d = 11 - n$, $3 \leq n \leq 8$, one finds a maximally noncompact form of E_n, denoted $E_{n,n}$. The maximally noncompact form of a Lie group of rank n has n more noncompact generators than compact generators. Thus, for example, $E_{7,7}$ has 133 generators of which 63 are compact and 70 are noncompact. A compact generator generates a circle, whereas a noncompact generator generates an infinite line. E_n are standard exceptional groups that appear in Cartan's classification of simple Lie algebras for $n = 6, 7, 8$. The definition for $n < 6$ can be obtained by extrapolation of the Dynkin diagrams. This gives the identifications (listing the maximally noncompact forms)[15]

$$E_{5,5} = SO(5,5), \quad E_{4,4} = SL(5,\mathbb{R}), \quad E_{3,3} = SL(3,\mathbb{R}) \times SL(2,\mathbb{R}). \qquad (8.145)$$

These noncompact Lie groups describe global symmetries of the classical low-energy supergravity theories. However, as was discussed already for the

---

14 Chapter 9 describes compactification spaces that (unlike tori) break some or all of the supersymmetries.
15 The compact forms of the same sequence of exceptional groups was encountered in the study of type I′ superstrings in Chapter 6.

case of the $E_{1,1} = SL(2, \mathbb{R})$ symmetry of type IIB superstring theory in ten dimensions, they are broken to infinite discrete symmetry groups by quantum and string-theoretic corrections. The correct statement for superstring theory/M-theory is that, for M-theory on $\mathbb{R}^d \times T^n$ or (equivalently) type IIB superstring theory on $\mathbb{R}^d \times T^{n-1}$, the resulting moduli space is invariant under an infinite discrete *U-duality* group. The group, denoted $E_n(\mathbb{Z})$, is a maximal discrete subgroup of the noncompact $E_{n,n}$ symmetry group of the corresponding supergravity theory.

The U-duality groups are generated by the Weyl subgroup of $E_{n,n}$ plus discrete shifts of axion-like fields. The subgroup $SL(n, \mathbb{Z}) \subset E_n(\mathbb{Z})$ can be understood as the geometric duality (modular group) of T^n in the M-theory picture. In other words, they correspond to disconnected components of the diffeomorphism group. The subgroup $SO(n-1, n-1; \mathbb{Z}) \subset E_n(\mathbb{Z})$ is the T-duality group of type IIB superstring theory compactified on T^{n-1}. These two subgroups intertwine nontrivially to generate the entire $E_n(\mathbb{Z})$ U-duality group. For example, in the $n = 3$ case the duality group is

$$E_3(\mathbb{Z}) = SL(3, \mathbb{Z}) \times SL(2, \mathbb{Z}). \tag{8.146}$$

The $SL(3, \mathbb{Z})$ factor is geometric from the M-theory viewpoint, and an

$$SO(2, 2; \mathbb{Z}) = SL(2, \mathbb{Z}) \times SL(2, \mathbb{Z}) \tag{8.147}$$

subgroup is the type IIB T-duality group. Clearly, $E_3(\mathbb{Z})$ is the smallest group containing both of these.

Toroidally compactified M-theory (or type II superstring theory) has a moduli space analogous to the Narain moduli space of the toroidally compactified heterotic string described in Chapter 7. Let H_n denote the maximal compact subgroup of $E_{n,n}$. For example, $H_6 = USp(8)$, $H_7 = SU(8)$ and $H_8 = Spin(16)$. Then one can define a homogeneous space

$$\mathcal{M}_n^0 = E_{n,n}/H_n. \tag{8.148}$$

This is directly relevant to the physics in that the scalar fields in the supergravity theory are defined by a sigma model on this coset space. Note that all the coset generators are noncompact. It is essential that they all be the same so that the kinetic terms of the scalar fields all have the same sign. The number of scalar fields is the dimension of the coset space $d_n = \dim \mathcal{M}_n^0$. For example, in three, four and five dimensions the number of scalars is

$$d_3 = \dim E_8 - \dim Spin(16) = 248 - 120 = 128, \tag{8.149}$$

$$d_4 = \dim E_7 - \dim SU(8) = 133 - 63 = 70, \tag{8.150}$$

$$d_5 = \dim E_6 - \dim USp(8) = 78 - 36 = 42. \tag{8.151}$$

The discrete duality-group identifications must still be accounted for, and this gives the moduli space

$$\mathcal{M}_n = \mathcal{M}_n^0 / E_n(\mathbb{Z}). \tag{8.152}$$

A nongeometric duality of M-theory

String theory possesses certain features, such as T-duality, that go beyond ones classical geometric intuition. This section shows that the same is true for M-theory by constructing an analogous duality transformation. There is a geometric understanding of the $SL(n,\mathbb{Z})$ subgroup of $E_n(\mathbb{Z})$ that comes from considering M-theory on $\mathbb{R}^{11-n} \times T^n$, since it is the modular group of T^n. But what does the rest of $E_n(\mathbb{Z})$ imply? To address this question it suffices to consider the first nontrivial case to which it applies, which is $n = 3$. In this case the U-duality group is $E_3(\mathbb{Z}) = SL(3,\mathbb{Z}) \times SL(2,\mathbb{Z})$. From the M-theory viewpoint the first factor is geometric and the second factor is not. So the question boils down to understanding the implication of the $SL(2,\mathbb{Z})$ duality in the M-theory construction. Specifically, we want to understand the nontrivial $\tau \to -1/\tau$ element of this group.

To keep the story as simple as possible, let us choose the T^3 to be rectangular with radii R_1, R_2, R_3, that is, $g_{ij} \sim R_i^2 \delta_{ij}$, and assume that $C_{123} = 0$. Choosing R_3 to correspond to the "eleventh" dimension makes contact with the type IIA theory on a torus with radii R_1 and R_2. In this set-up, the stringy duality of M-theory corresponds to simultaneous T-duality transformations of the type IIA theory for both of the circles. This T-duality gives a mapping to an equivalent point in the moduli space for which

$$R_i \to R_i' = \frac{\ell_s^2}{R_i} = \frac{\ell_p^3}{R_3 R_i} \quad i = 1, 2, \tag{8.153}$$

with ℓ_s unchanged. The derivation of this formula has used $\ell_p^3 = R_3 \ell_s^2$, which relates the 11-dimensional Planck scale ℓ_p to the ten-dimensional string scale ℓ_s. Under a T-duality the string coupling constant also transforms. The rule is that the coupling of the effective theory, which is eight-dimensional in this case, is invariant:

$$\frac{1}{g_8^2} = 4\pi^2 \frac{R_1 R_2}{g_s^2} = 4\pi^2 \frac{R_1' R_2'}{(g_s')^2}. \tag{8.154}$$

Thus

$$g_s' = \frac{g_s \ell_s^2}{R_1 R_2}. \tag{8.155}$$

What does this imply for the radius of the eleventh dimension R_3? Using $R_3 = g_s \ell_s \to R_3' = g_s' \ell_s'$,

$$R_3' = \frac{g_s \ell_s^3}{R_1 R_2} = \frac{\ell_p^3}{R_1 R_2}. \tag{8.156}$$

However, the 11-dimensional Planck length also transforms, because

$$\ell_p^3 = g_s \ell_s^3 \to (\ell_p')^3 = g_s' \ell_s^3 \tag{8.157}$$

implies that

$$(\ell_p')^3 = \frac{g_s \ell_s^5}{R_1 R_2} = \frac{\ell_p^6}{R_1 R_2 R_3}. \tag{8.158}$$

The perturbative type IIA description is only applicable for $R_3 \ll R_1, R_2$. However, even though T-duality was originally discovered in perturbation theory, it is supposed to be an exact nonperturbative property. Therefore, this duality mapping should be valid as an exact symmetry of M-theory without any restriction on the radii. Another duality is an interchange of circles, such as $R_3 \leftrightarrow R_1$. This corresponds to the nonperturbative S-duality of the type IIB superstring theory. Combining these dualities gives the desired stringy duality of M-theory on T^3, namely

$$R_1 \to \frac{\ell_p^3}{R_2 R_3}, \tag{8.159}$$

and cyclic permutations, accompanied by

$$\ell_p^3 \to \frac{\ell_p^6}{R_1 R_2 R_3}. \tag{8.160}$$

This basic stringy duality of M-theory, combined with the geometric ones, generates the entire U-duality group in every dimension. It is a property of quantum M-theory that goes beyond what can be understood from the effective 11-dimensional supergravity theory, which is geometrical.

EXERCISES

EXERCISE 8.7
Verify Eqs (8.131) and (8.132).

8.4 M-theory dualities

SOLUTION

Since the M-theory metric and the type IIB metric are related by

$$g^{(M)} = \beta^2 g^{(B)},$$

masses are related according to

$$M_{11} = \beta M_B.$$

Matching the mass of an M2-brane wrapped on the torus with a Kaluza–Klein excitation on the type IIB circle therefore gives

$$A_M T_{M2} = \beta \frac{1}{R_B}.$$

Similarly, using Eq. (8.128) for the mass of a Kaluza–Klein excitation on the torus, and equating it to the mass of a wrapped type IIB string gives

$$\frac{1}{R_{11} \mathrm{Im}\tau_M} = \beta(2\pi R_B T_{F1}).$$

Multiplying these equations together, using $A_M = (2\pi R_{11})^2 \mathrm{Im}\tau_M$, gives

$$\beta^2 = \frac{R_B A_M T_{M2}}{2\pi R_B T_{F1} R_{11} \mathrm{Im}\tau_M} = \frac{2\pi R_{11} T_{M2}}{T_{F1}},$$

which is Eq. (8.131). Taking the quotient of the same two equations, using $g_s = (\mathrm{Im}\tau_M)^{-1}$, gives

$$\frac{g_s^2}{R_B^2 T_{F1}} = T_{M2}(2\pi R_{11})^3,$$

which is Eq. (8.132). □

EXERCISE 8.8

Identifying type IIB superstring theory compactified on a circle and M-theory compactified on a torus, match the tensions of the nine-dimensional 4-branes.

SOLUTION

A (p,q) type IIB 5-brane wrapped on the circle is identified with an M5-brane wrapping a geodesic (p,q) cycle of the torus. Equating the resulting tensions gives

$$2\pi R_B \beta^5 T_{(p,q)} = L_{(p,q)} T_{M5},$$

where $L_{(p,q)} = 2\pi R_{11}|p - q\tau_M|$. We can check that the resulting D5-brane tension in ten dimensions agrees with the result quoted in Chapter 6. Indeed, setting $p = 1$ and $q = 0$, we obtain

$$T_{\text{D5}} = \frac{R_{11}}{R_{\text{B}}} \beta^{-5} T_{\text{M5}} = \frac{T_{\text{F1}}^3}{(2\pi)^2 g_s} = \frac{1}{(2\pi)^5 \ell_s^6 g_s},$$

which is the T_{D5} derived in Chapter 6. Therefore,

$$T_{(p,q)} = |p - q\tau_{\text{M}}| T_{\text{D5}}.$$

In particular, the NS5-brane tension is obtained by setting $q = 1$ and $p = 0$

$$T_{\text{NS5}} = |\tau_{\text{M}}| T_{\text{D5}}.$$

The standard result is obtained by setting $\tau_{\text{M}} = i/g_s$. □

EXERCISE 8.9
Verify that the three groups (8.145) are maximally noncompact.

SOLUTION

The group $SO(5,5)$ has dimension equal to 45, just like its compact form $SO(10)$. Its maximal compact subgroup is $SO(5) \times SO(5)$, which has dimension equal to 20. Thus, there are 25 noncompact generators and 20 compact generators. Since the rank of $SO(5,5)$, which is five, agrees with $25 - 20$, it is maximally noncompact. In the case of $SL(5, \mathbb{R})$, which is a noncompact form of $SU(5)$, the rank is four and the dimension is 24. The maximal compact subgroup is $SO(5)$, which has dimension equal to ten. Thus there are 14 noncompact generators and ten compact generators, and once again the difference is equal to the rank. This reasoning generalizes to $SL(n, \mathbb{R})$, which has $(n-1)(n+2)/2$ noncompact generators, $(n-1)n/2$ compact generators and rank $n-1$. The group $SL(3, \mathbb{R}) \times SL(2, \mathbb{R})$ is maximally noncompact, because each of its factors is. □

EXERCISE 8.10
Find a physical interpretation of Eqs (8.159) and (8.160).

SOLUTION

Equation (8.159) implies that

$$\frac{1}{R_1} \to (2\pi R_2)(2\pi R_3) T_{\text{M2}}.$$

Thus it interchanges a Kaluza–Klein excitation of the first circle with an

M2-brane wrapped on the second and third circles. The circles can be permuted, so it follows that these six 0-brane excitations belong to the $(\mathbf{3}, \mathbf{2})$ representation of the $SL(3, \mathbb{Z}) \times SL(2, \mathbb{Z})$ U-duality group.

Equation (8.160) implies that

$$T_{\mathrm{M2}} \to (2\pi R_1)(2\pi R_2)(2\pi R_3) T_{\mathrm{M5}}.$$

Therefore, it interchanges an unwrapped M2-brane with an M5-brane wrapped on the T^3. Thus these two 2-branes (from the eight-dimensional viewpoint) belong to the $(\mathbf{1}, \mathbf{2})$ representation of the U-duality group. □

Homework Problems

Problem 8.1
Derive the bosonic equations of motion of 11-dimensional supergravity.

Problem 8.2
Show that a particular solution of the bosonic equations of motion of 11-dimensional supergravity, called the Freund–Rubin solution, is given by a product space-time geometry $AdS_4 \times S^7$ with

$$F_4 = M\varepsilon_4,$$

where ε_4 is the volume form on AdS_4, and M is a free parameter with the dimensions of mass.[16] AdS_4 denotes four-dimensional anti-de Sitter space, which is a maximally symmetric space of negative curvature, with Ricci tensor

$$R_{\mu\nu} = -(M_4)^2 g_{\mu\nu} \qquad \mu, \nu = 0, 1, 2, 3.$$

The seven-sphere has Ricci tensor

$$R_{ij} = (M_7)^2 g_{ij} \qquad i, j = 4, 5, \ldots, 10.$$

What are the masses M_4 and M_7 in terms of the mass parameter M?

Problem 8.3
Derive Eq. (8.69) and transform the bosonic part of the type IIA supergravity action in ten dimensions from the string frame to the Einstein frame.

[16] Actually, in the quantum theory it has to be an integer multiple of a basic unit.

Problem 8.4
Derive the redefinitions of C_1, C_3, F_2 and F_4 that are required to display a factor of $e^{-2\Phi}$ in the terms S_R and S_{CS} of the type IIA action given in Eqs (8.41) and (8.42).

Problem 8.5
Show that S_{CS} in Eq.(8.42) is invariant under a $U(1)$ gauge transformation even though it contains F_4 rather than $\tilde{F}_4$.

Problem 8.6
Consider the type IIB bosonic supergravity action in ten dimensions given in Eq. (8.53). Setting $C_0 = 0$, perform the transformations $\Phi \to -\Phi$ and $g_{\mu\nu} \to e^{-\Phi}g_{\mu\nu}$. What theory do you obtain, and what does the result imply? How should the transformations be generalized when $C_0 \neq 0$?

Problem 8.7
Verify that the actions in Eqs (8.73) and (8.81) map into one another under the transformations (8.88) and (8.89).

Problem 8.8
Verify that the supersymmetry transformations of the fermi fields in the heterotic and type I theories map into one another to leading order in fermi fields under an S-duality transformation, if λ and χ are suitably rescaled.

Problem 8.9
Consider the Euclidean Taub–NUT metric (8.110). Show that there is no singularity at $r = 0$ by showing that the metric takes the following form near the origin:

$$ds^2 = d\rho^2 + \frac{\rho^2}{4}(d\theta^2 + d\phi^2 + d\psi^2 - 2\cos\theta\, d\phi\, d\psi)$$

with $\psi \sim \psi + 4\pi$, and that this corresponds to a metric on flat four-dimensional Euclidean space. Hint: let $\psi = \phi + 2y/R$.

Problem 8.10
Consider the ten-dimensional type IIA metric for a KK5-brane

$$ds = -dt^2 + \sum_{i=1}^{5} dx_i^2 + ds_{TN}^2,$$

where ds_{TN}^2 is given in Eqs (8.110) and (8.111).

(i) Use the rules presented in Section 6.4 to deduce the type IIB solution that results from a T-duality transformation in the y direction.
(ii) What is the type IIB interpretation of the result?
(iii) Verify that the tension of the type IIB solution supports this interpretation.

PROBLEM 8.11

In the presence of an M5-brane the 11-dimensional F_4 satisfies the Bianchi identity

$$dF_4 = \delta_W,$$

where δ_W is a delta function with support on the M5-brane world volume. How must the equation of motion of F_4 be modified in order to be compatible with this Bianchi identity? What does this imply for the field content on the M5-brane world volume? Hint: Consult Exercise 5.10.

PROBLEM 8.12

Verify that a type IIA D2-brane is an unwrapped M2-brane by showing that $T_{D2} = T_{M2}$. Do the same for the NS5-brane and the M5-brane. Verify that a wrapped M5-brane corresponds to a D4-brane.

PROBLEM 8.13

Verify Eqs (8.137) and (8.139).

PROBLEM 8.14

Verify that Eq. (8.139) implies that the Dirac quantization condition is satisfied if the M2- and M5-brane each carry one unit of charge and saturate the BPS bound.

PROBLEM 8.15

Derive Eqs (8.140), (8.141) and (8.142).

9
String geometry

Since critical superstring theories are ten-dimensional and M-theory is 11-dimensional, something needs to be done to make contact with the four-dimensional space-time geometry of everyday experience. Two main approaches are being pursued.[1]

Kaluza–Klein compactification

The approach with a much longer history is *Kaluza–Klein compactification*. In this approach the extra dimensions form a compact manifold of size l_c. Such dimensions are essentially invisible for observations carried out at energy $E \ll 1/l_c$. Nonetheless, the details of their topology have a profound influence on the spectrum and symmetries that are present at low energies in the effective four-dimensional theory. This chapter explores promising geometries for these extra dimensions. The main emphasis is on *Calabi–Yau* manifolds, but there is also some discussion of other manifolds of *special holonomy*. While compact Calabi–Yau manifolds are the most straightforward possibility, modern developments in nonperturbative string theory have shown that noncompact Calabi–Yau manifolds are also important. An example of a noncompact Calabi–Yau manifold, specifically the *conifold*, is discussed in this chapter as well as in Chapter 10.

Brane-world scenario

A second way to deal with the extra dimensions is the *brane-world* scenario. In this approach the four dimensions of everyday experience are identified with a defect embedded in a higher-dimensional space-time. This defect

---

[1] Some mathematical background material is provided in an appendix at the end of this chapter. Readers not familiar with the basics of topology and geometry may wish to study it first.

is typically given by a collection of coincident or intersecting branes. The basic fact (discussed in Chapter 6) that makes this approach promising is the observation that Yang–Mills gauge fields, like those of the standard model, are associated with the zero modes of open strings, and therefore they reside on the world volume of D-branes.

A variant of the Kaluza–Klein idea that is often used in brane-world scenarios is based on the observation that the extra dimensions could be much larger than one might otherwise conclude if the geometry is warped in a suitable fashion. In a *warped compactification* the overall scale of the four-dimensional Minkowski space-time geometry depends on the coordinates of the compact dimensions. This chapter concentrates on the more traditional Kaluza–Klein approach, where the geometry is a product of an internal manifold and an external manifold. Warped geometries and their use for brane-world constructions are discussed in Chapter 10.

Motivation

The only manifolds describing extra dimensions that have been discussed so far in this book are a circle and products of circles (or tori). Also, a $\mathbb{Z}_2$ orbifold of a circle has appeared a couple of times. If any of the five ten-dimensional superstring theories is compactified to four dimensions on a six-torus, then the resulting theory is very far from being phenomenologically acceptable, since no supersymmetry is broken. This means that there is $\mathcal{N} = 4$ or $\mathcal{N} = 8$ supersymmetry in four dimensions, depending on which ten-dimensional theory is compactified. This chapter explores possibilities that are phenomenologically much more attractive, such as orbifolds, Calabi–Yau manifolds and exceptional-holonomy manifolds. Compactification on these spaces leads to vacua with less supersymmetry in four dimensions.

In order to make contact with particle phenomenology, there are various properties of the $D = 4$ theory that one would like:

- The Yang–Mills gauge group $SU(3) \times SU(2) \times U(1)$, which is the gauge group of the standard model.
- An interesting class of $D = 4$ supersymmetric extensions of the standard model have $\mathcal{N} = 1$ supersymmetry at high energy. This supersymmetry must be broken at some scale, which could be as low as a TeV, to make contact with the physics observed at low energies. $\mathcal{N} = 1$ supersymmetry imposes restrictions on the theory that make calculations easier. Yet these restrictions are not so strong as to make the theory unrealistic, as happens in models with $\mathcal{N} \geq 2$.

At sufficiently high energy, supersymmetry in ten or 11 dimensions should be manifest. The issue being considered here is whether at energies that are low compared to the compactification scale, where there is an effective four-dimensional theory, there should be $\mathcal{N} = 1$ supersymmetry. One intriguing piece of evidence for this is that supersymmetry ensures that the three gauge couplings of the standard model unify at about 10^{16} GeV suggesting *supersymmetric grand unification* at this energy.

A technical advantage of supersymmetry, which appeared in the discussion of dualities in Chapter 8, and is utilized in Chapter 11 in the context of black hole physics, is that supersymmetry often makes it possible to extrapolate results from weak coupling to strong coupling, thereby providing information about strongly coupled theories. Supersymmetric theories are easier to solve than their nonsupersymmetric counterparts. The constraints imposed by supersymmetry lead to first-order equations, which are easier to solve than the second-order equations of motion. For the type of backgrounds considered here a solution to the supersymmetry constraints that satisfies the Bianchi identity for the three-form field strength is always a solution to the equations of motion, though the converse is not true.

If the ten-dimensional heterotic string is compactified on an internal manifold M, one wants to know when this gives $\mathcal{N} = 1$ supersymmetry in four dimensions. Given a certain set of assumptions, it is proved in Section 9.3 that the internal manifold must be a *Calabi–Yau three-fold*.

A first glance at Calabi–Yau manifolds

Calabi–Yau manifolds are complex manifolds, and they exist in any even dimension. More precisely, a *Calabi–Yau n-fold* is a Kähler manifold in n complex dimensions with $SU(n)$ holonomy. The only examples in two (real) dimensions are the complex plane $\mathbb{C}$ and the two-torus T^2. Any Riemann surface, other than a torus, is not Calabi–Yau. In four dimensions there are two compact examples, the K3 manifold and the four-torus T^4, as well as noncompact examples such as $\mathbb{C}^2$ and $\mathbb{C} \times T^2$. The cases of greatest interest are Calabi–Yau three-folds, which have six real (or three complex) dimensions. In contrast to the lower-dimensional cases there are many thousands of Calabi–Yau three-folds, and it is an open question whether this number is even finite. Compactification on a Calabi–Yau three-fold breaks 3/4 of the original supersymmetry. Thus, Calabi–Yau compactification of the heterotic string results in $\mathcal{N}=1$ supersymmetry in four dimensions, while for the type II superstring theories it gives $\mathcal{N}=2$.

Conifold transitions and supersymmetric cycles

Nonperturbative effects in the string coupling constant need to be included for the four-dimensional low-energy theory resulting from Calabi–Yau compactifications to be consistent. For example, massless states coming from branes wrapping *supersymmetric cycles* need to be included in the low-energy effective action, as otherwise the metric is singular and the action is inconsistent. This is discussed in Section 9.8.

Mirror symmetry

Compactifications on Calabi–Yau manifolds have an interesting property that is related to T-duality, which is a characteristic feature of the toroidal compactifications described in Chapters 6 and 7. This chapter shows that certain toroidal compactifications also have another remarkable property, namely invariance under interchange of the shape and size of the torus. This is the simplest example of a symmetry known as *mirror symmetry*, which is a property of more general Calabi–Yau manifolds. This property, discussed in Section 9.9, implies that two distinct Calabi–Yau manifolds, which typically have different topologies, can be physically equivalent. More precisely, type IIA superstring theory compactified on a Calabi–Yau manifold M is equivalent to type IIB superstring theory compactified on the mirror Calabi–Yau manifold W.[2] Evidence for mirror symmetry is given in Fig. 9.1. Some progress towards a proof of mirror symmetry is discussed in Section 9.9.

Exceptional-holonomy manifolds

Calabi–Yau manifolds have been discussed a great deal since 1985. More recently, other consistent backgrounds of string theory have been investigated, partly motivated by the string dualities discussed in Chapter 8. The most important examples, discussed in Section 9.12, are manifolds of G_2 and $Spin(7)$ holonomy. G_2 manifolds are seven-dimensional and break 7/8 of the supersymmetry, while $Spin(7)$ manifolds are eight-dimensional and break 15/16 of the supersymmetry. Calabi–Yau four-folds, which are also eight-dimensional, break 7/8 of the supersymmetry. They are discussed in the context of *flux compactifications* in Chapter 10.

[2] Even though it is called a symmetry, mirror symmetry is really a duality that relates pairs of Calabi–Yau manifolds.

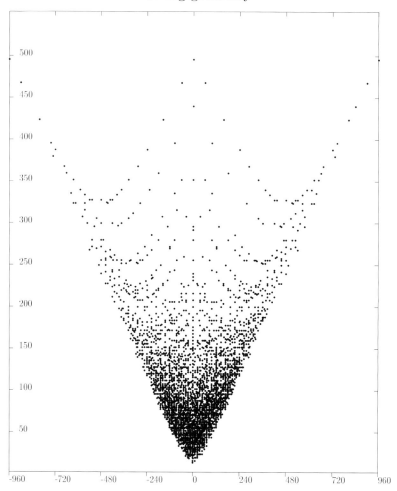

Fig. 9.1. This figure shows a plot of the sum $h^{1,1} + h^{2,1}$ against the Euler number $\chi = 2(h^{1,1} - h^{2,1})$ for a certain class of Calabi–Yau manifolds. The near-perfect symmetry of the diagram illustrates mirror symmetry, which is discussed in Section 9.7.

9.1 Orbifolds

Before discussing Calabi–Yau manifolds, let us consider a mathematically simpler class of compactification spaces called *orbifolds*. Sometimes it is convenient to know the explicit form of the metric of the internal space, which for almost all Calabi–Yau manifolds is not known,[3] not even for the

---

[3] Exceptions include tori and the complex plane.

four-dimensional manifold K3. Orbifolds include certain singular limits of Calabi–Yau manifolds for which the metric is known explicitly.

Suppose that X is a smooth manifold with a discrete isometry group G. One can then form the quotient space X/G. A point in the quotient space corresponds to an orbit of points in X consisting of a point and all of its images under the action of the isometry group. If nontrivial group elements leave points of X invariant, the quotient space has singularities. General relativity is ill-defined on singular spaces. However, it turns out that strings propagate consistently on spaces with orbifold singularities, provided so-called *twisted sectors* are taken into account. (Twisted sectors will be defined below). At nonsingular points, the orbifold X/G is locally indistinguishable from the original manifold X. Therefore, it is natural to induce local structures, such as the metric, to nonsingular regions of the orbifold. It is assumed here that the orbifold group action acts only on spatial dimensions. When the time direction is involved, new phenomena, such as closed time-like curves, can result.

Some simple examples

Compact examples

A circle is obtained by identifying points on the real line according to $x \sim x + 2\pi R$. The simplest example of an orbifold is the interval $S_1/\mathbb{Z}_2$ resulting after the identification of the circle coordinate $x \to -x$. This identification transforms a circle into an interval as shown in Fig. 9.2. This orbifold plays a crucial role in connection with the strong-coupling limit of the $E_8 \times E_8$ heterotic string, as discussed in Chapter 8.

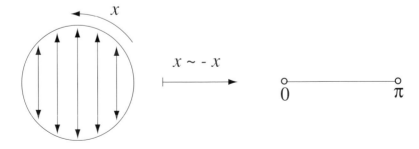

Fig. 9.2. The simplest example of an orbifold is the interval $S_1/\mathbb{Z}_2$.

Noncompact examples

A simple noncompact example of an orbifold results from considering the complex plane $\mathbb{C}$, described by a local coordinate z in the usual way, and the isometry given by the transformation

$$z \to -z. \tag{9.1}$$

This operation squares to one, and therefore it generates the two-element group $\mathbb{Z}_2$. The orbifold $\mathbb{C}/\mathbb{Z}_2$ is defined by identifying points that are in the same orbit of the group action, that is, by identifying z and $-z$. Roughly speaking, this operation divides the complex plane into two half-planes. More precisely, the orbifold corresponds to taking the upper half-plane and identifying the left and right halves of the boundary (the real axis) according to the group action. As depicted in Fig. 9.3, the resulting space is a cone.

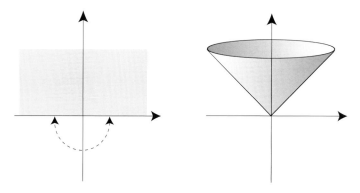

Fig. 9.3. To construct the orbifold $\mathbb{C}/\mathbb{Z}_2$ the complex plane is divided into two parts and identified along the real axis ($z \sim -z$). The resulting orbifold is a cone.

This orbifold is smooth except for a conical singularity at the point $(0,0)$, because this is the fixed point of the group action. One consequence of the conical singularity is that the circumference of a circle of radius R, centered at the origin, is πR and the conical deficit angle is π. An obvious generalization is the orbifold $\mathbb{C}/\mathbb{Z}_N$, where the group is generated by a rotation by $2\pi/N$. In this case there is again a singularity at the origin and the conical deficit angle is $2\pi(N-1)/N$. This type of singularity is an A_N singularity. It is included in the more general ADE classification of singularities, which is discussed in Sections 9.11 and 9.12.

The example $\mathbb{C}/\mathbb{Z}_2$ illustrates the following general statement: points that are invariant (or fixed) under some nontrivial group element map to singular points of the quotient space. Because of the singularities, these quotient spaces are not manifolds (which, by definition, are smooth), and they are called orbifolds instead. Not every discrete group action has fixed points. For example, the group $\mathbb{Z}$ generated by a translation $z \to z + a$ gives rise to the quotient space $\mathbb{C}/\mathbb{Z}$, which is a cylinder. Since there are no fixed points, the cylinder is a smooth manifold, and it would not be called an orbifold. When there are two such periods, whose ratio is not real, the quotient space $\mathbb{C}/(\mathbb{Z} \times \mathbb{Z})$ is a smooth torus.

The spectrum of states

What kind of physical states occur in the spectrum of free strings that live on an orbifold background geometry? In general, there are two types of states.

- The most obvious class of states, called *untwisted states*, are those that exist on X and are invariant under the group G. In other words, the Hilbert space of string states on X can be projected onto the subspace of G-invariant states. A string state Ψ on X corresponds to an orbifold string state on X/G if

$$g\Psi = \Psi, \quad \text{for all } g \in G. \tag{9.2}$$

For a finite group G, one can start with any state on X, Ψ_0, and construct a G-invariant state Ψ by superposing all the images $g\Psi_0$.

- There is a second class of physical string states on orbifolds whose existence depends on the fact that strings are extended objects. These states, called *twisted states*, are obtained in the following way. In a theory of closed strings, which is what is assumed here, strings must start and end at the same point, that is, $X^\mu(\sigma + 2\pi) = X^\mu(\sigma)$. A string that connects a point of X to one of its G images would not be an allowed configuration on X, but it maps to an allowed closed-string configuration on X/G. Mathematically, the condition is

$$X^\mu(\sigma + 2\pi) = gX^\mu(\sigma), \tag{9.3}$$

for some $g \in G$. The untwisted states correspond to $g = 1$. Twisted states are new closed-string states that appear after orbifolding. In general, there are various twisted sectors, labeled by the group element used to make the identification of the ends. More precisely, it is the conjugacy classes of G that give distinct twisted sectors. This distinction only matters if G is nonabelian.

In the example $\mathbb{C}/\mathbb{Z}_2$ it is clear that the twisted string states enclose the singular point of the orbifold. This is a generic feature of orbifolds.

In the quantum spectrum, the individual twisted-sector quantum states of the string are localized at the orbifold singularities that the classical configurations enclose. This is clear for low-lying excitations, at least, since the strings shrink to small size.

Orbifolds and supersymmetry breaking

String theories on an orbifold X/G generically have less unbroken supersymmetry than on X, which makes them phenomenologically more attractive. Let us examine how this works for a certain class of noncompact orbifolds that are a generalization of the example described above, namely orbifolds of the form $\mathbb{C}^n/\mathbb{Z}_N$. The conclusions concerning supersymmetry breaking are also applicable to compact orbifolds of the form $T^{2n}/\mathbb{Z}_N$.

The orbifold $\mathbb{C}^n/\mathbb{Z}_N$

Let us parametrize $\mathbb{C}^n$ by coordinates $(z^1, \ldots, z^n)$, and define a generator g of $\mathbb{Z}_N$ by a simultaneous rotation of each of the planes

$$g : z^a \to e^{i\phi^a} z^a, \quad a = 1, \ldots, n, \tag{9.4}$$

where the ϕ^a are integer multiples of $2\pi/N$, so that $g^N = 1$. The example of the cone corresponds to $n = 1$, $N = 2$ and $\phi^1 = \pi$.

Unbroken supersymmetries are the components of the original supercharge Q_α that are invariant under the group action. Since the group action in this example is a rotation, and the supercharge is a spinor, we have to examine how a spinor transforms under this rotation. The weights of spinor representations of a rotation generator in $2n$ dimensions have the form $(\pm\frac{1}{2}, \pm\frac{1}{2}, \ldots, \pm\frac{1}{2})$, a total of 2^n states. This corresponds to dividing the exponents by two in Eq. (9.4), which accounts for the familiar fact that a spinor reverses sign under a 2π rotation. An irreducible spinor representation of $Spin(2n)$ has dimension 2^{n-1}. An even number of $-$ weights gives one spinor representation and an odd number gives the other one. Under the same rotation considered above

$$g : Q_\alpha \to \exp\left(i \sum_{a=1}^{n} \varepsilon_\alpha^a \phi^a\right) Q_\alpha, \tag{9.5}$$

where ε_α is a spinor weight. Suppose, for example, that the ϕ^a are chosen so that

$$\frac{1}{2\pi} \sum_{a=1}^{n} \phi^a = 0 \mod N. \tag{9.6}$$

Then, in general, the only components of Q_α that are invariant under g are those whose weights ε_α have the same sign for all n components, since then $\sum \varepsilon_\alpha^a \phi^a = 0$. In special cases, other components may also be invariant. For each value of α for which the supercharge is not invariant, the amount of unbroken supersymmetry is cut in half. Thus, if there is invariance for only one value of α, the fraction of the supersymmetry that is unbroken is 2^{1-n}. This chapter shows that the same fraction of supersymmetry is preserved by compactification on a Calabi–Yau n-fold. In fact, some orbifolds of this type are singular limits of smooth Calabi–Yau manifolds.

9.2 Calabi–Yau manifolds: mathematical properties

Definition of Calabi–Yau manifolds

By definition, a Calabi–Yau n-fold is a Kähler manifold having n complex dimensions and vanishing first Chern class

$$c_1 = \frac{1}{2\pi}[\mathcal{R}] = 0. \tag{9.7}$$

A theorem, conjectured by Calabi and proved by Yau, states that any compact Kähler manifold with $c_1 = 0$ admits a Kähler metric of $SU(n)$ holonomy. As we will see below a manifold with $SU(n)$ holonomy admits a spinor field which is covariantly constant and as a result is necessarily Ricci flat. This theorem is only valid for compact manifolds. In order for it to be valid in the noncompact case, additional boundary conditions at infinity need to be imposed. As a result, metrics of $SU(n)$ holonomy correspond precisely to Kähler manifolds of vanishing first Chern class.

We will motivate the above theorem by showing that the existence of a covariantly constant spinor implies that the background is Kähler and has $c_1 = 0$. A fundamental theorem states that a compact Kähler manifold has $c_1 = 0$ if and only if the manifold admits a nowhere vanishing holomorphic n-form Ω. In local coordinates

$$\Omega(z^1, z^2, \ldots, z^n) = f(z^1, z^2, \ldots, z^n) dz^1 \wedge dz^2 \cdots \wedge dz^n. \tag{9.8}$$

In section 9.5 we will establish the vanishing of c_1 by explicitly constructing Ω in backgrounds of $SU(n)$ holonomy.

Hodge numbers of a Calabi–Yau n-fold

Betti numbers are fundamental topological numbers associated with a manifold.[4] The Betti number b_p is the dimension of the pth de Rham cohomology

---
4 There is more discussion of this background material in the appendix of this chapter.

of the manifold M, $H^p(M)$, which is defined in the appendix. When the manifold has a metric, Betti numbers count the number of linearly independent harmonic p-forms on the manifold. For Kähler manifolds the Betti numbers can be decomposed in terms of *Hodge numbers* $h^{p,q}$, which count the number of harmonic (p,q)-forms on the manifold

$$b_k = \sum_{p=0}^{k} h^{p,k-p}. \tag{9.9}$$

Hodge diamond

A Calabi–Yau n-fold is characterized by the values of its Hodge numbers. This is not a complete characterization, since inequivalent Calabi–Yau manifolds sometimes have the same Hodge numbers. There are symmetries and dualities relating different Hodge numbers, so only a small subset of these numbers is independent. The Hodge numbers of a Calabi–Yau n-fold satisfy the relation

$$h^{p,0} = h^{n-p,0}. \tag{9.10}$$

This follows from the fact that the spaces $H^p(M)$ and $H^{n-p}(M)$ are isomorphic, as can be proved by contracting a closed $(p,0)$-form with the complex conjugate of the holomorphic $(n,0)$-form and using the metric to make a closed $(0, n-p)$-form. Complex conjugation gives the relation

$$h^{p,q} = h^{q,p}, \tag{9.11}$$

and Poincaré duality gives the additional relation

$$h^{p,q} = h^{n-q,n-p}. \tag{9.12}$$

Any compact connected Kähler complex manifold has $h^{0,0} = 1$, corresponding to constant functions. A *simply-connected* manifold has vanishing *fundamental group* (first homotopy group), and therefore vanishing first homology group. As a result,[5]

$$h^{1,0} = h^{0,1} = 0. \tag{9.13}$$

In the important case of $n = 3$ the complete cohomological description of Calabi–Yau manifolds only requires specifying $h^{1,1}$ and $h^{2,1}$. The full set of Hodge numbers can be displayed in the Hodge diamond

---

[5] Aside from tori, the Calabi–Yau manifolds that are considered here are simply connected. Calabi–Yau manifolds that are not simply connected can then be constructed by modding out by discrete freely acting isometry groups. In all cases of interest, these groups are finite, and thus the resulting Calabi–Yau manifold still satisfies Eq. (9.13).

9.2 Calabi–Yau manifolds: mathematical properties

$$
\begin{array}{ccccccc}
 & & & h^{3,3} & & & \\
 & & h^{3,2} & & h^{2,3} & & \\
 & h^{3,1} & & h^{2,2} & & h^{1,3} & \\
h^{3,0} & & h^{2,1} & & h^{1,2} & & h^{0,3} \\
 & h^{2,0} & & h^{1,1} & & h^{0,2} & \\
 & & h^{1,0} & & h^{0,1} & & \\
 & & & h^{0,0} & & &
\end{array}
=
\begin{array}{ccccccc}
 & & & 1 & & & \\
 & & 0 & & 0 & & \\
 & 0 & & h^{1,1} & & 0 & \\
1 & & h^{2,1} & & h^{2,1} & & 1 \\
 & 0 & & h^{1,1} & & 0 & \\
 & & 0 & & 0 & & \\
 & & & 1 & & &
\end{array}
\quad (9.14)
$$

Using the relations discussed above, one finds that the Euler characteristic of the Calabi–Yau three-fold is given by

$$\chi = \sum_{p=0}^{6}(-1)^p b_p = 2(h^{1,1} - h^{2,1}). \quad (9.15)$$

In Chapter 10 compactifications of M-theory on Calabi–Yau four-folds are discussed. This corresponds to the case $n = 4$. These manifolds are characterized in terms of three independent Hodge numbers $h^{1,1}, h^{1,3}, h^{1,2}$. The Hodge diamond takes the form

$$
\begin{array}{ccccccccc}
 & & & & 1 & & & & \\
 & & & 0 & & 0 & & & \\
 & & 0 & & h^{1,1} & & 0 & & \\
 & 0 & & h^{2,1} & & h^{2,1} & & 0 & \\
1 & & h^{3,1} & & h^{2,2} & & h^{3,1} & & 1 \\
 & 0 & & h^{2,1} & & h^{2,1} & & 0 & \\
 & & 0 & & h^{1,1} & & 0 & & \\
 & & & 0 & & 0 & & & \\
 & & & & 1 & & & &
\end{array}
\quad (9.16)
$$

For Calabi–Yau four-folds there is an additional relation between the Hodge numbers, which will not be derived here, namely

$$h^{2,2} = 2(22 + 2h^{1,1} + 2h^{1,3} - h^{1,2}). \quad (9.17)$$

As a result, only three of the Hodge numbers can be varied independently. The Euler number can therefore be written as

$$\chi = \sum_{p=0}^{8}(-1)^p b_p = 6(8 + h^{1,1} + h^{3,1} - h^{2,1}). \quad (9.18)$$

9.3 Examples of Calabi–Yau manifolds

Calabi–Yau one-folds

The simplest examples of Calabi–Yau manifolds have one complex dimension.

Noncompact example: $\mathbb{C}$

A simple noncompact example is the complex plane $\mathbb{C}$ described in terms of the coordinates $(z, \bar{z})$. It can be described in terms of a flat metric

$$ds^2 = |dz|^2, \qquad (9.19)$$

and the holomorphic one-form is

$$\Omega = dz. \qquad (9.20)$$

Compact example: T^2

The only compact Calabi–Yau one-fold is the two-torus T^2, which can be described with a flat metric and can be thought of as a parallelogram with opposite sides identified. This simple example is discussed in Sections 9.5 and 9.9 in order to introduce concepts, such as mirror symmetry, that can be generalized to higher dimensions.

Calabi–Yau two-folds

Noncompact examples

Some simple examples of noncompact Calabi–Yau two-folds, which have two complex dimensions, can be obtained as products of the previous two manifolds: $\mathbb{C}^2 = \mathbb{C} \times \mathbb{C}$, $\mathbb{C} \times T^2$.

Compact examples: T^4, K3

Requiring a covariantly constant spinor is very restrictive in four real dimensions. In fact, K3 and T^4 are the only two examples of four-dimensional compact Kähler manifolds for which they exist. As a result, these manifolds are the only examples of Calabi–Yau two-folds. If one requires the holonomy to be $SU(2)$, and not a subgroup, then only K3 survives. By contrast, there are very many (possibly infinitely many) Calabi–Yau three-folds. Since K3 and T^4 are Calabi–Yau manifolds, they admit a Ricci-flat Kähler metric. Moreover, since $SU(2) = Sp(1)$, it turns out that they are also hyper-Kähler.[6] The explicit form of the Ricci-flat metric of a smooth K3

---

[6] In general, a $4n$-dimensional manifold of $Sp(n)$ holonomy is called hyper-Kähler. The notation $USp(2n)$ is also used for the same group when one wants to emphasize that the compact form is being used. Both notations are used in this book.

Orbifold limit of K3

A singular limit of K3, which is often used in string theory, is an orbifold of the T^4. This has the advantage that it can be made completely explicit. Consider the square T^4 constructed by taking $\mathbb{C}^2$ and imposing the following four discrete identifications:

$$z^a \sim z^a + 1 \qquad z^a \sim z^a + i, \qquad a = 1, 2. \tag{9.21}$$

There is a $\mathbb{Z}_2$ isometry group generated by

$$\mathcal{I} : (z^1, z^2) \to (-z^1, -z^2). \tag{9.22}$$

This $\mathbb{Z}_2$ action has 16 fixed points, for which each of the z^a takes one of the following four values

$$0, \ \frac{1}{2}, \ \frac{i}{2}, \ \frac{1+i}{2}. \tag{9.23}$$

Therefore, the orbifold $T^4/\mathbb{Z}_2$ has 16 singularities. These singularities can be repaired by a mathematical operation called *blowing up* the singularities of the orbifold.

Blowing up the singularities

The singular points of the orbifold described above can be "repaired" by the insertion of an *Eguchi–Hanson space*. The way to do this is to excise a small ball of radius a around each of the fixed points. The boundary of each ball is $S^3/\mathbb{Z}_2$ since opposite points on the sphere are identified, according to Eq. (9.22). One excises each ball and replaces it by a smooth noncompact Ricci-flat Kähler manifold whose boundary is $S^3/\mathbb{Z}_2$. The unique manifold that has an $S^3/\mathbb{Z}_2$ boundary and all the requisite properties to replace each of the 16 excised balls is an Eguchi–Hanson space. The metric of the Eguchi–Hanson space is

$$ds^2 = \Delta^{-1} dr^2 + \frac{1}{4} r^2 \Delta (d\psi + \cos\theta d\phi)^2 + \frac{1}{4} r^2 d\Omega_2^2, \tag{9.24}$$

with $\Delta = 1 - (a/r)^4$ and $d\Omega_2^2 = d\theta^2 + \sin^2\theta d\phi^2$. The radial coordinate is in the range $a \leq r \leq \infty$, where a is an arbitrary constant and ψ has period 2π.

Repairing the singularities in this manner gives a manifold with the desired topology, but the metric has to be smoothed out to give a true Calabi–Yau geometry. The orbifold then corresponds to the limit $a \to 0$. The nonzero

Hodge numbers of the Eguchi–Hanson space are $h^{0,0} = h^{1,1} = h^{2,2} = 1$. Moreover, the $(1,1)$-form is anti-self-dual and is given by

$$J = \frac{1}{2} r dr \wedge (d\psi + \cos\theta d\phi) - \frac{1}{4} r^2 \sin\theta d\theta \wedge d\phi, \tag{9.25}$$

as you are asked to verify in a homework problem. In terms of the complex coordinates

$$z_1 = r\cos(\theta/2) \exp\left[\frac{i}{2}(\psi + \phi)\right] \quad \text{and} \quad z_2 = r\sin(\theta/2) \exp\left[\frac{i}{2}(\psi - \phi)\right], \tag{9.26}$$

the metric is Kähler with Kähler potential

$$\mathcal{K} = \log\left[\frac{r^2 \exp(r^4 + a^4)^{1/2}}{a^2 + (r^4 + a^4)^{1/2}}\right]. \tag{9.27}$$

Hodge numbers of K3

The cohomology of K3 can be computed by combining the contributions of the T^4 and the Eguchi–Hanson spaces. The result obtained in this way remains correct after the metric has been smoothed out.

The Eguchi–Hanson spaces contribute a total of 16 generators to $H^{1,1}$, one for each of the 16 spaces used to blow up the singularities. Moreover, on the T^4 the following four representatives of $H^{1,1}$ cohomology classes survive the $\mathbb{Z}_2$ identifications:

$$dz^1 \wedge d\bar{z}^1, \quad dz^2 \wedge d\bar{z}^2, \quad dz^1 \wedge d\bar{z}^2, \quad dz^2 \wedge d\bar{z}^1. \tag{9.28}$$

This gives in total $h^{1,1} = 20$. In addition, there is one $H^{2,0}$ class represented by $dz^1 \wedge dz^2$ and one $H^{0,2}$ class represented by $d\bar{z}^1 \wedge d\bar{z}^2$. As a result, the Hodge numbers of K3 are given by the Hodge diamond

$$\begin{array}{ccccc} & & 1 & & \\ & 0 & & 0 & \\ 1 & & 20 & & 1 \\ & 0 & & 0 & \\ & & 1 & & \end{array} \tag{9.29}$$

Thus, the nonzero Betti numbers of K3 are $b_0 = b_4 = 1$, $b_2 = 22$, and the Euler characteristic is $\chi = 24$. The 22 nontrivial harmonic two-forms consist of three self-dual forms ($b_2^+ = 3$) and 19 anti-self-dual forms ($b_2^- = 19$).

Calabi–Yau n-folds

The complete classification of Calabi–Yau n-folds for $n > 2$ is an unsolved problem, and it is not even clear that the number of compact Calabi–Yau three-folds is finite. Many examples have been constructed. Here we mention a few of them.

Submanifolds of complex projective spaces

Examples of a Calabi–Yau n-folds can be constructed as a submanifold of $\mathbb{C}P^{n+1}$ for all $n > 1$. *Complex projective space*, $\mathbb{C}P^n$, sometimes just denoted P^n, is a compact manifold with n complex dimensions. It can be constructed by taking $\mathbb{C}^{n+1}/\{0\}$, that is the set of $(z^1, z^2, \ldots, z^{n+1})$ where the z^i are not all zero and making the identifications

$$(z^1, z^2, \ldots, z^{n+1}) \sim (\lambda z^1, \lambda z^2, \ldots, \lambda z^{n+1}), \tag{9.30}$$

for any nonzero complex $\lambda \neq 0$. Thus, lines[7] in $\mathbb{C}^{n+1}$ correspond to points in $\mathbb{C}P^n$.

$\mathbb{C}P^n$ is a Kähler manifold, but it is not a Calabi–Yau manifold. The simplest example is $\mathbb{C}P^1$, which is topologically the two-sphere S^2. Obviously, it does not admit a Ricci-flat metric. The standard metric of $\mathbb{C}P^n$, called the *Fubini–Study metric*, is constructed as follows. First one covers the manifold by $n+1$ open sets given by $z^a \neq 0$. Then on each open set one introduces local coordinates. For example, on the open set with $z^{n+1} \neq 0$, one defines $w^a = z^a/z^{n+1}$, with $a = 1, \ldots, n$. Then one introduces the Kähler potential (for this open set)

$$K = \log\left(1 + \sum_{a=1}^{n} |w^a|^2\right). \tag{9.31}$$

This determines the metric by formulas given in the appendix. A crucial requirement is that the analogous formulas for the Kähler potential on the other open sets differ from this one by Kähler transformations. You are asked to verify this in a homework problem.

Examples of Calabi–Yau manifolds can be obtained as subspaces of complex projective spaces. Specifically, let G be a *homogenous polynomial of degree* k in the coordinates z^a of $\mathbb{C}^{n+2}$, that is,

$$G(\lambda z^1, \ldots, \lambda z^{n+2}) = \lambda^k G(z^1, \ldots, z^{n+2}). \tag{9.32}$$

The submanifold of $\mathbb{C}P^{n+1}$ defined by

$$G(z^1, \ldots, z^{n+2}) = 0 \tag{9.33}$$

---
[7] A *line* in a complex manifold has one complex dimension.

is a compact Kähler manifold with n complex dimensions. This submanifold has vanishing first Chern class for $k = n+2$. One way of obtaining this result is to explicitly compute $c_1(X)$. To do so note that $c_1(X)$ can be expressed through the volume form since X is Kähler. As a volume form on X one can use the pullback of the $(n-1)$-power of the Kähler form of CP^{n+1}. Another way of obtaining this result is to use the *adjunction formula* of algebraic geometry, which implies

$$c_1(X) \sim [k - (n+2)] \, c_1(\mathbb{C}P^{n+1}). \tag{9.34}$$

This vanishes for $k = n + 2$.

- In the case of $n = 2$ (quartic polynomials in $\mathbb{C}P^3$) one obtains K3 manifolds. As an example consider

$$\sum_{a=1}^{4} (z^a)^4 = 0, \tag{9.35}$$

  as a quartic equation representing K3. Different choices of quartic polynomials give K3 manifolds that are diffeomorphic to each other but have different complex structures. Deformations of Calabi–Yau manifolds, in particular deformations of the complex structure, are discussed in Section 9.5.

- In the case of $n = 3$ this construction describes the *quintic hypersurface* in $\mathbb{C}P^4$. This manifold can be described by the polynomial

$$\sum_{a=1}^{5} (z^a)^5 = 0, \tag{9.36}$$

  or a more general polynomial of degree five in five variables. This manifold has the Hodge numbers

$$h^{1,1} = 1 \quad \text{and} \quad h^{2,1} = 101, \tag{9.37}$$

  which gives an Euler number of $\chi = -200$. Varying the coefficients of the quintic polynomial corresponds again to complex-structure deformations.

  The manifold defined by Eq. (9.36) can be covered by five open sets for which $z^a \neq 0$, $a = 1, \ldots, 5$. On the first open set, for example, one can define local coordinates $w^a = z^a/z^1$, $a = 2, 3, 4, 5$. These satisfy

$$\sum_{a=2}^{5} (w^a)^4 dw^a = 0. \tag{9.38}$$

9.3 Examples of Calabi–Yau manifolds

In terms of these coordinates the holomorphic three-form is given by

$$\Omega = \frac{dw^2 \wedge dw^3 \wedge dw^4}{(w^5)^4}. \tag{9.39}$$

Note that Eq. (9.39) seems to single out one of the coordinates. However, taking Eq. (9.38) into account one sees that the four coordinates w^a, $a = 2, \ldots, 5$, are treated democratically.

Weighted complex projective space: $W\mathbb{C}P^n_{k_1 \cdots k_{n+1}}$

One generalization entails replacing $\mathbb{C}P^n$ by *weighted complex projective space* $W\mathbb{C}P^n_{k_1 \cdots k_{n+1}}$. This n-dimensional complex space is defined by starting with $\mathbb{C}^{n+1}$ and making the identifications[8]

$$(\lambda^{k_1} z^1, \lambda^{k_2} z^2, \ldots, \lambda^{k_{n+1}} z^{n+1}) \sim \lambda^N (z^1, z^2, \ldots, z^{n+1}), \tag{9.40}$$

where $k_1, \ldots, k_{n+1}$ are positive integers, and N is their least common multiple. Further generalizations consist of products of such spaces with dimensions n_i. One can impose m polynomial constraint equations that respect the scaling properties of the coordinates. Generically, this produces a space with $\sum n_i - m$ complex dimensions. Then one has to compute the first Chern class, which is not so easy in general. Still this procedure has been automated, and several thousand inequivalent Calabi–Yau three-folds have been obtained. Other powerful techniques, based on *toric geometry*, which is not discussed in this book, have produced additional examples. Despite all this effort, the classification of Calabi–Yau three-folds is not yet complete.

Exercises

Exercise 9.1
Show that up to normalization $\int J \wedge J \wedge J$ is the volume of a compact six-dimensional Kähler manifold. Consider first the case of two real dimensions.

Solution
Here $J = ig_{a\bar{b}} dz^a \wedge d\bar{z}^{\bar{b}}$ is the Kähler form, which is discussed in the appendix. This result has to be true because $h^{3,3} = 1$ for a compact Kähler manifold in three complex dimensions. $J \wedge J \wedge J$, which is a $(3,3)$-form, must be

---

[8] Note that the λs have exponents and the zs have superscripts.

proportional to the volume form (up to an exact form), since it is closed but not exact. Still, it is instructive to demonstrate this explicitly. So let us do that now.

For one complex dimension (or two real dimensions) the Kähler form is $J = ig_{z\bar{z}}dz \wedge d\bar{z}$, where $z = x + iy$. The metric components then take the form

$$g_{xx} = g_{yy} = 2g_{z\bar{z}}, \quad g_{xy} = 0.$$

The Kähler form describes the volume, $V = \int J$, since

$$J = ig_{z\bar{z}}dz \wedge d\bar{z} = 2g_{z\bar{z}}dx \wedge dy = \sqrt{g}dx \wedge dy.$$

This argument generalizes to n complex dimensions, where $J = ig_{a\bar{b}}dz^a \wedge d\bar{z}^{\bar{b}}$. Setting $z^a = x^a + iy^a$ and using $\sqrt{g} = 2^n \det g_{a\bar{b}}$, one obtains for $n = 3$

$$\frac{1}{6}J \wedge J \wedge J = \sqrt{g}dx^1 \wedge \cdots \wedge dy^3,$$

which is the volume form. Thus,

$$V = \frac{1}{6}\int J \wedge J \wedge J.$$

□

EXERCISE 9.2

Consider the orbifold $T^2 \times T^2 / \mathbb{Z}_3$, where $\mathbb{Z}_3$ acts on the coordinates of $T^2 \times T^2$ by $(z^1, z^2) \to (\omega z^1, \omega^{-1}z^2)$, where $\omega = \exp(2\pi i/3)$ is a third root of unity, and (z^1, z^2) are the coordinates of the two tori. Compute the cohomology of M, including the contribution coming from the fixed points. Compare the result to the cohomology of K3.

SOLUTION

In order for the $\mathbb{Z}_3$ transformation to be a symmetry, let us choose the complex structure of the tori such that the periods are

$$z^a \sim z^a + 1 \sim z^a + e^{\pi i/3} \quad a = 1, 2.$$

The $\mathbb{Z}_3$ action has nine fixed points where each of the z^a takes one of the following three values:

$$0, \quad \frac{1}{\sqrt{3}}e^{\pi i/6}, \quad \frac{2}{\sqrt{3}}e^{\pi i/6}.$$

9.3 Examples of Calabi–Yau manifolds

The cohomology of the orbifold has two contributions: one from the harmonic forms of $T^2 \times T^2$ that are invariant under action $\mathbb{Z}_3$. The other one comes from the fixed points.

The $\mathbb{Z}_3$-invariant harmonic forms are:

$$1, \quad dz^1 \wedge dz^2, \quad d\bar{z}^1 \wedge d\bar{z}^2, \quad dz^1 \wedge d\bar{z}^1, \quad dz^2 \wedge d\bar{z}^2, \quad dz^1 \wedge dz^2 \wedge d\bar{z}^1 \wedge d\bar{z}^2.$$

Each of the nine singularities has a $\mathbb{C}P^1 \times \mathbb{C}P^1$ blow-up whose boundary is $S^3/\mathbb{Z}_3$. Each of these contributes two two-cycles or $h^{1,1} = 2$. The two two-cycles intersect at one point. Thus, the nonvanishing Hodge numbers of the orbifold are

$$h^{2,2} = h^{0,0} = h^{2,0} = h^{0,2} = 1, \qquad h^{1,1} = 2 + 9 \times 2 = 20,$$

while the other Hodge numbers vanish. These numbers are the same as those for K3. This orbifold is a singular limit of a smooth K3, like the $\mathbb{Z}_2$ orbifold considered in the text. $\mathbb{Z}_4$ and $\mathbb{Z}_6$ orbifolds also give singular K3 s.
□

EXERCISE 9.3

Consider $\mathbb{C}^2/G$, where G is the subgroup of $SU(2)$ generated by $(z^1, z^2) \to (\omega z^1, \omega^{-1} z^2)$ and $(z^1, z^2) \to (-z^2, z^1)$ with $\omega^{2n} = 1$. Show that in terms of variables invariant under the action of G the resulting (singular) space can be described by[9]

$$x^{n+1} + xy^2 + z^2 = 0.$$

SOLUTION

The variables

$$x = (z^1 z^2)^2, \quad y = \frac{i}{2}((z^1)^{2n} + (z^2)^{2n}), \quad z = \frac{1}{2}((z^1)^{2n} - (z^2)^{2n}) z^1 z^2$$

are invariant under the action of G. Thus

$$x^{n+1} = (z^1 z^2)^{2n+2},$$

$$xy^2 = -\frac{1}{4}((z^1)^{4n} + (z^2)^{4n} + 2(z^1 z^2)^{2n})(z^1 z^2)^2,$$

$$z^2 = \frac{1}{4}((z^1)^{4n} + (z^2)^{4n} - 2(z^1 z^2)^{2n})(z^1 z^2)^2.$$

[9] The singularity of this space is called a D_{n+2} singularity, because the blown-up geometry has intersection numbers encoded in the D_{n+2} Dynkin diagram. *Intersection number* is defined in Section 9.6, and the Dynkin diagram is explained in Section 9.11.

This leads to the desired equation

$$x^{n+1} + xy^2 + z^2 = 0.$$

□

9.4 Calabi–Yau compactifications of the heterotic string

Calabi–Yau compactifications of ten-dimensional heterotic string theories give theories in four-dimensional space-time with $\mathcal{N} = 1$ supersymmetry.[10] In other words, 3/4 of the original 16 supersymmetries are broken. As mentioned in the introduction, the motivation for this is the appealing, though unproved, possibility that this much supersymmetry extends down to the TeV scale in the real world. Another motivation for considering these compactifications is that it is rather easy to embed the standard-model gauge group, or a grand-unification gauge group, inside one of the two E_8 groups of the $E_8 \times E_8$ heterotic string theory.

Ansatz for the $D = 10$ space-time geometry

Let us assume that the ten-dimensional space-time M_{10} of the heterotic string theory decomposes into a product of a noncompact four-dimensional space-time M_4 and a six-dimensional internal manifold M, which is small and compact

$$M_{10} = M_4 \times M. \tag{9.41}$$

Previously, ten-dimensional coordinates were labeled by a Greek index and denoted x^μ. Now, the symbol x^M denotes coordinates of M_{10}, while x^μ denotes coordinates of M_4 and y^m denotes coordinates of the six-dimensional space M. This index rule is summarized by $M = (\mu, m)$. Generalizations of the ansatz in Eq. (9.41) are discussed in Chapter 10.

Maximally symmetric solutions

Let us consider solutions in which M_4 is maximally symmetric, that is, a homogeneous and isotropic four-dimensional space-time. Symmetries alone imply that the Riemann tensor of M_4 can be expressed in terms of its metric according to

$$R_{\mu\nu\rho\lambda} = \frac{R}{12}(g_{\mu\rho}g_{\nu\lambda} - g_{\mu\lambda}g_{\nu\rho}), \tag{9.42}$$

---

[10] This amount of supersymmetry is unbroken to every order in perturbation theory. In some cases it is broken by nonperturbative effects.

where the scalar curvature $R = g^{\mu\rho} g^{\nu\lambda} R_{\mu\nu\rho\lambda}$ is a constant. It is proportional to the four-dimensional cosmological constant. Maximal symmetry restricts the space-time M_4 to be either Minkowski ($R = 0$), AdS ($R < 0$) or dS ($R > 0$). The assumption of maximal symmetry along M_4 also requires the following components of the NS–NS three-form field strength H and the Yang–Mills field strength to vanish

$$H_{\mu\nu\rho} = H_{\mu\nu p} = H_{\mu n p} = 0 \quad \text{and} \quad F_{\mu\nu} = F_{\mu n} = 0. \tag{9.43}$$

In this chapter it is furthermore *assumed* that the internal three-form field strength H_{mnp} vanishes and the dilaton Φ is constant. These assumptions, made for simplicity, give rise to the backgrounds described in this chapter. Backgrounds with nonzero internal H-field and a nonconstant dilaton are discussed in Chapter 10.

Conditions for unbroken supersymmetry

The constraints that $\mathcal{N} = 1$ supersymmetry imposes on the vacuum arise in the following way. Each of the supersymmetry charges Q_α generates an infinitesimal transformation of all the fields with an associated infinitesimal parameter ε_α. Unbroken supersymmetries leave a particular background invariant. This is the classical version of the statement that the vacuum state is annihilated by the charges. The invariance of the bosonic fields is trivial, because each term in the supersymmetry variation of a bosonic field contains at least one fermionic field, but fermionic fields vanish in a classical background. Therefore, the only nontrivial conditions come from the fermionic variations

$$\delta_\varepsilon(\text{fermionic fields}) = 0. \tag{9.44}$$

In fact, for exactly this reason, only the bosonic parts of fermionic supersymmetry transformations were presented in Chapter 8. If the expectation values for the fermions still vanish after performing a supersymmetry variation, then one obtains a solution of the bosonic equations of motion that preserves supersymmetry for the type of backgrounds considered here. In fact, as is shown in Exercise 9.4, a solution to the supersymmetry constraints is always a solution to the equations of motion, while the converse is not necessarily true. Here we are applying this result for theories with local supersymmetry. This can be done if we impose the Bianchi identity satisfied by the three-form H as an additional constraint. In order to obtain unbroken $\mathcal{N} = 1$ supersymmetry, Eq. (9.44) needs to hold for four linearly independent choices of ε forming a four-component Majorana spinor (or equivalently a two-component Weyl spinor and its complex conjugate).

The supergravity approximation to heterotic string theory was described in Section 8.1. In particular, the bosonic part of the ten-dimensional action was presented. The full supergravity approximation also contains terms involving fermionic fields, which are incorporated in such a way that the theory has $\mathcal{N} = 1$ local supersymmetry (16 fermionic symmetries). As described in Section 8.1, the bosonic terms of the supersymmetry transformations of the fermionic fields can be written in the form[11]

$$\delta \Psi_M = \nabla_M \varepsilon - \tfrac{1}{4} \mathbf{H}_M \varepsilon,$$

$$\delta \lambda = -\tfrac{1}{2} \slashed{\partial} \Phi \varepsilon + \tfrac{1}{4} \mathbf{H} \varepsilon, \qquad (9.45)$$

$$\delta \chi = -\tfrac{1}{2} \mathbf{F} \varepsilon,$$

in the string frame. In addition, the three-form field strength H satisfies

$$dH = \frac{\alpha'}{4} \left[\mathrm{tr}(R \wedge R) - \mathrm{tr}(F \wedge F) \right]. \qquad (9.46)$$

The left-hand side is exact. Therefore, the cohomology classes of $\mathrm{tr}(R \wedge R)$ and $\mathrm{tr}(F \wedge F)$ have to be the same. In compactifications with branes, this condition can be modified by additional contributions.

Since the H-flux is assumed to vanish, the supersymmetry transformation of the gravitino simplifies,

$$\delta \Psi_M = \nabla_M \varepsilon. \qquad (9.47)$$

For an unbroken supersymmetry this must vanish, and therefore there should exist a nontrivial solution to the *Killing spinor equation*

$$\nabla_M \varepsilon = 0. \qquad (9.48)$$

This equation means that ε is a covariantly constant spinor.

$\mathcal{N} = 1$ supersymmetry implies that one such spinor should exist. Since the manifold M_{10} is a direct product, the covariantly constant spinor ε can be decomposed into a product structure

$$\varepsilon(x, y) = \zeta(x) \otimes \eta(y), \qquad (9.49)$$

or a sum of such terms. The properties of these spinors and the form of the decomposition are discussed in the next section. In making such decompositions of anticommuting (Grassmann-odd) spinors, it is always understood

---

[11] The notation introduced in Section 8.1 is simplified here according to $\tilde{\mathbf{H}}^{(3)} \to \mathbf{H}$ and $\tilde{H}_3 \to H$. Also, the fermionic variables that had tildes there are written here without tildes.

9.4 Calabi–Yau compactifications of the heterotic string

that the space-time components $\zeta(x)$ are anticommuting (Grassmann odd), while the internal components $\eta(y)$ are commuting (Grassmann even).

Properties of the external space

Let us consider the external components of Eq. (9.48) for which the index takes value $M = \mu$. The existence of a covariantly constant spinor $\zeta(x)$ on M_4, satisfying

$$\nabla_\mu \zeta = 0, \tag{9.50}$$

implies that the curvature scalar R in Eq. (9.42) vanishes, and hence M_4 is Minkowski space-time. This follows from

$$[\nabla_\mu, \nabla_\nu]\zeta = \frac{1}{4} R_{\mu\nu\rho\sigma} \Gamma^{\rho\sigma} \zeta = 0 \tag{9.51}$$

and the assumption of maximal symmetry (9.42). The details are shown in Exercises 9.6 and 9.7. Then ζ is actually constant, not just covariantly constant, and it is the infinitesimal transformation parameter of an unbroken global supersymmetry in four dimensions. This is a nontrivial result inasmuch as unbroken supersymmetry does not necessarily imply a vanishing cosmological constant by itself. AdS spaces can also be supersymmetrical, a fact that plays a crucial role in Chapter 12. However, this result does not solve the cosmological constant problem. The question that needs to be answered in order to make contact with the real world is whether the cosmological constant can vanish, or at least be extremely small, when supersymmetry is broken. The present result has nothing to say about this, since it is derived by requiring unbroken supersymmetry. To summarize: supersymmetry constrains the external space to be four-dimensional Minkowski space.

Properties of the internal manifold

Let us now consider the restrictions coming from the internal components $M = m$ of Eq. (9.48). The existence of a spinor that satisfies

$$\nabla_m \eta = 0, \tag{9.52}$$

and therefore is covariantly constant on M, leads to the *integrability condition*

$$[\nabla_m, \nabla_n]\eta = \frac{1}{4} R_{mnpq} \Gamma^{pq} \eta = 0. \tag{9.53}$$

This implies that the metric on the internal manifold M is Ricci-flat (see Exercises 9.6 and 9.7)

$$R_{mn} = 0. \tag{9.54}$$

However, in contrast to the external space-time, where maximal symmetry is assumed, it does not mean that M is flat, since the Riemann tensor can still be nonzero.

Holonomy and unbroken supersymmetry

For an orientable six-dimensional spin manifold,[12] the main case of interest here, parallel transport of a spinor η around a closed curve generically gives a rotation by a $Spin(6) = SU(4)$ matrix. This is the generic *holonomy group*.[13] A real spinor on such a manifold has eight components, but the eight components can be decomposed into two irreducible $SU(4)$ representations

$$\mathbf{8} = \mathbf{4} \oplus \bar{\mathbf{4}}, \tag{9.55}$$

where the $\mathbf{4}$ and $\bar{\mathbf{4}}$ represent spinors of opposite chirality, which are complex conjugates of one another. Thus, a spinor of definite chirality has four complex components.

A spinor that is covariantly constant remains unchanged after being parallel transported around a closed curve. The existence of such a spinor is required if some supersymmetry is to remain unbroken; see Eq. (9.48). The largest subgroup of $SU(4)$ for which a spinor of definite chirality can be invariant is $SU(3)$. The reason is that the $\mathbf{4}$ has an $SU(3)$ decomposition

$$\mathbf{4} = \mathbf{3} \oplus \mathbf{1}, \tag{9.56}$$

and the singlet is invariant under $SU(3)$ transformations. Since the condition for $\mathcal{N} = 1$ unbroken supersymmetry in four dimensions is equivalent to the existence of a covariantly constant spinor on the internal six-dimensional manifold, it follows that the manifold should have $SU(3)$ holonomy.

The supersymmetry charge of the heterotic string in ten dimensions is a Majorana–Weyl spinor with 16 real components, which form an irreducible representation of $Spin(9,1)$. Group theoretically, this decomposes with respect to an $SL(2,\mathbb{C}) \times SU(4)$ subgroup as[14]

$$\mathbf{16} = (\mathbf{2}, \mathbf{4}) \oplus (\bar{\mathbf{2}}, \bar{\mathbf{4}}). \tag{9.57}$$

---

[12] A spin manifold is a manifold on which spinors can be defined, that is, it admits spinors.
[13] More information about holonomy and spinors is given in the appendix.
[14] The other 16-dimensional spinor, which is not a supersymmetry of the heterotic string, then has the decomposition $\mathbf{16} = (\mathbf{2}, \bar{\mathbf{4}}) + (\bar{\mathbf{2}}, \mathbf{4})$.

Here $SL(2, \mathbb{C})$ is the four-dimensional Lorentz group, so **2** and $\bar{\mathbf{2}}$ correspond to positive- and negative-chirality Weyl spinors. On a manifold of $SU(3)$ holonomy only the singlet pieces of the **4** and the $\bar{\mathbf{4}}$ in Eq. (9.56) lead to covariantly constant spinors. Denoting them by fields $\eta_\pm(y)$, the covariantly constant spinor ε can be decomposed into a sum of two terms

$$\varepsilon(x, y) = \zeta_+ \otimes \eta_+(y) + \zeta_- \otimes \eta_-(y), \tag{9.58}$$

where $\zeta_\pm$ are two-component *constant* Weyl spinors on M_4. Note that

$$\eta_- = \eta_+^* \quad \text{and} \quad \zeta_- = \zeta_+^*, \tag{9.59}$$

since ε is assumed to be in a Majorana basis.

A representation of the gamma matrices that is convenient for this $10 = 4 + 6$ split is

$$\Gamma_\mu = \gamma_\mu \otimes 1 \quad \text{and} \quad \Gamma_m = \gamma_5 \otimes \gamma_m, \tag{9.60}$$

where γ_μ and γ_m are the gamma matrices of M_4 and M, respectively, and γ_5 is the usual four-dimensional chirality operator

$$\gamma_5 = -i\gamma_0\gamma_1\gamma_2\gamma_3, \tag{9.61}$$

which satisfies $\gamma_5^2 = 1$ and anticommutes with the other four γ_μs.

Internal Dirac matrices

The 8×8 Dirac matrices on the internal space M can be chosen to be antisymmetric. A possible choice of the six antisymmetric matrices satisfying $\{\gamma_i, \gamma_j\} = 2\delta_{ij}$ is

$$\sigma_2 \otimes 1 \otimes \sigma_{1,3} \quad \sigma_{1,3} \otimes \sigma_2 \otimes 1 \quad 1 \otimes \sigma_{1,3} \otimes \sigma_2. \tag{9.62}$$

One can then define a seventh antisymmetric matrix that anticommutes with all of these six as $\gamma_7 = i\gamma_1 \ldots \gamma_6$ or

$$\gamma_7 = \sigma_2 \otimes \sigma_2 \otimes \sigma_2. \tag{9.63}$$

The chirality projection operators are

$$P_\pm = (1 \pm \gamma_7)/2. \tag{9.64}$$

In terms of the matrices defined above, one defines matrices $\gamma_m = e_m^i \gamma_i$ in a real basis or γ_a and $\gamma_{\bar{a}}$ in a complex basis.

Kähler form and complex structure

Now let us consider possible fermion bilinears constructed from η_+ and η_-. Since these spinors are covariantly constant they can be normalized according to

$$\eta_+^\dagger \eta_+ = \eta_-^\dagger \eta_- = 1. \tag{9.65}$$

Next, define the tensor

$$J_m{}^n = i\eta_+^\dagger \gamma_m{}^n \eta_+ = -i\eta_-^\dagger \gamma_m{}^n \eta_-, \tag{9.66}$$

which by using the Fierz transformation formula (given in the appendix of Chapter 10) satisfies

$$J_m{}^n J_n{}^p = -\delta_m{}^p. \tag{9.67}$$

As a result, the manifold is almost complex, and J is the almost complex structure.

Since the spinors $\eta_\pm$ and the metric are covariantly constant, the almost complex structure is also covariantly constant, that is

$$\nabla_m J_n{}^p = 0. \tag{9.68}$$

This implies that the almost complex structure satisfies the condition that it is a complex structure, since it satisfies

$$N^p{}_{mn} = 0, \tag{9.69}$$

where $N^p{}_{mn}$ is the Nijenhuis tensor (see the appendix and Exercise A.4). So one can introduce local complex coordinates z^a and $\bar{z}^a$ in terms of which

$$J_a{}^b = i\delta_a{}^b, \quad J_{\bar{a}}{}^{\bar{b}} = -i\delta_{\bar{a}}{}^{\bar{b}} \quad \text{and} \quad J_a{}^{\bar{b}} = J_{\bar{a}}{}^b = 0. \tag{9.70}$$

Note that

$$g_{mn} = J_m{}^k J_n{}^l g_{kn}, \tag{9.71}$$

which together with Eq. (9.70) implies that the metric is hermitian with respect to the almost complex structure. Moreover, Eq. (9.71) implies that the quantity

$$J_{mn} = J_m{}^k g_{kn}, \tag{9.72}$$

is antisymmetric and as a result defines a two-form

$$J = \frac{1}{2} J_{mn} dx^m \wedge dx^n. \tag{9.73}$$

The components of J are related to the metric according to

$$J_{a\bar{b}} = ig_{a\bar{b}}. \tag{9.74}$$

9.4 Calabi–Yau compactifications of the heterotic string

One important property of J is that it is closed, since

$$dJ = \partial J + \bar{\partial} J = i\partial_a g_{b\bar{c}} dz^a \wedge dz^b \wedge dz^{\bar{c}} + i\partial_{\bar{a}} g_{b\bar{c}} dz^{\bar{a}} \wedge dz^b \wedge dz^{\bar{c}} = 0. \quad (9.75)$$

To see this, note that the metric is covariantly constant and take into account that we are using a torsion-free connection. As a result, the background is Kähler, and J is the Kähler form.

Holomorphic three-form

Let us now consider possible fermion bilinears, starting with ones that are bilinear in η_-. Remembering that η is Grassmann even, one can see that the bilinears $\eta_-^T \gamma_a \eta_-$ and $\eta_-^T \gamma_{ab} \eta_-$ vanish by symmetry. Also, the bilinear $\eta_-^T \eta_-$ vanishes by chirality. The only nonzero possibility, consistent with both chirality and symmetry, is

$$\Omega_{abc} = \eta_-^T \gamma_{abc} \eta_-. \quad (9.76)$$

This can be used to define a nowhere-vanishing $(3,0)$-form

$$\Omega = \frac{1}{6} \Omega_{abc} dz^a \wedge dz^b \wedge dz^c. \quad (9.77)$$

- Let us now show that Ω is closed. Since η and the metric are covariantly constant, it satisfies $\nabla_{\bar{d}} \Omega_{abc} = 0$. The connection terms vanish for a Kähler manifold, and therefore one deduces that $\bar{\partial}\Omega = 0$. It is obvious that $\partial\Omega = 0$, since there are only three holomorphic dimensions. Thus, Ω is closed, $d\Omega = (\partial + \bar{\partial})\Omega = 0$. The fact that $\bar{\partial}\Omega = 0$ implies that the coefficients Ω_{abc} are holomorphic.
- On the other hand, Ω is not exact. This can be understood as a consequence of the fact $\Omega \wedge \bar{\Omega}$ is proportional to the volume form, which has a nonzero integral over M (see Exercise 9.8). Therefore, $\Omega \wedge \bar{\Omega}$ is not exact, which implies that Ω is not exact. A Calabi–Yau manifold has $h^{3,0} = 1$, and Ω is a representative of the unique $(3,0)$ cohomology class. Other representatives differ by a nonzero multiplicative constant.

The existence of a holomorphic $(3,0)$-form implies that the manifold has a vanishing first Chern class. Indeed, since the holomorphic indices take three values, Ω_{abc} must be proportional to the Levi–Civita symbol ε_{abc}:

$$\Omega_{abc} = f(z) \varepsilon_{abc}, \quad (9.78)$$

with $f(z)$ a nowhere vanishing holomorphic function of z_1, z_2 and z_3. This implies that the norm of Ω, defined according to

$$||\Omega||^2 = \frac{1}{3!} \Omega_{abc} \bar{\Omega}^{abc}, \quad (9.79)$$

satisfies
$$\sqrt{g} = \frac{|f|^2}{||\Omega||^2}, \qquad (9.80)$$

where g denotes the determinant of the metric. As a result, the Ricci form is given by

$$\mathcal{R} = i\partial\bar{\partial}\log\sqrt{g} = -i\partial\bar{\partial}\log||\Omega||^2. \qquad (9.81)$$

Since $\log||\Omega||^2$ is a coordinate scalar, and therefore globally defined, this implies that $\mathcal{R}$ is exact and $c_1 = 0$. Since the internal spaces are Kähler manifolds with a vanishing first Chern class, they are by definition Calabi–Yau manifolds.

To summarize, assuming $H = 0$ and a constant dilaton, the requirement of unbroken $\mathcal{N} = 1$ supersymmetry of the heterotic string compactified to four dimensions implies that the internal manifold is Kähler and has a vanishing first Chern class. In other words, it is a Calabi–Yau three-fold. Such a manifold admits a unique Ricci-flat metric. The Ricci-flat metric is generally selected in the supergravity approximation (analyzed here), while stringy corrections can deform it to a metric that is not Ricci-flat.[15] The advantage of this formulation is that Kähler manifolds with vanishing first Chern class can be constructed using various methods (some of which are presented in Section 9.3). However, backgrounds in which only the holonomy is specified, which in the present case is $SU(n)$, are extremely difficult to deal with.

EXERCISES

EXERCISE 9.4
Given a theory with $\mathcal{N} = 1$ global supersymmetry, show that a supersymmetric state is a zero-energy solution to the equations of motion.

SOLUTION

A supersymmetric state $|\Psi\rangle$ is annihilated by one or more supercharges

$$Q|\Psi\rangle = 0.$$

---

[15] However, the known corrections to the metric can be absorbed in field redefinitions, so that the metric becomes Ricci-flat again.

For an $\mathcal{N} = 1$ supersymmetric theory there is no central charge, and we can write the Hamiltonian as
$$H = Q^\dagger Q,$$
which is positive definite. A supersymmetric state satisfies
$$H|\Psi\rangle = 0,$$
and therefore it gives a zero-energy solution of the equations of motion. The converse is not true, since there are positive-energy solutions of the equations of motion that are not supersymmetric. In classical terms, this result means that a field configuration satisfying $\delta_\varepsilon \psi = 0$, for all the fermi fields, as discussed in the text, is a solution of the classical field equations. □

EXERCISE 9.5
Prove that $\eta_\pm$ in Eqs (9.59) are Weyl spinors of opposite chirality, that is, γ_7 has eigenvalues ± 1.

SOLUTION

Using $\gamma_7 \equiv i\gamma_1\gamma_2\gamma_3\gamma_4\gamma_5\gamma_6$, one finds $\gamma_7^2 = 1$. This is manifest for the representation presented in Eq. (9.63). We can then define $P_\pm \equiv \frac{1}{2}(1 \pm \gamma_7)$, and $\eta_\pm \equiv P_\pm \eta$. Therefore,
$$\gamma_7 \eta_+ = \eta_+ \quad \text{and} \quad \gamma_7 \eta_- = -\eta_-.$$
In terms of holomorphic and antiholomorphic indices
$$\gamma_7 = (1 - \gamma_{\bar{1}}\gamma_1)(1 - \gamma_{\bar{2}}\gamma_2)(1 - \gamma_{\bar{3}}\gamma_3) = -(1 - \gamma_1\gamma_{\bar{1}})(1 - \gamma_2\gamma_{\bar{2}})(1 - \gamma_3\gamma_{\bar{3}}),$$
so the conditions $\gamma_a \eta_+ = 0$ and $\gamma_{\bar{a}} \eta_- = 0$ also give the same results. □

EXERCISE 9.6
Derive the identity $[\nabla_m, \nabla_n]\eta = \frac{1}{4} R_{mnpq} \Gamma^{pq} \eta$ used in Eq. (9.53).

SOLUTION

Using Eq. (9.60) and the definition of the covariant derivative in the appendix,
$$\nabla_n \eta = \partial_n \eta + \frac{1}{4}\omega_{npq}\gamma^{pq}\eta,$$
where ω_{npq} are the components of the spin connection. Thus
$$[\nabla_m, \nabla_n]\eta = [\partial_m + \frac{1}{4}\omega_{mpq}\gamma^{pq}, \partial_n + \frac{1}{4}\omega_{nrs}\gamma^{rs}]\eta.$$

In writing this one has used the fact that Christoffel-connection terms of the form $(\Gamma^p_{mn} - \Gamma^p_{nm})\partial_p \eta$ cancel by symmetry. The commutator above gives

$$\frac{1}{4}(\partial_m \omega_{nrs} - \partial_n \omega_{mrs})\gamma^{rs}\eta + \frac{1}{16}\omega_{mpq}\omega_{nrs}[\gamma^{pq}, \gamma^{rs}]\eta,$$

which simplifies to

$$\frac{1}{4}(\partial_m \omega_{nrs} - \partial_n \omega_{mrs} + \omega_{mrp}\omega_n{}^p{}_s - \omega_{nrp}\omega_m{}^p{}_s)\gamma^{rs}\eta = \frac{1}{4}R_{mnrs}\gamma^{rs}\eta,$$

where we have used

$$[\gamma_{rs}, \gamma^{pq}] = -8\delta^{[p}_{[r}\gamma_{s]}{}^{q]}.$$

$\square$

EXERCISE 9.7
Prove that $R_{mnpq}\gamma^{pq}\eta = 0$ implies that $R_{mn} = 0$.

SOLUTION
Multiplying the above equation with γ^n gives

$$\gamma^n \gamma^{pq} R_{mnpq}\eta = 0.$$

Using the gamma matrix identity

$$\gamma^n \gamma^{pq} = \gamma^{npq} + g^{np}\gamma^q - g^{nq}\gamma^p$$

and the equation

$$\gamma^{npq} R_{mnpq} = \gamma^{npq} R_{m[npq]} = 0,$$

which is the Bianchi identity, we get the expression

$$2g^{nq}\gamma^p R_{mnpq}\eta = 0.$$

This implies $\gamma^p R_{mp}\eta = 0$. If $\eta = \eta_+$ is positive chirality, for example, this gives

$$i\eta_-^T \gamma_q \gamma^p \eta_+ R_{mp} = J_q{}^p R_{mp} = 0.$$

J is invertible, so this implies that $R_{mp} = 0$, and thus the manifold is Ricci-flat. $\square$

EXERCISE 9.8
Show that $\Omega \wedge \overline{\Omega}$ is proportional to the volume form of the Calabi–Yau three-fold that we derived in Exercise 9.1.

SOLUTION

As in the case of Exercise 9.1, this is a nontrivial closed $(3,3)$-form, so this has to be true (up to an exact form) by uniqueness. Still, it is instructive to examine the explicit formulas and determine the normalization. By definition
$$\Omega = \frac{1}{6}\Omega_{a_1 a_2 a_3} dz^{a_1} \wedge dz^{a_2} \wedge dz^{a_3},$$
where $\Omega_{a_1 a_2 a_3} = \eta_-^T \gamma_{a_1 a_2 a_3} \eta_-$. Thus $\Omega \wedge \overline{\Omega}$ becomes
$$\frac{1}{36} dz^{a_1} \wedge dz^{a_2} \wedge dz^{a_3} \wedge d\bar{z}^{\bar{b}_1} \wedge d\bar{z}^{\bar{b}_2} \wedge d\bar{z}^{\bar{b}_3} \Omega_{a_1 a_2 a_3} \overline{\Omega}_{\bar{b}_1 \bar{b}_2 \bar{b}_3}$$
$$= -\frac{i}{36} J \wedge J \wedge J (\Omega_{a_1 a_2 a_3} \overline{\Omega}_{\bar{b}_1 \bar{b}_2 \bar{b}_3} g^{a_1 \bar{b}_1} g^{a_2 \bar{b}_2} g^{a_3 \bar{b}_3}).$$

Since $\frac{1}{6} J \wedge J \wedge J = dV$ is the volume form, we only need to prove that the extra factor is a constant. Because of the properties $\nabla_m \Omega_{abc} = 0$ and $\nabla_m g^{a\bar{b}} = 0$, we have
$$\nabla_m \|\Omega\|^2 = 0$$
where
$$\|\Omega\|^2 = \frac{1}{6} g^{a_1 \bar{b}_1} g^{a_2 \bar{b}_2} g^{a_3 \bar{b}_3} \Omega_{a_1 a_2 a_3} \overline{\Omega}_{\bar{b}_1 \bar{b}_2 \bar{b}_3}.$$

$\|\Omega\|^2$ is a scalar, and thus it is a constant. It follows that $\Omega \wedge \overline{\Omega}$ is $-i\|\Omega\|^2 dV$.
□

9.5 Deformations of Calabi–Yau manifolds

Calabi–Yau manifolds with specified Hodge numbers are not unique. Some of them are smoothly related by deformations of the parameters characterizing their shapes and sizes, which are called *moduli*. Often the entire moduli space of manifolds is referred to as a single Calabi–Yau space, even though it is really a continuously infinite family of manifolds. This interpretation was implicitly assumed earlier in raising the question whether or not there is a finite number of Calabi–Yau manifolds. There can also be more than one Calabi–Yau manifold of given Hodge numbers that are topologically distinct, with disjoint moduli spaces, since the Hodge numbers do not give a full characterization of the topology. On the other hand, when one goes beyond the supergravity approximation, it is sometimes possible to identify smooth topology-changing transitions, such as the conifold transition described in Section 9.8, which can even change the Hodge numbers.

This section and the next one explain how the moduli parametrize the space of possible choices of undetermined expectation values of massless scalar fields in four dimensions. They are undetermined because the effective potential does not depend on them, at least in the leading supergravity approximation. A very important property of the moduli space of Calabi–Yau three-folds is that it is the product of two factors, one describing the complex-structure moduli and one describing the Kähler-structure moduli.

Let us now consider the spectrum of fluctuations about a given Calabi–Yau manifold with fixed Hodge numbers. Some of these fluctuations come from metric deformations, while others are obtained from deformations of antisymmetric tensor fields.

Antisymmetric tensor-field deformations

As discussed in Chapter 8, the low-energy effective actions for string theories contain various p-form fields with kinetic terms proportional to

$$\int d^{10}x \sqrt{-g}\, |\, F_p\, |^2, \qquad (9.82)$$

where $F_p = dA_{p-1}$. An example of such a field is the type IIA or type IIB three-form $H_3 = dB_2$. The equation of motion of this field is[16]

$$\Delta B_{p-1} = d \star dB_{p-1} = 0. \qquad (9.83)$$

If we compactify to four dimensions on a product space $M_4 \times M$, where M is a Calabi–Yau three-fold, then the space-time metric is a sum of a four-dimensional piece and a six-dimensional piece. Therefore, the Laplacian is also a sum of two pieces

$$\Delta = \Delta_4 + \Delta_6, \qquad (9.84)$$

and the number of massless four-dimensional fields is given by the number of zero modes of the internal Laplacian Δ_6. These zero modes are counted by the Betti numbers b_p. The ten-dimensional field B_2, for example, can give rise to four-dimensional fields that are two-forms, one-forms and zero-forms. The number of these fields is summarized in the following table:

B_{MN}	$B_{\mu\nu}$	$B_{\mu n}$	B_{mn}
p-form in 4D	$p = 2$	$p = 1$	$p = 0$
# of fields in 4D	$b_0 = 1$	$b_1 = 0$	$b_2 = h^{1,1}$

[16] This assumes other terms vanish or can be neglected.

9.5 Deformations of Calabi–Yau manifolds

The b_2 scalar fields in this example are moduli originating from the B field. More generally, a p-form field gives rise to b_p moduli fields.

Metric deformations

The zero modes of the ten-dimensional metric (or graviton field) give rise to the four-dimensional metric $g_{\mu\nu}$ and a set of massless scalar fields originating from the internal components of the metric g_{mn}. In Calabi–Yau compactifications no massless vector fields are generated from the metric since $b_1 = 0$. A closely related fact is that Calabi–Yau three-folds have no continuous isometry groups.

The fluctuations of the metric on the internal space are analyzed by performing a small variation

$$g_{mn} \to g_{mn} + \delta g_{mn}, \qquad (9.85)$$

and then demanding that the new background still satisfies the Calabi–Yau conditions. In particular, one requires

$$R_{mn}(g + \delta g) = 0. \qquad (9.86)$$

This leads to differential equations for δg, and the number of solutions counts the number of ways the background metric can be deformed while preserving supersymmetry and the topology. The coefficients of these independent solutions are the moduli. They are the expectation values of massless scalar fields, called the *moduli fields*. These moduli parametrize changes of the size and shape of the internal Calabi–Yau manifold but not its topology.

A simple example: the torus

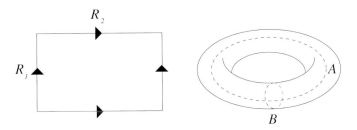

Fig. 9.4. A rectangular torus can be constructed by identifying opposite sides of a rectangle.

Consider the rectangular torus $T^2 = S^1 \times S^1$ displayed in Fig. 9.4. This torus is described by a flat metric. As discussed in Exercise 9.9, it is convenient to describe a torus using two complex parameters τ and ρ, which in the present case are related to the two radii by

$$\tau = i\frac{R_2}{R_1} \quad \text{and} \quad \rho = iR_1 R_2. \tag{9.87}$$

The shape, or complex structure, of the torus is described by τ, while the size is described by ρ. As a result, two transformations can be performed so that the torus remains a torus. A complex-structure deformation changes τ, while a Kähler-structure deformation changes ρ. These deformations are illustrated in Fig. 9.5.

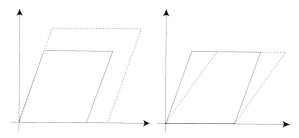

Fig. 9.5. Kähler structure deformations and complex structure deformations correspond to changing the size and shape of a torus respectively.

Recall that the holomorphic one-form on a torus is given by

$$\Omega = dz. \tag{9.88}$$

The complex-structure parameter τ can then be written as the quotient of the two periods

$$\tau = \frac{\int_A \Omega}{\int_B \Omega}, \tag{9.89}$$

where A and B are the cycles shown in Fig. 9.4. This definition is generalized to Calabi–Yau three-folds in the next section. The rectangular torus is not the most general torus. There can be an angle θ as shown in Fig. 9.6. When τ has a real part, mirror symmetry[17] only makes sense if ρ has a real part as well. The imaginary part of ρ then describes the volume, while the real part descends from the B field, as explained in Exercise 7.8.

Deformations of Calabi–Yau three-folds

In order to analyze the metric deformations of Calabi–Yau three-folds, let us use the strategy outlined in the introduction of this section and require that

---

[17] The mirror symmetry transformation that interchanges τ and ρ is discussed in Section 9.9.

9.5 Deformations of Calabi–Yau manifolds

g_{mn} and $g_{mn} + \delta g_{mn}$ both satisfy the Calabi–Yau conditions. In particular, they describe Ricci-flat backgrounds so that

$$R_{mn}(g) = 0 \quad \text{and} \quad R_{mn}(g + \delta g) = 0. \tag{9.90}$$

Some metric deformations only describe coordinate changes and are not of interest. To eliminate them one fixes the gauge

$$\nabla^m \delta g_{mn} = \frac{1}{2} \nabla_n \delta g_m^m, \tag{9.91}$$

where $\delta g_m^m = g^{mp} \delta g_{mp}$. Expanding the second equation in (9.90) to linear order in δg and using the Ricci-flatness of g leads to

$$\nabla^k \nabla_k \delta g_{mn} + 2 R_m{}^p{}_n{}^q \delta g_{pq} = 0. \tag{9.92}$$

This equation is known as the Lichnerowicz equation, which you are asked to verify in Problem 9.7. Using the properties of the index structure of the metric and Riemann tensor of Kähler manifolds, one finds that the equations for the mixed components $\delta g_{a\bar{b}}$ and the pure components δg_{ab} decouple.

Consider the infinitesimal $(1,1)$-form

$$\delta g_{a\bar{b}} dz^a \wedge d\bar{z}^{\bar{b}}, \tag{9.93}$$

which is harmonic if (9.92) is satisfied, as you are asked to verify in Problem 9.8. We imagine that after the variation $g + \delta g$ is a Kähler metric, which in classical geometry should be positive definite. The Kähler metric defines the Kähler form $J = i g_{a\bar{b}} dz^a \wedge d\bar{z}^{\bar{b}}$, and positivity of the metric is equivalent to

$$\int_{M_r} \underbrace{J \wedge \cdots \wedge J}_{r-\text{times}} > 0, \quad r = 1, 2, 3, \tag{9.94}$$

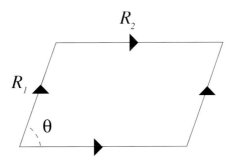

Fig. 9.6. The shape of a torus is described by a complex-structure parameter τ. The angle θ is the phase of τ.

where M_r is any complex r-dimensional submanifold of the Calabi–Yau three-fold. The subset of metric deformations that lead to a Kähler form satisfying Eq. (9.94) is called the *Kähler cone*. This space is a cone since if J satisfies (9.94), so does rJ for any positive number r, as illustrated in Fig. 9.7.

Fig. 9.7. The deformations of the Kähler form that satisfy Eq. (9.94) build the Kähler cone.

The five ten-dimensional superstring theories each contain an NS–NS two-form B. After compactification on a Calabi–Yau three-fold, the internal $(1,1)$-form $B_{a\bar{b}}$ has $h^{1,1}$ zero modes, so that this many additional massless scalar fields arise in four dimensions. The real closed two-form B combines with the Kähler form J to give the *complexified Kähler form*

$$\mathcal{J} = B + iJ. \tag{9.95}$$

The variations of this form give rise to $h^{1,1}$ massless complex scalar fields in four dimensions. Thus, while the Kähler form is real from a geometric viewpoint, it is effectively complex in the string theory setting, generalizing the complexification of the ρ parameter of the torus. This procedure is called the *complexification* of the Kähler cone. For M-theory compactifications, discussed later, there is no two-form B, and so the Kähler form, as well as the corresponding moduli space, is not complexified.

The purely holomorphic and antiholomorphic metric components g_{ab} and $g_{\bar{a}\bar{b}}$ are zero. However, one can consider varying to nonzero values, thereby changing the complex structure. With each such variation one can associate the complex $(2,1)$-form

$$\Omega_{abc} g^{c\bar{d}} \delta g_{\bar{d}\bar{e}} dz^a \wedge dz^b \wedge d\bar{z}^{\bar{e}}. \tag{9.96}$$

This is harmonic if (9.90) is satisfied. The precise relation to complex-structure variations is explained in Section 9.6.

9.6 Special geometry

The mathematics that is needed to describe Calabi–Yau moduli spaces, known as *special geometry*, is described in this section.

The metric on moduli space

The moduli space has a natural metric defined on it[18], which is given as a sum of two pieces. The first piece corresponds to deformations of the complex structure and the second to deformations of the complexified Kähler form

$$ds^2 = \frac{1}{2V} \int g^{a\bar{b}} g^{c\bar{d}} \left[\delta g_{ac} \delta g_{\bar{b}\bar{d}} + (\delta g_{a\bar{d}} \delta g_{c\bar{b}} - \delta B_{a\bar{d}} \delta B_{c\bar{b}}) \right] \sqrt{g}\, d^6 x, \quad (9.97)$$

where V is the volume of the Calabi–Yau manifold M. The fact that the metric splits into two pieces in this way implies that the geometry of the moduli space has a product structure (at least locally)

$$\mathcal{M}(M) = \mathcal{M}^{2,1}(M) \times \mathcal{M}^{1,1}(M). \quad (9.98)$$

Each of these factors has an interesting geometric structure of its own described below.

The complex-structure moduli space
The Kähler potential

Let us begin with the space of complex-structure deformations of the metric. First we define a set of $(2,1)$-forms according to

$$\chi_\alpha = \frac{1}{2} (\chi_\alpha)_{ab\bar{c}} dz^a \wedge dz^b \wedge d\bar{z}^{\bar{c}} \quad \text{with} \quad (\chi_\alpha)_{ab\bar{c}} = -\frac{1}{2} \Omega_{ab}{}^{\bar{d}} \frac{\partial g_{\bar{c}\bar{d}}}{\partial t^\alpha}, \quad (9.99)$$

where t^α, with $\alpha = 1, \ldots, h^{2,1}$ are local coordinates for the complex-structure moduli space. Indices are raised and lowered with the hermitian metric, so that $\Omega_{ab}{}^{\bar{d}} = g^{c\bar{d}} \Omega_{abc}$, for example. As in Eq. (9.96), these forms are harmonic. These relations can be inverted to show that under a deformation of the complex structure the metric components change according to

$$\delta g_{\bar{a}\bar{b}} = -\frac{1}{\|\Omega\|^2} \overline{\Omega}_{\bar{a}}{}^{cd} (\chi_\alpha)_{cd\bar{b}} \delta t^\alpha, \quad \text{where} \quad \|\Omega\|^2 = \frac{1}{6} \Omega_{abc} \overline{\Omega}^{abc}. \quad (9.100)$$

---

[18] The metric on the moduli space, which is a metric on the parameter space describing deformations of Calabi–Yau manifolds, should not be confused with the Calabi–Yau metric.

Writing the metric on moduli space as

$$ds^2 = 2G_{\alpha\bar{\beta}}\delta t^\alpha \delta \bar{t}^{\bar{\beta}}, \tag{9.101}$$

and using Eqs (9.97) and Eq. (9.100) for $\delta g_{\bar{a}\bar{b}}$, one finds that the metric on moduli space is

$$G_{\alpha\bar{\beta}}\delta t^\alpha \delta \bar{t}^{\bar{\beta}} = -\left(\frac{i \int \chi_\alpha \wedge \bar{\chi}_{\bar{\beta}}}{i \int \Omega \wedge \bar{\Omega}} \right) \delta t^\alpha \delta \bar{t}^{\bar{\beta}}. \tag{9.102}$$

Under a change in complex structure the holomorphic $(3,0)$-form Ω becomes a linear combination of a $(3,0)$-form and $(2,1)$-forms, since dz becomes a linear combination of dz and $d\bar{z}$. More precisely,

$$\partial_\alpha \Omega = K_\alpha \Omega + \chi_\alpha, \tag{9.103}$$

where $\partial_\alpha = \partial/\partial t^\alpha$ and K_α depends on the coordinates t^α but not on the coordinates of the Calabi–Yau manifold M. The concrete form of K_α is determined below. Moreover, the χ_α are precisely the $(2,1)$-forms defined in (9.99). Exercise 9.10 verifies Eq. (9.103).

Combining Eqs (9.102) and (9.103) and recalling that $G_{\alpha\bar{\beta}} = \partial_\alpha \partial_{\bar{\beta}} \mathcal{K}$, one sees that the metric on the complex-structure moduli space is Kähler with Kähler potential given by

$$\mathcal{K}^{2,1} = -\log\left(i \int \Omega \wedge \bar{\Omega} \right). \tag{9.104}$$

Exercise 9.9 considers the simple example of a two-dimensional torus and shows that the Kähler potential is given by Eq. (9.104) for $\Omega = dz$.

Special coordinates

In order to describe the complex-structure moduli space in more detail, let us introduce a basis of three-cycles A^I, B_J, with $I, J = 0, \ldots, h^{2,1}$, chosen such that their *intersection numbers* are

$$A^I \cap B_J = -B_J \cap A^I = \delta^I_J \quad \text{and} \quad A^I \cap A^J = B_I \cap B_J = 0. \tag{9.105}$$

The dual cohomology basis is denoted by (α_I, β^I). Then

$$\int_{A^J} \alpha_I = \int \alpha_I \wedge \beta^J = \delta^J_I \quad \text{and} \quad \int_{B_J} \beta^I = \int \beta^I \wedge \alpha_J = -\delta^I_J. \tag{9.106}$$

The group of transformations that preserves these properties is the symplectic modular group $Sp(2h^{2,1} + 2; \mathbb{Z})$.

9.6 Special geometry

In analogy with the torus example, we can define coordinates X^I on the moduli space by using the A periods of the holomorphic three-form

$$X^I = \int_{A^I} \Omega \quad \text{with} \quad I = 0, \ldots, h^{2,1}. \tag{9.107}$$

The number of coordinates defined this way is one more than the number of moduli fields. However, the coordinates X^I are only defined up to a complex rescaling, since the holomorphic three-form has this much nonuniqueness. To take account of this factor consider the quotient[19]

$$t^\alpha = \frac{X^\alpha}{X^0} \quad \text{with} \quad \alpha = 1, \ldots, h^{2,1}, \tag{9.108}$$

where the index α now excludes the value 0. This gives the right number of coordinates to describe the complex-structure moduli. Since the X^I give the right number of coordinates to span the moduli space, the B periods

$$F_I = \int_{B_I} \Omega \tag{9.109}$$

must be functions of the X, that is, $F_I = F_I(X)$. It follows that

$$\Omega = X^I \alpha_I - F_I(X) \beta^I. \tag{9.110}$$

A simple consequence of Eq. (9.103) is

$$\int \Omega \wedge \partial_I \Omega = 0, \tag{9.111}$$

which implies

$$F_I = X^J \frac{\partial F_J}{\partial X^I} = \frac{1}{2} \frac{\partial}{\partial X^I} \left(X^J F_J \right), \tag{9.112}$$

or, equivalently,

$$F_I = \frac{\partial F}{\partial X^I} \quad \text{where} \quad F = \frac{1}{2} X^I F_I. \tag{9.113}$$

As a result, all of the B periods are expressed as derivatives of a single function F called the *prepotential*. Moreover, since

$$2F = X^I \frac{\partial F}{\partial X^I}, \tag{9.114}$$

F is homogeneous of degree two, which means that if we rescale the coordinates by a factor λ

$$F(\lambda X) = \lambda^2 F(X). \tag{9.115}$$

[19] As usual in this type of construction, these coordinates parametrize the open set $X^0 \neq 0$.

Since the prepotential is defined only up to an overall scaling, strictly speaking it is not a function but rather a *section of a line bundle* over the moduli space.

The prepotential determines the metric on moduli space. Using the general rule for closed three-forms α and β

$$\int_M \alpha \wedge \beta = -\sum_I \left(\int_{A^I} \alpha \int_{B_I} \beta - \int_{A^I} \beta \int_{B_I} \alpha \right), \qquad (9.116)$$

the Kähler potential (9.104) can be rewritten in the form

$$e^{-\mathcal{K}^{2,1}} = -i \sum_{I=0}^{h^{2,1}} \left(X^I \bar{F}_I - \bar{X}^I F_I \right), \qquad (9.117)$$

as you are asked to verify in a homework problem. As a result, the Kähler potential is completely determined by the prepotential F, which is a holomorphic homogeneous function of degree two. This type of geometry is called *special geometry*.

An important consequence of the product structure (9.98) of the moduli space is that the complex-structure prepotential F is exact in α'. Indeed, the α' expansion is an expansion in terms of the Calabi–Yau volume V, which belongs to $\mathcal{M}^{1,1}(M)$, and it is independent of position in $\mathcal{M}^{2,1}(M)$, that is, the complex structure.[20] When combined with mirror symmetry, this important fact provides insight into an infinite series of stringy α' corrections involving the Kähler-structure moduli using a classical geometric computation involving the complex-structure moduli space only.

The Kähler transformations

The holomorphic three-form Ω is only determined up to a function f, which can depend on the moduli space coordinates X^I but not on the Calabi–Yau coordinates, that is, the transformation

$$\Omega \to e^{f(X)} \Omega \qquad (9.118)$$

should not lead to new physics. This transformation does not change the Kähler metric, since under Eq. (9.118)

$$\mathcal{K}^{2,1} \to \mathcal{K}^{2,1} - f(X) - \bar{f}(\bar{X}), \qquad (9.119)$$

which is a Kähler transformation that leaves the Kähler metric invariant.

---

[20] Since V and α' are the only scales in the problem, the only dimensionless quantity containing α' is $(\alpha')^3/V$. So if one knows the full V dependence, one also knows the full α' dependence.

9.6 Special geometry

Equations (9.103) and (9.104) determine K_α to be

$$K_\alpha = -\partial_\alpha \mathcal{K}^{2,1}. \tag{9.120}$$

One can then introduce the covariant derivatives

$$\mathcal{D}_\alpha = \partial_\alpha + \partial_\alpha \mathcal{K}^{2,1}, \tag{9.121}$$

and write

$$\chi_\alpha = \mathcal{D}_\alpha \Omega, \tag{9.122}$$

which now transforms as $\chi_\alpha \to e^{f(X)} \chi_\alpha$.

The Kähler-structure moduli space

The Kähler potential

An inner product on the space of $(1,1)$ cohomology classes is defined by

$$G(\rho, \sigma) = \frac{1}{2V} \int_M \rho_{a\bar{d}} \sigma_{\bar{b}c} g^{a\bar{b}} g^{c\bar{d}} \sqrt{g} d^6 x = \frac{1}{2V} \int_M \rho \wedge \star \sigma, \tag{9.123}$$

where $\star$ denotes the Hodge-star operator on the Calabi–Yau, and ρ and σ are real $(1,1)$-forms. Let us now define the cubic form

$$\kappa(\rho, \sigma, \tau) = \int_M \rho \wedge \sigma \wedge \tau, \tag{9.124}$$

and recall from Exercise 9.1 that $\kappa(J, J, J) = 6V$. Using the identity

$$\star \sigma = -J \wedge \sigma + \frac{1}{4V} \kappa(\sigma, J, J) J \wedge J \tag{9.125}$$

the metric can be rewritten in the form

$$G(\rho, \sigma) = -\frac{1}{2V} \kappa(\rho, \sigma, J) + \frac{1}{8V^2} \kappa(\rho, J, J) \kappa(\sigma, J, J). \tag{9.126}$$

If we denote by e_α a real basis of harmonic $(1,1)$-forms, then we can expand

$$\mathcal{J} = B + iJ = w^\alpha e_\alpha \quad \text{with} \quad \alpha = 1, \ldots, h^{1,1}. \tag{9.127}$$

The metric on moduli space is then

$$G_{\alpha\bar{\beta}} = \frac{1}{2} G(e_\alpha, e_\beta) = \frac{\partial}{\partial w^\alpha} \frac{\partial}{\partial \bar{w}^{\bar\beta}} \mathcal{K}^{1,1}, \tag{9.128}$$

where

$$e^{-\mathcal{K}^{1,1}} = \frac{4}{3} \int J \wedge J \wedge J = 8V. \tag{9.129}$$

A change in the normalization of the right-hand side would correspond to shifting the Kähler potential by an inconsequential constant. These equations show that the space spanned by w^α is a Kähler manifold and the Kähler potential is given by the logarithm of the volume of the Calabi–Yau.

We also define the *intersection numbers*

$$\kappa_{\alpha\beta\gamma} = \kappa(e_\alpha, e_\beta, e_\gamma) = \int e_\alpha \wedge e_\beta \wedge e_\gamma \tag{9.130}$$

and use them to form

$$G(w) = \frac{1}{6} \frac{\kappa_{\alpha\beta\gamma} w^\alpha w^\beta w^\gamma}{w^0} = \frac{1}{6w^0} \int \mathcal{J} \wedge \mathcal{J} \wedge \mathcal{J}, \tag{9.131}$$

which is analogous to the prepotential for the complex-structure moduli space. Here we have introduced one additional coordinate, namely w^0, in order to make $G(w)$ a homogeneous function of degree two. Then we find

$$e^{-\mathcal{K}^{1,1}} = i \sum_{A=0}^{h^{1,1}} \left(w^A \frac{\partial \bar{G}}{\partial \bar{w}^A} - \bar{w}^A \frac{\partial G}{\partial w^A} \right), \tag{9.132}$$

where now the new coordinate w^0 is included in the sum. In Eq. (9.132) it is understood that the right-hand side is evaluated at $w^0 = 1$. A homework problem asks you to verify that Eq. (9.132) agrees with Eq. (9.129).

The form of the prepotential

To leading order the prepotential is given by Eq. (9.131). However, note that the size of the Calabi–Yau belongs to $\mathcal{M}^{1,1}(M)$ and as a result α' corrections are possible. So Eq. (9.131) is only a leading-order result. However, the corrections are not completely arbitrary, because they are constrained by the symmetry. First note that the real part of w^α is determined by B, which has a gauge transformation. This leads to a *Peccei–Quinn* symmetry given by shifts of the fields by constants ε^α

$$\delta w^\alpha = \varepsilon^\alpha. \tag{9.133}$$

Together with the fact that $G(w)$ is homogeneous of degree two, this implies that perturbative corrections take the form

$$G(w) = \frac{\kappa_{ABC} w^A w^B w^C}{w^0} + i\mathcal{Y}(w^0)^2, \tag{9.134}$$

where $\mathcal{Y}$ is a constant. Note that the coefficient of $(w^0)^2$ is taken to be purely imaginary. Any real contribution is trivial since it does not affect the

Kähler potential. The result, which was derived by using mirror symmetry, is

$$\mathcal{Y} = \frac{\zeta(3)}{2(2\pi)^3}\chi(M), \qquad (9.135)$$

where $\chi(M) = 2(h^{1,1} - h^{2,1})$ is the Euler characteristic of the manifold.

Nonperturbatively, the situation changes, since the Peccei–Quinn symmetries are broken and corrections depending on w^α become possible. It turns out that sums of exponentially suppressed contributions of the form

$$\exp\left(-\frac{c_\alpha w^\alpha}{\alpha' w^0}\right), \qquad (9.136)$$

where c_α are constants, are generated. These corrections arise due to instantons, as is discussed in Section 9.8.

EXERCISES

EXERCISE 9.9

Use the definition (9.97) to show that the metric on the complex-structure moduli space of a two-dimensional torus is Kähler with Kähler potential given by

$$\mathcal{K} = -\log\left(i\int \Omega \wedge \overline{\Omega}\right) \quad \text{and} \quad \Omega = dz. \qquad (9.137)$$

SOLUTION

As we saw in Exercise **7.8**, a two-torus compactification, with complex-structure modulus $\tau = \tau_1 + i\tau_2$, can be described by a metric of the form

$$g = \frac{1}{\tau_2}\begin{pmatrix} \tau_1^2 + \tau_2^2 & \tau_1 \\ \tau_1 & 1 \end{pmatrix}.$$

Here we are setting $B = 0$ and $\sqrt{\det g} = 1$, since we are interested in complex-structure deformations. The torus metric then takes the form

$$ds^2 = \frac{1}{\tau_2}\left[\left(\tau_1^2 + \tau_2^2\right)dx^2 + 2\tau_1 dxdy + dy^2\right] = 2g_{z\bar{z}}dzd\bar{z},$$

where we have introduced a complex coordinate defined by

$$dz = dy + \tau dx \quad \text{and} \quad g_{z\bar{z}} = \frac{1}{2\tau_2}.$$

For these choices the Kähler potential is

$$\mathcal{K} = -\log\left(i\int dz \wedge d\bar{z}\right) = -\log(2\tau_2).$$

This gives the metric

$$G_{\tau\bar{\tau}} = \partial_\tau \partial_{\bar{\tau}} \mathcal{K} = \frac{1}{4\tau_2^2}.$$

Under a change in complex structure $\tau \to \tau + d\tau$ the metric components change by

$$\delta g_{zz} = \frac{d\tau}{2\tau_2^2} \quad \text{and} \quad \delta g_{\bar{z}\bar{z}} = \frac{d\bar{\tau}}{2\tau_2^2}.$$

Using the definition of the metric on moduli space (9.97) we find the moduli-space metric

$$ds^2 = 2G_{\tau\bar{\tau}}d\tau d\bar{\tau} = \frac{1}{2V}\int (g^{z\bar{z}})^2 \delta g_{zz} \delta g_{\bar{z}\bar{z}} \sqrt{g}d^2x = \frac{d\tau d\bar{\tau}}{2\tau_2^2}$$

in agreement with the computation based on the Kähler potential. $\square$

EXERCISE 9.10
Prove that $\partial_\alpha \Omega = \mathcal{K}_\alpha \Omega + \chi_\alpha$, where the χ_α are the $(2,1)$-forms defined in Eq. (9.99).

SOLUTION
By definition

$$\Omega = \frac{1}{6}\Omega_{abc} dz^a \wedge dz^b \wedge dz^c,$$

so the derivative gives

$$\partial_\alpha \Omega = \frac{1}{6}\frac{\partial \Omega_{abc}}{\partial t^\alpha} dz^a \wedge dz^b \wedge dz^c + \frac{1}{2}\Omega_{abc} dz^a \wedge dz^b \wedge \frac{\partial(dz^c)}{\partial t^\alpha}.$$

The first term is a $(3,0)$-form, while the derivative of dz^c is partly a $(1,0)$-form and partly a $(0,1)$-form. Since the exterior derivative d is independent of t^α, $\partial \Omega / \partial t^\alpha$ is closed, and hence

$$\partial \Omega / \partial t^\alpha \in H^{(3,0)} \oplus H^{(2,1)}.$$

Now we are going to show that the $(2,1)$-form here is exactly the χ_α in Eq. (9.99). By Taylor expansion we have

$$z^c(t^\alpha + \delta t^\alpha) = z^c(t^\alpha) + M_\alpha^c \delta t^\alpha,$$

which implies that

$$\frac{\partial(dz^c)}{\partial t^\alpha} = dM_\alpha^c = \frac{\partial M_\alpha^c}{\partial z^d} dz^d + \frac{\partial M_\alpha^c}{\partial \bar{z}^{\bar{d}}} dz^{\bar{d}}.$$

Therefore, the $(2,1)$-form is equal to

$$\frac{1}{2} \Omega_{abc} \frac{\partial M_\alpha^c}{\partial \bar{z}^{\bar{d}}} dz^a \wedge dz^b \wedge dz^{\bar{d}}.$$

We want to show that this is equal to

$$\chi_\alpha = -\frac{1}{4} \Omega_{abc} g^{c\bar{e}} \left(\frac{\partial g_{\bar{d}\bar{e}}}{\partial t^\alpha} \right) dz^a \wedge dz^b \wedge dz^{\bar{d}}.$$

Therefore, we need to show that

$$\frac{\partial M_\alpha^c}{\partial \bar{z}^{\bar{d}}} = -\frac{1}{2} g^{c\bar{e}} \left(\frac{\partial g_{\bar{d}\bar{e}}}{\partial t^\alpha} \right).$$

Differentiating the hermitian metric $ds^2 = 2 g_{a\bar{b}} dz^a d\bar{z}^{\bar{b}}$ in the same way that we did the holomorphic three-form gives the desired result

$$\frac{\partial g_{\bar{d}\bar{e}}}{\partial t^\alpha} = -2 g_{c\bar{e}} \frac{\partial M_\alpha^c}{\partial \bar{z}^{\bar{d}}}.$$

□

9.7 Type IIA and type IIB on Calabi–Yau three-folds

The compactification of type IIA or type IIB superstring theory on a Calabi–Yau three-fold M leads to a four-dimensional theory with $\mathcal{N} = 2$ supersymmetry. The metric perturbations and other scalar zero modes lead to moduli fields that belong to $\mathcal{N} = 2$ supermultiplets. These supermultiplets can be either vector multiplets or hypermultiplets, since these are the only massless $\mathcal{N} = 2$ supermultiplets that contain scalar fields.

$D = 4$, $\mathcal{N} = 2$ *supermultiplets*

Massless four-dimensional supermultiplets have a structure that is easily derived from the superalgebra by an analysis that corresponds to the massless analog of that presented in Exercise **8.2**. The physical states are labeled by their helicities, which are Lorentz-invariant quantities for massless states. For $\mathcal{N}$-extended supersymmetry the multiplet is determined by the maximal helicity with the rest of the states having multiplicities given by binomial coefficients. When the multiplet is not TCP self-conjugate, one must also adjoin the conjugate multiplet.[21]

In the case of $\mathcal{N} = 2$ this implies that the supermultiplet with maximal helicity 2 also has two helicity 3/2 states, and one helicity 1 state. Adding the TCP conjugate multiplet (with the opposite helicities) gives the $\mathcal{N} = 2$ *supergravity multiplet*, which contains one graviton, two gravitinos and one graviphoton. If the maximal helicity is 1, and one again adds the TCP conjugate, the same reasoning gives the $\mathcal{N} = 2$ *vector multiplet*, which contains one vector, two gauginos and two scalars. Finally, the multiplet with maximal helicity 1/2, called a *hypermultiplet* contains two spin 1/2 fields and four scalars. In each of these three cases there is a total of four bosonic and four fermionic degrees of freedom.

Type IIA

When the type IIA theory is compactified on a Calabi–Yau three-fold M, the resulting four-dimensional theory contains $h^{1,1}$ abelian vector multiplets and $h^{2,1} + 1$ hypermultiplets. The scalar fields in these multiplets parametrize the moduli spaces. There is no mixing between the two sets of moduli, so the moduli space can be expressed in the product form

$$\mathcal{M}^{1,1}(M) \times \mathcal{M}^{2,1}(M). \tag{9.138}$$

Each vector multiplet contains two real scalar fields, so the dimension of $\mathcal{M}^{1,1}(M)$ is $2h^{1,1}$. In fact, this space has a naturally induced geometry that promotes it into a special-Kähler manifold (with a holomorphic prepotential). Each hypermultiplet contains four real scalar fields, so the dimension of $\mathcal{M}^{2,1}(M)$ is $4(h^{2,1} + 1)$. This manifold turns out to be of a special type called *quaternionic Kähler*.[22] These geometric properties are inevitable consequences of the structure of the interaction of vector multiplets and hypermultiplets with $\mathcal{N} = 2$ supergravity. The massless field content of the compactified type IIA theory is explored in Exercise 9.12.

Type IIB

Compactification of the type IIB theory on a Calabi–Yau three-fold W yields $h^{2,1}$ abelian vector multiplets and $h^{1,1} + 1$ hypermultiplets. The correspond-

---

21 The only self-conjugate multiplets in four dimensions are the $\mathcal{N} = 4$ vector multiplet and the $\mathcal{N} = 8$ supergravity multiplet.
22 Note that quaternionic Kähler manifolds are not Kähler. The definition is given in the appendix.

ing moduli space takes the form

$$\mathcal{M}^{1,1}(W) \times \mathcal{M}^{2,1}(W). \tag{9.139}$$

In this case the situation is the opposite to type IIA, in that $\mathcal{M}^{2,1}(W)$ is special Kähler and $\mathcal{M}^{1,1}(W)$ is quaternionic Kähler. The massless field content of the compactified type IIB theory is explored in Exercise 9.13.

For each of the type II theories the dilaton belongs to the *universal hypermultiplet*, which explains the extra hypermultiplet in each case. This scalar is complex because there is a second scalar, an *axion a*, which is the four-dimensional dual of the two-form $B_{\mu\nu}$ ($dB = \star da$). The complex-structure moduli, being associated with complex $(2,1)$-forms, are naturally complex. The $h^{1,1}$ Kähler moduli are complex due to the B-field contribution in the complexified Kähler form $(J_{a\bar{b}} + iB_{a\bar{b}})$ as in the case of the heterotic string.

EXERCISES

EXERCISE 9.11

Explain the origin of the massless scalar fields in five dimensions that are obtained by compactifying M-theory on a Calabi–Yau three-fold.

SOLUTION

The massless fields in 11 dimensions are

$$\{G_{MN}, A_{MNP}, \Psi_M\}.$$

Let us decompose the indices of these fields in a $SU(3)$ covariant way, $M = (\mu, i, \bar{i})$. The fields belong to the following five-dimensional supermultiplets:

$$\text{gravity multiplet}: \quad G_{\mu\nu}, A_{ijk}, A_{\mu\nu\rho}, \text{ fermions}$$

$$h^{1,1} \quad \text{vector multiplets}: \quad A_{\mu j \bar{k}}, G_{j\bar{k}}, \text{ fermions}$$

$$h^{2,1} \quad \text{hypermultiplets}: \quad A_{ij\bar{k}}, G_{jk}, \text{ fermions}.$$

A five-dimensional duality transformation allows one to replace $A_{\mu\nu\rho}$ by a real scalar field. The gravity multiplet has eight bosonic and eight fermionic degrees of freedom. The other supermultiplets each have four bosonic and four fermionic degrees of freedom. The total number of massless scalar fields is

$$4h^{2,1} + h^{1,1} + 3.$$

□

EXERCISE 9.12

Consider the type IIA theory compactified on a Calabi–Yau three-fold. Explain the ten-dimensional origin of the massless fields in four dimensions.

SOLUTION

The massless fields in ten dimensions are

$$\{G_{MN}, B_{MN}, \Phi, C_M, C_{MNP}, \Psi_M^{(+)}, \Psi_M^{(-)}, \Psi^{(+)}, \Psi^{(-)}\},$$

where $\Psi_M^{(+)}$, $\Psi_M^{(-)}$ are the two Majorana–Weyl gravitinos of opposite chirality, while $\Psi^{(+)}$, $\Psi^{(-)}$ are the two Majorana–Weyl dilatinos. Writing indices in a $SU(3)$ covariant way, $M = (\mu, i, \bar{i})$, we can arrange the fields in $\mathcal{N} = 2$ supermultiplets:

$$\text{gravity multiplet}: \quad G_{\mu\nu}, \Psi_\mu, \widetilde{\Psi}_\mu, C_\mu$$

$$h^{1,1} \text{ vector multiplets}: \quad C_{\mu i\bar{j}}, G_{i\bar{j}}, B_{i\bar{j}}, \text{fermions}$$

$$h^{2,1} \text{ hypermultiplets}: \quad C_{ij\bar{k}}, G_{ij}, \text{fermions}$$

$$\text{universal hypermultiplet}: \quad C_{ijk}, \Phi, B_{\mu\nu}, \text{fermions}.$$

$B_{\mu\nu}$ is dual to a scalar field. Since the fields $C_{ij\bar{k}}$, G_{ij}, C_{ijk} are complex, the number of the massless scalar fields is $2h^{1,1} + 4h^{2,1} + 4$. There are $h^{1,1} + 1$ massless vector fields.

□

EXERCISE 9.13

Consider the type IIB theory compactified on a Calabi–Yau three-fold. Explain the ten-dimensional origin of the massless fields in four dimensions.

SOLUTION

The massless fields in ten dimensions are

$$\{G_{MN}, B_{MN}, \Phi, C, C_{MN}, C_{MNPQ}, \Psi_M^{(+)}, \widetilde{\Psi}_M^{(+)}, \Psi^{(-)}, \widetilde{\Psi}^{(-)}\}.$$

Let us use the same $SU(3)$ covariant notation as in the previous exercise. Compactification on a Calabi–Yau three-fold again gives $\mathcal{N} = 2$, $D = 4$ supersymmetry. The fields belong to the following supermultiplets:

$$\text{gravity multiplet}: \quad G_{\mu\nu}, \Psi_\mu, \widetilde{\Psi}_\mu, C_{\mu ijk}$$

$$h^{2,1} \quad \text{vector supermultiplets}: \quad C_{\mu ij\bar{k}}, G_{ij}, \text{fermions}$$

$$h^{1,1} \quad \text{hypermultiplets}: \quad C_{\mu\nu i\bar{j}}, G_{i\bar{j}}, B_{i\bar{j}}, C_{i\bar{j}}, \text{fermions}$$

$$\text{universal hypermultiplet}: \quad \Phi, C, B_{\mu\nu}, C_{\mu\nu}, \text{fermions}.$$

Now taking into account that G_{ij} is complex and that the four-form C has a self-duality constraint on its field strength, the total number of the massless scalar fields is $2h^{2,1} + 4(h^{1,1} + 1)$. The total number of massless vector fields is $h^{2,1} + 1$. □

9.8 Nonperturbative effects in Calabi–Yau compactifications

Until now we have discussed perturbative aspects of Calabi–Yau compactification that were understood prior to the second superstring revolution. This section and the following ones discuss some nonperturbative aspects of Calabi–Yau compactifications that were discovered during and after the second superstring revolution.

The conifold singularity

In addition to their nonuniqueness, one of the main problems with Calabi–Yau compactifications is that their moduli spaces contain singularities, that is, points in which the classical description breaks down. By analyzing a particular example of such a singularity, the *conifold singularity*, it became clear that the classical low-energy effective action description breaks down. Nonperturbative effects due to branes wrapping vanishing (or degenerating) cycles have to be taken into account.

To be concrete, let us consider the type IIB theory compactified on a Calabi–Yau three-fold. As we have seen in the previous section, the moduli space $\mathcal{M}^{2,1}(M)$ can be described in terms of homogeneous special coordinates X^I. A conifold singularity appears when one of the coordinates, say

$$X^1 = \int_{A^1} \Omega, \tag{9.140}$$

vanishes. The period of Ω over A^1 goes to zero, and therefore A^1 is called a *vanishing cycle*. At these points the metric on moduli space generically becomes singular. Indeed, the subspace $X^1 = 0$ has complex codimension 1, which is just a point if $h^{2,1} = 1$, and so it can be encircled by a closed loop. Upon continuation around such a loop the basis of three-cycles comes back to itself only up to an $Sp(2;\mathbb{Z})$ monodromy transformation. In general, the monodromy is

$$X^1 \to X^1 \quad \text{and} \quad F_1 \to F_1 + X^1. \tag{9.141}$$

This implies that near the conifold singularity

$$F_1(X^1) = \text{const} + \frac{1}{2\pi i} X^1 \log X^1. \tag{9.142}$$

In the simplest case one can assume that the other periods transform trivially. This result implies that near the conifold singularity the Kähler potential in Eq. (9.117) takes the form

$$\mathcal{K}^{2,1} \sim \log(|X^1|^2 \log |X^1|^2). \tag{9.143}$$

It follows that the metric $G_{1\bar{1}} = \frac{\partial^2 \mathcal{K}}{\partial X^1 \partial \overline{X^1}}$ is singular at $X^1 = 0$. This is a real singularity, and not merely a coordinate singularity, since the scalar curvature diverges, as you are asked to verify in a homework problem.

The singularity of the moduli space occurs for the following reason. The Calabi–Yau compactification is a description in terms of the low-energy effective action in which the massive fields have been integrated out. At the conifold singularity certain massive states become massless, and an inconsistency appears when such fields have been integrated out. The particular states that become massless at the singularity arise from D3-branes wrapping certain three-cycles, called *special Lagrangian cycles*, which are explained in the next section. Near the conifold singularity these states becomes light, and it is no longer consistent to exclude them from the low-energy effective action.

Supersymmetric cycles

This section explains how to calculate nonperturbative effects due to Euclideanized branes wrapping *supersymmetric cycles*. The world volume of a Euclideanized p-brane has $p + 1$ spatial dimensions, and it only exists for an instant of time. Note that only the world-volume time, and not the time in the physical Minkowski space, is Euclideanized. If a Euclideanized p-brane can wrap a $(p + 1)$-cycle in such a way that some supersymmetry is preserved, then the corresponding cycle is called supersymmetric. This gives a

9.8 Nonperturbative effects in Calabi–Yau compactifications

contribution to the path integral that represents a nonperturbative instanton correction to the theory. More precisely, fundamental-string instantons give contributions that are nonperturbative in α', whereas D-branes and NS5-branes give contributions that are also nonperturbative in g_s.[23] If the internal manifold is a Calabi–Yau three-fold, the values of p for which there are nontrivial $(p+1)$-cycles are $p = -1, 1, 2, 3, 5$.[24]

As was discussed in Chapter 6, type IIA superstring theory contains even-dimensional BPS D-branes, whereas the type IIB theory contains odd-dimensional BPS D-branes. Each of these D-branes carries a conserved R–R charge. So, in addition to fundamental strings wrapping a two-cycle and NS5-branes wrapping the entire manifold, one can consider wrapping D2-branes on a three-cycle in the IIA case. Similarly, one can wrap D1, D3 and D5-branes, as well as D-instantons, in the IIB case. These configurations give nonperturbative instanton contributions to the moduli-space geometry, that need to be included in order for string theory to be consistent. As explained in Section 9.9, these effects are crucial for understanding fundamental properties of string theory, such as *mirror symmetry*. There are different types of supersymmetric cycles in the context of Calabi–Yau compactifications, which we now discuss.[25]

Special Lagrangian submanifolds

For Calabi–Yau compactification of M-theory, which gives a five-dimensional low-energy theory, the possible instanton configurations arise from M2-branes wrapping three-cycles and M5-branes wrapping the entire Calabi–Yau manifold. Let us first consider a Euclidean M2-brane, which has a three-dimensional world volume. The goal is to examine the conditions under which a Euclidean membrane wrapping a three-cycle of the Calabi–Yau manifold corresponds to a stationary point of the path-integral-preserving supersymmetry. Once this is achieved, the next step is to determine the corresponding nonperturbative contribution to the low-energy five-dimensional effective action.

The M2-brane in 11 dimensions has a world-volume action, with global supersymmetry and local κ symmetry, whose form is similar to the actions for fundamental superstrings and D-branes described in Chapters 5 and 6. As in the other examples, in flat space-time this action is invariant under

---

[23] The g_s dependence is contained in the tension factor that multiplies the world-volume actions.
[24] A p-brane with $p = -1$ is the D-instanton of the type IIB theory.
[25] A vanishing potential for the tensor fields is assumed here. The generalization to a nonvanishing potential is known.

global supersymmetry

$$\delta_\varepsilon \Theta = \varepsilon \quad \text{and} \quad \delta_\varepsilon X^M = i\bar{\varepsilon}\Gamma^M \Theta, \qquad (9.144)$$

where $X^M(\sigma^\alpha)$, with $M = 0, \ldots, 10$, describes the membrane configuration in space-time. Θ is a 32-component Majorana spinor, and ε is a constant infinitesimal Majorana spinor. However, the question arises how much of this supersymmetry survives if a Euclideanized M2-brane wraps a three-cycle of the compactification. The M2-brane is also invariant under fermionic local κ symmetry, which acts on the fields according to

$$\delta_\kappa \Theta = 2P_+ \kappa(\sigma) \quad \text{and} \quad \delta_\kappa X^M = 2i\bar{\Theta}\Gamma^M P_+ \kappa(\sigma), \qquad (9.145)$$

where κ is an infinitesimal 32-component Majorana spinor, and $P_\pm$ are orthogonal projection operators defined by

$$P_\pm = \frac{1}{2}\left(1 \pm \frac{i}{6}\varepsilon^{\alpha\beta\gamma}\partial_\alpha X^M \partial_\beta X^N \partial_\gamma X^P \Gamma_{MNP}\right). \qquad (9.146)$$

The key to the analysis is the observation that a specific configuration $X^M(\sigma^\alpha)$ (and $\Theta = 0$) preserves the supersymmetry corresponding to a particular ε transformation, if this transformation can be compensated by a κ transformation. In other words, there should exist a $\kappa(\sigma)$ such that

$$\delta_\varepsilon \Theta + \delta_\kappa \Theta = \varepsilon + 2P_+ \kappa(\sigma) = 0. \qquad (9.147)$$

By acting with P_- this implies

$$P_- \varepsilon = 0. \qquad (9.148)$$

This equation is equivalent to the following two conditions:[26]

- The 11 coordinates X^M consist of X^a and $X^{\bar{a}}$, which refer to Calabi–Yau coordinates, and X^μ, which is the coordinate in five-dimensional space-time. In the supersymmetric instanton solution, $X^\mu = X_0^\mu$ is a constant, and the nontrivial embedding involves the other coordinates. The first condition is

$$\partial_{[\alpha} X^a \partial_{\beta]} X^{\bar{b}} J_{a\bar{b}} = 0. \qquad (9.149)$$

The meaning of this equation is that the pullback of the Kähler form of the Calabi–Yau three-fold to the membrane world volume vanishes.

- The second condition is[27]

$$\partial_\alpha X^a \partial_\beta X^b \partial_\gamma X^c \Omega_{abc} = e^{-i\varphi} e^{\mathcal{K}} \varepsilon_{\alpha\beta\gamma}. \qquad (9.150)$$

---

[26] The equivalence of Eq. (9.148) and the conditions (9.149) and (9.150) is proved in Exercise 9.15.
[27] $\varepsilon_{\alpha\beta\gamma}$ is understood to be a tensor here. Otherwise a factor of $\sqrt{G}$, where $G_{\alpha\beta}$ is the induced metric, would appear.

9.8 Nonperturbative effects in Calabi–Yau compactifications

The meaning of this equation is that the pullback of the holomorphic $(3,0)$-form Ω of the Calabi–Yau manifold to the membrane world volume is proportional to the membrane volume element. The complex-conjugate equation implies the same thing for the $(0,3)$-antiholomorphic form $\bar\Omega$. The phase φ is a constant that simply reflects an arbitrariness in the definition of Ω. The factor $e^{\mathcal{K}}$, where $\mathcal{K}$ is given by

$$\mathcal{K} = \frac{1}{2}(\mathcal{K}^{1,1} - \mathcal{K}^{2,1}), \tag{9.151}$$

is a convenient normalization factor. The term $\mathcal{K}^{2,1}$ is a function of the complex moduli belonging to $h^{2,1}$ hypermultiplets. $\mathcal{K}^{1,1}$ is a function of the real moduli belonging to $h^{1,1}$ vector supermultiplets.

The supersymmetric three-cycle conditions (9.149) and (9.150) define a *special Lagrangian submanifold*. When these conditions are satisfied, there exists a nonzero covariantly constant spinor of the form $\varepsilon = P_+ \eta$. Thus, the conclusion is that a Euclidean M2-brane wrapping a special Lagrangian submanifold of the Calabi–Yau three-fold gives a supersymmetric instanton contribution to the five-dimensional low-energy effective theory.

The conditions (9.149) and (9.150) imply that the membrane has minimized its volume. In order to derive a bound for the volume of the membrane consider

$$\int_\Sigma \varepsilon^\dagger P_-^\dagger P_- \varepsilon \, d^3\sigma \geq 0, \tag{9.152}$$

where Σ is the membrane world volume. Since

$$P_-^\dagger P_- = P_- P_- = P_-, \tag{9.153}$$

the inequality becomes

$$2\mathcal{V} \geq e^{-\mathcal{K}}\left(e^{i\varphi}\int_\Sigma \Omega + e^{-i\varphi}\int_\Sigma \bar\Omega\right), \tag{9.154}$$

where φ is a phase which can be adjusted so that we obtain

$$\mathcal{V} \geq e^{-\mathcal{K}}\left|\int_\Sigma \Omega\right|. \tag{9.155}$$

The bound is saturated whenever the membrane wraps a supersymmetric cycle $\mathcal{C}$, in which case

$$\mathcal{V} = e^{-\mathcal{K}}\left|\int_\mathcal{C} \Omega\right|. \tag{9.156}$$

Type IIA or type IIB superstring theory, compactified on a Calabi–Yau three-fold, also has supersymmetric cycles, which can be determined in a similar fashion. As in the case of M-theory, the type IIA theory receives instanton contributions associated with a D2-brane wrapping a special Lagrangian manifold. These contributions have a coupling constant dependence of the form $\exp(-1/g_s)$, because the D2-brane tension is proportional to $1/g_s$.

Black-hole mass formula

When the type IIB theory is compactified on a Calabi–Yau three-fold, four-dimensional supersymmetric black holes can be realized by wrapping D3-branes on special Lagrangian three-cycles. In the present case the bound for the mass of the black holes takes the form

$$M \geq e^{\mathcal{K}^{2,1}/2} \left| \int_{\mathcal{C}} \Omega \right| = e^{\mathcal{K}^{2,1}/2} \left| \int_M \Omega \wedge \Gamma \right|, \qquad (9.157)$$

where Γ is the three-form that is Poincaré dual to the cycle $\mathcal{C}$. Here we are assuming that the mass distribution on the D3-brane is uniform. Letting

$$\Gamma = q^I \alpha_I - p_I \beta^I, \qquad (9.158)$$

we can introduce special coordinates and use the expansion (9.110) to obtain the BPS bound

$$M \geq e^{\mathcal{K}^{2,1}/2} \left| p_I X^I - q^I F_I \right|. \qquad (9.159)$$

For BPS states the inequality is saturated, and the mass is equal to the absolute value of the central charge Z in the supersymmetry algebra. Thus Eq. (9.157) is also a formula for $|Z|$. As a result, BPS states become massless when a cycle shrinks to zero size. The above expression relating the central charge to the special coordinates plays a crucial role in the discussion of the attractor mechanism for black holes which will be presented in chapter 11.

Holomorphic cycles

In the case of type II theories other supersymmetric cycles also can contribute. For example, some supersymmetry can be preserved if a Euclidean type IIA string world sheet wraps a *holomorphic cycle*. This means that the embedding satisfies

$$\bar{\partial} X^a = 0 \quad \text{and} \quad \partial X^{\bar{a}} = 0, \qquad (9.160)$$

in addition to $X^\mu = X_0^\mu$. Thus, the complex structure of the Euclideanized string world sheet is aligned with that of the Calabi–Yau manifold. In this case, one says that it is *holomorphically embedded*. Recall that the type IIA theory corresponds to M-theory compactified on a circle. Therefore, from the M-theory viewpoint this example corresponds to a solution on $M_4 \times S^1 \times M$

9.8 Nonperturbative effects in Calabi–Yau compactifications

in which a Euclidean M2-brane wraps the circle and a holomorphic two-cycle of the Calabi–Yau.

EXERCISES

EXERCISE 9.14
Show that the submanifold $X = \overline{X}$ is a supersymmetric three-cycle inside the Calabi–Yau three-fold given by a quintic hypersurface in $\mathbb{C}P^4$.

SOLUTION

To prove the above statement, we should first check that the pullback of the Kähler form is zero. This is trivial in this case, because $X \to \overline{X}$ under the transformation $J \to -J$. On the other hand, the pullback of J onto the fixed surface $X = \overline{X}$ should give $J \to J$, so the pullback of J is zero.

Now let us consider the second condition, and compute the pullback of the holomorphic three-form. The equation for a quintic hypersurface in $\mathbb{C}P^4$ discussed in Section 9.3 is

$$\sum_{m=1}^{5}(X^m)^5 = 0.$$

Defining inhomogeneous coordinates $Y^k = X^k/X^5$, with $k = 1, 2, 3, 4$, on the open set $X^5 \neq 0$, the holomorphic three-form can be written as

$$\Omega = \frac{dY^1 \wedge dY^2 \wedge dY^3}{(Y^4)^4}.$$

The norm of Ω is

$$\|\Omega\|^2 = \frac{1}{6}\Omega_{abc}\overline{\Omega}^{abc} = \frac{1}{\hat{g}|Y^4|^8},$$

where $\hat{g} = \det g_{a\bar{b}}$. Using Eqs (9.104) and (9.129), as well as Exercise 9.8, one has

$$e^{-\mathcal{K}^{2,1}} = i\int \Omega \wedge \overline{\Omega} = V\|\Omega\|^2 = \frac{1}{8}e^{-\mathcal{K}^{1,1}}\|\Omega\|^2$$

which implies that

$$\|\Omega\|^2 = 8e^{2\mathcal{K}},$$

where $\mathcal{K} = \frac{1}{2}(\mathcal{K}^{1,1} - \mathcal{K}^{2,1})$. It follows that

$$\hat{g} = \frac{e^{-2\mathcal{K}}}{8|Y^4|^8}.$$

The pullback of the metric gives

$$h_{\alpha\beta} = 2\partial_\alpha Y^a g_{a\bar{b}} \partial_\beta Y^{\bar{b}}$$

so

$$\sqrt{h} = \sqrt{8\hat{g}} |\det(\partial Y)| = |\det(\partial Y)| \frac{e^{-\mathcal{K}}}{|Y^4|^4}.$$

Now we can calculate the pullback of the holomorphic $(3,0)$-form

$$\partial_\alpha Y^a \partial_\beta Y^b \partial_\gamma Y^c \Omega_{abc} = \frac{\varepsilon_{abc} \partial_\alpha Y^a \partial_\beta Y^b \partial_\gamma Y^c}{(Y^4)^4} = e^{-i\phi} e^{\mathcal{K}} \sqrt{h} \, \varepsilon_{\alpha\beta\gamma},$$

which is what we wanted to show. □

EXERCISE 9.15

Derive the equivalence between Eq. (9.148) and Eqs (9.149) and (9.150). For M-theory on $M_5 \times M$, where M is a Calabi–Yau three-fold, the M-theory spinor ε has the decomposition

$$\varepsilon = \lambda \otimes \eta_+ + \lambda^* \otimes \eta_-,$$

where λ is a spinor on M_5, and $\eta_\pm$ are Weyl spinors on the Calabi–Yau manifold. So the condition (9.148) takes the form

$$\left(1 - \frac{i}{6} \varepsilon^{\alpha\beta\gamma} \partial_\alpha X^m \partial_\beta X^n \partial_\gamma X^p \gamma_{mnp}\right) \left(e^{-i\theta} \eta_+ + \text{c.c.}\right) = 0,$$

where m, n, p label real coordinates of the internal Calabi–Yau manifold. Let us focus on the η_+ terms and take account of the complex-conjugate terms at the end of the calculation.

The formula can be simplified by using complex coordinates X^a and $\bar{X}^{\bar{a}}$, as in the text, and the conditions $\gamma_a \eta_+ = 0$. This implies that $\gamma_{abc} \eta_+ = 0$ and $\gamma_{ab\bar{c}} \eta_+ = 0$. The nonzero terms are

$$\gamma_{a\bar{b}\bar{c}} \eta_+ = -2i J_{a[\bar{b}} \gamma_{\bar{c}]} \eta_+$$

and

$$\gamma_{\bar{a}\bar{b}\bar{c}} \eta_+ = e^{-\mathcal{K}} \bar{\Omega}_{\bar{a}\bar{b}\bar{c}} \eta_-.$$

The first of these relations follows from the $\{\gamma_a, \gamma_{\bar{b}}\} = 2g_{a\bar{b}}$ and $J_{a\bar{b}} = ig_{a\bar{b}}$. The second one is an immediate consequence of the complex conjugate of

$\Omega_{abc} = e^{-\mathcal{K}} \eta_-^T \gamma_{abc} \eta_-$ and $\eta_+^T \eta_- = 1$. The dependence on $\mathcal{K}$ reflects a choice of normalization of Ω. The arbitrary phase θ could have been absorbed into η_+ earlier, but then it would reappear in this equation reflecting an arbitrariness in the phase of Ω.

Now we can write the above condition as

$$e^{-i\theta}\eta_+ + \frac{i}{6} e^{i\theta} \varepsilon^{\alpha\beta\gamma} \partial_\alpha X^a \partial_\beta X^b \partial_\gamma X^c e^{-\mathcal{K}} \Omega_{abc} \eta_+$$

$$-e^{-i\theta} \varepsilon^{\alpha\beta\gamma} \partial_\alpha X^a \partial_\beta X^{\bar{b}} \partial_\gamma X^{\bar{c}} J_{a\bar{b}} \gamma_{\bar{c}} \eta_+ + \text{c.c.} = 0.$$

Because η_-, $\gamma_{\bar{a}}\eta_-$, η_+, $\gamma_a \eta_+$ are linearly independent, this is equivalent to the following two conditions:

$$\varepsilon^{\alpha\beta\gamma} \partial_\alpha X^a \partial_\beta X^{\bar{b}} \partial_\gamma X^{\bar{c}} J_{a\bar{b}} = 0$$

and

$$e^{-i\theta} + \frac{i}{6} e^{i\theta} \varepsilon^{\alpha\beta\gamma} \partial_\alpha X^a \partial_\beta X^b \partial_\gamma X^c e^{-\mathcal{K}} \Omega_{abc} = 0.$$

Because the first equation is satisfied for all $\bar{c}$, we have

$$\partial_{[\alpha} X^a \partial_{\beta]} X^{\bar{b}} J_{a\bar{b}} = 0,$$

which is exactly Eq. (9.149). The second equation can be written as

$$\partial_\alpha X^a \partial_\beta X^b \partial_\gamma X^c \Omega_{abc} = -ie^{-2i\theta} e^{\mathcal{K}} \varepsilon_{\alpha\beta\gamma}.$$

Setting $e^{-i\varphi} = -ie^{-2i\theta}$ gives Eq. (9.150). □

9.9 Mirror symmetry

As T-duality illustrated, the geometry probed by point particles is different from the geometry probed by strings. In string geometry a circle of radius R can be equivalent to a circle of radius α'/R, providing a simple example of the surprising properties of string geometry. A similar phenomenon for Calabi–Yau three-folds, called *mirror symmetry*, is the subject of this section.

The mirror map associates with almost[28] any Calabi–Yau three-fold M another Calabi–Yau three-fold W such that

$$H^{p,q}(M) = H^{3-p,q}(W). \tag{9.161}$$

This conjecture implies, in particular, that $h^{1,1}(M) = h^{2,1}(W)$ and *vice*

---

[28] In the few cases where this fails, there still is a mirror, but it is not a Calabi–Yau manifold. However, it is just as good for string theory compactification purposes. This happens, for example, when M has $h^{2,1} = 0$, since any Calabi–Yau manifold W has $h^{11} \geq 1$.

versa. An early indication of mirror symmetry was that the space of thousands of string theory vacua appears to be self-dual in the sense that if a Calabi–Yau manifold with Hodge numbers $(h^{1,1}, h^{2,1})$ exists, then another Calabi–Yau manifold with flipped Hodge numbers $(h^{2,1}, h^{1,1})$ also exists. The set of vacua considered were known to be only a sample, so perfect matching was not expected. In fact, a few examples in this set had no candidate mirror partners. This was shown in Fig. 9.1.

These observations lead to the conjecture that the type IIA superstring theory compactified on M is exactly equivalent to the type IIB superstring theory compactified on W. This implies, in particular, an identification of the moduli spaces:

$$\mathcal{M}^{1,1}(M) = \mathcal{M}^{2,1}(W) \quad \text{and} \quad \mathcal{M}^{1,1}(W) = \mathcal{M}^{2,1}(M). \tag{9.162}$$

This is a highly nontrivial statement about how strings see the geometry of Calabi–Yau manifolds, since M and W are in general completely different from the classical geometry point of view. Indeed, even the most basic topology of the two manifolds is different, since the Euler characteristics are related by

$$\chi(M) = -\chi(W). \tag{9.163}$$

Nonetheless, the mirror symmetry conjecture implies that the type IIA theory compactified on M and the type IIB theory compactified on W are dual descriptions of the same physics, as they give rise to isomorphic string theories. A second, and genuinely different, possibility is given by the type IIA theory compactified on W, which (by mirror symmetry) is equivalent to the type IIB theory compactified on M.

Mirror symmetry is a very powerful tool for understanding string geometry. To see this note that the prepotential of the type IIB vector multiplets is independent of the Kähler moduli and the dilaton. As a result, its dependence on α' and g_s is exact. Mirror symmetry maps the complex-structure moduli space of type IIB compactified on W to the Kähler-structure moduli space of type IIA on the mirror M. The type IIA side does receive corrections in α'. As a result, a purely classical result is mapped to an (in general) infinite series of quantum corrections. In other words, a classical computation of the periods of Ω in W is mapped to a problem of counting holomorphic curves in M. Both sides should be exact to all orders in g_s, since the IIA dilaton is not part of $\mathcal{M}^{1,1}(M)$ and the IIB dilaton is not part of $\mathcal{M}^{2,1}(W)$.

Let us start by discussing mirror symmetry for a circle and a torus. These simple examples illustrate the basic ideas.

9.9 Mirror symmetry 413

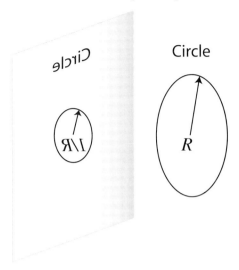

Fig. 9.8. T-duality transforms a circle of radius R into a circle of radius $1/R$. This duality is probably the origin of mirror symmetry.

The circle

The simplest example of mirror symmetry has already been discussed extensively in this book. It is T-duality. Chapter 6 showed that, when the bosonic string is compactified on a circle of radius R, the perturbative string spectrum is given by

$$\alpha' M^2 = \alpha' \left[\left(\frac{K}{R} \right)^2 + \left(\frac{WR}{\alpha'} \right)^2 \right] + 2N_\mathrm{L} + 2N_\mathrm{R} - 4, \qquad (9.164)$$

with

$$N_\mathrm{R} - N_\mathrm{L} = WK. \qquad (9.165)$$

These equations are invariant under interchange of W and K, provided that one simultaneously sends $R \to \alpha'/R$ as illustrated in Fig. 9.8. This turns out to be exactly true for the full interacting string theory, at least perturbatively.

The torus

One can also illustrate mirror symmetry for the two-torus $T^2 = S^1 \times S^1$, where the first circle has radius R_1 and the second circle has radius R_2. This torus may be regarded as an S^1 fibration over S^1. It is characterized by complex-structure and Kähler-structure parameters

$$\tau = i\frac{R_2}{R_1} \quad \text{and} \quad \rho = iR_1 R_2, \tag{9.166}$$

as in Section 9.5. Performing a T-duality on the fiber circle sends $R_1 \to 1/R_1$ (for $\alpha' = 1$), and as a result the moduli fields of the resulting mirror torus are

$$\tilde{\tau} = iR_1 R_2 \quad \text{and} \quad \tilde{\rho} = i\frac{R_2}{R_1}. \tag{9.167}$$

This shows that under the mirror map the complex-structure and Kähler-structure parameters have been interchanged, just as in the case of the Calabi–Yau three-fold.

T^3 *fibrations*

An approach to understanding mirror symmetry, which is based on T-duality, was proposed by Strominger, Yau and Zaslow (SYZ). If mirror symmetry holds, then a necessary requirement is that the spectrum of BPS states for the type IIA theory on M and type IIB on W must be the same. Verifying this would not constitute a complete proof, but it would give strong support to the mirror-symmetry conjecture. That is often the best that can be done for duality conjectures in string theory.

The BPS states to be compared arise from D-branes wrapping supersymmetric cycles of the Calabi–Yau. In the case of the type IIA theory, Dp-branes, with $p = 0, 2, 4, 6$, can wrap even-dimensional cycles of the Calabi–Yau. However, since only BPS states can be compared reliably, only supersymmetric cycles should be considered. In the simplest case one only considers the D0-brane, whose moduli space is the whole Calabi–Yau M, since the D0-brane can be located at any point in M. In the type IIB theory the BPS spectrum of wrapped D-branes arises entirely from D3-branes wrapping special Lagrangian three-cycles.

Since mirror symmetry relates the special Lagrangian three-cycle of W to the whole Calabi–Yau manifold M, its properties are very constrained. First, the D3-brane moduli space has to have three complex dimensions. Three real moduli are provided by the transverse position of the D3-brane. The remaining three moduli are obtained by assuming that mirror symmetry is implemented by three T-dualities. D0-branes are mapped to D3-branes under the action of three T-dualities. After performing the three T-dualities, three flat $U(1)$ gauge fields appear in the directions of the D3-brane. These are associated with the isometries of three circles which form a three-torus.

As a result, W is a T^3 fibration over a base B. By definition, a Calabi–Yau manifold is a T^3 fibration if it can be described by a three-dimensional base space B, with a three-torus above each point of B assembled so as to make a smooth Calabi–Yau manifold. A T^3 fibration is more general than a T^3 fiber bundle in that isolated T^3 fibers are allowed to be singular, which means that one or more of their cycles degenerate. Turning the argument around, M must also be a T^3 fibration. Mirror symmetry is a fiber-wise T-duality on all of the three directions of the T^3. A simple example of a fiber bundle is depicted in Fig. 9.9.

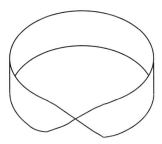

Fig. 9.9. A Moebius strip is an example of a nontrivial fiber bundle. It is a line segment fibered over a circle S^1. Calabi–Yau three-folds that have a mirror are conjectured to be T^3 fibrations over a base B. In contrast to the simple example of the Moebius strip, some of the T^3 fibers are allowed to be singular.

Since the number of T-dualities is odd, even forms and odd forms are interchanged. As a result, the $(1,1)$ and $(2,1)$ cohomologies are interchanged, as is expected from mirror symmetry. Moreover, there exists a holomorphic three-form on W, which implies that W is Calabi–Yau. The three T-dualities, of course, also interchange type IIA and type IIB.

The argument given above probably contains the essence of the proof of mirror symmetry. A note of caution is required though. We already pointed out that there are Calabi–Yau manifolds whose mirrors are not Calabi–Yau, so a complete proof would need to account for that. The T-duality rules and the condition that a supersymmetric three-cycle has to be special Lagrangian are statements that hold to leading order in α', while the full description of the mirror W requires, in general, a whole series of α' corrections.

9.10 Heterotic string theory on Calabi–Yau three-folds

As was discussed earlier, the fact that dH is an exact four-form implies that $\text{tr}(R \wedge R)$ and $\text{tr}(F \wedge F) = \frac{1}{30}\text{Tr}(F \wedge F)$ must belong to the same cohomology class. The curvature two-form R takes values in the Lie algebra of the holonomy group, which is $SU(3)$ in the case of Calabi–Yau compactification. Specializing to the case of the $E_8 \times E_8$ heterotic string theory, F takes values in the $E_8 \times E_8$ Lie algebra. The characteristic class $\text{tr}(R \wedge R)$ is nontrivial, and so it is necessary that gauge fields take nontrivial background values in the compact directions.

The easiest way – but certainly not the only one – to satisfy the cohomology constraint is for the field strengths associated with an $SU(3)$ subgroup of the gauge group to take background values that are equal to those of the curvature form while the other field strengths have zero background value. More fundamentally, the Yang–Mills potentials A can be identified with the potentials that give the curvature, namely the spin connections. This method of satisfying the constraint is referred to as *embedding the spin connection in the gauge group*. There are many ways of embedding $SU(3)$ inside $E_8 \times E_8$ and not all of them would work. The embedding is restricted by the requirement that the cohomology class of $\text{tr}(F \wedge F)$ gives exactly the class of $\text{tr}(R \wedge R)$ and not just some multiple of it. The embedding that satisfies this requirement is one in which the $SU(3)$ goes entirely into one E_8 factor in such a way that its commutant is E_6. In other words, $E_8 \supset E_6 \times SU(3)$. Thus, for this embedding, the unbroken gauge symmetry of the effective four-dimensional theory is $E_6 \times E_8$.

This specific scenario is not realistic for a variety of reasons, but it does have some intriguing features that one could hope to preserve in a better set-up. For one thing, E_6 is a group that has been proposed for grand unification. In that context, the gauge bosons belong to the adjoint **78** and chiral fermions are assigned to the **27**, which is a complex representation. This representation might also be used for Higgs fields. Clearly, these representations give a lot of extra fields beyond what is observed, so additional measures are required to lift them to high mass or else eliminate them altogether.

The presence of the second unbroken E_8 also needs to be addressed. The important observation is that all fields that participate in standard-model interactions must carry nontrivial standard-model quantum numbers. But the massless fields belonging to the adjoint of the second E_8 are all E_6 singlets. Fields that belong to nontrivial representations of both E_8s first occur for masses comparable to the string scale. Thus, if the string scale is comparable to the Planck scale, the existence of light fields carrying nontrivial quantum numbers of the second E_8 could only be detected by gravitational-strength interactions. These fields comprise the *hidden sector*. A hidden sector could actually be useful. Assuming that the hidden sector has a mass gap, perhaps due to confinement, one intriguing possibility is that hidden-sector particles comprise a component of the dark matter. It has also been suggested that gaugino condensation in the hidden sector could be the origin of supersymmetry breaking.

The adjoint of E_8, the **248**, is reducible with respect to the $E_6 \times SU(3)$ subgroup, with the decomposition

$$\mathbf{248} = (\mathbf{78}, \mathbf{1}) + (\mathbf{1}, \mathbf{8}) + (\mathbf{27}, \mathbf{3}) + (\overline{\mathbf{27}}, \overline{\mathbf{3}}). \tag{9.168}$$

The massless spectrum in four dimensions can now be determined. There are massless vector supermultiplets in the adjoint of $E_6 \times E_8$, since this is the unbroken gauge symmetry. In addition, there are $h^{1,1}$ chiral supermultiplets containing (complexified) Kähler moduli and $h^{2,1}$ chiral supermultiplets containing complex-structure moduli. These chiral supermultiplets are all singlets of the gauge group, since the ten-dimensional graviton is a singlet.

Let us now explain the origin of chiral matter, which belongs to chiral supermultiplets. It is easiest to focus on the origin of the scalars and invoke supersymmetry to infer that the corresponding massless fermions must also be present. For this purpose let us denote the components of the gauge fields as follows:

$$A_M = (A_\mu, A_a, A_{\bar{a}}). \tag{9.169}$$

Now let us look for the zero modes of A_a, which give massless scalars in four-dimensional space-time. As explained above, the corresponding fermions are chiral. The subscript a labels a quantity that transforms as a **3** of the holonomy $SU(3)$. However, the embedding of the spin connection in the gauge group means that this $SU(3)$ is identified with the $SU(3)$ in the decomposition of the gauge group. Therefore, the components of A_a belonging to the $(\overline{\mathbf{27}}, \overline{\mathbf{3}})$ term in the decomposition can be written in the form $A_{a, \bar{s}\bar{b}}$, where $\bar{s}$ labels the components of the $\overline{\mathbf{27}}$ and $\bar{b}$ labels the components of the $\overline{\mathbf{3}}$. This can be regarded as a $(1,1)$-form taking values in the $\overline{\mathbf{27}}$. However, a $(1,1)$-form has $h^{1,1}$ zero modes. Thus, we conclude that there are $h^{1,1}$ massless chiral supermultiplets belonging to the $\overline{\mathbf{27}}$ of E_6. The next case to consider is $A_{a,sb}$. To recast this as a differential form, one uses the inverse Kähler metric and the antiholomorphic $(0,3)$-form to define

$$A_{a\bar{d}\bar{e}s} = A_{a,sb} g^{b\bar{c}} \overline{\Omega}_{\bar{c}\bar{d}\bar{e}}. \tag{9.170}$$

This is a **27**-valued $(1,2)$-form. It then follows that there are $h^{2,1}$ massless chiral supermultiplets belonging to the **27** of E_6.

As an exercise in group theory, let us explore how the reasoning above is modified if the background gauge fields take values in $SU(4)$ or $SU(5)$

rather than $SU(3)$. In the first case, the appropriate embedding would be $E_8 \supset SO(10) \times SU(4)$, so that the unbroken gauge symmetry would be $SO(10) \times E_8$, and the decomposition of the adjoint would be

$$\mathbf{248} = (\mathbf{45}, \mathbf{1}) + (\mathbf{1}, \mathbf{15}) + (\mathbf{10}, \mathbf{6}) + (\mathbf{16}, \mathbf{4}) + (\overline{\mathbf{16}}, \overline{\mathbf{4}}). \quad (9.171)$$

This could lead to a supersymmetric $SO(10)$ grand-unified theory with generations of chiral matter in the $\mathbf{16}$, antigenerations in the $\overline{\mathbf{16}}$ and Higgs fields in the $\mathbf{10}$. This is certainly an intriguing possibility. In the $SU(5)$ case, the embedding would be $E_8 \supset SU(5) \times SU(5)$, so that the unbroken gauge symmetry would be $SU(5) \times E_8$. This could lead to a massless field content suitable for a supersymmetric $SU(5)$ grand-unified theory.

As a matter of fact, there are more complicated constructions in which these possibilities are realized. For the gauge fields to take values in $SU(4)$ or $SU(5)$, rather than $SU(3)$, requires more complicated ways of solving the topological constraints than simply embedding the holonomy group in the gauge group. The existence of solutions is guaranteed by a theorem of Uhlenbeck and Yau, though the details are beyond the scope of this book. For these more general embeddings there is no longer a simple relation between the Hodge numbers and the number of generations.

Starting from a Calabi–Yau compactification scenario that leads to a supersymmetric grand-unified theory, there are still a number of other issues that need to be addressed. These include breaking the gauge symmetry to the standard-model gauge symmetry and breaking the residual supersymmetry. If the Calabi–Yau space is not simply connected, as happens for certain quotient-space constructions, there is an elegant possibility. Wilson lines $W_i = \exp(\oint_{\gamma_i} A)$ can be introduced along the noncontractible loops γ_i without changing the field strengths. The unbroken gauge symmetry is then reduced to the subgroup that commutes with these Wilson lines. This can break the gauge group to $SU(3) \times SU(2) \times U(1)^n$, where $n = 3$ for the E_6 case, $n = 2$ for the $SO(10)$ case and $n = 1$ for the $SU(5)$ case. Experimentalists are on the lookout for heavy Z bosons, which would correspond to extra $U(1)$ factors.

9.11 K3 compactifications and more string dualities

Compactifications of string theory that lead to a four-dimensional spacetime are of interest for making contact with the real world. However, it is also possible to construct other consistent compactifications, which can also be of theoretical interest. This section considers a particularly interesting class of four-dimensional compact manifolds, namely Calabi–Yau two-folds.

As discussed earlier, the only Calabi–Yau two-fold with $SU(2)$ holonomy is the K3 manifold. It can be used to compactify superstring theories to six dimensions, M-theory to seven dimensions or F-theory to eight dimensions.

Compactification of M-theory on K3

M-theory has a consistent vacuum of the form $M_7 \times K3$, where M_7 represents seven-dimensional Minkowski space-time. The compactification breaks half of the supersymmetries, so the resulting vacuum has 16 unbroken supersymmetries. The moduli of the seven-dimensional theory have two potential sources. One source is the moduli-space of K3 manifolds, itself, which is manifested as zero modes of the metric tensor on K3. The other source is from zero modes of antisymmetric-tensor gauge fields. However, the only such field in M-theory is a three-form, and the third cohomology of K3 is trivial. Therefore, the three-form does not contribute any moduli in seven dimensions, and the moduli space of the compactified theory is precisely the moduli space of K3 manifolds.

Moduli space of K3

Let us count the moduli of K3. Kähler-structure deformations are given by closed $(1,1)$-forms,[29] so their number in the case of K3 is $h^{1,1} = 20$. Complex-structure deformations in the case of K3 correspond to coefficients for the variations

$$\delta g_{ab} \sim \Omega_{ac} g^{c\bar{d}} \omega_{b\bar{d}} + (a \leftrightarrow b), \tag{9.172}$$

where Ω is the holomorphic two-form and $\omega_{b\bar{d}}$ is a closed $(1,1)$-form. This variation vanishes if ω is the Kähler form, as you are asked to verify in a homework problem. Thus, there are 38 real (19 complex) complex-structure moduli. Combined with the 20 Kähler moduli this gives a 58-dimensional moduli space of K3 manifolds.

This moduli space is itself an orbifold. The result, worked out by mathematicians, is $\mathbb{R}^+ \times \mathcal{M}_{19,3}$, where

$$\mathcal{M}_{19,3} = \mathcal{M}^0_{19,3} / O(19,3; \mathbb{Z}) \tag{9.173}$$

and

$$\mathcal{M}^0_{19,3} = \frac{O(19,3; \mathbb{R})}{O(19, \mathbb{R}) \times O(3, \mathbb{R})}. \tag{9.174}$$

The $\mathbb{R}^+$ factor corresponds to the overall volume modulus, and the factor $\mathcal{M}_{19,3}$ describes a space of dimension $19 \times 3 = 57$, as required. In contrast

---

[29] This is true for any Calabi–Yau n-fold.

to the case of Calabi–Yau three-folds, the dependence on Kähler moduli and complex-structure moduli does not factorize. The singularities of the moduli space correspond to singular limits of the K3 manifold. Typically, one or more two-cycles of the K3 manifold degenerate (that is, collapse to a point) at these loci. In fact, the $\mathbb{Z}_2$ orbifold described in Section 9.3 is such a limit in which 16 nonintersecting two-cycles degenerate.

The proof that this is the right moduli space is based on the observation that the coset space characterizes the alignment of the 19 anti-self-dual and three self-dual two-forms in the space of two forms. Rather than trying to explain this carefully, let us confirm this structure by physical arguments.

Dual description of M-theory on $M_7 \times$ K3

The seven-dimensional theory obtained in this way has exactly the same massless spectrum, the same amount of supersymmetry, and the same moduli space as is obtained by compactifying (either) heterotic string theory on a three-torus. Recall that in Chapter 7 it was shown that the moduli space of the heterotic string compactified on T^n is $\mathcal{M}_{16+n,n} \times \mathbb{R}^+$, where

$$\mathcal{M}_{16+n,n} = \mathcal{M}^0_{16+n,n}/O(16+n,n;\mathbb{Z}) \tag{9.175}$$

and

$$\mathcal{M}^0_{16+n,n} = \frac{O(16+n,n;\mathbb{R})}{O(16+n,\mathbb{R}) \times O(n,\mathbb{R})}. \tag{9.176}$$

Therefore, it is natural to conjecture, following Witten, that heterotic string theory on a three-torus is dual to M-theory on K3.

In the heterotic description, the $\mathbb{R}^+$ modulus is associated with the string coupling constant, which is the vacuum expectation value of $\exp(\Phi)$, where Φ is the dilaton. Since this corresponds to the K3 volume in the M-theory description, one reaches the following interesting conclusion: the heterotic-string coupling constant corresponds to the K3 volume, and thus the strong-coupling limit of heterotic string theory compactified on a three-torus corresponds to the limit in which the volume of the K3 becomes infinite. Thus, this limit gives 11-dimensional M-theory! This is the same strong-coupling limit as was obtained in Chapter 8 for ten-dimensional type IIA superstring theory at strong coupling. The difference is that in one case the size of a K3 manifold becomes infinite and in the other the size of a circle becomes infinite.

An important field in the heterotic theory is the two-form B, whose field strength H includes Chern–Simon terms so that dH is proportional to $\text{tr} R^2 - \text{tr} F^2$. In the seven-dimensional K3 reduction of M-theory considered here, the B field arises as a dual description of A_3. The field A_3 also gives rise to 22 $U(1)$ gauge fields in seven dimensions, as required by the duality. The Chern–Simons 11-form gives seven-dimensional couplings of the B field to these gauge fields of the form required to account for the $\text{tr} F^2$ term in the dH equation. To account for the $\text{tr} R^2$ terms it is necessary to add higher-dimension terms to the M-theory action of the form $\int A_3 \wedge X_8$, where X_8 is quartic in curvature two-forms. Such terms, with exactly the required structure, have been derived by several different arguments. These include anomaly cancellation at boundaries as well as various dualities to string theories.

Matching BPS branes

As a further test of the proposed duality, one can compare BPS branes in seven dimensions. One interesting example is obtained by wrapping the M5-brane on the K3 manifold. This leaves a string in the seven noncompact dimensions. The only candidate for a counterpart in the heterotic theory is the heterotic string itself! To decide whether this is reasonable, recall that in the bosonic description of the heterotic string compactified on T^n there are $16+n$ left-moving bosonic coordinates and n right-moving bosonic coordinates. To understand this from the point of view of the M5-brane, the first step is to identify the field content of its world-volume theory. This is a tensor supermultiplet in six dimensions, whose bosonic degrees of freedom consist of five scalars, representing transverse excitations in 11 dimensions, and a two-form potential with an anti-self-dual three-form field strength.[30] This anti-self-dual three-form F_3 gives zero modes that can be expanded as a sum of terms

$$F_3 = \sum_{i=1}^{3} \partial_- X^i \omega_+^i + \sum_{i=1}^{19} \partial_+ X^i \omega_-^i, \qquad (9.177)$$

where $\omega_\pm^i$ denote the self-dual and anti-self-dual two-forms of K3, and $\partial_\pm X^i$ correspond to the left-movers and right-movers on the string world sheet. Since the latter are self-dual and anti-self-dual, respectively, all terms in this formula are anti-self-dual. In addition, the heterotic string has five more physical scalars, with both left-moving and right-moving components, describing transverse excitations in the noncompact dimensions. These are provided by the five scalars of the tensor multiplet.

Recall that the dimensions of a charged p-brane and its magnetic dual p'-brane are related in D dimensions by

$$p + p' = D - 4. \qquad (9.178)$$

[30] This field has three physical degrees of freedom, so the multiplet contains eight bosons and eight fermions, as is always the case for maximally supersymmetric branes.

For example, in 11 dimensions, the M5-brane is the magnetic dual of the M2-brane. It follows that in the compactified theory, the string obtained by wrapping the M5-brane on K3 is the magnetic dual of an unwrapped M2-brane. In the ten-dimensional heterotic string theory, on the other hand, the magnetic dual of a fundamental string (F1-brane) is the NS5-brane. After compactification on T^3, the magnetic dual of an unwrapped heterotic string is a fully wrapped NS5-brane. Thus, the heterotic NS5-brane wrapped on the three-torus corresponds to an unwrapped M2-brane.

The matching of tensions implies that

$$T_{F1} = T_{M5} V_{K3} \quad \text{and} \quad T_{NS5} V_{T^3} = T_{M2} \qquad (9.179)$$

or

$$\frac{1}{\ell_s^2} \sim \frac{V_{K3}}{\ell_p^6} \quad \text{and} \quad \frac{V_{T^3}}{g_s^2 \ell_s^6} \sim \frac{1}{\ell_p^3}, \qquad (9.180)$$

where the $\sim$ means that numerical factors are omitted. Combining these two relations gives the dimensionless relation

$$g_s \left(V_{T^3}/\ell_s^3\right)^{-1/2} \sim \left(V_{K3}/\ell_p^4\right)^{3/4}. \qquad (9.181)$$

The left-hand side of this relation is precisely the seven-dimensional heterotic-string coupling constant. This quantifies the earlier claim that $g_s \to \infty$ corresponds to $V_{K3} \to \infty$.

Nonabelian gauge symmetry

It is interesting to check how nonabelian gauge symmetries that arise in the heterotic string theory are understood from the M-theory point of view. We learned in Chapter 7 that the generic $U(1)^{22}$ abelian gauge symmetry of the heterotic string compactified on T^3 is enhanced to nonabelian symmetry at singularities of the Narain moduli space, which exist due to the modding out by the discrete factor $SO(19, 3; \mathbb{Z})$. It was demonstrated in examples that at such loci certain spin-one particles that are charged with respect to the $U(1)$s and massive away from the singular loci become massless to complete the nonabelian gauge multiplet. The nonabelian gauge groups that appear in this way are always of the type $A_n = SU(n+1)$, $D_n = SO(2n)$, E_6, E_7, E_8, or semisimple groups with these groups as factors. The ADE groups in the Cartan classification are the simple Lie groups with the property that all of their simple roots have the same length. Such Lie groups are called *simply-laced*. Given the duality that we have found, these results should be explainable in terms of M-theory on K3.

Generically, K3 compactification of M-theory gives 22 $U(1)$ gauge fields

9.11 K3 compactifications and more string dualities

in seven dimensions. These one-forms arise as coefficients in an expansion of the M-theory three-form A_3 in terms of the 22 linearly independent harmonic two-forms of K3. The three gauge fields associated with the self-dual two-forms correspond to those that arise from right-movers in the heterotic description and belong to the supergravity multiplet. Similarly, the 19 gauge fields associated with the anti-self-dual two-forms correspond to those that arise from left-movers in the heterotic description and belong to the vector supermultiplets.

The singularities of the Narain moduli space correspond to singularities of the K3 moduli space. So we need to understand why there should be nonabelian gauge symmetry at these loci. Each of these singular loci of the K3 moduli space correspond to degenerations of a specific set of two-cycles of the K3 surface. When this happens, wrapped M2-branes on these cycles give rise to new massless modes in seven dimensions. In particular, these should provide the charged spin-one gauge fields for the appropriate nonabelian gauge group.

The way to tell what group appears is as follows. The set of two-cycles that degenerate at a particular singular locus of the moduli space has a matrix of intersection numbers, which can be represented diagrammatically by associating a node with each degenerating cycle and by connecting the nodes by a line for each intersection of the two cycles. Two distinct cycles of K3 intersect either once or not at all, so the number of lines connecting any two nodes is either one or zero.

The diagrams obtained in this way look exactly like Dynkin diagrams, which are used to describe Lie groups. However, the meaning is entirely different. The nodes of Dynkin diagrams denote positive simple roots, whose number is equal to the rank of the Lie group, and the number of lines connecting a pair of nodes represents the angle between the two roots. For example, no lines represents $\pi/4$ and one line represents $2\pi/3$. For simply-laced Lie groups these are the only two cases that occur.

Mathematicians observed long ago that the intersection diagrams of degenerating two-cycles of K3 have an *ADE* classification, but it was completely mysterious what, if anything, this has to do with Lie groups. M-theory provides a beautiful answer. The diagram describing the degeneration of the K3 is identical to the Dynkin diagram that describes the resulting nonabelian gauge symmetry in seven dimensions. The *ADE* Dynkin diagrams are shown in Fig. 9.10. The simplest example is when a single two-cycle degenerates. This is represented by a single node and no lines, which is the Dynkin diagram for $SU(2)$. This case was examined in detail from the heterotic perspective in Chapter 7. A somewhat more complicated example is the degeneration corresponding to the $T^4/\mathbb{Z}_2$ orbifold discussed in Section 9.3. In this case 16 nonintersecting two-cycles degenerate, which gives $[SU(2)]^{16}$ gauge symmetry (in addition to six $U(1)$ factors). Similarly, the $\mathbb{Z}_3$ orbifold considered in Exercise 9.2 gives $[SU(3)]^9$ gauge symmetry (in addition to four $U(1)$ factors). The number of $U(1)$ factors is determined by requiring that the total rank is 22.

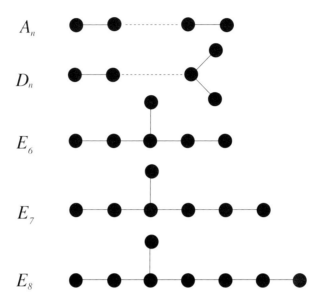

Fig. 9.10. The Dynkin diagrams of the simply-laced Lie algebras.

Type IIA superstring theory on K3

Compactification of the type IIA theory on K3 gives a nonchiral theory with 16 unbroken supersymmetries in six dimensions. This example is closely related to the preceding one, because type IIA superstring theory corresponds to M-theory compactified on a circle. Compactifying the seven-dimensional theory of the previous section on a circle, this suggests that the type IIA theory on K3 should be dual to the heterotic theory on T^4. A minimal spinor in six dimensions has eight components, so this is an $\mathcal{N} = 2$ theory from the six-dimensional viewpoint. Left–right symmetry of the type IIA theory implies that the two six-dimensional supercharges have opposite chirality, which agrees with what one obtains in the heterotic description.

Let us examine the spectrum of massless scalars (moduli) in six dimensions from the type IIA perspective. As in the M-theory case, the metric tensor gives 58 moduli. In addition to this, the dilaton gives one modulus and the two-form B_2 gives 22 moduli, since $b_2(K3) = 22$. The R–R fields C_1 and C_3 do not provide any scalar zero modes, since $b_1 = b_3 = 0$. Thus, the total number of moduli is 81. The heterotic string compactified on T^4 also has an 81-dimensional moduli space, obtained in Chapter 7,

$$\mathbb{R}^+ \times \mathcal{M}_{20,4}. \tag{9.182}$$

Thus, this should also be what one obtains from compactifying the type IIA superstring theory on K3. The $\mathbb{R}^+$ factor corresponds to the heterotic dilaton or the type IIA dilaton, so these two fields need to be related by the duality.

We saw above that the 58 geometric moduli contain 38 complex-structure moduli and 20 Kähler-structure moduli. Of the 22 moduli coming from B_2 the 20 associated with $(1,1)$-forms naturally combine with the 20 geometric Kähler-structure moduli to give 20 complexified Kähler-structure moduli, just as in the case of Calabi–Yau compactification described earlier. Altogether the 80-dimensional space $\mathcal{M}_{20,4}$ is parametrized by 20 complex Kähler-structure moduli and 20 complex-structure moduli. There is a mirror description of the type IIA theory compactified on K3, which is given by type IIA theory compactified on a mirror K3 in which the Kähler-structure moduli and complex-structure moduli are interchanged. While this is analogous to what we found for Calabi–Yau three-fold compactification, there are also some significant differences. For one thing, the two sets of moduli are incorporated in a single moduli space rather than a product of two separate spaces. Also, type IIA is related to type IIA, whereas in the Calabi–Yau three-fold case type IIA was related to type IIB. In that case, we used the SYZ argument to show that, when the Calabi–Yau has a T^3 fibration, this could be understood in terms of T-duality along the fibers. The corresponding statement now is that, when K3 has a T^2 fibration, the mirror description can be deduced by a T-duality along the fibers. The reason type IIA is related to type IIA is that this is an even number (two) of T-duality transformations.

Let us now investigate the relationship between the two dilatons, or equivalently the two string coupling constants, by matching branes. The analysis is very similar to that considered for the previous duality. For the purpose of this argument, let us denote the string coupling and string scale of the type IIA theory by g_A and ℓ_A and those of the heterotic theory by g_H and ℓ_H. Equating tensions of the type IIA NS5-brane wrapped on K3 and the heterotic string as well as the heterotic NS5-brane wrapped on T^4 and the type IIA string gives the relations

$$\frac{1}{\ell_H^2} \sim \frac{V_{K3}}{g_A^2 \ell_A^6} \quad \text{and} \quad \frac{V_{T^4}}{g_H^2 \ell_H^6} \sim \frac{1}{\ell_A^2}. \tag{9.183}$$

Let us now define six-dimensional string coupling constants by

$$g_{6H}^2 = g_H^2 \left(V_{T^4}/\ell_H^4\right)^{-1} \quad \text{and} \quad g_{6A}^2 = g_A^2 \left(V_{K3}/\ell_A^4\right)^{-1}. \tag{9.184}$$

Then these relations can be combined to give

$$g_{6H}^2 \sim g_{6A}^{-2}. \tag{9.185}$$

This means that the relation between the two six-dimensional theories is an S-duality that relates weak coupling and strong coupling, just like the duality relating the two $SO(32)$ superstring theories in ten dimensions.

Type IIB superstring theory on K3

Compactification of type IIB superstring theory on K3 gives a chiral theory with 16 unbroken supersymmetries in six dimensions. The two six-dimensional supercharges have the same chirality. The massless sector in six dimensions consists of a chiral $\mathcal{N} = 2$ supergravity multiplet coupled to 21 tensor multiplets. This is the unique number of tensor multiplets for which anomaly cancellation is achieved. The chiral $\mathcal{N} = 2$ supergravity has a $USp(4) \approx SO(5)$ R symmetry, and there is an $SO(21)$ symmetry that rotates the tensor multiplets. In fact, in the supergravity approximation, these combine into a noncompact $SO(21,5)$ symmetry. However, as always happens in string theory, this gets broken by string and quantum corrections to the discrete duality subgroup $SO(21,5;\mathbb{Z})$.

The gravity multiplet contains five self-dual three-form field strengths, while each of the tensor multiplets contains one anti-self-dual three-form field strength and five scalars. This is the same multiplet that appears on the world volume of an M5-brane, discussed a moment ago. It is the only massless matter multiplet that exists for chiral $\mathcal{N} = 2$ supersymmetry in six dimensions. Most of the three-form field strengths come from the self-dual five-form in ten dimensions as a consequence of the fact that K3 has three self-dual two-forms ($b_2^+ = 3$) and 19 anti-self-dual two-forms ($b_2^- = 19$). The additional two self-dual and anti-self-dual three-forms are provided by $F_3 = dC_2$ and $H_3 = dB_2$. The $5 \times 21 = 105$ scalar fields arise as follows: 58 from the metric, 1 from the dilaton Φ, 1 from C_0, 22 from B_2, 22 from C_2, and 1 from C_4.

The symmetries and the moduli counting described above suggest that the moduli space for K3 compactification of the type IIB theory should be $\mathcal{M}_{21,5}$. The natural question is whether this has a dual heterotic string interpretation. The closest heterotic counterpart is given by toroidal compactification to five dimensions, for which the moduli space is

$$\mathbb{R}^+ \times \mathcal{M}_{21,5}. \tag{9.186}$$

The extra modulus, corresponding to the $\mathbb{R}^+$ factor, is provided by the heterotic dilaton. Therefore, it is tempting to identify the heterotic string theory compactified to five dimensions on T^5 with the type IIB superstring compactified to five dimensions on K3 $\times S^1$. In this duality the heterotic-string coupling constant corresponds to the radius of the type IIB circle. Thus, the strong coupling limit of the toroidally compactified heterotic string theory in five dimensions gives the K3 compactified type IIB string in six dimensions. The relationship is analogous to that between the type IIA theory in ten dimensions and M-theory in 11 dimensions.

This picture can be tested by matching branes, as in the previous examples. However, the analysis is more complicated this time. The essential fact is that in five dimensions both constructions give 26 $U(1)$ gauge fields, with five of them belonging to the supergravity multiplet and 21 belonging to vector multiplets. Thus, point particles can carry 26 distinct electric charges. Their magnetic duals, which are strings, can also carry 26 distinct string charges. By matching the BPS formulas for their tensions one can deduce how to map parameters between the two dual descriptions and verify that, when the heterotic string coupling becomes large, the type IIB circle decompactifies.

Compactification of F-theory on K3

Type IIB superstring theory admits a class of nonperturbative compactifications, first described by Vafa, that go by the name of *F-theory*. The dilaton is not constant in these compactifications, and there are regions in which it is large. Therefore, since the value of the dilaton field determines the string coupling constant, these solutions cannot be studied using perturbation theory (except in special limits that correspond to orientifolds). This is the sense in which F-theory solutions are nonperturbative.

The crucial fact that F-theory exploits is the nonperturbative $SL(2, \mathbb{Z})$ symmetry of type IIB superstring theory in ten-dimensional Minkowski space-time. Recall that the R–R zero-form potential C_0 and the dilaton Φ can be combined into a complex field

$$\tau = C_0 + ie^{-\Phi}, \tag{9.187}$$

which transforms nonlinearly under $SL(2,\mathbb{Z})$ transformations in the same way as the modular parameter of a torus:

$$\tau \to \frac{a\tau + b}{c\tau + d}. \tag{9.188}$$

The two two-forms B_2 and C_2 transform as a doublet at the same time, while C_4 and the Einstein-frame metric are invariant.

F-theory compactifications involve 7-branes, which end up filling the d noncompact space-time dimensions and wrapping $(8-d)$-cycles in the compact dimensions. Therefore, before explaining F-theory, it is necessary to discuss the classification and basic properties of 7-branes. 7-branes in ten dimensions are codimension two, and so they can be enclosed by a circle, just as is the case for a point particle in three dimensions and a string in four dimensions. Just as in those cases, the presence of the brane creates a deficit angle in the orthogonal plane that is proportional to the tension of the brane. Thus, a small circle of radius R, centered on the core of the brane, has a circumference $(2\pi - \phi)R$, where ϕ is the deficit angle. In fact, this property is the key to searching for cosmic strings that might stretch across the sky.

The fact that fields must be single-valued requires that, when they are analytically continued around a circle that encloses a 7-brane, they return to their original values up to an $SL(2,\mathbb{Z})$ transformation. The reason for this is that $SL(2,\mathbb{Z})$ is a discrete gauge symmetry, so that the configuration space is the naive field space modded out by this gauge group. So the requirement stated above means that fields should be single-valued on this quotient space. The field τ, in particular, can have a nontrivial *monodromy transformation* like that in Eq. (9.188). Other fields, such as B_2 and C_2, must transform at the same time, of course.

Since 7-branes are characterized by their monodromy, which is an $SL(2,\mathbb{Z})$ transformation, there is an infinite number of different types. In the case of a D7-brane, the monodromy is $\tau \to \tau + 1$. This implies that $2\pi C_0$ is an angular coordinate in the plane perpendicular to the brane. More precisely, the 7-brane is characterized by the conjugacy class of its monodromy. If there is another 7-brane present the path used for the monodromy could circle the other 7-brane then circle the 7-brane of interest, and finally circle the other 7-brane in the opposite direction. This gives a monodromy described by a different element of $SL(2,\mathbb{Z})$ that belongs to the same conjugacy class and is physically equivalent. The conjugacy classes are characterized by a pair of coprime integers (p,q). This is interpreted physically as labelling the type

9.11 K3 compactifications and more string dualities

of IIB string that can end on the 7-brane. In this nomenclature, a D7-brane is a $(1,0)$ 7-brane, since a fundamental string can end on it.

Let us examine the type IIB equations of motion in the supergravity approximation. The relevant part of the type IIB action, described in Exercise **8.3**, is

$$\frac{1}{2}\int \sqrt{-g}\left(R - g^{\mu\nu}\frac{\partial_\mu\tau\partial_\nu\bar\tau}{(\text{Im}\tau)^2}\right)d^{10}x. \tag{9.189}$$

To describe a 7-brane, let us look for solutions that are independent of the eight dimensions along the brane, which has a flat Lorentzian metric, and parametrize the perpendicular plane as the complex plane with a local coordinate $z = re^{i\theta}$. The idea is that the brane should be localized at the origin of the z-plane. Now let us look for a solution to the equations of motion in the gauge in which the metric in this plane is conformally flat

$$ds^2 = e^{A(r,\theta)}(dr^2 + r^2d\theta^2) - (dx^0)^2 + (dx^1)^2 + \ldots + (dx^7)^2. \tag{9.190}$$

Just as in the case of the string world sheet, the conformal factor cancels out of the τ kinetic term. Therefore, its equation of motion is the same as in flat space. The τ equation of motion is satisfied if τ is a holomorphic function $\tau(z)$, as you are asked to verify in a homework problem.

The *elliptic modular function* $j(\tau)$ gives a one-to-one holomorphic map of the fundamental region of $SL(2,\mathbb{Z})$ onto the entire complex plane. It is invariant under $SL(2,\mathbb{Z})$ modular transformations, and it has a series expansion of the form

$$j(\tau) = \sum_{n=-1}^{\infty} c_n e^{2\pi i n\tau} \tag{9.191}$$

with $c_{-1} = 1$. Its leading asymptotic behavior for $\text{Im}\,\tau \to +\infty$ is given by the first term

$$j(\tau) \sim e^{-2\pi i\tau}. \tag{9.192}$$

If we choose the holomorphic function $\tau(z)$ to be given by

$$j\big(\tau(z)\big) = Cz, \tag{9.193}$$

where C is a constant, then for large z

$$\tau(z) \sim -\frac{1}{2\pi i}\log z. \tag{9.194}$$

This exhibits the desired monodromy $\tau \to \tau - 1$ as one encircles the 7-brane.[31]

---

[31] To get $\tau \to \tau + 1$ instead, one could replace z by $\bar z$, which corresponds to replacing the brane by an antibrane.

The tension of the 7-brane is given by

$$T_7 = \frac{1}{2}\int d^2x \frac{\vec{\partial}\tau \cdot \vec{\partial}\bar{\tau}}{(\operatorname{Im}\tau)^2} = \frac{1}{2}\int d^2x \frac{\partial\tau\bar{\partial}\bar{\tau} + \bar{\partial}\tau\partial\bar{\tau}}{(\operatorname{Im}\tau)^2}. \quad (9.195)$$

Now let us evaluate this for the solution proposed in Eq. (9.193). Since τ is holomorphic

$$T_7 = \frac{1}{2}\int d^2x \frac{\partial\tau\bar{\partial}\bar{\tau}}{(\operatorname{Im}\tau)^2} = \frac{1}{2}\int_{\mathcal{F}}\frac{d^2\tau}{(\operatorname{Im}\tau)^2} = \frac{\pi}{6}. \quad (9.196)$$

This has used the fact that the inverse image of the complex plane is the fundamental region $\mathcal{F}$. The volume of the moduli space was evaluated in Exercise **3.9**.

The integrand in Eq. (9.196) is the energy density that acts as a source for the gravitational field in the Einstein equation

$$R_{00} - \frac{1}{2}g_{00}R = -\frac{1}{2}g_{00}e^{-A}\frac{\partial\tau\bar{\partial}\bar{\tau}}{(\operatorname{Im}\tau)^2}. \quad (9.197)$$

Evaluating the curvature for the metric in Eq. (9.190), one obtains the equation

$$\partial\bar{\partial}A = -\frac{1}{2}\frac{\partial\tau\bar{\partial}\bar{\tau}}{(\tau-\bar{\tau})^2} = \partial\bar{\partial}\log\operatorname{Im}\tau. \quad (9.198)$$

The energy density is concentrated within a string-scale distance of the origin, where the supergravity equations aren't reliable. The total energy is reliable because of supersymmetry (saturation of the BPS bound), however. So, to good approximation, we can take $A = \alpha\log r$ and use $\nabla^2\log r = 2\pi\delta^2(\vec{x})$ to approximate the energy density by a delta function at the core. Doing this, one then matches the integrals of the two sides to determine $\alpha = -1/6$. This gives a result that is correct for large r, namely

$$A \sim -\frac{1}{6}\log r. \quad (9.199)$$

By the change of variables $\rho = r^{11/12}$ this brings the two-dimensional metric to the asymptotic form

$$ds^2 \sim d\rho^2 + \rho^2\left(\frac{11}{12}d\theta\right)^2, \quad (9.200)$$

which shows that there is a deficit angle of $\pi/6$ in the Einstein frame.

A more accurate solution, applicable for multiple 7-branes at positions z_i, $i = 1,\ldots,N$, can be constructed as follows. The general solution of Eq. (9.198) is

$$e^A = |f(z)|^2 \operatorname{Im}\tau \quad (9.201)$$

where $f(z)$ is holomorphic. This function is determined by requiring modular invariance and $r^{-1/6}$ singularities at the cores of 7-branes. The result is

$$f(z) = [\eta(\tau)]^2 \prod_{i=1}^{N} (z - z_i)^{-1/12}. \tag{9.202}$$

The Dedekind η function is

$$\eta(\tau) = q^{1/24} \prod_{n=1}^{\infty} (1 - q^n), \tag{9.203}$$

where

$$q = e^{2\pi i \tau}. \tag{9.204}$$

Under a modular transformation the Dedekind η function transforms as

$$\eta(-1/\tau) = \sqrt{-i\tau}\, \eta(\tau). \tag{9.205}$$

Thus, $|\eta(\tau)|^4 \mathrm{Im}\,\tau$ is modular invariant.

Since all 7-branes are related by modular transformations that leave the Einstein-frame metric invariant, it follows that in Einstein frame they all have a deficit angle of $\pi/6$. Suppose that 7-branes (of various types) are localized at (finite) points on the transverse space such that the total deficit angle is

$$\sum \phi_i = 4\pi. \tag{9.206}$$

Then the transverse space acquires the topology of a sphere with its curvature localized at the positions of the 7-branes, and the z-plane is better described as a projective space $\mathbb{C}P^1$. Since every deficit angle is $\pi/6$, Eq. (9.206) requires that there are a total of 24 7-branes. However, the choice of which types of 7-branes to use, and how to position them, is not completely arbitrary. For one thing, it is necessary that the monodromy associated with a circle that encloses all of them should be trivial, since the circle can be contracted to a point on the other side of the sphere without crossing any 7-branes.

The τ parameter is well defined up to an $SL(2,\mathbb{Z})$ transformation everywhere except at the positions of the 7-branes, where it becomes singular. A nicer way of expressing this is to say that one can associate a torus with complex-structure modulus $\tau(z)$ with each point in the z-plane. This gives a T^2 fibration with base space $\mathbb{C}P^1$, where the 24 singular fibers correspond to the positions of the 7-branes. Such a T^2 fibration is also called an *elliptic fibration*. Only the complex structure of the torus is specified by the modulus τ. Its size (or Kähler structure) is not a dynamical degree of freedom. Recall that the type IIB theory can be obtained by compactifying M-theory on a torus and letting the area of the torus shrink to zero. In this limit the modular parameter of the torus gives the τ parameter of the type IIB theory. Therefore, the best interpretation is that the torus in the F-theory construction has zero area.

A nice way of describing the complex structure of a torus is by an algebraic equation of the form

$$y^2 = x^3 + ax + b. \tag{9.207}$$

This describes the torus as a submanifold of $\mathbb{C}^2$, which is parametrized by complex numbers x and y. The constants a and b determine the complex structure τ of the torus. There is no metric information here, so the area is unspecified. The torus degenerates, that is, τ is ill-defined, whenever the discriminant of this cubic vanishes. This happens for

$$27a^3 - 4b^2 = 0. \tag{9.208}$$

Thus, the positions of the 7-branes correspond to the solutions of this equation. To ensure that $z = \infty$ is not a solution, we require that a^3 and b^2 are polynomials of the same degree.

Since there should be 24 7-branes, the equation should have 24 solutions. Thus, $a = f_8(z)$ and $b = f_{12}(z)$, where f_n denoted a polynomial of degree n. The total space can be interpreted as a K3 manifold that admits a T^2 fibration. The only peculiar feature is that the fibers have zero area. Let us now count the number of moduli associated with this construction. The polynomials f_8 and f_{12} have arbitrary coefficients, which contribute $9 + 13 = 22$ complex moduli. However, four of these are unphysical because of the freedom of an $SL(2, \mathbb{C})$ transformation of the z-plane and a rescaling $f_8 \to \lambda^2 f_8$, $f_{12} \to \lambda^3 f_{12}$. This leaves 18 complex moduli. In addition there is one real modulus (a Kähler modulus) that corresponds to the size of the $\mathbb{C}P^1$ base space. The complex moduli parametrize the positions of the 7-branes (modulo $SL(2,\mathbb{C})$) in the z-plane. The fact that there are fewer than 21 such moduli shows that the positions of the 7-branes (as well as their monodromies) is not completely arbitrary.

Remarkably, there is a dual theory that has the same properties. The heterotic string theory compactified on a torus to eight dimensions has 16 unbroken supersymmetries and the moduli space

$$\mathbb{R}^+ \times \mathcal{M}_{18,2}. \tag{9.209}$$

The real modulus is the string coupling constant, which therefore corresponds to the area of the $\mathbb{C}P^1$ in the F-theory construction. The second factor has 18×2 real moduli or 18 complex moduli. In fact, mathematicians knew before the discovery of F-theory that this is the moduli space of elliptically fibered K3 manifolds. Thus, F-theory compactified on an elliptically fibered K3 (with section) is conjectured to be dual to the heterotic string theory compactified on T^2.

This duality can be related to the others, and so it constitutes one more link in a consistent web of dualities. For example, if one compactifies on another circle, and uses the duality between type IIB on a circle and M-theory on a torus, this torus becomes identified with the F-theory fiber torus, which now has finite area. Then one recovers the duality between M-theory on K3 and the heterotic string on T^3 for the special case of elliptically fibered K3s.

The F-theory construction described above is the simplest example of a large class of possibilities. More generally, F-theory on an elliptically fibered Calabi–Yau n-fold (with section) gives a solution for $(12 - 2n)$-dimensional Minkowski space-time. For example, using elliptically fibered Calabi–Yau four-folds one can obtain four-dimensional F-theory vacua with $\mathcal{N} = 1$ supersymmetry. It is an interesting challenge to identify duality relations between such constructions and other ones that can give $\mathcal{N} = 1$, such as the heterotic string compactified on a Calabi–Yau three-fold.

9.12 Manifolds with G_2 and $Spin(7)$ holonomy

Since the emergence of string dualities and the discovery of M-theory, special-holonomy manifolds have received considerable attention. Manifolds of $SU(3)$ holonomy have already been discussed at length. 7-manifolds with G_2 holonomy and 8-manifolds with $Spin(7)$ holonomy are also of interest for a number of reasons. They constitute the *exceptional-holonomy manifolds*. We refer to them simply as G_2 manifolds and $Spin(7)$ manifolds, respectively.

G_2 manifolds

Suppose that M-theory compactified to four dimensions on a 7-manifold M_7,

$$M_{11} = M_4 \times M_7, \tag{9.210}$$

gives rise to $\mathcal{N} = 1$ supersymmetry in four dimensions. An analysis of the supersymmetry constraints, along the lines studied for Calabi–Yau three-folds, constrains M_7 to have G_2 holonomy. In such a compactification to flat $D = 4$ Minkowski space-time, there should exist *one* spinor (with four independent components) satisfying

$$\delta \psi_M = \nabla_M \varepsilon = 0. \tag{9.211}$$

The background geometry is then $\mathbb{R}^{3,1} \times M_7$, where M_7 has G_2 holonomy, and ε is the covariantly constant spinor of the G_2 manifold tensored with a constant spinor of $\mathbb{R}^{3,1}$. As in the case of Calabi–Yau three-folds, Eq. (9.211) implies that M_7 is Ricci flat. Of course, it cannot be Kähler, or even complex, since it has an odd dimension. Let us now examine why Eq. (9.211) implies that M_7 has G_2 holonomy.

The exceptional group G_2

G_2 can be defined as the subgroup of the $SO(7)$ rotation group that preserves the form

$$\varphi = dy^{123} + dy^{145} + dy^{167} + dy^{246} - dy^{257} - dy^{347} - dy^{356}, \tag{9.212}$$

where

$$dy^{ijk} = dy^i \wedge dy^j \wedge dy^k, \tag{9.213}$$

and y^i are the coordinates of $\mathbb{R}^7$. G_2 is the smallest of the five exceptional simple Lie groups $(G_2, F_4, E_6, E_7, E_8)$, and it has dimension 14 and rank 2. Its Dynkin diagram is given in Fig. 9.11. Let us describe its embedding in $Spin(7)$, the covering group of $SO(7)$, by giving the decomposition of three representations of $Spin(7)$, the vector **7**, the spinor **8** and the adjoint **21**:

- Adjoint representation: decomposes under G_2 as **21 = 14 + 7**.
- The vector representation is irreducible **7 = 7**.
- The spinor representation decomposes as **8 = 7 + 1**.

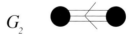

Fig. 9.11. The G_2 Dynkin diagram.

The singlet in the spinor representation precisely corresponds to the covariantly constant spinor in Eq. (9.211) and this decomposition is the reason why G_2 compactifications preserve 1/8 of the original supersymmetry, leading to an $\mathcal{N} = 1$ theory in four dimensions in the case of M-theory. While

Calabi–Yau three-folds are characterized by the existence of a nowhere vanishing covariantly constant holomorphic three-form, a G_2 manifold is characterized by a covariantly constant real three-form Φ, known as the *associative calibration*

$$\Phi = \frac{1}{6}\Phi_{abc} e^a \wedge e^b \wedge e^c, \quad (9.214)$$

where e^a are the seven-beins of the manifold. The Hodge dual four-form $\star\Phi$ is known as the coassociative calibration.

A simple compact example

Smooth G_2 manifolds were first constructed by resolving the singularities of orbifolds. A simple example is the orbifold T^7/Γ, where T^7 is the flat seven-torus and Γ is a finite group of isometries preserving the calibration Eq. (9.212) generated by

$$\alpha : (y^1, \ldots, y^7) \to (y^1, y^2, y^3, -y^4, -y^5, -y^6, -y^7), \quad (9.215)$$

$$\beta : (y^1, \ldots, y^7) \to (y^1, -y^2, -y^3, y^4, y^5, 1/2 - y^6, -y^7), \quad (9.216)$$

$$\gamma : (y^1, \ldots, y^7) \to (-y^1, y^2, -y^3, y^4, 1/2 - y^5, y^6, 1/2 - y^7). \quad (9.217)$$

In a homework problem you are asked to verify that α, β, γ have the following properties: (1) they preserve the calibration, (2) $\alpha^2 = \beta^2 = \gamma^2 = 1$, (3) the three generators commute. The group Γ is isomorphic to $\mathbb{Z}_2^3$. The fixed points of α (and similarly for β and γ) are 16 copies of T^3, while (β, γ) act freely on the fixed-point set of α (similarly for the fixed-point set of β and γ). The singularities of this orbifold can be blown up in a similar way discussed in Section 9.1 for $K3$, that is, by cutting out a ball $B^4/\mathbb{Z}_2$ around each singularity and replacing it with an Eguchi–Hanson space. The result is a smooth G_2 manifold.

Supersymmetric cycles in G_2 manifolds

As in the case of Calabi–Yau three-folds, supersymmetric cycles in G_2 manifolds play a crucial role in describing nonperturbative effects. Supersymmetric three-cycles can be defined for G_2 manifolds in a similar manner as for Calabi–Yau three-folds in Section 9.8. A supersymmetric three-cycle is a configuration that solves the equation

$$P_-\epsilon = \frac{1}{2}\left(1 - \frac{i}{6}\varepsilon^{\alpha\beta\gamma}\partial_\alpha X^M \partial_\beta X^N \partial_\gamma X^P \Gamma_{MNP}\right)\epsilon = 0, \quad (9.218)$$

where now the spinor ϵ lives in seven dimensions. Here $\alpha, \beta, \ldots$ are indices on the cycle while $M, N, \ldots$ are $D = 11$ indices. By a similar calculation to that in Exercise 9.15, one can verify that the defining equation for a supersymmetric three-cycle is

$$\partial_{[\alpha} X^a \partial_\beta X^b \partial_{\gamma]} X^c \Phi_{abc} = \varepsilon_{\alpha\beta\gamma}. \qquad (9.219)$$

This means that the pullback of the three-form onto the cycle is proportional to the volume form. A G_2 manifold can also have supersymmetric four-cycles, which solve the equation

$$P_-\epsilon = \frac{1}{2}\left(1 - \frac{i}{4!}\varepsilon^{\alpha\beta\gamma\sigma}\partial_\alpha X^M \partial_\beta X^N \partial_\gamma X^Q \partial_\sigma X^P \Gamma_{MNPQ}\right)\epsilon = 0. \qquad (9.220)$$

The solution has the same form as Eq. (9.219) with the associative calibration replaced by the dual coassociative calibration $\star\Phi$. Both type of cycles break 1/2 of the original supersymmetry.

Obviously, there is interest in the phenomenological implications of M-theory compactifications on G_2 manifolds, because these give $\mathcal{N} = 1$ theories in four dimensions. Let us mention a few topics in this active area of research.

G_2 manifolds and strongly coupled gauge theories

Compactification of M-theory on a smooth G_2 manifold does not lead to chiral matter or nonabelian gauge symmetry. The reason is that M-theory is a nonchiral theory and compactification on a smooth manifold cannot lead to a chiral theory. A chiral theory can only be obtained if singularities or other defects, where chiral fermions live, are included. Singularities arise, for example, when a supersymmetric cycle shrinks to zero size.

M-theory compactification on a G_2 manifold with a conical singularity leads to interesting strongly coupled gauge theories, which have been investigated in some detail. The local structure of a conical singularity is described by a metric of the form

$$ds^2 = dr^2 + r^2 d\Omega_{n-1}^2. \qquad (9.221)$$

Here r denotes a radial coordinate and $d\Omega_{n-1}^2$ is the metric of some compact manifold Y. In general, this metric describes an n-dimensional space X that has a singularity at $r = 0$ unless $d\Omega_{n-1}^2$ is the metric of the unit sphere, S^{n-1}. An example is a lens space $S^3/\mathbb{Z}_{N+1}$, which corresponds to an A_N singularity.

Singularities can give rise to nonabelian gauge groups in the low-energy effective action. Recall from Chapter 8 that M-theory compactified on K3 is dual to the heterotic string on T^3, and that there is enhanced gauge symmetry at the singularities of K3, which have an *ADE* classification.

9.12 Manifolds with G_2 and Spin(7) holonomy

Invoking this duality for fibered manifolds, there should be a duality between compactification of heterotic theories on Calabi–Yau manifolds with a T^3 fibration and M-theory on G_2 manifolds with a K3 fibration.

In order to obtain four-dimensional theories with nonabelian gauge symmetry, one strategy is to embed ADE singularities in G_2 manifolds. In general, the singularities of four-dimensional manifolds can be described as $\mathbb{C}^2/\Gamma$, where Γ is a subgroup of the holonomy group $SO(4)$. The points that are left invariant by Γ then correspond to the singularities. The holonomy group of K3 is $SU(2)$, and as a result Γ has to be a subgroup of $SU(2)$ to give unbroken supersymmetry. The finite subgroups of $SU(2)$ also have an ADE classification consisting of two infinite series (A_n, $n = 1, 2, \ldots$ and D_k, $k = 4, 5 \ldots$) and three exceptional subgroups (E_6, E_7 and E_8). So for example, the generators for the two infinite series can be represented according to

$$\begin{pmatrix} e^{2\pi i/n} & 0 \\ 0 & e^{-2\pi i/n} \end{pmatrix}, \tag{9.222}$$

for the A_n series. Meanwhile D_k has two generators given by

$$\begin{pmatrix} e^{\pi i/(k-2)} & 0 \\ 0 & e^{-\pi i/(k-2)} \end{pmatrix} \quad \text{and} \quad \begin{pmatrix} 0 & i \\ i & 0 \end{pmatrix}. \tag{9.223}$$

In the heterotic/M-theory duality discussed in Section 9.11, the heterotic string gets an enhanced symmetry group whenever the K3 becomes singular. In general, M-theory compactified on a background of the form $\mathbb{R}^4/\Gamma_{ADE} \times \mathbb{R}^{6,1}$ gives rise to a Yang–Mills theory with the corresponding ADE gauge group, near the singularity. Embedding four-dimensional singular spaces into G_2 manifolds, M-theory compactification can therefore give rise to nonabelian gauge groups in four dimensions.

G_2 manifolds and intersecting D6-brane models

Another area where G_2 manifolds play an important role is *intersecting* D6-brane models.[32] Recall that Section 8.3 showed that N parallel D6-branes in the type IIA theory are interpreted in M-theory as a multi-center Taub–NUT metric times a flat seven-dimensional Minkowski space-time. Half of the supersymmetry is preserved by a stack of parallel branes. If they are not parallel, the amount of supersymmetry preserved depends on types of rotations that relate the branes. Any configuration preserving at least one supersymmetry is described by a special-holonomy manifold from the M-theory perspective. If the position of the branes is such that they can

[32] This is one of the constructions used in attempts to obtain realistic models.

be interpreted in M-theory as a seven-manifold on which one covariantly constant real spinor can be defined times flat four-dimensional Minkowski space-time, then this is a G_2 holonomy configuration.

For parallel D6-branes, the 7-manifold with G_2 holonomy is a direct product of the multi-center Taub–NUT metric times $\mathbb{R}^3$, as you are asked to verify in a homework problem. As discussed in Chapter 8, certain type IIA fields, such as the dilaton and the $U(1)$ gauge field, lift to pure geometry in 11 dimensions. From the M-theory perspective, strings stretched between two D6-branes have an interpretation as membranes wrapping one of the $n(n+1)/2$ holomorphic embeddings of S^2 in multi-center Taub–NUT, as shown in Fig. 9.12. When two D6-branes come close to each other, these strings become massless, resulting in nonabelian gauge symmetry. Without entering into the details, let us mention that chiral matter can be realized when D6-branes intersect at appropriate angles, because the GSO projection removes massless fermions of one chirality. This leads to interesting models with some realistic features.

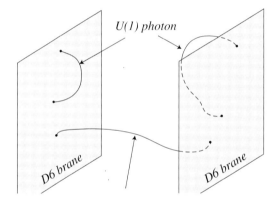

Fig. 9.12. Strings stretched between two D6-branes can be interpreted as membranes wrapping a holomorphically embedded S^2 in a multi-center Taub–NUT geometry.

$Spin(7)$ *manifolds*

Eight-dimensional manifolds of $Spin(7)$ holonomy are of interest in the study of string dualities including connections to strongly coupled gauge theories. Compactification of M-theory on a $Spin(7)$ manifold gives a theory with $\mathcal{N} = 1$ supersymmetry in three dimensions. The supercharge has two components, so 1/16 of the original supersymmetry is preserved. This is less supersymmetry than the minimal amount for a Lorentz-invariant supersymmetric theory in four dimensions. Witten has speculated that the existence of such a three-dimensional theory might indicate the existence of a theory in four dimensions with no supersymmetry that upon circle compactification develops an $\mathcal{N} = 1$ supersymmetry in three dimensions. This is one of many speculations that have been considered in attempts to explain why the observed cosmological constant is so tiny.

$Spin(7)$ is the subgroup of $Spin(8)$ that leaves invariant the self-dual four-form

$$\begin{aligned}\Omega = & \ dy^{1234} + dy^{1256} + dy^{1278} + dy^{1357} - dy^{1368} - dy^{1458} - dy^{1467} - \\ & \ dy^{2358} - dy^{2367} - dy^{2457} + dy^{2468} + dy^{3456} + dy^{3478} + dy^{5678},\end{aligned}$$

where

$$dy^{ijkl} = dy^i \wedge dy^j \wedge dy^k \wedge dy^l, \qquad (9.224)$$

and y_i with $i = 1, \ldots, 8$ are the coordinates of $\mathbb{R}^8$. This 21-dimensional Lie group is compact and simply-connected.

The decomposition of the adjoint is $\mathbf{28} = \mathbf{21} + \mathbf{7}$. $Spin(8)$ has three eight-dimensional representations: the fundamental and two spinors, which are sometimes denoted $\mathbf{8}_v$, $\mathbf{8}_s$ and $\mathbf{8}_c$. Because of the triality of $Spin(8)$, discussed in Chapter 5, it is possible to embed $Spin(7)$ inside $Spin(8)$ such that one spinor decomposes as $\mathbf{8}_c = \mathbf{7} + \mathbf{1}$, while the $\mathbf{8}_v$ and $\mathbf{8}_s$ both reduce to the spinor $\mathbf{8}$ of the $Spin(7)$ subgroup. By choosing such an embedding, the $Spin(7)$ holonomy preserves 1/16 of the original supersymmetry corresponding to the singlet in the decomposition of the two $Spin(8)$ spinors.

Examples of compact $Spin(7)$ manifolds can be obtained, as in the G_2 case, as the blow-ups of orbifolds. The simplest example starts with an orbifold $T^8/\mathbb{Z}_2^4$. $Spin(7)$ manifolds are not Kähler in general. As in the G_2 case, it is interesting to consider manifolds with singularities, which can lead to strongly coupled gauge theories.

EXERCISES

EXERCISE 9.16
Verify that the calibration (9.212) is invariant under 14 linearly independent combinations of the 21 rotation generators of $\mathbb{R}^7$.

Solution

An infinitesimal rotation has the form $R_{ij} = \delta_{ij} + a_{ij}$, where a_{ij} is infinitesimal, and $a_{ij} = -a_{ji}$. This acts on the coordinates by $y'^i = R_{ij}y^j$. Now plug this into the three-form (9.212) and keep only the linear terms in a. Requiring the three-form to be invariant results in the equations

$$a_{14} + a_{36} + a_{27} = 0, \quad a_{15} + a_{73} + a_{26} = 0,$$

$$a_{16} + a_{43} + a_{52} = 0, \quad a_{17} + a_{35} + a_{42} = 0,$$

$$a_{76} + a_{54} + a_{32} = 0, \quad a_{12} + a_{74} + a_{65} = 0,$$

$$a_{13} + a_{57} + a_{64} = 0.$$

These seven constraints leave $21 - 7 = 14$ linearly independent rotations under which the calibration is invariant. This construction ensures that they generate a group. □

Appendix: Some basic geometry and topology

This appendix summarizes some basic geometry and topology needed in this chapter as well as other chapters of this book. This summary is very limited, so we refer the reader to GSW as well as some excellent review articles for a more detailed discussion. The mathematically inclined reader may prefer to consult the math literature for a more rigorous approach.

Real manifolds

What is a manifold?

A real d-dimensional manifold is a space which locally looks like Euclidean space $\mathbb{R}^d$. More precisely, a real manifold of dimension d is defined by introducing a covering with open sets on which local coordinate systems are introduced. Each of these coordinate systems provides a homeomorphism between the open set and a region in $\mathbb{R}^d$. The manifold is constructed by pasting together the open sets. In regions where two open sets overlap, the two sets of local coordinates are related by smooth transition functions. Some simple examples of manifolds are as follows:

- $\mathbb{R}^d$ and $\mathbb{C}^d$ are examples of noncompact manifolds.

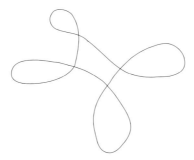

Fig. 9.13. This is not a one-dimensional manifold, because the intersection points are singularities.

- The n-sphere $\sum_{i=1}^{n+1}(x_i^2) = 1$ is an example of a compact manifold. The case $n = 0$ corresponds to two points at $x = \pm 1$, $n = 1$ is a circle and $n = 2$ is a sphere. In contrast to the one-dimensional noncompact manifold $\mathbb{R}^1$, the compact manifold S^1 needs two open sets to be constructed.
- The space displayed in Fig. 9.13 is not a one-dimensional manifold since there is no neighborhood of the cross over points that looks like $\mathbb{R}^1$.

Homology and cohomology

Many topological aspects of real manifolds can be studied with the help of homology and cohomology groups. In the following let us assume that M is a compact d-dimensional manifold with no boundary.

A p-form A_p is an antisymmetric tensor of rank p. The components of A_p are

$$A_p = \frac{1}{p!} A_{\mu_1 \cdots \mu_p} dx^{\mu_1} \wedge \cdots \wedge dx^{\mu_p}, \qquad (9.225)$$

where $\wedge$ denotes the wedge product (an antisymmetrized tensor product). From a mathematician's viewpoint, these p-forms are the natural quantities to define on a manifold, since they are invariant under diffeomorphisms and therefore do not depend on the choice of coordinate system. The possible values of p are $p = 0, 1, \ldots, d$.

The *exterior derivative* d gives a linear map from the space of p-forms into the space of $(p+1)$-forms given by

$$dA_p = \frac{1}{p!} \partial_{\mu_1} A_{\mu_2 \cdots \mu_{p+1}} dx^{\mu_1} \wedge \cdots \wedge dx^{\mu_{p+1}}. \qquad (9.226)$$

A crucial property that follows from this definition is that the operator d is nilpotent, which means that $d^2 = 0$. This can be illustrated by applying d^2

to a zero form

$$ddA_0 = d\left(\frac{\partial A_0}{\partial x^\mu}dx^\mu\right) = \frac{\partial^2 A_0}{\partial x^\mu \partial x^\nu}dx^\mu \wedge dx^\nu, \tag{9.227}$$

which vanishes due to antisymmetry of the wedge product. A p-form is called *closed* if

$$dA_p = 0, \tag{9.228}$$

and *exact* if there exists a globally defined $(p-1)$-form A_{p-1} such that

$$A_p = dA_{p-1}. \tag{9.229}$$

A closed p-form can always be written locally in the form dA_{p-1}, but this may not be possible globally. In other words, a closed form need not be exact, though an exact form is always closed.

Let us denote the space of closed p-forms on M by $C^p(M)$ and the space of exact p-forms on M by $Z^p(M)$. Then the pth *de Rham cohomology group* $H^p(M)$ is defined to be the quotient space

$$H^p(M) = C^p(M)/Z^p(M). \tag{9.230}$$

$H^p(M)$ is the space of closed forms in which two forms which differ by an exact form are considered to be equivalent. The dimension of $H^p(M)$ is called the *Betti number*. Betti numbers are very basic topological invariants characterizing a manifold. The Betti numbers of S^2 and T^2 are described in Fig. 9.14. Another especially important topological invariant of a manifold is the *Euler characteristic*, which can be expressed as an alternating sum of Betti numbers

$$\chi(M) = \sum_{i=0}^{d}(-1)^i b_i(M). \tag{9.231}$$

The Betti numbers of a manifold also give the dimensions of the *homology groups*, which are defined in a similar way to the cohomology groups. The analog of the exterior derivative d is the boundary operator δ, which acts on submanifolds of M. Thus, if N is a submanifold of M, then δN is its boundary. This operator associates with every submanifold its boundary with signs that take account of the orientation. The boundary operator is also nilpotent, as the boundary of a boundary is zero. Therefore, it can be used to define homology groups of M in the same way that the exterior derivative was used to define cohomology groups of M. Arbitrary linear combinations of submanifolds of dimension p are called p-chains. Here again, to be more precise, one should say what type of coefficients is used to form

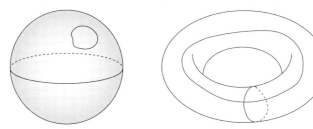

Fig. 9.14. The Betti numbers b_p count the number of p-cycles which are not boundaries. For the sphere all one-cycles can be contracted to a point and the Betti numbers are $b_0 = b_2 = 1$ and $b_1 = 0$. The torus supports nontrivial one-cycles and as a result the Betti numbers are $b_0 = b_2 = 1$ and $b_1 = 2$.

the linear combinations. A chain that has no boundary is called *closed*, and a chain that is a boundary is called *exact*. A closed chain z_p, also called a *cycle*, satisfies

$$\delta z_p = 0. \tag{9.232}$$

The simplicial homology group $H_p(M)$ is defined to consist of equivalence classes of p-cycles. Two p-cycles are equivalent if and only if their difference is a boundary.

Poincaré duality

A fundamental theorem is Stokes' theorem. Given a real manifold M, let A be an arbitrary p-form and let N be an arbitrary $(p+1)$-chain. Then Stokes' theorem states

$$\int_N dA = \int_{\delta N} A. \tag{9.233}$$

This formula provides an isomorphism between $H^p(M)$ and $H_{d-p}(M)$ that is called *Poincaré duality*. To every closed p-form A there corresponds a $(d-p)$-cycle N with the property

$$\int_M A \wedge B = \int_N B, \tag{9.234}$$

for all closed $(d-p)$-forms B. The fact that the left-hand side only depends on the cohomology class of A and the right-hand side only depends on the homology class of N is an immediate consequence of Stokes' theorem and the fact that M has no boundary. Poincaré duality allows us to determine the Betti numbers of a manifold by counting the nontrivial cycles of the manifold. For example, S^N has Betti numbers $b_0 = 1, b_1 = 0, \ldots, b_N = 1$.

Riemannian geometry

Metric tensor

The manifolds described so far are entirely characterized by their topology. Next, we consider manifolds endowed with a metric. If the metric is positive definite, the manifold is called a Riemannian manifold. If it has indefinite signature, as in the case of general relativity, it is called a pseudo-Riemannian manifold. In either case the metric is a symmetric tensor characterized by an infinitesimal line element

$$ds^2 = g_{\mu\nu}(x) dx^\mu dx^\nu, \qquad (9.235)$$

which allows one to compute the length of a curve by integration. The line element itself is coordinate independent. This fact allows one to compute how the metric components $g_{\mu\nu}(x)$ transform under general coordinate transformations (diffeomorphisms).

The metric tensor can be expressed in terms of the *frame*. This consists of d linearly independent one-forms e^α that are defined locally on M. In terms of a basis of one-forms

$$e^\alpha = e^\alpha_\mu dx^\mu. \qquad (9.236)$$

The components e^α_μ form a matrix called the *vielbein*. Let $\eta^{\alpha\beta}$ and $\eta_{\alpha\beta}$ denote the flat metric whose only nonzero entries are ± 1 on the diagonal. In the Riemannian case (Euclidean signature) η is the unit matrix. In the Lorentzian case, there is one -1 corresponding to the time direction. The metric tensor is given in terms of the frame by

$$g = \eta_{\alpha\beta} e^\alpha \otimes e^\beta. \qquad (9.237)$$

In terms of components this corresponds to

$$g_{\mu\nu} = \eta_{\alpha\beta} e^\alpha_\mu e^\beta_\nu. \qquad (9.238)$$

The inverse vielbein and metric are denoted e^μ_α and $g^{\mu\nu}$.

Harmonic forms

The metric is needed to define the Laplace operator acting on p-forms on a d-dimensional space given by

$$\Delta_p = d^\dagger d + d d^\dagger = (d + d^\dagger)^2, \qquad (9.239)$$

where

$$d^\dagger = (-1)^{dp+d+1} \star d \star \qquad (9.240)$$

for Euclidean signature, and there is an extra minus sign for Lorentzian signature. The Hodge $\star$-operator acting on p-forms is defined as

$$\star(dx^{\mu_1}\wedge\cdots\wedge dx^{\mu_p}) = \frac{\varepsilon^{\mu_1\cdots\mu_p\mu_{p+1}\cdots\mu_d}}{(d-p)!|g|^{1/2}}g_{\mu_{p+1}\nu_{p+1}}\cdots g_{\mu_d\nu_d}dx^{\nu_{p+1}}\wedge\cdots\wedge dx^{\nu_d}. \tag{9.241}$$

The Levi–Civita symbol ε transforms as a tensor density, while $\varepsilon/|g|^{1/2}$ is a tensor. A p-form A is said to be *harmonic* if and only if

$$\Delta_p A = 0. \tag{9.242}$$

Harmonic p-forms are in one-to-one correspondence with the elements of the group $H^p(M)$. Indeed, from the definition of the Laplace operator it follows that if A_p is harmonic

$$(dd^\dagger + d^\dagger d)A_p = 0, \tag{9.243}$$

and as a result

$$(A_p, (dd^\dagger + d^\dagger d)A_p) = 0 \Rightarrow (d^\dagger A_p, d^\dagger A_p) + (dA_p, dA_p) = 0. \tag{9.244}$$

Using a positive-definite scalar product it follows that A_p is closed and co-closed. The Hodge theorem states that on a compact manifold that has a positive definite metric a p-form has a unique decomposition into harmonic, exact and co-exact pieces

$$A_p = A_p^{\rm h} + dA_{p-1}^{\rm e.} + d^\dagger A_{p-1}^{\rm c.e.}. \tag{9.245}$$

As a result, a closed form can always be written in the form

$$A_p = A_p^{\rm h} + dA_{p-1}^{\rm e.}. \tag{9.246}$$

Since the Hodge dual turns a closed p-form into a co-closed $(d-p)$-form and *vice versa*, it follows that the Hodge dual provides an isomorphism between the space of harmonic p-forms and the space of harmonic $(d-p)$-forms. Therefore,

$$b_p = b_{d-p}. \tag{9.247}$$

The connection

Another fundamental geometric concept is the *connection*. There are actually two of them: the *affine connection* and the *spin connection*, though they are related (via the vielbein). Connections are not tensors, though the arbitrariness in their definitions corresponds to adding a tensor. Also, they are used in forming covariant derivatives, which are constructed so that they map tensors to tensors. The expressions for the connections can be deduced from the fundamental requirement that the vielbein is covariantly constant

$$\nabla_\mu e_\nu^\alpha = \partial_\mu e_\nu^\alpha - \Gamma^\rho_{\mu\nu} e_\rho^\alpha + \omega_\mu{}^\alpha{}_\beta e_\nu^\beta = 0. \tag{9.248}$$

This equation determines the affine connection Γ and the spin connection ω up to a contribution characterized by a *torsion tensor*, which is described in Chapter 10. The affine connection, for example, is given by the Levi–Civita connection plus a torsion contribution

$$\Gamma^\rho_{\mu\nu} = \left\{{\rho \atop \mu\nu}\right\} + K_{\mu\nu}{}^\rho, \tag{9.249}$$

where the Levi–Civita connection is

$$\left\{{\rho \atop \mu\nu}\right\} = \frac{1}{2} g^{\rho\lambda} (\partial_\mu g_{\nu\lambda} + \partial_\nu g_{\mu\lambda} - \partial_\lambda g_{\mu\nu}), \tag{9.250}$$

and K is called the contortion tensor. The formula for the spin connection, given by solving Eq. (9.248), is

$$\omega_\mu{}^\alpha{}_\beta = -e^\nu_\beta (\partial_\mu e_\nu^\alpha - \Gamma^\lambda_{\mu\nu} e_\lambda^\alpha). \tag{9.251}$$

Curvature tensors

The curvature tensor can be constructed from either the affine connection Γ or the spin connection ω. Let us follow the latter route. The spin connection is a Lie-algebra valued one-form $\omega^\alpha{}_\beta = \omega_\mu{}^\alpha{}_\beta dx^\mu$. The algebra in question is $SO(d)$, or a noncompact form of $SO(d)$ in the case of indefinite signature. Thus, it can be regarded as a Yang–Mills gauge field. The curvature two-form is just the corresponding field strength,

$$R^\alpha{}_\beta = d\omega^\alpha{}_\beta + \omega^\alpha{}_\gamma \wedge \omega^\gamma{}_\beta, \tag{9.252}$$

which in matrix notation becomes

$$R = d\omega + \omega \wedge \omega. \tag{9.253}$$

Its components have two base-space and two tangent-space indices $R_{\mu\nu}{}^\alpha{}_\beta$. One can move indices up and down and convert indices from early Greek to late Greek by contracting with metrics, vielbeins and their inverses. In particular, one can form $R^\mu{}_{\nu\rho\lambda}$, which coincides with the *Riemann curvature tensor* that is usually constructed from the affine connection. Contracting a pair of indices gives the *Ricci tensor*

$$R_{\nu\lambda} = R^\mu{}_{\nu\mu\lambda}, \tag{9.254}$$

Appendix: Some basic geometry and topology 447

and one more contraction gives the *scalar curvature*

$$R = g^{\mu\nu} R_{\mu\nu}. \tag{9.255}$$

Holonomy groups

The holonomy group of a Riemannian manifold M of dimension d describes the way various objects transform under parallel transport around closed curves. The objects that are parallel transported can be tensors or spinors. For spin manifolds (that is, manifolds that admit spinors), spinors are the most informative. The reason is that the most general transformation of a vector is a rotation, which is an element of $SO(d)$.[33] The corresponding transformation of a spinor, on the other hand, is an element of the covering group $Spin(d)$. So let us suppose that a spinor is parallel transported around a closed curve. As a result, the spinor is rotated from its original orientation

$$\varepsilon \to U\varepsilon, \tag{9.256}$$

where U is an element of $Spin(d)$ in the spinor representation appropriate to ε. Now imagine taking several consecutive paths each time leaving and returning to the same point. The result for the spinor after two paths is, for example,

$$\varepsilon \to U_1 U_2 \varepsilon = U_3 \varepsilon. \tag{9.257}$$

As a result, the U matrices build a group, called the *holonomy group* $\mathcal{H}(M)$.

The generic holonomy group of a Riemannian manifold M of real dimension d that admits spinors is $Spin(d)$. Now one can consider different special classes of manifolds in which $\mathcal{H}(M)$ is only a subgroup of $Spin(d)$. Such manifolds are called manifolds of *special holonomy*.

- $\mathcal{H} \subseteq U(d/2)$ if and only if M is Kähler.
- $\mathcal{H} \subseteq SU(d/2)$ if and only if M is Calabi–Yau.
- $\mathcal{H} \subseteq Sp(d/4)$ if and only if M is hyper-Kähler.
- $\mathcal{H} \subseteq Sp(d/4) \cdot Sp(1)$ if and only if M is quaternionic Kähler.

In the first two cases d must be a multiple of two, and in the last two cases it must be a multiple of four. Kähler manifolds and Calabi–Yau manifolds are discussed later in this appendix. Hyper-Kähler and quaternionic Kähler manifolds will not be considered further. There are two other cases of special holonomy. In seven dimensions the exceptional Lie group G_2 is

---

[33] Reflections are avoided by assuming that the manifold is oriented.

a possible holonomy group, and in eight dimensions $Spin(7)$ is a possible holonomy group. The G_2 case is of possible physical interest in the context of compactifying M-theory to four dimensions.

Complex manifolds

A complex manifold of complex dimension n is a special case of a real manifold of dimension $d = 2n$. It is defined in an analogous manner using complex local coordinate systems. In this case the transition functions are required to be biholomorphic, which means that they and their inverses are both holomorphic. Let us denote complex local coordinates by z^a ($a = 1, \ldots, n$) and their complex conjugates $\bar{z}^{\bar{a}}$.

A complex manifold admits a tensor J, with one covariant and one contravariant index, which in complex coordinates has components

$$J_a{}^b = i\delta_a{}^b, \qquad J_{\bar{a}}{}^{\bar{b}} = -i\delta_{\bar{a}}{}^{\bar{b}}, \qquad J_a{}^{\bar{b}} = J_{\bar{a}}{}^b = 0. \tag{9.258}$$

These equations are preserved by a holomorphic change of variables, so they describe a globally well-defined tensor.

Sometimes one is given a real manifold M in $2n$ dimensions, and one wishes to determine whether it is a complex manifold. The first requirement is the existence of a tensor, $J^m{}_n$, called an *almost complex structure*, that satisfies

$$J_m{}^n J_n{}^p = -\delta_m{}^p. \tag{9.259}$$

This equation is preserved by a smooth change of coordinates. The second condition is that the almost complex structure is a *complex structure*. The obstruction to this is given by a tensor, called the *Nijenhuis tensor*

$$N^p{}_{mn} = J_m{}^q \partial_{[q} J_{n]}{}^p - J_n{}^q \partial_{[q} J_{m]}{}^p. \tag{9.260}$$

When this tensor is identically zero, J is a complex structure. Then it is possible to choose complex coordinates in every open set that defines the real manifold M such that J takes the values given in Eq. (9.258) and the transition functions are holomorphic.

On a complex manifold one can define (p, q)-forms as having p holomorphic and q antiholomorphic indices

$$A_{p,q} = \frac{1}{p!q!} A_{a_1 \cdots a_p \bar{b}_1 \cdots \bar{b}_q} dz^{a_1} \wedge \cdots \wedge dz^{a_p} \wedge d\bar{z}^{\bar{b}_1} \wedge \cdots \wedge d\bar{z}^{\bar{b}_q}. \tag{9.261}$$

The real exterior derivative can be decomposed into holomorphic and antiholomorphic pieces

$$d = \partial + \bar{\partial} \tag{9.262}$$

with

$$\partial = dz^a \frac{\partial}{\partial z^a} \quad \text{and} \quad \bar{\partial} = d\bar{z}^{\bar{a}} \frac{\partial}{\partial \bar{z}^{\bar{a}}}. \tag{9.263}$$

Then ∂ and $\bar{\partial}$, which are called *Dolbeault operators*, map (p,q)-forms to $(p+1,q)$-forms and $(p,q+1)$-forms, respectively. Each of these exterior derivatives is nilpotent

$$\partial^2 = \bar{\partial}^2 = 0, \tag{9.264}$$

and they anticommute

$$\partial\bar{\partial} + \bar{\partial}\partial = 0. \tag{9.265}$$

Complex geometry

Let us now consider a complex Riemannian manifold. In terms of the complex local coordinates, the metric tensor is given by

$$ds^2 = g_{ab}dz^a dz^b + g_{a\bar{b}}dz^a d\bar{z}^{\bar{b}} + g_{\bar{a}b}d\bar{z}^{\bar{a}} dz^b + g_{\bar{a}\bar{b}}d\bar{z}^{\bar{a}} d\bar{z}^{\bar{b}}. \tag{9.266}$$

The reality of the metric implies that $g_{\bar{a}\bar{b}}$ is the complex conjugate of g_{ab} and that $g_{a\bar{b}}$ is the complex conjugate of $g_{\bar{a}b}$. A *hermitian manifold* is a special case of a complex Riemannian manifold, which is characterized by the conditions

$$g_{ab} = g_{\bar{a}\bar{b}} = 0. \tag{9.267}$$

These conditions are preserved under holomorphic changes of variables, so they are globally well defined.

The Dolbeault cohomology group $H^{p,q}_{\bar{\partial}}(M)$ of a hermitian manifold M consists of equivalence classes of $\bar{\partial}$-closed (p,q)-forms. Two such forms are equivalent if and only if they differ by a $\bar{\partial}$-exact (p,q)-form. The dimension of $H^{p,q}_{\bar{\partial}}(M)$ is called the *Hodge number* $h^{p,q}$. We can define the Laplacians

$$\Delta_\partial = \partial\partial^\dagger + \partial^\dagger\partial \quad \text{and} \quad \Delta_{\bar{\partial}} = \bar{\partial}\bar{\partial}^\dagger + \bar{\partial}^\dagger\bar{\partial}. \tag{9.268}$$

A *Kähler manifold* is defined to be a hermitian manifold on which the Kähler form

$$J = ig_{a\bar{b}}dz^a \wedge d\bar{z}^{\bar{b}} \tag{9.269}$$

is closed

$$dJ = 0. \tag{9.270}$$

It follows that the metric on these manifolds satisfies $\partial_a g_{b\bar{c}} = \partial_b g_{a\bar{c}}$, as well as the complex conjugate relation, and therefore it can be written locally in the form

$$g_{a\bar{b}} = \frac{\partial}{\partial z^a} \frac{\partial}{\partial \bar{z}^b} \mathcal{K}(z, \bar{z}), \quad (9.271)$$

where $\mathcal{K}(z, \bar{z})$ is called the *Kähler potential*. Thus,

$$J = i\partial\bar{\partial}\mathcal{K}.$$

The Kähler potential is only defined up to the addition of arbitrary holomorphic and antiholomorphic functions $f(z)$ and $\bar{f}(\bar{z})$, since

$$\tilde{\mathcal{K}}(z, \bar{z}) = \mathcal{K}(z, \bar{z}) + f(z) + \bar{f}(\bar{z}) \quad (9.272)$$

leads to the same metric. In fact, there are such relations on the overlaps of open sets.

On Kähler manifolds the various Laplacians all become identical

$$\Delta_d = 2\Delta_{\bar{\partial}} = 2\Delta_{\partial}. \quad (9.273)$$

The various possible choices of cohomology groups (based on d, ∂ and $\bar{\partial}$) each have a unique harmonic representative of the corresponding type, as in the real case described earlier. Therefore, in the case of Kähler manifolds, it follows that they are all identical

$$H^{p,q}_{\bar{\partial}}(M) = H^{p,q}_{\partial}(M) = H^{p,q}(M). \quad (9.274)$$

As a consequence, the Hodge and the Betti numbers are related by

$$b_k = \sum_{p=0}^{k} h^{p,k-p}. \quad (9.275)$$

If ω is a (p,q)-form on a Kähler manifold with n complex dimensions, then the complex conjugate form $\omega^\star$ is a (q,p)-form. It follows that

$$h^{p,q} = h^{q,p}. \quad (9.276)$$

Similarly, if ω is a (p,q)-form, then $\star\omega$ is a $(n-p, n-q)$-form. This implies that

$$h^{n-p,n-q} = h^{p,q}. \quad (9.277)$$

One way of understanding this result is to focus on the harmonic representatives of the cohomology classes, which are both closed and co-closed. As in the case of real manifolds, the Hodge dual of a closed form is co-closed and *vice versa*, so the Hodge dual of a harmonic form is harmonic.

In terms of complex local coordinates, only the mixed components of the

Ricci tensor are nonvanishing for a hermitian manifold. Therefore, one can define a $(1,1)$-form, called the *Ricci form*, by

$$\mathcal{R} = iR_{a\bar{b}}dz^a \wedge d\bar{z}^b. \tag{9.278}$$

For a hermitian manifold, the exterior derivative of the Ricci form is proportional to the torsion. However, for a Kähler manifold the torsion vanishes, and therefore the Ricci form is closed $d\mathcal{R} = 0$. It is therefore a representative of a cohomology class belonging to $H^{1,1}(M)$. This class is called the *first Chern class*

$$c_1 = \frac{1}{2\pi}[\mathcal{R}]. \tag{9.279}$$

EXERCISES

EXERCISE **A.1**
Use Stokes' theorem to verify Poincaré duality.

SOLUTION

Consider a form $A \in H^p(M)$. It can be expanded in a basis $\{w^i\}$, so that $A = \alpha_i w^i$. Consider also a form $B \in H^{d-p}(M)$, which is expanded in a basis $\{v^j\}$ as $B = \beta_j v^j$. Therefore,

$$\int_M A \wedge B = \alpha_i \beta_j \int_M w^i \wedge v^j \equiv \alpha_i \beta_j m^{ij}.$$

Now we define the dual basis to $\{v^j\}$ as $\{Z_j\}$, which are $(d-p)$-cycles that satisfy

$$\int_{Z_j} v^i = \delta^i_j.$$

According to Stokes' theorem, we can integrate B over the $(d-p)$-cycle $N = \alpha_i m^{ij} Z_j$, to get

$$\int_N B = \int_{\alpha_i m^{ij} Z_j} \beta_\gamma v^\gamma = \alpha_i \beta_\gamma m^{ij} \delta^\gamma_j = \alpha_i \beta_j m^{ij} = \int_M A \wedge B.$$

It follows that, for any $A \in H^p(M)$, there is a corresponding $N \in H_{d-p}(M)$. This implies Poincaré duality

$$H^p(M) \approx H_{d-p}(M)$$

Exercise A.2

Consider the complex plane with coordinate $z = x + iy$ and the standard flat Euclidean metric ($ds^2 = dx^2 + dy^2$). Compute $\star dz$ and $\star d\bar{z}$.

Solution

Because we have a Euclidean metric, it is easy to check
$$\star dx = dy \quad \text{and} \quad \star dy = -dx,$$
where we have used $\varepsilon^{xy} = -\varepsilon^{yx} = 1$. Thus
$$\star dz = -i dz \quad \text{and} \quad \star d\bar{z} = i d\bar{z}.$$

Exercise A.3

If ∇ is a torsion-free connection, which means that $\Gamma^p_{mn} = \Gamma^p_{nm}$, show that Eq. (9.260) is equivalent to
$$N^p{}_{mn} = J^q{}_m \nabla_q J^p{}_n - J^q{}_n \nabla_q J^p{}_m - J^p{}_q \nabla_m J^q{}_n + J^p{}_q \nabla_n J^q{}_m.$$

Solution

By definition

$$J^q{}_m \nabla_q J^p{}_n + J^p{}_q \nabla_n J^q{}_m - J^q{}_n \nabla_q J^p{}_m - J^p{}_q \nabla_m J^q{}_n$$

$$= J^q{}_m (\partial_q J^p{}_n + \Gamma_{q\lambda}{}^p J^\lambda{}_n - \Gamma_{qn}{}^\lambda J^p{}_\lambda)$$

$$+ J^p{}_q (\partial_n J^q{}_m + \Gamma_{n\lambda}{}^q J^\lambda{}_m - \Gamma_{nm}{}^\lambda J^q{}_\lambda) - (n \leftrightarrow m).$$

Because $J^q{}_m \Gamma_{q\lambda}{}^p J^\lambda{}_n$ and $J^p{}_q \Gamma_{nm}{}^\lambda J^q{}_\lambda$ are symmetric in (n,m), if the connection is torsion-free, these terms cancel. To see the cancellation of

$$J^p{}_q \Gamma_{n\lambda}{}^q J^\lambda{}_m - J^q{}_m \Gamma_{qn}{}^\lambda J^p{}_\lambda,$$

we only need to exchange the index λ and q of the first term. So all the affine connection terms cancel out, and the expression simplifies to

$$N^p{}_{mn} = J^q{}_m \partial_q J^p{}_n + J^p{}_q \partial_n J^q{}_m - (n \leftrightarrow m),$$

which is what we wanted to show.

Homework Problems

Problem 9.1
By considering the orbifold limit in Section 9.3 explain why the 22 harmonic two-forms of K3 consist of three self-dual forms and 19 anti-self-dual forms.

Problem 9.2
Show that the Eguchi–Hanson space defined by Eq. (9.24) is Ricci flat and Kähler and that the Kähler form is anti-self-dual.

Problem 9.3
Show that the curvature two-form of S^2 using the Fubini–Study metric is

$$R = -2\frac{dz \wedge d\bar{z}}{(1+z\bar{z})^2}.$$

Using this result compute the Chern class and the Chern number (that is, the integral of the Chern class over S^2) for the tangent bundle of S^2.

Problem 9.4
Show that the Kähler potential for $\mathbb{C}P^n$ given in Eq. (9.31) undergoes a Kähler transformation when one changes from one set of local coordinates to another one. Construct the Fubini–Study metric.

Problem 9.5
Show that $\mathcal{K} = -\log(\int J)$ is the Kähler potential for the Kähler-structure modulus of a two-dimensional torus.

Problem 9.6
Consider a two-dimensional torus characterized by two complex parameters τ and ρ (that is, an angle θ is also allowed). Show that T-duality interchanges the complex-structure and Kähler parameters, as mentioned in Section 9.5, and that the spectrum is invariant under this interchange.

Problem 9.7
Verify the Lichnerowicz equation discussed in Section 9.5:

$$\nabla^k \nabla_k \delta g_{mn} + 2R_m{}^p{}_n{}^q \delta g_{pq} = 0.$$

Hint: use $R_{mn} = 0$ and the gauge condition in Eq. (9.91).

Problem 9.8
Use (9.92) to show that (9.93) and (9.96) are harmonic.

Problem 9.9
Check the result for the Kähler potential Eq. (9.117).

Problem 9.10
Show that Eq. (9.132) agrees with Eq. (9.129).

Problem 9.11
Compute the scalar curvature of the conifold metric in Eq. (9.143), and show that it diverges at $X^1 = 0$. Thus, the conifold singularity is a real singularity in the moduli space.

Problem 9.12
Show that the operators in Eq. (9.146) are projection operators.

Problem 9.13
Consider the $E_8 \times E_8$ heterotic string compactified on a six-dimensional orbifold

$$\frac{T^2 \times T^2 \times T^2}{\mathbb{Z}_4},$$

where $\mathbb{Z}_4$ acts on the complex coordinates (z_1, z_2, z_3) of the three tori, as $(z_1, z_2, z_3) \to (iz_1, iz_2, -z_3)$. Identify the spin connection with the gauge connection of one of the E_8s to find the spectrum of massless modes and gauge symmetries in four dimensions.

Problem 9.14
Verify that Eq. (9.172) vanishes if J is the Kähler form.

Problem 9.15
As mentioned in Section 9.11, compactification of the type IIB theory on K3 leads to a chiral theory with $\mathcal{N} = 2$ supersymmetry in six dimensions. Since this theory is chiral, it potentially contains gravitational anomalies. Using the explicit form of the anomaly characteristic classes discussed in Chapter 5, show that anomaly cancellation requires that the massless sector contain 21 matter multiplets (called tensor multiplets) in addition to the supergravity multiplet.

Problem 9.16
Consider the second term in the action (9.189) restricted to two dimensions described by a complex variable z. Form the equation of motion of the field τ and show that it is satisfied by any holomorphic function $\tau(z)$.

Problem 9.17
Consider a Calabi–Yau three-fold given as an elliptically fibered manifold over $\mathbb{C}P^1 \times \mathbb{C}P^1$

$$y^2 = x^3 + f(z_1, z_2)x + g(z_1, z_2),$$

where z_1, z_2 represent the two $\mathbb{C}P^1$s and f, g are polynomials in f in (z_1, z_2).

(i) What is the degree of the polynomials f and g? Hint: write down the holomorphic three-form and insist that it has no zeros or poles at infinity.
(ii) Compute the number of independent complex structure deformations of this Calabi–Yau. What do you obtain for the Hodge number $h^{2,1}$?
(iii) How many Kähler deformations do you find, and what does this imply for $h^{1,1}$?

Problem 9.18
Verify properties (i)–(iii) for the G_2 orbifold T^7/Γ defined in Section 9.12. Show that the blow-up of each fixed point gives 12 copies of T^3.

Problem 9.19
Verify that the solution to the constraint equation for a supersymmetric three-cycle in a G_2 manifold Eq. (9.218) is given by Eq. (9.219). Repeat the calculation for the supersymmetric four-cycle.

Problem 9.20
Show that the direct product of the multi-center Taub–NUT metric discussed in Section 8.3 with flat $\mathbb{R}^3$ corresponds to a 7-manifold with G_2 holonomy.

Problem 9.21
Find the conditions, analogous to those in Exercise 9.16, defining the $Spin(7)$ action that leaves invariant the four-form (9.224). Verify that there are the correct number of conditions.

10
Flux compactifications

Moduli-space problem

The previous chapter described Calabi–Yau compactification for a product manifold $M_4 \times M$. When the ten-dimensional heterotic string is compactified on such a manifold the resulting low-energy effective action has $\mathcal{N} = 1$ supersymmetry, which makes it phenomenologically attractive in a number of respects. Certain specific Calabi–Yau compactifications even lead to three-generation models.

An unrealistic feature of these models is that they contain massless scalars with undetermined vacuum expectation values (vevs). Therefore, they do not make specific predictions for many physical quantities such as coupling constants. These scalar fields are called moduli fields, since their vevs are moduli for which there is no potential in the low-energy four-dimensional effective action. This moduli-space problem or moduli-stabilization problem has been recognized, but not emphasized, in the traditional string theory literature. This situation changed with the discovery of string dualities and recognition of the key role that branes play in string theory.

As discussed in Chapter 8, the moduli-space problem already arises for simple circle compactification of $D = 11$ supergravity, where the size of the circle is a modulus, dual to the vev of the type IIA dilaton, which is undetermined. A similar problem, in a more complicated setting, appears for the volume of the compact space in conventional Calabi–Yau compactifications of any superstring theory. In this case the size of the internal manifold cannot be determined.

Warped compactifications

Recently, string theorists have understood how to generate a potential that can stabilize the moduli fields. This requires compactifying string theory

Flux compactifications

on a new type of background geometry, a *warped geometry*.[1] Warped compactifications also provide interesting models for superstring and M-theory cosmology. Furthermore, they are relevant to the duality between string theory and gauge theory discussed in Chapter 12.

In a warped geometry, background values for certain tensor fields are nonvanishing, so that associated fluxes thread cycles of the internal manifold. An n-form potential A with an $(n+1)$-form field strength $F = dA$ gives a *magnetic flux* of the form[2]

$$\int_{\gamma_{n+1}} F, \tag{10.1}$$

that depends only on the homology of the cycle γ_{n+1}. Similarly, in D dimensions the same field gives an *electric flux*

$$\int_{\gamma_{D-n-1}} \star F, \tag{10.2}$$

where the star indicates the Hodge dual in D dimensions. This flux depends only on the homology of the cycle γ_{D-n-1}.

Flux quantization

This chapter explores the implications of flux compactifications for the moduli-space problem, and it presents recent developments in this active area of research. The fluxes involved are strongly constrained. This is important if one hopes to make predictions for physical parameters such as the masses of quarks and leptons. The form of the n-form tensor fields that solve the equations of motion is derived, and the important question of which of these preserve supersymmetry and which do not is explored.

In addition to the equations of motion, a second type of constraint arises from flux-quantization conditions. Section 10.5 shows that when branes are the source of the fluxes, the quantization is simple to understand: the flux (suitably normalized) through a cycle surrounding the branes is the number of enclosed branes, which is an integer. For manifolds of nontrivial homology, there can be integrally quantized fluxes through nontrivial cycles even when there are no brane sources, as is explained in Section 10.1. In such cases, the quantization is a consequence of the generalized Dirac quantization condition explained in Chapter 6. In special cases, there can be an offset by some fraction in the flux quantization rule due to effects induced by curvature.

---

[1] Warped geometries have been known for a long time, but their role in the moduli-stabilization problem was only recognized in the 1990s.
[2] It is a matter of convention which flux is called *magnetic* and which flux is called *electric*.

This happens in M-theory, for example, due to higher-order quantum gravity corrections to the $D = 11$ supergravity action, as is explained in Section 10.5.

Flux compactifications

Let us begin by considering compactifications of M-theory on manifolds that are conformally Calabi–Yau four-folds. For these compactifications, the metric differs from a Calabi–Yau metric by a conformal factor. Even though these models are phenomenologically unrealistic, since they lead to three-dimensional Minkowski space-time, in some cases they are related to $\mathcal{N} = 1$ theories in four dimensions. This relatively simple class of models illustrates many of the main features of flux compactifications. More complicated examples, such as type IIB and heterotic flux compactifications, are discussed next. In the latter case nonzero fluxes require that the internal compactification manifolds are non-Kähler but still complex. It is convenient to describe them using a connection with torsion.

The dilaton and the radial modulus

Two examples of moduli are the dilaton, whose value determines the string coupling constant, and the radial modulus, whose value determines the size of the internal manifold. Classical analysis that neglects string loop and instanton corrections is justified when the coupling constant is small enough. Similarly, a supergravity approximation to string theory is justified when the size of the internal manifold is large compared to the string scale. When there is no potential that fixes these two moduli, as is the case in the absence of fluxes, these moduli can be tuned so that these approximations are arbitrarily good. Therefore, even though compactifications without fluxes are unrealistic, at least one can be confident that the formulas have a regime of validity. This is less obvious for flux compactifications with a stabilized dilaton and radial modulus, but it will be shown that the supergravity approximation has a regime of validity for flux compactifications of M-theory on manifolds that are conformally Calabi–Yau four-folds.

More generally, moduli fields are stabilized *dynamically* in flux compactifications. While this is certainly what one wants, it also raises new challenges. How can one be sure that a classical supergravity approximation has any validity at all, once the value of the radial modulus and the dilaton are stabilized? There is generally a trade-off between the number of moduli that are stabilized and the amount of mathematical control that one has. This poses a challenge, since in a realistic model *all* moduli should be stabilized.

Some models are known in which all moduli are fixed, and a supergravity approximation still can be justified. In these models the fluxes take integer values N, which can be arbitrarily large in such a way that the supergravity description is valid in the large N limit.

The string theory landscape

Even though flux compactifications can stabilize the moduli fields appearing in string theory compactifications, there is another troubling issue. Flux compactifications typically give very many possible vacua, since the fluxes can take many different discrete values, and there is no known criterion for choosing among them. These vacua can be regarded as extrema of some potential, which describes the *string theory landscape*. Section 10.6 discusses one approach to addressing this problem, which is to accept the large degeneracy and to characterize certain general features of typical vacua using a statistical approach.

Fluxes and dual gauge theories

Chapter 12 shows that superstring theories in certain backgrounds, which typically involve nonzero fluxes, have a dual gauge-theory description. The simplest examples involve conformally invariant gauge theories. However, there are also models that provide dual supergravity descriptions of *confining* supersymmetric gauge theories. Section 10.2 describes a flux model that is dual to a confining gauge theory in the context of the type IIB theory, the *Klebanov–Strassler* (KS) model. The dual gauge theory aspects of this model are discussed in Chapter 12.

Brane-world scenarios

An alternative to the usual Kaluza–Klein compactification method of hiding extra dimensions, called the *brane-world* scenario, is described in Section 10.2. One of the goals of this approach is to solve the *gauge hierarchy problem*, that is, to explain why gravity is so much weaker than the other forces. The basic idea is that the visible Universe is a 3-brane, on which the standard model fields are confined, embedded in a higher dimension space-time. Extra dimensions have yet to be observed experimentally, of course, but in this set-up it is not out of the question that this could be possible.[3] While

---

[3] The search for extra dimensions is one of the goals of the Large Hadron Collider (LHC) at CERN, which is scheduled to start operating in 2007.

the standard model fields are confined to the 3-brane, gravity propagates in all $4+n$ dimensions. Section 10.2 shows that the hierarchy problem can be solved if the $(4+n)$-dimensional background geometry is not factorizable, that is, if it involves a warp factor, like those of string theory flux compactifications. In fact, flux compactifications of string theory give a warp factor in the geometry, which could provide a solution to the hierarchy problem. This is an alternative to the more usual approach to the hierarchy problem based on supersymmetry broken at the weak scale.

Fluxes and superstring cosmology

The Standard Big Bang model of cosmology (SBB) is the currently accepted theory that explains many features of the Universe such as the existence of the *cosmic microwave background* (CMB). The CMB accounts for most of the radiation in the Universe. This radiation is nearly isotropic and has the form of a black-body spectrum. However, there are small irregularities in this radiation that can only be explained if, before the period described by the SBB, the Universe underwent a period of rapid expansion, called *inflation*. This provides the initial conditions for the SBB theory. Different models of inflation have been proposed, but inflation ultimately needs to be derived from a fundamental theory, such as string theory. This is currently a very active area of research in the context of flux compactifications, and it is described towards the end of this chapter. Cosmology could provide one of the most spectacular ways to verify string theory, since strings of cosmic size, called *cosmic strings*, could potentially be produced.

10.1 Flux compactifications and Calabi–Yau four-folds

In the traditional string theory literature, compactifications to fewer than four noncompact space-time dimensions were not considered to be of much interest, since the real world has four dimensions. However, this situation changed with the discovery of the string dualities described in Chapters 8 and 9. In particular, it was realized that M-theory compactifications on conformally Calabi–Yau four-folds, which are discussed in this section, are closely related to certain F-theory compactifications to four dimensions. Since the three-dimensional theories have $\mathcal{N}=2$ supersymmetry, which means that there are four conserved supercharges, they closely resemble four-dimensional theories with $\mathcal{N}=1$ supersymmetry.

Recall that Exercise 9.4 argued that a supersymmetric solution to a theory with global $\mathcal{N}=1$ supersymmetry is a zero-energy solution of the equa-

10.1 Flux compactifications and Calabi–Yau four-folds

tions of motion. By solving the first-order supersymmetry constraints one obtains solutions to the second-order equations of motion, and thus a consistent string-theory background. One has to be careful when generalizing this to theories with local supersymmetry, since solving the Killing spinor equations does not automatically ensure that a solution to the full equations of motion. This section shows that the supersymmetry constraints for flux compactifications, together with the Bianchi identity, yield a solution to the equations of motion, which can be derived from a potential for the moduli, and that this potential describes the stabilization of these moduli.

M-theory on Calabi–Yau four-folds

The bosonic part of the action for 11-dimensional supergravity, presented in Chapter 8, is

$$2\kappa_{11}^2 S = \int d^{11}x \sqrt{-G} \left(R - \frac{1}{2}|F_4|^2 \right) - \frac{1}{6} \int A_3 \wedge F_4 \wedge F_4. \tag{10.3}$$

The only fermionic field is the gravitino and a supersymmetric configuration is a nontrivial solution to the Killing spinor equation

$$\delta \Psi_M = \nabla_M \varepsilon + \frac{1}{12} \left(\Gamma_M \mathbf{F}^{(4)} - 3 \mathbf{F}^{(4)}_M \right) \varepsilon = 0. \tag{10.4}$$

The notation is the same as in Section 8.1. This equation needs to be solved for some nontrivial spinor ε and leads to constraints on the background metric as well as the four-form field strength. In Chapter 9 a similar analysis of the supersymmetry constraints for the heterotic string was presented. However, there the three-form tensor field was set to zero for simplicity.

In general, it is inconsistent to set the fluxes to zero, unless additional simplifying assumptions (or truncations) are made. This section shows that vanishing fluxes are inconsistent for most M-theory compactifications on eight manifolds due to the effects of quantum corrections to the action Eq. (10.3).

Warped geometry

Let us now construct flux compactifications of M-theory to three-dimensional Minkowski space-time preserving $\mathcal{N} = 2$ supersymmetry.[4] The most general ansatz for the metric that is compatible with maximal symmetry and Poincaré invariance of the three-dimensional space-time is a *warped metric*. This means that the space-time is not a direct product of an external space-time with an internal manifold. Rather, a scalar function depending on the

---

[4] A similar analysis can be performed to obtain models with $\mathcal{N} = 1$ supersymmetry.

coordinates of the internal dimensions $\Delta(y)$ is included. The explicit form for the metric ansatz is

$$ds^2 = \underbrace{\Delta(y)^{-1}\eta_{\mu\nu}dx^\mu dx^\nu}_{\text{3D flat space-time}} + \underbrace{\Delta(y)^{1/2}g_{mn}(y)dy^m dy^n}_{\text{8D internal manifold}}, \quad (10.5)$$

where x^μ are the coordinates of the three-dimensional Minkowski space-time M_3 and y^m are the coordinates of the internal Euclidean eight-manifold M. In the following we consider the case in which the internal manifold is a Calabi–Yau four-fold, which results in $\mathcal{N} = 2$ supersymmetry in three dimensions. The scalar function $\Delta(y)$ is called the *warp factor*. The powers of the warp factor in Eq. (10.5) have been chosen for later convenience.

In general, a warp factor can have a dramatic influence on the properties of the geometry. Consider the example of a torus, which can be described by the flat metric

$$ds^2 = d\theta^2 + d\varphi^2 \quad \text{with} \quad 0 \leq \theta \leq \pi, \quad 0 \leq \varphi \leq 2\pi. \quad (10.6)$$

By including a suitable warp factor, the torus turns into a sphere

$$ds^2 = d\theta^2 + \sin^2\theta d\varphi^2, \quad (10.7)$$

leading to topology change. Moreover, once the warp factor is included, it is no longer clear that the space-time splits into external and internal components. However, this section shows (for flux compactifications of M-theory on Calabi–Yau four-folds) that the effects of the warp factor are subleading in the regime in which the size of the four-fold is large. In this regime, one can use the properties of Calabi–Yau manifolds discussed in Chapter 9.

Decomposition of Dirac matrices

To work out the dimensional reduction of Eq. (10.4), the 11-dimensional Dirac matrices need to be decomposed. The decomposition that is required for the $11 = 3 + 8$ split is

$$\Gamma_\mu = \Delta^{-1/2}(\gamma_\mu \otimes \gamma_9) \quad \text{and} \quad \Gamma_m = \Delta^{1/4}(1 \otimes \gamma_m), \quad (10.8)$$

where γ_μ are the 2×2 Dirac matrices of M_3. Concretely, they can be represented by

$$\gamma_0 = i\sigma_1, \quad \gamma_1 = \sigma_2 \quad \text{and} \quad \gamma_2 = \sigma_3, \quad (10.9)$$

where the σ's are the Pauli matrices

$$\sigma_1 = \begin{pmatrix} 0 & 1 \\ 1 & 0 \end{pmatrix}, \quad \sigma_2 = \begin{pmatrix} 0 & -i \\ i & 0 \end{pmatrix} \quad \text{and} \quad \sigma_3 = \begin{pmatrix} 1 & 0 \\ 0 & -1 \end{pmatrix}. \tag{10.10}$$

Moreover, γ_m are the 16×16 gamma matrices of M and γ_9 is the eight-dimensional chirality operator that satisfies $\gamma_9^2 = 1$ and anticommutes with the other eight γ_m's. It is both possible and convenient to choose a representation in which the γ_m and γ_9 are real symmetric matrices. In a tangent-space basis one can choose the eight 16×16 Dirac matrices on the internal space M to be

$$\begin{array}{cc} \sigma_2 \otimes \sigma_2 \otimes 1 \otimes \sigma_{1,3}, & \sigma_2 \otimes \sigma_{1,3} \otimes \sigma_2 \otimes 1, \\ \sigma_2 \otimes 1 \otimes \sigma_{1,3} \otimes \sigma_2, & \sigma_{1,3} \otimes 1 \otimes 1 \otimes 1 \end{array} . \tag{10.11}$$

Then one can define a ninth symmetric matrix that anticommutes with all of these eight as

$$\gamma_9 = \gamma_1 \ldots \gamma_8 = \sigma_2 \otimes \sigma_2 \otimes \sigma_2 \otimes \sigma_2, \tag{10.12}$$

from which the chirality projection operators

$$P_\pm = (1 \pm \gamma_9)/2 \tag{10.13}$$

are constructed.

Decomposition of the spinor

The 11-dimensional Majorana spinor ε decomposes into a sum of two terms of the form

$$\varepsilon(x, y) = \zeta(x) \otimes \eta(y) + \zeta^*(x) \otimes \eta^*(y), \tag{10.14}$$

where ζ is a two-component anticommuting spinor in three dimensions, while η is a commuting 16-component spinor in eight dimensions. A theory with $\mathcal{N} = 2$ supersymmetry in three dimensions has two linearly independent Majorana–Weyl spinors η_1, η_2 on M, which have been combined into a complex spinor in Eq. (10.14). In general, these two spinors do not need to have the same chirality. However, for Calabi–Yau four-folds the spinor on the internal manifold is complex and Weyl

$$\eta = \eta_1 + i\eta_2 \quad \text{with} \quad (\gamma_9 - 1)\eta = 0. \tag{10.15}$$

This sign choice for the eigenvalue of γ_9, which is just a convention, is called positive chirality. The two real spinors η_1, η_2 correspond to the two singlets in the decomposition of the $\mathbf{8}_c$ representation of $Spin(8)$ to $SU(4)$, the holonomy group of a Calabi–Yau four-fold,

$$\mathbf{8}_c \to \mathbf{6} \oplus \mathbf{1} \oplus \mathbf{1}. \tag{10.16}$$

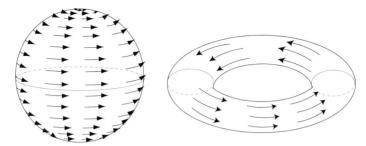

Fig. 10.1. This figure illustrates the Poincaré–Hopf index theorem. A continuous vector field on a sphere must have at least two zeros, which in this case are located at the north and south poles, since the Euler characteristic is 2. On the other hand, a vector field on a torus can be nonvanishing everywhere since $\chi = 0$.

Nonchiral spinors

If η_1 and η_2 have opposite chirality the complex spinor $\eta = \eta_1 + i\eta_2$ is nonchiral. The two spinors of opposite chirality define a vector field on the internal manifold

$$v_a = \eta_1^\dagger \gamma_a \eta_2. \qquad (10.17)$$

Requiring this vector to be nonvanishing leads to an interesting class of solutions. Indeed, the *Poincaré–Hopf index theorem* of algebraic topology states that the number of zeros of a continuous vector field must be at least equal to the absolute value of the Euler characteristic χ of the background geometry. As a result, a nowhere vanishing vector field only exists for manifolds with $\chi = 0$. An example of this theorem is illustrated in Fig. 10.1. Flux backgrounds representing M5-branes filling the three-dimensional space-time and wrapping supersymmetric three-cycles on the internal space are examples of this type of geometries. Moreover, once the spinor is nonchiral, compactifications to AdS$_3$ spaces become possible. Compactifications to AdS space are considered in Chapter 12, so the discussion in this chapter is restricted to spinors of positive chirality. It will turn out that AdS$_3$ is not a solution in this case.

Solving the supersymmetry constraints

The constraints that follow from Eq. (10.4) are influenced by the warp-factor dependence of the metric. As was pointed out in Chapter 8, there is a relation between the covariant derivatives of a spinor with respect to a pair of metrics that differ by a conformal transformation. In particular, in the present case, the internal and external components of the metric are rescaled with a different power of the warp factor and the vielbeins are given

10.1 Flux compactifications and Calabi–Yau four-folds

by $E^\alpha_\mu = \Delta^{-1/2} e^\alpha_\mu$ and $E^\alpha_m = \Delta^{1/4} e^\alpha_m$. This leads to

$$\nabla_\mu \varepsilon \to \nabla_\mu \varepsilon - \tfrac{1}{4} \Delta^{-7/4} \left(\gamma_\mu \otimes \gamma_9 \gamma^m \right) \partial_m \Delta \varepsilon,$$

$$\nabla_m \varepsilon \to \nabla_m \varepsilon + \tfrac{1}{8} \Delta^{-1} \partial_n \Delta \left(1 \otimes \gamma_m{}^n \right) \varepsilon. \tag{10.18}$$

For compactifications to maximally symmetric three-dimensional space-time, Poincaré invariance restricts the possible nonvanishing components of F_4 to

$$F_{mnpq}(y) \quad \text{and} \quad F_{\mu\nu\rho m} = \varepsilon_{\mu\nu\rho} f_m(y), \tag{10.19}$$

where $\varepsilon_{\mu\nu\rho}$ is the completely antisymmetric Levi–Civita tensor of M_3. Once the gamma matrices are decomposed as in Eq (10.8), the nonvanishing flux components take the form

$$\mathbf{F}^{(4)} = \Delta^{-1}(1 \otimes \mathbf{F}) + \Delta^{5/4}(1 \otimes \gamma_9 \mathbf{f}),$$

$$\mathbf{F}^{(4)}_\mu = \Delta^{3/4}(\gamma_\mu \otimes \mathbf{f}), \tag{10.20}$$

$$\mathbf{F}^{(4)}_m = -\Delta^{3/2} f_m(y)(1 \otimes \gamma_9) + \Delta^{-3/4}(1 \otimes \mathbf{F}_m),$$

where $\mathbf{F}$, $\mathbf{F}_m$ and $\mathbf{f}$ are defined like their ten-dimensional counterparts, but the tensor fields are now contracted with eight-dimensional Dirac matrices

$$\mathbf{F} = \frac{1}{24} F_{mnpq} \gamma^{mnpq}, \quad \mathbf{F}_m = \frac{1}{6} F_{mnpq} \gamma^{npq} \quad \text{and} \quad \mathbf{f} = \gamma^m f_m. \tag{10.21}$$

The gravitino supersymmetry transformation Eq. (10.4) has external and internal components depending on the value of the index M.

External component of the gravitino equation

Let us analyze the external component $\delta \Psi_\mu = 0$ first. In three-dimensional Minkowski space-time a covariantly constant spinor, satisfying

$$\nabla_\mu \zeta = 0, \tag{10.22}$$

can be found. As a result, the $\delta \Psi_\mu = 0$ equation becomes

$$\partial \!\!\!/ \Delta^{-3/2} \eta + \mathbf{f} \eta + \frac{1}{2} \Delta^{-9/4} \mathbf{F} \eta = 0, \tag{10.23}$$

which by projecting on the two chiralities using the projection operators $P_\pm$ leads to

$$\mathbf{F} \eta = 0, \tag{10.24}$$

and

$$f_m(y) = -\partial_m \Delta^{-3/2}. \tag{10.25}$$

The last of these equations provides a relation between the external flux component and the warp factor.

Internal component of the gravitino equation

After decomposing the gamma matrices and fluxes using Eqs (10.8) and (10.20), respectively, the internal component of the supersymmetry transformation $\delta\Psi_m = 0$ takes the form

$$\nabla_m \eta + \frac{1}{4}\Delta^{-1}\partial_m\Delta\,\eta - \frac{1}{4}\Delta^{-3/4}\mathbf{F}_m\eta = 0. \tag{10.26}$$

This equation leads to

$$\mathbf{F}_m\xi = 0 \quad \text{and} \quad \nabla_m\xi = 0, \tag{10.27}$$

where

$$\xi = \Delta^{1/4}\eta. \tag{10.28}$$

Since ξ is a nonvanishing covariantly constant complex spinor with definite chirality, the second expression in Eq. (10.27) states that the internal manifold M is conformal to a Calabi–Yau four-fold.

Conditions on the fluxes

The mathematical properties of Calabi–Yau four-folds are similar to those of three-folds, as discussed in Chapter 9. The covariantly constant spinor appearing in Eq. (10.27) can be used to define the almost complex structure of the internal manifold

$$J_a{}^b = -i\xi^\dagger \gamma_a{}^b \xi, \tag{10.29}$$

which has the same properties as for the Calabi–Yau three-fold case, as you are asked to verify in Problems 10.2, 10.3. Recall that the Dirac algebra for a Kähler manifold

$$\{\gamma^a, \gamma^b\} = 0, \quad \{\gamma^{\bar{a}}, \gamma^{\bar{b}}\} = 0, \quad \{\gamma^a, \gamma^{\bar{b}}\} = 2g^{a\bar{b}}, \tag{10.30}$$

can be interpreted as an algebra of raising and lowering operators. This is useful for evaluating the solution of Eq. (10.27). To see this rewrite Eq. (10.29) as

$$J_{a\bar{b}} = ig_{a\bar{b}} = -i\xi^\dagger \gamma_{a\bar{b}}\xi = -i\xi^\dagger(\gamma_a\gamma_{\bar{b}} - g_{a\bar{b}})\xi. \tag{10.31}$$

This implies

$$0 = \xi^\dagger \gamma_a \gamma_{\bar{b}} \xi = (\gamma_{\bar{a}}\xi)^\dagger(\gamma_{\bar{b}}\xi). \tag{10.32}$$

10.1 Flux compactifications and Calabi–Yau four-folds

By setting $\bar{a} = \bar{b}$ the previous equation implies that ξ is a highest-weight state that is annihilated by $\gamma_{\bar{a}}$,

$$\gamma_{\bar{a}}\xi = \gamma^a \xi = 0, \tag{10.33}$$

for all indices on the Calabi–Yau four-fold. Using this result, Exercise 10.3 shows that the first condition in Eq. (10.27) implies the vanishing of the following flux components:

$$F^{4,0} = F^{0,4} = F^{1,3} = F^{3,1} = 0, \tag{10.34}$$

and that the only nonvanishing component is $F \in H^{(2,2)}$, which must satisfy the *primitivity condition*

$$F_{a\bar{b}c\bar{d}}\, g^{c\bar{d}} = 0. \tag{10.35}$$

Since ξ has a definite chirality, F is self-dual on the Calabi–Yau four-fold, as is explained in Exercise 10.2. The self-duality implies that Eq. (10.35) can be written in the following form:[5]

$$F^{2,2} \wedge J = 0. \tag{10.36}$$

As a result of the above analysis, supersymmetry is unbroken if F lies in the primitive $(2,2)$ cohomology, that is,

$$F \in H^{(2,2)}_{\text{primitive}}(M). \tag{10.37}$$

In the following the general definition of primitive forms is given and their relevance in building the complete de Rham cohomology is discussed.

Primitive cohomology

Any harmonic (p,q)-form of a Kähler manifold can be expressed entirely in terms of primitive forms, a representation known as the *Lefschetz decomposition*. This construction closely resembles the Fock-space construction of angular momentum states $|j, m\rangle$ using raising and lowering operators $J_\pm$. Chapter 9 discussed the Hodge decomposition of the de Rham cohomology of a compact Kähler manifold. The Lefschetz decomposition is compatible with the Hodge decomposition, as is shown below.

On a compact Kähler manifold M of complex dimension d (and real dimension $2d$) with Kähler form J, one can define an $SU(2)$ action on harmonic

---

[5] Problem 10.5 asks you to verify that the primitivity condition is modified when the complex spinor on the internal manifold is nonchiral.

forms (and hence the de Rham cohomology) by

$$J_3 : G \to \tfrac{1}{2}(d-n)G,$$

$$J_- : G \to J \wedge G, \qquad (10.38)$$

$$J_+ : G \to \tfrac{1}{2(n-2)!} J^{p_1 p_2} G_{p_1 p_2 \ldots p_n} dx^{p_3} \wedge \cdots \wedge dx^{p_n},$$

where G is a harmonic n-form. Notice that J_- lowers the J_3 eigenvalue by one and as a result acts as a lowering operator while J_+ increases the value of J_3 by one and thus acts as a raising operator. Problem 10.6 asks you to verify that these operators satisfy an $SU(2)$ algebra.

As in the case of the angular momentum algebra, the space of harmonic forms can be classified according to their J_3 and J^2 eigenvalues, with basis states denoted by

$$|j, m\rangle \quad \text{with} \quad m = -j, -j+1, \ldots, j-1, j. \qquad (10.39)$$

Primitive forms are defined as highest-weight states that are annihilated by J_+, that is,

$$J_+ G_{\text{primitive}} = 0, \qquad (10.40)$$

and may be denoted by $|j, j\rangle$. All other states (or harmonic forms) can then be obtained by acting with lowering operators J_- on primitive forms. A primitive n-form also satisfies

$$J_-^{2j+1} G_{\text{primitive}} = 0 \quad \text{where} \quad j = \frac{d-n}{2}. \qquad (10.41)$$

Notice that the primitive forms in the middle-dimensional cohomology (that is, with $n = d$) correspond to $j = 0$. So they are singlets $|0, 0\rangle$ that are annihilated by both the raising and lowering operators

$$J_+ G = 0 \quad \text{or} \quad J_- G = 0. \qquad (10.42)$$

These two formulas correspond to Eqs (10.35) and (10.36), respectively. This discussion makes it clear that primitive forms can be used to construct any harmonic form and hence representatives of every de Rham cohomology class. Schematically, the Lefschetz decomposition is[6]

$$H^n(M) = \bigoplus_k J_-^k H^{n-2k}_{\text{primitive}}(M). \qquad (10.43)$$

---

[6] It would be more precise to write $\text{Harm}^n(M)$ instead of $H^n(M)$.

The Lefschetz decomposition is compatible with the Hodge decomposition, so that we can also write

$$H^{(p,q)}(M) = \bigoplus_k J_-^k H_{\text{primitive}}^{(p-k,q-k)}(M). \qquad (10.44)$$

In this way any harmonic (p,q)-form can be written in terms of primitive forms. If M is a Calabi–Yau four-fold, it follows from Eq. (10.41) that primitive (p,q)-forms satisfy

$$\underbrace{J \wedge \cdots \wedge J}_{5-p-q \text{ times}} \wedge F_{\text{primitive}}^{p,q} = 0. \qquad (10.45)$$

In the case of a Calabi–Yau four-fold, it is a useful fact that the Hodge $\star$ operator has the eigenvalue $(-1)^p$ on the primitive $(p, 4-p)$ cohomology (see Exercise 10.4). This is of relevance in Section 10.3.

Tadpole-cancellation condition

We have learned that unbroken supersymmetry requires that the internal flux components $F_{mnpq}(y)$ are given by a primitive $(2, 2)$-form, Eq. (10.37), and the external flux components $f_m(y)$ are determined in terms of the warp factor by Eq. (10.25). The equation that determines the warp factor follows from the equation of motion of the four-form field strength. Using self-duality, it would make the energy density $|F_4|^2$ proportional to the Laplacian of $\log \Delta$, which gives zero when integrated over the internal manifold. If this were the whole story, one would be forced to conclude that the flux vanishes, so that one is left with ordinary Calabi–Yau compactification of the type discussed in Chapter 9. However, quantum gravity corrections to 11-dimensional supergravity must be taken into account, and then nonzero flux is required for consistency. Let us explain how this works.

The action for 11-dimensional supergravity receives quantum corrections, denoted δS, coming from an eight-form X_8 that is quartic in the Riemann tensor

$$\delta S = -T_{\text{M2}} \int_M A_3 \wedge X_8, \qquad (10.46)$$

where

$$X_8 = \frac{1}{(2\pi)^4} \left[\frac{1}{192} \text{tr} R^4 - \frac{1}{768} (\text{tr} R^2)^2 \right]. \qquad (10.47)$$

This correction term was first derived by considering a one-loop scattering amplitude in type IIA string theory involving four gravitons $G_{\mu\nu}$ and one two-form tensor field $B_{\mu\nu}$. In the type IIA theory the correction takes a similar form as in M-theory, with the three-form A_3 replaced by the NS–NS

470 Flux compactifications

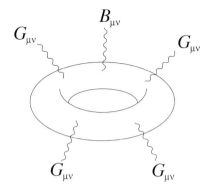

Fig. 10.2. The higher-order interaction in Eq. (10.46) can be determined by calculating a one-loop diagram in type IIA string theory, involving four gravitons and one NS–NS two-form field, whose result can then be lifted to M-theory.

two-form B_2. Since the result does not depend on the dilaton, it can be lifted to M-theory.

The δS term is also required for the cancellation of anomalies on boundaries of the 11-dimensional space-time, such as those that are present in the strongly coupled $E_8 \times E_8$ theory, which is also know as *heterotic M-theory*. This was discussed in Chapter 5. Together with the original $\int A_3 \wedge F_4 \wedge F_4$ term it gives the complete Chern–Simons part of the theory, so it is not just the leading term in some expansion. In fact, it is the only higher-derivative term that can contribute to the problem at hand in the large-volume limit.

The field strength satisfies the Bianchi identity

$$dF = 0. \tag{10.48}$$

Furthermore, the δS term contributes to the 11-dimensional equation of motion of the four-form field strength. Combining Eqs (10.3) and (10.46), the result is

$$d \star F_4 = -\frac{1}{2} F_4 \wedge F_4 - 2\kappa_{11}^2 T_{M2} X_8. \tag{10.49}$$

Using Eq. (10.25) this gives an equation for the warp factor

$$d \star_8 d \log \Delta = \frac{1}{3} F \wedge F + \frac{4}{3} \kappa_{11}^2 T_{M2} X_8. \tag{10.50}$$

Integrating this expression over the internal manifold leads to the *tadpole-cancellation condition*, as follows. The integral of the left-hand side vanishes, since it is exact, and (for the time being) it is assumed that no explicit delta function singularities are present. In other words, it is assumed that no

10.1 Flux compactifications and Calabi–Yau four-folds

space-filling M2-branes are present. To obtain the result of the X_8 integration, it is convenient to express the anomaly characteristic class X_8 in terms of the first and second Pontryagin forms of the internal manifold

$$P_1 = \frac{1}{(2\pi)^2}\left(-\frac{1}{2}\mathrm{tr} R^2\right) \quad \text{and} \quad P_2 = \frac{1}{(2\pi)^4}\left[-\frac{1}{4}\mathrm{tr} R^4 + \frac{1}{8}(\mathrm{tr} R^2)^2\right]. \tag{10.51}$$

This gives

$$X_8 = \frac{1}{192}(P_1^2 - 4P_2). \tag{10.52}$$

For complex manifolds the Pontryagin classes are related to the Chern classes by

$$P_1 = c_1^2 - 2c_2 \quad \text{and} \quad P_2 = c_2^2 - 2c_1 c_3 + 2c_4. \tag{10.53}$$

Thus

$$X_8 = \frac{1}{192}(c_1^4 - 4c_1^2 c_2 + 8c_1 c_3 - 8c_4). \tag{10.54}$$

Calabi–Yau manifolds have vanishing first Chern class, so the only nontrivial contribution comes from the fourth Chern class. This in turn is related to the Euler characteristic χ, so

$$\int_M X_8 = -\frac{1}{24}\int_M c_4 = -\frac{\chi}{24}. \tag{10.55}$$

Thus, Eq. (10.50) leads to the tadpole-cancellation condition

$$\frac{1}{4\kappa_{11}^2 T_{\mathrm{M2}}}\int_M F \wedge F = \frac{\chi}{24}. \tag{10.56}$$

Fluxes without sources

Using the last equation, it is possible to estimate the order of magnitude of the internal flux components. Expressing κ_{11}^2 and the M2-brane tension in terms of the 11-dimensional Planck length ℓ_p yields

$$4\kappa_{11}^2 T_{\mathrm{M2}} = 2(2\pi\ell_\mathrm{p})^6. \tag{10.57}$$

As a result, the order of magnitude of the fluxes is

$$F_{mnpq} \simeq O\left(\frac{\ell_\mathrm{p}^3}{\sqrt{v}}\right), \tag{10.58}$$

where v is the volume of the Calabi–Yau four-fold. Comparing this result with Eq. (10.50) shows that the warp factor satisfies $\log \Delta \sim \ell_\mathrm{p}^6/v^{3/4}$, or if

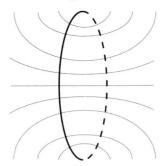

Fig. 10.3. According to Maxwell's theory, an electric current in a solenoid generates a magnetic field even though no monopoles, electric or magnetic, are present. The integral of the field strength and its dual over any closed surface in space vanishes. Similarly, nontrivial flux solutions exist in M-theory, even when no δ-function sources, corresponding to M2-branes or M5-branes, are present.

this is small

$$\Delta \simeq 1 + O\left(\frac{\ell_p^6}{v^{3/4}}\right). \tag{10.59}$$

In the approximation in which the size of the Calabi–Yau is very large, that is, when $\ell_p/v^{1/8} \to 0$, the background metric becomes unwarped.

This analysis shows that nontrivial flux solutions are possible even in the absence of explicit delta function sources for M2-branes or M5-branes, which would appear in the equation of motion and Bianchi identity for F_4. A rather similar situation appears in ordinary Maxwell theory, where a magnetic flux is generated by an electric current running through a loop even though there are no magnetic monopoles, as illustrated in Fig. 10.3.

According to Eq. (10.50), nonsingular solutions for the warp factor and the background geometry are possible even in the absence of explicit brane sources. In fact, a nonsingular background is necessary to justify rigorously the validity of the supergravity approximation everywhere in space-time. Nevertheless, the supergravity approximation is valid for singular solutions provided that the delta-function singularities are treated carefully.

Inclusion of M2-brane sources

If M2-branes filling the external Minkowski space are also present, an additional integer N (the number of M2-branes) appears on the left-hand side of Eq. (10.56), resulting in

$$N + \frac{1}{4\kappa_{11}^2 T_{M2}} \int_M F \wedge F = \frac{\chi}{24}. \tag{10.60}$$

Since F is self-dual, both terms on the left-hand side of this equation are positive. So if $\chi > 0$, there are supersymmetry preserving solutions with nonvanishing flux or M2-branes. The number of these solutions is finite, because of quantization constraints on the fluxes that are discussed in Section 10.6. For $\chi < 0$ there are no supersymmetric solutions.

Interactions of moduli fields

As discussed in Chapter 9, a Calabi–Yau four-fold has three independent Hodge numbers ($h^{1,1}$, $h^{2,1}$ and $h^{3,1}$), each of which gives the multiplicities of scalar fields in the lower-dimensional theory. The purpose here is to show that many of these fields can be stabilized by fluxes.

The $D = 3$ field content

The variations of the complex structure of a Calabi–Yau four-fold are parametrized by $h^{3,1}$ *complex* parameters T^I, the complex-structure moduli fields, which belong to chiral supermultiplets. Deformations of the Kähler structure give rise to $h^{1,1}$ *real* moduli K^A. Thus, the Kähler form is

$$J = \sum_{A=1}^{h^{1,1}} K^A e_A, \qquad (10.61)$$

where e_A is a basis of harmonic (1, 1)-forms. Together with the $h^{1,1}$ vectors arising from the three-form A_3 these give $h^{1,1}$ three-dimensional vector supermultiplets. Moreover, $h^{2,1}$ additional *complex* moduli N^I, belonging to chiral supermultiplets, arise from the three-form A_3. For simplicity of the presentation, the scalars N^I are ignored in the discussion that follows. The conditions for unbroken $\mathcal{N} = 2$ supersymmetry in three dimensions, described above, can be regarded as conditions that determine some of the scalar fields in terms of the fluxes. Let us therefore derive the three-dimensional interactions that account for these conditions. A more direct derivation, based on a Kaluza–Klein compactification, is given in Section 10.3.

In the absence of flux it is possible to make duality transformations that replace the vector multiplets by chiral multiplets. In particular, the vectors are replaced by scalars. Once this is done, the Kähler moduli are complexified. When fluxes are present there are nontrivial Chern–Simons terms. Nevertheless the duality transformation is still possible, but it becomes more complicated. Thus, we prefer to work with the real Kähler moduli.

Superpotential for complex-structure moduli

The complex-structure moduli T^I appear in chiral multiplets, and the interactions responsible for stabilizing them are encoded in the superpotential

$$W^{3,1}(T) = \frac{1}{2\pi} \int_M \Omega \wedge F, \qquad (10.62)$$

where Ω is the holomorphic four-form of the Calabi–Yau four-fold, and we have set $\kappa_{11} = 1$. There are several different methods to derive Eq. (10.62). The simplest method, which is the one used here, is to verify that this superpotential leads to the supersymmetry constraints Eq. (10.34). An alternative derivation is presented in Section 10.3, where it is shown that Eq. (10.62) arises from Kaluza–Klein compactification of M-theory on a manifold that is conformally Calabi–Yau four-fold.

In space-times with a vanishing cosmological constant, the conditions for unbroken supersymmetry are the vanishing of the superpotential and the vanishing of the Kähler covariant derivative of the superpotential, that is,

$$W^{3,1} = \mathcal{D}_I W^{3,1} = 0 \quad \text{with} \quad I = 1, \ldots, h^{3,1}, \qquad (10.63)$$

where $\mathcal{D}_I W^{3,1} = \partial_I W^{3,1} - W^{3,1} \partial_I \mathcal{K}^{3,1}$, and $\mathcal{K}^{3,1}$ is the Kähler potential on the complex-structure moduli space introduced in Section 9.6, namely

$$\mathcal{K}^{3,1} = -\log\left(\int_M \Omega \wedge \overline{\Omega}\right). \qquad (10.64)$$

The Kähler potential is now formulated in terms of the holomorphic four-form instead of the three-form used in Chapter 9. The condition $W^{3,1} = 0$ implies

$$F^{4,0} = F^{0,4} = 0. \qquad (10.65)$$

As in the three-fold case of Section 9.6, $\partial_I \Omega$ generates the $(3,1)$ cohomology so that the second condition in Eq. (10.63) imposes the constraint

$$F^{1,3} = F^{3,1} = 0. \qquad (10.66)$$

The form of the superpotential in Eq. (10.62) holds to all orders in perturbation theory, because of the standard *nonrenormalization theorem* for the superpotentials. This theorem, which is most familiar for $\mathcal{N} = 1$ theories in $D = 4$, also holds for $\mathcal{N} = 2$ theories in $D = 3$.[7] Supersymmetry

---

[7] The basic argument is that since the superpotential is a holomorphic function, the size of the internal manifold could only appear in the superpotential paired up with a corresponding axion. However, the superpotential cannot depend on this axion, as otherwise the axion shift symmetry would be violated. Correspondingly, the superpotential does not depend on the size of the internal manifold, and its form is not corrected in perturbation theory. Nonperturbative corrections are nevertheless allowed, as they violate the axion shift symmetry. For more details see GSW, Vol. II.

implies that the superpotential Eq. (10.62) generates a scalar potential for the complex-structure moduli fields, so that these fields are stabilized. This potential is discussed in Section 10.3.

Interactions of the Kähler moduli

The primitivity condition,

$$F^{2,2} \wedge J = 0, \tag{10.67}$$

is the equation that stabilizes the Kähler moduli. This condition can be derived from the *real* potential

$$W^{1,1}(K) = \int_M J \wedge J \wedge F, \tag{10.68}$$

where J is the Kähler form. This interaction is sometimes called a superpotential in the literature, but it is not a holomorphic function, so this name is somewhat misleading. Supersymmetry imposes the constraint

$$W^{1,1} = \partial_A W^{1,1} = 0 \quad \text{with} \quad A = 1, \ldots, h^{1,1}, \tag{10.69}$$

which leads to the primitivity condition. Section 10.3 shows that $W^{1,1}$ appears in the scalar potential for the moduli fields of M-theory compactified on a Calabi–Yau four-fold.

F-theory on Calabi–Yau four-folds

The M-theory compactifications on manifolds that are conformally Calabi–Yau four-folds are dual to certain F-theory compactifications on Calabi–Yau four-folds, which lead to four-dimensional space-times with $\mathcal{N} = 1$ supersymmetry. Thus, this dual formulation is more attractive from the phenomenological point of view. The F-theory backgrounds one is interested in are nonperturbative type IIB backgrounds in which the Calabi–Yau four-fold is elliptically fibered, as was discussed in Chapter 9.

To be concrete, the Calabi–Yau four-fold one is interested in can be described locally as a product of a Calabi–Yau three-fold times a torus.[8] The conditions on the four-form fluxes derived above correspond to conditions on three-form fluxes in the type IIB theory. Concretely, the relation between the F-theory four-form and type IIB three-form is

$$F_4 = \frac{1}{\tau - \bar{\tau}} \left(G_3^\star \wedge dz - G_3 \wedge d\bar{z} \right), \tag{10.70}$$

[8] Locally, this is always possible, except at singular fibers.

where
$$dz = d\sigma_1 + \tau d\sigma_2. \tag{10.71}$$

$\sigma_{1,2}$ are the coordinates parametrizing the torus, and τ is its complex structure, which in the type IIB theory is identified with the axion–dilaton field. Moreover, $G_3 = F_3 - \tau H_3$ is a combination of the type IIB R–R and NS–NS three-forms. In components, this implies that

$$F^{1,3} = \frac{1}{\tau - \bar\tau} \left[(G_3^\star)^{0,3} \wedge dz - (G_3)^{1,2} \wedge d\bar z \right], \tag{10.72}$$

$$F^{0,4} = -\frac{1}{\tau - \bar\tau} (G_3)^{0,3} \wedge d\bar z. \tag{10.73}$$

Imposing the M-theory supersymmetry constraints $F^{0,4} = F^{1,3} = 0$ leads to the supersymmetry constraints for the type IIB three-form

$$G_3 \in H^{(2,1)}, \tag{10.74}$$

while the remaining components of G_3 vanish. The next section shows that any harmonic (2, 1)-form on a Calabi–Yau three-fold with $h^{1,0} = 0$ is primitive. Therefore, primitivity is automatic if the background is a Calabi–Yau three-fold with nonvanishing Euler characteristic. Otherwise, it is an additional constraint that has to be imposed.

Many examples of M-theory and F-theory compactifications on Calabi–Yau four-fold have been constructed in the literature. A simple example is described by M-theory on $K3 \times K3$, which leads to a theory with $\mathcal{N} = 4$ supersymmetry in three dimensions. Other examples include orbifolds of $T^2 \times T^2 \times T^2 \times T^2$.

EXERCISES

EXERCISE 10.1
Explain the powers of Δ in Eq. (10.20).

SOLUTION

The powers of Δ in Eq. (10.20) are a straightforward consequence of the powers of Δ appearing in the gamma matrices in Eq. (10.8). □

10.1 Flux compactifications and Calabi–Yau four-folds

EXERCISE 10.2
Show that if the Killing spinor ξ has positive chirality, that is, if $\gamma_9 \xi = +\xi$, F is self-dual on the Calabi–Yau four-fold, as stated in the text. What happens if we reverse the chirality of ξ?

SOLUTION
Using the gamma-matrix identities listed in the appendix of this chapter it is possible to show that

$$\mathbf{F}_m \mathbf{F}^m \xi = -2\mathbf{F}^2 \xi - \frac{1}{12} F_{mnpq} \left(F^{mnpq} \mp \star F^{mnpq} \right) \xi,$$

where $\gamma_9 \xi = \pm \xi$. Since $\mathbf{F}_m \xi = \mathbf{F}\xi = 0$, it follows that

$$(F \mp \star F)^2 = 0.$$

This quantity is positive and therefore

$$F = \pm \star F \qquad \text{for} \qquad \gamma_9 \xi = \pm \xi.$$

Thus, positive-chirality spinors lead to a self-dual F. If the chirality is reversed, self-duality is replaced by anti-self-duality. $\square$

EXERCISE 10.3
Show that a harmonic four-form on a Calabi–Yau four-fold satisfying $\mathbf{F}_m \xi = 0$ belongs to $H^{(2,2)}_{\text{primitive}}$.

SOLUTION
In complex coordinates the condition $\mathbf{F}_m \xi = 0$ implies

$$F_{m\bar{a}\bar{b}\bar{c}} \gamma^{\bar{a}\bar{b}\bar{c}} \xi + 3 F_{m\bar{a}bc} \gamma^{\bar{a}bc} \xi = 0,$$

where m can be a holomorphic or antiholomorphic index. Each of these terms has to vanish separately:

- Using Eq. (10.33), the condition $F_{m\bar{a}\bar{b}\bar{c}} \gamma^{\bar{a}\bar{b}\bar{c}} \xi = 0$ implies

$$F_{m\bar{a}\bar{b}\bar{c}} \left\{ \gamma_{\bar{d}}, \gamma^{\bar{a}\bar{b}\bar{c}} \right\} \xi = 6 F_{m\bar{d}\bar{b}\bar{c}} \gamma^{\bar{b}\bar{c}} \xi = 0.$$

This in turn implies that

$$F_{m\bar{d}\bar{b}\bar{c}} \left[\gamma_{\bar{e}}, \gamma^{\bar{b}\bar{c}} \right] \xi = 4 F_{m\bar{d}\bar{e}\bar{c}} \gamma^{\bar{c}} \xi = 0,$$

which yields

$$F_{m\bar{d}\bar{e}\bar{c}} \left\{ \gamma_{\bar{f}}, \gamma^{\bar{c}} \right\} \xi = 2 F_{m\bar{d}\bar{e}\bar{f}} \xi = 0.$$

Since m can be holomorphic or antiholomorphic and F is real, this results in
$$F^{4,0} = F^{3,1} = F^{1,3} = F^{0,4} = 0.$$

- Applying the same reasoning as above, the condition $F_{m\bar{a}\bar{b}\bar{c}}\gamma^{\bar{a}\bar{b}c}\xi = 0$ implies that
$$F_{a\bar{b}c\bar{d}}g^{c\bar{d}} = 0.$$

Using the self-duality of F shown in Exercise 10.2 and the relation between J and the metric, this equation can be re-expressed as
$$F \wedge J = 0.$$

As a result, $F \in H^{(2,2)}_{\text{primitive}}$. □

Exercise 10.4
Show that a harmonic $(3, 1)$-form on a Calabi–Yau four-fold is anti-self-dual.

Solution

A harmonic $(3, 1)$-form
$$F^{3,1} = \frac{1}{6} F_{abc\bar{d}} dz^a \wedge dz^b \wedge dz^c \wedge dz^{\bar{d}}$$

satisfies
$$F_{abc\bar{d}} J^{c\bar{d}} = 0.$$

If this did not vanish, it would give a harmonic $(2, 0)$-form, but this does not exist on a Calabi–Yau four-fold. Using this equation and the explicit representation of the ε symbol,
$$\varepsilon_{abcd\bar{p}\bar{q}\bar{r}\bar{s}} = (g_{a\bar{p}} g_{b\bar{q}} g_{c\bar{r}} g_{d\bar{s}} \pm \text{permutations}),$$

it is easy to verify that $\star F^{3,1} = -F^{3,1}$. Note that this argument can be easily generalized to show that a primitive $(p, 4 - p)$-form satisfies
$$\star F^{(p, 4-p)} = (-1)^p F^{(p, 4-p)}.$$

□

Exercise 10.5
Show that the supersymmetry constraints in Eqs (10.63) and (10.69) lead to the flux constraints in Eqs (10.65)–(10.67).

Solution

In analogy to the three-fold case discussed in Chapter 9, the following formulas hold for four-folds:

$$\partial_I \Omega = K_I \Omega + \chi_I, \quad I = 1, ..., h^{3,1}$$

and

$$J = K^A e_A, \quad A = 1, ..., h^{1,1},$$

where χ_I and e_A describe bases of harmonic $(3,1)$-forms and $(1,1)$-forms, respectively. Since Ω is a $(4,0)$-form one obtains from Eq. (10.63)

$$\int_M \Omega \wedge F^{0,4} = 0 \quad \text{and} \quad \int_M \chi_I \wedge F^{1,3} = 0.$$

Since $h^{0,4} = 1$, the first constraint leads to $F^{0,4} = 0$. Since χ_I describes a basis of harmonic $(3,1)$-forms, $\star F^{3,1} = \sum_{I=1}^{h^{3,1}} A^I \chi_I$, which leads to

$$\int_M \star F^{3,1} \wedge F^{1,3} = \int_M \star (F^{1,3})^* \wedge F^{1,3} = \int_M |F^{1,3}|^2 \sqrt{g}\, d^8 x = 0,$$

as F is real. This leads to the flux constraint

$$F^{1,3} = F^{3,1} = F^{0,4} = F^{4,0} = 0.$$

Using $\partial_A W^{1,1} = 0$ and Eq. (10.68), one gets

$$\int e_A \wedge J \wedge F^{2,2} = 0.$$

Since $\star(J \wedge F^{2,2})$ is a harmonic $(1,1)$-form, we have

$$\star(J \wedge F^{2,2}) = \sum_{A=1}^{h^{1,1}} U^A e_A.$$

So the above constraint results in

$$\int_M \star(J \wedge F^{2,2}) \wedge (J \wedge F^{2,2}) = \int_M |J \wedge F^{2,2}|^2 \sqrt{g}\, d^8 x = 0,$$

which leads to the primitivity condition Eq. (10.67). Notice that the condition $W^{1,1} = 0$ is then satisfied, too. □

10.2 Flux compactifications of the type IIB theory

No-go theorems for warped compactifications of perturbative string theory date back as far as the 1980s. The arguments used then, based on low-energy supergravity approximations to string theory, were claimed to rule out warped compactifications to a Minkowski or a de Sitter space-time. If the internal spaces are compact and nonsingular, and no brane sources are included, the warp factor and fluxes are necessarily trivial in the leading supergravity approximation. These theorems were revisited in the 1990s when the contributions of branes and higher-order corrections to low-energy supergravity actions were understood better. These ingredients made it possible to evade the no-go theorems and to construct warped compactifications of the type IIB theory, which we will describe in detail below.

The no-go theorem

The no-go theorem states that if the type IIB theory is compactified on internal spaces that are compact and nonsingular, and no brane sources are included, the warp factor and fluxes are necessarily trivial in the leading supergravity approximation. This subsection shows how this result is derived and then it shows how sources invalidate the no-go theorem. A similar no-go theorem shows that compactifications to $D = 4$ de Sitter space-time do not solve the equations of motion (see Problem 10.8).

Type IIB action in the Einstein frame

For illustrative purposes, as well as concreteness, let us consider warped compactifications of the type IIB theory to four-dimensional Minkowski space-time M_4 on a compact manifold M. The ten-dimensional low-energy effective action for the type IIB theory was presented in Chapter 8. In the Einstein frame it takes the form[9]

$$S = \frac{1}{2\kappa^2} \int d^{10}x \sqrt{-G} \left[R - \frac{|\partial \tau|^2}{2(\operatorname{Im}\tau)^2} - \frac{|G_3|^2}{2\operatorname{Im}\tau} - \frac{|\tilde{F}_5|^2}{4} \right]$$

$$+ \frac{1}{8i\kappa^2} \int \frac{C_4 \wedge G_3 \wedge G_3^\star}{\operatorname{Im}\tau}, \tag{10.75}$$

where

$$G_3 = F_3 - \tau H_3, \tag{10.76}$$

[9] Recall that the Einstein-frame and string-frame metrics are related by $g^{\mathrm{E}}_{MN} = e^{-\Phi/2} g^{\mathrm{S}}_{MN}$.

and $F_3 = dC_2$, $H_3 = dB_2$. The R–R scalar C_0, which is sometimes called an axion, and the dilaton Φ are combined in the complex axion–dilaton field

$$\tau = C_0 + ie^{-\Phi}. \tag{10.77}$$

The only change in notation from that described in Section 8.1 is the use of M, N (rather than μ, ν) for ten-dimensional vector indices. As explained in that section,

$$\widetilde{F}_5 = \star_{10} \widetilde{F}_5 \tag{10.78}$$

has to be imposed as a constraint. Here $\star_{10}$ is the Hodge-star operator in ten dimensions. $|G_3|^2$ is defined by

$$|G_3|^2 = \frac{1}{3!} G^{M_1 N_1} G^{M_2 N_2} G^{M_3 N_3} G_{M_1 M_2 M_3} G^{\star}_{N_1 N_2 N_3}. \tag{10.79}$$

The equations of motion and their solution

To compactify the theory to four dimensions, let us consider a warped-metric ansatz of the form

$$ds_{10}^2 = \sum_{M,N=0}^{9} G_{MN} dx^M dx^N = e^{2A(y)} \underbrace{\eta_{\mu\nu} dx^\mu dx^\nu}_{4D} + e^{-2A(y)} \underbrace{g_{mn}(y) dy^m dy^n}_{6D}, \tag{10.80}$$

where x^μ denote the coordinates of four-dimensional Minkowski space-time, and y^m are local coordinates on M. Poincaré invariance implies that the warp factor $A(y)$ is allowed to depend on the coordinates of the internal manifold only.

Poincaré invariance and the Bianchi identities restrict the allowed components of the flux. The three-form flux G_3 is allowed to have components along M only, while the self-dual five-form flux $\widetilde{F}_5$ should take the form

$$\widetilde{F}_5 = (1 + \star_{10}) d\alpha \wedge dx^0 \wedge dx^1 \wedge dx^2 \wedge dx^3, \tag{10.81}$$

where $\alpha(y)$ is a function of the internal coordinates, which will turn out to be related to the warp factor $A(y)$.

The no-go theorem is derived by using the equations of motion following from the action Eq. (10.75). The ten-dimensional Einstein equation can be written in the form

$$R_{MN} = \kappa^2 \left(T_{MN} - \frac{1}{8} G_{MN} T \right), \tag{10.82}$$

where

$$T_{MN} = -\frac{2}{\sqrt{-G}} \frac{\delta S}{\delta G^{MN}} \tag{10.83}$$

is the energy–momentum tensor, and T is its trace. This equation has an external piece ($\mu\nu$) and an internal piece (mn), but the mixed piece vanishes trivially. The external piece takes the form[10]

$$R_{MN} = -\frac{1}{4} G_{MN} \left(\frac{1}{2 \operatorname{Im} \tau} |G_3|^2 + e^{-8A} |\partial \alpha|^2 \right) \qquad M, N = 0, 1, 2, 3. \tag{10.84}$$

Transforming to the metric $\eta_{\mu\nu}$ gives an equation determining the warp factor in terms of the fluxes

$$\Delta A = \frac{e^{4A}}{8 \operatorname{Im} \tau} |G_3|^2 + \frac{1}{4} e^{-8A} |\partial \alpha|^2, \tag{10.85}$$

or, equivalently

$$\Delta e^{4A} = \frac{e^{8A}}{2 \operatorname{Im} \tau} |G_3|^2 + e^{-4A} \left(|\partial \alpha|^2 + |\partial e^{4A}|^2 \right). \tag{10.86}$$

The no-go theorem is a simple consequence of this equation. If both sides are integrated over the internal manifold M, the left-hand side vanishes, because it is a total derivative. The right-hand side is a sum of positive-definite terms, which only vanishes if the individual terms vanish. As a result, one is left with constant A, α and vanishing G_3. The assumption of maximal symmetry would, in principle, allow an external space-time with a cosmological constant Λ, which for $\Lambda < 0$ results in AdS space-times while for $\Lambda > 0$ gives dS space-times. It turns out that the above no-go theorem can be generalized to include this cosmological constant. As you are asked to show in Problem 10.8, Λ appears with a positive coefficient on the right-hand side of Eq. (10.86). Using the same reasoning as above, one obtains another no-go theorem which excludes dS solutions in the absence of sources and/or singularities in the background geometry.

Flux-induced superpotentials

It turns out that brane sources can and do invalidate the no-go theorem. There is an energy–momentum tensor associated with these sources, which contributes to the right-hand side of Eq. (10.86) in the form

$$2\kappa^2 e^{2A} \mathcal{J}_{\text{loc}}, \tag{10.87}$$

where

$$\mathcal{J}_{\text{loc}} = \frac{1}{4} \left(\sum_{M=5}^{9} T_M{}^M - \sum_{M=0}^{3} T_M{}^M \right)_{\text{loc}}, \tag{10.88}$$

---

[10] Indices M, N are used (rather than μ, ν) to emphasize that this curvature is constructed using the metric G_{MN}.

and T^{loc} denotes the energy–momentum tensor associated with the local sources given by

$$T^{\text{loc}}_{MN} = -\frac{2}{\sqrt{-G}} \frac{\delta S_{\text{loc}}}{\delta G^{MN}}. \tag{10.89}$$

Here S_{loc} is the action describing the sources. For a Dp-brane wrapping a $(p-3)$-cycle Σ in M the relevant interactions are

$$S_{\text{loc}} = -\int_{\mathbb{R}^4 \times \Sigma} d^{p+1}\xi T_p \sqrt{-g} + \mu_p \int_{\mathbb{R}^4 \times \Sigma} C_{p+1}. \tag{10.90}$$

This is the action to leading order in α' and for the case of vanishing fluxes on the brane. This action was described in detail in Section 6.5. In order to describe D7-branes wrapped on four-cycles it is necessary to include the first α' correction given by the Chern–Simons term on the D7-brane

$$-\mu_3 \int_{\mathbb{R}^4 \times \Sigma} C_4 \wedge \frac{p_1(R)}{48}. \tag{10.91}$$

It turns out that Eq. (10.87) can contribute negative terms on the right-hand side of Eq. (10.86).

These sources also contribute to the Bianchi identity[11] for $\widetilde{F}_5$

$$d\widetilde{F}_5 = H_3 \wedge F_3 + 2\kappa^2 T_3 \rho_3. \tag{10.92}$$

Here ρ_3 is the D3 charge density from the localized sources and, as usual, it contains a delta function factor localized along the source.

Tadpole-cancellation condition

Integrating Eq. (10.92) over the internal manifold M leads to the type IIB tadpole-cancellation condition

$$\frac{1}{2\kappa^2 T_3} \int_M H_3 \wedge F_3 + Q_3 = 0, \tag{10.93}$$

where Q_3 is the total charge associated with ρ_3. As a result, nonvanishing Q_3 charges induce three-form expectation values. It is shown below that G_3 is imaginary self-dual. Therefore, three-form fluxes are only induced if Q_3 is negative. Problem 10.12 asks you to check that the D7-branes generate a negative contribution to the right-hand side of Eq. (10.86) by solving the equations of motion in the presence of branes.

A useful way of describing the type IIB solution is by lifting it to F-theory compactified on an elliptically fibered Calabi–Yau four-fold X. As explained in Section 9.3, the base of the fibration encodes the type IIB geometry while

---

11 Because of self-duality, this is the same as the equation of motion.

the fiber describes the behavior of the type IIB axion–dilaton τ. In this description, the tadpole-cancellation condition takes a form similar to that found for M-theory on a four-fold

$$\frac{\chi(X)}{24} = N_{D3} + \frac{1}{2\kappa^2 T_3} \int_M H_3 \wedge F_3, \tag{10.94}$$

where $\chi(X)$ is the Euler characteristic of the four-fold, and N_{D3} is the D3-brane charge present in the compactification.[12] The left-hand side of this equation can be interpreted as the negative of the D3-brane charge induced by curvature of the D7-branes. Thus, the equation is the condition for the total D3-brane charge from all sources to cancel.

Conditions on the fluxes

What conditions does the background satisfy? To answer this question there are several ways to proceed. One way is to solve the equations of motion previously described but now taking brane sources into account. Schematically, this is done by inserting the $\widetilde{F}_5$ flux of Eq. (10.81) into the Bianchi identity Eq. (10.92) and subtracting the result from the contracted Einstein equation Eq. (10.86), taking the energy–momentum tensor contribution from the brane sources into account. The resulting constraint is

$$\Delta\left(e^{4A} - \alpha\right) = \frac{1}{6\,\mathrm{Im}\tau} e^{8A} \mid iG_3 - \star G_3 \mid^2 + e^{-4A} \mid \partial(e^{4A} - \alpha) \mid^2 \tag{10.95}$$
$$+ 2\kappa^2 e^{2A} \left(\mathcal{J}_{\mathrm{loc}} - T_3 \rho_3^{\mathrm{loc}}\right).$$

Most localized sources satisfy the BPS-like bound

$$\mathcal{J}_{\mathrm{loc}} \geq T_3 \rho_3^{\mathrm{loc}}. \tag{10.96}$$

As a result, for the kinds of sources that are considered here, the solutions to the equations are characterized by the following conditions:

- The three-form field strength G_3 is imaginary self-dual,

$$\star G_3 = iG_3, \tag{10.97}$$

where the $\star$ denotes the Hodge dual in six dimensions. A solution to the imaginary self-dual condition is a harmonic form of type $(2,1) + (0,3)$. It is shown below that only the primitive part of the $(2,1)$ component is allowed in supersymmetric solutions.
- There is a relation between the warp factor and the four-form potential

$$e^{4A} = \alpha. \tag{10.98}$$

[12] This includes D3-branes and instantons on D7-branes.

10.2 Flux compactifications of the type IIB theory

- The sources saturate the BPS bound, that is,

$$\mathcal{J}_{\mathrm{loc}} = T_3 \rho_3^{\mathrm{loc}}. \tag{10.99}$$

This equation is satisfied by D3-branes, for example. Indeed, using the relevant terms in the world-volume action for the D3-brane in Eq. (10.90) shows

$$T_0{}^0 = T_1{}^1 = T_2{}^2 = T_3{}^3 = -T_3\rho_3 \quad \text{and} \quad T_m{}^m = 0. \tag{10.100}$$

This implies that the BPS inequality is not only satisfied but also saturated. On the other hand, anti-D3-branes satisfy the inequality but do not saturate it, since the left-hand side of Eq. (10.99) is still positive but the right-hand side has the opposite sign. A different way to saturate the bound is to use D7-branes wrapped on four-cycles and O3-planes. D5-branes wrapped on collapsing two-cycles satisfy, but do not saturate, the BPS bound.

The superpotential

The constraint Eq. (10.97) can be derived from a superpotential for the complex-structure moduli fields

$$W = \int_M \Omega \wedge G_3, \tag{10.101}$$

where Ω denotes the holomorphic three-form of the Calabi–Yau three-fold.

Let us derive the conditions for unbroken supersymmetry using the superpotential Eq. (10.101). For concreteness, consider the case of a Calabi–Yau manifold with a single Kähler modulus, which characterizes the size of the Calabi–Yau. Before turning on fluxes, there are massless fields describing the complex-structure moduli z^α ($\alpha = 1, \ldots, h^{2,1}$), the axion–dilaton τ and the superfield ρ containing the Kähler modulus.

As is explained in Exercise 10.6, the Kähler potential can be computed from the dimensional reduction of the ten-dimensional type IIB action by taking the Calabi–Yau manifold to be large. The result for the radial modulus ρ is

$$\mathcal{K}(\rho) = -3\, \log[-i(\rho - \bar\rho)]. \tag{10.102}$$

This should be added to the results for the axion–dilaton and complex-structure moduli, which are

$$\mathcal{K}(\tau) = -\log[-i(\tau - \bar\tau)] \quad \text{and} \quad \mathcal{K}(z^\alpha) = -\log\left(i \int_M \Omega \wedge \bar\Omega\right). \tag{10.103}$$

The total Kähler potential is given by

$$\mathcal{K} = \mathcal{K}(\rho) + \mathcal{K}(\tau) + \mathcal{K}(z^\alpha). \qquad (10.104)$$

Conditions for unbroken supersymmetry

Supersymmetry is unbroken if

$$\mathcal{D}_a W = \partial_a W + \partial_a \mathcal{K} W = 0, \qquad (10.105)$$

where $a = \rho, \tau, \alpha$ labels all the moduli superfields. In order to evaluate this condition, first note that the superpotential in Eq. (10.101) is independent of the radial modulus. As a result,

$$\mathcal{D}_\rho W = \partial_\rho \mathcal{K} W = -\left(\frac{3}{\rho - \bar\rho}\right) W = 0, \qquad (10.106)$$

which implies that supersymmetric solutions obey

$$W = 0. \qquad (10.107)$$

So the $(0,3)$ component of G_3 has to vanish. The equation

$$\mathcal{D}_\tau W = \frac{1}{\tau - \bar\tau} \int_M \Omega \wedge \overline{G}_3 = 0 \qquad (10.108)$$

implies that the $(3,0)$ component of G_3 has to vanish as well. The remaining conditions are

$$\mathcal{D}_\alpha W = \int_M \chi_\alpha \wedge G_3 = 0, \qquad (10.109)$$

where χ_α is a basis of harmonic $(2,1)$-forms introduced in Chapter 9. Since this condition holds for all harmonic $(2,1)$-forms, one concludes that supersymmetry is unbroken if

$$G_3 \in H^{(2,1)}(M). \qquad (10.110)$$

Remark on primitivity

Compact Calabi–Yau three-folds with a vanishing Euler characteristic satisfy $h^{1,0} = 0$. In this case any harmonic $(2,1)$-form is primitive. To see this, let us apply the Lefschetz decomposition to the present case. A harmonic $(2,1)$-form

$$\chi = \frac{1}{2}\chi_{ab\bar c}dz^a \wedge dz^b \wedge d\bar z^{\bar c} \qquad (10.111)$$

can be decomposed into a part parallel to J and an orthogonal part according to

$$\chi = v \wedge J + (\chi - v \wedge J) = \chi_\parallel + \chi_\perp, \qquad (10.112)$$

where
$$v = \frac{3}{2}\chi_{ap\bar{q}}J^{p\bar{q}}dz^a, \qquad (10.113)$$
which has been chosen so that
$$\chi_\perp \wedge J = 0. \qquad (10.114)$$

On the other hand, if such a one-form v exists, it is harmonic, which implies $h^{1,0} \neq 0$. As a result, $\chi = \chi_\perp$, and any harmonic $(2,1)$-form is primitive. Note that the vanishing of $h^{1,0}$ is required to prove that any harmonic $(2,1)$-form is primitive. On a six-torus $h^{1,0} \neq 0$ and there are harmonic $(2,1)$-forms that are not primitive. If $h^{1,0} \neq 0$ supersymmetry is unbroken if
$$G_3 \in H^{(2,1)}_{\text{primitive}}. \qquad (10.115)$$

Note that besides being primitive, the χ_α are also imaginary self-dual. The behavior of three-forms under the Hodge-star operation is displayed in the table. Expressing the Levi–Civita tensor in the form
$$\varepsilon_{abc\bar{p}\bar{q}\bar{r}} = -i\left(g_{a\bar{p}}g_{b\bar{q}}g_{c\bar{r}} \pm \text{permutations}\right) \qquad (10.116)$$
allows us to check these rules by the reasoning of Exercise 10.4. Then Eq. (10.110) agrees with the condition that G_3 is imaginary self-dual.

$(3,0)$	Ω	$\star\Omega = -i\Omega$
$(2,1)$	χ_α	$\star\chi_\alpha = i\chi_\alpha$
$(1,2)$	$\bar{\chi}_\alpha$	$\star\bar{\chi}_\alpha = -i\bar{\chi}_\alpha$
$(0,3)$	$\bar{\Omega}$	$\star\bar{\Omega} = i\bar{\Omega}$

An example: flux background on the conifold

As discussed in Chapter 9, different Calabi–Yau manifolds are connected by conifold transitions. At the connection points the Calabi–Yau manifolds degenerate. This section explores further the behavior of a Calabi–Yau manifold near a conifold singularity of its moduli space. By including these singular points it is possible to describe many, and possibly all, Calabi–Yau manifolds as part of a single connected web. In order to be able to include these singular points, it is necessary to understand how to smooth out the singularities. This can be done in two distinct ways, called *deformation* and *resolution*.

Conifold singularities occur commonly in the moduli spaces of compact Calabi–Yau spaces, but they are most conveniently analyzed in terms of

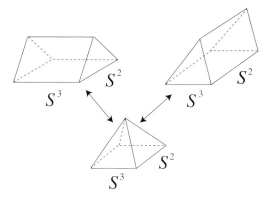

Fig. 10.4. The deformation and the resolution of the singular conifold near the singularity at the tip of the cone.

the noncompact Calabi–Yau space obtained by magnifying the region in the vicinity of a singularity of the three-fold. This noncompact Calabi–Yau space is called the conifold, and its geometry is given by a cone. This section describes the space-time geometry of the conifold, together with its smoothed out cousins, the deformed conifold and the resolved conifold. Type IIB superstring theory compactified on a *deformed conifold* is an interesting example of a flux compactification. It is the superstring dual of a confining gauge theory, which is described in Chapter 12. Here we settle for a supergravity analysis.

The conifold

At a conifold point a Calabi–Yau three-fold develops a conical singularity, which can be described as a hypersurface in $\mathbb{C}^4$ given by the quadratic equation

$$\sum_{A=1}^{4} (w^A)^2 = 0 \quad \text{for} \quad w^A \in \mathbb{C}^4. \tag{10.117}$$

This equation describes a surface that is smooth except at $w^A = 0$. It describes a cone with an $S^2 \times S^3$ base. To see that it is a cone note that if w^A solves Eq. (10.117) then so does λw^A, where λ is a complex constant. Letting $w^A = x^A + iy^A$, and introducing a new coordinate ρ, Eq. (10.117) can be recast as three real equations

$$\vec{x} \cdot \vec{x} - \frac{1}{2}\rho^2 = 0, \quad \vec{y} \cdot \vec{y} - \frac{1}{2}\rho^2 = 0, \quad \vec{x} \cdot \vec{y} = 0. \tag{10.118}$$

The first equation describes an S^3 with radius $\rho/\sqrt{2}$. Then the last two equations can be interpreted as describing an S^2 fibered over the S^3. It can be shown that a Ricci flat and Kähler metric on this space is given by a cone

$$ds^2 = dr^2 + r^2 d\Sigma^2, \tag{10.119}$$

where $r = \sqrt{3/2}\rho^{2/3}$ and $d\Sigma^2$ is the metric on the five-dimensional base, which has the topology $S^2 \times S^3$. Explicitly, the metric on the base can be written in terms of angular variables

$$d\Sigma^2 = \frac{1}{9}\left(2d\beta + \sum_{i=1}^{2} \cos\theta_i d\phi_i\right)^2 + \frac{1}{6}\sum_{i=1}^{2}\left(d\theta_i^2 + \sin^2\theta_i d\phi_i^2\right). \tag{10.120}$$

The range of the angular variables is

$$0 \leq \beta \leq 2\pi, \qquad 0 \leq \theta_i \leq \pi \quad \text{and} \quad 0 \leq \phi_i \leq 2\pi, \tag{10.121}$$

for $i = 1, 2$, while $0 \leq r < \infty$. This space has the isometry group $SU(2) \times SU(2) \times U(1)$.[13]

In order to describe this background in more detail, it is convenient to introduce the basis of one-forms

$$\begin{aligned}
g^1 &= \tfrac{1}{\sqrt{2}}(e^1 - e^3), & g^2 &= \tfrac{1}{\sqrt{2}}(e^2 - e^4), \\
g^3 &= \tfrac{1}{\sqrt{2}}(e^1 + e^3), & g^4 &= \tfrac{1}{\sqrt{2}}(e^2 + e^4), \\
g^5 &= e^5,
\end{aligned} \tag{10.122}$$

with

$$\begin{aligned}
e^1 &= -\sin\theta_1 d\phi_1, & e^2 &= d\theta_1, \\
e^3 &= \cos 2\beta \sin\theta_2 d\phi_2 - \sin 2\beta d\theta_2, \\
e^4 &= \sin 2\beta \sin\theta_2 d\phi_2 + \cos 2\beta d\theta_2, \\
e^5 &= 2d\beta + \cos\theta_1 d\phi_1 + \cos\theta_2 d\phi_2.
\end{aligned} \tag{10.123}$$

In terms of this basis the metric takes the form

$$d\Sigma^2 = \frac{1}{9}(g^5)^2 + \frac{1}{6}\sum_{i=1}^{4}(g^i)^2. \tag{10.124}$$

The conifold has a conical singularity at $r = 0$. In fact, this would also

[13] Compact Calabi–Yau three-folds do not have continuous isometry groups.

be true for any choice of the five-dimensional base space other than a five-sphere of unit radius. As was already mentioned, in the case of the conifold there are two ways of smoothing out the singularity at the tip of the cone, called deformation and resolution.

The deformed conifold

The deformation consists in replacing Eq. (10.117) by

$$\sum_{A=1}^{4}(w^A)^2 = z, \qquad (10.125)$$

where z is a nonzero complex constant. Since $w^A \in \mathbb{C}^4$ we can rescale these coordinates and assume that z is real and nonnegative. This defines a Calabi–Yau three-fold for any value of z. As a result, z spans a one-dimensional moduli space. At the singularity of the moduli space ($z = 0$) the manifold becomes singular (at $\rho = 0$).

For large r the deformed conifold geometry reduces to the singular conifold with $z = 0$, that is, it is a cone with an $S^2 \times S^3$ base. Moving from ∞ towards the origin, the S^2 and S^3 both shrink. Decomposing w^A into real and imaginary parts, as before, yields

$$z = \vec{x} \cdot \vec{x} - \vec{y} \cdot \vec{y}, \qquad (10.126)$$

and using the definition

$$\rho^2 = \vec{x} \cdot \vec{x} + \vec{y} \cdot \vec{y}, \qquad (10.127)$$

shows that the range of r is

$$z \leq \rho^2 < \infty. \qquad (10.128)$$

As a result, the singularity at the origin is avoided for $z > 0$. This shows that as ρ^2 gets close to z the S^2 disappears leaving just an S^3 with finite radius.

The resolved conifold

The second way of smoothing out the conifold singularity is called resolution. In this case as the apex of the cone is approached, it is the S^3 which shrinks to zero size, while the size of the S^2 remains nonvanishing. This is also called a small resolution, and the nonsingular space is called the resolved conifold.

In order to describe how this works, let us make a linear change of variables to recast the singular conifold in the form

$$\det \begin{pmatrix} X & U \\ V & Y \end{pmatrix} = 0. \tag{10.129}$$

Away from $(X, Y, U, V) = 0$ this space is equivalently described as the space

$$\begin{pmatrix} X & U \\ V & Y \end{pmatrix} \begin{pmatrix} \lambda_1 \\ \lambda_2 \end{pmatrix} = 0, \tag{10.130}$$

in which λ_1 and λ_2 don't both vanish. The solutions for λ_i are determined up to an overall multiplicative constant, that is,

$$(\lambda_1, \lambda_2) \simeq \lambda(\lambda_1, \lambda_2) \quad \text{with} \quad \lambda \in \mathbb{C}^\star. \tag{10.131}$$

As a result, the variables (X, Y, U, V) and (λ_1, λ_2) lie in $\mathbb{C}^4 \times \mathbb{C}P^1$ and satisfy the condition (10.130). This describes the resolved conifold, which is nonsingular. Why is the singularity removed? In order to answer this question note that for $(X, Y, U, V) \neq (0, 0, 0, 0)$ this space is the same as the singular conifold. However, at the point $(X, Y, U, V) = (0, 0, 0, 0)$ any solution for (λ_1, λ_2) is allowed. This space is $\mathbb{C}P^1$, which is a two-sphere.

Fluxes on the conifold

Let us now consider a flux background of the conifold geometry given by N space-time-filling D3-branes located at the tip of the conifold, as well as M D5-branes wrapped on the S^2 in the base of the deformed conifold and filling the four-dimensional space-time. These D5-branes are usually called *fractional D3-branes*.

This background can be constructed by starting with a set of M D5-branes, which give

$$\int_{S^3} F_3 = 4\pi^2 \alpha' M. \tag{10.132}$$

This can also be written as

$$F_3 = \frac{M\alpha'}{2} \omega_3 \quad \text{where} \quad \omega_3 = g^5 \wedge \omega_2, \tag{10.133}$$

and

$$\omega_2 = \frac{1}{2} \left(\sin \theta_1 d\theta_1 \wedge d\phi_1 - \sin \theta_2 d\theta_2 \wedge d\phi_2 \right). \tag{10.134}$$

In order to describe a supersymmetric background, the complex three-form G_3 should be an imaginary self-dual $(2, 1)$-form. This implies that an H_3 flux has to be included. Imaginary self-duality determines the H_3 flux to be

$$H_3 = \frac{3}{2r} g_s M \alpha' dr \wedge \omega_2, \tag{10.135}$$

where $g_s = e^\Phi$ is the string coupling constant, which is assumed to be constant, while the axion C_0 has been set to zero. Once H_3 and F_3 are both present, F_5 is determined by the Bianchi identity

$$d\tilde{F}_5 = H_3 \wedge F_3 + 2\kappa^2 T_3 \rho_3, \qquad (10.136)$$

to be

$$\tilde{F}_5 = (1 + \star_{10})\mathcal{F}, \qquad (10.137)$$

where

$$\mathcal{F} = \frac{1}{2}\pi(\alpha')^2 N_{\text{eff}}(r)\omega_2 \wedge \omega_3 \qquad (10.138)$$

and

$$N_{\text{eff}}(r) = N + \frac{3}{2\pi}g_s M^2 \log\left(\frac{r}{r_0}\right). \qquad (10.139)$$

Note that the total five-form flux is now radially dependent, with

$$\int_\Sigma \tilde{F}_5 = \frac{1}{2}(\alpha')^2 \pi N_{\text{eff}}(r). \qquad (10.140)$$

The geometry in this case is a *warped conifold*, where the metric has the form

$$ds_{10}^2 = e^{2A(r)}\eta_{\mu\nu}dx^\mu dx^\nu + e^{-2A(r)}(dr^2 + r^2 d\Sigma^2). \qquad (10.141)$$

The metric of the base, $d\Sigma^2$, is given in Eq. (10.120). The volume form for the metric in these coordinates is given by

$$\sqrt{-g} = \frac{1}{54}e^{-2A}r^5 \sin\theta_1 \sin\theta_2. \qquad (10.142)$$

Using this and

$$\omega_2 \wedge \omega_3 = -d\beta \wedge \sin\theta_1 d\theta_1 \wedge d\phi_1 \wedge \sin\theta_2 d\theta_2 \wedge d\phi_2, \qquad (10.143)$$

one obtains

$$\star(\omega_2 \wedge \omega_3) = 54 r^{-5} e^{8A} dr \wedge dx^0 \wedge dx^1 \wedge dx^2 \wedge dx^3. \qquad (10.144)$$

The warp factor is determined in terms of the five-form flux by Eq. (10.81), or equivalently

$$\star_{10}\mathcal{F} = d\alpha \wedge dx^0 \wedge dx^1 \wedge dx^2 \wedge dx^3, \qquad (10.145)$$

while $\alpha = \exp(4A)$ according to Eq. (10.98). Using the expression for the five-form flux in Eq. (10.138) this leads to the equation

$$d\alpha = 27\pi(\alpha')^2 \alpha^2 r^{-5} N_{\text{eff}}(r) dr. \qquad (10.146)$$

Integration then gives the warp factor

$$e^{-4A(r)} = \frac{27\pi(\alpha')^2}{4r^4}\left[g_sN + \frac{3}{2\pi}(g_sM)^2\log\left(\frac{r}{r_0}\right) + \frac{3}{8\pi}(g_sM)^2\right], \quad (10.147)$$

where r_0 is a constant of integration.

Problem 10.13 asks you to show that G_3 is primitive. This result implies that this is a supersymmetric background. Note that in this section we have used the constraints in Eqs (10.98) and (10.115), which were derived for compact spaces. However, these constraints can also be derived from the Killing spinor equations for type IIB, which are local. As a result, they also hold for noncompact spaces.

Warped space-times and the gauge hierarchy

The observation that Poincaré invariance allows space-times with extra dimensions that are warped products has interesting consequences for phenomenology. *Brane-world* scenarios are toy models based on the proposal that the observed four-dimensional world is confined to a brane embedded in a five-dimensional space-time.[14] In one version of this proposal, the fifth dimension is not curled up. Instead, it is infinitely extended. If we live on such a brane, why is there a four-dimensional Newtonian inverse-square law for gravity instead of a five-dimensional inverse-cube law? The answer is that the space-time is warped. Let's explore how this works.

Localizing gravity with warp factors

The action governing five-dimensional gravity with a cosmological constant Λ in the presence of a 3-brane is

$$S \sim \int d^5x \sqrt{-G}\,(R - 12\Lambda) - T\int d^4x\sqrt{-g}, \quad (10.148)$$

where T is the 3-brane tension, G_{MN} is the five-dimensional metric, and $g_{\mu\nu}$ is the induced four-dimensional metric of the brane. This action admits a solution of the equations of motion of the form

$$ds^2 = e^{-2A(x_5)}\eta_{\mu\nu}dx^\mu dx^\nu + dx_5^2, \quad (10.149)$$

with

$$A(x_5) = \sqrt{-\Lambda}|x_5|. \quad (10.150)$$

[14] There could be an additional compact five-dimensional space that is ignored in this discussion.

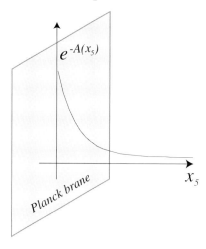

Fig. 10.5. Gravity is localized on the Planck brane due to the presence of a warp factor in the metric.

Here $-\infty \leq x_5 \leq \infty$ is infinite, and the brane is at $x_5 = 0$. Moreover, for a static solution it is necessary that the brane tension is related to the space-time cosmological constant Λ by

$$T = 12\sqrt{-\Lambda}, \tag{10.151}$$

which requires that the cosmological constant is negative. This geometry is locally anti-de Sitter (AdS_5), except that there is a discontinuity in derivatives of the metric at $x^5 = 0$. This discontinuity is determined by the delta function brane source using standard matching formulas of general relativity.

The metric (10.149) contains a warp factor, which has the interesting consequence that, even though the fifth dimension is infinitely extended, four-dimensional gravity is observed on the brane. This way of concealing an extra dimension is an alternative to compactification. Computing the normal modes of the five-dimensional graviton in this geometry, one finds that the zero mode, which is interpreted as the four-dimensional graviton, is localized in the vicinity of the brane and that G_4 controls its interactions. The effective four-dimensional Planck mass on the brane is given by

$$M_4^2 = M_5^3 \int dx_5 e^{-2\sqrt{-\Lambda}|x_5|}, \tag{10.152}$$

or in terms of Newton's constant

$$G_4 = G_5 \left(\int dx_5 e^{-2\sqrt{-\Lambda}|x_5|} \right)^{-1}. \tag{10.153}$$

10.2 Flux compactifications of the type IIB theory

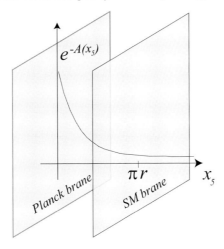

Fig. 10.6. On the SM brane the energy scales are redshifted due to the presence of the warp factor in the metric.

Large hierarchies from warp factors

If instead of one 3-brane, two parallel 3-branes are considered, the implications for phenomenology are even more interesting. In this construction the background geometry is again a warped product, but now the warp factor provides a natural way to solve the hierarchy problem.

Imagine that the 3-branes are again embedded in a five-dimensional space-time as shown Fig. 10.6. One brane is located at $x_5 = \pi r$, and called the *standard-model brane* (SM), while the other brane is located at $x_5 = 0$ and called the *Planck brane* (P). The action governing five-dimensional gravity coupled to the two branes is

$$S = \int d^5x \sqrt{-G}\,(R - 12\Lambda) - T_{\text{SM}} \int d^4x \sqrt{-g^{\text{SM}}} - T_{\text{P}} \int d^4x \sqrt{-g^{\text{P}}}, \tag{10.154}$$

where T_{SM} and T_{P} are the tensions of the two branes. The metric is again assumed to be a warped product

$$ds^2 = e^{-2A(x_5)} \eta_{\mu\nu} dx^\mu dx^\nu + dx_5^2 \tag{10.155}$$

in the interval $0 \le x_5 \le \pi r$.

The equations of motion are solved by a warp factor of the form

$$A(x_5) = \sqrt{-\Lambda}\,|x_5|, \tag{10.156}$$

as before, and

$$T_{\text{P}} = -T_{\text{SM}} = 12\sqrt{-\Lambda}. \tag{10.157}$$

Negative tension may seem disturbing. However, negative-tension branes can be realized in orientifold models and in F-theory compactifications. In this solution the metric is normalized so that it takes the form

$$g_{\mu\nu}^{\text{P}} = \eta_{\mu\nu}. \qquad (10.158)$$

on the Planck brane. Then, because of the warp factor, the SM brane metric is

$$g_{\mu\nu}^{\text{SM}} = e^{-2\pi r\sqrt{-\Lambda}} \eta_{\mu\nu}. \qquad (10.159)$$

This scale factor means that objects with energy E at the Planck brane are red-shifted on the SM brane, and appear as objects with energy $e^{-\pi r\sqrt{-\Lambda}}E$. By choosing the separation scale r suitably, one can arrange for this factor to be of order 10^{-16}, so as to find TeV scale physics on the SM brane by starting with Planck-scale physics on the Planck brane. This is an interesting proposal (due to Randall and Sundrum) for solving the gauge hierarchy problem. This scenario has a number of remarkable implications. It becomes conceivable that phenomena that used to be relegated to ultra-high energy scales may be accessible at accelerator energies. Thus, Kaluza–Klein modes, fundamental strings, black holes, gravitational radiation could all be observable. The LHC experiments are preparing to search for all of these possibilities. Supersymmetry, which many view as more likely to be discovered, seems quite mundane by comparison. Not surprisingly, these ideas have attracted a lot of attention, and there is a large and rapidly growing literature on the subject. In the following, we settle for a brief sketch of how this scenario might be realized in string theory.

A large hierarchy on the deformed conifold

It is interesting that the above approach to solving the hierarchy problem appears naturally in string theory.[15] The branes that seem best suited to this purpose are the D3-branes in a type IIB orientifold or F-theory construction. One can imagine D3-branes placed at points on a compact internal manifold. To get a large hierarchy two sets of D3-branes would need to be separated by the distance r. This distance would then determine the size of the hierarchy. However, r is a modulus in the four-dimensional theory, since the D3-brane coordinates have no potential. In the following we will see that one can obtain a warped background generating a large and stable hierarchy by using the flux backgrounds discussed at the beginning of this section.

To be concrete, one can consider the deformed conifold geometry. Locally,

---

15 So does supersymmetry.

near the tip of the cone, the flux solution is similar to the one described in the previous section. Globally, however, the background solution must be changed, since we are interested in a compact solution. The conifold solution presented in the previous section is noncompact with r unbounded. This can be interpreted as a singular limit of a compact manifold in which one of the cycles degenerates to infinite size.

Let us assume that there are M units of F_3 flux through an A-cycle and $-K$ units of H_3 flux through a B-cycle, that is,

$$\frac{1}{2\pi\alpha'} \int_A F_3 = 2\pi M \quad \text{and} \quad \frac{1}{2\pi\alpha'} \int_B H_3 = -2\pi K. \tag{10.160}$$

Using Poincaré duality, the superpotential can then be written as

$$W = \int G_3 \wedge \Omega = (2\pi)^2 \alpha' \left(M \int_B \Omega - K\tau \int_A \Omega \right), \tag{10.161}$$

The complex coordinate describing the cycle collapsing at the tip of the conifold is

$$z = \int_A \Omega. \tag{10.162}$$

The discussion of special geometry in Section 9.6 explained that the dual coordinates, that is, the coordinates defining periods of the B-cycles, are functions of the periods of the A-cycles. More concretely, since we are describing a conifold singularity, we can invoke the result derived in Section 9.8 that

$$\int_B \Omega = \frac{z}{2\pi i} \log z + \text{holomorphic}. \tag{10.163}$$

Using these results, the Kähler covariant derivative of the superpotential can be rewritten in the form[16]

$$\mathcal{D}_z W \simeq (2\pi)^2 \alpha' \left(\frac{M}{2\pi i} \log z - i \frac{K}{g_s} + \ldots \right) \tag{10.164}$$

in the limit in which K/g_s is large. The equation $\mathcal{D}_z W = 0$ is solved by

$$z \simeq e^{-2\pi K/M g_s}. \tag{10.165}$$

Thus, one obtains a large hierarchy of scales if, for example, $M = 1$ and $K/g_s = 5$. It is assumed that the dilaton is frozen in this solution.

The solution for the warp factor can be estimated in the following way. As

---

[16] This assumes $z \ll 1$, which is the case of interest.

498   Flux compactifications

will be discussed in more detail in Chapter 12, close to a set of N D3-branes
the space-time metric takes the form

$$ds^2 = \left(\frac{r}{R}\right)^2 |d\vec{x}|^2 + \left(\frac{R}{r}\right)^2 (dr^2 + r^2 d\Omega_5^2) \quad \text{with} \quad R^4 = 4\pi g_s N (\alpha')^2, \tag{10.166}$$

where r is the distance from the D3-brane, which is located at $r \approx 0$. We would like to estimate the warp factor close to the D3-brane. Since the background is the deformed conifold, r has a minimal value determined by the deformation parameter z according to

$$r_{\min} \simeq \rho_{\min}^{2/3} = z^{1/3} \simeq e^{-2\pi K/3Mg_s}, \tag{10.167}$$

showing that the warp factor approaches a small and positive value close to the D3-brane.

EXERCISES

EXERCISE 10.6
Show that in a Calabi–Yau three-fold compactification of type IIB superstring theory the Kähler potential for the radial modulus, the axion–dilaton modulus and the complex-structure moduli is given by

$$\mathcal{K} = -3\log\left[-i(\rho - \bar{\rho})\right] - \log[-i(\tau - \bar{\tau})] - \log\left(i \int_M \Omega \wedge \bar{\Omega}\right).$$

SOLUTION

The part of the Kähler potential depending on the complex-structure moduli (the last term) was derived in Chapter 9. The way to derive the contribution from the radial modulus ρ and the axion–dilaton modulus τ is to consider the action on a background of the form

$$ds^2 = e^{-6u(x)} \underbrace{g_{\mu\nu} dx^\mu dx^\nu}_{4D} + e^{2u(x)} \underbrace{g_{mn} dy^m dy^n}_{CY_3}.$$

Here $u(x)$ parametrizes the volume of the Calabi–Yau three-fold. The power of $e^{u(x)}$ in the first term has been chosen to give a canonically normalized Einstein term in four dimensions.

The supersymmetric partner of the radial modulus is another axion b, which descends from the four-form according to

$$C_{\mu\nu pq} = a_{\mu\nu} J_{pq},$$

where J is the Kähler form. In four dimensions the two-form a can be dualized to a scalar b according to

$$da = e^{-8u(x)} \star db.$$

Setting

$$\rho = \frac{b}{\sqrt{2}} + ie^{4u},$$

the resulting low-energy effective action is

$$S = \frac{1}{2\kappa_4^2} \int d^4x \sqrt{-g} \left(R - \frac{1}{2} \frac{\partial_\mu \tau \partial^\mu \bar\tau}{(\mathrm{Im}\,\tau)^2} - \frac{3}{2} \frac{\partial_\mu \rho \partial^\mu \bar\rho}{(\mathrm{Im}\,\rho)^2} \right).$$

Here the four-dimensional gravitational coupling constant is given by $\kappa_4^2 = \kappa_{10}^2/\mathcal{V}$, where $\mathcal{V}$ is the volume of the Calabi–Yau three-fold computed using the metric g_{mn}. The kinetic terms for τ and ρ correspond to the first two terms in the Kähler potential $\mathcal{K}$. □

10.3 Moduli stabilization

The important fact about compactifications with flux is that there is a non-trivial scalar potential for the moduli fields.[17] This should not be surprising, since the background flux modifies the equations that determine the geometry. The complete scalar potential V for the moduli fields can be obtained from the superpotential and the Kähler potential by a standard supergravity formula, as was discussed earlier, or by a direct Kaluza–Klein compactification, as is done here.

Scalar potential for M-theory

In the following the scalar potential for flux compactifications of M-theory on a Calabi–Yau four-fold is derived from the low-energy expansion of the action Eq. (10.3) on the warped geometry described by Eq. (10.5). This further illustrates that the constraints derived from $W^{3,1}$ in Eq. (10.62) stabilize the complex-structure moduli, while the equations derived from $W^{1,1}$ in Eq. (10.68) stabilize the Kähler moduli.

---

[17] Calling these fields moduli in this setting is a bit of an oxymoron, since moduli are defined to have no potential. However, this has become standard usage.

As you are asked to check in Problem **10.18**, fluxes generate a scalar potential for the moduli

$$V(T, K) = \frac{1}{4\mathcal{V}^3}\left(\int_M F \wedge \star F - \frac{1}{6}T_{\mathrm{M2}}\chi\right), \qquad (10.168)$$

where we set $\kappa_{11} = 1$, as in Section 10.1. The terms that contribute to the potential originate from the internal component of the flux while the f_m term has been dropped, because it gives a subleading contribution in the large-volume limit.

Since F is a four-form it lies in the middle-dimensional cohomology of the Calabi–Yau four-fold. According to Eq. (10.44) the $(2,2)$-component of the four-form flux has the Lefschetz decomposition

$$F^{2,2} = F^{2,2}_\mathrm{o} + J \wedge F^{1,1}_\mathrm{o} + J \wedge J F^{0,0}_\mathrm{o}, \qquad (10.169)$$

where the subindex o indicates that the flux is primitive. As was shown in Eq. (10.67), only the primitive term, that is, the first term, is nonzero for a supersymmetric solution. However, here all terms are included in order to allow for the possibility of supersymmetry breaking. Since the first and third terms are self-dual, and the second term is anti-self-dual,

$$\star F^{2,2} = F^{2,2} - 2J \wedge F^{1,1}_\mathrm{o}, \qquad (10.170)$$

where $\star$ denotes the Hodge dual on the internal manifold. It follows from Exercise 10.4 that

$$\star F^{4,0} = F^{4,0} \qquad \text{and} \qquad \star F^{3,1} = -F^{3,1}, \qquad (10.171)$$

and similarly for the $(0,4)$ and $(1,3)$ components, since F is real. Taking the previous two equations into account, the total four-form flux satisfies

$$\star F = F - 2F^{3,1} - 2F^{1,3} - 2J \wedge F^{1,1}_\mathrm{o}. \qquad (10.172)$$

Therefore, after taking the wedge product with F, the kinetic term for the flux appearing in Eq. (10.168) can be rewritten in the form

$$\int_M F \wedge \star F = \int_M F \wedge F - 4\int_M F^{3,1} \wedge F^{1,3} - 2\int_M J \wedge F^{1,1}_\mathrm{o} \wedge J \wedge F^{1,1}_\mathrm{o}. \qquad (10.173)$$

All other terms vanish by orthogonality relations given by the Hodge decomposition and the Lefschetz decomposition. Inserting this into the scalar potential Eq. (10.168), we realize that the first term on the right-hand side of Eq. (10.173) cancels due to the tadpole-cancellation condition Eq. (10.60) with $N = 0$. As a result, only the anti-self-dual part of F contributes to the scalar potential.

10.3 Moduli stabilization

Supersymmetry-breaking solutions

The preceding results imply the existence of supersymmetry-breaking solutions of the equations of motion. Indeed, any flux satisfying

$$F = \star F \quad \text{and} \quad F \notin H^{(2,2)}_{\text{primitive}} \qquad (10.174)$$

solves the equations of motion and breaks supersymmetry. Fluxes of the form

$$F \sim \Omega \quad \text{or} \quad F \sim J \wedge J \qquad (10.175)$$

provide examples. Moreover, since these flux components do not appear in the scalar potential they do not generate a cosmological constant.

The scalar potential

The second term on the right-hand side of Eq. (10.173) can be rewritten according to

$$\int_M F^{3,1} \wedge F^{1,3} = -e^{\mathcal{K}^{3,1}} G^{I\bar{J}} \mathcal{D}_I W^{3,1} \mathcal{D}_{\bar{J}} \overline{W}^{3,1}, \qquad (10.176)$$

and as a result yields a scalar potential for the complex-structure moduli. This result is obtained by expanding $F^{3,1}$ in a basis of $(3,1)$-forms. The explicit calculation is rather similar to Exercise 10.5. Analogously, the last term on the right-hand side of Eq. (10.173) generates a potential for the Kähler-structure moduli

$$\int_M J \wedge F_0^{1,1} \wedge J \wedge F_0^{1,1} = -\mathcal{V}^{-1} G^{AB} \mathcal{D}_A W^{1,1} \mathcal{D}_B W^{1,1}, \qquad (10.177)$$

where[18]

$$\mathcal{D}_A = \partial_A - \frac{1}{2} \partial_A \mathcal{K}^{1,1} \quad \text{with} \quad \mathcal{K}^{1,1} = -3 \log \mathcal{V}, \qquad (10.178)$$

and G^{AB} is the inverse of the metric G_{AB}

$$G_{AB} = -\frac{1}{2} \partial_A \partial_B \log \mathcal{V}. \qquad (10.179)$$

Here $\mathcal{V} = \frac{1}{24} \int J \wedge J \wedge J \wedge J$ is the Calabi–Yau volume. In total, the scalar potential becomes

$$V(T, K) = e^{\mathcal{K}} G^{I\bar{J}} \mathcal{D}_I W^{3,1} \mathcal{D}_{\bar{J}} \overline{W}^{3,1} + \frac{1}{2} \mathcal{V}^{-4} G^{AB} \mathcal{D}_A W^{1,1} \mathcal{D}_B W^{1,1}, \qquad (10.180)$$

---

[18] Note that $\mathcal{K}^{1,1}$ is not a Kähler potential, since it is function of real fields. Nevertheless, it has some similar properties.

where $\mathcal{K} = \mathcal{K}^{3,1} + \mathcal{K}^{1,1}$. This potential is manifestly nonnegative, which shows that compactifications to AdS$_3$ spaces cannot be obtained in this way.

The radial modulus

Note that not all of the moduli need contribute to the potential Eq. (10.180). For example, it does not depend on the radial modulus, which characterizes the overall volume of the compact manifold M. Therefore, this modulus is not stabilized. The reason for this is that the conditions for unbroken supersymmetry in Eqs (10.65), (10.66) and (10.67), and also the conditions for the existence of supersymmetry breaking solutions in Eq. (10.174), are invariant under the rescaling of the volume by a constant. While this may seem disappointing, it is also quite fortunate. This freedom means that the volume can be chosen sufficiently large to justify the approximations that have been made. At sufficiently large volume, most of the higher-derivative terms of M-theory can be dropped. The situation, of course, changes once nonperturbative effects are included. It is expected that such effects stabilize the radial modulus and that the calculations made remain valid when the flux quantum is large. This is not specific to the M-theory compactifications discussed in this section, but holds for most of the flux compactifications discussed in the literature. Very few models have been constructed in which all moduli are stabilized without nonperturbative effects.

The scalar potential for type IIB

The scalar potential for type IIB compactified on a Calabi–Yau three-fold follows from a standard supergravity formula. In Section 10.2 the formulas for the superpotential W and Kähler potential $\mathcal{K}$ were presented. Given these potentials, $\mathcal{N} = 1$ supergravity determines the scalar potential in terms of these quantities[19]

$$V = e^{\mathcal{K}} \left(G^{a\bar{b}} \mathcal{D}_a W \overline{\mathcal{D}_b W} - 3|W|^2 \right), \tag{10.181}$$

where $G_{a\bar{b}} = \partial_a \partial_{\bar{b}} \mathcal{K}$ is the metric on moduli space, with a, b labelling all the superfields, and $G^{a\bar{b}}$ is its inverse. Moreover, $\mathcal{D}_a = \partial_a + \partial_a \mathcal{K}$.

As it should be, this scalar potential is invariant under the Kähler transformation

$$\mathcal{K}(z, \bar{z}) \to \mathcal{K}(z, \bar{z}) - f(z) - \bar{f}(\bar{z}), \tag{10.182}$$

---

[19] This compactification gives $\mathcal{N} = 2$ supersymmetry, but an $\mathcal{N} = 1$ formalism is still applicable. Moreover, one only has $\mathcal{N} = 1$ when orientifold planes are included.

since the superpotential transforms according to

$$W(z) \to e^{f(z)} W(z). \tag{10.183}$$

This transformation is a consequence of the linear dependence of W on Ω and the behavior of the holomorphic three-form under Kähler transformations. Here z refers to the moduli fields and $f(z)$ is a holomorphic function of these fields. The four-dimensional gravitational constant (or Planck length) κ_4 has been set to one in the above formulas.

A simple calculation shows that this potential does not depend on the radial modulus (except as an overall factor). Using the result for the Kähler potential for ρ derived in exercise **10.6**, one finds

$$G^{\rho\bar\rho} \mathcal{D}_\rho W \mathcal{D}_{\bar\rho} \overline{W} - 3|W|^2 = 0. \tag{10.184}$$

As a result, the scalar potential is of the *no-scale* type

$$V = e^{\mathcal{K}} \sum_{i,j \ne \rho} G^{i\bar j} \mathcal{D}_i W \mathcal{D}_{\bar j} \overline{W}, \tag{10.185}$$

where i, j label all the fields excluding ρ. At the minimum of the potential

$$\mathcal{D}_i W = 0, \tag{10.186}$$

which implies $V = 0$ even though supersymmetry is broken in general, since

$$\mathcal{D}_\rho W \ne 0. \tag{10.187}$$

These solutions have the interesting property that $V = 0$ at the minimum of the potential, so that the cosmological constant vanishes at the same time supersymmetry that is broken. Even though this may seem encouraging for achieving the goal of breaking supersymmetry without generating a large vacuum energy density, it does not constitute a solution of the cosmological constant problem. There is no reason to believe that this result continues to hold when α' and g_s corrections are included. In the next section we will see that nonperturbative corrections to W depending on ρ can generate a nonvanishing cosmological constant.

Moduli stabilization by nonperturbative effects

The type IIB no-go theorem excludes the possibility of compactification to four-dimensional de Sitter (dS) space, or more generally to a space with a positive cosmological constant. This section shows that this conclusion can be circumvented when nonperturbative effects are taken into account. This is of interest, since the Universe appears to have a small positive cosmological constant.

The basic idea is to stabilize all moduli of the type IIB compactification and to break the no-scale structure by adding nonperturbative corrections to the superpotential. These contributions are combined in such a way that supersymmetry is not broken. This leads to an AdS vacuum with a negative vacuum energy density. Then one adds anti-D3-branes that break the supersymmetry and give a positive vacuum energy density.

In the simplest case, there is only one exponential correction to the superpotential, but in general there may be multiple exponentials. The corrections to the Kähler potential can be ignored in the large-volume limit. The Kähler potential for the radial modulus is then equal to its tree-level expression. Assuming that all other modes are massive and can be integrated out, one is left with an effective theory for the radial modulus. In the following we assume that the only Kähler modulus is the size, while the complex structure and the dilaton become massive due to the presence of fluxes.

The superpotential is assumed to be given by the tree-level result W_0 together with an exponential generated by nonperturbative effects

$$W = W_0 + Ae^{ia\rho}. \qquad (10.188)$$

One source of nonperturbative effects is instantons arising from Euclidean D3-branes wrapping four-cycles. These give a contribution to the superpotential of the form

$$W_{\text{inst}} = T(z^\alpha)e^{2\pi i\rho}, \qquad (10.189)$$

where $T(z^\alpha)$ is the one-loop determinant that is a function of the complex-structure moduli, and ρ is the radial modulus. Another possible source for such corrections is gluino condensation in the world-volume gauge theory of D7-branes, which might be present and wrapped around internal four-cycles.

The coefficient a is a constant that depends on the specific source of the nonperturbative effects. For simplicity, we assume that a, A and W_0 are real and that the axion vanishes. At the supersymmetric minimum all Kähler covariant derivatives of the superpotential vanish including $\mathcal{D}_\rho W = 0$. Using the Kähler potential in Eq. (10.102), Exercise 10.7 shows that the effective potential

$$V = e^{\mathcal{K}} \left(G^{\rho\bar\rho}\mathcal{D}_\rho W \mathcal{D}_{\bar\rho}\overline{W} - 3|W|^2 \right) \qquad (10.190)$$

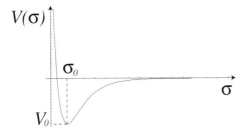

Fig. 10.7. Form of the potential as a function of the radial modulus. The values of the potential and the size depend on the values used for A, a and W_0. The figure displays a minimum at which the potential is negative leading to an AdS vacuum.

has a minimum that is given by

$$V_0 = -\frac{a^2 A^2}{6\sigma_0} e^{-2a\sigma_0}. \tag{10.191}$$

Here σ_0 is the value of σ in the radial modulus $\rho = i\sigma$ at the minimum of the potential. Since this potential is negative, the only maximally symmetric space-time allowed by such a supersymmetric compactification is AdS space-time.

One can break supersymmetry explicitly by adding anti-D3-branes. This gives an additional term in the scalar potential of the form

$$\delta V = \frac{D}{\sigma^2}, \tag{10.192}$$

where D is proportional to the number of anti-D3-branes.

It can be chosen to make the vacuum energy density positive, so that a compactification to dS space becomes possible. Including the anti-D3-brane contribution results in the scalar potential

$$V(\sigma) = \frac{aAe^{-a\sigma}}{2\sigma^2}\left(\frac{1}{3}\sigma aAe^{-a\sigma} + W_0 + Ae^{-a\sigma}\right) + \frac{D}{\sigma^2}. \tag{10.193}$$

The form of $V(\sigma)$ is displayed in Fig. 10.8. It shows that a vacuum with a positive cosmological constant can be obtained. Strictly speaking, the vacua obtained in this way are only metastable. However, the lifetime could be extremely long.

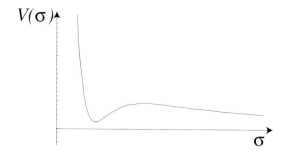

Fig. 10.8. Form of the potential as a function of the radial modulus after taking anti-D3-branes into account. The figure displays a minimum at which the potential is positive leading to a de Sitter vacuum.

EXERCISES

EXERCISE 10.7
Derive the extremum of the potential in Eq. (10.191).

SOLUTION

The only solution is supersymmetric, so let us assume it from the outset. Using
$$W = W_0 + Ae^{ia\rho},$$
the solution for $\rho = i\sigma$ in the ground state, which we denote by σ_0, is the solution of
$$\mathcal{D}_\rho W = \partial_\rho W + \partial_\rho \mathcal{K}\, W = 0 \quad \text{with} \quad \mathcal{K} = -3\log\left[-i(\rho - \bar\rho)\right].$$
This gives
$$W_0 = -A\left(\frac{2}{3}a\sigma_0 + 1\right)e^{-a\sigma_0},$$
or
$$W = -\frac{2}{3}Aa\sigma_0 e^{-a\sigma_0}.$$
So the minimum of the potential
$$V = e^{\mathcal{K}}\left(G^{\rho\bar\rho}\mathcal{D}_\rho W \mathcal{D}_{\bar\rho}\overline{W} - 3|W|^2\right)$$

is
$$V_0 = -\frac{a^2 A^2}{6\sigma_0} e^{-2a\sigma_0},$$
in agreement with Eq. (10.191). □

EXERCISE 10.8
Show that the potential Eq. (10.181) can be expressed entirely in terms of the Kähler-transformation invariant combination
$$\widetilde{\mathcal{K}} = \mathcal{K} + \log |W|^2.$$

SOLUTION

Using this definition, Eq. (10.181) is equal to
$$V = e^{\widetilde{\mathcal{K}}} \left(G^{a\bar{b}} \frac{\mathcal{D}_a W}{W} \frac{\overline{\mathcal{D}_b W}}{\overline{W}} - 3 \right).$$
However,
$$G_{a\bar{b}} = \partial_a \partial_{\bar{b}} \mathcal{K} = \partial_a \partial_{\bar{b}} \widetilde{\mathcal{K}},$$
and thus the inverse metric $G^{a\bar{b}}$ only depends on $\widetilde{\mathcal{K}}$. Also,
$$\frac{\mathcal{D}_a W}{W} = \partial_a \log W + \partial_a \mathcal{K} = \partial_a \widetilde{\mathcal{K}}.$$
Therefore,
$$V = e^{\widetilde{\mathcal{K}}} \left(G^{a\bar{b}} \partial_a \widetilde{\mathcal{K}} \partial_{\bar{b}} \widetilde{\mathcal{K}} - 3 \right)$$
only depends on $\widetilde{\mathcal{K}}$. □

EXERCISE 10.9
Use dimensional analysis to restore the factors of κ_4 in the scalar potential. Discuss the limit $\kappa_4 \to 0$.

SOLUTION

W has dimension three, $\mathcal{K}$ has dimension two and the scalar potential V has dimension four. Therefore, restoring the powers of κ_4, Eq. (10.181) takes the form
$$\kappa_4^4 V = e^{\kappa_4^2 \mathcal{K}} \left(\kappa_4^4 G^{a\bar{b}} \mathcal{D}_a W \overline{\mathcal{D}_b W} - 3\kappa_4^6 |W|^2 \right),$$

where $\mathcal{D}_a W = \partial_a W + \kappa_4^2 \partial_a \mathcal{K} W$. Thus

$$V = e^{\kappa_4^2 \mathcal{K}} \left(G^{a\bar{b}} \mathcal{D}_a W \mathcal{D}_{\bar{b}} \overline{W} - 3\kappa_4^2 |W|^2 \right).$$

For small κ_4,

$$V = G^{a\bar{b}} \partial_a W \partial_{\bar{b}} \overline{W} + \mathcal{O}(\kappa_4^2).$$

As expected, one finds the global supersymmetry formula plus corrections proportional to Newton's constant. □

10.4 Fluxes, torsion and heterotic strings

This section explores compactifications of the weakly coupled heterotic string in the presence of a nonzero three-form field H.[20] A nonvanishing H flux has two implications for the background geometry. First, the background geometry becomes a warped product, like that discussed in the previous sections. The second consequence of nonvanishing H is that its contributions to the various equations can be given a geometric interpretation as torsion of the internal manifold. If the gauge fields are not excited, heterotic supergravity is a truncation of either type II supergravity theory. Therefore, some of the analysis in this section applies to those cases and *vice versa*.

Warped geometry

As in the previous sections, when H flux is included, the space-time is no longer a direct-product space of the form $M_{10} = M_4 \times M$. (For simplicity, in the following we assume that the external space-time is four-dimensional.) Analysis of the heterotic supersymmetry transformation laws will show that a warp factor $e^{2D(y)}$ must be included in the metric in order to provide a consistent solution. In the *Einstein frame*, let us write the background metric for the warped compactification in the form

$$ds^2 = e^{2D(y)}(\underbrace{g_{\mu\nu}(x)dx^\mu dx^\nu}_{\text{4D}} + \underbrace{g_{mn}(y)dy^m dy^n}_{\text{6D}}) \qquad (10.194)$$

As before, x denotes the coordinates of the external space, y the internal coordinates, the indices μ, ν label the coordinates of the external space and m, n label the coordinates of the internal space.

The function $D(y)$ depends only on the internal coordinates. It will be shown that supersymmetry can be satisfied when there is nonzero H flux provided that

$$D(y) = \Phi(y), \qquad (10.195)$$

[20] The index on H_3 is suppressed.

where Φ is the dilaton field. In the case without H flux, the dilaton is constant, so the geometry is a direct product in the Einstein frame. When $\partial_m \Phi \neq 0$, it becomes a warped product. This warp factor is exactly the one that converts the Einstein frame to the *string frame*. So the geometry actually is a direct product with respect to the string-frame metric even when there is nonzero H flux. Since the internal space is compact and the dilaton field $\Phi(y)$ is nonsingular (in the absence of NS5-branes), the dilaton is bounded. Therefore, shifting by a constant can make the coupling arbitrarily weak, so that perturbation theory is justified.

Torsion

The use of a connection with torsion is natural, since the three-form H is part of the supergravity multiplet. The torsion two-form T^α is defined in terms of the frame and spin-connection one-forms by[21]

$$T^\alpha = de^\alpha + \omega^\alpha{}_\gamma \wedge e^\gamma, \tag{10.196}$$

which can be written in terms of connection coefficients Γ^r_{mn} according to

$$T^\alpha = \Gamma^r_{mn} e^\alpha{}_r dx^m \wedge dx^n, \tag{10.197}$$

Since torsion is a tensor, it has intrinsic geometric meaning. A connection is torsion-free if it is symmetric in its lower indices.

In defining the geometry one is free to choose what torsion tensor to include in the connection as one pleases. A connection, which is not a tensor, can always be redefined by a tensor, and in this way the torsion is changed. In particular, one can choose to use the Christoffel connection, which has no torsion. The use of a connection with torsion has the geometric consequences described below. However, you are never required to use such a connection. In flux compactifications of the heterotic string there is a natural choice, since by incorporating the three-form flux in the connection, in the way described below, one is able to define a covariantly constant spinor.

Geometrically, torsion measures the failure of infinitesimal parallelograms, defined by the parallel transport of a pair of vectors, to close. Parallel transport for the case in which the torsion vanishes is illustrated in Fig. 10.9 and a case in which it does not vanish is illustrated in Fig. 10.10.

As a simple example consider the Euclidean metric $ds^2 = dx^2 + dy^2$ on the two-dimensional plane $\mathbb{R}^2$. If parallel transport is defined in the usual

---

[21] There are other meanings of the word torsion that should not be confused with the one introduced here.

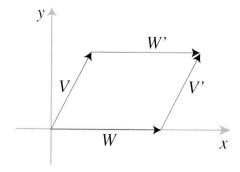

Fig. 10.9. The vectors V and W are parallel transported to V' and W' using a torsion-free connection. The resulting parallelogram closes.

sense of elementary geometry, the Christoffel connection vanishes in cartesian coordinates. However, any connection compatible with the flat metric is allowed. This means one can choose any connection that respects angles and distances or equivalently which leaves the metric covariantly constant. In the present case this means that one can choose any spin connection one-form taking values in the Lie algebra of the two-dimensional rotation group, so

$$\omega_{\alpha\beta} = h\varepsilon_{\alpha\beta}, \qquad (10.198)$$

where h can be any one-form. Parallel transport of a vector now leads to a (would-be) parallelogram that fails to close, as indicated in Fig. 10.10. Mathematically, this means that $\nabla_V W - \nabla_W V \neq [V, W]$.

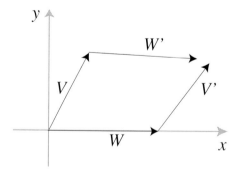

Fig. 10.10. The vectors V and W are parallel transported to V' and W' using a connection that has torsion. The resulting parallelogram fails to close.

Conditions for unbroken supersymmetry

The goal of this subsection is to derive the supersymmetry constraints for compactifications of the heterotic string to maximally symmetric four-dimensional space-time allowing for nonzero H flux. As was explained in Section 9.4, a supersymmetric configuration is one for which a spinor ε exists that satisfies

$$\delta\Psi_M = \nabla_M \varepsilon - \tfrac{1}{4}\mathbf{H}_M \varepsilon = 0,$$

$$\delta\lambda = -\tfrac{1}{2}\slashed{\partial}\Phi\varepsilon + \tfrac{1}{4}\mathbf{H}\varepsilon = 0, \qquad (10.199)$$

$$\delta\chi = -\tfrac{1}{2}\mathbf{F}\varepsilon = 0,$$

in the notation of Section 8.1. A very convenient fact is that these formulas are written in the string frame. Therefore, the warp factor is already taken into account, and they can be analyzed using a space-time that is a direct product of external and internal spaces, just as in Chapter 9. As before, Φ is the dilaton, F is the nonabelian Yang–Mills field strength and H is the three-form field strength satisfying the Bianchi identity

$$dH = \frac{\alpha'}{4}\left[\mathrm{tr}(R\wedge R) - \mathrm{tr}(F\wedge F)\right]. \qquad (10.200)$$

Poincaré invariance of the external space-time requires some components to vanish

$$H_{\mu\nu\rho} = H_{\mu\nu p} = H_{\mu n p} = 0 \quad \text{and} \quad F_{\mu\nu} = F_{\mu n} = 0. \qquad (10.201)$$

The nonvanishing fields can depend on the coordinates of the internal manifold only.

One class of consistent solutions of Eq. (10.199) has a vanishing three-form and a constant dilaton. These solutions are the conventional Calabi–Yau compactifications described in Chapter 9. Now let us consider solutions with

$$H_{mnp} \neq 0 \quad \text{and} \quad \partial_m \Phi \neq 0. \qquad (10.202)$$

The supersymmetry transformation of the gravitino can be rewritten conveniently in terms of a covariant derivative with torsion. To understand this, recall that

$$\nabla_M \varepsilon = \partial_M \varepsilon + \frac{1}{4}\omega_{MAB}\Gamma^{AB}\varepsilon. \qquad (10.203)$$

This result is written for tangent-space indices A, B and base-space indices

M, N, P of the ten-dimensional space-time. In the ten-dimensional theory, the supersymmetry variation of the gravitino can be written as

$$\tilde{\nabla}_M \varepsilon = (\nabla_M - \frac{1}{8} H_{MAB} \Gamma^{AB}) \varepsilon, \qquad (10.204)$$

where ∇_M is the torsion-free connection, since this combination appears in the supersymmetry transformation of the gravitino field. Here the derivative $\tilde{\nabla}_M$ is defined with respect to a connection with torsion. The three-form flux shifts the spin connection according to

$$\tilde{\omega}^A{}_B = \omega^A{}_B - \frac{1}{2} H_M{}^A{}_B dx^M. \qquad (10.205)$$

Using Eq. (10.196) one sees that the three-form flux represents an additional contribution to the torsion one-form

$$\tilde{T}^A = T^A + \frac{1}{2} H^A{}_{MN} dx^M \wedge dx^N. \qquad (10.206)$$

We will choose $T^A = 0$ so that $\tilde{T}^A$ is given by the three-form flux.

The supersymmetry parameter and gamma matrices decompose into internal and external pieces

$$\varepsilon(x, y) = \zeta_+(x) \otimes \eta_+(y) + \zeta_-(x) \otimes \eta_-(y), \qquad (10.207)$$

where $\zeta_\pm$ are Weyl spinors on M_4 and $\eta_\pm$ are Weyl spinors on M that satisfy

$$\zeta_- = \zeta_+^\star \quad \text{and} \quad \eta_- = \eta_+^\star. \qquad (10.208)$$

The gamma matrices split as

$$\Gamma_\mu = \gamma_\mu \otimes 1 \quad \text{and} \quad \Gamma_m = \gamma_5 \otimes \gamma_m. \qquad (10.209)$$

The conditions (10.199) have several components. From the external component of the gravitino transformation one obtains

$$\delta \psi_\mu = \nabla_\mu \zeta_+ = 0, \qquad (10.210)$$

which implies that $R = 0$. Here R is the scalar curvature of the external space, which by maximal symmetry is a constant. Even though solutions with a negative cosmological constant, that is, AdS compactifications, can be compatible with supersymmetry, only Minkowski-space compactifications are possible in the present set-up. This part of the analysis is unaffected by the H flux and is the same as in Chapter 9.

The internal component of the gravitino supersymmetry condition requires the existence of H-covariant spinors $\eta_\pm$ with

$$\tilde{\nabla}_m \eta_\pm = (\nabla_m - \frac{1}{8} H_{mnp} \gamma^{np}) \eta_\pm = 0 \qquad (10.211)$$

for a supersymmetric solution. Eq. (10.211) implies that the scalar quantity $\eta_+^\dagger \eta_+$ is a constant, and so once again it can be normalized so that $\eta_+^\dagger \eta_+ = 1$. In terms of this spinor, one can define the tensor

$$J_m{}^n = i\eta_+^\dagger \gamma_m{}^n \eta_+ = -i\eta_-^\dagger \gamma_m{}^n \eta_-. \tag{10.212}$$

Moreover, using Fierz transformations, it is possible to show that

$$J_m{}^n J_n{}^p = -\delta_m{}^p. \tag{10.213}$$

Thus, the background geometry is almost complex, and J is an almost complex structure. This implies that the metric has the property

$$g_{mn} = J_m{}^k J_n{}^l g_{kl}, \tag{10.214}$$

and that the quantity

$$J_{mn} = J_m{}^k g_{kn} \tag{10.215}$$

is antisymmetric. As a result, it can be used to define a two-form

$$J = \frac{1}{2} J_{mn} dx^m \wedge dx^n, \tag{10.216}$$

which is sometimes called the *fundamental form*. It should not be confused with the Kähler form.

The tensor $J_n{}^p$ is covariantly constant with respect to the connection $\widetilde{\nabla}$ with torsion,

$$\widetilde{\nabla}_m J_n{}^p = \nabla_m J_n{}^p - \frac{1}{2} H_{sm}{}^p J_n{}^s - \frac{1}{2} H^s{}_{mn} J_s{}^p = 0. \tag{10.217}$$

Again, it is understood that ∇ uses the Christoffel connection. Using this result, it follows that the Nijenhuis tensor, defined in the appendix of chapter 9, vanishes (see Exercise 10.10). As a result, J is a complex structure, and the manifold is complex. So one can introduce local complex coordinates z^a and $\bar{z}^{\bar{a}}$ in terms of which

$$J_a{}^b = i\delta_a{}^b, \qquad J_{\bar{a}}{}^{\bar{b}} = -i\delta_{\bar{a}}{}^{\bar{b}} \qquad \text{and} \qquad J_a{}^{\bar{b}} = J_{\bar{a}}{}^b = 0. \tag{10.218}$$

The metric is hermitian with respect to J, since combining Eqs (10.214) and (10.218) implies that the metric has only mixed components $g_{a\bar{b}}$. The fundamental form J is then related to the metric by

$$J_{a\bar{b}} = ig_{a\bar{b}}. \tag{10.219}$$

Inserting the relation between the fundamental form and the metric into Eq. (10.217) gives

$$H = i(\partial - \bar{\partial})J. \tag{10.220}$$

By definition $dJ = 0$ for a Kähler manifold. As a result, backgrounds with nonvanishing H are non-Kähler.

Let us consider the implications of the dilatino equation in Eq. (10.199). Evaluating it in complex coordinates and using $\gamma^{\bar{a}}\eta_+ = \gamma^a \eta_- = 0$, one finds that

$$\partial_a \Phi = -\frac{1}{2} H_{ab\bar{c}} g^{b\bar{c}} \qquad (10.221)$$

and the complex-conjugate relation. This relation implies the existence of a unique nowhere-vanishing holomorphic three-form Ω. This three-form is given by

$$\Omega = e^{-2\Phi} \eta_-^T \gamma_{abc} \eta_- dz^a \wedge dz^b \wedge dz^c. \qquad (10.222)$$

Using Eq. (10.221), Exercise 10.11 shows that Ω is holomorphic, that is,

$$\bar{\partial}\Omega = 0. \qquad (10.223)$$

Note that the norm of Ω, defined by

$$||\Omega||^2 = \Omega_{a_1 a_2 a_3} \bar{\Omega}_{\bar{b}_1 \bar{b}_2 \bar{b}_3} g^{a_1 \bar{b}_1} g^{a_2 \bar{b}_2} g^{a_3 \bar{b}_3}, \qquad (10.224)$$

is related to the dilaton by

$$||\Omega||^2 = e^{-4(\Phi + \Phi_0)}, \qquad (10.225)$$

where Φ_0 is an arbitrary constant.

The existence of the holomorphic $(3,0)$-form implies the vanishing of the first Chern class, that is, $c_1 = 0$. Together with Eq. (10.211) this implies that the background has $SU(3)$ holonomy. However, since the internal manifolds are not Kähler they cannot be Calabi–Yau. Note that even though the background is not Kähler, it still satisfies the weaker condition

$$d\left(e^{-2\Phi} J \wedge J\right) = 0, \qquad (10.226)$$

which means that the background is conformally balanced.

The vanishing of the supersymmetry variation of the gluino, $\mathbf{F}\varepsilon = 0$, implies that

$$(F_{ab}\gamma^{ab} + F_{\bar{a}\bar{b}}\gamma^{\bar{a}\bar{b}} + 2F_{a\bar{b}}\gamma^{a\bar{b}})\eta = 0 \qquad (10.227)$$

and hence that the gauge field satisfies

$$g^{a\bar{b}} F_{a\bar{b}} = F_{ab} = F_{\bar{a}\bar{b}} = 0, \qquad (10.228)$$

which is called the hermitian Yang–Mills equation.

Once a solution for the hermitian Yang–Mills field has been found, the fundamental form is constrained to satisfy the differential equation

$$i\partial\bar{\partial}J = \frac{\alpha'}{8}\left[\text{tr}(R \wedge R) - \text{tr}(F \wedge F)\right], \qquad (10.229)$$

which is a consequence of the anomaly cancellation condition.

To summarize, supersymmetry is unbroken if the external space-time is Minkowski and the internal space satisfies the following conditions:

- It is complex and hermitian.
- There exists a holomorphic $(3,0)$-form Ω that is related to the fundamental form by the condition that the background is conformally balanced, that is,

$$d(||\Omega||J \wedge J) = 0. \qquad (10.230)$$

- The gauge field satisfies the hermitian Yang–Mills condition.
- The fundamental form satisfies the differential equation in Eq. (10.229).

These are the only conditions that have to be imposed. Once a solution of the above constraints has been found, H and Φ are determined by the data of the geometry according to

$$H = i(\partial - \bar{\partial})J \quad \text{and} \quad \Phi = \Phi_0 - \frac{1}{2}\log||\Omega||. \qquad (10.231)$$

There exist six-dimensional compact internal spaces that solve the above constraints and lead to interesting phenomenological models in four dimensions. However, they lie beyond the scope of this book. In the following we describe a simpler example in which the internal space is four-dimensional.

Conformal K3

Four-dimensional internal spaces for heterotic-string backgrounds with torsion can be constructed by considering an ansatz of the form of a direct product in the string-frame, as before, with

$$g_{mn}(y) = e^{2D(y)}g_{mn}^{\text{K3}}(y), \qquad (10.232)$$

where $g_{mn}^{\text{K3}}(y)$ represents the (unknown) metric of K3, and $g_{mn}(y)$ is the internal part of the string-frame metric. In this four-dimensional example, the internal manifold is given by a conformal factor times a Calabi–Yau manifold.

In this background the dilatino and gravitino supersymmetry conditions can be written in the form

$$(\partial_m \Phi + \frac{1}{2}\partial_m h)\gamma^m \eta = 0 \tag{10.233}$$

and

$$\nabla_m \eta + \frac{1}{4}\partial_n h \; \gamma_m{}^n \eta = 0. \tag{10.234}$$

Here $dh = \star H$ is the one-form dual to H in four dimensions and the Hodge-star operator is defined with respect to the metric g_{mn}. The first equation implies that

$$\Phi(y) = -\frac{1}{2}h(y) + \text{const.} \tag{10.235}$$

In other words, the flux is given in terms of the dilaton by $H = -2 \star d\Phi$. In terms of the metric rescaled by the factor e^{2D}, Eq. (10.234) takes the form

$$\widetilde{\nabla}_m \eta + \frac{1}{2}\partial_n D \; \gamma_m{}^n \eta + \frac{1}{4}\partial_n h \; \gamma_m{}^n \eta = 0. \tag{10.236}$$

Therefore, for the choice

$$D(y) = \Phi(y) \tag{10.237}$$

one finds $\widetilde{\nabla}_m \eta = 0$. This is just the Killing spinor equation required to define a Calabi–Yau manifold. Since K3 is the only Calabi–Yau manifold in four dimensions, one is justified in identifying the rescaled metric with the K3 metric.

The conformal factor and the dilaton are constrained by the Bianchi identity for the H flux

$$d \star d\Phi = -\frac{\alpha'}{8}\left[\text{tr}(R \wedge R) - \text{tr}(F \wedge F)\right]. \tag{10.238}$$

Solutions can be found if the right-hand side is exact. The conditions

$$F_{\bar{a}\bar{b}} = F_{ab} = g^{a\bar{b}}F_{a\bar{b}} = 0 \tag{10.239}$$

are conformally invariant. Therefore, they only need to be solved for K3.

EXERCISES

EXERCISE 10.10
Show that the backgrounds described in Section 10.4 are complex.

Solution

In order to prove that the manifold is complex one computes the Nijenhuis tensor, which was defined in the appendix of Chapter 9 to be

$$N_{mn}{}^p = J_m{}^q J_{[n}{}^p{}_{,q]} - J_n{}^q J_{[m}{}^p{}_{,q]}.$$

Eq. (10.217) implies that the Nijenhuis tensor takes the form

$$N_{mnp} = \frac{1}{2}\left(H_{mnp} - 3J_{[m}{}^q J_n{}^s H_{p]qs}\right)$$

Identities for Dirac matrices, which are listed in the appendix of this chapter, imply

$$\begin{aligned}J_{[m}{}^p J_{n]}{}^q &= \tfrac{1}{4} g^{pr} g^{qs} (J \wedge J)_{mrns} + \tfrac{1}{2} J_{mn} J^{pq} \\ &= \tfrac{1}{2} \eta^\dagger \gamma^{pq}{}_{mn} \eta - \tfrac{1}{2} \eta^\dagger \gamma^{pq} \eta\, \eta^\dagger \gamma_{mn} \eta\, ,\end{aligned}$$

where the last line has used the six-dimensional identity

$$\frac{1}{2}(J \wedge J) = *J.$$

As a result, one obtains

$$\begin{aligned}N_{mnp} &= -\tfrac{1}{12} \eta^\dagger_+ \left\{ H, \gamma_{mnp} + 3i\gamma_{[m} J_{np]} \right\} \eta_+ \\ &= -\tfrac{1}{12} \eta^\dagger_+ \left[\partial\!\!\!/\Phi, \gamma_{mnp} + 3i\gamma_{[m} J_{np]} \right] \eta_+ \\ &= 0.\end{aligned}$$

This proves that the manifold is complex. □

Exercise 10.11
Prove that Ω in Eq. (10.222) is holomorphic.

Solution

A holomorphic three-form is a $\bar\partial$ closed form of type $(3,0)$. In order to prove that Ω is holomorphic, we compute $\bar\partial \Omega$. We start by computing its covariant derivative.

The covariant derivative (defined with respect to the Christoffel connection) acting on the tensor Ω is

$$\nabla_{\bar k} \Omega_{abc} = \partial_{\bar k} \Omega_{abc} - 3\Gamma^p_{\bar k [a} \Omega_{bc]p} = \partial_{\bar k} \Omega_{abc} - \Gamma^p_{\bar k p} \Omega_{abc}.$$

Using the definition of the Christoffel connection and expanding Eq. (10.220) in components implies

$$\Gamma^p_{\bar k p} = g^{p \bar q} \partial_{[\bar k} g_{\bar q] p} = \frac{1}{2} H_{\bar k p \bar q} g^{p \bar q} = \partial_{\bar k} \Phi.$$

518   Flux compactifications

As a result,
$$\nabla_{\bar{k}}\Omega_{abc} = \partial_{\bar{k}}\Omega_{abc} - \partial_{\bar{k}}\Phi\Omega_{abc}.$$

On the other hand, using the definition of Ω, one obtains
$$\nabla_{\bar{k}}\Omega_{abc} = -\partial_{\bar{k}}\Phi\Omega_{abc}.$$

Indeed, to see this last relation, use
$$\nabla_{\bar{k}}\Omega_{abc} = \nabla_{\bar{k}}\left(e^{-2\Phi}\eta_-^T\gamma_{abc}\eta_-\right) = -2\partial_{\bar{k}}\Phi\Omega_{abc} + 2e^{-2\Phi}\eta_-^T\gamma_{abc}\nabla_{\bar{k}}\eta_-.$$

Using Eq. (10.211), this is equal to
$$-2\partial_{\bar{k}}\Phi\Omega_{abc} + \frac{1}{2}H_{\bar{k}n\bar{p}}g^{n\bar{p}}\Omega_{abc} = -\partial_{\bar{k}}\Phi\Omega_{abc}.$$

This implies that Ω is holomorphic.   □

10.5 The strongly coupled heterotic string

This feature is generic and is not special to the type IIB theory. It also applies to the heterotic theory. The subject of moduli stabilization in the strongly coupled heterotic string is still relatively unexplored and an active area of current research.

A natural way to describe the strongly coupled $E_8 \times E_8$ heterotic string theory is in terms of M-theory. This formulation, called *heterotic M-theory*, was introduced in Chapter 8. Recall that it has a space-time geometry $\mathbb{R}^{10} \times S^1/\mathbb{Z}_2$. The quotient space $S^1/\mathbb{Z}_2$ can be regarded as a line interval that arises when the $E_8 \times E_8$ heterotic string is strongly coupled, with a length equal to $g_s\ell_s$. The gauge fields of the two E_8 gauge groups live on the two ten-dimensional boundaries of the resulting 11-dimensional space-time. This section explores some phenomenological implications of fluxes in heterotic M-theory and briefly describes moduli stabilization in the context of the strongly coupled theory. For heterotic M-theory compactified on a Calabi–Yau three-fold, the four-form field strength F_4 does not vanish if higher-order terms in $\kappa^{2/3}$ are taken into account. The Yang–Mills fields act as magnetic sources in the Bianchi-identity for F_4 and therefore an F_4 of order $\kappa^{2/3}$ is required for consistency. As in the previous sections, a warped geometry again plays a crucial role in heterotic M-theory compactifications.

One rather intriguing result is that, in heterotic M-theory, Newton's constant is bounded from below by an expression that is close to the correct

Newton's constant from the $D = 10$ heterotic string

As was shown in Chapter 8, the leading terms of the ten-dimensional effective action for the heterotic string in the string frame are

$$L_{\text{eff}} = \int d^{10}x \sqrt{-G} e^{-2\Phi} \left(\frac{4}{\alpha'^4} R - \frac{1}{\alpha'^3} \text{tr}|F|^2 \right) + \ldots. \tag{10.240}$$

If this theory is compactified on a Calabi–Yau manifold with volume $\mathcal{V}$, the resulting four-dimensional low-energy effective action takes the form

$$L_{\text{eff}} = \int d^4x \, \mathcal{V} \sqrt{-G} e^{-2\Phi} \left(\frac{4}{\alpha'^4} R - \frac{1}{\alpha'^3} \text{tr}|F|^2 \right) + \ldots. \tag{10.241}$$

In the supergravity approximation, the volume of the Calabi–Yau manifold is assumed to be large $\mathcal{V} > \alpha'^3$. Thus, the value of Newton's constant from the previous formula is

$$G_4 = \frac{e^{2\Phi} \alpha'^4}{64\pi \mathcal{V}}. \tag{10.242}$$

The value of the unification gauge coupling constant is

$$\alpha_U = \frac{e^{2\Phi} \alpha'^3}{16\pi \mathcal{V}}. \tag{10.243}$$

The previous two formulas lead to an expression for Newton's constant in terms of these variables

$$G_4 = \frac{1}{4} \alpha_U \alpha'. \tag{10.244}$$

If one assumes that the string is weakly coupled, then $e^{2\Phi} \ll 1$, and the volume of the Calabi–Yau is bounded from above

$$\mathcal{V} \ll \frac{\alpha'^3}{16\pi \alpha_U}. \tag{10.245}$$

In heterotic-string compactifications of the type described in Chapter 9, the size of the compactification manifold gives a bound on the unification scale. Thus, for a Calabi–Yau manifold that can be characterized by a single length scale, the volume satisfies $\mathcal{V} \approx M_U^{-6}$. Inserting this value into Eq. (10.245) and Eq. (10.244) one obtains a lower bound for Newton's constant[22]

$$G_4 > \frac{\alpha_{\rm U}^{4/3}}{M_{\rm U}^2},\qquad(10.246)$$

which is too large by a significant factor. The lesson is that by insisting on perturbative control, one obtains unrealistic values for the four-dimensional Newton's constant.

Newton's constant from heterotic M-theory

This situation can be improved in the context of the strongly coupled heterotic string. At strong coupling, the corrections to the predicted value of Newton's constant are closer to the phenomenologically interesting regime. If simultaneously the Calabi–Yau volume is large then the successful weak-coupling prediction for the gauge coupling constants is not ruined. Let us illustrate how fluxes at strong coupling can lead to the right prediction for G_4 in the example of the strongly coupled $E_8 \times E_8$ heterotic string, as described in terms of heterotic M-theory.[23]

The terms of interest in the action for heterotic M-theory are

$$L = \frac{1}{2\kappa_{11}^2}\int_{M_{11}} d^{11}x\sqrt{g}R - \sum_i \frac{1}{8\pi(4\pi\kappa_{11}^2)^{2/3}}\int_{M_i^{10}} d^{10}x\sqrt{g}|F_i|^2,\qquad(10.247)$$

where $i = 1, 2$ labels the gauge fields of the two different E_8 gauge groups, and κ_{11} is the 11-dimensional gravitational constant as usual. If this theory is compactified on a Calabi–Yau manifold with volume $\mathcal{V}$ times an interval $S_1/\mathbb{Z}_2$ of length πd, one can read off the value of Newton's constant and the gauge couplings to be

$$G_4 = \frac{\kappa_{11}^2}{8\pi^2\mathcal{V}d} \qquad \text{and} \qquad \alpha_{\rm U} = \frac{(4\pi\kappa_{11}^2)^{2/3}}{2\mathcal{V}}.\qquad(10.248)$$

These formulas show that, if $\alpha_{\rm U}$ and $M_{\rm U}$ are made small enough, then Newton's constant G_4 can be made small by taking d to be large enough.

The length of the interval d cannot be arbitrarily large, because there is a value of order $(\mathcal{V}/\kappa_{11})^{2/3}$, beyond which one of the two E_8's is driven to infinite coupling. To derive this bound, the concrete form of the supergravity background needs to be worked out. This was done by Witten by solving the constraint following from the gravitino supersymmetry transformation.

---

22 Typical values are $\alpha_{\rm U} \sim 1/25$ and $M_{\rm U} \sim 2\times 10^{16}{\rm GeV}$, whereas $G_4 = m_{\rm p}^{-2}$ and $m_{\rm p} \sim 10^{19}{\rm GeV}$.
23 A similar conclusion can be drawn for the strongly coupled $SO(32)$ heterotic string theory, whose strong-coupling limit is given by the weakly coupled ten-dimensional type I superstring theory.

In this background the metric is warped and the fluxes are nonvanishing due to the Bianchi identity

$$(dF)_{11IJKL} = -\frac{3\sqrt{2}}{2\pi}\left(\frac{\kappa_{11}}{4\pi}\right)^{2/3}[\text{tr}F_{[IJ}F_{KL]} - \frac{1}{2}\text{tr}R_{[IJ}R_{KL]}]\delta(x^{11}). \quad (10.249)$$

The delta-function singularity on the right-hand side of this equation comes from the boundaries or $\mathbb{Z}_2$-fixed planes, and it requires the fluxes F_4 to be nonvanishing. This Bianchi identity reproduces the right Bianchi identity for the perturbative heterotic string in the weakly coupled limit (in which the length of the interval goes to zero). As a side remark, one can see from Eq. (10.249) that, when higher-order corrections are taken into account, fluxes no longer obey the ordinary Dirac quantization condition. Namely, in the appropriate normalization, the Bianchi identity implies that fluxes are half-integer quantized,

$$[F_4/2\pi] = \lambda(F) - \lambda(R)/2, \quad (10.250)$$

where λ describes the first Pontryagin class, which is an integer. Also, F refers to the E_8 bundle and R refers to the tangent bundle.

Requiring that the infinite coupling regime be avoided gives a lower bound on Newton's constant, which (up to a numerical factor) is

$$G_4 \geq \frac{\alpha_U^2}{M_U^2}. \quad (10.251)$$

This bound is about an order of magnitude weaker than what was derived from the weakly coupled heterotic string at the beginning of this section. Inclusion of numerical factors, such as $16\pi^2$, gives a bound that is close to the correct value. Moreover, the bound can be weakened further if one chooses a Calabi–Yau manifold that is much smaller in some directions than in others, so that its size is not well characterized by a single scale.

Moduli stabilization

Moduli stabilization in the context of the heterotic string has not been explored in detail. It is, of course, desirable to see if a potential for the interval length d can be generated and to make sure that the resulting value for the interval is in agreement with the value of Newton's constant. Without entering into details, let us only mention that such a potential can be derived from nonperturbative effects in a similar manner as was done for the type IIB theory. The nonperturbative effects come from open M2-brane instantons that wrap the length of the interval (as illustrated in Fig. 10.11) and gluino condensation on the hidden boundary. Both effects combine in such a way that the length of the interval is stabilized in a phenomenologically interesting regime.

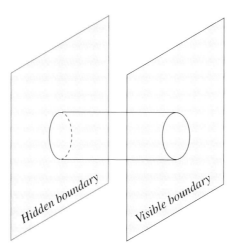

Fig. 10.11. Open M2-brane instantons stretching between both boundaries together with gluino condensation generate a potential for the interval length.

10.6 The landscape

One of the goals of string theory is to derive the standard model of elementary particles from first principles and to compute as many of its parameters as possible. The dream of a unique consistent quantum vacuum capable of making these predictions evaporated when it was discovered that there are several consistent superstring vacua in ten dimensions. Soon it became evident that the situation is even more complicated, because continua of supersymmetric vacua exist parametrized by the dilaton and other moduli. These vacua are unrealistic because they contain massless scalars, the moduli fields, and they have unbroken supersymmetry. Until supersymmetry is broken, one cannot answer the question of why the value of the cosmological constant is incredibly small but nonzero. This problem has been addressed in the recent string theory literature in the context of flux compactifications.

The anthropic principle

One approach proposed in the literature argues that there is a large number of nonsupersymmetric vacua so that the typical spacing between adjacent values for the cosmological constant is smaller than the observed value. In this case, it is reasonable that some vacua should have approximately the observed value. Moreover, a significantly larger value than is observed would not lead to galaxy formation and the development of life in the Universe, so our existence ensures that a small value was chosen. In these discussions, the possible string theory vacua are viewed as the local minima of a very complicated potential function with many peaks and valleys. This function is visualized as a *landscape*. This picture is based on an intuition derived from nonrelativistic quantum mechanics. This intuition surely breaks down if the scale of the peaks and valleys approaches the string scale or the Planck scale, as it is based on the use of the low energy effective actions that can be derived from string theory. However, it provides a valid description if it is smaller than those scales by a factor that can be made arbitrarily large.

The statistical approach

Motivated by the existence of this enormous number of vacua, a statistical analysis of their properties has been proposed. Consider the type IIB flux vacua discussed in Section 10.2, where the minima of the potential are described by isolated points. In the statistical approach, ensembles of randomly chosen systems are picked and specific quantities of interest are studied. Rather than studying individual vacua, the overall distribution of vacua on the moduli space is analyzed. Important examples of quantities that can be analyzed statistically are the cosmological constant and the supersymmetry breaking scale. These studies are motivating string theorists to rethink the concept of *naturalness* in quantum field theory. If the multiplicity of vacua can compensate for small numbers such as the ratio of the weak scale to the Planck scale, then it could undermine one of the arguments for low-energy supersymmetry breaking.

In order to study the number and distribution of type IIB flux vacua, the ensemble is built from the low-energy effective theories with flux described by the superpotential of Eq. (10.101) and subject to the tadpole-cancellation condition Eq. (10.94). It is rather important in this approach that the number of vacua that is found is finite. Fortunately, this seems to be a consequence of the constraints given by the tadpole-cancellation condition, which provides a bound on the possible fluxes. Additional constraints come from supersymmetry and duality symmetries as is discussed below. The number of vacua, with all moduli stabilized, is finite for this class of examples, but this might not be true in general.

Counting of vacua

Let us now describe the counting of supersymmetric type IIB flux vacua discussed in Section 10.2. Recall that in these vacua the three-form $G_3 =$

$F_3 - \tau H_3$ is nonvanishing. Since the three-forms F_3 and H_3 are harmonic, they are fully characterized by their periods on a basis of three-cycles

$$N_{\text{RR}}^\alpha = \eta^{\alpha\beta}\int_{\Sigma_\beta} F_3 \quad \text{and} \quad N_{\text{NS}}^\alpha = \eta^{\alpha\beta}\int_{\Sigma_\beta} H_3. \qquad (10.252)$$

Here $\eta_{\alpha\beta}$ is the intersection matrix of three-cycles and $\eta^{\alpha\beta}$ is its inverse. Recall that (for suitable normalizations) these N's are integers as a consequence of the generalized Dirac quantization condition. In this notation the tadpole-cancellation condition Eq. (10.94) gives the following constraint on the fluxes

$$0 \leq \eta_{\alpha\beta} N_{\text{RR}}^\alpha N_{\text{NS}}^\beta \leq L, \qquad (10.253)$$

where

$$L = \chi/24 - N_{D3}. \qquad (10.254)$$

Here χ is the Euler characteristic of the 3-fold and N_{D3} is a positive integer describing the total R–R charge, as in Eq. (10.94).

Using Eq. (10.101), the superpotential can be written in terms of the periods of the holomorphic three-form

$$\Pi_\alpha = \int_{\Sigma_\alpha} \Omega, \qquad (10.255)$$

as

$$W = (N_{\text{RR}}^\alpha - \tau N_{\text{NS}}^\alpha)\Pi_\alpha = N \cdot \Pi. \qquad (10.256)$$

A supersymmetric flux vacuum is determined by the flux quanta N^α and solves the equation

$$\mathcal{D}_i W = 0, \qquad (10.257)$$

where $W = 0$ corresponds to Minkowski space and $W \neq 0$ corresponds to AdS space.

A simple example

The simplest examples of flux compactifications are orientifolds, such as $T^6/\mathbb{Z}_2$. As an example, let us count the flux vacua for the simple toy model of a rigid Calabi–Yau with no complex-structure moduli, $b_3 = 2$ and periods $\Pi_1 = 1$ and $\Pi_2 = i$. The Kähler moduli are ignored as these moduli fields are fixed by nonperturbative effects and therefore can be ignored in a perturbative description. This simple example illustrates all the features of more realistic six-dimensional examples. It has no geometric moduli at all, only the axion–dilaton modulus τ, which can be viewed as the complex-structure modulus of a torus.

10.6 The landscape

The superpotential takes the simple form

$$W = N \cdot \Pi = A\tau + B, \quad (10.258)$$

with coefficients

$$A = -(N_{\text{NS}}^1 + iN_{\text{NS}}^2) = a_1 + ia_2, \quad (10.259)$$

$$B = N_{\text{RR}}^1 + iN_{\text{RR}}^2 = b_1 + ib_2. \quad (10.260)$$

Using Eq. (10.103), the condition Eq. (10.257) gives

$$D_\tau W = \partial_\tau W + \partial_\tau K W = \partial_\tau W - \frac{1}{\tau - \bar{\tau}} W = -\frac{A\bar{\tau} + B}{\tau - \bar{\tau}} = 0. \quad (10.261)$$

This determines the τ-parameter of the axion–dilaton to be

$$\tau = -\bar{B}/\bar{A}. \quad (10.262)$$

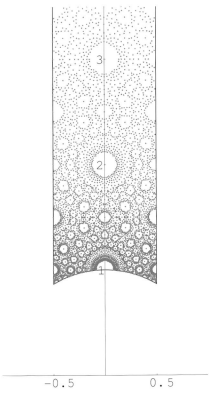

Fig. 10.12. Values of τ in the fundamental region of $SL(2,\mathbb{Z})$ for a rigid Calabi–Yau manifold with $L = 150$.

526   Flux compactifications

One final restriction on the vacua comes from the $SL(2,\mathbb{Z})$ duality symmetry of the type IIB theory. This symmetry allows one to restrict the value of the integers appearing in the previous formula to $a_2 = 0$ and $0 \leq b_1 < a_1$, which then implies that $a_1 b_2 \leq L$. For each choice of L, the values of τ that correspond to allowed choices of the fluxes can be computed using Eq. (10.262). A scatter plot of these values for the choice $L = 150$ is shown in Fig. 10.12. This figure shows that, at particular points, such as $\tau = ni$, there are holes. At the center of these holes there is a large degeneracy of vacua. For example, there are 240 vacua for $\tau = 2i$. So one concludes from this simple toy example that the statistical analysis provides the information where vacua with certain properties can be found in the moduli space. With these techniques it is possible to compute the distribution function of vacua on the moduli space of string compactifications and such an analysis can be generalized to the nonsupersymmetric case. However, this is beyond the scope of this book. On the more speculative side, it has been proposed that the landscape can be described in terms of a *wave function of the Universe*, providing an alternative way of thinking about the issue of how to choose among the many different flux vacua. This subject is an active area of current string theory research.

10.7 Fluxes and cosmology

Superstring theory and M-theory have implications for cosmology, some of which are addressed in this section. The main conceptual issues arise when the classical space-time description derived from general relativity breaks down, and the curvature of space-time diverges. This happens at the beginning of the Universe in the SBB, when the classical space-time becomes singular and the energy density becomes infinite. Here, one might hope that string theory smoothes out the singularity, due to the finite size of the string, so that there could be a sensible cosmology before the Big Bang. When the curvature of space-time and the string coupling become large, the perturbative formulation of string theory becomes unreliable, and one needs to turn to other techniques, such as the Matrix-theory proposal for M-theory,[24] which is an interesting area of current research.

Some basic cosmology

Before discussing string-theory cosmology, some basic features of the standard model of cosmology, including its successes and shortcomings, are pre-

[24] Matrix theory is introduced in Chapter 12.

10.7 Fluxes and cosmology

sented. The next two subsections are intended to present a basic "tool kit" of cosmology for the string-theory student. The interested student should consult cosmology textbooks for a more detailed and complete explanation.

The perfect-fluid description

Let us consider four-dimensional general relativity in the presence of a *perfect fluid*, which describes the energy content of the Universe. By definition, a perfect fluid is described in terms of a stress-energy tensor that is a smoothly varying function of position and is isotropic in the local rest frame. The perfect-fluid description is suggested by the fact that the matter and radiation distribution of the Universe looks remarkably homogeneous and isotropic on very large cosmological scales. For instance, most of the radiation contained in the Universe is accounted for by the *cosmic microwave background* (CMB), which is isotropic up to tiny fluctuations of order 10^{-5} once the dipole moment due to the motion of the Sun and Earth is subtracted. Furthermore, galaxy surveys indicate a homogeneous distribution at scales greater than 100 Mpc (1 pc = 3.086×10^{16}m). The energy–momentum tensor of a perfect fluid takes the form

$$T_{00} = \rho, \qquad T_{ij} = p g_{ij}. \tag{10.263}$$

This tensor is characterized by three quantities: the mass-energy density ρ, the pressure p and the spatial components of the metric g_{ij}. In addition, it is generally assumed that there is a simple relation between the mass-energy density ρ and pressure p given by the *equation of state*

$$p = w\rho, \tag{10.264}$$

where w is a constant that depends on whether the Universe is dominated by relativistic particles (termed radiation), nonrelativistic particles (collectively called matter) or vacuum energy. Some of the cosmologically relevant gravitating sources are listed in Table 10.1.

type of fluid	w	$\rho \sim a^{-3(w+1)}$	$a(t) \sim t^{2/3(w+1)}$
radiation	1/3	$1/a^4$	$t^{1/2}$
matter	0	$1/a^3$	$t^{2/3}$
vacuum energy	-1	const.	$e^{\sqrt{\Lambda/3}\,t}$

Table 10.1: *Cosmologically most relevant gravitating sources. The time dependence of the scale factor a is given for $k = 0$.*

Flux compactifications

Friedmann–Robertson–Walker Universe

The homogeneity and isotropy of the $D = 4$ space-time uniquely determines the metric to be of the following *Friedmann–Robertson–Walker* (FRW) type

$$ds^2 = -dt^2 + a^2(t)\left(\frac{dr^2}{1 - kr^2} + r^2(d\theta^2 + \sin^2\theta d\phi^2)\right). \tag{10.265}$$

The only functional freedom remaining in this metric is the time-dependent scale-factor $a(t)$ which determines the radial size of the Universe. It is determined by the Einstein equations

$$G_{\mu\nu} = R_{\mu\nu} - \frac{1}{2}g_{\mu\nu}R = 8\pi G T_{\mu\nu} - \Lambda g_{\mu\nu}, \tag{10.266}$$

and therefore by the dynamics of the theory. Here G denotes Newton's constant. A cosmological constant has been included in this equation, since recent astronomical observations indicate that it has a positive (nonvanishing) value $\Lambda = 10^{-120} M_P^4 = (10^{-3} \text{eV})^4$. In addition, the metric is characterized by the discrete parameter k, which characterizes the spatial curvature[25]

$$R_{\text{curv}} = a|k|^{-1/2}. \tag{10.267}$$

It takes the values $-1, 0, 1$ depending on whether there is enough gravitating energy in the Universe to render it *closed, flat or open*. The precise definition of these terms is given below. For the flat case, $k = 0$, the time-dependence of the scale factor for various cosmic fluids is displayed in Table 10.1.

Friedmann and acceleration equations

The Einstein field equations, which determine $a(t)$, reduce for the FRW ansatz to the Friedmann and acceleration equations, respectively

$$H^2 = \frac{1}{3M_P^2}\rho_{\text{tot}} - \frac{k}{a^2} + \frac{\Lambda}{3}, \tag{10.268}$$

$$\frac{\ddot{a}}{a} = -\frac{1}{6M_P^2}(\rho_{\text{tot}} + 3p_{\text{tot}}) + \frac{\Lambda}{3}, \tag{10.269}$$

where

$$H(t) = \dot{a}(t)/a(t) \tag{10.270}$$

defines the Hubble parameter, which determines the rate of expansion of the Universe. Furthermore,

$$\rho_{\text{tot}} = \sum_i \rho_i, \qquad p_{\text{tot}} = \sum_i p_i \tag{10.271}$$

---

[25] In these conventions r is dimensionless and $a(t)$ is a length. For $k = 0$, $\sqrt{-g} = a^3$.

are the total energy density and pressure, while $M_{\rm P} = (8\pi G)^{-1/2}$ denotes the reduced Planck mass.[26] The index i labels different contributing fluids, as listed in Table 10.1. Sometimes the cosmological constant is regarded as a time-independent contribution to the energy density and pressure of the vacuum $\rho_{vac} = -p_{vac} = M_p^2 \Lambda$. It does not appear explicitly in the previous equations.

Open, flat and closed Universes

It follows from the Friedmann equation Eq. (10.268) that (for $\Lambda = 0$) the Universe is flat, $k = 0$, when the energy density equals the critical density

$$\rho_c = 3H^2 M_{\rm P}^2. \tag{10.272}$$

This is a time-dependent function that at present has the value $\rho_{c,0} = 1.7 \times 10^{-29} {\rm g/cm}^3$.

It is customary to define the energy density of the various fluids that are present in units of ρ_c by introducing the *density parameter* $\Omega_i = \rho_i/\rho_c$ for the ith fluid. In terms of the sum over all such contributions, $\Omega = \sum_i \Omega_i = \rho_{\rm tot}/\rho_c$, the Friedmann equation takes the simple form

$$\Omega - 1 = \frac{k}{a^2 H^2} - \frac{\Lambda}{3H^2} \ . \tag{10.273}$$

This illustrates that there is a simple relation between the curvature k and the deviation from the critical density ρ_c. The classification of cosmological models as open (infinite), flat or closed (finite), which is summarized in Table 10.2, follows from this equation.[27]

ρ	Ω	spatial curvature k	type of Universe
$< \rho_c$	< 1	-1	open
$= \rho_c$	$= 1$	0	flat
$> \rho_c$	> 1	1	closed

Table 10.2: *The classification of cosmological models.*

The Friedmann and acceleration equations imply the *continuity or fluid equation*, which expresses energy conservation

$$\dot{\rho}_{\rm tot} + 3H(\rho_{\rm tot} + p_{\rm tot}) = 0 \ . \tag{10.274}$$

---

[26] The reduced Planck mass has a numerical value $M_{\rm P} = 2.436 \times 10^{18}$ GeV and differs by a factor $\sqrt{8\pi}$ from the alternative definition $m_{\rm p} = 1.22 \times 10^{19}$ GeV.

[27] The value of Λ has been absorbed into Ω in this table.

If there is a single cosmic fluid, with equation of state given by Eq. (10.264), one obtains from here the following dependence of ρ on the FRW scale-factor

$$\rho \sim \frac{1}{a^{3(w+1)}} \,. \qquad (10.275)$$

This relation, valid for any value of k, is displayed in Table 10.1 for the most important cosmic fluids. The acceleration equation implies that $\ddot{a} < 0$ for fluids with $\rho + 3p > 0$, and hence the associated FRW cosmologies describe decelerating Universes. Under the general assumption that the energy density ρ is positive, one can show that a FRW cosmology implies an initial singularity. This forms the basis for the SBB model of cosmology in which a FRW Universe starts from an initial it Big-Bang singularity.

The SBB model of cosmology

Let us now briefly summarize the successes and remaining puzzles of the SBB model of cosmology. In the cosmological time period starting at the time of *nucleosynthesis*, when protons and neutrons bound together to form atomic nuclei (mostly of hydrogen and helium), the SBB model is very well confirmed by three main observations. These are

- *The Hubble redshift law*: by extrapolation of the measured velocities of galaxies of the nearby galaxy cluster, Hubble made the bold conjecture that the Universe is undergoing a uniform expansion, so that galaxies that are separated by a distance L recede from one another with a velocity $v = H_0 L$, where H_0 is the present Hubble parameter. This relation and deviations from it are well understood.
- *Nucleosynthesis*: the relative abundance of the light elements, such as 75% H, 24% ^{3}He and smaller fractions of Deuterium and ^{4}He, is explained by the theory of nucleosynthesis and constitutes the earliest observational confirmation of the SBB model.
- *The cosmic microwave background* (CMB): most of the radiation contained in the Universe at present is nearly isotropic and has the form of a blackbody spectrum with temperature about $2.7\,°$K. It is known as the Cosmic Microwave Background (CMB). The discovery of this radiation in 1964 by Penzias and Wilson constitutes one of the great triumphs of the SBB model, which predicts a black-body distribution for the CMB. The measurement of the CMB's temperature fluctuations, $\delta T/T$, whose spatial variation is decomposed into a power spectrum, provides information on the energy-density fluctuations $\delta\rho/\rho$ in the early Universe. This is important for understanding the potential microscopic origin of the observed large-scale structure of the Universe.

However, puzzles still remain in the SBB model. Some of the most important ones are

- *The horizon problem*: the observed CMB is isotropic. However, when we follow the evolution of the Universe backwards in time according to the SBB model the sky decomposes into lots of causally disconnected patches. It needs to be explained why opposite points in the sky look so similar even though they cannot have been in causal contact since the Big Bang.
- *The flatness problem*: observation shows that $\Omega = \rho_{\text{tot}}/\rho_{\text{c}} \simeq 1$ at the current epoch. From the SBB evolution one finds that the comoving Hubble length $1/(aH)$ increases in time. Hence the Friedmann equation Eq. (10.273) shows that Ω would have to be fine-tuned to a value extremely close to one at earlier times in order to comply with present observation.
- *Unwanted relics*: the SBB model does not explain why some relics, that could in principle be abundant, are so rare. Examples of such relics are magnetic monopoles, which would be produced when the gauge group of a grand-unified theory is broken to a smaller group. Other examples are domain walls, cosmic strings or the gravitino. Perhaps not all of these objects exist, but some of them probably do. The presence of unwanted relics would be dramatic, since some of them could quickly dominate the evolution of the Universe.
- *The origin of the CMB anisotropies*: the SBB does not explain the observed CMB anisotropies occurring at a relative magnitude of about 10^{-5}.

These four puzzles are successfully addressed by an *inflationary phase* in the early Universe (taking place prior to the Big Bang), as discussed in the next section. There are more puzzles, which may or may not be connected to inflation, such as

- *Dark matter*: rotation curves of galaxies and the application of the virial theorem to the dynamics of clusters of galaxies indicate that there must be some form of invisible matter, called *dark matter*, which clusters around galaxies and is responsible for explaining the large-scale structure of the Universe. This dark matter should be predominantly *cold*, meaning that it is composed of particles that were nonrelativistic at the time of decoupling with no significant random motion.
- *Dark energy*: measurements of high red-shift Type I supernovas imply that our Universe is undergoing an accelerated expansion in the present epoch. A positive $\ddot{a}$ requires an unusual equation of state with sources of negative pressure appearing in the energy–momentum tensor, as the inequality $\rho + 3p < 0$ needs to be satisfied. The presence of a positive

cosmological constant on the right-hand side of the acceleration equation Eq. (10.269) would give such a repulsive force.
- *Why four dimensions?*: Critical M-theory or string theory predicts 11 or ten dimensions, respectively. The answer to the question of why we only observe four large dimensions might be provided within the context of cosmology.

These last three problems seem to require new physics beyond the SBB for their solution. For example, supersymmetry can provide viable dark matter candidates such as the lightest supersymmetric partner of the standard model particles (LSP). A thorough understanding of quantum gravity may be required to solve the latter two questions. On the other hand, as is discussed in the next subsection, there is a simple mechanism within the FRW cosmology framework that solves the first set of four puzzles.

Basics of inflation

Inflationary cosmology was introduced in the 1980s to solve some of the previously mentioned problems of the SBB model. This theory does not replace the SBB model, rather it describes an era in the evolution of our Universe prior to the Big Bang, without destroying any of its successes.

Definition of inflation

Very generally, a period of *inflation* is defined as a period in which the Universe is accelerating and thus the scale factor satisfies

$$\ddot{a}(t) > 0. \tag{10.276}$$

Equivalently, this condition can be rephrased as

$$\frac{d}{dt}\left(\frac{1}{aH}\right) < 0. \tag{10.277}$$

This equation states that the *comoving*[28] Hubble length $1/aH$, which is the most important characteristic scale of an expanding Universe, decreases in time. From the acceleration equation Eq. (10.269), one finds that inflation implies

$$\rho_{\text{tot}} + 3p_{\text{tot}} < 0, \tag{10.278}$$

so that, assuming $\rho > 0$, the effective pressure of the material driving the expansion has to be negative. Scalar (spin-0) particles have this property, as is discussed next.

---

[28] In general, a comoving point is defined as a point moving with the expansion of the Universe, that is, a point with vanishing momentum density.

The inflaton

The scalar particles used to construct different inflationary models are called *inflatons*. When there is just one such inflaton, it is described by the Lagrangian

$$\mathcal{L} = -\frac{1}{2}g^{\mu\nu}\partial_\mu\phi\partial_\nu\phi - V(\phi), \tag{10.279}$$

where ϕ is the inflaton and $V(\phi)$ is its potential. Different inflationary models are described by different potentials, which ultimately should be derived from a fundamental theory, such as string theory. The components of the energy–momentum tensor following from Eqs (10.279), (10.83) and (10.263) determine the expressions for the density and pressure to be

$$\rho_\phi = \frac{1}{2}\dot\phi^2 + V(\phi), \tag{10.280}$$

$$p_\phi = \frac{1}{2}\dot\phi^2 - V(\phi). \tag{10.281}$$

Here spatial gradients are assumed to be negligible, so that ϕ can be regarded to be a function of t only.

We conclude from this that inflation takes place as long as $\dot\phi^2 < V(\phi)$, which is generally the case for potentials that are flat enough. Neglecting k, Λ and other forms of matter, these expressions can be substituted into the Friedmann equation Eq. (10.268) and the continuity equation Eq. (10.274) to get the equations of motion

$$H^2 = \frac{1}{3M_\mathrm{P}^2}[V(\phi) + \frac{1}{2}\dot\phi^2] \tag{10.282}$$

and

$$\ddot\phi + 3H\dot\phi = -\frac{dV}{d\phi}. \tag{10.283}$$

One observes that the field equation for the inflaton looks like a harmonic oscillator with a friction term given by the Hubble parameter. Different models of inflation can be obtained by solving these two equations for a variety of potentials $V(\phi)$. Some examples are discussed below. Before doing so, let us first explain why inflation solves some of the problems not explained within the context of the SBB model.

Solution to some problems of the SBB model

From the form of the Friedmann equation, it becomes evident why inflation can solve some of the unanswered questions of the SBB model. According to Eq. (10.277), the comoving Hubble length decreases in time during inflation, and this is just what is needed to solve the flatness problem. Whereas usually Ω is driven away from 1, the opposite happens during inflation, as we can see from Eq. (10.273) (the Friedmann equation), with the cosmological constant term set to zero or absorbed into Ω. The curvature term become negligible once the comoving Hubble length increases. Hence, if inflation lasts for a long enough time, it brings Ω very close to 1 without the necessity for fine-tuning Ω. The horizon problem is solved as the distance between comoving points gets drastically stretched during inflation. This allows the entire present observable Universe to lie within a region that was well inside the Hubble radius before inflation. Since the Hubble radius is a good proxy for the particle horizon size, that is, the size over which massless particles can causally influence each other, the whole currently observable Universe could have been causally connected before inflation. Likewise, this stretching dilutes the density of any undesired relic particles, provided they are produced before the inflationary era.

Different inflationary models

Cosmologists have considered a large number of models and studied their inflationary behavior. The models studied in the literature can be classified according to three independent criteria.

- *Initial conditions for inflation*: many inflationary models are based on the assumption that the Universe was in a state of thermal equilibrium with a very high temperature at the beginning of inflation. The inflaton was at the minimum of its temperature dependent effective potential $V(\phi, T)$. The main idea of *chaotic inflation* is to study all possible initial conditions for the Universe including those where the Universe is outside of thermal equilibrium and the scalar is no longer at its minimum.
- *Behavior of the model during inflation*: there are various possibilities for the time dependence of the scale factor $a(t)$. Power law inflation is one example that is discussed next.
- *End of inflation*: there are basically two possibilities for ending the inflationary era, *slow roll* or a *phase transition*. In the first type of model the inflaton is a slowly evolving (or "rolling") field, which at the end of inflation becomes faster and faster. Phase transition models contain at least two scalar fields. One of the fields becomes tachyonic at the end of inflation, which generally signals an instability, where a phase transition takes place. *Hybrid inflation* is an example. This type of inflation is

of particular interest in recent attempts to make contact between string theory and inflation.

Power-law inflation

It is hard to find the exact solution of Eqs (10.282) and (10.283) for a generic inflaton potential $V(\phi)$, so approximations or numerical studies have to be made. However, there is one known analytic solution called *power-law inflation*. For power-law inflation the potential is

$$V(\phi) = V_0 \exp\left(-\sqrt{\frac{2}{p}}\frac{\phi}{M_P}\right), \tag{10.284}$$

where V_0 and p are constants. The scale factor and inflaton that solve the spatially flat equations of motion are

$$a(t) = a_0 t^p, \tag{10.285}$$

$$\phi(t) = \sqrt{2p} M_P \log\left(\sqrt{\frac{V_0}{p(3p-1)}}\frac{t}{M_P}\right). \tag{10.286}$$

The scale factor is inflationary as long as $p > 1$.

Slow-roll approximation

As stated above, finding exact solutions to Eqs (10.282) and (10.283) is difficult, so approximations need to be made. The so-called *slow roll approximation* neglects one term in each equation

$$H^2 \approx \frac{V(\phi)}{3M_P^2}, \tag{10.287}$$

$$3H\dot\phi \approx -V'(\phi), \tag{10.288}$$

where primes are derivatives with respect to the inflaton. A *necessary* condition for the slow-roll approximation to be valid is that the two *slow-roll parameters* ε and η are small

$$\varepsilon(\phi) = \frac{1}{2} M_P^2 (V'/V)^2 \ll 1, \tag{10.289}$$

$$|\eta(\phi)| = M_P^2\, |V''/V| \ll 1. \tag{10.290}$$

The parameter ε is positive by definition, but the absolute value is required on the left-hand side of the second equation, since η can be negative. Obtaining a solution to the slow-roll conditions is *sufficient* to achieve inflation, but not *necessary*. This can be seen by rewriting the condition for inflation Eq. (10.276) as

$$\frac{\ddot{a}}{a} = \dot{H} + H^2 > 0, \tag{10.291}$$

where $a > 0$ needs to be taken into account. This is obviously satisfied for $\dot{H} > 0$. From the Friedman and acceleration equations this requires in $p_\phi < \rho_\phi$, which is not satisfied for the scalar field described by Eqs (10.280), (10.281). If $\dot{H} < 0$, then the following inequality has to be satisfied

$$-\frac{\dot{H}}{H^2} < 1. \tag{10.292}$$

This can be rewritten in terms of ε using the slow-roll approximation

$$-\frac{\dot{H}}{H^2} \approx \frac{M_P^2}{2}\left(\frac{V'}{V}\right)^2 = \varepsilon. \tag{10.293}$$

By the slow-roll approximation, $\varepsilon \ll 1$, we observe that this condition leads to $\ddot{a} > 0$ and inflation. The second restriction $\eta \ll 1$ guarantees the friction term dominates in Eq. (10.283) so that inflation lasts long enough. The above conditions provide a straightforward method to check if a particular potential is inflationary. For the simple example of $V(\phi) = m^2\phi^2/2$, the slow-roll approximation holds for $\phi^2 > 2M_P^2$, and inflation ends once the scalar field gets so close to the minimum that the slow-roll conditions break down.

Exit from inflation

From the previous discussion, one concludes that the slow-roll conditions provide a way to characterize the *exit from inflation*. The inflationary process comes to an end when the approximations break down, which happens for a value of ϕ for which $\varepsilon(\phi) = 1$. A simple calculation shows that, for power-law inflation, the slow-roll parameters are given by constants

$$\varepsilon = \eta/2 = 1/p, \tag{10.294}$$

so that inflation never ends. In principle, this is a problem. One way of solving it could be provided by embedding this model into string theory, where additional dynamics might provide an end to the inflationary era.

Hybrid inflation

An inflationary model that has played a role in recent string-cosmology developments, called *hybrid inflation*, was constructed in the early 1990s. This model is based on two scalar fields: the inflaton ψ, whose potential is flat and satisfies the slow-roll conditions, and another scalar ϕ, whose mass depends on the inflaton field. Inflation ends in this model not because the slow-roll approximation breaks down, but because the field ϕ becomes tachyonic, that is, its mass squared becomes negative. This signals an instability, where a phase transition takes place. During this phase transition topological defects, such as *cosmic strings*[29], can be formed. The explicit form of the potential for hybrid inflation is

$$V(\phi, \psi) = a(\psi^2 - 1)\phi^2 + b\phi^4 + c, \qquad (10.295)$$

where a, b, c are positive constants. From the form of $V(\phi, \psi)$, one easily observes that, for $\psi^2 > 1$, the field ϕ has a positive mass squared, it becomes massless at $\psi = 1$ and ϕ is tachyonic for $\psi^2 < 1$. Since ϕ is driven to zero for $\psi > 1$, the potential in the ψ direction is flat and satisfies the slow-roll conditions, so that ψ is identified with the inflaton, while ϕ is called the *tachyon*. As discussed in the next section, precisely such a tachyon appears in brane–antibrane inflation, which is how hybrid inflation makes its appearance in string theory. After inflation $\psi^2 < 1$, ϕ acquires a vev and ψ becomes massive.

Number of e-foldings

There are various model-dependent quantities that can be compared with cosmological observations, and which can eventually be used to rule out some of the inflationary models. The amount of inflation that occurs after time t is characterized by the ratio of the scale factors at time t and at the end of inflation. This ratio determines *number of e-foldings* $N(t)$

$$N(t) = \log\left(\frac{a(t_{\text{end}})}{a(t)}\right), \qquad (10.296)$$

where t_{end} is the time when inflation ends. This quantity measures the amount of inflation that remains to take place at any given time t. Using the slow-roll approximation, N can be conveniently rewritten in terms of the inflaton and its potential

$$N(t) = \int_t^{t_{\text{end}}} \frac{\dot{a}}{a} dt = \int_t^{t_{\text{end}}} H dt \approx \frac{1}{M_{\text{P}}^2} \int_{\phi_{\text{end}}}^{\phi} \frac{V}{V'} d\phi. \qquad (10.297)$$

Here ϕ_{end} is the value of the inflaton at the end of inflation, which satisfies $\epsilon(\phi_{\text{end}}) = 1$ when inflation ends through a breakdown of the slow-roll approximation. To solve the flatness and horizon problems, the number of

---

[29] The existence of cosmic strings would be extraordinary, as a direct experimental evidence of string theory would be provided. This subject is nevertheless beyond the scope of this book.

e-foldings has to be larger than 60, a criterion that can be used to rule out some inflationary models.

Gravitational waves and density perturbations

Inflation not only explains the homogeneity and isotropy of the Universe, but it also predicts the *spectrum of gravitational waves* (also called *tensor perturbations*) as well as the *density perturbations* (also called *scalar perturbations*) of the CMB. Density perturbations create anisotropies in the CMB and are responsible for the formation and clustering of galaxies. The size of these irregularities depends on the energy scale at which inflation takes place. The observed scalar perturbations are in excellent agreement with the predictions of inflation. Gravitational waves do not affect the formation of galaxies but lead to polarization of the CMB, which is beginning to show up in the WMAP (Wilkinson Microwave Anisotropy Probe) satellite experiment and will be measured better in future missions.

Without entering into much detail, let us mention that such fluctuations in the energy density of the Universe can be explained in the context of inflation as originating from the quantum fluctuations of the inflaton. Inflation produces density perturbations at every scale. The amplitude of these perturbations depends on the form of the inflaton potential V. More precisely, the spectrum for density perturbations $\delta_H(k) \sim \delta\rho/\rho$ and gravitational waves $A_G(k)$ are given by the expressions

$$\delta_H(k) = \sqrt{\frac{512\pi}{75}} \frac{V^{2/3}}{M_P^3 V'}\bigg|_{k=aH}, \tag{10.298}$$

$$A_G(k) = \sqrt{\frac{32}{75}} \frac{V^{1/2}}{M_P^2}\bigg|_{k=aH}. \tag{10.299}$$

Here k is the comoving wave number, appearing because the fluctuations are typically analyzed in a Fourier expansion into comoving modes $\delta\phi = \Sigma\delta\phi_k e^{ikx}$. The right-hand side of these equations is to be evaluated at a particular time during inflation for which $k = aH$, which for a given k corresponds to a particular value of ϕ.

Comparison with cosmological data

Cosmological data lead to $\delta_H = 1.91 \times 10^{-5}$, provided that $A_G \ll \delta_H$. To compare with observational data, it is useful to express the spectrum in terms of observable quantities and to make a power-law approximation

$$\delta_H(k) \approx k^{n-1}, \qquad A_G^2(k) \approx k^{n_G}. \tag{10.300}$$

Here n and n_G are called the spectral indices for scalar and tensor perturbations, respectively

$$n - 1 = \frac{d \ln \delta_H^2}{d \ln k}, \qquad n_G = \frac{d \ln A_G^2}{d \ln k}. \tag{10.301}$$

The spectral indices can be expressed in terms of the slow-roll parameters

$$n = 1 - 6\varepsilon + 2\eta, \tag{10.302}$$

$$n_G = -2\varepsilon, \tag{10.303}$$

which shows that, in the slow-roll approximation, the spectrum is almost scale invariant $n \approx 1$. Because spectral indices are measurable quantities, we can use these relations to gain information about the inflaton potential. Recent results from WMAP indicate that $n \sim .95$.

Fluxes and inflation

The embedding of inflation into string theory is difficult in conventional Calabi–Yau compactification. Even though such compactifications contain many scalar fields that could potentially serve as inflatons, namely the moduli fields, these fields are generically either massless or have a potential with a runaway behavior, which makes their interpretation as inflatons rather difficult. This situation has changed quite a bit with the development of a better nonperturbative understanding of string theory and flux compactifications.

Brane–brane inflation

One of the first attempts to embed inflation into string theory (developed in the late 1990s) makes use of D-branes. In this approach a pair of D-branes is considered and the inflaton is identified with the scalar field describing the separation of the branes, that is, it is the lowest mode of the open string that connects the two D-branes. If supersymmetry is preserved, there is no net force between the branes and no potential for the inflaton. This has been verified by a one-loop string amplitude calculation, which is not presented here. The intuitive argument is that, for a BPS brane configuration, the gravitational attraction between the branes is compensated by the repulsive Coulomb forces between the two branes coming from various NS–NS and R–R fields. However, when supersymmetry is broken (in a certain way), there is a net attractive force between the branes. This leads to a potential for the inflaton field.

Even though this was the first proposal that demonstrated the possibility of making connections between string theory or brane physics and inflation, the concrete model had some problems, such as a drastic fine tuning required to reproduce the experimental values of the density perturbations or the lack of a satisfactory explanation for the end of inflation.

Brane–antibrane inflation

Some of these problems were solved in the context of brane–antibrane inflation. Consider instead a D3/anti-D3 system located at specific points of a Calabi–Yau three-fold. For a D3/anti-D3 system supersymmetry is broken, and there is a net attractive force between the branes and antibranes, whose explicit form is given by the potential (for a large distance between the brane and the antibrane)

$$V(r) = 2T_3 \left(1 - \frac{1}{2\pi^3}\frac{T_3}{M_{10}^8 r^4}\right), \tag{10.304}$$

where M_{10} is the ten-dimensional Planck mass, T_3 is the D3-brane tension and r is the separation between the brane and the antibrane. One can write this potential in terms of the canonically normalized scalar $\phi = T_3^{1/2} r$, where it takes the form

$$V(\phi) = 2T_3 \left(1 - \frac{1}{2\pi^3}\frac{T_3^3}{M_{10}^8 \phi^4}\right). \tag{10.305}$$

Using this potential, one can compute the slow-roll parameters appearing in Eqs (10.289) and (10.290)

$$\epsilon = \frac{1}{2} M_P^2 (V'/V)^2 \sim \frac{L^6}{r^{10}}. \tag{10.306}$$

$$\eta = M_P^2 (V''/V) \sim \frac{L^6}{r^6}. \tag{10.307}$$

M_P is the four-dimensional Planck mass appearing in Eq. (10.290) which is related to the ten-dimensional Planck mass by $M_P^2 = L^6 M_{10}^8$. Here L^6 approximately represents the volume of the Calabi–Yau three-fold. The D3 and anti-D3-branes are localized at specific points on the Calabi–Yau manifold, that is, they cannot be separated by more than L. As a result, it is not possible to achieve $|\eta| \ll 1$, as needed for slow-roll inflation. Different proposals for solving this problem have been presented in the literature, such as D3- and anti-D3-branes in a warped geometry (this is discussed next), branes at angles or collisions of multiple branes [30].

---

30 A more recent proposal is to give up the slow roll condition.

Inflation and fluxes

In the previous treatment of the D3/anti-D3 system the size L was treated as a constant. However, it is known that in string theory the size of the internal manifold is a modulus. The potential (in the four-dimensional Einstein frame) for this field (again for a D3/anti-D3-brane distance large as compared to the string scale) is

$$V(\phi, L) \approx \frac{2T_3}{L^{12}}. \tag{10.308}$$

This potential is very steep for small L. As a result, treating L as a dynamical field causes the Calabi–Yau size to become large too fast to realize slow-roll inflation. This issue could in principle be avoided if the radial modulus of the internal manifold is stabilized. In Section 10.3 a mechanism was described to stabilize the radial modulus of a D3/anti-D3 system in terms of fluxes and nonperturbative effects. The stabilization of the radial modulus using nonperturbative corrections to the superpotential does not solve this problem (unless some degree of fine tuning is allowed), but it puts it into a new perspective.

As discussed in Section 10.3, to analyze the stabilization of the moduli of a D3/anti-D3 system on an internal warped geometry the scalar potential for the radial modulus of the internal manifold and the scalars describing the positions of the branes need to be derived. $\mathcal{N} = 1$ supersymmetry dictates that this potential is determined in terms of a Kähler potential and a superpotential.

Consider first a single D3-brane position modulus ϕ and the radial modulus of the internal space ρ. The Kähler potential is given by

$$\mathcal{K}(\rho, \bar{\rho}, \phi, \bar{\phi}) = -3 \log \left[\rho + \bar{\rho} - k(\phi, \bar{\phi}) \right]. \tag{10.309}$$

Here the real part of ρ is related to the size L by

$$2L = \rho + \bar{\rho} - k(\phi, \bar{\phi}), \tag{10.310}$$

while the imaginary part of ρ is the axion χ. Furthermore, $k(\phi, \bar{\phi})$ is the canonical Kähler potential for the inter-brane distance, which is given by $k(\phi, \bar{\phi}) = \phi \bar{\phi}$.

The other quantity that determines the form of the low-energy effective action is the superpotential W. As explained in Section 10.3, W takes the form

$$W(\rho) = W_0 + A e^{-a\rho}. \tag{10.311}$$

Here W_0 is the perturbative superpotential

$$W_0 = \int G_3 \wedge \Omega, \qquad (10.312)$$

where Ω is the holomorphic $(3,0)$ form. The exponential contribution depending on ρ comes from nonperturbative effects, as discussed in Chapter 10. Further contributions to the scalar potential involving the radial modulus come from corrections to the Kähler potential, which will be ignored in the following. The complete form of these corrections is not known at present.

The above results for the Kähler potential and the superpotential can be used to compute the scalar potential for the Calabi–Yau volume and the brane position, which is determined by supersymmetry

$$V = e^{\mathcal{K}} \left(G^{a\bar{b}} D_a W D_{\bar{b}} \bar{W} - 3|W|^2 \right). \qquad (10.313)$$

Using the previous expressions for the Kähler potential and the superpotential the potential takes the form

$$V = \frac{1}{6L} \left(|\partial_\rho W|^2 - \frac{3}{2L}(\bar{W} \partial_\rho W + W \partial_{\bar{\rho}} \bar{W}) \right) + \left(\frac{|\partial_\rho W|^2}{12L^2} \right) \phi \bar{\phi}. \qquad (10.314)$$

As explained in Chapter 10, including the effects of the anti-D3-brane gives an additional term in the potential

$$V = \frac{1}{6L} \left(|\partial_\rho W|^2 - \frac{3}{2L}(\bar{W} \partial_\rho W + W \partial_{\bar{\rho}} \bar{W}) \right) + \left(\frac{|\partial_\rho W|^2}{12L^2} \right) \phi \bar{\phi} + \frac{D}{(2L)^2}, \qquad (10.315)$$

where D is a positive constant. This potential can be expanded about a minimum in which $\rho = \rho_c$, and $\phi = 0$. After transforming to a canonically normalized field $\varphi = \phi/\sqrt{3/(\rho + \bar{\rho})}$, the potential can be written in the form

$$V = \frac{V_0(\rho_c)}{(1 - \varphi\bar{\varphi}/3)^2} \approx V_0(\rho_c) \left(1 + \frac{2}{3} \varphi \bar{\varphi} \right). \qquad (10.316)$$

This potential leads to a slow-roll parameter $\eta = 2/3$, which again indicates that no slow-roll inflation can be described in this scenario, at least not in an obvious manner. Allowing a certain amount of fine tuning of the interbrane distance would obviously solve this problem. As previously mentioned, other alternatives based on inflation are currently explored in the literature. Other approaches aim to propose an alternative to inflation such as brane gases, time-dependent warped geometries, models based on Matrix theory or models that make a connection to the dS/CFT correspondence. It is fair to say that, even though it is an exciting prospect, the application of string theory to cosmology is still at its early stages.

Hybrid inflation and exit from inflation

One very attractive aspect of D3/anti-D3-brane inflation is that it provides a natural mechanism to end inflation based on the hybrid inflation mechanism previously discussed. The potential for the interbrane distance discussed so far is valid for distances that are large compared to the string scale. Since the force between the D3-brane and the anti-D3-brane is attractive, the branes collide and annihilate with one another. This process is described in terms of an additional field T, which corresponds to the tachyon of hybrid inflation. For large brane separation, this field is massive. It becomes massless once the branes come sufficiently close to one another and tachyonic when they annihilate. The form of the potential describing this process is the same as the potential for hybrid inflation previously discussed:

$$V(\phi, T) = a \left((\phi/\ell_s)^2 - b\right) T^2 + cT^4 + V(\phi), \qquad (10.317)$$

where a, b and c are positive constants. The collision of branes results in the production of strings of cosmic size, which are called *cosmic strings*. Even though they are not an inevitable prediction, the discovery of such objects would be a spectacular way to verify string theory. Further progress in string cosmology, together with more observational data, may someday provide direct evidence of string theory.

Appendix: Dirac matrix identities

This appendix lists various identities satisfied by Dirac matrices. These have been used in this chapter to analyze the conditions for unbroken supersymmetry of flux compactifications.

$$[\gamma_m, \gamma^r] = 2\gamma_m{}^r \qquad \{\gamma_m, \gamma^r\} = 2\delta_m{}^r$$

$$[\gamma_{mn}, \gamma^r] = -4\delta_{[m}{}^r \gamma_{n]} \qquad \{\gamma_{mn}, \gamma^r\} = 2\gamma_{mn}{}^r$$

$$[\gamma_{mnp}, \gamma^r] = 2\gamma_{mnp}{}^r \qquad \{\gamma_{mnp}, \gamma^r\} = 6\delta_{[m}{}^r \gamma_{np]}$$

$$[\gamma_{mnpq}, \gamma^r] = -8\delta_{[m}{}^r \gamma_{npq]} \qquad \{\gamma_{mnpq}, \gamma^r\} = 2\gamma_{mnpq}{}^r$$

$$[\gamma_{mnpqk}, \gamma^r] = 2\gamma_{mnpqk}{}^r \qquad \{\gamma_{mnpqk}, \gamma^r\} = 10\delta_{[m}{}^r \gamma_{npqk]}$$

$$[\gamma_{mn}, \gamma^{rs}] = -8\delta_{[m}^{[r}\gamma_{n]}^{s]} \qquad \{\gamma_{mn}, \gamma^{rs}\} = 2\gamma_{mn}{}^{rs} - 4\delta_{[mn]}{}^{rs}$$
$$[\gamma_{mnp}, \gamma^{rs}] = 12\delta_{[m}^{[r}\gamma_{np]}^{s]} \qquad \{\gamma_{mnp}, \gamma^{rs}\} = 2\gamma_{mnp}{}^{rs} - 12\delta_{[mn}{}^{rs}\gamma_{p]}$$
$$[\gamma_{mnpq}, \gamma^{rs}] = -16\delta_{[m}^{[r}\gamma_{npq]}^{s]} \qquad \{\gamma_{mnpq}, \gamma^{rs}\} = 2\gamma_{mnpq}{}^{rs} - 24\delta_{[mn}{}^{rs}\gamma_{pq]}$$
$$[\gamma_{mnpqk}, \gamma^{rs}] = 20\delta_{[m}^{[r}\gamma_{npqk]}^{s]} \qquad \{\gamma_{mnpqk}, \gamma^{rs}\} = 2\gamma_{mnpqk}{}^{rs} - 40\delta_{[mn}{}^{rs}\gamma_{pqk]}$$

$$[\gamma_{mnp}, \gamma^{rst}] = 2\gamma_{mnp}{}^{rst} - 36\delta_{[mn}^{[rs}\gamma_{p]}^{t]}$$
$$[\gamma_{mnpq}, \gamma^{rst}] = -24\delta_{[m}^{[r}\gamma_{npq]}^{st]} + 48\delta_{[mnp}{}^{rst}\gamma_{q]}$$
$$[\gamma_{mnpqk}, \gamma^{rst}] = 2\gamma_{mnpqk}{}^{rst} - 120\delta_{[mn}^{[rs}\gamma_{pqk]}^{t]}$$

$$\{\gamma_{mnp}, \gamma^{rst}\} = 18\delta_{[m}^{[r}\gamma_{np]}^{st]} - 12\delta_{[mnp]}{}^{rst}$$
$$\{\gamma_{mnpq}, \gamma^{rst}\} = 2\gamma_{mnpq}{}^{rst} - 72\delta_{[mn}^{[rs}\gamma_{pq]}^{t]}$$
$$\{\gamma_{mnpqk}, \gamma^{rst}\} = 30\delta_{[m}^{[r}\gamma_{npqk]}^{st]} - 120\delta_{[mnp}{}^{rst}\gamma_{qk]}$$

$$[\gamma_{mnpq}, \gamma^{rstu}] = -32\delta_{[m}^{[r}\gamma_{npq]}^{stu]} + 192\delta_{[mnp}^{[rst}\gamma_{q]}^{u]}$$
$$[\gamma_{mnpqk}, \gamma^{rstu}] = 40\delta_{[m}^{[r}\gamma_{npqk]}^{stu]} - 480\delta_{[mnp}^{[rst}\gamma_{qk]}^{u]}$$

$$\{\gamma_{mnpq}, \gamma^{rstu}\} = 2\gamma_{mnpq}{}^{rstu} - 144\delta_{[mn}^{[rs}\gamma_{pq]}^{tu]} + 48\delta_{[mnpq]}{}^{rstu}$$
$$\{\gamma_{mnpqk}, \gamma^{rstu}\} = 2\gamma_{mnpqk}{}^{rstu} - 240\delta_{[mn}^{[rs}\gamma_{pqk]}^{tu]} + 240\delta_{[mnpq}{}^{rstu}\gamma_{k]}$$

$$[\gamma_{mnpqk}, \gamma^{rstuv}] = 2\gamma_{mnpqk}{}^{rstuv} - 400\delta_{[mn}^{[rs}\gamma_{pqk]}^{tuv]} + 1200\delta_{[mnpq}^{[rstu}\gamma_{k]}^{v]}$$
$$\{\gamma_{mnpqk}, \gamma^{rstuv}\} = 50\delta_{[m}^{[r}\gamma_{npqk]}^{stuv]} - 1200\delta_{[mnp}^{[rst}\gamma_{qk]}^{uv]} + 240\delta_{[mnpqk]}{}^{rstuv}$$

In general,

$$\left.\begin{array}{l}[\gamma_{m_1...m_p}, \gamma^{n_1...n_q}] \quad pq \text{ odd} \\ \{\gamma_{m_1...m_p}, \gamma^{n_1...n_q}\} \quad pq \text{ even}\end{array}\right\} = 2\gamma_{m_1...m_p}{}^{n_1...n_q}$$

$$-\frac{2p!q!}{2!(p-2)!(q-2)!}\delta_{[m_1 m_2}{}^{[n_1 n_2}\gamma_{m_3...m_p]}{}^{n_3...n_q]}$$

$$+\frac{2p!q!}{4!(p-4)!(q-4)!}\delta_{[m_1...m_4}{}^{[n_1...n_4}\gamma_{m_5...m_p]}{}^{n_5...n_q]}$$

$$-\ldots$$

and

$$\left.\begin{array}{l}[\gamma_{m_1...m_p}, \gamma^{n_1...n_q}] \quad pq \text{ even} \\ \{\gamma_{m_1...m_p}, \gamma^{n_1...n_q}\} \quad pq \text{ odd}\end{array}\right\} = \frac{(-1)^{p-1}2p!q!}{1!(p-1)!(q-1)!}\delta_{[m_1}{}^{[n_1}\gamma_{m_2...m_p]}{}^{n_2...n_q]}$$

$$-\frac{(-1)^{p-1}2p!q!}{3!(p-3)!(q-3)!}\delta_{[m_1 m_2 m_3}{}^{[n_1 n_2 n_3}\gamma_{m_4...m_p]}{}^{n_4...n_q]}$$

$$+\ldots$$

The Fierz transformation identity for commuting spinors is

$$\chi\bar{\psi} = \frac{1}{2^{[d/2]}} \sum_{p=0}^{d} \frac{1}{p!} \gamma^{m_p...m_1}\bar{\psi}\gamma_{m_1...m_p}\chi. \tag{10.318}$$

In the case of anticommuting spinors there is an additional minus sign.

Homework Problems

Problem 10.1
Show that covariant derivatives with respect to conformally transformed metrics $\hat{g}_{MN} = \Omega^2 g_{MN}$ are related by

$$\hat{\nabla}_M \eta = \nabla_M \eta + \frac{1}{2}\Omega^{-1}\nabla_N \Omega \Gamma_M{}^N \eta.$$

Use this result to derive Eq. (10.18).

Problem 10.2
Re-express the supersymmetry transformation Eq. (10.26) in terms of the rescaled spinor $\xi = \Delta^{1/4}\eta$. Use this equation to show that the almost

Flux compactifications

complex structure defined by Eq. (10.29) is covariantly constant

$$\nabla_p J_m{}^n = 0,$$

where ∇_p is defined with respect to the metric g_{mn} appearing in Eq. (10.5).

Problem 10.3
Use the Fierz identity Eq. (10.318) to show that the almost complex structure given in Eq. (10.29) satisfies $J^2 = -1$.

Problem 10.4
Consider a flux compactification of M-theory on an eight manifold to three-dimensional Minkowski space-time. Suppose that two Majorana–Weyl spinors of opposite chirality ξ_+, ξ_- on the eight-dimensional internal manifold can be found

$$P_\pm \xi = \frac{1}{2}(1 \pm \gamma^9)\xi = \xi_\pm,$$

so that the 8D spinor $\xi = \xi_+ + \xi_-$ is nonchiral. Assuming that the internal flux component is self-dual, show that, after an appropriate rescaling of the spinor, the internal component of the gravitino supersymmetry transformation takes the form

$$\nabla_m \xi_+ - \frac{1}{4}\Delta^{-3/4} \mathbf{F}_m \xi_- = 0, \qquad \nabla_m \xi_- = 0.$$

Problem 10.5
Consider M-theory compactified on an eight manifold with a nonchiral complex spinor on the internal space. Recall that Eq. (10.17) showed that a nonvanishing vector field can be constructed.

(i) Use the Fierz identity (10.318) to show that Eq. (10.17) implies that the vector field relates the two (real) spinors of opposite chirality

$$\eta_1 = v^a \gamma_a \eta_2.$$

(ii) Use part (i) and the result of Problem 10.4 to show that the primitivity condition Eq. (10.36) is modified to

$$F \wedge J + \star dv = 0,$$

where v has been rescaled by a constant.

Problem 10.6
Show that the operations J_3, J_+, J_- in Eq. (10.38) define an $SU(2)$ algebra.

PROBLEM **10.7**
Verify Eq. (10.226), which shows that the flux backgrounds for the weakly coupled heterotic string in Section 10.4 are conformally balanced.

PROBLEM **10.8**
Show that, in the absence of sources or singularities in the background geometry, type IIB theories compactified to four dimensions do not admit dS space-times as solutions to the equations of motion. In other words, repeat the computation that led to Eq. (10.86) by allowing a cosmological constant Λ in external space-time.

PROBLEM **10.9**
Assuming a constant dilaton, show that the scalar potential of type IIB theory compactified on a Calabi–Yau three-fold in the presence of fluxes is given by
$$V = e^{\mathcal{K}} \left(G^{a\bar{b}} \mathcal{D}_a W \mathcal{D}_{\bar{b}} \overline{W} - 3|W|^2 \right),$$
where
$$W = \int_M \Omega \wedge G_3.$$
Here a, b label all the holomorphic moduli. You can assume that the Kähler potential is given by Eq. (10.104).

PROBLEM **10.10**
Show that, in a Calabi–Yau four-fold compactification of M-theory, the stationary points of
$$|Z(\gamma)|^2 = \frac{|\int_\gamma \Omega|^2}{\int \Omega \wedge \bar{\Omega}}$$
are given by the points in moduli space where $|Z(\gamma)|^2 = 0$, or if $|Z(\gamma)|^2 \neq 0$, then F has to satisfy $F^{1,3} = F^{3,1} = 0$. In the above expression γ is the Poincaré dual cycle to the four-form F. A related result is derived in Chapter 11 in the context of the attractor mechanism for black holes.

PROBLEM **10.11**
Show that the Christoffel connection does not transform as a tensor under coordinate transformations, but that torsion transforms as a tensor.

PROBLEM **10.12**
Show that D7-branes give a negative contribution to the right-hand side of

Problem 10.13

Show that the Kähler form J of the singular conifold described in Section 10.2 can be written in terms of a basis of one-forms according to

$$J = \frac{2}{3} dr \wedge g_5 + \frac{1}{3}\left(e^2 \wedge e^1 + e^3 \wedge e^4\right)$$

Deduce that G_3, given by Eqs (10.133) and (10.135), is primitive.

Problem 10.14

For the heterotic string with torsion there is an identity of the form

$$\star_6 H = -e^{-a\Phi} d(e^{a\Phi} J).$$

Derive the value of the parameter a for which this is true.

Problem 10.15

Verify Eqs (10.170), (10.173), (10.176) and (10.177) for flux compactifications of M-theory on a Calabi–Yau four-fold.

Problem 10.16

Fill in the details of the Kaluza–Klein compactification to derive the scalar potential Eq. (10.168).

Problem 10.17

Derive the formula for Newton's constant in the context of the strongly coupled heterotic string Eq. (10.248).

Problem 10.18

Derive the result Eq. (10.221) from the dilatino equation Eq. (10.199).

Problem 10.19

Show that the Einstein field equations that determine $a(t)$ reduce for the FRW ansatz to the Friedmann and acceleration equations.

Problem 10.20

When the slow-roll parameters satisfy Eqs (10.289) and (10.290), show that it is consistent to neglect the corresponding two terms in the FRW equations.

11
Black holes in string theory

Black holes are a fascinating research area for many reasons. On the one hand, they appear to be a very important constituent of our Universe. There are super-massive black holes with masses ranging from a million to a billion solar masses at the centers of most galaxies. The example of M31 is pictured in Fig. 11.1. Much smaller black holes are formed as remnants of certain supernovas.

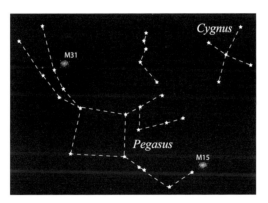

Fig. 11.1. The nuclei of many galaxies, including M31, are quite violent places, and the existence of supermassive black holes is frequently postulated to explain them. M15, on the other hand, is one of the most densely packed globular clusters known in the Milky Way galaxy. The core of this cluster has undergone a core collapse, and it has a central density cusp with an enormous number of stars surrounding what may be a central black hole.

From the theoretical point of view, black holes provide an intriguing arena in which to explore the challenges posed by the reconciliation of general relativity and quantum mechanics. Since string theory purports to provide a consistent quantum theory of gravity, it should be able to address these challenges. In fact, some of the most fascinating developments in string theory concern quantum-mechanical aspects of black-hole physics. These are the subject of this chapter.

The action for *general relativity* (GR) in D dimensions without any sources is given by the Einstein–Hilbert action[1]

$$S = \frac{1}{16\pi G_D} \int d^D x \sqrt{-g} R, \tag{11.1}$$

where G_D is the D-dimensional Newton gravitational constant. The classical equation of motion is the vanishing of the Einstein tensor

$$G_{\mu\nu} = R_{\mu\nu} - \frac{1}{2} g_{\mu\nu} R = 0, \tag{11.2}$$

or, equivalently (for $D > 2$), $R_{\mu\nu} = 0$. Thus, the solutions are Ricci-flat space-times. Straightforward generalizations are provided by adding electromagnetic fields, spinor fields or tensor fields of various sorts, such as those that appear in supergravity theories. Some of the most interesting solutions describe black holes. They have singularities at which certain curvature invariants diverge. In most cases these singularities are shielded by an *event horizon*, which is a hypersurface separating those space-time points that are connected to infinity by a time-like path from those that are not. The conjecture that space-time singularities should always be surrounded by a horizon in physically allowed solutions is known as the *cosmic censorship conjecture*.[2] Classically, black holes are stable objects, whose mass can only increase as matter (or radiation) crosses the horizon and becomes trapped forever. Quantum mechanically, black holes have thermodynamic properties, and they can decay by the emission of thermal radiation.

Challenges posed by black holes

A long list of challenges is presented by black holes. Some of them have been addressed by string theory already, while others remain active areas of research. Here are some of the most important ones:

- Does the existence of black holes and branes imply that quantum mechanics must break down and that pure quantum states can evolve into mixed states? The fact that this superficially appears to be the case is known as the *information loss puzzle*. String theory is constructed as a quantum theory, and therefore the answer is expected to be "no." In fact, various arguments have been constructed that point quite strongly in that

---

[1] See the Appendix of Chapter 9 for a brief review of Riemannian geometry.
[2] This is a modern version of the conjecture. Originally, the conjecture was that, starting from "good" initial conditions, general relativity never generates naked singularities.

direction. However, a complete resolution of the information loss puzzle undoubtedly requires understanding how string theory makes sense of the singularity, where quantum gravity effects become very important. So it is fair to say that this is still an open question.

- Can string theory elucidate the thermodynamic description of black holes? Does black-hole entropy have a microscopic explanation in terms of a large degeneracy of quantum states? One of the most important achievements of string theory in recent times (starting with work of Strominger and Vafa) is the construction of examples that provide an affirmative answer to this question. This chapter describes explicit string solutions for which a microscopic derivation of the Bekenstein–Hawking entropy is known.

- Are there black-hole solutions that correspond to single microstates rather than thermodynamic ensembles? If so, do they have a singularity and a horizon? Or do these properties arise from thermodynamic averaging? These questions are currently under discussion. However, since the answers are not yet clear, they will not be addressed further in this chapter.

- What, if anything, renders black-hole singularities harmless in string theory? In some cases, as illustrated by the analysis of the conifold in Chapter 9, the singularity can be "lifted" once nonperturbative states are taken into account. One natural question is whether string theory can elucidate the status of the cosmic censorship conjecture?

- Does string theory forbid the appearance of closed time-like curves? Such causality-violating solutions can be constructed. There needs to be a good explanation why such solutions should or should not be rejected as unphysical. It may be that they only occur when sources have unphysical properties.

- What generalizations of black-hole solutions exist in dimensions $D > 4$? The case of five dimensions is discussed extensively in this chapter, and explicit supersymmetric black-hole solutions are presented. Black holes fall into two categories: (1) *large black holes* that have finite-area horizons in the supergravity approximation; (2) *small black holes* that have horizons of zero area, and hence a *naked singularity*, in the supergravity approximation. The small black holes acquire horizons of finite area when stringy corrections to the supergravity approximation are taken into account. It seems that large supersymmetric black holes only arise for $D \leq 5$. This is one reason why there has been a lot of interest in the $D = 5$ case. Another reason is that nonspherical horizon topologies become possible for $D > 4$. The example of $D = 5$ black rings will be described.

Chapter 12 describes black p-brane solutions. Black branes are higher-

dimension generalizations of black-hole solutions. These solutions play an important role in the context of the AdS/CFT correspondence.
- A recent speculative suggestion is that black holes might be copiously produced at particle accelerators, such the LHC.[3] This prediction hinges on the possibility of lowering the scale at which gravity becomes strong in suitably *warped backgrounds*, such as those discussed in Chapter 10. The scale might even be as low as the TeV scale. If correct, this would provide one way of testing string theory at particle accelerators, which would be quite fantastic.

11.1 Black holes in general relativity

In order to introduce the reader to some basic notions of black-hole physics, let us begin with the simplest black-hole solutions of general relativity in four dimensions, which are the *Schwarzschild* and *Reissner–Nordström* black holes. The latter black hole is a generalization of the Schwarzschild solution that is electrically charged. Another generalization, known as the *Kerr* black hole, is a black hole with angular momentum. Certain black holes with angular momentum are considered in Section 11.3.

Schwarzschild black hole

The Schwarzschild solution in spherical coordinates

For a spherically symmetric mass distribution of mass M in four space-time dimensions, there is a unique solution to the vacuum Einstein's equations

$$R_{\mu\nu} = 0, \tag{11.3}$$

that describes the geometry outside of the mass distribution.[4] In four dimensions it is given by the Schwarzschild black-hole metric, which in Schwarzschild coordinates (t, r, θ, ϕ) is

$$ds^2 = g_{\mu\nu}dx^\mu dx^\nu = -\left(1 - \frac{r_{\rm H}}{r}\right)dt^2 + \left(1 - \frac{r_{\rm H}}{r}\right)^{-1}dr^2 + r^2 d\Omega_2^2, \tag{11.4}$$

where

$$r_{\rm H} = 2G_4 M. \tag{11.5}$$

---

[3] The LHC is the *Large Hadron Collider* at CERN, which is scheduled to start operating in 2007.
[4] The statement that the Schwarzschild black hole is the unique vacuum solution of Einstein's equations in four dimensions with spherical symmetry. Its time independence is known as Birkhoff's theorem.

Here r_H is known as the *Schwarzschild radius*, and G_4 is Newton's constant. The metric describing the unit two-sphere is

$$d\Omega_2^2 = d\theta^2 + \sin^2\theta d\phi^2. \tag{11.6}$$

The Schwarzschild metric only depends on the total mass M (which is both inertial and gravitational), and it reduces to the Minkowski metric as $M \to 0$. Note that t is a time-like coordinate for $r > r_H$ and a space-like coordinate for $r < r_H$, while the reverse is true for r. The surface $r = r_H$, called the *event horizon*, separates the previous two regions. This metric is *stationary* in the sense that the metric components are independent of the Schwarzschild time coordinate t, so that $\partial/\partial t$ is a Killing vector. This Killing vector is time-like outside the horizon, null on the horizon, and space-like inside the horizon.

It becomes clear that M has the interpretation of a mass by considering the weak field limit, that is, the asymptotic $r \to \infty$ behavior of Eq. (11.4). In this limit we should recover Newtonian gravity.[5] The Newtonian potential Φ in these stationary coordinates can be read off from the tt component of the metric

$$g_{tt} \sim -(1 + 2\Phi). \tag{11.7}$$

As a result, in the case of the Schwarzschild black hole,

$$\Phi = -\frac{MG_4}{r}, \tag{11.8}$$

so that it becomes clear that the parameter M is the black-hole mass.

Schwarzschild black hole in D dimensions

The four-dimensional Schwarzschild metric (11.4) can be generalized to D dimensions, where it takes the form

$$ds^2 = -h dt^2 + h^{-1} dr^2 + r^2 d\Omega_{D-2}^2, \tag{11.9}$$

with

$$h = 1 - \left(\frac{r_H}{r}\right)^{D-3} \tag{11.10}$$

and

$$r_H^{D-3} = \frac{16\pi M G_D}{(D-2)\Omega_{D-2}}. \tag{11.11}$$

---

[5] This is nicely illustrated by considering a massive test particle moving in the curved background. This is a homework problem.

Here Ω_n is the volume of a unit n-sphere, namely[6]

$$\Omega_n = \frac{2\pi^{(n+1)/2}}{\Gamma\left(\frac{n+1}{2}\right)}. \tag{11.12}$$

For large r, this again determines the Newton potential and therefore the black-hole mass M.

The singularities

As can be seen from Eq. (11.4), the coefficients of the metric become singular at $r = 0$ and also at the Schwarzschild radius $r = r_{\rm H}$. In general, a singularity in a metric component could be a coordinate-dependent phenomenon. In order to determine whether a physical singularity is present, coordinate-independent quantities, that is, scalars, should be analyzed. Such a scalar quantity should involve the Riemann tensor. For example, the $D = 4$ Schwarzschild solution yields, after a straightforward calculation,

$$R^{\mu\nu\rho\sigma} R_{\mu\nu\rho\sigma} = \frac{12 r_{\rm H}^2}{r^6}. \tag{11.13}$$

This is evidence that the singularity at the horizon $r = r_{\rm H}$ is only a coordinate singularity, as we will prove shortly, while it proves that a physical singularity is located at $r = 0$.

For objects that are not black holes, the behavior of the solution at the point $r = 0$ is of no physical relevance, since these objects have a mass distribution of finite size, and there is no horizon or singularity. The metric describing the sun, for example, is perfectly well defined at $r = 0$. If, however, the mass is concentrated inside the Schwarzschild radius, then the singularity at $r = 0$ becomes relevant, and the resulting solution is called a Schwarzschild black hole.

In general relativity, it is common practice to set Newton's constant equal to unity, $G_4 = 1$, as a choice of length scale. We prefer not to do so, both because we are interested in Newton's constant in various space-time dimensions, and because the string scale, rather than Newton's constant, is the natural length scale in string theory. G_4, and more generally G_D, are related to the string scale, the string coupling, and a $(10 - D)$-dimensional compactification volume V by $G_D = G_{10}/V$ and $G_{10} = 8\pi^6 g_{\rm s}^2 \ell_{\rm s}^8$.

Schwarzschild solution in Kruskal–Szekeres coordinates

There are other coordinate systems in which the Schwarzschild solution does not even have a coordinate singularity at the horizon. One such coordinate

---

[6] This can be derived by computing $\int \exp(-r^2)\, d^{n+1}x$ in spherical coordinates and comparing to the answer computed in Cartesian coordinates.

system, called the *Kruskal–Szekeres* coordinate system, is related to the Schwarzschild coordinates previously introduced by

$$u = \left(\frac{r}{r_\text{H}} - 1\right)^{1/2} e^{r/2r_\text{H}} \cosh\left(\frac{t}{2r_\text{H}}\right), \qquad (11.14)$$

$$v = \left(\frac{r}{r_\text{H}} - 1\right)^{1/2} e^{r/2r_\text{H}} \sinh\left(\frac{t}{2r_\text{H}}\right). \qquad (11.15)$$

In these coordinates the metric takes the form

$$ds^2 = \frac{4r_\text{H}^3}{r} e^{-r/r_\text{H}} \left(-dv^2 + du^2\right) + r^2 d\Omega_2^2. \qquad (11.16)$$

Note that, from Eqs (11.14) and (11.15), it follows that

$$u^2 - v^2 = \left(\frac{r}{r_\text{H}} - 1\right) e^{r/r_\text{H}}. \qquad (11.17)$$

Different regions of space-time determined by this metric are represented in the *Kruskal diagram* shown in Fig. 11.2. Equation (11.17) shows that the event horizon $r = r_\text{H}$ corresponds to $u = \pm v$, which is represented by a pair of solid lines in Fig. 11.2. Equation (11.17) also shows that $v^2 < u^2$ when $r > r_\text{H}$. The metric in the u, v coordinates can be analytically extended to the region in between the horizon and the singularity. In these coordinates the curvature singularity at $r = 0$ corresponds to the hyperbola $v^2 - u^2 = 1$. This is a pair of space-like curves represented by dashed lines in Fig. 11.2. Thus the space-time is well defined for

$$-\infty < u < +\infty \qquad \text{and} \qquad v^2 < u^2 + 1. \qquad (11.18)$$

As can be seen from Eq. (11.16), the singularity at the horizon is no longer present in these coordinates.

The Schwarzschild geometry in Kruskal–Szekeres coordinates displays more space-time regions than those represented by the original Schwarzschild coordinates, which are only good for $r > r_\text{H}$. The additional regions are unphysical in the sense that a physical black hole that forms by collapse would only have the future singularity (with $u > 0$) and not the past one (with $u < 0$). The latter behaves like a time-reversed black hole and is sometimes called a *white hole*.

The Kruskal–Szekeres coordinates have the additional advantage that geodesics take a very simple form. The equation $ds = 0$ is satisfied by lines with the property $du = \pm dv$ (and fixed position on the two-sphere). This means that null geodesics are 45° lines in Fig. 11.2.

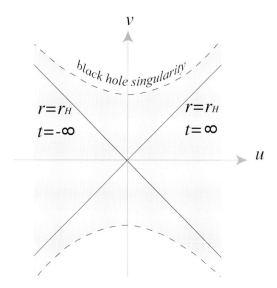

Fig. 11.2. The Schwarzschild black hole in Kruskal–Szekeres coordinates. The solid lines correspond to the horizon, while the dashed lines correspond to the singularity. The shaded region describes the part of the diagram in which the Kruskal–Szekeres coordinates are well defined.

For $|u| > |v|$,
$$t = r_{\rm H} \log\left(\frac{u+v}{u-v}\right), \qquad (11.19)$$

and so the horizon maps to $t = \pm\infty$. It takes an infinite amount of Schwarzschild time to reach the horizon, which reflects the fact that the horizon is infinitely redshifted for an asymptotic observer. From Fig. 11.2 one can infer that light rays emitted by a source situated inside the black hole, which means inside the horizon but outside the singularity, never escape to the region outside the black hole. This is the reason why the surface $r = r_{\rm H}$ is called the *event horizon*. In general, such event horizons are *null hypersurfaces*, which means that vectors n_μ normal to these surfaces satisfy $n^2 = 0$. In the case at hand, the horizon is a two-sphere of radius $r_{\rm H}$ times a null line. In Fig. 11.2, only the null line is shown. It is customary to say that the horizon is S^2 and leave the null line implicit.[7] In particular, it follows from Eq. (11.5) that the area of the event horizon is

$$A = 4\pi r_{\rm H}^2 = 16\pi (MG_4)^2. \qquad (11.20)$$

---

[7] There is a theorem to the effect that S^2 is the only possible horizon topology for a black hole in four dimensions. We will see later that there are other possibilities, besides a sphere, in higher dimensions.

Reissner–Nordström black hole

Reissner–Nordström metric in spherical coordinates

The generalization of the Schwarzschild black hole to one with electric charge Q, but no angular momentum, is called the *Reissner–Nordström black hole*. Charged black holes play a very special role in string theory, because in some cases they are supersymmetric. Thus, by the usual BPS-type reasoning, they can provide information about string theory at strong coupling. In four dimensions the metric of a Reissner–Nordström black hole can be written in the form

$$ds^2 = -\Delta\, dt^2 + \Delta^{-1} dr^2 + r^2 d\Omega_2^2, \tag{11.21}$$

where

$$\Delta = 1 - \frac{2MG_4}{r} + \frac{Q^2 G_4}{r^2}. \tag{11.22}$$

This metric is a solution to Einstein's equations in the presence of an electric field

$$G_{\mu\nu} = R_{\mu\nu} - \frac{1}{2} R g_{\mu\nu} = 8\pi G_4 T_{\mu\nu}, \tag{11.23}$$

where $T_{\mu\nu}$ is in general the energy–momentum tensor for this field

$$T_{\mu\nu} = F_{\mu\rho} F_\nu{}^\rho - \frac{1}{4} g_{\mu\nu} F_{\rho\sigma} F^{\rho\sigma}. \tag{11.24}$$

Since the problem has spherical symmetry, the only nonvanishing component of the $U(1)$ electric field strength is given by the radial component of the electric field E_r

$$F_{tr} = E_r = \frac{Q}{r^2}, \tag{11.25}$$

as is verified in Exercise 11.1. The Reissner–Nordström metric can be generalized to include magnetic charges as well as electric charges, which results in a nonvanishing component $F_{\theta\phi}$. This generalization is described in Exercise 11.2.

Singularities

The metric components in Eq. (11.21) are singular for three values of r. The dependence of the function $\Delta(r)$ which illustrates these singularities is shown in Fig. 11.3. There is a physical curvature singularity at $r = 0$, which can be verified by computing again the scalar $R^{\mu\nu\rho\sigma} R_{\mu\nu\rho\sigma}$. In addition, the factor g_{tt} in the metric vanishes for

$$r = r_\pm = MG_4 \pm \sqrt{(MG_4)^2 - Q^2 G_4}, \tag{11.26}$$

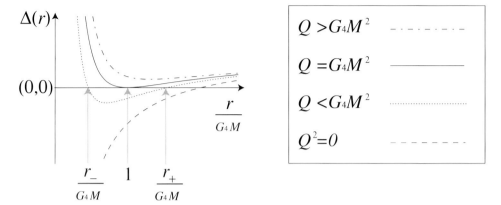

Fig. 11.3. Plots of the function $\Delta(r)$ for the Reissner–Nordström black hole

which are referred to as the *inner horizon* and the *outer horizon*. The outer horizon, $r = r_+$, is the event horizon in this case. Note that it is only present if

$$M\sqrt{G_4} \geq |Q|. \tag{11.27}$$

If this bound is not satisfied, then the metric has a naked singularity at $r = 0$ that is unshielded by a horizon. According to the *cosmic censorship conjecture*, naked singularities should never be produced in physical processes, so that these solutions would be unphysical.

Extremal Reissner–Nordström black hole for $D = 4$

In the limiting case

$$r_\pm = MG_4 \quad \text{or} \quad M\sqrt{G_4} = |Q| \tag{11.28}$$

the black hole is called *extremal*, and it has the maximal charge that is allowed given its mass, as follows from the bound (11.27). When the Reissner–Nordström solution arises as a solution of a supersymmetric theory, the saturation of this bound is often equivalent to the saturation of a BPS bound, which then implies that the extremal Reissner–Nordström black-hole solution has some unbroken supersymmetry.

The metric of an extremal Reissner–Nordström black hole takes the form

$$ds^2 = -\left(1 - \frac{r_0}{r}\right)^2 dt^2 + \left(1 - \frac{r_0}{r}\right)^{-2} dr^2 + r^2 d\Omega_2^2, \tag{11.29}$$

where $r_0 = MG_4$. Let us shift the definition of r by letting $\tilde{r} = r - r_0$ and then dropping the tilde. After some simple algebra this leaves

$$ds^2 = -\left(1 + \frac{r_0}{r}\right)^{-2} dt^2 + \left(1 + \frac{r_0}{r}\right)^2 \left(dr^2 + r^2 d\Omega_2^2\right). \tag{11.30}$$

This is a convenient form of the extremal Reissner–Nordström metric in which the horizon is located at $r = 0$. As in the Schwarzschild case, the space-time is regular at the horizon, which is again only a coordinate singularity. In the near-horizon limit, where $r \approx 0$, the geometry approaches $AdS_2 \times S^2$, as is shown in Exercise 11.3.

Extremal Reissner–Nordström black hole for $D = 5$

Reissner–Nordström black holes have straightforward generalizations to other space-time dimensions. As pointed out in the introduction, an extremal Reissner–Nordström black hole in $D = 5$ is of interest in connection with the microscopic derivation of the black-hole entropy. Its metric can be written in a form similar to Eq. (11.29)

$$ds^2 = -\left[1 - \left(\frac{r_0}{r}\right)^2\right]^2 dt^2 + \left[1 - \left(\frac{r_0}{r}\right)^2\right]^{-2} dr^2 + r^2 d\Omega_3^2. \tag{11.31}$$

Alternatively, one can define $\tilde{r} = \sqrt{r^2 - r_0^2}$ and then drop the tilde to obtain a form analogous to Eq. (11.30)

$$ds_5^2 = -\left[1 + \left(\frac{r_0}{r}\right)^2\right]^{-2} dt^2 + \left[1 + \left(\frac{r_0}{r}\right)^2\right] \left(dr^2 + r^2 d\Omega_3^2\right). \tag{11.32}$$

Using this expression, it is easy to see that the horizon at $r = 0$ has radius r_0, and therefore and its area is

$$A = \Omega_3 r_0^3 = 2\pi^2 r_0^3. \tag{11.33}$$

Comparing with Eq. (11.11), the mass and charge (suitably normalized) of this black hole are

$$M = \frac{Q}{\sqrt{G_5}} = \frac{3\pi r_0^2}{4G_5}. \tag{11.34}$$

EXERCISES

EXERCISE 11.1
The Reissner–Nordström black hole discussed in Section 11.1 is a solution

of Einstein–Maxwell theory

$$S = \int d^4x \sqrt{-g} \left(\frac{1}{2\kappa^2} R - \frac{1}{4} F_{\mu\nu} F^{\mu\nu} \right).$$

The equation of motion and Bianchi identity for the gauge field are

$$\nabla_\mu F^{\mu\nu} = \frac{1}{\sqrt{-g}} \partial_\mu \left(\sqrt{-g} F^{\mu\nu} \right) = 0,$$

$$\epsilon^{\mu\nu\rho\sigma} \partial_\nu F_{\rho\sigma} = 0.$$

Find the most general solution for the gauge field that solves these equations for the spherically symmetric background

$$ds^2 = -e^{2A(r)} dt^2 + e^{2B(r)} dr^2 + r^2 d\Omega_2^2.$$

SOLUTION

Since $F_{\mu\nu}$ is static and spherically symmetric, there are only two independent nonvanishing components for the field strength, $F_{tr}(r, \theta, \phi)$ and $F_{\theta\phi}(r, \theta, \phi)$. For the particular metric of this exercise

$$\sqrt{-g} = e^{A+B} r^2 \sin\theta.$$

The nontrivial equation of motion for the electric field is

$$\partial_r \left(\sqrt{-g} F^{rt} \right) = \partial_r \left(e^{A+B} r^2 \sin\theta \cdot (-e^{-2A-2B} F_{rt}) \right)$$

$$= \partial_r \left(e^{-A-B} r^2 \sin\theta F_{tr} \right) = 0.$$

The most general static solution of this equation is

$$F_{tr} = e^{A+B} \frac{q(\theta, \phi)}{r^2}.$$

The Bianchi identity $\varepsilon^{\mu\nu\rho\sigma} \partial_\nu F_{\rho\sigma} = 0$, leads to additional constraints, $\partial_\theta F_{tr} = 0 = \partial_\phi F_{tr}$, so that

$$F_{tr} = e^{A+B} \frac{q}{r^2},$$

where q is constant. For the values of A and B given in Eq. (11.21), this reduces to Eq. (11.25). These values also solve the Einstein equation (11.23).

The equations of motion for the magnetic field takes the form

$$\partial_\theta \left(e^{A+B} r^2 \sin\theta F^{\theta\phi} \right) = 0,$$

$$\partial_\phi \left(e^{A+B} r^2 \sin\theta F^{\phi\theta} \right) = 0.$$

11.1 Black holes in general relativity

The solution to these equations is

$$F_{\theta\phi} = p(r,t)\sin\theta.$$

Taking into account the Bianchi identity, $\partial_r F_{\theta\phi} = \partial_t F_{\theta\phi} = 0$, one obtains

$$F_{\theta\phi} = p\sin\theta,$$

where p is a constant. This field can then be inserted in the Einstein equation to determine the functions A and B. □

EXERCISE 11.2
Show that the parameters q and p in the previous exercise are electric and magnetic charges.

SOLUTION

As discussed in Chapter 8, magnetic and electric charge are given by

$$Q_{\text{mag}} = \frac{1}{4\pi}\int F = \frac{1}{4\pi}\int_0^\pi d\theta \int_0^{2\pi} d\phi\, F_{\theta\phi}$$

and

$$Q_{\text{el}} = \frac{1}{4\pi}\int \star F = \frac{1}{4\pi}\int_0^\pi d\theta \int_0^{2\pi} d\phi\, (\star F)_{\theta\phi}.$$

Inserting $F_{\theta\phi} = p\sin\theta$ in the first integral gives $Q_{\text{mag}} = p$. To evaluate the electric charge it is necessary to compute the dual of the electric field:

$$(\star F)_{\theta\phi} = \sqrt{-g}F^{rt} = e^{A+B}r^2\sin\theta e^{-2(A+B)}F_{tr} = q\sin\theta.$$

Thus $Q_{\text{el}} = q$.

EXERCISE 11.3
Show that the near-horizon geometry of a $D=4$ extremal Reissner–Nordström black hole is $AdS_2 \times S^2$.

SOLUTION

Near the horizon $r \approx 0$. In this limit Eq. (11.30) becomes

$$ds^2 = -\left(\frac{r_0}{r}\right)^{-2}dt^2 + \left(\frac{r_0}{r}\right)^2 dr^2 + r_0^2 d\Omega_2^2.$$

Setting $\tilde{r} = r_0^2/r$, and dropping the tilde,

$$ds^2 = \left(\frac{r_0}{r}\right)^2(-dt^2 + dr^2) + r_0^2 d\Omega_2^2.$$

This gives a constant negative curvature in the r and t directions, which is AdS_2. Similarly, in the angular directions one has a sphere, with constant positive curvature. In each case the radius of curvature is r_0. As a result, the geometry in the near-horizon limit is $AdS_2 \times S^2$. This is also known as the Bertotti–Robinson metric. $\square$

11.2 Black-hole thermodynamics

Entropy and temperature

Classical black holes behave like thermodynamical objects characterized by a temperature and an entropy. The microscopic quantum origin of these features is addressed in Section 11.4. For now, let us consider the thermodynamic description, that is, the macroscopic description of black holes.

Given a static metric, such as the $D = 4$ Schwarzschild metric Eq. (11.4), there is an elementary method of computing the temperature. The key point to recall is that a system that has a temperature $T = \beta^{-1}$ is periodic in Euclideanized time $\tau = it$ with period β. A simple way to understand this fact is to recall that a thermodynamic partition function is given by

$$Z = \mathrm{Tr}\left(e^{-\beta H}\right),$$

where H is the Hamiltonian of the system. Since quantum mechanical evolution by a time interval t is given by e^{-iHt}, the trace corresponds to imposing a periodicity β in Euclidean time.

The way to determine the temperature of a black hole is to consider its analytic continuation to Euclidean time and then to examine the periodicity of this coordinate. This period is determined by requiring that the Euclideanized metric is regular at the horizon. This may sound like a cookbook recipe, but it is by far the easiest way to carry out the computation. It can be confirmed in a variety of ways, for example by showing that a black hole emits blackbody radiation at the computed temperature.

In order to examine the vicinity of the horizon, let us define ρ by

$$r = r_\mathrm{H}(1 + \rho^2), \tag{11.35}$$

and expand the Euclideanized version of the Schwarzschild metric Eq. (11.4) about $\rho = 0$. This gives

$$ds^2 \approx 4r_\mathrm{H}^2 \left(d\rho^2 + \rho^2\left(\frac{d\tau}{2r_\mathrm{H}}\right)^2 + \frac{1}{4}d\Omega_2^2\right). \tag{11.36}$$

The first two terms describe a flat plane in polar coordinates provided that the period of τ is

$$\beta = 4\pi r_H = 8\pi M G_4. \tag{11.37}$$

Thus the temperature of the Schwarzschild black hole is $T = 1/(8\pi M G_4)$. Since the temperature decreases as M increases, the specific heat is negative.

Very massive black holes are accurately described by classical solutions of Einstein's theory of general relativity. Classically, black holes are stable and black, which means that nothing can ever escape from inside the horizon. Thus the mass can only increase as matter falls through the horizon. If one takes the thermodynamic interpretation of black holes into account, the analogy suggests that

$$dM = TdS, \tag{11.38}$$

where M is the mass of the black hole, T is its temperature and S is the black hole's entropy. The black-hole entropy should be taken into account in the *second law of thermodynamics*,

$$dS/dt \geq 0. \tag{11.39}$$

The entropy of black holes added to the entropy of their surroundings always has to increase with time.

For a Schwarzschild black hole, $\beta = 1/T = 8\pi M G_4$. Requiring that $S \to 0$ as $M \to 0$, to fix an integration constant, one obtains

$$S = 4\pi M^2 G_4. \tag{11.40}$$

Bekenstein–Hawking entropy formula

From Eq. (11.4) it follows that the area A of the event horizon of a Schwarzschild black hole is given by

$$A = 4\pi r_H^2 = 16\pi (MG_4)^2, \tag{11.41}$$

so the entropy can be written in the form

$$S = \frac{A}{4G_4}. \tag{11.42}$$

This is one-quarter of the area of the horizon measured in units of the Planck length. This relation, known as the *Bekenstein–Hawking (BH) entropy formula*, appears to be universally valid (for any black hole in any dimension), at least when A is sufficiently large. For an arbitrary (not necessarily Schwarzschild) black hole in D dimensions, the formula becomes

$$S = \frac{A}{4G_D}, \tag{11.43}$$

where A is the volume (usually called the area) of the $(D-2)$-dimensional horizon. According to this formula, the entropy of a $D=4$ Reissner–Nordström black hole is

$$S = \pi r_+^2 / G_4. \tag{11.44}$$

For an extremal Reissner–Nordström black hole this is $S = \pi M^2 G_4$.

Hawking radiation

When an object has a finite temperature, it emits thermal radiation, which for a black hole would suggest that its mass should decrease in time. This contradicts the known classical behavior, namely that the mass can only increase, discussed earlier. This paradox led Hawking to consider quantum corrections to the classical description. He argued that the gravitational fields at the horizon are strong enough for quantum mechanical pair production in the vicinity of the horizon to lead to the emission of thermal radiation. Roughly speaking, one particle in the virtual pair falls into the hole, and the other one is emitted as a physical on-shell particle. For large black holes, this can be demonstrated reliably using quantum field theory in a classical curved space-time background geometry. Since gravity is treated classically, the black hole has to be big for this analysis to be reliable. In this way, Hawking argued that a black hole emits radiation, and as a consequence it loses mass. The outgoing radiation is thermal, when back-reaction can be neglected. Thus, the black hole behaves as if it were a black body with the temperature computed earlier. The classical statement that nothing can escape from a black hole is undermined by quantum effects. The fact that the entropy of the black hole decreases when Hawking radiation is emitted is consistent with the second law of thermodynamics when the entropy of the emitted radiation is taken into account.

Pure states and mixed states

Hawking has argued that quantum mechanics breaks down when gravity is taken into account. First, by a semi-classical analysis, he argued that black holes emit thermal radiation at a temperature determined by the parameters of the black hole (mass, charge, and angular momentum). Such radiation has no correlations, and therefore is in a mixed state, characterized by a density matrix. On the other hand, a collapsing shell of matter that forms a black hole can be in a pure quantum state. Thus, he argued, pure states can evolve into mixed states in a quantum theory of gravity. This contradicts the basic tenet of unitary evolution in quantum mechanics, and it is referred to as *information loss* or *loss of quantum coherence*.

The AdS/CFT conjecture, described in Chapter 12, certainly would appear to contradict this reasoning, since the AdS space in which black holes can form is dual to a unitary conformal field theory. Thus string solutions, at least ones that are asymptotically AdS, probably provide counterexamples to Hawking's claim. That said, it should be admitted that it is an extremely subtle matter to explain in detail where Hawking's argument for information loss breaks down. This question has been discussed extensively in the literature, but it is not yet completely settled.

EXERCISE 11.4

Show that the temperature of a $D = 4$ Reissner–Nordström black hole is

$$T = \frac{\sqrt{(MG_4)^2 - Q^2 G_4}}{2\pi r_+^2}.$$

What happens to this temperature in the extremal limit?

SOLUTION

Using the same reasoning as in Section 11.2, we set $r = r_+(1+\rho^2)$ and expand the Euclideanized metric of the Reissner–Nordström black hole about $\rho = 0$ to get

$$ds^2 = \frac{4r_+^3}{r_+ - r_-} \left[d\rho^2 + \rho^2 \left(\frac{(r_+ - r_-)d\tau}{2r_+^2} \right)^2 + \frac{r_+ - r_-}{4r_+} d\Omega^2 \right].$$

The value of β that follows from this expression is

$$\beta = \frac{4\pi r_+^2}{r_+ - r_-},$$

which leads to a temperature

$$T = \frac{r_+ - r_-}{4\pi r_+^2} = \frac{\sqrt{(MG_4)^2 - Q^2 G_4}}{2\pi r_+^2}.$$

In the extremal limit, $M\sqrt{G_4} = |Q|$, this gives a vanishing temperature $T = 0$. □

EXERCISE 11.5

Estimate the Schwarzschild radius, temperature, and entropy of a one solar mass Schwarzschild black hole. Estimate its lifetime due to the emission of Hawking radiation. The sun has a mass of $M = 2.0 \times 10^{33} g$.

SOLUTION

Reinstating $\hbar$, c and k_B by dimensional analysis, in order to express these quantities in ordinary units, gives

$$r_H = \frac{2G_4 M}{c^2} \sim 3.0 \times 10^3 \mathrm{m}, \quad T = \frac{\hbar c^3}{8\pi M G_4 k_B} \sim 6.0 \times 10^{-8} \mathrm{K},$$

$$S = \frac{A}{4G_4} = \frac{\pi R^2 c^3}{G_4 \hbar} \sim 1.0 \times 10^{77}, \quad \Delta t \sim \frac{G_4^2 M^3}{\alpha \hbar c^4} \sim 10^{66} \mathrm{years}.$$

The value of the coefficient α is about 10^{-3}. □

11.3 Black holes in string theory

This section considers supersymmetric (and hence extremal) black holes that have finite entropy in the supergravity approximation. These include three-charge black holes in five dimensions and four-charge black holes in four dimensions, which can be interpreted as approximations to solutions of toroidally compactified string theory. For this class of compactifications, finite-horizon-area black-hole solutions that are asymptotically flat only exist in the supergravity approximation in four and five dimensions. The reason for this can be explained by referring to the extremal Reissner–Nordström solutions given in Eqs (11.30) and (11.32). In each case the coefficient of dr^2 takes the form

$$g_{rr} = \left(1 + (r_0/r)^{D-3}\right)^{\frac{2}{D-3}}. \tag{11.45}$$

This behaves near the horizon ($r = 0$) like $g_{rr} \sim (r_0/r)^2$, which is necessary to obtain a finite horizon radius and area. It appears that constructions obtained by string-theory or M-theory compactification always give an outer exponent that is a positive integer. This can only correspond to $2/(D-3)$ if $D = 4$ or $D = 5$. In the multi-charge examples that are discussed in this section, the expression $1 + (r_0/r)^{D-3}$ is replaced by a product of factors that can have different radii, but the same conclusion still applies.

For all other supersymmetric black holes, including any supersymmetric solution for $D > 5$, the horizon has zero radius in the supergravity approximation. To obtain a nonzero radius in these cases, it is necessary to include stringy corrections, that is, corrections to the Einstein–Hilbert action that are higher order in the curvature tensor. This is discussed in Section 11.6.

Extremal three-charge black holes for $D = 5$

The simplest nontrivial example for which the entropy can be calculated involves supersymmetric black holes in five dimensions that carry three different kinds of charges. These can be studied in the context of compactifications of the type IIB superstring theory on a five-torus T^5. The analysis is carried out in the approximation that five of the ten dimensions of the IIB theory are sufficiently small and the black holes are sufficiently large so that a five-dimensional supergravity analysis can be used.

$\mathcal{N} = 8$ supergravity for $D = 5$

The supergravity theory in question is $\mathcal{N} = 8$ supergravity in five dimensions. This contains a number of one-form and two-form gauge fields. However, by duality transformations, the two-forms can be replaced by one-forms. Once this is done, the resulting theory contains 27 $U(1)$ gauge fields. Furthermore, the theory has a noncompact $E_{6,6}$ global U-duality symmetry.[8] The 27 $U(1)$ s belong to the fundamental **27** representation of this group. Therefore, a charged black hole in this theory can carry 27 different types of electric charges. Some of these electric charges can be realized by wrapping branes and exciting Kaluza–Klein excitations. A specific example is discussed below.

The black-hole solution

Three-charge black holes in five dimensions can be obtained by taking Q_1 D1-branes wrapped on an S^1 of radius R inside the T^5, Q_5 D5-branes wrapped on the $T^5 = T^4 \times S^1$, and n units of Kaluza–Klein momentum along the same circle. Each of these objects breaks half of the supersymmetry, so altogether 7/8 of the supersymmetry is broken, and one is left with solutions that have four conserved supercharges. Other equivalent string-theoretic constructions of these black-hole solutions are related to the one considered here by U-duality transformations. Some examples are given later.

There are a variety of ways to analyze this system. One of them is in terms of a five-dimensional gauge theory. Since the Q_1 D1-branes are embedded inside the Q_5 D5-branes, this configuration can be described entirely in terms of the $U(Q_5)$ world-volume gauge theory of the D5-branes. In this description a D-string wound on a circle is described by a $U(Q_5)$ instanton that is localized in the other four directions. So, altogether, there are Q_1 such instantons. The Kaluza–Klein momentum can also be described as excitations in this gauge theory.

---

[8] In the supergravity approximation it is a continuous symmetry.

The five-dimensional metric describing this black-hole can be obtained from the ten-dimensional type IIB theory by wrapping the corresponding branes as described above, or it can be constructed directly. In either case, the resulting metric can be written in Einstein frame in the form

$$ds^2 = -\lambda^{-2/3} dt^2 + \lambda^{1/3}\left(dr^2 + r^2 d\Omega_3^2\right), \tag{11.46}$$

where

$$\lambda = \prod_{i=1}^{3}\left[1 + \left(\frac{r_i}{r}\right)^2\right]. \tag{11.47}$$

The relation between the parameters r_i and the charges Q_i is derived below. This solution describes an extremal three-charge black hole with a vanishing temperature $T = 0$. Note that this formula reduces to the extremal Reissner–Nordström black-hole metric given in Eq. (11.32) in the special case $r_1 = r_2 = r_3$, that is, when the three charges are equal. The dilaton is a constant, so there is a globally well defined string coupling constant g_s. Thus, the string-frame metric differs from the one given above only by a constant factor.

The horizon of the black hole in Eq. (11.46) is located at $r = 0$, and its area is

$$A = 2\pi^2 r_1 r_2 r_3. \tag{11.48}$$

This vanishes when any of the three charges vanishes, which is the reason that three charges have been considered in the first place. Put differently, one needs to break 7/8 of the supersymmetry in order to form a horizon that has finite area in the supergravity approximation, and this requires introducing three different kinds of excitations. When there is only one or two nonzero charges, there still is a horizon of finite area, but its dependence on the string scale is such that its area vanishes in the supergravity approximation. For the supergravity approximation to string theory to be valid, it is necessary that the geometry is slowly varying at the string scale. This requires $r_i \gg \ell_s$.

The black hole mass

Using Eq. (11.11) one can read off the mass of the black hole M to be

$$M = M_1 + M_2 + M_3 \quad \text{where,} \quad M_i = \frac{\pi r_i^2}{4G_5}. \tag{11.49}$$

The fact that the masses are additive in this way is a consequence of the form of the metric. However, this had to be the case, because the BPS condition is satisfied, and the charges are additive.

To express the result for r_i in terms of ten-dimensional quantities, the value of G_5 needs to be determined. Letting $(2\pi)^4 V$ denote the volume of the T^4 and R be the radius of the S^1 one obtains

$$G_5 = \frac{G_{10}}{(2\pi)^5 RV}. \quad (11.50)$$

As explained in Chapter 8, $G_{10} = 8\pi^6 g_s^2 \ell_s^8$ is the 10-dimensional Newton constant in string units. Putting these facts together gives the relation

$$r_i^2 = \frac{g_s^2 \ell_s^8}{RV} M_i. \quad (11.51)$$

The masses M_i can be computed at weak string coupling using string-theoretic considerations, namely the formulas for the mass of winding and momentum modes derived in Chapter 7. In the string frame, the masses are

$$M_1 = 2\pi R T_{D1} Q_1 = \frac{Q_1 R}{g_s \ell_s^2},$$

$$M_2 = (2\pi)^5 RV T_{D5} Q_5 = \frac{Q_5 RV}{g_s \ell_s^6}, \quad (11.52)$$

$$M_3 = \frac{n}{R}.$$

Here Q_1 and Q_5 are the numbers of wrapped D1-branes and D5-branes, respectively, and hence the values of the corresponding charges. Similarly, n is the integer that specifies momentum on the circle.

The quantities T_{D1} and T_{D5} are the tensions of a single D1-brane and D5-brane given in Chapter 6. Using these relations, the conditions $r_i^2 \gg \ell_s^2$ become

$$g_s Q_1 \gg \frac{V}{\ell_s^4}, \quad g_s Q_5 \gg 1, \quad g_s^2 n \gg \frac{R^2 V}{\ell_s^6}. \quad (11.53)$$

If R and V are of order string scale, and one wants $g_s \ll 1$, so as to be in the perturbative string theory regime, then all three charges must be large. Since the effective expansion parameters in string perturbation theory are actually $g_s Q_1$ and $g_s Q_5$,[9] this takes one out of the perturbative regime. On the other hand, when the couplings are small, the mass and the spectrum of excitations can be computed by string-theoretic considerations.

The crucial fact is that supersymmetry allows us to extrapolate certain properties from weak coupling to strong coupling reliably, so that results that are obtained in the two limits can be compared meaningfully. The property of this type that is of most interest is the number of quantum

---

[9] These correspond to the 't Hooft couplings in the corresponding large-N world-volume gauge theories.

states. It is computed in weakly coupled string theory and compared to the classical entropy, which is meaningful for strong coupling. This type of reasoning could break down if short supermultiplets join up to give a long supermultiplet. Strictly speaking, the quantity that can be continued safely from weak coupling to strong coupling is an index, which typically counts the number of bosonic states minus the number of fermionic states, whereas the entropy is the logarithm of the sum of these numbers. Usually, this distinction can be ignored.

The entropy

Using Eqs (11.48), (11.50) and (11.51), one finds that the entropy is

$$S = \frac{A}{4G_5} = \frac{2\pi g_s \ell_s^4}{\sqrt{RV}} \sqrt{M_1 M_2 M_3}. \tag{11.54}$$

Using the relations in (11.52) to re-express this in terms of the charges, one obtains the elegant result

$$S = 2\pi \sqrt{Q_1 Q_5 n}. \tag{11.55}$$

As was mentioned earlier, there are 27 possible electric charges, and this is the result when only a specific three of them are nonzero. The charges transform as a **27** representation of the noncompact $E_{6,6}$ symmetry group of $\mathcal{N} = 8$ supergravity in $D = 5$. The entropy should be invariant under this symmetry group.[10] In other words, there should be an $E_{6,6}$ invariant Δ that is cubic in the 27 electric charges such that the entropy takes the form

$$S = 2\pi \sqrt{\Delta}. \tag{11.56}$$

The invariant Δ generalizes the factor $Q_1 Q_5 n$ appearing in Eq. (11.55).

The construction of the cubic invariant is relatively simple in this case. The **27** representation is also an irreducible representation of the maximal compact subgroup $USp(8)$. That group has a unique cubic invariant, which therefore must also be the $E_{6,6}$ invariant. In the case of five dimensions the central charge matrix Z_{AB} is a real antisymmetric 8×8 matrix that is also symplectic traceless. This means that, given a symplectic matrix Ω_{AB}, one has $\text{tr}(\Omega Z) = 0$.[11] This is one real condition, so Z contains 27 independent real charges, as expected. The unique cubic invariant with manifest $USp(8)$

---

[10] Stringy corrections to the formula need only be invariant under the discrete $E_6(\mathbb{Z})$ U-duality subgroup.

[11] We can choose Ω_{AB} to be the antisymmetric matrix whose nonzero matrix elements with $A < B$ are $\Omega_{12} = \Omega_{34} = \Omega_{56} = \Omega_{78} = 1$. A symplectic matrix A satisfies $A^T \Omega A = \Omega$.

symmetry is

$$\Delta = -\frac{1}{48}\text{tr}(\Omega Z \Omega Z \Omega Z), \tag{11.57}$$

where the normalization is chosen for later convenience.

By a transformation of the form, $Z \to A^T Z A$, where A is a symplectic matrix,[12] the matrix Z_{AB} can be brought to a canonical form in which its only nonzero entries for $A < B$ are $Z_{12} = x_1$, $Z_{34} = x_2$, $Z_{56} = x_3$, $Z_{78} = x_4$, where $\sum x_i = 0$ and the x_i s are real. A symmetric way of writing this is

$$x_1 = Q_1 - Q_2 - Q_3, \qquad x_2 = -Q_1 + Q_2 - Q_3,$$

$$x_3 = -Q_1 - Q_2 + Q_3, \qquad x_4 = Q_1 + Q_2 + Q_3. \tag{11.58}$$

If one evaluates Δ for these choices, one finds the desired result:

$$\Delta = \frac{1}{24}\sum x_i^3 = Q_1 Q_2 Q_3. \tag{11.59}$$

Thus, up to a change of basis, the three-charge solution is completely general.

Duality and other black-hole configurations

Three-charge supersymmetric black holes in five dimensions have been described above as D1-D5-P bound states in the toroidally compactified type IIB theory. Here D1 refers to the Q_1 D1-branes wrapped on a y^1 circle, D5 refers to the Q_5 D5-branes wrapped on the $y^1 \cdots y^5$ torus, and P refers to the n units of Kaluza–Klein momentum on the y^1 circle. Using the various possible S and T dualities that exist for type II theories, this brane configuration can be related to various dual configurations describing black holes that have an entropy given by Eq. (11.55), with the corresponding charges of the dual brane configuration. For example, an S-duality transformation replaces the D1-branes by F1-branes (fundamental strings) and the D5-branes by NS5-branes. The Kaluza–Klein momenta P are unaffected. Alternatively, a T-duality transformation along the y^1 direction maps the type IIB configuration to a type IIA configuration with the D1-branes mapping to D0-branes and the D5-branes mapping to D4-branes. Moreover, the Kaluza–Klein momentum maps to an F1-brane wrapped n times on the dual y^1 circle. Further T dualities give a host of other equivalent type IIA and type IIB configurations. Exercise 11.6 works out an example of such a dual description.

---

[12] This is appropriate, because $USp(8)$ is the automorphism group of the $\mathcal{N} = 8$, $D = 5$ supersymmetry algebra.

M-theory interpretation

Starting from the Type IIA configuration described above, one can carry out two more T-duality transformations along the y^2 and y^3 directions to obtain a type IIA configuration consisting of Q_1 D2-branes wrapped on y^2 and y^3, Q_5 D2-branes wrapped on y^4 and y^5 and n fundamental strings wrapped on y^1. This configuration can be interpreted at strong coupling as M-theory compactified on a 6-torus. Calling the M-theory circle coordinate y^6, the n fundamental type IIA strings are then identified as n M2-branes wrapped on the y^1 and y^6 circles. The two sets of D2-branes are then identified as sets of M2-branes. Altogether, there are three sets of M2-branes wrapped on three orthogonal tori. This is a satisfying picture in that it puts the three sources of charges on a symmetrical footing, which nicely accounts for their symmetrical appearance in the entropy formula. The verification that this brane configuration gives the same entropy as before is a homework problem.

Nonextremal three-charge black holes for $D=5$

The extremal three-charge black-hole solutions in five dimensions given above have nonextremal generalizations, which describe nonsupersymmetric black holes with finite temperature. These black holes are described by the metric

$$ds^2 = -h\, \lambda^{-2/3} dt^2 + \lambda^{1/3} \left(\frac{dr^2}{h} + r^2 d\Omega_3^2 \right), \tag{11.60}$$

where

$$h = 1 - \frac{r_0^2}{r^2} \tag{11.61}$$

and

$$\lambda = \prod_{i=1}^{3} \left[1 + \left(\frac{r_i}{r} \right)^2 \right] \quad \text{with} \quad r_i^2 = r_0^2 \sinh^2 \alpha_i, \quad i = 1, 2, 3. \tag{11.62}$$

This reduces to the extremal metric in Eq. (11.46) in the limit $r_0 \to 0$ with r_i held fixed. Moreover, the limit $\alpha_i \to 0$ with r_0 held fixed gives the Schwarzschild metric in five dimensions given in Eq. (11.9).

The mass of this black hole can be read off using the same rules as before resulting in

$$M = \frac{\pi r_0^2}{8 G_5} \left(\cosh 2\alpha_1 + \cosh 2\alpha_2 + \cosh 2\alpha_3 \right). \tag{11.63}$$

The inclusion of the factor h in the metric shifts the position of the event horizon from $r = 0$ to $r = r_0$. At the horizon the factor λ takes the value

$$\lambda(r_0) = \prod_{i=1}^{3} \cosh^2 \alpha_i. \tag{11.64}$$

The radial size of the horizon is

$$r_H = r_0 \left[\lambda(r_0)\right]^{1/6}, \tag{11.65}$$

and thus the area of the horizon is

$$A = 2\pi^2 r_H^3 = 2\pi^2 r_0^3 \cosh \alpha_1 \cosh \alpha_2 \cosh \alpha_3. \tag{11.66}$$

The entropy is then given by

$$S = \frac{A}{4G_5} = \frac{2\pi r_0^3 V_6}{\ell_p^9} \cosh \alpha_1 \cosh \alpha_2 \cosh \alpha_3. \tag{11.67}$$

To convert to string units, one would replace ℓ_p^9 by $g_s^2 \ell_s^9$.

These formulas have a suggestive interpretation. Let us imagine that, in addition to there being Q_1 D1-branes wrapping the y^1 circle, there are also $\overline{Q}_1$ anti-D1-branes wrapping the same circle. Similarly, anti-D5-branes and right-moving Kaluza–Klein excitations can be introduced. Then the net charge in each case is

$$\widehat{Q}_i = Q_i - \overline{Q}_i \quad i = 1, 2, 3. \tag{11.68}$$

The three types of electric charge have $\widehat{Q}_i \sim \sinh 2\alpha_i$. By identifying $Q_i \sim \exp(2\alpha_i)$ and $\overline{Q}_i \sim \exp(-2\alpha_i)$ one interprets the net charge as a difference of brane and antibrane contributions and the expression for $M_i \sim \cosh 2\alpha_i$ as the sum of brane and antibrane contributions. This also allows the entropy in Eq. (11.67) to be rewritten in form

$$S = \frac{A}{4G_5} = 2\pi \prod_{i=1}^{3} (\sqrt{Q_i} + \sqrt{\overline{Q}_i}), \tag{11.69}$$

which is a nice generalization of Eq. (11.55).

Rotating supersymmetric black holes for $D = 5$

In five dimensions it is possible for a three-charge black hole to rotate and still be supersymmetric.[13] This is not possible in four dimensions, where all

---

[13] A rotating time-independent black-hole solution in four dimensions is known as a *Kerr black hole*. The solution under consideration here is quite different from that one.

rotating black holes, even extremal ones, are not supersymmetric. The key is to note that the rotation group in five dimensions is $SO(4) \sim SU(2) \times SU(2)$. Supersymmetry requires restricting the rotation to one of the two $SU(2)$ factors, which corresponds to simultaneous rotation, with equal angular momentum, in two orthogonal planes. There are more general ways in which a five-dimensional black hole can rotate, of course, but this is the only one that is supersymmetric. It preserves 1/8 of the original 32 supersymmetries, just like the previous examples. To describe this case, let us introduce angular coordinates as follows:

$$x^1 = r \cos\theta \cos\psi, \quad x^2 = r \cos\theta \sin\psi, \tag{11.70}$$

$$x^3 = r \sin\theta \cos\phi, \quad x^4 = r \sin\theta \sin\phi. \tag{11.71}$$

Then

$$dx^i dx^i = dr^2 + r^2 d\Omega_3^2 \tag{11.72}$$

describes Euclidean space for

$$d\Omega_3^2 = d\theta^2 + \sin^2\theta d\phi^2 + \cos^2\theta d\psi^2, \quad 0 \le \theta \le \pi/2, \quad 0 \le \phi, \psi \le 2\pi. \tag{11.73}$$

The metric of the desired supersymmetric rotating black hole is a relatively simple generalization of Eq. (11.46)

$$ds^2 = -\lambda^{-2/3} \left(dt - \frac{a}{r^2} \sin^2\theta d\phi + \frac{a}{r^2} \cos^2\theta d\psi \right)^2 + \lambda^{1/3} \left(dr^2 + r^2 d\Omega_3^2 \right), \tag{11.74}$$

where λ is again given by Eq. (11.47). This metric describes simultaneous rotation in the 12 and 34 planes. The parameter a is related to $J_{12} = J_{34} = J$ by

$$J = \frac{\pi a}{4G_5}. \tag{11.75}$$

The area of the horizon at $r = 0$, and hence the entropy, is computed in Exercise 11.7 and shown to be

$$S = \frac{A}{4G_5} = 2\pi \sqrt{Q_1 Q_5 n - J^2}. \tag{11.76}$$

Extremal four-charge black holes for $D = 4$

The metric and entropy

The construction of supersymmetric black holes in four dimensions is quite similar to the five-dimensional case. Before proposing a specific brane realization, let us write down the metric and explore its properties. The analog of Eq. (11.46) is

$$ds^2 = -\lambda^{-1/2} dt^2 + \lambda^{1/2} \left(dr^2 + r^2 d\Omega_2^2 \right),\tag{11.77}$$

where

$$\lambda = \prod_{i=1}^{4} \left(1 + \frac{r_i}{r}\right).\tag{11.78}$$

This reduces to Eq. (11.30) when all four r_i are equal. We can read off the mass of the black hole from the large distance behavior of g_{tt} using Eqs (11.7) and (11.8). The result is

$$M = \sum_{i=1}^{4} M_i \quad \text{with} \quad M_i = \frac{r_i}{4G_4}.\tag{11.79}$$

The area of the horizon, which is located at $r = 0$, is

$$A = 4\pi \sqrt{r_1 r_2 r_3 r_4}.\tag{11.80}$$

Putting these facts together, the resulting entropy is

$$S = \frac{A}{4G_4} = 16\pi G_4 \sqrt{M_1 M_2 M_3 M_4}.\tag{11.81}$$

Type IIA brane construction

It still remains to relate the four masses to four electric (or magnetic) charges. This requires some sort of brane construction involving four types of branes or excitations. To be specific, let us consider the type IIA theory compactified on a six torus that is a product of six circles with coordinates $y^1, \ldots, y^6$ and radii $R_1, \ldots, R_6$. A brane configuration that preserves 1/8 of the $\mathcal{N} = 8$ supersymmetry, and therefore is suitable, is the following: Q_1 D2-branes wrapped on the y^1 and y^6 circles, Q_2 D6-branes wrapped on all six circles, Q_3 NS5-branes wrapped on the first five circles, and Q_4 units of Kaluza–Klein momentum on the first circle. The masses that correspond to these types of excitations are

$$\begin{aligned} M_1 &= (2\pi R_1)(2\pi R_6) T_{D2} Q_1 = \tfrac{1}{g_s \ell_s^3}(R_1 R_6) Q_1, \\ M_2 &= (2\pi R_1) \cdots (2\pi R_6) T_{D6} Q_2 = \tfrac{1}{g_s \ell_s^7}(R_1 \cdots R_6) Q_2, \\ M_3 &= (2\pi R_1) \cdots (2\pi R_5) T_{NS5} Q_3 = \tfrac{1}{g_s^2 \ell_s^6}(R_1 \cdots R_5) Q_3, \\ M_4 &= \tfrac{1}{R_1} Q_4. \end{aligned} \tag{11.82}$$

Inserting these masses into the entropy formula given above and using

$$G_4 = \frac{G_{10}}{(2\pi R_1)\cdots(2\pi R_6)} = \frac{g_s^2 \ell_s^8}{8 R_1 \cdots R_6} \tag{11.83}$$

yields the final formula for the entropy

$$S = 2\pi\sqrt{Q_1 Q_2 Q_3 Q_4}. \tag{11.84}$$

This result bears a striking resemblance to Eq. (11.55).

Dual brane configurations

As in the five-dimensional three-charge case, there are many other equivalent brane configurations that are related by various S- and T-duality transformations, and Eq. (11.84) applies to all of them. For example, a T-duality along directions 1,2,3 gives a type IIB configuration. Following this by an S-duality gives a type IIB configuration that has Q_1 D3-branes wrapping directions 2,3,6, Q_2 D3-branes wrapping directions 4,5,6, Q_3 D5-branes wrapping directions 1–5, and Q_4 D1-branes wrapping direction 1. A further T-duality along direction 6 gives a type IIA configuration consisting of three sets of D2-branes wrapping orthogonal two-tori and a set of D6-branes wrapping the entire 6-torus.

$\mathcal{N} = 8$ supergravity in $D = 4$

The effective four-dimensional theory in this case is $\mathcal{N} = 8$ supergravity, which has a noncompact $E_{7,7}$ duality group. This is a continuous symmetry in the supergravity approximation, though it is broken to the infinite discrete U-duality group $E_7(\mathbb{Z})$ by string theory corrections. Since we are working in the supergravity approximation, the entropy of extremal black holes should be invariant under the continuous symmetry group. Writing the entropy in the form $S = 2\pi\sqrt{\Delta}$, we found $\Delta = Q_1 Q_2 Q_3 Q_4$ for a certain specific four-charge black hole in Eq. (11.84). We can use group theory to figure out how this should generalize.

$\mathcal{N} = 8$ supergravity has 28 $U(1)$ gauge fields. There are therefore 28 distinct electric and magnetic charges that a black hole can carry. These charges form a **56** representation of the $E_{7,7}$ duality group. There is a unique $E_{7,7}$-symmetric quartic invariant that can be constructed out of these charges, and the product $Q_1 Q_2 Q_3 Q_4$ corresponds to a special case of that invariant. The way this works is as follows: The matrix of central charges is an 8×8 complex antisymmetric matrix

$$Z_{AB} = q_{AB} + i p_{AB}, \tag{11.85}$$

where the q_{AB} denote the 28 electric charges and the p_{AB} denote the 28 magnetic charges. The invariant Δ is a quartic expression in these central charges. The $E_{7,7}$ duality group has an $SU(8)$ subgroup, which can be made manifest. Subscripts A, B label an **8** and superscripts label an $\bar{\mathbf{8}}$ of $SU(8)$. Thus the complex conjugate of the central charge is denoted $\bar{Z}^{AB}$. Now consider the formula

$$\Delta = \operatorname{tr}(Z\bar{Z}Z\bar{Z}) - \frac{1}{4}\left(\operatorname{tr}Z\bar{Z}\right)^2 + 4(\operatorname{Pf}Z + \operatorname{Pf}\bar{Z}), \qquad (11.86)$$

where the Pfaffian is

$$\operatorname{Pf}Z = \frac{1}{2^4 \cdot 4!}\varepsilon^{ABCDEFGH}Z_{AB}Z_{CD}Z_{EF}Z_{GH}. \qquad (11.87)$$

Each of the terms in Eq. (11.86) has manifest $SU(8)$ symmetry. The claim is that this particular combination is the unique one (up to normalization) for which this extends to $E_{7,7}$ symmetry.

By a transformation of the form $Z \to U^T Z U$, where U is a unitary matrix,[14] Z can be brought to a canonical form in which the only nonzero entries (with $A < B$) are $z_1 = Z_{12}$, $z_2 = Z_{34}$, $z_3 = Z_{56}$, $z_4 = Z_{78}$. The z_is are complex, in general. In this basis one has

$$\Delta = 2\sum|z_i|^4 - \left(\sum|z_i|^2\right)^2 + 8\operatorname{Re}(z_1 z_2 z_3 z_4). \qquad (11.88)$$

As a matter of fact, by an $SU(8)$ transformation, it is possible to remove three phases. So, for example, all four z_i could be chosen to have the same phase or else three of the z_i could be chosen to be real. Thus, the five-charge case discussed below, is the generating solution for the arbitrary case in the same sense that the three-charge solution was in five dimensions.

To make contact with the four-charge black hole considered previously, all four z_i can be chosen to be real in order to give four electric charges. For the specific choices

$$z_1 = \frac{1}{4}(Q_1 + Q_2 + Q_3 + Q_4), \qquad z_2 = \frac{1}{4}(Q_1 + Q_2 - Q_3 - Q_4),$$

$$z_3 = \frac{1}{4}(Q_1 - Q_2 + Q_3 - Q_4), \qquad z_4 = \frac{1}{4}(Q_1 - Q_2 - Q_3 + Q_4), \qquad (11.89)$$

one finds after some algebra that $\Delta = Q_1 Q_2 Q_3 Q_4$ in agreement with what we found earlier by other methods.

---

14 This is appropriate because $U(8)$ is the automorphism group of the $\mathcal{N} = 8$, $D = 4$ supersymmetry algebra.

Five-charge configuration

It is possible to add P_1 D0-branes to the D2-D2-D2-D6 configuration described above without breaking any additional supersymmetry. The resulting 5-charge configuration differs from the configurations considered so far in an important respect. Namely, a D0-brane and a wrapped D6-brane are mutually nonlocal. In other words, they are electric and magnetic with respect to the same gauge field. Let us not attempt to write down the solution that describes such a black hole. It is given by an $E_{7,7}$ transformation of the solution that we presented. Rather, let us simply note that the E_7 quartic invariant can be evaluated for all possible choices of electric and magnetic charges, so it is simply a matter of reading off what it gives. To do this we should simply replace $Q_1 \to Q_1 + iP_1$ in each of the four z_is and re-evaluate Δ. After some algebra one finds that Eq. (11.88) gives

$$\Delta = Q_1 Q_2 Q_3 Q_4 - \frac{1}{4} P_1^2 Q_1^2. \tag{11.90}$$

Thus

$$S = 2\pi \sqrt{Q_1 Q_2 Q_3 Q_4 - \frac{1}{4} P_1^2 Q_1^2}. \tag{11.91}$$

If one chooses to make the more common convention of calling D0-branes electrically charged and D6-branes magnetically charged, then we should make an electric–magnetic duality transformation, which amounts to renaming the charges as follows: $Q_1 = P_0$ and $P_1 = -Q_0$. Written this way, the entropy takes the form

$$S = 2\pi \sqrt{P_0 Q_2 Q_3 Q_4 - \frac{1}{4} P_0^2 Q_0^2}. \tag{11.92}$$

The 4d/5d connection

The astute reader may have noticed a resemblance between the entropy of a rotating black hole in five dimensions, given in Eq. (11.76), and the four-dimensional entropy describing a four-charge black hole Eq. (11.92). Specifically, the two formulas agree if one sets $P_0 = 1$ and makes the identification $J = Q_0/2$. This turns out to be more than a coincidence. Without going into the mathematical details, let us explain qualitatively how this comes about.

Since the four-dimensional black hole has $P_0 = 1$, there is one D6-brane. In Chapter 8 it was explained that a D6-brane of the type IIA theory is a higher-dimensional analog of a Kaluza–Klein monopole. This means that, from the 11-dimensional M-theory perspective, the four dimensions transverse to the brane form the Taub–NUT geometry

$$ds^2_{\text{TN}} = \left(1 + \frac{R}{2r}\right)(dr^2 + r^2 d\Omega_2^2)$$

$$+ \left(1 + \frac{R}{2r}\right)^{-1}(dy + R\sin^2(\theta/2)\, d\phi)^2. \tag{11.93}$$

This geometry can be visualized as analogous to a cigar with the D6-brane localized to the region near the tip. Far from the tip of the cigar, the geometry looks like $\mathbb{R}^3 \times S^1$, where the circle is the M-theory circle, which in type IIA units has radius $R = g_s \ell_s$. The fact that the number of D0-branes is Q_0 means that there are Q_0 units of momentum around the M-theory circle. On the other hand, near the tip of the cigar the geometry would be nonsingular and look like $\mathbb{R}^4$ if there were no other branes in the problem. However, their presence makes the story more complicated.

Now consider the strong-coupling limit of the previous four-dimensional picture. In this limit the radius of the M-theory circle approaches infinity, and the Taub–NUT geometry approaches flat $\mathbb{R}^4$ far from the origin. However, near the origin there is a five-dimensional black hole. The Q_0 units of momentum around the M-theory circle are still present, but now as angular momentum $J_{12} = J_{34} = Q_0/2$ about the origin, which was the tip of the cigar. In the limit one is left with a five-dimensional black hole with M2-brane charges Q_1, Q_2, Q_3 and $J = J_{12} = J_{34}$. Since the entropy does not depend on the string coupling constant, which controls the size of the M-theory circle, its value must be the same for the four- and five-dimensional black holes, which is what we found.

EXERCISES

EXERCISE 11.6
Verify that the D0-D4-F1 realization of the black hole discussed in Section 11.3 has entropy given by Eq. (11.55), with the charges replaced by the charges of the dual configuration.

SOLUTION
Consider Q_0 D0-branes, Q_4 D4-branes wrapping the T^4, which has a volume

$(2\pi)^4 V$, and Q_1 F1-branes wrapping the y^1 circle, which has radius R. This leads to the masses

$$M_1 = \frac{Q_0}{g_s \ell_s}, \quad M_2 = \frac{Q_4}{(2\pi)^4 g_s \ell_s^5}(2\pi)^4 V, \quad M_3 = \frac{Q_1}{2\pi \ell_s^2} 2\pi R.$$

Inserting this into the expression for the entropy Eq. (11.54) gives

$$S = \frac{A}{4G_5} = \frac{2\pi g_s \ell_s^4}{\sqrt{RV}}\sqrt{M_1 M_2 M_3} = 2\pi \sqrt{Q_0 Q_4 Q_1}.$$

Comparison of this formula with Eq. (11.55) shows that the D0-D4-F1 system gives the same entropy for $Q_0 = Q_1$, $Q_4 = Q_5$ and $Q_1 = n$, which is what we wanted to show.

Note that the various dualities that relate the different brane descriptions of the black hole do not change the five-dimensional metric except by an overall constant factor. Such a factor has no bearing on the computation of the entropy, which is dimensionless. □

EXERCISE 11.7
Compute the area of the horizon of the rotating black hole described in Section 11.3 and deduce its entropy.

SOLUTION

In the near-horizon limit $r \approx 0$ and constant t the metric Eq. (11.74) reduces to

$$ds^2 = R^2 d\Omega_3^2 - (a/R^2)^2(\cos^2\theta d\psi - \sin^2\theta d\phi)^2$$

$$= R^2 d\theta^2 + R^2(\cos\theta \sin\theta)^2(d\phi + d\psi)^2 + (R^2 - (a/R^2)^2)(\cos^2\theta d\psi - \sin^2\theta d\phi)^2,$$

where

$$R^2 = (r_1 r_2 r_3)^{2/3}.$$

The easiest way to compute the area of the horizon described by this metric is to define the orthonormal one-forms

$$e_1 = R d\theta,$$

$$e_2 = R \cos\theta \sin\theta (d\phi + d\psi),$$

$$e_3 = \sqrt{R^2 - (a/R^2)^2}(\cos^2\theta d\psi - \sin^2\theta d\phi).$$

11.3 Black holes in string theory

Then the area of the horizon obtained from this metric is given by

$$A = \int e_1 \wedge e_2 \wedge e_3 = R^2 \sqrt{R^2 - (a/R^2)^2} \int \cos\theta \sin\theta d\theta \wedge d\phi \wedge d\psi$$

$$= 2\pi^2 \sqrt{(r_1 r_2 r_3)^2 - a^2}.$$

Using this result,

$$S = \frac{A}{4G_5} = 2\pi \sqrt{Q_1 Q_5 n - J^2},$$

where the angular momentum J is related to the parameter a by

$$J = \frac{\pi a}{4G_5}.$$

If $a > r_1 r_2 r_3$, the black hole is *over-rotating*, and the geometry has a naked singularity, at least in the supergravity approximation. □

EXERCISE 11.8

Consider the dual configuration of the $D = 4$ extremal four-charge black hole described in Section 11.3. Show that this gives the entropy in Eq. (11.84), with the charges replaced by the charges of the dual configuration.

SOLUTION

The dual configuration has three sets of D2-branes and one set of D6-branes. The associated masses are

$$M_1 = (2\pi R_2)(2\pi R_3) T_{D2} = \frac{R_2 R_3}{g_s \ell_s^3} Q_1, \quad M_2 = \frac{R_4 R_5}{g_s \ell_s^3} Q_2,$$

$$M_3 = \frac{R_1 R_6}{g_s \ell_s^3} Q_3, \quad M_4 = (2\pi)^6 (R_1 \cdots R_6) T_{D6} P_0 = \frac{R_1 \cdots R_6}{g_s \ell_s^7} P_0.$$

Therefore, the entropy is

$$S = \frac{A}{4G_4} = 16\pi \frac{g_s^2 \ell_s^8}{8 R_1 \cdots R_6} \sqrt{M_1 M_2 M_3 M_4} = 2\pi \sqrt{Q_1 Q_2 Q_3 P_0},$$

which reproduces Eq. (11.84). □

EXERCISE 11.9

Construct the nonextremal generalization of the four-charge black hole by analogy with the construction given for nonextremal black holes in five dimensions. Interpret the masses, charges, and entropy in terms of branes and antibranes, as was done in the five-dimensional case.

SOLUTION

The formulas analogous to Eqs (11.60) – (11.62) for $D = 4$ black holes take the form

$$ds^2 = -\lambda^{-1/2}\left(1 - \frac{r_0}{r}\right) dt^2 + \lambda^{1/2}\left(\left(1 - \frac{r_0}{r}\right)^{-1} dr^2 + r^2 d\Omega_2^2\right),$$

$$\lambda = \prod_{i=1}^{4}\left(1 + \frac{r_0 \sinh^2 \alpha_i}{r}\right).$$

We can then extract the values of the masses in the usual way,

$$M = \frac{r_0}{4G_4}\sum_{i=1}^{4}\sinh^2 \alpha_i + \frac{r_0}{2G_4} = \frac{r_0}{8G_4}\sum_{i=1}^{4}\cosh 2\alpha_i,$$

which gives

$$M_i = \frac{r_0 \cosh 2\alpha_i}{8G_4}.$$

The outer horizon, located at $r = r_0$, has an area

$$A = 4\pi r_0^2 \prod_{i=1}^{4} \cosh \alpha_i.$$

This result can be interpreted as signaling the presence of $\overline{Q}_i$ antibranes in addition to the Q_i branes. Identifying $\overline{Q}_i \sim \exp(-2\alpha_i)$ and $Q_i \sim \exp(2\alpha_i)$, we see that the result for the mass comes from the sum of the masses of the branes and the antibranes, while the net charge comes from the difference

$$\widehat{Q}_i = Q_i - \overline{Q}_i.$$

The result for the entropy can then be written in terms of these charges

$$S = \frac{A}{4G_4} = 2\pi \prod_{i=1}^{4}(\sqrt{Q_i} + \sqrt{\overline{Q}_i}).$$

□

11.4 Statistical derivation of the entropy
Extremal black holes

Now let us turn to the microscopical derivation of the entropy of the three-charge supersymmetric black hole in five dimensions. The four-charge supersymmetric black hole in four dimensions can be analyzed in a similar manner, but that is left as a homework problem. The derivation was first given by Strominger and Vafa in the context of type IIB compactifications on $K3 \times S^1$. The discussion that follows analyzes the somewhat simpler case of the toroidal compactification described in Section 11.3. The analysis can be carried out either for the D1-D5-P system or for the S-dual F1-NS5-P system. For definiteness, the discussion that follows refers to the former set-up.

The fact that there are Q_1 units of charge associated with D1-branes means that there are Q_1 windings of D1-branes around the circle. However, the way this is achieved has not been specified. The two extreme possibilities are (1) there are Q_1 D1-branes each of which wraps around the circle once and (2) there is a single D1-brane that wraps around the circle Q_1 times. Altogether, the distinct possibilities correspond to the partitions of Q_1. When there is more than one D1-brane, it is important that they form a bound state in order to give a single black hole. The Q_5 units of D5-brane charge also can be realized in various ways. In all cases, one wants the D1-D5-P system to form a bound state, so that one ends up with a localized object in the noncompact dimensions.

The low-energy physics of these bound states is described by an orbifold conformal field theory that is defined on the circle of radius R. The fields in the conformal field theory correspond to the zero modes of open strings that connect the D1-branes to the D5-branes. There are $Q_1 Q_5$ distinct such strings, since each strand of D1-branes can connect to each strand of D5-branes. That is the picture locally. However, imagine displacing this (small) connecting string repeatedly around the circle. If there is a single multiply wound D1-brane and a single multiply wound D5-brane (along the circle), and if Q_1 and Q_5 have no common factors, then one must go around the circle $Q_1 Q_5$ times to get back to where one started. Thus, the excitations of this system are the same as what one gets from having a single string wound around the circle $Q_1 Q_5$ times. Since this string is localized in the noncompact dimensions, the only bosonic zero modes in its world-volume theory correspond to its position in the four transverse compact dimensions. Since the system is supersymmetric, there must therefore be four boson and four fermion zero modes on the string world volume.

The system described above can be represented as an orbifold conformal field theory that is obtained by taking the tensor product of $Q_1 Q_5$ theories describing singly wound strings and then modding out by all of their $(Q_1 Q_5)!$ permutations. This orbifold theory has many twisted sectors,[15] and

---

15 They are given by the conjugacy classes of the permutation group $S_{Q_1 Q_5}$.

just one of them corresponds to a single string wound Q_1Q_5 times. However, this sector gives more low-energy degrees of freedom than any of the other sectors, all of which involve multiple strings. Excitations of shorter strings have higher energy, which suppresses them entropically. Therefore, one obtains an excellent approximation to the entropy by only counting the excitations of this long string, which is what we will do.

In view of the preceding, let us consider a single string wound Q_1Q_5 times around a circle of radius R that is only allowed to oscillate in four transverse directions. The question to be answered is how many different ways are there of constructing a *supersymmetric* excitations of energy n/R. The string can have left-moving and right-moving excitations, and the level-matching condition is $N_L - N_R = nW$, where the winding number is

$$W = Q_1Q_5. \tag{11.94}$$

Supersymmetry requires that either N_L or N_R vanishes, since then that sector contributes a short (supersymmetric) representation in the tensor product of left-movers and right-movers that gives the physical states of the closed string. Whether N_L or N_R should vanish is determined by the sign of nW.

If N_m^i and n_m^i denote excitation numbers of the four transverse bosonic and fermionic oscillators, respectively, then evaluation of N_L or N_R gives

$$|nW| = \sum_{i=1}^{4} \sum_{m=1}^{\infty} m(N_m^i + n_m^i). \tag{11.95}$$

The degeneracy $d(Q_1, Q_5, n)$ is then given by N_0 times the number of choices for N_m^i and n_m^i that gives $|nQ_1Q_5|$. The factor N_0 denotes the degeneracy of the left-moving or right-moving ground state, which is always 16 for a type II string. However, multiplicative numerical factors turn out to be completely negligible.

The degeneracy is given by the coefficient of w^{nW} in the generating function

$$G(w) = N_0 \prod_{m=1}^{\infty} \left(\frac{1 + w^m}{1 - w^m}\right)^4. \tag{11.96}$$

The numerator takes account of the four fermions and the denominator takes account of the four bosons. To be precise, in this formula the fermions are taken here to be in the R sector. The NS sector would give an equal contribution (after GSO projection).

The degeneracy is evaluated for large nW by representing it as a contour integral and using a saddle-point evaluation, as was described in Chapter 2.

11.4 Statistical derivation of the entropy

It is already clear that the answer is a function of $N = nW = nQ_1Q_5$ only. However, while this is true for the single-string sector under consideration, it is not true for the subdominant multiple-string configurations that are not being considered. Evaluation of the degeneracy for large N requires knowing the behavior of G near $w = 1$. This is obtained using the Jacobi theta function identity

$$\theta_4(0|\tau) = \frac{1}{\sqrt{-i\tau}} \theta_2(0|-1/\tau), \tag{11.97}$$

where $w = e^{i\pi\tau}$, for the representations

$$\theta_4(0|\tau) = \prod_{m=1}^{\infty} \left(\frac{1-w^m}{1+w^m} \right) \tag{11.98}$$

$$\theta_2(0|\tau) = \sum_{n=-\infty}^{\infty} w^{(n-1/2)^2}. \tag{11.99}$$

This implies that as $w \to 1$

$$G(w) \to \left(-\frac{\log w}{\pi} \right)^2 \exp\left(-\frac{\pi^2}{\log w} \right). \tag{11.100}$$

Then writing the degeneracy in the form

$$d(Q_1, Q_5, n) = \frac{1}{2\pi i} \oint \frac{G(w) dw}{w^{N+1}}, \tag{11.101}$$

and using a saddle-point approximation, one finds for large N that

$$d(Q_1, Q_5, n) \sim (Q_1 Q_5 n)^{-7/4} \exp\left(2\pi \sqrt{Q_1 Q_5 n} \right), \tag{11.102}$$

and as a result the microcanonical black-hole entropy is given by

$$S = \log d \sim 2\pi \sqrt{Q_1 Q_5 n} - \frac{7}{4} \log(Q_1 Q_5 n) + \dots \tag{11.103}$$

Remarkably, the leading term in Eq. (11.103) reproduces the result obtained earlier in Eq. (11.55) by computing the area of the horizon in the supergravity approximation. The exponential factor in the degeneracy factor is multiplied by a power of $Q_1 Q_5 n$, and the first correction to the entropy formula is proportional to the logarithm of this factor. This term is a stringy correction to the entropy computed in the supergravity approximation. For the particular black hole considered here, the leading correction to the BH entropy formula is proportional to $\log(A/G_D)$. That seems to be the rule quite generally. However, in contrast to the famous factor of $1/4$ in the leading term, $A/4G_D$, the coefficient of the logarithm is not universal.

Nonextremal black holes

It is natural to try to extend this analysis to nonextremal black holes. The goal would be to reproduce the BH entropy formula by counting microstates. However, in this case the black holes are no longer supersymmetric, and the entropy formula is not guaranteed to extrapolate from weak coupling to strong coupling without corrections. Because of this lack of control, the result has not been derived in the general case using controlled approximations. What has been done successfully, in a mathematically controlled way, is to compare the results for *nearly extremal* black holes for which the nonextremality can be treated as a perturbation.

Let us consider the nonextremal $D = 5$ black holes described in Section 11.3 in the special case that the only antibranes are $\bar{n}$ Kaluza–Klein excitations. In this case, the macroscopic entropy formula Eq. (11.69) becomes

$$S = 2\pi(\sqrt{Q_1 Q_5 n} + \sqrt{Q_1 Q_5 \bar{n}}). \tag{11.104}$$

The interpretation in terms of the world-volume theory of a string of winding number $W = Q_1 Q_5$ is that the equations $N_L = nQ_1 Q_5$ and $N_R = 0$, which were appropriate in the extremal case are now replaced by

$$N_L = nQ_1 Q_5 \quad \text{and} \quad N_R = \bar{n} Q_1 Q_5. \tag{11.105}$$

In this case the degeneracy of states contains both a left-moving and a right-moving factor

$$d \sim \exp(2\pi\sqrt{N_L} + 2\pi\sqrt{N_R}). \tag{11.106}$$

Taking the logarithm of both sides gives the microscopic entropy

$$S = 2\pi(\sqrt{N_L} + \sqrt{N_R}), \tag{11.107}$$

in exact agreement with the macroscopic formula! Surely this is better agreement than one had any right to expect. At the very least, the approximations that have been made require $\bar{n} \ll n$. We will not describe the precise requirements for the approximations to be justified. Suffice it to say, there is some region for which they are justified, but the result that one obtains turns out to give agreement in an even larger region. It would be nice if one could understand why this happened.

Hawking radiation

The nonextremal black holes have a finite temperature and decay by the emission of Hawking radiation. The brane picture makes the instability clear: it can be interpreted as brane–antibrane annihilation. Specifically, for the set-up in the preceding subsection, where there are both left-moving and right-moving Kaluza–Klein excitations, they can collide to give a massless closed-string state, which is then emitted from the black hole. The calculation has been carried out for $\bar{n} \ll n$ with the conclusion that the decay rate as a function of frequency is

$$d\Gamma(\omega) = \frac{A}{e^{\omega/T} - 1} \frac{d^4 k}{(2\pi)^4}, \qquad (11.108)$$

where the temperature is

$$T = \frac{2\sqrt{\bar{n}}}{\pi R}. \qquad (11.109)$$

If one considers D1-brane anti-D1-brane or D5-brane/anti-D5-brane annihilations instead, then a different viewpoint is convenient. When a brane and an antibrane coincide, their common world volume contains a tachyonic mode that arises as the lowest mode of the open string that connects the brane to the antibrane. This tachyon signals an instability of the worldvolume theory, which results in the emission of closed-string radiation as in the previous discussion. In fact, one can test this reasoning by using Witten's string field theory to describe the open string. Sen has argued persuasively that this theory gives a potential for the tachyon field, and that the decay corresponds to sliding down this potential from a local maximum to a local minimum, that is, tachyon condensation. Furthermore, the value of the potential at the minimum should be lower than its value at the maximum by exactly twice the brane tension. Thus the world-volume tachyon rolling to the minimum of its potential precisely corresponds to brane–antibrane annihilation. This results in the emission of closed-string quanta. In the black-hole setting considered here, these quanta comprise the Hawking radiation. This prediction for the gap between the maximum and minimum of the potential has been tested numerically in Witten's bosonic string field theory, and it has been verified to high precision. Moreover, it has recently been derived analytically.

11.5 The attractor mechanism

Moduli fields

As has been discussed in previous sections, black holes can appear when a superstring theory or M-theory is compactified to lower dimensions and when branes are wrapped on nontrivial cycles of the compact manifold. The compactified theories typically have many moduli of the type considered in Chapters 9 and 10. These moduli appear as part of the black-hole solutions, which turn out to exist for generic values of these moduli at infinity, that is, far from the black hole where the geometry is essentially flat. As a result, there is the dangerous possibility that the entropy of the black hole may depend on parameters that are continuous, namely the moduli fields at infinity, and not only on discrete black-hole charges. This would be a problem, since the number of microstates with given charges is an integer, that should not depend on parameters that can be varied continuously. It should only depend on quantities that take discrete values, such as electric/magnetic charges and angular momenta.[16]

The attractor mechanism

In order to resolve this puzzle, one has to realize that the entropy of a black hole is determined by the behavior of the solution at the horizon of the black hole and not at infinity. The obvious way to reconcile this with the observations in the preceding paragraph is for the moduli fields to vary with the radius in such a way that their values at the horizon of the black hole are completely determined by the discrete quantities, such as the charges, regardless of their values at infinity. In other words, the radial dependence of these moduli is determined by differential equations whose solutions flow to definite values at the horizon, regardless of their boundary values at infinity. This solution is called an *attractor* and its existence is the essence of the *attractor mechanism*. The existence of an attractor is necessary for a microscopic description of the black-hole entropy to be possible.

The attractor equations arise from combining laws of black-hole physics with properties of the internal compactification manifolds. To be specific, this section gives the derivation of the attractor equations for type IIB superstring theory compactified on Calabi–Yau three-folds. A crucial ingredient in these cases is *special geometry*, a tool used to describe the relevant moduli spaces that was introduced in Chapter 9.

Black holes in type IIB Calabi–Yau compactifications

As discussed in Chapter 9, when type IIB superstring theory is compactified on a Calabi–Yau three-fold M, the resulting theory in four dimensions has $\mathcal{N} = 2$ supersymmetry. The four-dimensional theory is $\mathcal{N} = 2$ supergravity coupled to $h^{2,1}$ abelian vector multiplets and $h^{1,1} + 1$ hypermultiplets.

---

[16] The entropy formulas given earlier depend on integers only, though they are not logarithms of integers. The reason, of course, is that the formulas are not exact.

11.5 The attractor mechanism

The vector multiplets contain the complex-structure moduli, while the hypermultiplets contain the Kähler moduli and the dilaton. The following discussion focuses on the fields contained in vector multiplets, since the entropy does not depend on the hypermultiplets, at least in the supergravity approximation, as will become clear.

Brief review of special geometry

An $\mathcal{N} = 2$ vector multiplet contains a complex scalar, a gauge field and a pair of Majorana (or Weyl) fermions. The moduli space describing the scalars is $h^{2,1}$-dimensional and is a *special-Kähler* manifold. The Kähler potential for the complex-structure moduli space is

$$\mathcal{K} = -\log\left(i \int_M \Omega \wedge \overline{\Omega}\right), \tag{11.110}$$

where Ω is the holomorphic three-form of the Calabi–Yau manifold, as usual. In this set up a black hole can be realized by wrapping a set of D3-branes on a special Lagrangian three-cycle $\mathcal{C}$. In order to describe this, let us introduce the Poincaré dual three-form to $\mathcal{C}$, which we denote by Γ.

This black hole carries electric and magnetic charges with respect to the $h^{2,1}$ $U(1)$ gauge fields originating from the ten-dimensional type IIB self-dual five-form F_5 as well as the graviphoton belonging to the $\mathcal{N} = 2$ supergravity multiplet. In order to describe the charges, let us introduce a basis of three-cycles A^I, B_J (with $I, J = 1, \ldots, h^{2,1} + 1$), which can be chosen such that the intersection numbers are

$$A^I \cap B_J = -B_J \cap A^I = \delta^I_J \quad \text{and} \quad A^I \cap A^J = B_I \cap B_J = 0. \tag{11.111}$$

The Poincaré dual three-forms are denoted α^I and β_I. The group of transformations that preserves these properties is the symplectic modular group $Sp(2h^{2,1} + 2; \mathbb{Z})$. The symplectic coordinates introduced in Chapter 9 are

$$X^I = e^{\mathcal{K}/2} \int_{A^I} \Omega \quad \text{and} \quad F_I = e^{\mathcal{K}/2} \int_{B_I} \Omega. \tag{11.112}$$

Recall that the definition of Ω can be rescaled by a factor that is independent of the manifold coordinates and that this corresponds to a rescaling of the homogeneous coordinates X^I. This freedom has been used to include the factors of $e^{\mathcal{K}/2}$, which will be convenient later.

The electric and magnetic charges, q_I and p^I, that result in four dimensions are encoded in the homology class $\mathcal{C} = p^I B_I - q_I A^I$ or the equivalent cohomology class $\Gamma = p^I \alpha_I - q_I \beta^I$. Thus, in terms of a canonical homology basis A^I, B_I, one can write

$$\int_{A^I} \Gamma = \int_M \Gamma \wedge \beta^I = p^I \quad \text{and} \quad \int_{B_I} \Gamma = \int_M \Gamma \wedge \alpha_I = q_I. \quad (11.113)$$

The central charge, which is determined by the charges, is given by

$$Z(\Gamma) = e^{i\alpha}|Z| = e^{\mathcal{K}/2} \int_M \Gamma \wedge \Omega = e^{\mathcal{K}/2} \int_C \Omega. \quad (11.114)$$

This expression for the central charge can be derived from the $\mathcal{N} = 2$ supersymmetry algebra, as was shown in Chapter 9. it can be re-expressed as follows:

$$Z(\Gamma) = e^{\mathcal{K}/2} \sum_I \left(\int_{A^I} \Gamma \int_{B_I} \Omega - \int_{B_I} \Gamma \int_{A^I} \Omega \right) = p^I F_I - q_I X^I. \quad (11.115)$$

The attractor equations and dyonic black holes

Let us now show that the complex-structure moduli fields at the horizon are determined by the charges of the black hole, independent of the values of these fields at infinity. In order to illustrate this, we will derive the differential equations satisfied by the complex-structure moduli fields for the case of four-dimensional spherically symmetric supersymmetric black holes. These conditions restrict the space-time metric to be of the form

$$ds^2 = -e^{2U(r)}dt^2 + e^{-2U(r)}d\vec{x} \cdot d\vec{x}, \quad (11.116)$$

where $\vec{x} = (x_1, x_2, x_3)$ and $r = |\vec{x}|$ is the radial distance and $r = 0$ is the event horizon. Note that this requires using a coordinate system that is singular at the horizon like the one in Eq. (11.77), for example. Let us also assume that the holomorphic complex-structure moduli fields t^α only depend on the radial coordinate, so that $t^\alpha = t^\alpha(r)$, with $\alpha = 1, \ldots, h^{2,1}$. Recall that these coordinates are related to the homogeneous coordinates X^I introduced above by $t^\alpha = X^\alpha/X^0$. It is convenient to introduce the variable $\tau = 1/r$. Then $\tau = 0$ corresponds to spatial infinity, while $\tau = \infty$ corresponds to the horizon of the black hole.

The first-order differential equations satisfied by $U(\tau)$ and $t^\alpha(\tau)$ can be derived by solving the conditions for unbroken supersymmetry

$$\delta\psi_\mu = \delta\lambda^\alpha = 0, \quad (11.117)$$

where ψ_μ is the gravitino, and λ^a represents the gauginos. These equations

11.5 The attractor mechanism

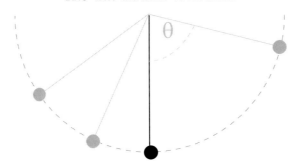

Fig. 11.4. The pendulum with a dissipative force acting on it evolves towards $\theta = 0$ independently of the initial conditions.

imply a set of first-order differential equations[17]

$$\frac{dU(\tau)}{d\tau} = -e^{U(\tau)}|Z|, \tag{11.118}$$

$$\frac{dt^\alpha(\tau)}{d\tau} = -2e^{U(\tau)}G^{\alpha\bar{\beta}}\partial_{\bar{\beta}}|Z|. \tag{11.119}$$

Recall that $G^{\alpha\bar{\beta}}$ is the inverse of $G_{\alpha\bar{\beta}} = \partial_\alpha \partial_{\bar{\beta}} \mathcal{K}$. In this form the conditions for unbroken supersymmetry can be interpreted as differential equations describing a dynamical system with τ playing the role of time.

The physical scenario described by these equations has a nice analogy with dynamical systems. Consider, for example, a pendulum with a dissipative force acting on it. In general, the final position of the pendulum is independent of its initial position and velocity. The point at $\theta = 0$ in Fig. 11.4 represents the attractor in this simple example. Solving the equations in the near-horizon limit is then equivalent to solving the late-time behavior of the dynamical system. It will turn out that the horizon represents an attractor, that is, a point (or surface) in the phase space to which the system evolves after a long period of time. This means that the moduli approach fixed values at the horizon that are independent of the initial conditions.

Solution of the attractor equations

In order to solve Eqs (11.118) and (11.119) explicitly near the horizon, let us first note that these differential equations can be written in the alternative equivalent form

$$2\frac{d}{d\tau}\left[e^{-U(\tau)+\mathcal{K}/2}\mathrm{Im}\left(e^{-i\alpha}\Omega\right)\right] \sim -\Gamma. \tag{11.120}$$

---

[17] The derivations are given in hep-th/9807087. Since Eq. (11.114) is homogeneous of degree one in the X s, $Z(t^\alpha)$ means $(X^0)^{-1}Z(X^I)$.

The $\sim$ symbol means that two sides are cohomologous. In other words, they are allowed to differ by an exact three-form, though this freedom can be eliminated by choosing harmonic representatives. Equation (11.118) is one real equation and (11.119) is $h^{2,1}$ complex equations, making $2h^{2,1} + 1$ real equations altogether. Equation (11.120) can be projected along each of the classes of H^3, and there are $2h^{2,1} + 2$ of these. So, if it really is equivalent to Eqs (11.118) and (11.119), it is necessary that there is a redundancy among these equations.

Consider integrating Eq. (11.120) over the A cycles and the B cycles. Using Eqs (11.112) and (11.113) this gives

$$2\frac{d}{d\tau}\left[e^{-U(\tau)}\text{Im}\left(e^{-i\alpha}X^I\right)\right] = -p^I \qquad (11.121)$$

and

$$2\frac{d}{d\tau}\left[e^{-U(\tau)}\text{Im}\left(e^{-i\alpha}F_I\right)\right] = -q_I. \qquad (11.122)$$

Contracting the first equation with q_I and subtracting the second equation contracted with p^I gives

$$2\frac{d}{d\tau}\left[e^{-U(\tau)}\text{Im}\left(e^{-i\alpha}(q_I X^I - p^I F_I)\right)\right] = 0. \qquad (11.123)$$

However, Eqs (11.114) and (11.115) imply that

$$e^{-i\alpha}(q_I X^I - p^I F_I) = -|Z|, \qquad (11.124)$$

so that Eq. (11.123) is automatically satisfied. This is the required redundancy that leaves $2h^{2,1} + 1$ nontrivial equations.

Equation (11.118) can be obtained from Eq. (11.120) by projecting both sides on $e^{-i\alpha}e^{\mathcal{K}/2}\Omega$. This means taking the wedge product with this three-form and then integrating over the manifold. The derivation of Eq. (11.118) by this method is given in Exercise 11.10. To deduce the complex equations in Eq. (11.119), one should project along $e^{-i\alpha}e^{\mathcal{K}/2}D_\alpha\Omega$. Together with the previous result, this extracts the full information content of Eq. (11.120). The derivation of Eq. (11.119) is left as a homework problem.

The differential equation (11.120) can be integrated, since Γ does not depend on τ. Its expansion in a real cohomology basis only depends on the electric and magnetic charges carried by the black hole.[18] The result is

$$2e^{-U(\tau)+\mathcal{K}/2}\text{Im}\left(e^{-i\alpha}\Omega\right) \sim -\Gamma\tau + 2\left[e^{-U(\tau)+\mathcal{K}/2}\text{Im}\left(e^{-i\alpha}\Omega\right)\right]_{\tau=0}. \qquad (11.125)$$

This equation yields implicitly the solution for the moduli fields $t^\alpha = t^\alpha(\tau)$.

---

[18] Of course, its Hodge decomposition depends on the complex structure and thus on τ, but this is not relevant to the argument.

Equation (11.119) implies that

$$\frac{d|Z|}{d\tau} = \frac{dt^\alpha(\tau)}{d\tau}\partial_\alpha|Z| + \frac{dt^{\bar\alpha}(\tau)}{d\tau}\partial_{\bar\alpha}|Z| = -4e^U G^{\alpha\bar\beta}\partial_\alpha|Z|\partial_{\bar\beta}|Z| \leq 0. \quad (11.126)$$

As a result, $|Z|$ is a monotonically decreasing function of τ converging to a minimum. The fixed point is then determined by

$$\frac{d|Z|}{d\tau} \to 0 \quad \text{as} \quad \tau \to \infty. \quad (11.127)$$

In order to solve for the moduli fields near the horizon, we assume that the central charge has a nonvanishing value $Z = Z_\star \neq 0$ at the fixed point. Therefore, Eq. (11.118) can be integrated to give, for large τ,

$$\tau^{-1}e^{-U(\tau)} \to |Z_\star|. \quad (11.128)$$

Substituting into the metric, this implies that the near-horizon geometry is $AdS_2 \times S^2$, just as in Exercise 11.3,

$$ds^2 \to -\frac{r^2}{|Z_\star|^2}dt^2 + |Z_\star|^2\frac{dr^2}{r^2} + |Z_\star|^2(d\theta^2 + \sin^2\theta d\phi^2), \quad (11.129)$$

and it determines the area of the horizon to be

$$A = 4\pi|Z_\star|^2. \quad (11.130)$$

In the near-horizon limit Eq. (11.125) can be solved giving rise to the *attractor equation*, which is a determining equation for the complex-structure moduli in the near-horizon limit. In this limit Eq. (11.125) implies that

$$2e^{\mathcal{K}/2}\text{Im}\left(\overline{Z}_\star\Omega\right) \sim -\Gamma \quad (11.131)$$

at the horizon. This implies that

$$\Gamma = \Gamma_{(3,0)} + \Gamma_{(0,3)} \quad (11.132)$$

at the horizon, that is, the only nonvanishing terms in the Hodge decomposition of Γ are $(3,0)$ and $(0,3)$, while the $(1,2)$ and $(2,1)$ parts vanish. This is a property of the fixed-point, and it need not be true away from the horizon. Therefore, the attractor mechanism can be viewed as a method to determine Ω at the horizon in terms of the charges of the black hole.

The attractor condition (11.131) and the charges defined in Eq. (11.113) give the alternative formulas[19]

$$p^I = -2\text{Im}\left(\overline{Z}X^I\right) \quad \text{and} \quad q_I = -2\text{Im}\left(\overline{Z}F_I\right). \quad (11.133)$$

This form of the attractor equations is used in the following sections.

---

[19] These equations often appear with plus signs. Clearly, the signs depend on conventions that have been made along the way.

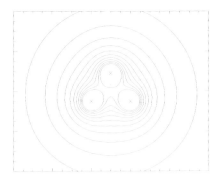

Fig. 11.5. Lines with constant τ of the 3-center solution with identical charges.

Multi-center solutions

There are stationary multi-black holes solutions that are known as multi-center solutions. The reason these exist is that, when each of the black holes preserves the same supersymmetry charge, this supersymmetry is an unbroken symmetry of the multi-black hole system. In this case, the BPS condition result in a *no-force condition*, which means that the total force acting on each of the black holes due to the presence of the others exactly cancels, so that each of them can remain at rest. The various attractive and repulsive forces due to gravity, scalar fields, and gauge fields are guaranteed to cancel out due to supersymmetry. This is true even though the field configurations are much more complicated than they are for a single black hole.

The attractor equations can be generalized to the case where are black-hole horizons, with charges encoded in harmonic three-forms Γ_p, at different points $\vec{x}_p$. In the special case where all of the component black-holes have the same charges, the flow parameter τ is naturally defined to be

$$\tau = \sum_p \frac{1}{|\vec{x} - \vec{x}_p|}. \qquad (11.134)$$

Surfaces with constant τ in the 3-center case are displayed in Fig. 11.5. In general, the charges are not identical. In order to describe such a solution, known as a *multi-center solution*, one has to consider a slightly generalized metric of the form

$$ds^2 = -e^{2U}(dt + \omega_i dx^i)^2 + e^{-2U} d\vec{x} \cdot d\vec{x}. \qquad (11.135)$$

11.5 The attractor mechanism

The attractor equations can then be shown to take the form

$$H = 2e^{-U}\text{Im}\left(e^{-i\alpha}e^{\mathcal{K}/2}\Omega\right),$$
$$\star d\omega = \int_M dH \wedge H,$$
(11.136)

where $H(\vec{x})$ is a harmonic function of the space-like coordinates (as well as a differential form in the compact dimensions), $\star$ is the Hodge star operator on $\mathbb{R}^3$ and α is the phase of $Z(\sum \Gamma_p)$. The first of these equations is the generalization of Eq. (11.125), while the second one has no counterpart in the one-center case. Since each of the horizons is an attractor, the flow equation in this case is called a *split attractor flow*.

If we have N centers in asymptotically flat space-time, integration gives

$$H = -\sum_{p=1}^{N} \Gamma_p \frac{1}{|\vec{x} - \vec{x}_p|} + 2\text{Im}\left(e^{-i\alpha}e^{\mathcal{K}/2}\Omega\right)_{r=\infty}.$$
(11.137)

Acting with the operator $d\star$ on the second equation in (11.136) gives the condition

$$\int_M \Delta H \wedge H = 0.$$
(11.138)

Using

$$\Delta \frac{1}{|\vec{x} - \vec{x}_p|} = -4\pi\delta^{(3)}(\vec{x} - \vec{x}_p),$$
(11.139)

one then obtains

$$\sum_{q=1}^{N} \frac{1}{|\vec{x}_p - \vec{x}_q|} \int_M \Gamma_p \wedge \Gamma_q = 2\text{Im}\left[e^{-i\alpha}Z(\Gamma_p)\right]_{r=\infty}.$$
(11.140)

It can be shown that a multi-center solution exists as long as the this equation is satisfied. It determines the position of the charges. So, for example, in the two-center case the separation of the horizons is determined by

$$|\vec{x}_1 - \vec{x}_2| = \frac{\int_M \Gamma_1 \wedge \Gamma_2}{2\text{Im}\left[e^{-i\alpha}Z(\Gamma_1)\right]_{r=\infty}}.$$
(11.141)

Black rings

In four dimensions there is a theorem to the effect that the topology of a black-hole event horizon is necessarily that of a two-sphere. It therefore came as a surprise when people realized that there are more possibilities in higher dimensions. In all of the five-dimensional examples discussed so far, the horizon has S^3 topology. However, there are also asymptotically flat supersymmetric solutions in five dimensions for which the topology of the horizon is $S^1 \times S^2$. In fact, there are so many solutions of this type that they are not uniquely determined by their mass, charges and angular momenta, in contrast to black holes. These solutions are called *black rings*. As you might guess, rotation is required to support this topology. These solutions can be found by considering $\mathcal{N} = 2$ supergravity coupled to a set of vector multiplets in five dimensions. This can be realized by compactifying M-theory on a Calabi–Yau three-fold, as was discussed in Chapter 9. This is a conceptually beautiful subject, but the formulas tend to get a bit complicated. So we will just list the essential results without the derivations.[20]

In order to present the supersymmetric black-ring solutions, let us first describe the most general solutions with unbroken supersymmetry. The scalars in the vector multiplets are real and denoted by Y^A. The BPS equations are then solved by

$$ds_5^2 = -f^{-2}(dt + \omega)^2 + f ds_X^2, \tag{11.142}$$

where

$$ds_X^2 = \sum_{m,n=1}^{4} h_{mn} dx^m dx^n. \tag{11.143}$$

Here X is a four-dimensional hyperkähler space with metric h_{mn}, ω is a one-form on X and f is a scalar function depending on the coordinates of X. The $U(1)$ field strength two-forms F^A in the vector multiplets are determined by

$$F^A = d\left[f^{-1}Y^A(dt + \omega)\right] + \Theta^A, \tag{11.144}$$

where Θ^A are closed self-dual two-forms on X, that is,

$$\Theta^A = \star_4 \Theta^A. \tag{11.145}$$

Moreover, supersymmetry implies that ω and f are determined by

$$d\omega + \star_4 d\omega = -f Y_A \Theta^A, \tag{11.146}$$

and

$$\nabla^2(fY_A) = 3 D_{ABC} \Theta^B \Theta^C, \tag{11.147}$$

where D_{ABC} are the intersection numbers of two-forms (or dual four-cycles) describing the geometry of the Y^A moduli space. This is the most general solution preserving supersymmetry in five dimensions. So, for example, the five-dimensional three-charge black holes and rotating black holes described

---

[20] For further details see hep-th/0504126.

11.5 The attractor mechanism

in Section 11.3 are special cases of this solution, as you are asked to check in a homework problem.

In order to obtain an example of a black-ring solution, it is sufficient to consider the case of one modulus, that is, $A = 1$ and $D_{111} = 1$. The space X is taken to be Taub–NUT with metric

$$ds_X^2 = H^0 d\vec{x} \cdot d\vec{x} + \frac{1}{H^0}(dx^5 + \omega^0)^2 \quad \text{and} \quad d\omega^0 = \star_3 dH^0, \quad (11.148)$$

where x^5 has period 4π. The solution is then formulated in terms of a set of two-center harmonic functions H defined by

$$H^0 = \frac{4}{R_{\text{TN}}^2} + \frac{1}{|\vec{x}|} \qquad H_0 = -\frac{q_0}{L} + \frac{q_0}{|\vec{x} - \vec{x}_0|}, \quad (11.149)$$

and

$$H^1 = \frac{p^1}{|\vec{x} - \vec{x}_0|}, \qquad H_1 = 1 + \frac{q_1}{|\vec{x} - \vec{x}_0|}, \quad (11.150)$$

where $\vec{x}_0 = (0, 0, L)$ and (p^1, q_1, q_0) are constants. When compactified to four dimensions, this background is a bound state of one D6-brane located at $|\vec{x}| = 0$ and a black hole with D4-D2-D0 brane charges (p^1, q_1, q_0).

Using these harmonic functions, Eq. (11.145) is solved by

$$\Theta^1 = d\left[\frac{H^1}{H^0}(dx^5 + \omega^0)\right] + \star_3 dH^1, \quad (11.151)$$

while the solution to Eq. (11.147) is provided by

$$f = H_1 + 3\frac{(H^1)^2}{H^0}. \quad (11.152)$$

Moreover,

$$\omega = -\left[H_0 + 2\frac{(H^1)^3}{(H^0)^2} + \frac{H_1 H^1}{H^0}\right](dx^5 + \omega^0) + \omega^{(4)}, \quad (11.153)$$

where $\omega^{(4)}$ is the solution of

$$d\omega^{(4)} = H_I \star_3 dH^I - H^I \star_3 dH_I. \quad (11.154)$$

The black-ring solution is then obtained by taking the limit $R_{\text{TN}} \to \infty$. This is the same sort of limit considered earlier when we discussed the 4d/5d connection relating a rotating black hole in five dimensions to one with suitable charges in four dimensions.

The five-dimensional metric can then be written in the form

$$ds_5^2 = G_{\mu\nu}^{(4)} dx^\mu dx^\nu + \lambda(dx^5 - A_\mu dx^\mu)^2, \quad (11.155)$$

where $G^{(4)}_{\mu\nu}$ is the four-dimensional metric, λ is a scalar and A_μ is a $U(1)$ gauge field. The four-dimensional metric satisfies the two-center attractor equations.

EXERCISES

EXERCISE 11.10
Deduce Eq. (11.118) by projecting both sides of Eq. (11.120) on $e^{-i\alpha+\mathcal{K}/2}\Omega$.

SOLUTION
Equation (11.120) is equivalent to

$$\frac{d}{d\tau}\left[e^{-U}\left(e^{-i\alpha+\mathcal{K}/2}\Omega - e^{i\alpha+\mathcal{K}/2}\overline{\Omega}\right)\right] \sim -i\Gamma.$$

Taking the wedge product with $e^{-i\alpha+\mathcal{K}/2}\Omega$, only the second term on the left contributes since $\Omega \wedge \Omega = \Omega \wedge \frac{d}{d\tau}\Omega = 0$. Thus

$$-\frac{d}{d\tau}\left(e^{-U}\right)e^{\mathcal{K}}\Omega \wedge \overline{\Omega} - e^{-U}e^{-i\alpha+\mathcal{K}/2}\Omega \wedge \frac{d}{d\tau}\left(e^{i\alpha+\mathcal{K}/2}\overline{\Omega}\right)$$

$$\sim -ie^{-i\alpha+\mathcal{K}/2}\Omega \wedge \Gamma.$$

The imaginary part of this equation is now integrated over the manifold. A useful identity that implies that the integral of the second term is real is

$$\int e^{i\alpha+\mathcal{K}/2}\overline{\Omega} \wedge \frac{d}{d\tau}\left(e^{-i\alpha+\mathcal{K}/2}\Omega\right) = \int e^{-i\alpha+\mathcal{K}/2}\Omega \wedge \frac{d}{d\tau}\left(e^{i\alpha+\mathcal{K}/2}\overline{\Omega}\right),$$

which is derived by differentiating Eq. (11.110) written in the form

$$\int \left(e^{-i\alpha+\mathcal{K}/2}\Omega\right) \wedge \left(e^{i\alpha+\mathcal{K}/2}\overline{\Omega}\right) = -i.$$

In this way, one obtains

$$i\frac{d}{d\tau}\left(e^{-U}\right)e^{\mathcal{K}}\int \Omega \wedge \overline{\Omega} = -\frac{1}{2}e^{\mathcal{K}/2}\int \left(e^{-i\alpha}\Omega + e^{i\alpha}\overline{\Omega}\right) \wedge \Gamma.$$

Using Eq. (11.110) to simplify the left-hand side and Eq. (11.114) to simplify the right-hand side, one obtains

$$\frac{d}{d\tau}\left(e^{-U}\right) = |Z|,$$

11.6 Small BPS black holes in four dimensions

Section 11.4 showed how the counting of microscopic degrees of freedom reproduces the BH entropy of certain supersymmetric black holes. A crucial requirement for this agreement is that Dp-branes wrapped on cycles of the internal manifold excite enough different charges to give a solution with a nonvanishing classical black-hole horizon.

Black holes can also be created using fundamental strings and their excitations without invoking solitonic branes. The string spectrum consists of an infinite tower of states with arbitrarily large masses. For sufficiently high excitations, or sufficiently large coupling constant, gravitational collapse becomes unavoidable. This implies that the Hilbert space of string excitations should contain black holes. This opens up the interesting possibility that certain string excitations admit an alternative interpretation as black holes. In this section we discuss evidence that black holes are an alternative description of certain elementary string excitations. The evidence again follows from comparing the black-hole entropy obtained by counting microscopic quantum states to the macroscopic black-hole entropy described by geometry.

The difficulty in making a black hole out of perturbative string excitations is that an elementary string states do not excite all four types of charges in the λ factor of the metric in Eq. (11.77). Therefore, the area of the horizon would vanish in the supergravity approximation, leaving a null singularity at the origin. For large string excitation number N, the entropy is proportional to $\sqrt{N}$, and the area of the horizon is $A \sim \sqrt{N} \ell_p^2$ where ℓ_p is the four-dimensional Planck length. Even though the area of the horizon is large in Planck units, it is of order one in string units, which explains why the supergravity approximation gives zero. Therefore, α' corrections to the supergravity approximation are important for obtaining a horizon of finite radius that shields the singularity, as required by the cosmic censorship conjecture.

Microstate counting

As a specific example, let us consider the heterotic string compactified on a six-torus to four dimensions, which was discussed in Chapter 7.[21] This gives

---

[21] The techniques discussed in this section are of more general applicability than this specific example.

28 $U(1)$ gauge fields.[22] These transform as a vector of the $O(22,6;\mathbb{Z})$ duality group. 22 of the gauge fields belong to 22 vector multiplets, while the other 6 belong to the supergravity multiplet. The allowed charges of these gauge fields are given by sites of the Narain lattice, as described in Chapter 7. Since this is an even lattice, a charge vector in this lattice squares to an even integer. In other words, since the charges are encoded in the internal momenta (p_R, p_L), where p_L has 22 components and p_R has six components,

$$p_R^2 - p_L^2 = 2N. \tag{11.156}$$

The mass formula for these states is

$$\frac{1}{4}\alpha' M^2 = \frac{1}{2}p_R^2 + N_R = \frac{1}{2}p_L^2 + N_L - 1, \tag{11.157}$$

where N_L and N_R are the usual oscillator excitation numbers.

Dabholkar–Harvey states

Most states with masses given by Eq. (11.157) are unstable, but the BPS states are stable. The BPS states, that is, the state belonging to short supermultiplets for which the mass saturates the BPS bound, have $N_R = 0$. In this case

$$\alpha' M^2 = 2p_R^2, \tag{11.158}$$

while N_L is arbitrary. This results in a whole tower of stable states, which are sometimes called *Dabholkar–Harvey* states. For these states the level-matching condition reduces to

$$N_L - 1 = N. \tag{11.159}$$

For example, if there is winding number w and Kaluza–Klein excitation number n on one cycle of the torus, then $N = |nw|$. In general, the degeneracy of states for large N is given by

$$d_N \approx \exp\left(4\pi\sqrt{N}\right), \tag{11.160}$$

resulting in a leading contribution to the black-hole entropy given by

$$S = \log d_N \approx 4\pi\sqrt{N}. \tag{11.161}$$

---

[22] We assume generic positions in the moduli space so that there is no enhanced gauge symmetry.

Counting states

Let us now compute the corrections to Eq. (11.161). The degeneracy d_N denotes the number of ways that the 24 left-moving bosonic oscillators can give $N_L = N + 1$ units of excitation. This can be encoded in a partition function

$$Z(\beta) = \sum d_N e^{-\beta N} = \frac{16}{\Delta(q)}, \tag{11.162}$$

where

$$q = e^{-\beta} = e^{2\pi i \tau}, \tag{11.163}$$

and the factor of 16 is the degeneracy of right-moving ground states. The factor $\Delta(q)$ is related to the Dedekind η function by

$$\Delta(q) = \eta(\tau)^{24} = q \prod_{n=1}^{\infty} (1 - q^n)^{24}. \tag{11.164}$$

The large-N degeneracy depends on the value of this function as $q \to 1$ or $\beta \to 0$. Under a modular transformation the Dedekind η function transforms as

$$\eta(-1/\tau) = \sqrt{-i\tau}\,\eta(\tau). \tag{11.165}$$

As a result,

$$\Delta(e^{-\beta}) = \left(\frac{\beta}{2\pi}\right)^{-12} \Delta(e^{-4\pi^2/\beta}), \tag{11.166}$$

which, by using $\Delta(q) \approx q$ for small q, gives the estimate

$$\Delta(e^{-\beta}) \approx \left(\frac{\beta}{2\pi}\right)^{-12} e^{-4\pi^2/\beta}. \tag{11.167}$$

This result is extremely accurate, since all corrections are exponentially suppressed.

Now one can compute d_N, as in earlier chapters.

$$d_N = \frac{1}{2\pi i} \oint Z(\beta) \frac{dq}{q^{N+1}} = \frac{1}{2\pi i} \oint \frac{16}{\Delta(q)} \frac{dq}{q^{N+1}}. \tag{11.168}$$

Using Eq. (11.166), this can be approximated for large N by

$$d_N \approx 16\, \hat{I}_{13}(4\pi\sqrt{N}), \tag{11.169}$$

where

$$\hat{I}_\nu(z) = \frac{1}{2\pi i} \int_{\varepsilon-i\infty}^{\varepsilon+i\infty} (t/2\pi)^{-\nu-1} e^{t+z^2/4t} dt \tag{11.170}$$

is a modified Bessel function. This formula includes all inverse powers of N, but it does not include terms that are exponentially suppressed for large N. A saddle-point estimate for large N gives

$$S = \log d_N \approx 4\pi\sqrt{N} - \frac{27}{2}\log\sqrt{N} + \frac{15}{2}\log 2 + \ldots \qquad (11.171)$$

This shows that the leading-order entropy of the black hole obtained by counting microstates is proportional to the mass, $S \sim M$. We could try to compare this to the corresponding macroscopic black-hole solution, but the black hole constructed of perturbative Dabholkar–Harvey states only excites two of the four charges that are needed to get a nonvanishing area of the event horizon. So the result is zero in the supergravity approximation. This is the best that one could hope for in this approximation, because if the area were nonzero, the entropy would be proportional to M^2.

So how can we construct a macroscopic black hole that reproduces the entropy (11.169)? The resolution lies in realizing that elementary string states become heavy enough to form black holes at large coupling. As a result, one should expect that string-theoretic corrections to supergravity, such as terms in the action that are higher order in the curvature, modify the macroscopic geometry and the associated entropy, yielding a nonvanishing result.

Macroscopic entropy

The preceding analysis gave a very accurate result for the degeneracy of states d_N of a certain class of supersymmetric black holes. Remarkably, this formula has been reproduced precisely from a dual macroscopic analysis. The crucial point is that the supergravity approximation is inadequate for this problem, and one must include higher-order terms in the string effective action. In general, this is a hopelessly difficult problem. However, in the case at hand, it turns out that the relevant higher-order terms can be computed.

In order to compute these corrections, it is more convenient to work with the type IIA string theory compactified on $K3 \times T^2$ instead of the heterotic string on T^6. According to a duality discussed in Chapter 9, this is an equivalent theory. In this description the machinery of special geometry is applicable. The quantum gravity corrections are then encoded in corrections to the prepotential. No closed expression for these corrections is known in general, but luckily in this case there is $\mathcal{N} = 4$ supersymmetry. When there is this much supersymmetry, a nonrenormalization theorem implies that only the first correction to the prepotential is nonvanishing, and this is enough to reproduce the microscopic entropy discussed in the previous section. A key ingredient in the analysis is the attractor mechanism.

Type IIA superstring theory on $K3 \times T^2$ has $\mathcal{N} = 4$ supersymmetry in four dimensions, but the attractor mechanism analysis is carried out most conveniently using the $\mathcal{N} = 2$ complex special geometry formalism. It is still applicable when there are additional supersymmetries. When one goes beyond the supergravity approximation and includes higher-genus contributions to the effective action, the holomorphic prepotential $F(X^I)$ generalizes to a function

$$F(X^I, W^2) = \sum_{h=0}^{\infty} F_h(X^I) W^{2h}, \qquad (11.172)$$

where h denotes the genus and W is a chiral superfield that appears in the description of the $\mathcal{N} = 2$ supergravity multiplet.[23] The first component of W is the anti-self-dual part of the graviphoton field strength. The *graviphoton* is the $U(1)$ gauge field contained in the $\mathcal{N} = 2$ supergravity multiplet. The prepotential satisfies the homogeneity equation

$$X^I \partial_I F(X^I, W^2) + W \partial_W F(X^I, W^2) = 2 F(X^I, W^2), \qquad (11.173)$$

which generalizes the formula presented in Chapter 9. Topological string theory techniques, which are not described in this book, enable one to compute the coefficients of terms in the effective action of the form

$$\int d^4x d^4\theta W^{2h} F_h(X^I), \qquad (11.174)$$

which is exactly what is required.

When terms of higher-order than the Einstein–Hilbert term contribute to the action in a significant way, the BH entropy formula is no longer correct. The appropriate generalization has been worked out by Wald. Wald's formula (see Problem 11.15) is applied to the R^2 corrected action in the present case.

The attractor equations that determine the moduli in terms of the charges, and make the central charge extremal, are[24]

$$\begin{aligned} p^I &= \mathrm{Re}\,(CX^I) \\ q_I &= \mathrm{Re}\,(CF_I), \end{aligned} \qquad (11.175)$$

---

[23] Since we do not wish to describe this formalism, as well as other issues, the argument presented here is sketchy. The reader is referred to hep-th/0507014 for further details.

[24] The coordinates X^I, F_I in this section and those in section 11.5 differ by a rescaling of the holomorphic three-form Ω by a factor $2i\bar{Z}/C$, where C is an arbitrary field introduced here for bookkeeping purposes.

where p^I denote magnetic charges and q_I denote electric charges as before. Moreover, in the conventions that are usually used, the graviphoton field strength at the horizon takes the value

$$C^2 W^2 = 256. \tag{11.176}$$

After taking the corrections into account, it can be shown that the black-hole entropy is

$$S = \frac{\pi i}{2}\left(q_I \overline{CX}^I - p^I \overline{CF}_I\right) + \frac{\pi}{2}\operatorname{Im}\left(C^3 \partial_C F\right). \tag{11.177}$$

The first term in this equation agrees with the attractor value $S = \pi |Z_\star|^2$ (for $G_4 = 1$) derived in the previous section when one takes account the rescaling mentioned in the footnote. The second term is a string theory correction.

The first equation in (11.175) is solved by writing

$$CX^I = p^I + \frac{i}{\pi}\phi^I. \tag{11.178}$$

In order to solve the second equation, we define

$$\mathcal{F}(\phi, p) = -\pi \operatorname{Im} F(p^I + \frac{i}{\pi}\phi^I, 256). \tag{11.179}$$

Using this definition,

$$q_I = \frac{1}{2}\left(CF_I + \overline{CF}_I\right) = -\frac{\partial}{\partial \phi^I}\mathcal{F}(\phi, p), \tag{11.180}$$

where we have used

$$\frac{\partial}{\partial \phi^I} = \frac{i}{\pi C}\frac{\partial}{\partial X^I} - \frac{i}{\pi \overline{C}}\frac{\partial}{\partial \overline{X}^I}. \tag{11.181}$$

The homogeneity relation for the prepotential then implies

$$C\partial_C F\left(X^I, \frac{256}{C^2}\right) = X^I \frac{\partial}{\partial X^I} F - 2F. \tag{11.182}$$

As a result, the corrected entropy can be written in the form

$$S(p, q) = \mathcal{F}(\phi, p) - \phi^I \frac{\partial}{\partial \phi^I}\mathcal{F}(\phi, p). \tag{11.183}$$

In other words, the entropy of the black hole is the Legendre transform of $\mathcal{F}$ with respect to ϕ^I. So it is more convenient to specify the ϕ^I, which play the role of chemical potentials, rather than the electric charges q_I.

11.6 Small BPS black holes in four dimensions

For the reasons just described, it is natural to consider a mixed ensemble with the partition function

$$\mathcal{Z}(\phi^I, p^I) = e^{\mathcal{F}(\phi^I, p^I)} = \sum_{q^I} \Omega(q_I, p^I) e^{-\phi^I q_I}, \qquad (11.184)$$

which is microcanonical with respect to the magnetic charges p^I and canonical with respect to the electric charges q_I. Moreover, $\Omega(q_I, p^I)$ are the black-hole degeneracies, and $\log \Omega$ is the microcanonical entropy. The black-hole entropy is then obtained according to

$$S(q,p) = \log \Omega(q,p). \qquad (11.185)$$

The inverse transform is (formally)

$$\Omega(q_I, p^I) = \int e^{\mathcal{F}(\phi^I, p^I) + \phi^I q_I} \prod d\phi^I, \qquad (11.186)$$

which, in principle, allows one to obtain the microscopic black-hole degeneracies by using amplitudes computed by topological string theory.

Heterotic compactification on T^6

In the special case of the heterotic string on T^6, one can use these results by going to the S-dual description in terms of the type IIA theory on $K3 \times T^2$. In this description the Kaluza–Klein modes and winding modes map to D4-branes wrapped on the K3 and D0-branes. The D0-branes are electrically charged with respect to one gauge field and the D4-branes are magnetically charged with respect to another one. Thus, only two charges, q_0 and p^1 say, are nonzero.

The prepotential is particularly simple in this case. Since this theory has an $\mathcal{N} = 4$ supersymmetry, the only nonvanishing contributions to the prepotential are F_0 and F_1. For F_0 one takes the tree-level amplitude given by

$$F_0 = -\frac{1}{2} C_{ab} X^a X^b \left(\frac{X^1}{X^0} \right), \qquad a, b = 2, \ldots, 23 \qquad (11.187)$$

where C_{ab} is the intersection matrix of two-cycles on K3, and

$$\tau = \tau_1 + i\tau_2 = X^1/X^0 \qquad (11.188)$$

is the Kähler modulus of the torus.

The only additional contribution is F_1. Schematically, this term can be obtained by taking the ten-dimensional interaction $\int B \wedge Y_8$ and compactifying on $K3 \times T^2$. In the type IIA description there is an $SL(2, \mathbb{Z})$ T-duality symmetry associated with the T^2 factor, which corresponds to an $SL(2, \mathbb{Z})$

S-duality symmetry of the dual heterotic string theory in four dimensions. The modular parameter of this symmetry is τ, and it transforms nonlinearly under $SL(2,\mathbb{Z})$ transformations in the usual way. Its real part, τ_1, which is an axion-like field, arises from a duality transformation of the two-form B in four dimensions. Accordingly, the ten-dimensional interaction gives rise to a four-dimensional term of the form

$$\frac{1}{8\pi}\int \tau_1 \left(\mathrm{tr} R \wedge R - \mathrm{tr} F \wedge F\right). \tag{11.189}$$

The normalization is fixed by the requirement that this should be well defined up to a multiple of 2π when τ_1 is shifted by an integer, since such shifts are part of the $SL(2,\mathbb{Z})$ group. To get the rest of the group working, specifically the transformation $\tau \to -1/\tau$, it is necessary to add higher-order terms by the replacement

$$\tau = \frac{1}{2\pi i}\log q \to \frac{24}{2\pi i}\log \eta(\tau) = \frac{1}{2\pi i}\log \Delta(q). \tag{11.190}$$

In the heterotic viewpoint, the corrections given by this substitution have the interpretation as instanton contributions due to Euclideanized NS5-branes wrapping the six-torus.

It follows that the S-duality invariant and supersymmetric completion of the $\mathrm{tr} R \wedge R$ term is[25]

$$\frac{1}{16\pi^2}\mathrm{Im}\int \log \Delta(q)\,\mathrm{tr}\left[(R - iR^\star)\wedge(R - iR^\star)\right]. \tag{11.191}$$

The factor involving the curvatures is part of $\int d^4\theta W^2$, and its coefficient determines F_1 to be

$$F_1 = \frac{i}{128\pi}\log \Delta(q). \tag{11.192}$$

This shows that F_1 is independent of the K3 moduli. Moreover, $F_h = 0$ for $h > 1$. As a result, one finds that the prepotential for this case takes a particularly simple form, namely

$$F(X,W^2) = -\frac{1}{2}C_{ab}X^a X^b\left(\frac{X^1}{X^0}\right) - \frac{W^2}{128\pi i}\log \Delta(q). \tag{11.193}$$

Using these formulas one can solve the attractor equations and the Legendre transformation obtaining

$$\phi_0 = -2\pi\sqrt{\frac{p^1}{q_0}}. \tag{11.194}$$

---

[25] In terms of two-forms, $R^\star$ is defined by a duality transformation of the Lorentz indices $(R^\star)^{mn} = \frac{1}{2}\varepsilon^{mn}{}_{pq}R^{pq}$.

One then reproduces the desired entropy formula

$$S \sim \log\left(16\hat{I}_{13}(4\pi\sqrt{p^1 q_0})\right). \tag{11.195}$$

The analysis described above was restricted to supersymmetric black holes. However, the analysis can be extended to the entropy of black holes that are extremal, but not necessarily supersymmetric. Specifically, the entropy given by Wald's formula is given by extremizing an entropy function with respect to moduli fields as well as electric fields at the horizon. This implies that the attractor mechanism is very general: if the entropy function depends on a specific modulus, that modulus is fixed at the horizon. If it does not depend on a modulus, the entropy does not depend on it either.

HOMEWORK PROBLEMS

PROBLEM 11.1
Consider motion of a massive particle in an arbitrary $D = 4$ space-time. The Newtonian limit can be obtained when the curvature of the space-time is small and the velocity is small $v \ll 1$. Expand the space-time metric about flat Minkowski space, $g_{\mu\nu} = \eta_{\mu\nu} + \tilde{g}_{\mu\nu}$ with $|\tilde{g}_{\mu\nu}| \ll 1$, to show that the Newtonian potential Φ is related to the metric by $\Phi = -\tilde{g}_{tt}/2$.

PROBLEM 11.2
Verify that the metric in Eq. (11.9) has a vanishing Ricci tensor, so that D-dimensional Schwarzschild black hole is a solution to Einstein's equations.

PROBLEM 11.3
Derive Eq. (11.11).

PROBLEM 11.4
Re-express the metric in Eq. (11.9) in a higher-dimensional generalization of Kruskal–Szekeres coordinates and verify that there is no singularity at the horizon.

PROBLEM 11.5
Calculate the temperature of the nonextremal black hole (11.60). What happens in the limit $r_0 \to 0$?

Problem 11.6
By similar reasoning to Exercise 11.6, show that the entropy of the three-charge extremal $D = 5$ black hole is given correctly by M-theory on $T^6 = T^2 \times T^2 \times T^2$ with Q_1 M2-branes wrapping the first T^2, Q_2 M2-branes wrapping the second T^2 and Q_3 M2-branes wrapping the third T^2.

Problem 11.7
Verify that Eq. (11.90) follows from Eq. (11.88).

Problem 11.8
Deduce Eq. (11.119) by projecting both sides of Eq. (11.120) on $e^{-i\alpha+\mathcal{K}/2}D_a\Omega$ and using reasoning similar to that in Exercise 11.12. Warning: this is a difficult problem.

Problem 11.9
Show that the Kähler potential in Eq. (11.110) can be recast in the form

$$\mathcal{K} = -\log[2\,\mathrm{Im}(\overline{X}^I F_I)].$$

What form does this equation take when re-expressed in terms of $t^\alpha = X^\alpha/X^0$ and $\widetilde{F}(t^\alpha) = (X^0)^{-2} F(X^I)$?

Problem 11.10
Show that the five-dimensional three-charge black hole with rotation discussed in Section 11.3 solves Eqs (11.144) to (11.154).

Problem 11.11
Show that the horizon of the black-ring solution described by Eqs (11.148) to (11.154) has the topology $S^1 \times S^2$. What is the area of the horizon and what is the entropy of the corresponding black hole?

Problem 11.12
The Dedekind η function can be represented in the form

$$\eta(\tau) = q^{1/24} \prod_{n=1}^{\infty}(1-q^n) = \sum_{n=-\infty}^{\infty}(-1)^n q^{\frac{3}{2}(n-1/6)^2}.$$

Use the Poisson resummation formula and this representation of the η function to verify the modular transformation (11.165).

Problem 11.13
Verify the result for the partition function (11.162).

Problem 11.14
Derive Eq. (11.194).

Problem 11.15
Wald's formula determines the entropy of a $D = 4$ black hole when the effective action contains terms of higher order in the curvature tensor. Denoting the effective Lagrangian density by $\mathcal{L}$, Wald's formula expresses the entropy as an integral over the horizon of the black hole

$$S = 2\pi \int_{S^2} \varepsilon_{\mu\nu} \varepsilon_{\rho\lambda} \frac{\partial \mathcal{L}}{\partial R_{\mu\nu\rho\lambda}} d^2\Omega.$$

Verify that Wald's formula gives the usual BH entropy formula when only the Einstein–Hilbert term is present.

Problem 11.16
Perform microstate counting to obtain the entropy of the nonextremal three-charge black hole given in Eq. (11.69).

Problem 11.17
Perform microstate counting to obtain the entropy of the $D = 5$ rotating black hole given in Eq. (11.76). Also, derive the entropy formula given in Eq. (11.84).

12
Gauge theory/string theory dualities

Many remarkable dualities relating string theories and M-theory have been described in previous chapters. However, this is far from the whole story. There is an entirely new class of dualities that relates conventional (non-gravitational) quantum field theories to string theories and M-theory.

There are three main areas in which such a gauge theory/string theory duality emerged around the mid to late 1990s that are described in this chapter:

- Matrix theory
- Anti-de Sitter/conformal field theory (AdS/CFT) correspondence
- Geometric transitions

Historically, string theory was introduced in the 1960s to describe hadrons (particles made of quarks and gluons that experience strong interactions). Strings would bind quarks and anti-quarks together to build a meson, as depicted in Fig. 12.1 or three quarks to make a baryon. As this approach was developed, it gradually became clear that critical string theory requires the presence of a spin 2 particle in the string's spectrum. This ruled out critical string theory as a theory of hadrons, but it led to string theory becoming a candidate for a quantum theory of gravity. Also, QCD emerged as the theory of the strong interaction. The idea that there should be some other string theory that gives a dual description of QCD was still widely held, but it was unclear how to construct it. Given this history, the discovery of the dualities described in this chapter is quite surprising. String theory and M-theory were believed to be fundamentally different from theories based on local fields, but here are precise equivalences between them, at least for certain background geometries. In fact, it seems possible that every nonabelian gauge theory has a dual description as a quantum gravity theory. To the extent that this is true, it answers the question whether quantum mechanics breaks down when gravity is taken into account with a resounding *no*, because the dual field theories are quantum theories with unitary evolution.

Fig. 12.1. A meson can be viewed as a quark and an antiquark held together by a string.

The methods introduced in this chapter can be used to study the infrared limits of various quantum field theories. Realistic models of QCD, for example, should be able to explain confinement and chiral-symmetry breaking. These properties are not present in models such as $\mathcal{N} = 4$ super Yang–Mills theory due to the large amount of unbroken supersymmetry. There is a variety of ways to break these symmetries so as to get richer models, in both the AdS/CFT and geometric transition approaches. In this setting, phenomena such as confinement and chiral-symmetry breaking can be understood.

Matrix theory

With the discovery of the string dualities described in Chapter 8, it became a challenge to understand M-theory beyond the leading $D = 11$ supergravity approximation. Unlike ten-dimensional superstring theories, there is no massless dilaton, and therefore there is no dimensionless coupling constant on which to base a perturbation expansion. In short, 11-dimensional supergravity is not renormalizable. Of course, ten-dimensional supergravity theories are also not renormalizable, but superstring theory allows us to do better. So one of the most fundamental goals of modern string theory research is to understand better what M-theory is. An early success was a quantum description of M-theory in a flat 11-dimensional space-time background, called Matrix theory. This theory is discussed in Section 12.2. Its fundamental degrees of freedom are D0-branes instead of strings. The generalization to toroidal space-time backgrounds is also described. Matrix theory is formulated in a noncovariant way, and it is difficult to use for explicit computations, so the quest for a simpler formulation of Matrix theory or a variant of it is an important goal of current string theory research. Nevertheless, the theory is correct, and it has passed some rather nontrivial tests that are described in Section 12.2.

AdS/CFT duality

By considering collections of coincident M-branes or D-branes, one finds a space-time geometry that has the features discussed in Chapter 10. The branes are sources of flux and curvature, and a warped geometry results. In certain limits the gauge theory on the world-volume of the branes describes precisely the same physics as string theory or M-theory in the warped geometry created by the branes. In this way one is led to a host of remarkable gauge theory/string theory dualities.

In their most straightforward realization, AdS/CFT dualities relate type IIB superstring theory or M-theory in space-time geometries that are asymptotically anti-de Sitter (AdS) times a compact space to conformally invariant field theories.[1] Anti-de Sitter space is a maximally symmetric space-time with a negative cosmological constant. Even though it is spatially infinite in extent, one can define a boundary at infinity. For reasons to be explained, the space-time manifold of the conformal field theory (CFT) is associated with this boundary of the AdS space. Therefore, these are *holographic dualities*. The name is meant to reflect the similarity to ordinary holography, which records three-dimensional images on two-dimensional emulsions.

The conjectured AdS/CFT correspondences are dualities in the usual sense: when one description is weakly coupled, the dual description is strongly coupled. Thus, assuming that the conjecture is correct, it allows the use of weak-coupling perturbative methods in one theory to learn nontrivial facts about the strongly coupled dual theory. Just as Matrix theory can be regarded as defining quantum M-theory in certain space-time backgrounds, a possible point of view is that the AdS/CFT dualities serve to complete the quantum definitions of string theories and M-theory for another class of space-time backgrounds. Ideally, one would like to have a background-independent definition of these quantum theories, but that does not exist yet. Even so, what has been achieved is really quite remarkable.

The AdS/CFT conjecture emerged from considering the space-time geometry in the vicinity of a large number (N) of coincident p-branes. The three basic examples of AdS/CFT duality, which have maximal supersymmetry (32 supercharges), correspond to taking the p-branes to be either M2-branes, D3-branes, or M5-branes. The corresponding world-volume theories (in three, four, or six dimensions) have superconformal symmetry, and therefore they are superconformal field theories (SCFT). In each case the dual M-theory or string-theory geometry is the product of an anti-de Sitter space-time and a sphere:

---

[1] The conformal group in D dimensions was defined in Chapter 3.

- SCFT on N M2-branes $\leftrightarrow$ M-theory on $AdS_4 \times S^7$,
- SCFT on N M5-branes $\leftrightarrow$ M-theory on $AdS_7 \times S^4$,
- SCFT on N D3-branes $\leftrightarrow$ type IIB on $AdS_5 \times S^5$.

The background in each case has nonvanishing antisymmetric tensor gauge fields with N units of flux threading the sphere. This is clearly required by Gauss's law, since the sphere surrounds the p-branes, each of which carries one unit of the appropriate type of charge. Because it is the case that is best understood, the duality based on coincident D3-branes in type IIB superstring theory is described in greatest detail.

The AdS/CFT correspondence has various extensions and generalizations. One natural direction to explore is the possibility of a dS/CFT correspondence. Such a correspondence is much less well understood, however, since theories in a de Sitter space-time cannot be supersymmetric. dS/CFT duality relates string theory on a D-dimensional de Sitter space to a Euclidean conformal field theory on a $(D-1)$-dimensional sphere. One of the motivations for such a conjecture is the observational evidence for a small positive cosmological constant, which suggests that the Universe is approaching a de Sitter cosmology in the far future. Such a correspondence might also have relevance for the very early Universe. Instead of the usual M-branes and D-branes, the extended objects that are required in this context are Euclidean objects, called *S-branes*, which are discussed in Section 12.1.

Geometric transitions and topological strings

Geometric transitions were originally discovered in the context of the type IIA string theory, but there is a mirror type IIB version of this duality and an M-theory interpretation of this transition. The basic idea of this approach is to construct an $\mathcal{N} = 1$ supersymmetric, confining gauge theory by wrapping D5-branes on topologically nontrivial two-cycles of a Calabi-Yau manifold. The open string excitations on the D5-branes define a supersymmetric gauge theory. If moduli are varied so that the two-cycles shrink to zero size, the theory undergoes a geometric transition to a closed-string sector in which the D-branes disappear and fluxes emerge. Many quantities of the gauge theory, in particular the superpotential, can be computed in terms of fluxes integrated over suitable cycles. This is the subject of Section 12.6.

12.1 Black-brane solutions in string theory and M-theory

In order to discuss the above-mentioned dualities, let us start by introducing black p-brane solutions, which are higher-dimensional counterparts of four-dimensional classical black-hole solutions. The relevant equations of motion are those that are obtained from the actions describing the low-energy limits of superstring theory and M-theory, which were discussed in Chapter 8. Higher-order corrections to these equations are not important in this context.

Black-hole solutions in four dimensions are point-like and (in the absence of angular momentum) have $SO(3)$ rotational symmetry. There is also an $\mathbb{R}$ symmetry associated with time-translation invariance. In higher dimensions, $D > 4$, it is also possible to obtain solutions that describe the geometry and other fields associated with black p-branes, which are p-dimensional extended objects surrounded by an event horizon. If the theory is initially formulated in $D = d + 1$ dimensions, the presence of an extremal p-brane breaks the Lorentz symmetry

$$SO(d,1) \to SO(d-p) \times SO(p,1). \tag{12.1}$$

The first factor describes the rotational symmetry transverse to the brane and the second factor describes the Lorentz symmetry along the brane. There are also translational symmetries along the brane that enlarge the Lorentz symmetry to a Poincaré symmetry. Moreover, we are mainly interested in cases that have Killing spinors and preserve supersymmetries. In fact, the main focus here is on higher-dimensional analogs of extremal and near-extremal Reissner–Nordström black holes. In the nonextremal case the Lorentz symmetry along the brane is broken to a subgroup.

The black M2-brane and M5-brane solutions of 11-dimensional supergravity are considered first. Then we present the black Dp-brane solutions of the type II supergravity theories.

Extremal black M-branes

Chapter 8 presented the bosonic part of the 11-dimensional supergravity action, which is

$$2\kappa_{11}^2 S = \int d^{11}x \sqrt{-G} \left(R - \frac{1}{2}|F_4|^2 \right) - \frac{1}{6} \int A_3 \wedge F_4 \wedge F_4. \tag{12.2}$$

Varying the metric and the three-form gives the field equations that we wish to solve. Alternatively, as discussed in previous chapters, one can obtain supersymmetric solutions by solving the Killing spinor equation

$$\delta \Psi_M = \nabla_M \varepsilon + \frac{1}{12} \left(\Gamma_M \mathbf{F}^{(4)} - 3\mathbf{F}^{(4)}_M \right) \varepsilon = 0. \tag{12.3}$$

There are two types of solutions corresponding to the two types of BPS branes in M-theory, M2-branes and M5-branes, since these are the objects that are electric and magnetic sources of the four-form flux.

Extremal black M2-brane

The supersymmetric (or BPS) M2-brane solution of the Killing spinor equation should have $SO(2,1) \times SO(8)$ symmetry. The metric takes the form[2]

$$ds^2 = H^{-2/3} dx \cdot dx + H^{1/3} dy \cdot dy, \tag{12.4}$$

while the four-form flux has the form

$$F_4 = dx^0 \wedge dx^1 \wedge dx^2 \wedge dH^{-1}. \tag{12.5}$$

Since this has nonzero time components F_{0ijk}, it is an called electric field strength.[3]

The symbol $dx \cdot dx$ represents the three-dimensional Minkowski metric along the brane, while $dy \cdot dy$ represents the Euclidean metric for the eight dimensions perpendicular to the brane. Denoting by r the radial coordinate in the transverse space, that is, $r = |\vec{y}|$, it turns out that the Killing spinor equation is solved if H solves the 8-dimensional Laplace equation.[4] Thus, one of the solutions is

$$H = 1 + \frac{r_2^6}{r^6}, \tag{12.6}$$

where

$$r_2^6 = 32\pi^2 N_2 \ell_p^6 \tag{12.7}$$

and N_2 is the number of M2-branes. This describes a source at $r = 0$, which is the black M2-brane horizon. The strength of the source is proportional to the M2-brane charge and hence the number of M2-branes, as is checked in Exercise 12.1.

This solution describes the fields created by a set of flat coincident M2-branes in the supergravity approximation. The sources are the charge and energy density of the M2-branes. It is a straightforward analog of the extremal Reissner–Nordström black hole described in Chapter 11.

Extremal black M5-brane

The magnetic dual of the preceding solution is the black M5-brane describing the field configuration sourced by N_5 coincident flat M5-branes in 11 dimensions. The BPS M5-brane solution must have $SO(5,1) \times SO(5)$ symmetry. Therefore, the metric takes the form

$$ds^2 = H^{-1/3} dx \cdot dx + H^{2/3} dy \cdot dy, \tag{12.8}$$

---

[2] The precise form of this solution is verified in Problem 12.1.
[3] F_{0ijk} and F_{ijkl} are called the electric and magnetic components of F_4, respectively.
[4] This structure is quite common. It also appeared at several points in Chapters 10 and 11.

where now $dx \cdot dx$ is the six-dimensional Lorentz metric, and $dy \cdot dy = dr^2 + r^2 d\Omega_4^2$ is the five-dimensional Euclidean metric. As before, the powers of H are chosen such that a supersymmetric solution is obtained if H solves Laplace's equation (this time in five dimensions), so that

$$H = 1 + \frac{r_5^3}{r^3}, \quad (12.9)$$

where

$$r_5^3 = \pi N_5 \ell_p^3. \quad (12.10)$$

The four-form flux in this case is magnetic

$$F_4 = \star \left(dx^0 \wedge dx^1 \wedge \ldots \wedge dx^5 \wedge dH^{-1} \right), \quad (12.11)$$

as expected for the black M5-brane solution.

Near-horizon limits

The extremal M2-brane solution has a horizon at $r = 0$. Let us write the perpendicular part of the metric in spherical coordinates

$$dy \cdot dy = dr^2 + r^2 d\Omega_7^2. \quad (12.12)$$

Then as $r \to 0$, the coefficient of $d\Omega_7^2$ has a finite limit

$$r^2 H^{1/3} \to r_2^2. \quad (12.13)$$

Therefore, r_2 is the radius of horizon, which in this case has topology $S^7 \times \mathbb{R}^2$ times a null line. The 11-dimensional near-horizon geometry is

$$ds^2 \sim (r/r_2)^4 dx \cdot dx + (r_2/r)^2 dr^2 + r_2^2 d\Omega_7^2. \quad (12.14)$$

The first two terms describe four-dimensional anti-de Sitter space, so altogether the near-horizon geometry of this extremal black M2-brane is $AdS_4 \times S^7$.[5]

Anti-de Sitter space in $(d+1)$ dimensions

To understand the near-horizon geometry, let us describe $(d+1)$-dimensional anti-de Sitter space (AdS_{d+1}) of radius R by the metric

$$ds^2 = R^2 \frac{dx \cdot dx + dz^2}{z^2}, \quad (12.15)$$

---

5 You are asked to construct this solution in Problem 8.2.

where $dx \cdot dx$ represents the metric of d-dimensional Minkowski space-time. The first two terms in the near-horizon M2-brane geometry (12.14) are brought to this form by the change of variables

$$z = \frac{r_2^3}{2r^2}, \qquad (12.16)$$

which gives $R = r_2/2$. The AdS_4 radius is half the S^7 radius.

The horizon of the extremal M5-brane solution is again at $r = 0$, and it has the topology $S^4 \times \mathbb{R}^5$ times a null line, where the spherical factor has radius r_5. The near-horizon geometry in this case is

$$ds^2 \sim (r/r_5)dx \cdot dx + (r_5/r)^2 dr^2 + r_5^2 d\Omega_4^2. \qquad (12.17)$$

The change of variables $r = 4r_5^3/z^2$ shows that the first two terms again describe an anti-de Sitter space with $R = 2r_5$, so the near-horizon geometry is that of $AdS_7 \times S^4$. The AdS_7 radius is twice the S^4 radius.

The $AdS_4 \times S^7$ and $AdS_7 \times S^4$ geometries discussed above have been obtained as the near-horizon geometries of a collection of coincident M2-branes and M5-branes, respectively, embedded in an asymptotically Minkowski space-time. However, they have a more far-reaching significance than that. They are exact BPS solutions of M-theory. Not only do they solve the equations of motion of 11-dimensional supergravity, but they solve the equations of M-theory including all the (mostly unknown) higher-order corrections to the low-energy effective action. This result is established by arguing that all higher-order corrections necessarily give vanishing corrections to the equations of motion when evaluated in these backgrounds as a consequence of their high symmetry. The same is true for the $AdS_5 \times S^5$ solution of the type IIB theory discussed below.

Extremal black D-branes

The construction of extremal black D-brane solutions can be carried out in the same way as that of black M-brane solutions. In this case the action that is required is a type II supergravity action. However, if the goal is to construct a black Dp-brane, then the only one R–R field, C_{p+1} with field strength $F_{p+2} = dC_{p+1}$, needs to be included in the action. Also, the NS–NS two-form vanishes, and therefore it can be dropped. Thus, the required string-frame action, which can be read off from Section 8.1, is

$$S^{(p)} = \frac{1}{2\kappa^2} \int \sqrt{-g}\left[e^{-2\Phi}\left(R + 4(\partial\Phi)^2\right) - \frac{1}{2}|F_{p+2}|^2\right]d^{10}x. \qquad (12.18)$$

When p is even this is a type IIA action and when p is odd it is a type IIB action. In the special case $p = 3$, the constraint $F_5 = \star F_5$ has to be imposed and an extra factor of $1/2$ should be inserted in the F_5 kinetic term. While the problem is conceptually the same as the black M-brane problems, and the solutions are very similar, there is one significant difference. This is the presence of the dilaton field Φ. The solution has a spatially varying dilaton field for all values of p except for $p = 3$.

The extremal black Dp-brane solution has the metric

$$ds^2 = H_p^{-1/2} dx \cdot dx + H_p^{1/2} dy \cdot dy, \qquad (12.19)$$

where the harmonic function H_p is given by

$$H_p(r) = 1 + \left(\frac{r_p}{r}\right)^{7-p}. \qquad (12.20)$$

As before, $dx \cdot dx$ is the $(p+1)$-dimensional Lorentz metric along the brane and $dy \cdot dy = dr^2 + r^2 d\Omega_{8-p}^2$ is the Euclidean metric in the $9-p$ perpendicular directions. The dilaton is given by

$$e^\Phi = g_s H_p^{(3-p)/4}. \qquad (12.21)$$

Problem 12.15 asks you to verify these formulas.

Since $H_p \to 1$ as $r \to \infty$, the dilaton approaches a constant. Thus, the parameter g_s is the string coupling constant at infinity. This formula displays the important fact that the dilaton is a constant for $p = 3$ only. If $p < 3$, the coupling becomes large for $r \to 0$, which puts the system in a nonperturbative regime, where the solution is unreliable.

The R–R field strength is

$$F_{p+2} = dH_p^{-1} \wedge dx^0 \wedge dx^1 \wedge \ldots \wedge dx^p, \qquad (12.22)$$

which is realized for the R–R potential

$$C_{01\ldots p} = H_p(r)^{-1} - 1. \qquad (12.23)$$

This can be rewritten in the form

$$F_{p+2} = Q \star \omega_{8-p}, \qquad (12.24)$$

where Q is the D-brane charge and ω_n is the volume form for a unit n-sphere. This form ensures that $\star F$ integrates over the sphere to give the charge, as required by Gauss's law. In the special case of $p = 3$ this should be replaced by

$$F_5 = Q(\omega_5 + \star \omega_5) \qquad (12.25)$$

12.1 Black-brane solutions in string theory and M-theory

in order to incorporate self-duality. Using the formula $G_{10} = 8\pi^6 g_s^2 \ell_s^8$ from Chapter 8 and

$$T_{Dp} = (2\pi)^{-p} \ell_s^{-(p+1)} g_s^{-1}, \tag{12.26}$$

from Chapter 6, one obtains

$$(r_p/\ell_s)^{7-p} = (2\sqrt{\pi})^{5-p} \Gamma\left(\frac{7-p}{2}\right) g_s N. \tag{12.27}$$

The extremal black D3-brane

In the special case of $p = 3$ the formulas above give a constant dilaton. In this case, letting $r_3 = R$, Eq. (12.27) takes the form

$$R^4 = 4\pi g_s N \alpha'^2. \tag{12.28}$$

Furthermore, the near-horizon limit of the metric takes the form

$$ds^2 \sim (r/R)^2 dx \cdot dx + (R/r)^2 dr^2 + R^2 d\Omega_5^2. \tag{12.29}$$

The change of variables $z = R^2/r$ brings this to the form

$$ds^2 \sim R^2 \frac{dx \cdot dx + dz^2}{z^2} + R^2 d\Omega_5^2. \tag{12.30}$$

This shows that the near-horizon geometry is $AdS_5 \times S^5$, where both factors have radius R.

Nonextremal black D-branes

The extremal black D-brane solutions, which describe the geometry and other fields generated by a set of coincident D-branes, are supersymmetric. However, the equations of motion following from the action Eq. (12.18) also have nonsupersymmetric charged solutions, which are called *nonextremal black p-branes* (see Problem 12.6). We only consider $p < 7$ here, since the other cases are somewhat special and not relevant to the discussion in the remainder of this chapter.[6] For $p < 7$ the line element is given by

$$ds^2 = -\Delta_+(r)\Delta_-(r)^{-1/2} dt^2 + \Delta_-(r)^{1/2} dx^i dx^i$$

$$+ \Delta_+(r)^{-1} \Delta_-(r)^\gamma dr^2 + r^2 \Delta_-(r)^{\gamma+1} d\Omega_{8-p}^2, \tag{12.31}$$

---

[6] 7-branes have a conical deficit angle at their core, like point particles in $D = 3$. Their geometry is discussed in Chapter 9 in connection with F-theory. 8-branes are domain walls in ten dimensions that divide the space-time into disjoint regions and 9-branes are space-time-filling.

where x^i, $i = 1, \ldots, p$, describes the spatial coordinates along the brane,

$$\gamma = -\frac{1}{2} - \frac{5-p}{7-p} \qquad (12.32)$$

and

$$\Delta_\pm(r) = 1 - \left(\frac{r_\pm}{r}\right)^{7-p}. \qquad (12.33)$$

The dilaton and electric field are given by

$$e^\Phi = g_s \Delta_-(r)^{(p-3)/4} \qquad (12.34)$$

and

$$\star F_{p+2} = Q \omega_{8-p}, \qquad (12.35)$$

respectively. Here, ω_n is the volume form for the unit n-sphere, so that $\int \omega_n = \Omega_n$ and $N = Q\Omega_{8-p}$, the R–R charge of the brane, is an integer.

Nonextremal black D3-branes

The case $p = 3$ is again special because the brane is self-dual. In this case one has

$$F_5 = Q(\omega_5 + \star \omega_5). \qquad (12.36)$$

Also, when $p = 3$ it follows from Eqs (12.32) and (12.34) that the dilaton is constant and $\gamma = -1$.

To recover the extremal solutions with $r_+ = r_-$, discussed in the previous section, one should make a change of radial coordinate like in the previous chapter for extremal Reissner–Nordström black holes. Namely, define $\tilde{r}$ by

$$\tilde{r}^{7-p} = r^{7-p} - r_+^{7-p}, \qquad (12.37)$$

so that in this new coordinate the horizon is at $\tilde{r} = 0$.

Mass and charge of the solutions

The solutions given above are two-parameter families of solutions labeled by r_+ and r_-. These radii are in turn related to the mass per unit p-volume T and the charge per unit volume Q, as in the case of Reissner–Nordström black holes. These solutions have an event horizon at $r = r_+$ and an inner horizon at $r = r_-$. The mass per unit volume and charge of the black Dp-brane are related to the radii $r_\pm$ by

$$T = \frac{\Omega_{8-p}}{2\kappa_{10}^2}\left[(8-p)r_+^{7-p} - r_-^{7-p}\right] \qquad (12.38)$$

and
$$Q = \frac{(7-p)}{2}(r_+ r_-)^{(7-p)/2}, \tag{12.39}$$
respectively. The charge is determined by the asymptotic behavior of the gauge field in the usual way. The mass density is determined by the asymptotic behavior of the geometry by a standard prescription of general relativity known as the ADM mass formula.

As in the case of the Reissner–Nordström black hole, the singularity at $r = 0$ is shielded by the horizon provided that $r_+ > 0$. In the case of Dp-branes, the inequality $r_+ \geq r_-$ is equivalent to the Bogomolny bound
$$T \geq N T_{\mathrm{D}p}, \tag{12.40}$$
where $T_{\mathrm{D}p}$ is the tension of a single BPS D-brane, given in Chapter 6. Thus, extremal black Dp-branes saturate the bound and are supersymmetric.

Gregory–Laflamme instability

Under certain circumstances, nonextremal black p-branes can be unstable to break up into black branes of lower dimension. This instability is caused by the different shapes that horizons of black branes can have in string theory. Accordingly, there are different entropies. Indeed, in four dimensions event horizons are always spherically symmetric, but in higher dimensions the event horizons can have different topologies, as discussed in Chapter 11. A higher-dimensional black brane decays into lower-dimensional branes, if in the process the entropy increases. This is the basic idea of the Gregory–Laflamme instability.

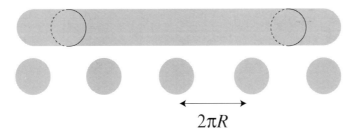

Fig. 12.2. A black string breaks into black holes if the entropy becomes larger in this process.

Let us illustrate the idea with the example illustrated in Fig. 12.2. Imagine that one considers an uncharged five-dimensional black string, which is given by the product of a four-dimensional Schwarzschild solution times the real line. Imagine wrapping this string on a circle of radius R. Denoting the

Schwarzschild radius by r_1, the mass is proportional to $r_1 R$ and the entropy is $S_1 \sim r_1^2 R$. Now consider a black hole that is localized on the circle. This can be constructed by starting with a periodic array with spacing $2\pi R$ on the covering space. Denoting the radius of this black hole by r_0, its mass is proportional to r_0^2 and the entropy is $S_0 \sim r_0^3$. Now let us equate the masses. We then obtain the following relation for the entropies

$$\frac{S_1}{S_0} = k \frac{r_0}{R}, \tag{12.41}$$

where k is a numerical constant that is not relevant for the present discussion. Thus, holding the mass fixed, $S_1 > S_0$ for $R < R_c$ and $S_1 < S_0$ for $R > R_c$, where R_c is the critical radius at which the entropies are equal.

Intuitively, the lower-entropy configuration is unstable and decays into the higher-entropy configuration. In other words, for large enough radius, the black string decays into an array of black holes. More generally, a long enough segment of black string must break. This is somewhat like a QCD string, which can break if a quark–antiquark pair is formed at the endpoints. The existence of this instability has been confirmed by studying the world-volume theory of the black string and showing that it develops a tachyonic mode for $R > R_c$.

S-branes

The black-brane solutions that have been described so far are static solutions of the low-energy effective action of string theory. However, since space and time appear on an equal footing in relativity, one should also be able to construct time-dependent solutions. This is the case, and there are solutions called *S-branes* that are quite similar to Dp-branes. They satisfy Dirichlet boundary conditions, but now in the time direction. Like conventional Dp-branes, they have a perturbative interpretation as hyperplanes on which strings can end, and they can be obtained as solutions of the equations of motion.

S-branes in field theory

The simplest example of an S-brane can be found in a four-dimensional field theory with one scalar field ϕ and the potential

$$V(\phi) = (\phi^2 - a^2)^2, \tag{12.42}$$

where ϕ is real. There are two classical minima located at

$$\phi = \phi_\pm = \pm a. \tag{12.43}$$

12.1 Black-brane solutions in string theory and M-theory

A time-dependent configuration is constructed by choosing initial conditions

$$\phi(\vec{x},0) = 0 \quad \text{and} \quad \dot{\phi}(\vec{x},0) = v. \tag{12.44}$$

If v is chosen to be positive, then after a sufficient amount of time, that is, for $t \to +\infty$, the scalar approaches the ϕ_+ minimum. The time-reversed process would start at $\phi = \phi_-$ for $t = -\infty$ and then evolve to $\phi(\vec{x},0) = 0$. Altogether, for the desired solution, $\phi \to \pm a$ for $t \to \pm\infty$. This solution is called a space-like brane, or S-brane, to contrast it with a D-brane which would take $\phi \to \pm a$ for $x \to \pm\infty$.

S-branes couple to tensor fields, pretty much as D-branes do. For example, an S0-brane, which is defined to have one spatial dimension, couples in four dimensions to an electromagnetic field. The corresponding Maxwell equations are

$$dF = 0 \quad \text{and} \quad d^\dagger F = dz\delta(t)\delta(x)\delta(y). \tag{12.45}$$

This corresponds to an S0-brane extended in the z-direction. Note the $\delta(t)$ on the right-hand side of the second equation. This describes the fact that S-branes are localized in time and underscores the difference from D-branes. The solution of Maxwell equations in this case is

$$F = \text{Re}\left(dz \wedge d\frac{1}{\sqrt{t^2 - x^2 - y^2 - i\epsilon}}\right). \tag{12.46}$$

S-branes in string theory

There are several ways of describing S-branes in string theory. In perturbation theory they can be represented as branes with Dirichlet boundary conditions in the time direction. They can also be obtained as Euclidean analogs of the black-brane solutions. The simplest example of such a construction is an S0-brane solution in four dimensions, which can be obtained from the Schwarzschild solution by analytic continuation. The $SO(3)$ radial symmetry of the black-hole solution is replaced by the hyperbolic symmetry $SO(2,1)$.

Starting with

$$ds^2 = -\left(1 - \frac{2M}{r}\right)dt^2 + \left(1 - \frac{2M}{r}\right)^{-1}dr^2 + r^2(\sin^2\theta d\phi^2 + d\theta^2), \tag{12.47}$$

and transforming $t \to ir$, $r \to it$, $\theta \to i\theta$, $M \to iP$ yields

$$ds^2 = -\left(1 - \frac{2P}{t}\right)^{-1}dt^2 + \left(1 - \frac{2P}{t}\right)dr^2 + t^2 d\Sigma^2, \tag{12.48}$$

where
$$d\Sigma^2 = \sinh^2\theta d\phi^2 + d\theta^2 \qquad (12.49)$$

is the metric of the hyperbolic space in two dimensions. Making a change of coordinates given by

$$t = 2P\cosh^2(\eta/2), \qquad (12.50)$$

with $-\infty < \eta < \infty$, the metric takes the form

$$ds^2 = C^2(\eta)\left(-d\eta^2 + d\Sigma^2\right) + D^2(\eta)dr^2, \qquad (12.51)$$

with

$$C(\eta) = t(\eta) \quad \text{and} \quad D(\eta) = \tanh\left(\frac{\eta}{2}\right). \qquad (12.52)$$

The S0-brane is localized at the horizon where $\eta = 0$. Solutions for Sp-branes have also been constructed. They provide time-dependent backgrounds, which could play an interesting role in cosmology. However, they are difficult to study, since they are not supersymmetric. S-branes play a prominent role in the dS/CFT correspondence, which is an interesting analog of the AdS/CFT correspondence discussed in this chapter. Since the Universe has a positive cosmological constant, it is natural to search for such an analog. Defining a quantum theory of gravity in dS space is difficult, and it is not described in this book, but it is a promising direction to explore.

EXERCISES

EXERCISE 12.1
Relate the horizon radius r_2 in Eq. (12.6) to the number of M2-branes N_2.

SOLUTION

The rule for computing the mass of a black hole given in Chapter 11 needs to be slightly generalized. The generalization is to ignore the spatial dimensions along the brane and interpret the result as a tension (mass per unit volume). In the present case this gives

$$g_{00} \sim -1 + \frac{2}{3}(r_2/r)^6 = -1 + \frac{16\pi G_{11} N_2 T_{M2}}{9\Omega_7 r^6}.$$

Here, $G_{11} = 16\pi^7 \ell_p^9$ is Newton's constant in 11 dimensions, $\Omega_7 = \pi^4/3$ is the volume of a unit seven-sphere,[7] and $T_{M2} = (2\pi)^{-2}\ell_p^{-3}$ is the tension of an M2-brane. Putting these together gives

$$r_2^6 = 32\pi^2 N_2 \ell_p^6.$$

$\Box$

EXERCISE 12.2
Describe the zero-charge limit of the nonextremal black p-brane solution.

SOLUTION

In the special case of an uncharged black p-brane, which can be achieved by starting with an equal number of branes and antibranes, $r_- = 0$, so that $\Delta_- = 1$. Then the solution collapses to

$$ds^2 = -\Delta_+(r)dt^2 + dx^i dx^i + \Delta_+(r)^{-1}dr^2 + r^2 d\Omega_{8-p}^2,$$

which is the $(10-p)$-dimensional Schwarzschild metric times p-dimensional Euclidean space. In this case the dilaton is a constant, the tension of the brane is proportional to r_+^{7-p}/G_{10} and the entropy per unit p-volume is proportional to r_+^{8-p}/G_{10}. $\Box$

12.2 Matrix theory

The analysis of brane configurations and their near-horizon geometry leads to some extremely remarkable duality conjectures. Historically, the first one of these dualities was *Matrix theory*,[8] so let us begin our discussion of gauge/gravity dualities with this example.

As discussed in Chapter 8, M-theory is believed to be a consistent quantum theory of gravity in 11 dimensions. Although we do not have a precise formulation of quantum M-theory, several aspects are well understood:

- There are numerous dualities relating superstring theories to specific compactifications of M-theory.
- At low energies and large distances M-theory reduces to 11-dimensional supergravity.

Matrix theory constitutes an important step towards understanding quantum M-theory when all 11 dimensions are noncompact, and it has generalizations that characterize certain compactifications.

---
[7] As explained in Chapter 11, $\Omega_n = 2\pi^{(n+1)/2}\left[\Gamma((n+1)/2)\right]^{-1}$.
[8] The originators (Banks, Fischler, Shenker and Susskind) called it *M(atrix) theory*, since it relates to M-theory. We choose to omit the parentheses.

Matrix theory in the infinite-momentum frame

The Matrix-theory conjecture states that M-theory in the *infinite-momentum frame* is described by a specific supersymmetric matrix model. The only dynamical degrees of freedom (or *partons*) are identified as the D0-branes of type IIA superstring theory, so that the calculation of any physical quantity in M-theory can be reduced to a calculation in the Matrix-model quantum mechanics. Recall that type IIA superstring theory corresponds to M-theory compactified on a circle of radius R, and the D0-brane corresponds to the first Kaluza–Klein excitation of the massless fields (or supergraviton) of M-theory on this circle. A general Kaluza–Klein excitation is a point-like object whose 11-component of momentum is

$$p_{11} = \frac{N}{R}, \tag{12.53}$$

where N is an integer. From the ten-dimensional perspective, this is interpreted as a threshold bound state of N D0-branes. The term *infinite-momentum frame* refers to the limit in which p_{11} and N go to infinity.

Chapter 6 described the world-volume theories of various D-branes and collections of D-branes. In particular, the action describing a system of N D0-branes is ten-dimensional super Yang–Mills theory dimensionally reduced to 0+1 dimensions supplemented by higher-order corrections (of the Born–Infeld type). The claim, however, is that these higher-order terms do not contribute in the infinite-momentum frame, and therefore the bosonic part of the Lagrangian is given precisely by

$$\mathcal{L} = \frac{1}{2R} \text{Tr} \left(-(D_\tau X^i)^2 + \frac{1}{2}[X^i, X^j]^2 \right), \tag{12.54}$$

where $i = 1, \ldots, 9$ labels the transverse directions. This quantum mechanical system has a $U(N)$ gauge symmetry.

Matrix theory and DLCQ

In the original formulation of the conjecture, which relates M-theory to Matrix theory, an $N \to \infty$ limit was required. Later, a somewhat stronger version of the conjecture was formulated for finite N. This version states that the *discrete light-cone quantization* (DLCQ) of M-theory is exactly described by the $U(N)$ Matrix theory in Eq. (12.54) supplemented by the usual fermion terms. In the DLCQ approach, the circle is chosen to be in a null direction rather than space-like. For a null circle, the radius R has no invariant meaning, but the integer N does. The DLCQ predictions agree with the infinite-momentum frame ones in the limit $N \to \infty$.

To test this conjecture, some quantities can be computed in both M-theory and Matrix theory and can then be compared for finite N. This can be done by computing the effective action for the scattering of two (groups of) D0-branes. This conjecture has been verified up to two loops in the gauge theory, beyond which calculations in Matrix theory become very difficult, though they are well defined.

Super Yang–Mills action in $0+1$ dimensions

To compute the effective action for two D0-branes, the background-field method is used. This is a technique that allows a gauge choice to be made and quantum computations to be carried out without sacrificing manifest gauge invariance. The complete gauge-theory action is obtained from ten-dimensional super Yang–Mills theory dimensionally reduced to $0+1$ dimensions.

The Matrix-theory action

After covariant gauge fixing, the Lagrangian contains a $U(2)$ gauge field A_μ, a gauge-fixing term and ghost fields. The complete Lagrangian is

$$\mathcal{L} = \text{Tr}\left(\frac{1}{2g}F_{\mu\nu}^2 - i\bar{\psi}D\psi + \frac{1}{g}(\bar{D}^\mu A_\mu)^2\right) + \mathcal{L}_{\mathcal{G}}, \qquad (12.55)$$

where $F_{\mu\nu}$ is a $U(2)$ field strength with $\mu, \nu = 0, \ldots, 9$, ψ is a real 16-component spinor in the adjoint of $U(2)$ and $\mathcal{L}_{\mathcal{G}}$ is the ghost Lagrangian, whose explicit form is given in Problem 12.14. For the gauge-fixing term, it is convenient to use the background-field gauge condition

$$\bar{D}^\mu A_\mu = \partial^\mu A_\mu + [B^\mu, A_\mu], \qquad (12.56)$$

where B_μ is the background field. After dimensional reduction to 0+1 dimensions, the field strength and the derivative of the fermionic fields can be expressed in terms of the matrices X_i as

$$\begin{aligned} F_{0i} &= \partial_\tau X_i + [A, X_i], \\ F_{ij} &= [X_i, X_j], \\ D_\tau \psi &= \partial_\tau \psi + [A, \psi], \\ D_i \psi &= [X_i, \psi]. \end{aligned} \qquad (12.57)$$

Here, A denotes the zero component of the gauge field in Eq. (12.55). Setting $g = 2R$ in Eq. (12.55), we recover Eq. (12.54).

This action can be expanded around a classical background B^i by setting

$$X^i = B^i + \sqrt{g} Y^i, \qquad (12.58)$$

where Y^i represents the quantum degrees of freedom. For example, to describe the motion of two D0-branes on straight lines, one chooses the background fields

$$B^1 = i\frac{v\tau}{2}\sigma^3 \quad \text{and} \quad B^2 = i\frac{b}{2}\sigma^3. \qquad (12.59)$$

Here, v is the relative velocity of the two D0-branes, b is the impact parameter and σ^3 is a Pauli matrix. Furthermore, $B^i = 0$ for $i = 0$ and $i = 3,\ldots 9$. A convenient form of the action is written in terms of $U(2)$ generators by decomposing the fields in terms of Pauli matrices,

$$A = \frac{i}{2}\left(A_0 \mathbb{1} + A_a \sigma^a\right), \qquad (12.60)$$

and similarly for the fields X^i and ψ. The zero components of this decomposition describe the motion of the center of mass and are ignored in the following. The Lagrangian is now a sum of four terms

$$\mathcal{L} = \mathcal{L}_Y + \mathcal{L}_A + \mathcal{L}_\mathcal{G} + \mathcal{L}_{\text{fermi}}, \qquad (12.61)$$

whose explicit form is given in Problem 12.14.

The field content

Since A and the X^i are ten traceless 2×2 matrices, they give 30 bosonic fields. Defining

$$r^2 = b^2 + (v\tau)^2, \qquad (12.62)$$

one finds that the bosonic Lagrangians $\mathcal{L}_Y$ and $\mathcal{L}_A$ are described in terms of 16 bosons with mass-squared $m_\mathcal{B}^2 = r^2$, two bosons with $m_\mathcal{B}^2 = r^2 + 2v$, two bosons with $m_\mathcal{B}^2 = r^2 - 2v$ and ten massless bosons. All these fields are real. The ghost action is described in terms of two complex bosons with $m_\mathcal{G}^2 = r^2$ and one complex massless boson.

Feynman rules for Matrix theory

There are two possible approaches to compute the gauge-invariant background field effective action. The first one treats the background field exactly, so that this field enters in the propagators and vertices of the theory. To compute the effective action, one has to sum over all 1PI graphs without external lines. The second approach treats the background field perturbatively, so that it appears as external lines in the one-particle irreducible

(1PI) graphs of the theory. Here, the first approach is followed, and the background field is treated exactly.

We can now derive the Feynman rules. The explicit form of the vertices can be read off from the actions described in Problem 12.14, where cubic and quartic vertices appear. The concrete form of the propagators can be easily obtained, since the problem can be mapped onto the problem of finding the propagators for the one-dimensional harmonic oscillator. By definition, the propagators of the bosonic fields solve the equation

$$(-\partial_\tau^2 + \mu^2 + (v\tau)^2)\Delta_{\mathcal{B}}\left(\tau, \tau' | \mu^2 + (v\tau)^2\right) = \delta(\tau - \tau'), \tag{12.63}$$

where $\mu^2 = b^2$ or $b^2 \pm 2v$ depending on the type of boson that one is considering. This is nothing but the propagator for a one-dimensional harmonic oscillator, so that the propagators of all the bosonic fields take the form

$$\Delta_{\mathcal{B}}\left(\tau, \tau' | \mu^2 + (v\tau)^2\right)$$

$$= \int_0^\infty ds\, e^{-\mu^2 s} \sqrt{\frac{v}{2\pi \sinh 2sv}} \exp\left[-\frac{v}{2}\left(\frac{(\tau^2 + \tau'^2)\cosh 2sv - 2\tau\tau'}{\sinh 2sv}\right)\right]. \tag{12.64}$$

The propagator of the fermionic fields is the solution to the equation

$$(-\partial_\tau + m_{\mathcal{F}})\,\Delta_{\mathcal{F}}\left(\tau, \tau' | m_{\mathcal{F}}\right) = \delta(\tau - \tau'), \tag{12.65}$$

where $m_{\mathcal{F}} = v\tau\gamma_1 + b\gamma_2$ is the fermionic mass matrix. Using gamma matrix algebra, it is verified in Exercise 12.13 that the fermionic propagator can be expressed in terms of the bosonic propagator by

$$\Delta_{\mathcal{F}}(\tau, \tau' | m_{\mathcal{F}}) = (\partial_\tau + m_{\mathcal{F}})\,\Delta_{\mathcal{B}}\left(\tau, \tau' | r^2 - v\gamma_1\right). \tag{12.66}$$

This is a Dirac-like operator acting on a bosonic propagator of a particle with mass $r^2 - v\gamma_1$. Since Eq. (12.64) provides a closed expression for $\Delta_{\mathcal{B}}$, one can use Eq. (12.66) to obtain an exact expression for the fermionic propagator $\Delta_{\mathcal{F}}$. Diagonalizing the mass matrix we find that our theory contains eight real fermions with mass $m_{\mathcal{F}}^2 = r^2 + v$ and eight real fermions with $m_{\mathcal{F}}^2 = r^2 - v$. The third component of ψ is massless. The effective action can be derived using these Feynman rules. The one-loop effective action is considered first.

One-loop effective action

The one-loop effective action can be characterized by a potential $V(r)$, which is related to the phase shift δ in the scattering amplitude of the two D0-

branes by

$$\delta(b, v) = -\int d\tau \, V(b^2 + v^2\tau^2). \tag{12.67}$$

The phase shift is obtained from the determinants of the operators $-\partial_\tau^2 + M^2$ that originate from integrating out the massive degrees of freedom at one-loop. The result for the one-loop determinants is

$$\det{}^4(-\partial_\tau^2 + r^2 + v)\det{}^4(-\partial_\tau^2 + r^2 - v)$$

$$\det{}^{-1}(-\partial_\tau^2 + r^2 + 2v)\det{}^{-1}(-\partial_\tau^2 + r^2 - 2v)\det{}^{-6}(-\partial_\tau^2 + r^2), \tag{12.68}$$

where the first line is the fermionic contribution and the second line is the bosonic contribution. In a *proper-time* representation of the determinants the phase shift is written as

$$\delta = \int_0^\infty \frac{ds}{s} \frac{e^{-sb^2}}{\sinh sv} \left(3 - 4\cosh sv + \cosh 2sv\right). \tag{12.69}$$

The integrand can be expanded for large impact parameter, and one obtains to leading order in inverse powers of r

$$V(r) = \frac{15}{16} \frac{v^4}{r^7}. \tag{12.70}$$

As shown in the next section, this is precisely the result expected for a single supergraviton exchange in 11 dimensions. Therefore, $(0 + 1)$-dimensional Matrix theory seems to know about the propagation of massless modes in 11 dimensions! Of course, one would like to check if this agreement holds beyond one-loop order, so the two-loop effective action is computed next.

Two-loop effective action

Feynman diagrams that contribute

The two-loop effective action is given by the sum of all diagrams of the form contained in Fig. 12.3. The propagators for the fluctuations Y and the gauge field A are indicated by wavy lines, ghost propagators by dashed lines and the solid lines indicate the fermion propagators. The explicit expression is

$$\int d\tau \lambda_4 \Delta_1(\tau, \tau | m_1) \Delta_2(\tau, \tau | m_2) \tag{12.71}$$

for the diagram involving the quartic vertex λ_4, where Δ_1 and Δ_2 are the propagators of the corresponding particles with masses m_1 and m_2, respec-

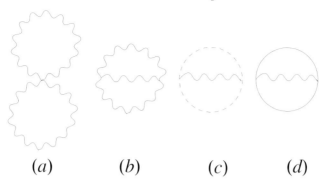

Fig. 12.3. Feynman diagrams that contribute to the two-loop effective action. The different types of lines represent different fields, as explained in the text.

tively. Similarly,

$$\int d\tau d\tau' \lambda_3^{(1)} \lambda_3^{(2)} \Delta_1(\tau,\tau'|m_1) \Delta_2(\tau,\tau'|m_2) \Delta_3(\tau,\tau'|m_3) \tag{12.72}$$

for the diagram involving the cubic vertices $\lambda_3^{(1)}$ and $\lambda_3^{(2)}$.

Massive states that contribute

Let us see what masses are involved in these diagrams. Equation (12.71) is well behaved when m_1 and m_2 are both different from zero. If $m_1 = 0$, it contributes

$$\int \frac{dp}{p^2} \tag{12.73}$$

to the relevant integrals. However, this expression vanishes in dimensional regularization. Dimensional regularization of ill-defined integrals is defined by requiring three properties: translation symmetry, dilatation symmetry and factorization. Invariance under dilatations imposes the condition that the integral Eq. (12.73) vanishes. Therefore, diagrams containing a quartic vertex only contribute when they involve two massive particles. A similar argument for Eq. (12.72) leads to the conclusion that exactly one massless state is present, as otherwise the corresponding diagram vanishes.

Nonrenormalization theorem for the v^4 term

Dimensional analysis of the two-loop effective action gives $g\mathcal{L}_2$, which has an expansion of the form

$$\mathcal{L}_2 = \alpha_0 \frac{1}{r^2} + \alpha_2 \frac{v^2}{r^6} + \alpha_4 \frac{v^4}{r^{10}} + \ldots \tag{12.74}$$

Odd powers in v in this series are missing because of time-reversal invariance. The α_is are numerical coefficients that are determined by computation of the Feynman diagrams. This is a cumbersome but straightforward calculation. Only the final results are quoted here. First, the coefficient of the v^4/r^{10} term, which appears at two loops in Matrix theory, turns out to be equal to zero once all the contributions coming from bosons and fermions are added up. The vanishing of this numerical coefficient is in agreement with the nonrenormalization theorem for the v^4 term appearing in Matrix theory and is required in order to have agreement with M-theory, as shown in the next section. So Matrix theory has passed the first two-loop test: the vanishing of the v^4/r^{10} term. However, this is only one term in the effective action for two D0-branes.

Dimensional analysis of the two-loop effective action

By dimensional analysis, described in Exercise 12.4, the allowed terms have a double expansion in v and r of the following form

$$g\mathcal{L} = \sum_{m=0}^{\infty} g^m \mathcal{L}_m = c_{00} v^2 + \sum_{m,n=1}^{\infty} c_{mn} g^m \frac{v^{2n+2}}{r^{3m+4n}}. \tag{12.75}$$

Specifically,

$$\begin{aligned}
\mathcal{L}_0 &= c_{00} v^2 \\
\mathcal{L}_1 &= c_{11} \frac{v^4}{r^7} + c_{12} \frac{v^6}{r^{11}} + c_{13} \frac{v^8}{r^{15}} + \ldots \\
\mathcal{L}_2 &= c_{21} \frac{v^4}{r^{10}} + c_{22} \frac{v^6}{r^{14}} + c_{23} \frac{v^8}{r^{18}} + \ldots \\
\mathcal{L}_3 &= c_{31} \frac{v^4}{r^{13}} + c_{32} \frac{v^6}{r^{17}} + c_{33} \frac{v^8}{r^{21}} + \ldots
\end{aligned} \tag{12.76}$$

The subscript on $\mathcal{L}_n$ labels the number of Matrix-theory loops. It has just been argued that the c_{12} vanishes. As a test of the conjectured duality, let us now explore how this result arises from the M-theory point of view.

Comparison with M-theory and more predictions

In this section M-theory amplitudes are computed and compared to the Matrix theory predictions described above.

Probe and source gravitons

The calculation is set up in such a way that, when two gravitons scatter, one of the gravitons is taken to be heavy and serves as the *source* graviton.

12.2 Matrix theory

The other graviton is light and is the *probe* graviton. The way this can make sense is for the two gravitons to have momenta $p_- = N_1/R$ and $p_- = N_2/R$, and for N_1 to be much larger than N_2, so that the first graviton is the source of the gravitational field. Note that the circle is null, as required for DLCQ.

The source graviton is taken to have vanishing transverse velocity. Its worldline is $x^- = x^i = 0$ and it produces the Aichelburg–Sexl metric

$$G_{\mu\nu} = \eta_{\mu\nu} + h_{\mu\nu}, \qquad (12.77)$$

where the only nonvanishing component of $h_{\mu\nu}$ is

$$h_{--} = \frac{2\kappa_{11}^2 p_-}{7\Omega_8 r^7}\delta(x^-) = \frac{15\pi N_1}{R M_p^9 r^7}\delta(x^-). \qquad (12.78)$$

Here, $\kappa_{11}^2 = 16\pi^5/M_p^9$, where M_p is the 11-dimensional Planck mass up to a convention-dependent numerical factor, and Ω_8 is the volume of the eight-sphere. This metric is obtained from the Schwarzschild metric by taking the limit of infinite boost in the + direction while the mass is taken to zero. The latter accounts for the absence of higher-order terms in $1/r$ or N_1 dependence. The source graviton is in a state of definite p_- so the average over the $x^- \in (0, 2\pi R)$ direction gives

$$h_{--} = \frac{15 N_1}{2 R^2 M_p^9 r^7}. \qquad (12.79)$$

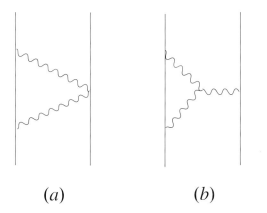

Fig. 12.4. Matrix theory Feynman diagrams. (a) illustrates a probe graviton (thin straight line) interacting with the metric of the source graviton (heavy straight line) at second order in perturbation theory. (b) illustrates a nonvanishing nonlinear correction to the metric of the source.

Action for probe graviton

The action of the probe graviton in this field is now determined. To find it, it is sufficient to consider the action for a massive scalar, since spin effects give a more rapid fall off with r.

$$S = -m \int d\tau \, (-G_{\mu\nu}\dot{x}^\mu \dot{x}^\nu)^{1/2} \qquad (12.80)$$

$$= -m \int d\tau \, (-2\dot{x}^- - v^2 - h_{--}\dot{x}^-\dot{x}^-)^{1/2} \,,$$

where the form of the Aichelburg-Sexl metric was used with $x^+ = \tau$. A dot denotes a τ derivative, and $v^2 = \dot{x}^i \dot{x}^i$. This action vanishes if $m \to 0$ with fixed velocities, but for the process being considered here it is p_- that is to be fixed. We therefore carry out a Legendre transformation on x^-:

$$p_- = m \frac{1 + h_{--}\dot{x}^-}{(-2\dot{x}^- - v^2 - h_{--}\dot{x}^-\dot{x}^-)^{1/2}} \,. \qquad (12.81)$$

The appropriate Lagrangian for x^i at fixed p_- is (minus) the Routhian,

$$\mathcal{L}'(p_-) = -\mathcal{R}(p_-) = \mathcal{L} - p_-\dot{x}^-(p_-). \qquad (12.82)$$

Equation (12.81) determines $\dot{x}^-(p_-)$; it is convenient before solving to take the limit $m \to 0$, where it reduces to $G_{\mu\nu}\dot{x}^\mu\dot{x}^\nu = 0$. Then

$$\dot{x}^- = \frac{\sqrt{1 - h_{--}v^2} - 1}{h_{--}} \,. \qquad (12.83)$$

In the $m \to 0$ limit at fixed $p_- = N_2/R$ the effective Lagrangian becomes

$$-p_-\dot{x}^-(p_-) = p_-\left\{\frac{v^2}{2} + \frac{h_{--}v^4}{8} + \frac{h_{--}^2 v^6}{16} + \ldots\right\}$$

$$= \frac{N_2}{2R}v^2 + \frac{15}{16}\frac{N_1 N_2}{R^3 M_{\rm p}^9}\frac{v^4}{r^7} + \frac{225}{64}\frac{N_1^2 N_2}{R^5 M_{\rm p}^{18}}\frac{v^6}{r^{14}} + \ldots \qquad (12.84)$$

In this formula the explicit dependence on $R^3 M_{\rm p}$ has been restored by dimensional analysis. The second and third terms correspond to the diagrams in Fig. 12.4.

What do we see from this expression?

- The v and r dependencies exactly match with the diagonal terms ($m = n$) appearing in the previous section, and the N dependence agrees with the leading large-N behavior N^{L+1}, where L is the number of loops.
- The coefficient of the v^4/r^7 term agrees with the one-loop Matrix-theory result.

- The absence of a two-loop term gv^4/r^{10} is in agreement with the previous Matrix-theory result. It reflects the existence of a nonrenormalization theorem for the v^4 term.

- There appears a new term with a coefficient 225/64 that should correspond to a two-loop term in Matrix theory. Can Matrix theory reproduce this two-loop coefficient? Computing the v^6/r^{14}-term in Matrix theory by extending the calculation of the two-loop v^4-term, precise agreement has been achieved.

Reproducing the N dependence

Next, the N dependence of this result needs to be reconstructed. Recall that we are considering the scattering of two D0-branes in Matrix theory. To get the right N_1 and N_2 dependence, one must consider the scattering of a group of N_1 D0-branes against N_2 D0-branes. One can easily reconstruct the N-dependence of this scattering process. In double line notation, every graph involving three index loops is of order N^3. Terms proportional to N_1^3 or N_2^3 would only involve one block (graviton) and so could not depend on r. Symmetry under the interchange of N_1 and N_2 determines that the $SU(2)$ result is multiplied by

$$\frac{N_1 N_2^2 + N_1^2 N_2}{2}, \qquad (12.85)$$

which agrees with the supergravity result for the terms of interest. Finally, restoring the dependence on M and R, the two-loop result of Matrix theory is precisely the result found in the supergravity calculation Eq. (12.84).

This highly nontrivial agreement is a strong test of the Matrix theory conjecture. As pointed out earlier, calculations become very difficult in the Matrix-theory picture at higher orders, but there is no reason to anticipate problems.

Matrix theory for toroidal compactifications

Let us now consider what happens when p of the transverse dimensions in the previous construction are taken to form a torus T^p. Requiring consistency with some of the dualities discussed in previous chapters provides some additional tests of the Matrix-theory conjecture.

In Chapter 8 it was argued that, when M-theory is compactified on a T^p, the resulting theory has a nonperturbative U-duality symmetry that is given by $E_p(\mathbb{Z})$. Let us recall the first few cases of this group:

$$E_2(\mathbb{Z}) = SL(2,\mathbb{Z}), \quad E_3(\mathbb{Z}) = SL(3,\mathbb{Z}) \times SL(2,\mathbb{Z}), \quad E_4(\mathbb{Z}) = SL(5,\mathbb{Z}). \tag{12.86}$$

It is an interesting test of the Matrix-theory proposal to see whether it can reproduce these symmetries. First of all, compactification on T^p gives a modular symmetry $SL(p,\mathbb{Z})$ as a straightforward geometric symmetry. So this gives the full result for $p = 2$, but only a subgroup of the desired symmetries for $p > 2$. Recall that Chapter 8 attributed the enhancement of $SL(p,\mathbb{Z})$ to $E_p(\mathbb{Z})$ to a nongeometric duality of M-theory. So the question is whether Matrix theory is smart enough to know about such nongeometric dualities.

Let us now consider the problem from the Matrix-theory side. To start with, we have a system of N D0-branes on a geometry containing a torus T^p. It is very convenient in this case to carry out T-duality transformations along all of the torus directions. That leads to a system in which the compact space consists of the dual torus $\widehat{T}^p$, which is wrapped by N Dp-branes. The world-volume theory of these Dp-branes is maximally supersymmetric Yang–Mills theory on the dual torus assuming that it remains true that higher-dimension corrections can be dropped in the infinite-momentum frame. Thus, it is a gauge theory in $p+1$ dimensions. So the question arises whether the required symmetry enhancement for $p > 2$ can be understood in terms of these gauge theories.

The first nontrivial case is $p = 3$. This is our old friend $\mathcal{N} = 4$ super Yang–Mills theory in $D = 4$, which features prominently in the remainder of this chapter. In the present setting it is compactified on T^3, which gives a geometric $SL(3,\mathbb{Z})$ duality group. However, as was discussed in Chapter 8, this gauge theory also has a nonperturbative $SL(2,\mathbb{Z})$ S-duality group. So the full duality group is $SL(3,\mathbb{Z}) \times SL(2,\mathbb{Z})$ exactly as desired.

Next, let us consider the case $p = 4$. This leads us to consider super Yang–Mills theory in $4 + 1$ dimensions. One may be tempted to reject this as nonrenormalizable, but let us proceed anyway. The duality group of the torus is $SL(4,\mathbb{Z})$, but the desired group is $SL(5,\mathbb{Z})$. The clue to what happens is given by the observation that the Yang–Mills coupling constant g_{YM} in five dimensions has the dimensions of a length. The claim is that this gauge theory generates a fifth spatial dimension, which is a circle, and the size of this circle is controlled by g_{YM}. This is reminiscent of how type IIA string theory grows an extra dimension at strong coupling.

In fact, we already know that this is true. The five-dimensional gauge theory in question is the world-volume theory of a set of coincident D4-branes in the type IIA theory. However, we know that, from the M-theory viewpoint, D4-branes are really M5-branes that wrap the extra spatial dimension that M-theory provides. Thus, the D4-brane system is better viewed as a set of M5-branes wrapping a T^5. The desired $SL(5, \mathbb{Z})$ duality group is then recognized to be the modular group of this torus. This six-dimensional world-volume theory is believed to be a well-defined quantum field theory. The reason its discovery was made relatively recently is that it is strongly coupled in the UV, and therefore it does not have a simple Lagrangian description.

The situation for $p > 4$ is even more challenging and has not been worked out in detail. However, it should be clear already that Matrix theory is capable of capturing a great deal of subtle physics. In fact, its validity can be deduced from the gauge theory/string theory dualities considered in the next section.

Exercises

Exercise 12.3
Show that the fermionic propagator can be expressed in terms of the bosonic propagator as indicated in Eq. (12.65).

Solution
Comparing Eqs (12.63), (12.65) and (12.66), we need to show that

$$(\partial_\tau - v\tau\gamma_1 - b\gamma_2)(\partial_\tau + v\tau\gamma_1 + b\gamma_2) = \partial_\tau^2 - r^2 + v\gamma_1,$$

where we have used $r^2 = b^2 + (v\tau)^2$ and $m_\mathcal{F} = v\tau\gamma_1 + b\gamma_2$. This follows from some simple gamma matrix algebra and the derivative acting on the τ term. Thus, one obtains the desired relation between the bosonic and fermionic propagators. □

Exercise 12.4
Show that the only terms in the Matrix theory effective action up to three loops are the terms appearing in Table (12.76).

Solution
The solution follows from dimensional analysis. Since the action is dimen-

sionless, it follows that the Lagrangian has dimension $[\mathcal{L}] = -1$. From the explicit form of the Lagrangian in Eq. (12.54), it then follows that $[R] = [g] = -3$, $[X^i] = -1$. Also, $[r] = [b] = -1$ and $[v] = -2$. It follows that $g^m v^{2n+2}/r^{3m+4n}$ has dimension -4 as required. Therefore, these dimensions lead to the expansion appearing in Table (12.76). Dimensional analysis determines the entire v and r dependence of the effective actions at each order in the perturbation expansion. Only the numerical coefficients need to be computed by evaluating Feynman diagrams. □

12.3 The AdS/CFT correspondence

The basic idea of the AdS/CFT duality and its generalizations is that string theory or M-theory in the near-horizon geometry of a collection of coincident D-branes or M-branes is equivalent to the low-energy world-volume theory of the corresponding branes. This section explains the AdS/CFT correspondence.

The D3-brane case

The conjecture

The AdS/CFT conjecture (for the case of D3-branes) is that type IIB superstring theory in the $AdS_5 \times S^5$ background described in Section 12.1 is dual to $\mathcal{N} = 4$, $D = 4$ super Yang–Mills theory with gauge group $SU(N)$. This string theory background corresponds to the ground state of the gauge theory, and excitations and interactions in one description correspond to excitations and interactions in the dual description.

D-brane world-volume theories were studied in considerable detail in Chapter 6. In the case of type II superstring theories we learned that the world-volume theory of N coincident BPS D-branes is a maximally supersymmetric $U(N)$ gauge theory. The formulas become complicated when terms that are higher order in α' or nontrivial background fields are taken into account. However, in the absence of background fields and at lowest order in α', the result is very simple: the low-energy effective action on the world volume of N coincident Dp-branes is given by the dimensional reduction of supersymmetric $U(N)$ gauge theory in ten dimensions to $p + 1$ dimensions. This theory is all that is required for the analysis that follows. The $U(1)$ subgroup of $U(N)$ decouples as a free theory and does not participate in the duality.[9] So the gauge group in the duality is really $SU(N)$, not $U(N)$. The distinction between the two is a subleading effect in the large-N limit.

The coupling constants

The dimensionless effective coupling of super Yang–Mills theory in $p+1$ dimensions is scale dependent. At an energy scale E, it is determined by dimensional analysis to be

$$g_{\text{eff}}^2(E) \sim g_{\text{YM}}^2 N E^{p-3}. \tag{12.87}$$

This coupling is small, so that perturbation theory applies, for large E (the UV) for $p < 3$ and for small E (the IR) for $p > 3$.

The special case $p = 3$ corresponds to $\mathcal{N} = 4$ super Yang–Mills theory in four dimensions, which is known to be a UV finite, conformally invariant field theory. In that case $g_{\text{eff}}^2(E)$ is independent of the scale E and corresponds to the 't Hooft coupling constant

$$\lambda = g_{\text{YM}}^2 N. \tag{12.88}$$

This is the combination that is held constant in the large-N expansion of the gauge theory discussed below.

The Yang–Mills coupling constant is the same as the open-string coupling constant, since the gauge fields are massless modes of open strings. Using the relation between open- and closed-string coupling constants, this gives the identification

$$g_{\text{YM}}^2 = 4\pi g_{\text{s}}. \tag{12.89}$$

Fortunately, the dilaton in Eq. (12.21) is constant, so there is no ambiguity in this identification. Indeed, this constancy reflects the fact that the coupling is energy independent. Combining this with the identity $R^4 = 4\pi g_{\text{s}} N \alpha'^2$, obtained in Eq. (12.28), gives the relation

$$R = \lambda^{1/4} \ell_{\text{s}}. \tag{12.90}$$

The last equation relates R/ℓ_{s}, which is the radius of both the S^5 and the AdS_5 in string units, to the 't Hooft coupling of the dual gauge theory. When the field theory is weakly coupled, the dual string theory geometry is strongly curved, which makes computations difficult. Conversely, when the string-theory geometry is weakly curved, and a supergravity approximation is justified, the dual gauge theory is strongly coupled.

---

9 More precisely, the $U(1)$ lives on the boundary and the $SU(N)$ lives in the bulk, which is why the $U(1)$ is not relevant.

Rank of the gauge group

Another important fact about the duality is that the rank of the gauge group corresponds to the five-form flux through the five-sphere

$$\int_{S^5} F_5 = N. \tag{12.91}$$

To understand this, recall that the extremal D3-brane construction started with N coincident D3-branes, which carry a total of N units of D3-brane charge. This charge is measured by enclosing the D3-branes with a five-sphere and computing the five-form flux. Thus, the parameter N, which labels the gauge-theory group, corresponds to the five-form flux in the dual type IIB description.

Symmetry matching

If the proposed correspondence is true, it is necessary that the two dual theories should have the same symmetry. This requirement is relatively easy to test, because the symmetry in each case is independent of the parameters λ and N. So it doesn't matter which theory is in a strongly coupled regime. In each case the complete symmetry is given by the superalgebra $PSU(2,2|4)$, as we explain below.[10] This supergroup group has a bosonic subgroup that is $SU(2,2) \times SU(4)$. In addition, it contains 32 fermionic generators that transform as $(4,4) + (\bar{4},\bar{4})$ under this group. This supergroup is described in more detail in Exercise 12.8.

Let us discuss the symmetry of the string theory solution first. The AdS_5 geometry has the isometry $SO(4,2)$ and the S^5 geometry has the isometry $SO(6)$. The theory has fermions that belong to spinor representations, so it is better to refer to the covering groups which are $SU(2,2)$ and $SU(4)$, respectively. Thus the bosonic subgroup of the supergroup is realized by the geometry. This background realizes all 32 supersymmetries of the type IIB superstring theory as vacuum symmetries. In other words, it has just as much supersymmetry as the ten-dimensional Minkowski vacuum, which corresponds to $R \to \infty$. The conserved supercharges transform as $(4,4) + (\bar{4},\bar{4})$ under $SU(2,2) \times SU(4)$ and combine with the space-time isometries to give $PSU(2,2|4)$.

Now let us turn to the symmetry of the dual $\mathcal{N} = 4$ super Yang–Mills theory. First of all, as we have already asserted, this is a conformally invariant field theory. This has been proved to be an exact property of the

---

[10] A superalgebra of the form $SU(m|n)$ has a bosonic subalgebra $SU(m) \times SU(n) \times U(1)$. When $m = n$ the $U(1)$ factor decouples from the rest of the algebra. The letter P indicates that this $U(1)$ factor is absent.

quantum theory, not just a feature of the classical field theory. This is a very special feature, which implies in particular, that there is an exact cancellation of ultraviolet divergences to all orders in perturbation theory, so that no renormalization scale needs to be introduced to define the theory. One is still free, however, to define the theory at a given energy scale by integrating out all degrees of freedom above that scale. However, since the theory is conformal, the effective coupling defined in this way is independent of the energy scale.

The $SU(4)$ symmetry arises as the global $SU(4)$ R symmetry of the dual $\mathcal{N} = 4$ super Yang–Mills theory. By definition, an R symmetry is a symmetry that does not commute with the supersymmetries. In particular, the four fermions of one chirality transform as a **4** and those of the opposite chirality transform as a **$\bar{4}$**, and the six scalar fields form a **6**. The linearly realized supersymmetries account for 16 fermionic symmetries. However, there are 16 additional nonlinearly realized fermionic symmetries. One way of discovering these is to compute the commutators of the linearly realized supersymmetries with the conformal transformations. Putting all this together, one is led to the desired superconformal algebra $PSU(2,2|4)$.

Large-N limit

The large-N limit, at fixed λ, is of particular interest. Large-N gauge-theory amplitudes have a convenient topological expansion. Specifically, using a double-line notation for adjoint $U(N)$ fields, and filling in the space between the lines so that propagators look like ribbons, the Feynman diagrams can be viewed as two-dimensional surfaces and assigned an Euler characteristic χ. As described in Exercise 12.7, the contribution of diagrams of genus g (or Euler characteristic $\chi = 2 - 2g$) to field-theory amplitudes scales for large N and fixed λ as N^χ. The proof uses Euler's theorem that a two-dimensional simplicial complex with V vertices, E edges, and F faces has

$$\chi = V - E + F. \tag{12.92}$$

Since g_s corresponds to λ/N, the $1/N$ expansion at fixed λ corresponds to the loop expansion of the dual string description.

Planar diagrams

The leading terms in the large-N fixed-λ expansion of the gauge theory define the planar (or genus 0) approximation. It is conjectured that $\mathcal{N} = 4$ super Yang–Mills theory is integrable in this approximation. There is quite a bit of circumstantial evidence for this conjecture, including the existence of an infinite number of conserved charges, but it has not yet been proved.

There is hope that exact analytic computations of correlation functions in the planar approximation may be possible some day. In any case, in the planar approximation they can be computed perturbatively in λ. There is also hope of carrying out exact tree-level calculations of the type IIB superstring theory in the $AdS_5 \times S^5$ background. According to the duality, this would predict the complete planar approximation to the gauge theory. Unfortunately, this computation also is not yet tractable with currently known methods. So the tests of the duality that have been carried out to date are more limited than this, but still very impressive.

Stringy corrections

The preceding discussion shows that, in the string-theory description, stringy effects are suppressed for $\lambda \gg 1$ (so that the radius R is much larger than the string length scale). Similarly, quantum corrections (given by string loops) are small when $N \gg 1$, provided that the limit is carried out at fixed λ. Geometrically, this means that R is much larger than the Planck length. To understand this, recall that Chapter 8 showed that the ten-dimensional Planck length is given by

$$\ell_p = g_s^{1/4} \ell_s. \tag{12.93}$$

Combining this with $\lambda = 4\pi g_s N$ gives

$$N = \frac{1}{4\pi}(R/\ell_p)^4. \tag{12.94}$$

The dictionary

Now that we have described the basic features of the D3-brane correspondence, let us summarize the conclusions for this case:[11]

- The integer N gives the rank of the gauge group, which corresponds to the flux of the five-form R–R gauge field threading the five-sphere.
- The Yang–Mills coupling constant g_{YM} is related to the string coupling constant by $g_{YM}^2 = 4\pi g_s$. The fact that g_{YM} does not depend on the energy scale corresponds to the fact the dilaton is a constant for the black D3-brane solution.
- The supergroup $PSU(2,2|4)$ is the isometry group of the superstring theory background, and it is also the superconformal symmetry group of the $\mathcal{N} = 4$ gauge theory. All of the generators correspond to Killing vectors and Killing spinors of the space-time geometry. In the gauge theory, some

[11] All of this can be generalized to other Dp-branes, but this is the simplest, most symmetrical, example for the reasons that have been explained.

of the operators generate the super-Poincaré subgroup, and the rest generate other conformal transformations. In particular, 16 of the fermionic operators generate linearly realized Poincaré supersymmetries and the other 16 generate superconformal symmetries.
- The common radius R of the AdS_5 and S^5 geometries is related to the 't Hooft parameter $\lambda = g_{\rm YM}^2 N$ of the gauge theory by $R = \lambda^{1/4} \ell_{\rm s}$.

Duality for M-branes

There are similar AdS/CFT conjectures for the two M-theory cases for which extremal black-brane solutions were constructed in Section 12.1. However, they have been explored in much less detail than the D3-brane case. There are at least three reasons for this: (1) computations are much more difficult in M-theory than in type IIB superstring theory; (2) the dual conformal field theories are much more elusive than the $\mathcal{N} = 4$ super Yang–Mills theory; (3) there is great interest in using AdS/CFT dualities to learn more about *four-dimensional* gauge theories.

The M2-brane conjecture

A stack of M2-branes has an $AdS_4 \times S^7$ near-horizon geometry, and M-theory for this geometry (with N units of $\star F_4$ flux through the sphere) is dual to a conformally invariant $SU(N)$ gauge theory in three dimensions. One significant difference from the type IIB superstring example, is that the M-theory background does not contain a dilaton field, and therefore there is no weak-coupling limit. Correspondingly, the three-dimensional conformal field theory does not have an adjustable coupling constant, and it is necessarily strongly coupled. As a result, it does not need to have a classical Lagrangian description. In fact, there does not appear to be one. Therefore, this three-dimensional CFT is much more difficult to analyze than $\mathcal{N} = 4$ super Yang–Mills theory.

CFT for the M2-brane case

One way of thinking about the three-dimensional CFT is as follows. The low-energy effective world-volume theory on a collection of N coincident D2-branes of type IIA superstring theory is a maximally supersymmetric $U(N)$ Yang–Mills theory in three dimensions. This theory is not conformal because the Yang–Mills coupling in three dimensions is dimensionful and introduces a scale. Recall that the type IIA coupling constant is proportional to the radius of a circular eleventh dimension. When this coupling becomes large, the gauge-theory coupling constant also becomes large. In view of

Eq. (12.87), this corresponds to a flow to the infrared in the gauge theory. It also corresponds to the radius of the circular eleventh dimension increasing giving an 11-dimensional M-theory geometry in the limit. Therefore, in the limit, the coupling becomes infinite, and one reaches the conformally-invariant fixed-point theory that describes a collection of coincident M2-branes in 11 dimensions. This theory should have an $SO(8)$ R symmetry corresponding to rotations in the eight dimensions that are transverse to the M2-branes in 11 dimensions.

The $AdS_4 \times S^7$ metric has the isometry group

$$SO(3,2) \times SO(8) \approx Sp(4) \times Spin(8). \qquad (12.95)$$

As before, the first factor is the symmetry of the AdS space, which corresponds to the conformal symmetry group of the dual gauge theory. Also, the second factor is the symmetry of the sphere, which corresponds to the R symmetry of the dual gauge theory. This solution is maximally supersymmetric, which means that there are 32 conserved supercharges. In the dual gauge theory 16 supersymmetries are realized linearly, and the other 16 are conformal supersymmetries. Including the supersymmetries, the complete isometry superalgebra is $OSp(8|4)$. This contains 32 fermionic generators (the supercharges) transforming as $(\mathbf{8}, \mathbf{4})$ under $Spin(8) \times Sp(4)$.

The M5-brane case

Similar remarks apply to the six-dimensional CFT associated with a stack of M5-branes that is dual to M-theory with an $AdS_7 \times S^4$ geometry. The $AdS_7 \times S^4$ metric has the isometry group

$$SO(6,2) \times SO(5) \approx Spin(6,2) \times USp(4). \qquad (12.96)$$

Including the supersymmetries, the complete isometry superalgebra in this case is $OSp(6,2|4)$. This superalgebra contains 32 fermionic generators transforming as $(\mathbf{8}, \mathbf{4})$ under $Spin(6,2) \times USp(4)$.

The problem of defining the conformal field theory on the M5-branes is more severe than in the M2-brane case. To define a field theory, a weak-coupling description in the UV is required. Unlike the M2-brane case, there is no such description in the M5-brane case, because it is a six-dimensional theory. Still, there must be a CFT associated with the M5-brane system. The problem is that we don't know how to describe it other than via the AdS/CFT duality.

12.3 The AdS/CFT correspondence

The structure of anti-de Sitter space

In Section 8.1, AdS_{d+1}, where $d = p+1$, has been described in Poincaré coordinates. In these coordinates, the AdS_{d+1} metric is given by

$$ds^2 = \frac{R^2}{z^2}\left((dx^2)_{d+1} + dz^2\right), \quad z \geq 0. \tag{12.97}$$

Recall that the boundary at spatial infinity ($r \to \infty$) corresponds to $z = 0$, since $z \sim R^2/r$. Similarly, the horizon at $r = 0$ corresponds to $z = \infty$.

From AdS to CAdS

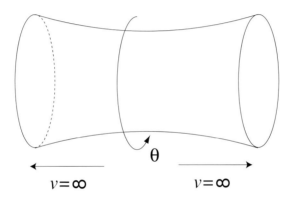

Fig. 12.5. AdS_{p+2} is a hyperboloid in $\mathbb{R}^{2,p+1}$ with a closed time-like curve in the θ direction.

Poincaré coordinates do not give a complete description of the Lorentzian AdS_{d+1} space-time. To understand this, it is useful to consider a hypersurface in a $(d+2)$-dimensional Lorentzian space of signature $(d, 2)$, describing a hyperboloid

$$y_1^2 + \ldots + y_d^2 - t_1^2 - t_2^2 = -R^2 = -1, \tag{12.98}$$

as depicted in Fig. 12.5. In the last step the radius R has been set equal to one, for convenience. This description makes the $SO(p+1, 2)$ symmetry manifest. The relation between the coordinates introduced here and the Poincaré coordinates given earlier is

$$(z, x^0, x^i) = \left((t_1 + y_d)^{-1}, t_2 z, y_i(t_1 + y_d)^{-1}\right). \tag{12.99}$$

To pass to spherical coordinates for both the ys and the ts, the notation

$$(y_1, \ldots, y_d) \to (v, \Omega_p) \quad \text{and} \quad (t_1, t_2) \to (\tau, \theta) \tag{12.100}$$

is introduced. In these coordinates the hypersurface is $v^2 - \tau^2 = -1$, and the metric on this surface is

$$ds^2 = \sum dy_i^2 - \sum dt_j^2 = \frac{dv^2}{1+v^2} + v^2 d\Omega_p^2 - (1+v^2)d\theta^2. \qquad (12.101)$$

Note that the time-like coordinate θ is periodic! This would imply that the conjugate energy eigenvalues are quantized as multiples of a basic unit. This is definitely *not* what type IIB superstring theory on $AdS_5 \times S^5$ gives. The energy quantization does hold for the supergravity modes and their Kaluza–Klein excitations, but it is not true for the stringy excitations.

CAdS/CFT correspondence

The solution to this problem is to replace the AdS space-time with its covering space CAdS. Therefore, strictly speaking, one should speak of *CAdS/CFT duality*, but that is not usually done. To describe the covering space, let us replace the circle parametrized by θ by a real line parametrized by t. This gives a global description of the desired space-time geometry. Letting $v = \tan \rho$, the metric becomes

$$ds^2 = \frac{1}{\cos^2 \rho}(d\rho^2 + \sin^2 \rho \, d\Omega_p^2 - dt^2). \qquad (12.102)$$

This has topology $B_{p+1} \times \mathbb{R}$ which can be visualized as a solid cylinder. The $\mathbb{R}$ factor, which is a real line, corresponds to the *global time coordinate t*, and B_{p+1} denotes a solid ball whose boundary is the sphere S^p. The boundary of the CAdS space-time at spatial infinity ($\rho = \pi/2$) is $S^p \times \mathbb{R}$. The Poincaré coordinates cover a subspace of the global space-time, called the *Poincaré patch*, as shown in Fig. 12.6. This diagram, which shows the global causal structure of the geometry, is called a *Penrose diagram*. All light rays travel at 45 degrees in a Penrose diagram.

When one uses the covering space CAdS to describe the bulk theory, the spatial coordinates of the dual gauge theory are naturally taken to form a sphere S^p. This does have a significant technical advantage: when the spatial coordinates form a sphere, the Hamiltonian has a discrete spectrum rather than a continuous one. This can be traced to the fact that the time coordinate in global coordinates differs from the time coordinate in the Poincaré patch. Thus, if P_0 denotes the Yang–Mills Hamiltonian appropriate to the Poincaré patch time coordinate, then

$$H = \frac{1}{2}(P_0 + K_0), \qquad (12.103)$$

is the Hamiltonian appropriate to global time, and it has a discrete spec-

12.3 The AdS/CFT correspondence   647

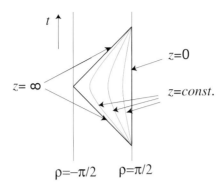

Fig. 12.6. This diagram shows how the Poincaré patch is embedded in global CAdS for the special case of AdS_2.

trum.[12] Geometrically, it is as though the geometry in Eq. (12.102) is creating a potential well so that the center point $\rho = 0$ is at a minimum.

It should also be noted that, in global coordinates, there is no horizon. The horizon is a feature of the description in terms of the coordinates of the Poincaré patch but not of the global space-time. One can see in Fig. 12.6 that there is nothing special about the horizon of the Poincaré patch in the global description. In case this sounds surprising, recall that even flat space-time appears to have a horizon for a uniformly accelerating observer, who sees a Rindler space.

Euclideanized AdS geometry

As discussed in the next subsection, it is useful to consider the Euclideanized AdS geometry (EAdS) to test the AdS/CFT correspondence more precisely. This can be obtained by Wick-rotating the t_2 coordinate:

$$y_1^2 + \ldots + y_d^2 - t_1^2 + t_2^2 = -1. \tag{12.104}$$

The symmetry is now $SO(d+1,1)$. This manifold should not be confused with Lorentzian signature de Sitter space, which would have $+1$ on the right-hand side. As before, EAdS can be described in Poincaré coordinates by

$$ds^2 = \frac{1}{z^2}(dz^2 + (dx^2)_d), \tag{12.105}$$

where now $(dx^2)_d = dx_1^2 + \ldots + dx_d^2$. Unlike the Lorentzian case, these coordinates describe the space globally. They give a description that is

---
[12] K_0 is one of the conformal group generators.

equivalent to the one given by the metric

$$ds^2 = d\rho^2 + \sinh^2 \rho \, d\Omega_d^2, \tag{12.106}$$

which is the analog of Eq. (12.102). Another equivalent metric in terms of $d+1$ coordinates u_i is

$$ds^2 = \frac{4 \sum du_i^2}{(1 - \sum u_i^2)^2}, \tag{12.107}$$

where $\sum u_i^2 \leq 1$. The latter form shows that the topology is that of a ball B_{d+1}, whose boundary is a sphere S^d. Thus the dual Euclideanized gauge theory should be compactified on a sphere — S^4 for our main example. In this case the $SO(5)$ subgroup of the $SO(5,1)$ conformal group is realized as the symmetry of the S^4.

Holographic duality

The notion of holography was first introduced into gravitational physics in the context of trying to encode the degrees of freedom of a black hole on its horizon, with roughly one degree of freedom (or Q-bit, to be more precise) per Planck area, as suggested by the Bekenstein–Hawking entropy formula. While that may be possible, it is not understood in detail how to do that; if it were, we would have a more general understanding of the microscopic origin of the entropy formula than currently exists.

Holography in AdS space

The holographic duality considered here is somewhat similar, except that it applies to the entire space-time rather than to a black hole. The AdS/CFT duality is *holographic* in the sense that the physics of the $(d+1)$-dimensional *bulk* – or even the ten or 11-dimensional bulk, if the sphere is taken into account – is encoded in the dual d-dimensional gauge theory.

How does the hologram work?

The basic idea is that d-dimensional x^μ coordinates of a point in the bulk correspond to the x^μ position in the field theory. The more subtle question is how the radial coordinate r or z is encoded in the gauge theory.

One approach to defining the gauge theory as a quantum theory is to define it as a function of an energy (or momentum) scale E, as discussed earlier. This should be interpreted (in the Wilsonian sense) to mean that fields with momenta above this scale have been integrated out. For theories that are conformal, the resulting effective theory at the scale E is independent of E, since there is no other scale to which E can be compared. Let us consider a scale transformation of the gauge theory $x^\mu \to ax^\mu$. Scale invariance implies that, if this is accompanied by a rescaling of the energy scale $E \to E/a$, this is a symmetry. Since x^μ in the gauge theory is identified with x^μ in the bulk, this scaling can be performed in the AdS metric Eq. (12.97). However, there we see that, when $x^\mu \to ax^\mu$ is accompanied by $z \to z/a$, this is a symmetry of the metric. Thus, we are led to the identification

$$E \sim 1/z \sim r. \tag{12.108}$$

This reasoning shows that the radial coordinate in the bulk corresponds to the energy scale in the dual gauge theory, but it does not establish the constant of proportionality. Dimensional analysis suggests that one should identify

$$E = kr\ell_s^{-2}, \tag{12.109}$$

where k is a dimensionless constant. For example, if one identifies the energy of a string stretched from the horizon at $r = 0$ to a point with radial coordinate r with the scale E, this would determine a value of k.[13] However, other analyses can lead to different constants of proportionality. Perhaps this reflects the ambiguity in defining the energy scale of the field theory in the first place. The important, and unambiguous, fact is that in any scheme for defining energy scales there is a correspondence of ratios $E_1/E_2 = z_2/z_1 = r_1/r_2$.

This identification of radial coordinate in the bulk with energy scale in the gauge theory is very striking, and one might wonder how it could be reconciled with any notion of locality in the bulk theory. It is difficult to define gauge-invariant local observables in a gravitational theory with diffeomorphism symmetry. So it is not clear that locality should be an exact principle for the bulk theory. However, one would expect it to hold for scales larger than the string scale, when local field theory in a fixed gravitational background is a reasonable approximation. One reason the holographic correspondence proposed here is not in manifest conflict with locality is the observation that changes in energy scale in the gauge theory are given by the renormalization group equation, which is local in the energy scale.

Given the holographic identification, one can ask where the dual gauge theory is located. If one regards the theory without any degrees of freedom integrated out as the most fundamental, this corresponds to $E \to \infty$. Then one can say that the dual gauge theory is located at the boundary $r = \infty$

---

[13] This energy is finite, and proportional to r, even though the proper distance is infinite, due to the compensating effect of the red-shift factor.

or $z = 0$. As one integrates out high-momentum degrees of freedom, it gets translated inwards toward the horizon.

The correspondence can be generalized to bulk theories that are only AdS asymptotically as $r \to \infty$. In such a case, the dual gauge theory is not conformal, but (according to the holographic principle) it should approach a conformal fixed point in the ultraviolet. It is often convenient to work in a small curvature regime of the bulk theory, where a supergravity approximation can be used. In cases where the dual gauge theory is not exactly conformal, the 't Hooft coupling constant may not be large, for a sufficiently large range of energies, for there to be a large fifth dimension in the dual description.

S-duality

Section 8.2 described the S-duality of type IIB superstring theory. In flat ten-dimensional space-time the S-duality group was shown to be $SL(2,\mathbb{Z})$. In particular, it was shown that the complex scalar field

$$\tau_{IIB} = C_0 + ie^{-\Phi}, \tag{12.110}$$

transforms as a modular parameter under $SL(2,\mathbb{Z})$ transformations.

Section 8.2 also described an $SL(2,\mathbb{Z})$ S-duality of $U(N)$ $\mathcal{N} = 4$ super Yang–Mills theory. As explained there, if one includes a topological θ term in the Lagrangian, one can define a complexified gauge theory coupling

$$\tau_{YM} = \frac{\theta}{2\pi} + \frac{4\pi i}{g_{YM}^2}. \tag{12.111}$$

Now one is led to an important implication of the AdS/CFT duality. Namely, the S-duality of the gauge theory is induced by the S-duality of the string theory. Since the AdS/CFT correspondence requires the S-duality of the gauge theory, any test of this S-duality is also a test of the correspondence. Conversely, the existing evidence in support of the S-duality of $\mathcal{N} = 4$ super Yang–Mills theory can be regarded as support for the AdS/CFT conjecture.

The identification $g_{YM}^2 = 4\pi g_s$, where $e^\Phi = g_s$ for the extremal D3-brane solution, naturally generalizes to

$$\tau_{YM} = \tau_{IIB}, \tag{12.112}$$

provided one makes the identification $\theta = 2\pi C_0$. The black D3-brane solution can be generalized to allow a constant nonzero value of C_0 without making any other changes, and this corresponds to adding a θ term to the dual gauge theory. The conclusion is that the value of the complex coupling in the gauge theory is identified with the vacuum expectation value of the complex scalar field of the string theory. Each of them is defined on a space that is identical to the moduli space of complex structures of a torus, which was described in Chapter 3 and utilized several times in Chapter 8.

A more precise correspondence

The tests of AdS/CFT duality described so far only required analyzing perturbations of $AdS_5 \times S^5$. Another successful test in this framework, which we have not explained, was to show that all the linearized supergravity states correspond to states in the dual gauge theory. However, there is more than this perturbative framework that needs to be understood to define a precise map between a CFT and its AdS dual, since this set-up is only sensitive to the supergravity states and their Kaluza–Klein excitations, but does not probe the underlying string-theory structure of the theory.

A conformal field theory does not have particle states or an S-matrix. The only physical observables, that is, well-defined and meaningful quantities, in a CFT are correlation functions of gauge-invariant operators. Thus, what is required is an explicit prescription for relating such correlation functions to computable quantities in the AdS string-theory background. These are very similar to ordinary S-matrix elements, with the definition suitably generalized to AdS boundary conditions at infinity. Since the dimension of the AdS exceeds that of the CFT by one, it is sensible that off-shell quantities in the CFT should correspond to on-shell quantities in AdS.

The gauge-invariant operators are defined at a point, which corresponds to perturbing the gauge theory in the ultraviolet. Therefore, according to the holographic energy/radius correspondence, the gauge theory should be considered to be at the AdS boundary.

It is technically easier to work with the Euclideanized conformal field theory and to relate its correlation functions to quantities in the Euclideanized anti-de Sitter geometry. So this is a good place to start. After that, we discuss the case of Lorentzian-signature case. The prescription requires a one-to-one correspondence of bulk fields ϕ and gauge-invariant operators $\mathcal{O}$ of the boundary CFT.

The path integral

Schematically the correspondence works as follows. Denoting boundary values of ϕ by ϕ_0, one computes the bulk-theory path integral with these boundary values to define a partition function

$$Z_{\text{string}}(\phi_0) = \int_{\phi_0} D\phi \, e^{-S_{\text{string}}}. \tag{12.113}$$

Then this is identified with the field theory expression

$$\left\langle \exp \int_{S^d} \phi_0 \mathcal{O} \right\rangle_{CFT}, \tag{12.114}$$

which is the generating function of correlation functions. To complete the explanation one should carefully explain how one removes the divergences that occur in these formal expressions. Even then, these formulas are hard to evaluate in practice, except in regimes where perturbation theory is applicable, as illustrated in Fig. 12.7. The Feynman rules require interaction vertices in the bulk and three types of propagators: bulk to bulk, bulk to boundary, and boundary to boundary.

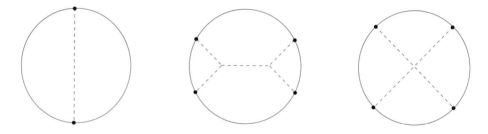

Fig. 12.7. Correlation functions in strongly coupled gauge theories can be calculated in terms of ordinary Feynman diagrams in the bulk theory with propagators that terminate on the boundary.

One of the bulk fields is the metric $g_{\mu\nu}$. It corresponds to the energy–momentum tensor $T_{\mu\nu}$ of the CFT, which is always a gauge-invariant operator. *For this reason the AdS/CFT correspondence always involves a gravitational theory for the bulk.* The asymptotic behavior of the metric as the boundary is approached is well defined up to a conformal rescaling ($g_{\mu\nu} \sim \lambda g_{\mu\nu}$.) Thus, ϕ_0 in this case denotes the boundary value of the conformal class of the metric. Another example for which the bulk field to boundary operator correspondence is known is the dilaton. The dilaton corresponds to the Lagrangian of the CFT, because a small change in the gauge coupling, which is dual to the string coupling determined by the dilaton, adds an operator proportional to the Lagrangian.[14]

---

[14] The analogous M-theory backgrounds do not have a dilaton, so maybe it is not surprising that one is unable to construct a Lagrangian for the dual CFT in those cases.

Anomalous dimensions

As an example, consider a scalar field ϕ of mass m in EAdS$_5$, whose metric is given in Eq. (12.105). The quadratic terms in the action are

$$S \sim \int d^4y\, dz\, [z^2(\partial_y \phi)^2 + z^2(\partial_z \phi)^2 + m^2 R^2 \phi^2]/z^5. \tag{12.115}$$

Considering five dimensions rather than ten implies a truncation to the lowest Kaluza–Klein mode on the five-sphere. The classical field equation derived from this action has two independent solutions that are given by Bessel functions. To decide which solutions are normalizable, we are particularly interested in the asymptotic behavior at the boundary, which corresponds to $z = 0$. The two solutions both give power behavior of the form $\phi \sim z^\alpha$, and the two values of α, determined by the equation of motion, are the roots of[15]

$$\alpha(\alpha - 4) = m^2 R^2, \tag{12.116}$$

which are

$$\alpha_\pm = 2 \pm \sqrt{4 + m^2 R^2}. \tag{12.117}$$

You are asked to derive this result from the equation of motion in Problem 12.8.

In defining the boundary value ϕ_0 that enters in Eq. (12.113), a singular factor must be removed. The reason for this is that the more singular solution goes as $\phi \sim z^{\alpha_-}$ near the boundary. Since the boundary theory is conformal, a conformal rescaling that is x independent is allowed. Thus, the regularized boundary value of the field is

$$\phi_0(x) = \lim_{z \to 0} z^{-\alpha_-} \phi(x, z). \tag{12.118}$$

As a result of this *renormalization*, the corresponding boundary operator acquires a scaling dimension. The naive scaling dimension would be four for a scalar field. However, one also has to take account of the scaling property of $z^{-\alpha_-}$, recalling that z scales like x. This contributes an anomalous dimension $-\alpha_- \geq 0$. As a result, one obtains the scaling dimension for the dual gauge theory operator

$$\Delta = 4 - \alpha_- = 2 + \sqrt{4 + m^2 R^2}. \tag{12.119}$$

For example, the dilaton has $m = 0$, which agrees with the fact that the SYM Lagrangian has $\Delta = 4$. The corresponding analysis for the graviton "predicts" that the stress tensor should have $\Delta = 4$, which is also correct.

[15] When the boundary theory has d dimensions, this generalizes to $\alpha(\alpha - d) = m^2 R^2$.

As an aside, and for completeness, let us note the following. The restriction to the α_- solution is correct for the $AdS_5 \times S^5$ case, but it is not always correct. In AdS_{d+1} space-time the mass-squared parameter m^2 can be negative without making the vacuum unstable, as it would in flat space-time. In fact, the stability bound, known as the *Breitenlohner–Freedman bound*, can be deduced from the d-dimensional generalization of Eq. (12.115). It is

$$m^2 R^2 > -d^2/4. \tag{12.120}$$

For $m^2 R^2 > 1 - d^2/4$ there is a unique admissible boundary condition, which is the case for all fields in the theory under consideration here. However, some theories have fields with masses in the window

$$1 - d^2/4 > m^2 R^2 > -d^2/4, \tag{12.121}$$

and then both solutions $\alpha_\pm$ are admissible, since they both satisfy the unitarity bound $\Delta > (d-2)/2$.

Lorentzian signature and the Hamiltonian

The CAdS/CFT duality for Lorentzian signature entails new issues. The boundary-value problem in this case no longer has unique solutions, because one can add normalizable (propagating) modes. Nonnormalizable bulk modes correspond to backgrounds that couple to gauge-invariant local operators of the boundary gauge theory, as in the Euclidean case. In addition, the normalizable modes of the Lorentzian case correspond to localized fluctuations of the gauge theory.

From the point of view of the dual CFT, there is a Hamiltonian that generates the time evolution. This is conceptually cleanest when the spatial dimensions form a p-dimensional sphere, as is natural in the global CAdS formulation. In this case, if one imagines expanding all fields in Fourier modes on the sphere, one has a quantum mechanical theory with an infinite number of degrees of freedom. However, there are only a finite number of states below any fixed energy.

The $AdS_5 \times S^5$ solution (for example) corresponds to the ground state of this Hamiltonian, whereas nonnormalizable modes in the bulk correspond to excited states of the gauge theory. Generic excited states of the gauge theory correspond to bulk geometries with string-scale curvatures for which the supergravity approximation is not valid. Smooth geometries correspond to highly excited states of the gauge theory with a smooth distribution of excitations. This correspondence has been worked out in complete detail for the half-BPS states. While this is far from the whole story, it is a highly instructive starting point. In this case the excitations of the CFT correspond to placing N free fermions in a harmonic oscillator potential. That study makes it clear how the energy eigenvalues of the fermions in the gauge theory encode the geometry of the dual string theory. Generically, unless N is very large and the eigenvalues are smoothly distributed, one obtains a turbulent quantum-foam-like geometry. The term *bubbling AdS* has been introduced in this context.

The modification of the bulk solution obtained by the addition of normalizable modes, on the other hand, corresponds to changing the Hamiltonian of the boundary CFT by the addition of relevant perturbations. Relevant perturbations are defined to be ones with dimension less than four, which are important in the IR and unimportant in the UV. For example, the addition of a mass term for one (or more) of the six scalar fields is a relevant perturbation of the gauge theory.

Chiral primary operators

An alternative to analyzing the gauge theory on S^p and using the Hamiltonian approach is to consider the gauge theory on $\mathbb{R}^p$, as is natural in the Poincaré patch description of the AdS space. In this case the physical observables are correlation functions of gauge-invariant operators. The gauge-invariant operators correspond to the various states in the Hamiltonian description by a state–operator correspondence that is a higher-dimensional analog of that described for two-dimensional conformal field theories in Chapter 3.

Testing the AdS/CFT correspondence in this set-up involves finding the correspondence between gauge-invariant operators in the gauge theory and particle states in the string theory. In each case these are classified by representations of the superconformal symmetry algebra. Such representations include three types: long, short, and ultrashort. As explained in Chapters 8 and 11, the $\mathcal{N} = 4$ supersymmetry algebra provides lower bounds (BPS bounds) on allowed masses or energies. If neither bound is saturated, the representation is long and all 16 of the linearly realized supersymmetry generators are effective in building up the multiplet structure. In this case the allowed helicities cover a range of eight units, since each charge can shift the helicity by one half. If one of the bounds is saturated and the other is not, the representation is called short and eight of the supersymmetry generators are effective. Then the helicities in the multiplet cover a range of four units. In the ultrashort case, both bounds are saturated, and the helicities cover a range of two units. The $\mathcal{N} = 4$ super Yang–Mills fields themselves have helicities ranging from -1 to $+1$ and form an ultrashort multiplet. However, they are not gauge-invariant operators.

In the string description short multiplets arise as the five-dimensional supergravity multiplet and all of its Kaluza–Klein excitations on the five-sphere. The harmonics on the five-sphere give $SU(4)$ irreducible representations denoted $(0, n, 0)$ in Dynkin notation. In $SO(6)$ language, these correspond to rank-n tensors that are totally symmetric and traceless. Clearly, the helicities range from -2 to $+2$ for these multiplets, since the five-sphere harmonic does not contribute to the helicity. All of the excited string states belong to long multiplets, which are much more difficult to analyze. However, it is possible to say something about a certain class of them in the plane-wave limit, as is done in Section 12.5.

Let us now consider some local operators in the gauge theory that belong to short multiplets. The $SU(N)$ super Yang–Mills fields are described as traceless $N \times N$ hermitian matrices. The way to form gauge-invariant combinations is to consider traces of various products. The quantities that are allowed inside the traces are the six scalars, four spinors, and Yang–Mills field strength, as well as arbitrary covariant derivatives of these fields. One can also consider products of such traces. However, it turns out that single-trace operators correspond to single-particle states and multi-trace operators correspond to multi-particle states in leading order. At higher orders in λ and $1/N$, there can be mixing between operators with differing numbers of traces.

A convenient way of characterizing a supermultiplet is by finding the *primary* operator of lowest dimension. By definition, this operator is annihilated by all of the conformal symmetries S_α and K_μ. The other operators in the supermultiplet are reached by commuting or anticommuting the primary operator with the super-Poincaré generators Q_α and P_μ. These operators are called *descendants* and are characterized by the fact that they can be expressed as Q acting on some operator. In the case of short multiplets, the primary operator is also annihilated by half of the Q supersymmetry generators. Such operators are called *chiral primary operators*.

As an example, consider the trace of a product of n scalar fields

$$\mathcal{O}^{I_1 I_2 \cdots I_n} = \mathrm{Tr}\left(\phi^{I_1} \phi^{I_2} \cdots \phi^{I_n}\right). \tag{12.122}$$

It turns out that if any of the indices are antisymmetrized this operator is a descendant. A commutator $[\phi_I, \phi_J]$ is a descendant field because it appears in the supersymmetry transformation of fermion fields. To understand this, recall that in ten dimensions $\delta\psi \sim F_{\mu\nu}\Gamma^{\mu\nu}\varepsilon$. On reduction to four dimensions $F_{IJ} \to [\phi_I, \phi_J]$.

The way to make a primary operator is to totally symmetrize all n indices and remove all traces to make a traceless symmetric tensor. However, this is not quite the whole story. These operators can be related to multi-trace operators when $n > N$. By a *multi-trace operator*, we mean an operator that is a product of operators of the form in Eq. (12.122). Thus, to state the final conclusion, these operators provide a complete list of single-trace chiral primary operators for $n = 2, 3, \ldots, N$. This rule reflects the fact that these are the orders of the independent Casimir invariants of $SU(N)$. This is explored further in Exercise 12.9.

These chiral primary operators form the $(0, n, 0)$ representation of $SU(4)$. In the large-N limit this matches perfectly with what one finds from Kaluza–Klein reduction on the five-sphere in the dual string-theory picture. It has been shown that the masses of these bulk scalar fields match the conformal dimensions of the chiral primary operators in the way required by the duality that was described earlier. It is interesting, though, that for finite N the Kaluza–Klein excitations with $n > N$ seem to be missing in the dual gauge-theory description. This is how it should be, however. The infinite tower of Kaluza–Klein excitations actually is truncated at N. The reason will be explained later.

Anomalies

In general, it is difficult to compare gauge theory and string theory correlation functions, because the AdS/CFT correspondence relates weak coupling to strong coupling. However, there are certain quantities that are controlled by anomalies that can be computed exactly enabling the comparison to be made. Let us describe an example.

The $SU(4)$ R symmetry is a chiral symmetry of $\mathcal{N} = 4$ super Yang–Mills theory. This is evident because left-handed and right-handed fermions belong to complex-conjugate representations ($\mathbf{4}$ and $\bar{\mathbf{4}}$). If one were to add $SU(4)$ gauge fields and make this symmetry into a local symmetry, one would obtain an inconsistent quantum theory, because the $SU(4)$ currents would acquire an anomalous divergence

$$(\nabla^\mu J_\mu)^a = \frac{N^2 - 1}{384\pi^2} i\, d^{abc} \epsilon^{\mu\nu\rho\lambda} F^b_{\mu\nu} F^c_{\rho\lambda}. \qquad (12.123)$$

Such $SU(4)$ gauge fields are not present in the super Yang–Mills theory, so there is no inconsistency. However, they do exist in the bulk theory, where they arise by the Kaluza–Klein mechanism as a consequence of the isometry of the five-sphere. The anomaly means that if one turns on nonzero field strengths for these gauge fields the bulk theory would no longer be gauge invariant. The associated anomaly can be computed from the bulk perspective and compared to the gauge theory anomaly described above as a nontrivial test of the AdS/CFT correspondence.

The way to see the anomaly in the string-theory description is to consider the Chern–Simons term in the low-energy effective five-dimensional action of the bulk theory

$$S_{\text{CS}} = \frac{iN^2}{96\pi^2} \int_{\text{AdS}_5} d^5x \left(d^{abc} \epsilon^{\mu\nu\rho\lambda\sigma} A^a_\mu \partial_\nu A^b_\rho \partial_\lambda A^c_\sigma + \ldots \right). \tag{12.124}$$

Under a gauge transformation $\delta A^a_\mu = \nabla_\mu \Lambda^a$, the Chern–Simons term changes by a boundary term

$$-\frac{iN^2}{384\pi^2} \int d^4x d^{abc} \epsilon^{\mu\nu\rho\lambda} \Lambda^a F^b_{\mu\nu} F^c_{\rho\lambda}. \tag{12.125}$$

Identifying this with $-\int d^4x \Lambda^a (\nabla^\mu J_\mu)^a$ in the dual gauge theory, one obtains exact agreement with the gauge theory calculation to leading order in large N. A more refined analysis, at one-loop order in the string theory, has been carried out. It shows that the factor really is $N^2 - 1$ rather than N^2. This agreement is a very nontrivial test of the AdS/CFT correspondence, since the two computations look completely different.

A similar anomaly analysis can be carried out for the conformal (or Weyl) anomaly that arises from coupling the gauge theory to gravity. Agreement is again found at leading order in large N.

Near-extremal black D3-brane

Nonextremal black D-branes solutions were presented in Section 12.1. Like nonextremal black-hole solutions, they have thermodynamic properties including temperature and entropy. This section explores these properties for the near-extremal black D3-brane and interprets them in the context of the AdS/CFT duality. The analysis is carried out for an asymptotically $AdS_5 \times S^5$ space-time, where the radius of each factor is R.

The metric for a nonextremal black D-brane is given in Eq. (12.31). Here, we specialize that formula to the case $p = 3$ and re-express it terms of the coordinate $z = R^2/r$, which was introduced earlier. It has a horizon at $z = z_0$ that encloses a singularity. Including the five-sphere, the metric in Poincaré coordinates for a near-extremal black 3-brane in an asymptotically $AdS_5 \times S^5$ space-time is

$$ds^2 = \frac{R^2}{z^2} \left(-f(z)dt^2 + d\vec{x} \cdot d\vec{x} + f^{-1}(z)dz^2 \right) + R^2 d\Omega_5^2, \tag{12.126}$$

where
$$f(z) = 1 - (z/z_0)^4. \tag{12.127}$$

This space-time approaches $AdS_5 \times S^5$ asymptotically at infinity, which in these coordinates is given by $z \to 0$.

The temperature

The temperature of this black 3-brane can be determined in the standard way. Specifically, one introduces a Euclidean time coordinate $\tau = it$ and requires that the periodicity of τ is such that there is no conical singularity at the horizon ($z = z_0$). Substituting $z = z_0 - \varepsilon$ in Eq. (12.126) and expanding in ε, one obtains

$$ds^2 = \frac{R^2}{z_0^2}\left(\frac{4\varepsilon}{z_0}d\tau^2 + dx^i dx^i + \frac{z_0}{4\epsilon}d\varepsilon^2\right) + R^2 d\Omega_5^2. \tag{12.128}$$

Now making the change of variables $\varepsilon = z_0 \rho^2/R^2$, one sees that ρ and $\theta = 2\tau/z_0$ parametrize a plane in polar coordinates. Thus, the required period of τ is $\beta = \pi z_0$. As usual, β is identified as the inverse temperature of the black D3-brane.

The entropy

The entropy of this black 3-brane per unit three-volume (as measured in the x^i coordinates) is given by the Bekenstein–Hawking formula (horizon area divided by $4G_{10}$), is

$$\frac{S}{V} = \frac{1}{4G_{10}}\left(\frac{R}{z_0}\right)^3 \cdot R^5 \Omega_5 = \frac{\pi^2}{2}N^2 T^3. \tag{12.129}$$

One can try to test this result with a dual CFT computation of the entropy carried out for the $\mathcal{N} = 4$ gauge theory at temperature T. However, exact agreement should not be expected. The preceding analysis is based on a supergravity approximation to the string-theory geometry. This is valid when the string is weakly coupled and the curvature is small, in other words for large N and large λ. The CFT computation can be carried out for small λ by simply adding up the contributions of the individual free fields. Since these are opposite limits, the results need not agree. Nonetheless, let us carry out the comparison. The small λ CFT computation gives

$$\frac{S}{V} = \frac{2\pi^2}{3}N^2 T^3. \tag{12.130}$$

The T^3 dependence was inevitable because the theory is conformal and there is no other scale. The N^2 factor is also obvious, because each of the fields in the CFT is an $N \times N$ matrix, and the entropy is proportional to the number of fields.[16] So only the numerical coefficient requires care. It is determined by adding the contributions of one vector, four spinors and six scalars in each supermultiplet.

The two preceding results differ by a factor of 4/3, but it was already emphasized that agreement should not be expected. One result corresponds to the limit $\lambda \to 0$ and the other one to the limit $\lambda \to \infty$. The results for these two limits suggest that there should be a formula for the entropy density of the large-N limit of the gauge theory of the form

$$\frac{S}{V} = c(\lambda) \frac{\pi^2}{2} N^2 T^3, \qquad (12.131)$$

where $\lambda = g_{\text{YM}}^2 N$ and $g_{\text{YM}}^2 = 4\pi g_s$, as before. The gravity calculation then implies that $c(\infty) = 1$, while the CFT calculation implies that $c(0) = 4/3$. It is conjectured that the function $c(\lambda)$ extrapolates smoothly between these two values. The complete function $c(\lambda)$ is not known yet, but the next-to-leading terms in the two limits have been computed and are given by

$$c(\lambda) = \frac{4}{3} - \frac{2\lambda}{\pi^2} + \ldots \quad \text{for small } \lambda, \qquad (12.132)$$

$$c(\lambda) = 1 + \frac{15\zeta(3)}{8\lambda^{3/2}} + \ldots \quad \text{for large } \lambda. \qquad (12.133)$$

Giant gravitons and the stringy exclusion principle

In Chapter 6 we discussed the Myers effect, in which a D0-brane in an electric four-form flux is polarized into a spherical D2-brane. A similar phenomenon can be realized in the present setting by considering a massless particle, such as a graviton, moving along a great circle of the S^5. These are BPS solutions that are included in the Kaluza–Klein spectrum discussed earlier. As the momentum of the particle is increased the effect of the background five-form flux becomes more important, and the particle becomes polarized into a sphere. What we are discussing here can be viewed as the polarization of a graviton by a five-form flux into a spherical D3-brane. Such a configuration is sometimes called a *giant graviton*.

Giant gravitons can occur inside the anti-de Sitter space, localized on the five-sphere, or inside the five-sphere, localized in the anti-de Sitter space. The two cases differ in one interesting respect. In the latter case the radius of the giant graviton is bounded by the radius R of the five-sphere. This fact, referred to as the *stringy exclusion principle*, implies that the tower of

---

[16] Since we are interested in large N, we do not distinguish between N^2 and $N^2 - 1$.

Kaluza–Klein excitations is actually cut off at N. This is a string-theoretic effect that is not visible in the supergravity approximation.

In the previous section, this stringy exclusion principle was anticipated by classifying the single-trace chiral primary operators in the dual gauge theory. What we are finding now is exactly what is required for agreement of the two pictures. This success of AdS/CFT duality is highly significant in that it is nonperturbative in the $1/N$ expansion.

Confinement/deconfinement phase transition

One can explore whether gauge theories are confining or not by evaluating Wilson loops. These are gauge-invariant operators, and thus physical observables, of the boundary gauge theory. Given a closed contour $\mathcal{C}$ in $\mathbb{R}^4$ and a representation D of the gauge group (usually chosen to be the fundamental), one defines the Wilson-loop operator

$$W(\mathcal{C}) = \text{Tr}\left(P \exp \oint_{\mathcal{C}} A\right). \tag{12.134}$$

Here, the gauge field A is a matrix of one forms in the representation D, and P denotes that the integral is path ordered. The choice of starting point for the path does not matter once the trace is taken. Physically, one can think of $\mathcal{C}$ as the world line of a heavy external quark. The usual assertion is that, for a square contour with sides of length L, one finds for large L that $W \sim \exp(-cL)$ for a nonconfining theory. This behavior is referred to as a *perimeter law*. On the other hand, $W \sim \exp(-cL^2)$ for a confining theory. This behavior is referred to as an *area law*.

In a conformally invariant theory, such as $\mathcal{N} = 4$ SYM theory, dimensional analysis together with a large-N limit requires that the potential for a quark-antiquark pair with separation L should be of the form $V(L) = v(\lambda)/L$. This is a nonconfining (Coulomb-like) behavior corresponding to a perimeter law.[17] Perturbation theory implies that, for small coupling, $v(\lambda)$ is proportional to λ.

The dual string picture can be used to derive the behavior of $v(\lambda)$ for large λ, the limit in which the AdS curvature is small compared to the string tension T. In other words, $TR^2 \sim \sqrt{\lambda} \gg 1$. In the string picture one views the contour $\mathcal{C}$ as the boundary of a string world sheet. Then, in the large tension limit, the Wilson loop is given to good approximation by

$$\langle W \rangle \sim \exp(-T \cdot \text{Area}). \tag{12.135}$$

[17] An area law would correspond to a linear potential $V \sim L$.

Here, the area is that of the minimal area surface (embedded in AdS_5) with boundary $\mathcal{C}$. The fact that this is an area might seem to contradict what was said previously. However, because of the curvature of the AdS space, this area actually grows linearly for large L, and one finds for large λ that $v(\lambda) \sim TR^2 \sim \sqrt{\lambda}$. One subtlety in this analysis is that the area of the world sheet is actually divergent, because the proper distance to the boundary of AdS is infinite. However, the divergent part has a universal behavior that can be subtracted as part of a consistent regularization procedure. Then the results asserted above can be obtained.

Compactification on a circle

An example of a confining gauge theory is pure Yang–Mills in three dimensions. We can make contact with that theory starting from $\mathcal{N} = 4$ SYM in two steps. The first step is to take one of the three spatial dimensions to be a circle of radius r_0. Then, for energies below $1/r_0$, the theory is effectively three-dimensional. The second step is to get rid of all massless particles other than the gauge fields. A convenient way to achieve this is to require the fermi fields to be antiperiodic on the circle, so that their masses are of order $1/r_0$. The bosons are given periodic boundary conditions. Even so, the scalars of the SYM-theory also get masses of order $1/r_0$ induced by radiative corrections. So do the three-dimensional scalars corresponding to the component of the gauge fields along the compact direction.

The next problem is to identify the bulk supergravity geometry that has this geometry for its boundary. A trick for finding the answer is to start with the black 3-brane solution and perform a double Wick rotation ($t \to iy$ and $x_3 \to it$) giving

$$ds^2 = \frac{R^2}{z^2}\Big(-dt^2 + f(z)dy^2 + dx_1^2 + dx_2^2 + f^{-1}(z)dz^2\Big), \quad (12.136)$$

where R is the AdS radius and

$$f(z) = 1 - (z/z_0)^4. \quad (12.137)$$

This solution does not have a horizon, and it is only defined for $z < z_0$. This means that z_0 is the end of space. As a result, the warp factor $(1/z)$ cannot go to zero. This changes the analysis of the Wilson loop asymptotics, and one concludes that there is an area law (confinement) in this case.

Recall that the y coordinate is periodic with period $2\pi r_0$. For the metric to be nonsingular for $z \to z_0$, one needs to take $z_0 = 2r_0$. This geometry is topologically $B_2 \times \mathbb{R}^{2,1}$, and its boundary has the topology $S^1 \times \mathbb{R}^{2,1}$. The

12.3 The AdS/CFT correspondence

y circle on the boundary (where $z = 0$) is the boundary of a disk whose interior corresponds to $z > 0$.

Finite temperature

The preceding compactified theory can be studied at finite temperature by Euclideanizing the time coordinate t and imposing periodicity β, as usual. The boundary topology of the finite temperature theory is $\mathbb{R}^2 \times S^1 \times S^1$, where the first circle is the spatial circle of radius r_0 and the second circle is the periodic Euclidean time of circumference β. There are two choices for the topology of the finite temperature bulk theory that could give this boundary. The one implied by the analysis given above is $\mathbb{R}^2 \times B_2 \times S^1$. An alternative possibility is $\mathbb{R}^2 \times S^1 \times B_2$. In the latter solution the disk has the time circle as its boundary. Since it has a perimeter law, it does not give confinement. In the bulk-boundary correspondence, one should include all possible bulk configurations that give a specified boundary configuration. In this case there are two of them. The Wilson-loop analysis for one case indicates confinement and for the other indicates deconfinement. So what should we conclude?

In a saddle-point approximation, the string-theory partition function has roughly the form

$$Z_{\text{string}} \sim e^{-S_1} + e^{-S_2}, \qquad (12.138)$$

where the two terms represent the contributions of the two possible bulk topologies. Actually, both S_1 and S_2 contain an infinite factor – the volume of the space-time. However, one can compute the difference $S_2 - S_1$, which is finite. It is a positive function of β and r_0 times $\beta^2 - r_0^2$. Therefore, one or the other is dominant as $N \to \infty$ depending on the ratio of β to r_0. This implies that, for large N, there is a phase transition, which is known as the Hawking–Page phase transition. The interpretation of this phase transition in the dual field theory is that the low-temperature phase (in which the first term dominates) exhibits confinement and a mass gap, whereas the high-temperature phase (in which the second term dominates) has deconfinement. In other words, there are physical unconfined quanta (gluons, etc.) carrying color quantum numbers.

This is roughly the same picture one expects for QCD. Even though the bulk theory that should be dual to QCD is not known, it is now reasonably clear that there should be a dual five-dimensional string theory that contains gravity. At low temperature, one bulk geometry dominates, and at high temperature there should be a different dual geometry. The quark-gluon deconfinement phase transition should be analogous to the Hawking–Page phase transition. In fact, there have been qualitative successes in accounting for data on high-energy collisions of large nuclei in the RHIC collider at Brookhaven National Laboratory using this type of a holographic model. More specifically, when heavy nuclei collide at high energies, it is believed that a quark-gluon plasma is formed, which quickly cools due to expansion. However, before it cools through the deconfinement phase transition temperature, high-energy partons (deconfined quarks or gluons) can travel through the plasma with an effective viscosity that one can try to deduce from the observations. This is an example of a parameter that has been estimated using the string theory/gauge theory duality. It is quite remarkable that members of the nuclear physics community are now becoming interested in black holes in five-dimensional anti-de Sitter space-time!

Proving the conjecture

We have presented many pieces of evidence that support the validity of the AdS/CFT conjecture. Some of them are highly nontrivial and very impressive. So one might wonder whether the construction of a proof that it is correct is a reasonable goal. One problem with this is that there is no other known way of giving a complete definition of string theory. We know pretty well how to define the perturbation expansion, and we know many facts about the nonperturbative physics in various string vacua. Certainly, the duality should reproduce, or at least not contradict, what is known. So a falsification of the conjecture would be straightforward.

Perhaps, the right attitude at this point is to assume that the conjecture is correct, as long as this does not lead to contradictions or paradoxes. Since, at least in four dimensions, the dual gauge theories are unambiguously defined, this means that the duality can be taken as the definition of string theory (or M-theory) for the class of background configurations where it applies. There is also room for further progress in precisely specifying how the duality map works.

One might hope that some completely independent fundamental formulation of string theory would be found some day. If this were to happen, then the goal of proving AdS/CFT would become better formulated. A more modest goal is to learn how to apply AdS/CFT ideas to a larger class of backgrounds.

To be specific, there is a lot of interest in space-times that are asymptotically de Sitter, rather than anti-de Sitter, mostly motivated by the astrophysical/cosmological observations that point to a positive cosmological constant. One idea has been to try to formulate dS/CFT dualities. Another idea that is being explored is to formulate dualities for asymptotically AdS space-times that have large regions within them that are nearly de Sitter.

EXERCISES

EXERCISE 12.5
Explain why the superpotential of an $\mathcal{N} = 1$ supersymmetric theory with an unbroken R symmetry should have R charge equal to two.

SOLUTION

The four Grassmann coordinates of $\mathcal{N} = 1$ superspace decompose into two chiral components θ_α and two antichiral components $\bar\theta_{\dot\alpha}$. The superfield formulation of $\mathcal{N} = 1$ gauge theories contains two types of terms, called D terms and F terms, where D terms are given by integrals over the full superspace and F terms are given by integrals over chiral superspace. The Lagrangian density has the general structure

$$\mathcal{L} = \mathcal{L}_K + \mathcal{L}_V + \mathcal{L}_W,$$

where the first term is a D term and the last two terms are F terms and their complex conjugates. For example,

$$\mathcal{L}_W = \int d^2\theta\, W + \int d^2\bar\theta\, \overline{W},$$

where the superpotential W is a holomorphic function of chiral superfields.

Under an R-symmetry transformation, the different components of a superfield carry different charges. One can assign uniform R-charge assignment to the entire superfield if one adopts the rule that θ carries R-charge $+1$ and $\bar\theta$ carries R-charge -1. These statements are a consequence of the commutation relations

$$[R, Q_\alpha] = Q_\alpha \quad \text{and} \quad [R, \overline{Q}_{\dot\alpha}] = -\overline{Q}_{\dot\alpha}.$$

Integration over θ is like differentiation, and therefore $d^2\theta$ carries R-charge -2. Therefore, for $\mathcal{L}_W$ to be R-invariant, it is necessary that the superpotential W have R-charge equal to 2. This entails finding suitable R-charge assignments for all the chiral superfields so that each term in the superpotential has R-charge equal to 2. □

EXERCISE 12.6

Formulate $\mathcal{N}=4$ super Yang–Mills theory in terms of $\mathcal{N}=1$ superfields.

SOLUTION

Expressed in terms of $\mathcal{N}=1$ superfields, the $\mathcal{N}=4$ theory can be recast as a theory with a vector superfield V and three adjoint chiral superfields Φ_i, $i = 1, 2, 3$. The Lagrangian consists of a kinetic term for the chiral superfields

$$\mathcal{L}_K = \int d^4\theta \sum_{i=1}^{3} \text{tr}\left(\Phi_i^\dagger e^V \Phi_i\right),$$

a kinetic term for the gauge superfields[18]

$$\mathcal{L}_V = \frac{1}{4g^2}\int d^2\theta\, \text{tr}(W^\alpha W_\alpha) + \text{h.c.},$$

and a superpotential term

$$\mathcal{L}_W = \int d^2\theta\, W + \text{h.c.}.$$

The first two terms are completely determined by the choice of gauge group and representations, so that all that remains is to specify the superpotential W.

The only parameter of the theory is the dimensionless coupling constant g. Therefore, the superpotential must be cubic in the chiral superfields, since these are the terms with dimensionless coefficients. By formulating the theory using $\mathcal{N}=1$ superfields, only one of the four supersymmetries is manifest and only a subgroup of the $SU(4)$ R-symmetry group can be made manifest. Specifically, an $SU(3) \times U(1)$ subgroup of the $SU(4)$ R-symmetry group is manifest in this formulation. The three chiral superfields transform as a **3** of the $SU(3)$. The $U(1)$ factor is the usual R symmetry of an $\mathcal{N}=1$ theory. From these considerations the superpotential is completely determined up to an overall normalization

$$W \sim \epsilon^{ijk} \text{tr}\left(\Phi_i \Phi_j \Phi_k\right) \sim \text{tr}(\Phi_1[\Phi_2, \Phi_3]).$$

The normalization is uniquely determined by the requirement of $\mathcal{N}=4$ supersymmetry or $SU(4)$ R symmetry. The correct result turns out to be

$$W = \sqrt{2}\,\text{tr}(\Phi_1[\Phi_2, \Phi_3]). \tag{12.139}$$

---

[18] The adjoint fields V, W^α, Φ_i are all written as matrices. For example $V = \sum V^a t^a$. The matrices t^a give a representation of the Lie algebra $[T^a, T^b] = if^{abc}T^c$ for which $\text{tr}(t^a t^b) = k\delta^{ab}$. The formulas are written for the fundamental representation of $SU(N)$, which has $k = 1$.

12.3 The AdS/CFT correspondence

Using the result of the previous exercise, one learns that the three chiral superfields have $U(1)$ R charge 2/3. Since they are chiral, their scaling dimension is $\Delta = 3R/2 = 1$, which receives no quantum corrections. □

EXERCISE 12.7
Consider a four-dimensional $U(N)$ gauge theory with only adjoint representation fields. Verify that a Feynman diagram with Euler characteristic χ (in the same sense as in Eq. (12.92) scales as N^χ for large N.

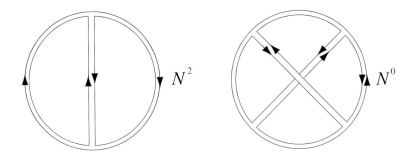

Fig. 12.8. Two-loop planar and nonplanar vacuum diagrams.

SOLUTION

In this case vertices, propagators and loops of the Feynman diagram correspond to the vertices, edges and faces of the two-dimensional surfaces, respectively. Each vertex contributes a factor $1/g^2$, each propagator contributes a factor g^2 and each loop contributes a factor N. So the Feynman diagram is proportional to

$$(g^2)^{E-V} N^F = (\lambda/N)^{E-V} N^F = N^\chi \lambda^{E-V}.$$

For fixed λ, the diagram scales as N^χ. A couple of examples are shown in Fig. 12.8. □

EXERCISE 12.8
Describe the Lie superalgebra for the supergroup $PSU(2,2|4)$.

SOLUTION

The discussion that follows is relevant to all four-dimensional superconformal groups $SU(2,2|N)$. There is a similar analysis for three-dimensional superconformal groups $OSp(N|4)$ and six-dimensional superconformal groups $OSp(6,2|2N)$.

Conformal groups in arbitrary dimension were described in Chapter 3. The generators consist of the Poincaré generators $M_{\mu\nu}$ and P_μ as well as the conformal generator K_μ and the dilatation generator D. Generators can be assigned conformal weights determined by their commutation with D. Thus, since

$$[D, M_{\mu\nu}] = 0, \quad [D, P_\mu] = -iP_\mu, \quad [D, K_\mu] = iK_\mu,$$

M and D have weight 0, P has weight 1 and K has weight -1. This immediately determines which terms can appear in a commutation relation. For example,

$$[P_\mu, K_\nu] = 2iM_{\mu\nu} - 2i\eta_{\mu\nu}D.$$

For N-extended supersymmetry in four dimensions, one adjoins supercharges Q_α^A and their complex conjugates $Q_{\dot\alpha A}$. An upper A labels an $\mathbf{N}$ of $SU(N)$ and a lower A labels the conjugate $\overline{\mathbf{N}}$ representation. Since $\{Q, Q\} \sim P$, Q has conformal weight $1/2$. It follows that the commutators $[Q, K]$ should generate weight $-1/2$ generators. These are the superconformal supercharges denoted $S_{\alpha A}$ and $S_{\dot\alpha}^A$. The $SU(N)$ group theory requires that $\{Q_\alpha^A, S_{\dot\beta}^B\} = 0$, but

$$\{Q_\alpha^A, S_{\beta B}\} = c_1 \sigma_{\alpha\beta}^{\mu\nu} \delta_B^A M_{\mu\nu} + c_2 \varepsilon_{\alpha\beta} R^A{}_B + c_3 \varepsilon_{\alpha\beta} \delta_B^A D.$$

In the special case of $N = 4$ one can impose $R^A{}_A = 0$, so that the R-symmetry group is $SU(4)$. The supergroup is then called $PSU(2,2|4)$. Given normalization conventions, the coefficients c_i can be determined by analyzing the Jacobi identities.

An alternative approach is to use the defining representation of the superalgebra in terms of supermatrices. In the case of $SU(M|N)$ they have $M + N$ rows and columns which can be written in block form

$$X = \begin{pmatrix} A & B \\ C & D \end{pmatrix},$$

where A is $M \times M$ hermitian, D is $N \times N$ hermitian and $B = C^\dagger$ is $M \times N$ fermionic. Also, the supertrace vanishes:

$$\text{Str} X = \text{tr} A - \text{tr} D = 0.$$

You are asked to explore this algebra in a homework problem. In this notation, the generators of the supergroup $PSU(2,2|4)$ are assembled in a supermatrix of the form

$$X = \begin{pmatrix} \frac{1}{2}M_{\mu\nu}(\sigma^{\mu\nu})^\alpha{}_\beta + D\delta^\alpha{}_\beta & K_\mu(\sigma^\mu)^\alpha{}_{\dot\beta} & S^\alpha{}_B \\ P_\mu(\sigma^\mu)^{\dot\alpha}{}_\beta & \frac{1}{2}M_{\mu\nu}(\sigma^{\mu\nu})^{\dot\alpha}{}_{\dot\beta} - D\delta^{\dot\alpha}{}_{\dot\beta} & Q^{\dot\alpha}{}_B \\ Q^A{}_\beta & S^A{}_{\dot\beta} & R^A{}_B \end{pmatrix}.$$

□

EXERCISE 12.9
Explain why the operators defined in Eq. (12.122) can be related to multi-trace operators for $n > N$.

SOLUTION
The operators

$$\mathcal{O}^{I_1 I_2 \cdots I_n} = \mathrm{Tr}\left(\phi^{I_1}\phi^{I_2}\cdots\phi^{I_n}\right) + \ldots$$

are totally symmetric and traceless, which ensures that these are chiral primary operators. For example,

$$\mathcal{O}^{IJ} = \mathrm{Tr}\left(\phi^I \phi^J\right) - \frac{1}{6}\delta^{IJ}\left(\phi^K \phi^K\right).$$

These conditions imply that $\mathcal{O}$ belongs to an irreducible representation of the $SU(4)$ R-symmetry group. The case $n = 1$ vanishes because the fields are $N \times N$ traceless hermitian matrices as appropriate for the Lie algebra of $SU(N)$. The claim is that for $n > N$ operators of this type can be re-expressed as products of traces up to terms that involve commutators, which can be ignored since they are descendent operators.

To make the argument, one can assume that the fields are commuting, since this only involves dropping commutators. Since each field has $N-1$ independent eigenvalues, there are $N-1$ algebraically independent symmetric monomials made from these eigenvalues. In the case of $SU(N)$ these independent monomials have order $n = 2, 3, \ldots, N$. This is the same reasoning by which one argues that these are the orders of the independent Casimir invariants of $SU(N)$.

□

12.4 Gauge/string duality for the conifold and generalizations

The duality between type IIB superstring theory in an $AdS_5 \times S^5$ background with N units of flux and $\mathcal{N} = 4$ super Yang–Mills theory with gauge group $SU(N)$ is the simplest example of a large class of string theory/gauge theory dualities. This section says a little bit about some other examples.

Other gauge/string dualities

In Section 12.1 we found that the metric for an extremal black D3-brane is

$$ds^2 = H_3^{-1/2} dx \cdot dx + H_3^{1/2}(dr^2 + r^2 d\Omega_5^2), \qquad (12.140)$$

where $H_3 = 1 + (R/r)^4$ and $dx \cdot dx$ is the four-dimensional Minkowski metric on the brane. The horizon is at $r = 0$, and the near-horizon geometry is $AdS_5 \times S^5$. There are also N units of five-form flux, and

$$R^4 = \lambda \ell_s^4, \qquad (12.141)$$

where $\lambda = g_{YM}^2 N = 4\pi g_s N$. In this section we wish to consider generalizations of this construction that are obtained by replacing the S^5 by other compact Einstein spaces.[19] In other words,

$$ds^2 = H_3^{-1/2} dx \cdot dx + H_3^{1/2}(dr^2 + r^2 ds_5^2), \qquad (12.142)$$

where ds_5^2 is the metric of the Einstein space.

Other $\mathcal{N} = 4$ examples

The six-dimensional metric $dr^2 + r^2 ds_5^2$ describes a cone. In fact, for any choice of ds_5^2, other than a unit five-sphere, there is a singularity at the tip of the cone, $r = 0$. The physical interpretation is that N D3-branes are localized at this conical singularity. Another example that is still maximally supersymmetric is obtained by making antipodal identifications of the five-sphere, which amounts to replacing it by the smooth space $\mathbb{R}P^5$. There are actually a few distinct ways to carry out this construction. Depending on this choice, the dual $\mathcal{N} = 4$ super Yang–Mills theory has a gauge group that is either $SO(2N)$, $SO(2N+1)$ or $USp(2N)$.

$\mathcal{N} = 2$ examples

Another class of possibilities is to replace the sphere by an orbifold of the sphere S^5/Γ, where Γ is a suitably chosen discrete group of $SU(2)$ such as $\mathbb{Z}_n$. In this case, half the supersymmetry is broken so that the dual gauge is an $\mathcal{N} = 2$ superconformal Yang–Mills theory. In such examples the gauge group is typically a product group $\prod SU(N_i)$. In addition to the vector multiplets it contains hypermultiplets that transform as bifundmentals of the form $(\mathbf{N_i}, \overline{\mathbf{N_j}})$. For example, in the case of $\mathbb{Z}_2$ one obtains the gauge group $SU(N) \times SU(N)$ with hypermultiplets that transform as $(\mathbf{N}, \overline{\mathbf{N}})$ and $(\overline{\mathbf{N}}, \mathbf{N})$.

---

[19] An *Einstein space* is one for which $R_{mn} = c g_{mn}$, that is, the Ricci tensor is proportional to the metric tensor.

$\mathcal{N} = 1$ examples

The five-dimensional space X_5 is called a *Sasaki–Einstein space*, if it is an Einstein space and if the six-dimensional cone over X_5 is a noncompact Calabi–Yau space (with a conical singularity). Then 3/4 of the supersymmetry is broken. Therefore, the dual gauge theory on the world-volume of the D3-branes should be an $\mathcal{N} = 1$ superconformal gauge theory. The formula for the AdS radius is modified to

$$R^4 = 4\pi\lambda\ell_s^4 \frac{\text{Vol}(S^5)}{\text{Vol}(X_5)}. \tag{12.143}$$

For this ratio to be meaningful, it is important to choose coordinates in which $R_{mn} = g_{mn}$ for the Sasaki–Einstein space.

$T^{1,1}$ and the conifold

Chapter 10 introduced a noncompact Calabi–Yau space, called the conifold, with this structure. Recall that it was defined as a hypersurface in $\mathbb{C}^4$ by the simple equation $\sum (w^A)^2 = 0$. In this case, the five-dimensional space X_5 is $T^{1,1} = SU(2) \times SU(2)/U(1)$, which has $SU(2) \times SU(2) \times U(1)$ isometry. $T^{1,1}$ is the simplest nontrivial case of a Sasaki–Einstein space. Its metric was given in Chapter 10. As was explained there, it has the topology $S^3 \times S^2$. This example is the simplest case of an infinite family of possible choices. This section explores this example in some detail, and then comments very briefly on the other ones. The bulk theory contains vector superfields that realize the $SU(2) \times SU(2)$ symmetry. There is also a $U(1)$ gauge field in the AdS_5 supergravity multiplet. As is always the case, these local symmetries of the bulk theory correspond to global symmetries of the dual gauge theory. In particular, the part coming from the supergravity multiplet, which is $U(1)$ in this case, is dual to the global R symmetry of the gauge theory. This R symmetry is contained in the superconformal algebra $SU(2, 2|1)$.

The dual gauge theory

Let us now describe the gauge theory in more detail. The $T^{1,1}$ space can be obtained by smoothing out the $\mathbb{Z}_2$ orbifold theory described above. This fact allows us to deduce that this is also an $SU(N) \times SU(N)$ gauge theory. Each $\mathcal{N} = 2$ hypermultiplet decomposes into two $\mathcal{N} = 1$ chiral supermultiplets. Thus, the gauge theory has two chiral superfields, denoted A_i, transforming under the gauge group as $(\mathbf{N}, \overline{\mathbf{N}})$ and two chiral superfields, denoted B_i, transforming as $(\overline{\mathbf{N}}, \mathbf{N})$. The A_i fields form a doublet of one $SU(2)$ symmetry and the B_i fields form a doublet of the other $SU(2)$. All four fields A_i and

B_i have R-charge equal to 1/2, a result that is determined by anomaly cancellation.

The superconformal symmetry group implies an inequality between the R-charge and the dimension

$$\Delta \geq 3R/2, \tag{12.144}$$

which is analogous to a BPS bound. The As and Bs are chiral fields that saturate this bound, and therefore their dimensions are determined by their R-charges. Thus, they have dimension $\Delta = 3/4$.[20] Since the naive dimension is 1, this means that these fields have anomalous dimension $\Delta_a = -1/4$. The superpotential, which is required to have dimension equal to three and R-charge equal to two, is of the form

$$W \sim \operatorname{tr}(A_1 B_1 A_2 B_2 - A_1 B_2 A_2 B_1). \tag{12.145}$$

This is the unique structure with these values that respects all the local and global symmetries.

A test of the duality

Many tests of this duality have been carried out successfully. Let us describe one of these results. To make a comparison to the bulk theory, one must form gauge-invariant operators. The appropriate ones in this case are

$$\mathcal{O}_k = \operatorname{tr}\left(A_{i_1} B_{j_1} \ldots A_{i_k} B_{j_k}\right). \tag{12.146}$$

These are chiral primary operators if the is and js are separately symmetrized. Then this operator belongs to the $(\mathbf{k+1}, \mathbf{k+1})$ representation of the global $SU(2) \times SU(2)$. The R-charge of these operators is k and the dimension is $\Delta_k = 3k/2$. The duality requires that the bulk theory should have corresponding scalar fields belonging to short supermultiplets with mass given by $m_k^2 = \Delta_k(\Delta_k - 4)$ as in Eq. (12.116). The Kaluza–Klein analysis of the scalar modes on $T^{1,1}$ gives precisely such states with the correct $SU(2) \times SU(2) \times U(1)$ quantum numbers.

Adding fractional D3-branes

The fun really starts when we add wrapped D5-branes to the previous construction. This breaks the AdS symmetry and take us out of the realm of conformal symmetry for the dual gauge theory. This is significant, because then many physically important phenomena appear. There are various ways

---

[20] These fields are allowed to violate the unitarity bound $\Delta \geq 1$, because that bound only applies to gauge-invariant operators.

12.4 Gauge/string duality for the conifold and generalizations   673

one can add D5-branes. They could wrap the entire $T^{1,1}$, its three-cycle or its two cycle. As in Section 10.2, we wish to consider the case in which M D5-branes wrap the two-cycle at the base of the deformed conifold geometry. As was explained there, when such fractional D3-branes are included, the geometry undergoes a logarithmic warping, but supersymmetry remains unbroken, since the relevant three-form flux is primitive.

The effect of the D5-branes on the gauge symmetry can be understood as follows. Suppose that there is only one of them and it is somehow held at a nonzero value $r = r_0$. Then it forms a domain wall in five dimensions separating the regions $r < r_0$ and $r > r_0$. As one crosses the domain wall, the gauge symmetry changes from $SU(N) \times SU(N)$ for $r < r_0$ to $SU(N) \times SU(N+1)$ for $r > r_0$. Iterating this M times, and letting $r_0 \to 0$, one deduces that, when there are M D5-branes, in addition to the usual N D3-branes, at the tip of the conifold the gauge symmetry is

$$SU(M+N) \times SU(N). \qquad (12.147)$$

More precisely, this is the gauge symmetry in the ultraviolet, which corresponds to large r. The four chiral superfields then belong to the representation $(\mathbf{M+N}, \overline{\mathbf{N}})$ and its complex conjugate. It was shown in Section 10.2 that the warp factor is modified to the form[21]

$$e^{-4A(r)} = \frac{L^4}{r^4}\log(r/r_s), \qquad L^2 = \frac{9g_s M \alpha'}{2\sqrt{2}}. \qquad (12.148)$$

This logarithmic warping of the AdS geometry implies that the dual $\mathcal{N} = 1$ gauge theory with $M > 0$ is no longer conformal, and therefore it has a nontrivial renormalization group flow. The details are described below.

R-symmetry breaking

One consequence of the addition of the fractional D3-branes is to break the $U(1)$ gauge symmetry of the bulk theory and the dual global $U(1)$ R symmetry of the gauge theory. These phenomena can be explored separately and shown to match as required.

In the gauge theory with $M > 0$ the $U(1)$ R-symmetry current J^μ develops an anomaly. By computing the appropriate one-loop triangle diagrams, as described in Exercise 12.8, one finds that

$$\partial_\mu J^\mu = \frac{M}{16\pi^2}\left(F^a_{\mu\nu}\widetilde{F}^{a\mu\nu} - G^b_{\mu\nu}\widetilde{G}^{b\mu\nu}\right), \qquad (12.149)$$

where $F^a_{\mu\nu}$ are $SU(M+N)$ field strengths and $G^b_{\mu\nu}$ are $SU(N)$ field strengths.

---
[21] Various constants that were presented in Section 10.2 are absorbed in the parameter r_s here.

The integrated expression is the same as what one gets by shifting a θ parameter by π/M, and therefore the R symmetry is broken to the discrete group $\mathbb{Z}_{2M}$.

In the bulk theory, the $U(1)$ gauge field eats a scalar field, thereby breaking the $U(1)$ gauge symmetry spontaneously. The scalar field that is eaten is the one that is dual to the operator $F^a_{\mu\nu}\widetilde{F}^{a\mu\nu} - G^b_{\mu\nu}\widetilde{G}^{b\mu\nu}$. In terms of the geometry, the $U(1)$ symmetry appears as symmetry under shifts of the coordinate β introduced in Section 10.2. However, after the fractional D3-branes are introduced, there is also a nonzero R–R potential C_2 that shifts by $C_2 \to C_2 + M\alpha'\varepsilon\omega_2$ when $\beta \to \beta + \varepsilon$. Integrating this over the S^2 inside $T^{1,1}$, one deduces that the symmetry is broken to discrete shifts by multiples of π/M, that is, $\mathbb{Z}_{2M}$.

The duality cascade

The warp factor in Eq. (12.148) contains the factor $\log(r/r_s)$. Using the rule that r is proportional to energy scale in the gauge theory, this corresponds to $\log(\mu/\Lambda)$, where μ is the running scale and Λ is the fundamental scale of the gauge theory. These formulas are well defined for large r and large μ. So the question arises as to what happens as these are decreased and approach the singularity. This is referred to as the *flow to the infrared*.

In the $\mathcal{N} = 4$ theory, we had the relation $g_{\text{YM}}^2 = 4\pi g_s$, which can be re-expressed as $\alpha = g_s$, where $\alpha = g_{\text{YM}}^2/4\pi$. The generalization of this to the $\mathcal{N} = 1$ theory described by the warped conifold is

$$\frac{1}{\alpha_1(\mu)} + \frac{1}{\alpha_2(\mu)} = \frac{1}{g_s}. \quad (12.150)$$

The index 1 refers to $SU(M+N)$ and the index 2 refers to $SU(N)$. The dilaton is a constant in the warped conifold geometry, so this implies the constancy of the left-hand side. This can be verified by computing the one-loop beta functions of the gauge theory. The difference of the inverse couplings exhibits the expected logarithmic running

$$\frac{1}{\alpha_1(\mu)} - \frac{1}{\alpha_2(\mu)} = \frac{3M}{\pi}\log(\mu/\Lambda) + \text{const.} \quad (12.151)$$

The coefficient on the right-hand side is easily computed in the gauge theory, and it has been verified in the dual string geometry.

These formulas show that, as μ decreases, which corresponds to decreasing r, $1/\alpha_1$ decreases and $1/\alpha_2$ increases. What happens when $1/\alpha_1 \to 0$? This question needs to be answered both in the framework of the gauge theory and the string theory. There is a beautiful answer in the gauge theory, known as *Seiberg duality*.

To cut a long story short, Seiberg showed that one can continue $\mathcal{N} = 1$ gauge theories of this type across the singularity, provided one replaces the gauge theory by a different one, called the *Seiberg dual*, on the other side! For an $SU(N_c)$ theory with N_c colors and $N_f > N_c$ flavors (meaning chiral superfields in the fundamental representation), the Seiberg dual is an $SU(N_f - N_c)$ gauge theory with N_f flavors.[22] In the present context, $N_c = M + N$ and $N_f = 2N$. Therefore, the $SU(M + N)$ gauge group gets replaced with an $SU(N - M)$ gauge group. Altogether, when the dust settles, one has an $SU(N) \times SU(N - M)$ gauge theory that is isomorphic to the theory one started with in the UV, with N replaced by $N - M$. This process repeats k times as one flows to the infrared so long as $N - kM$ remains positive, and then it ends. For example, if N is an integer multiple of M, the final gauge theory in the IR is $\mathcal{N} = 1$ $SU(M)$ gauge theory with no chiral matter. The renormalization group flow is plotted in Fig. 12.9.

Confinement

$\mathcal{N} = 1$ $SU(M)$ gauge theory with no chiral matter is well known to exhibit confinement and a mass gap. Also, it has a gaugino condensate that breaks the discrete R symmetry $\mathbb{Z}_{2M} \to \mathbb{Z}_2$. So the question arises how these features are manifested in the bulk string theory. The basic mechanism was already hinted at in Chapter 10. The naked singularity in the metric at r_s is removed because the conifold becomes a deformed conifold. Recall that this corresponds to a manifold given by an equation of the form $\sum (w^A)^2 = \varepsilon^2$. The parameter ε is related to r_s by $\varepsilon \sim r_s^{3/2}$. This smooths out the tip of the conifold and cuts off the space-time before one reaches a horizon. In other words, $r = 0$ is no longer part of the space-time.

---

22 There are also some other fields that are not relevant to the present discussion.

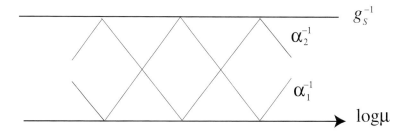

Fig. 12.9. The renormalization group flow of the duality cascade.

Sasaki–Einstein spaces

As stated earlier, Sasaki–Einstein spaces in five dimensions are defined to be Einstein spaces, whose cones are Calabi–Yau three-folds. Such manifolds can also be expressed as circle bundles over a four-dimensional Kähler base. In the case of $T^{1,1}$ the explicit metric in Chapter 10 shows that the base is $S^2 \times S^2$, and the metric has $SU(2) \times SU(2) \times U(1)$ isometry.

In 2004 an infinite family of new Sasaki–Einstein spaces, called $Y^{p,q}$, where p and q are coprime integers, were discovered. All of these spaces are topologically $S^2 \times S^3$, but their isometry is only $SU(2) \times U(1) \times U(1)$. Each of the $Y^{p,q}$ Sasaki–Einstein spaces give rise to a dual $\mathcal{N} = 1$ conformal field theory, all of which have been identified. Moreover, the phenomena discussed above, including a duality cascade when fractional D3-branes are present, occur for each of these theories. In 2005 an even larger set of Sasaki–Einstein spaces, denoted $L^{p,q,r}$, was constructed. These are also topologically $S^3 \times S^2$, but their isometry is further reduced to $U(1) \times U(1) \times U(1)$. One now has a rich set of examples with which to carry out many interesting studies of string theory/gauge theory duality.

EXERCISES

EXERCISE 12.10
Verify the R-symmetry anomaly Eq. (12.149).

SOLUTION

If a classical Lagrangian has a chiral $U(1)$ global symmetry, there can be an anomaly due to one-loop triangle diagrams, which have one $U(1)$ current and two gauge fields attached. Each chiral fermion circulating in the loop gives a contribution to the anomaly that is proportional to its $U(1)$ charge. In the case of a single gauge group G with field strength $F^a_{\mu\nu}$, the formula is

$$\partial_\mu J^\mu = \frac{K}{32\pi^2} F^a_{\mu\nu} \widetilde{F}^{a\mu\nu},$$

where $K = \sum n_m R_m$. Here, n_m is the number of chiral fermions with R-charge R_m. More precisely, in the case of $SU(N)$, each fundamental representation chiral fermion counts as $n_m = 1$ and each adjoint representation chiral fermion counts as $n_m = 2N$.

The chiral superfields A_i and B_i each have $R = 1/2$. The chiral fermions in these multiplets, which are coefficients of θ_α, therefore have R-charge $-1/2$. Similarly, the chiral gluinos have $R = 1$. Thus, in the case of the $SU(M+N)$ the total contribution is

$$K = 4N \cdot (-1/2) + 2(M+N) \cdot 1 = 2M$$

and in the case of $SU(N)$

$$K = 4(M+N) \cdot (-1/2) + 2N \cdot 1 = -2M.$$

There are the required results. □

12.5 Plane-wave space-times and their duals

As was explained earlier, the tree-level approximation to the type IIB superstring theory in an $AdS_5 \times S^5$ background, with N units of R–R flux through the five-sphere, corresponds to the planar approximation to the dual $\mathcal{N} = 4$ super Yang–Mills theory with an $SU(N)$ gauge group. Both sides of this duality, even for the planar/tree-level approximation, are difficult. With great effort, one can compute a few order in λ in the field theory and a few orders in α' in the string theory. However, these are expansions in opposite limits and cannot be compared.

Compared with superstring theory in flat space, there are two severe complications. One is that the background geometry causes the world-sheet theory to be a nonlinear system. Thus, solving classical string theory in this geometry is mathematically the same as solving a complicated interacting two-dimensional quantum field theory. The second difficulty is that the background includes nonzero R–R gauge fields, specifically the self-dual five-form field strength that threads the five-sphere with N units of flux. The RNS formalism is not capable of handling R–R backgrounds, so one is forced to use the GS formalism. This formalism is notoriously difficult to quantize, especially if one wants to keep the symmetries of the geometry manifest.

The type IIB plane-wave

There is a plane-wave limit of $AdS_5 \times S^5$ geometry that can be defined that gives a space-time of intermediate complexity between AdS and flat space-time, which is also maximally supersymmetric. In this geometry it is more difficult to define the duality, because there is not a well-defined dual gauge-theory. Instead, one has to consider limits of appropriately defined families of correlation functions. This can be done analytically, however, and the duality can be subjected to nontrivial tests. Let us discuss the string theory side of the story first.

The geometry is a product of two factors, one for the AdS space and one for the sphere:

$$ds^2(AdS_5) = R^2(-\cosh^2\rho\, dt^2 + d\rho^2 + \sinh^2\rho\, d\Omega_3^2), \quad (12.152)$$

$$ds^2(S^5) = R^2(\cos^2\theta\, d\phi^2 + d\theta^2 + \sin^2\theta\, d\tilde\Omega_3^2). \quad (12.153)$$

The appropriate limit to consider is an example of a type of limit proposed by Penrose and therefore called a *Penrose limit*. The idea is to blow up the neighborhood of a light-like trajectory in the space-time in such a way as to obtain a nontrivial limit. Specifically, we wish to focus on the neighborhood of a point moving around a circumference of the sphere with the speed of light.

In order to implement the desired Penrose limit, let us define new coordinates as follows:

$$r = R\sinh\rho, \quad y = R\sin\theta, \quad (12.154)$$

$$x^+ = t/\mu, \quad x^- = \mu R^2(\phi - t). \quad (12.155)$$

Here, μ is an arbitrary mass scale introduced so that $x^\pm$ have dimensions of length. By rescaling x^+ and x^-, μ could be set equal to one without loss of generality, but this won't be done. The coordinate x^- has period $2\pi\mu R^2$ as a consequence of its dependence on ϕ, and so the conjugate (angular) momentum is $P_- = J/\mu R^2$, where J is an integer. This integer is interpreted in the dual gauge theory as the R charge associated with an arbitrarily chosen $U(1)$ subgroup of the $SU(4)$ R symmetry.

Now consider the infinite-radius limit $R \to \infty$, holding $r, y, x^\pm$ fixed. This gives the plane-wave geometry[23]

$$ds^2 = 2dx^+ dx^- + g_{++}(x^I)(dx^+)^2 + \sum_{I=1}^{8} dx^I dx^I, \quad (12.156)$$

where

$$g_{++}(x^I) = -\mu^2(r^2 + y^2). \quad (12.157)$$

---

[23] This is a special case of a plane wave. The general definition allows $g_{++} = A_{IJ}(x^+)x^I x^J$. The term pp-wave is frequently used.

12.5 Plane-wave space-times and their duals

The radial coordinates r and y are defined by

$$r^2 = \sum_{1}^{4}(x^I)^2 \quad \text{and} \quad y^2 = \sum_{5}^{8}(x^I)^2. \tag{12.158}$$

Note that the limit $\mu \to 0$ gives flat ten-dimensional Minkowski space-time. The dimensionless statement is $\mu \alpha' P_- \to 0$.

As far as the space-time geometry is concerned, there is $SO(8)$ rotational symmetry in the eight transverse directions. However, this symmetry is broken to $SO(4) \times SO(4)$ by the R–R five-form field strength, which (in the limit) has the form

$$F_5 \sim \mu \, dx^+ \wedge (dx^1 \wedge dx^2 \wedge dx^3 \wedge dx^4 + dx^5 \wedge dx^6 \wedge dx^7 \wedge dx^8). \tag{12.159}$$

The limiting solution has just as much symmetry as the original one; it is still a maximally symmetric space-time. In fact, its supergroup of isometries is a Wigner–Inönü contraction of the original $PSU(2,2|4)$ supergroup.

The complicated $AdS_5 \times S^5$ GS world-sheet action, mentioned above, simplifies dramatically in the plane-wave limit, especially if one chooses light-cone gauge $x^+ = P_- \tau$.[24] For this choice, one finds that the action is a free two-dimensional field theory! The only modification of the flat-space light-cone gauge world-sheet action, described in Chapter 5, is that the eight world-sheet bosons x^I and the eight world-sheet fermions S^a are now massive, with mass μ. Thus, the x^I, for example, satisfy a two-dimensional Klein-Gordon equation, rather than a two-dimensional wave equation. To be explicit, the light-cone gauge world-sheet action takes the form

$$S = \frac{1}{2\pi\alpha'} \int d^2\sigma \left(\frac{1}{2}(\dot{x}^2 - x'^2 - \mu^2 x^2) + i\bar{S}(\rho \cdot \partial + \mu \Gamma_*)S \right). \tag{12.160}$$

The matrix $\Gamma_* = \Gamma_1 \Gamma_2 \Gamma_3 \Gamma_4 = \Gamma_5 \Gamma_6 \Gamma_7 \Gamma_8$, where these matrices Γ_8 are $SO(8)$ gamma matrices. Normally, the last product should also contain a factor Γ_9. That is not shown because the chiral spinor S satisfies $\Gamma_9 S = S$. Because of the matrix Γ_*, the fermion mass term, which arises from the coupling to F_5, breaks the symmetry from $SO(8)$ to $SO(4) \times SO(4)$, as expected. This action has as much supersymmetry as in the flat-space $\mu = 0$ limit.

It is easy to generalize the usual decomposition of the motion of a free string in harmonic oscillators to this case. Fourier analysis and quantization give harmonic oscillator operators satisfying

$$[a_m^I, a_n^{J\dagger}] = \delta^{IJ} \delta_{mn} \quad m, n \in \mathbb{Z}, \quad I, J = 1, 2, ..., 8. \tag{12.161}$$

---

[24] Many authors write $P^\pm$ rather that $P_\mp$. These are equivalent in flat space, where the former notation is usually used. However, the latter is more precise, since momenta are naturally one-forms.

The mass terms in the world-sheet action mix left-movers and right-movers. Therefore, it is convenient to allow mode numbers to run over all integers rather than to treat left-movers and right-movers separately. In the limit $\mu \to 0$, left-movers and right-movers would decouple and correspond to positive and negative indices. Note also that the zero modes are described by harmonic oscillators, rather than continuous momenta p_I. This reflects the fact that g_{++} acts like a confining quadratic potential restricting motion into the transverse directions.

The frequency of the nth oscillator is

$$\omega_n = \sqrt{1 + (n/\mu\alpha)^2}, \tag{12.162}$$

where $\alpha = \alpha' P_-$. Then the light-cone Hamiltonian, which describes evolution in τ (and hence x^+) is

$$H_{\ell c} = \mu \sum_{n=-\infty}^{\infty} \sum_{I=1}^{8} \omega_n a_n^{I\dagger} a_n^I + \text{ fermions}. \tag{12.163}$$

The eigenvalues of this Hamiltonian give the allowed values of P_+. The zero-point energies of the bosons and fermions cancel, so no regularization is required.

The Fock space is constrained by

$$\sum_{n=-\infty}^{\infty} \sum_{I=1}^{8} n\, a_n^{I\dagger} a_n^I + \text{ fermions} = 0, \tag{12.164}$$

which generalizes the usual level-matching condition. This constraint arises as a consequence of translation symmetry of the spatial world-sheet coordinate.

The dual gauge theory limit

Let us now consider the implications for the dual gauge theory. The Penrose limit $R \to \infty$ corresponds to $J, N \to \infty$ with finite

$$\lambda' = g_{\text{YM}}^2 N/J^2, \tag{12.165}$$

which is the loop expansion parameter introduced by Berenstein, Maldacena, and Nastase (BMN). By definition, BMN operators are the class of gauge-invariant operators of the gauge theory that survive, with finite anomalous dimension, in the Penrose/BMN limit.

The key duality formula relates the anomalous-dimension operator Δ_a of a BMN operator to the light-cone gauge energy of the corresponding state in the plane-wave string theory

$$\Delta_a = \Delta - J \quad \leftrightarrow \quad P_+ = H_{\ell c}. \tag{12.166}$$

Here, Δ denotes the dimension and J is the $U(1)$ R charge. Note that both of these become infinite in the limit under consideration, but that the difference remains finite for BMN operators. Viewed in terms of global coordinates, so that the dual gauge theory is defined on S^3 rather than R^3, Δ_a can be alternatively interpreted as an energy. For half-BPS states, which correspond to short representations, the anomalous dimension Δ_a vanishes. The BMN operators, by contrast, are not BPS, but they are kept sufficiently close to BPS operators so that the anomalous dimension remains finite in the limit. These operators are characterized by having an R charge J that scales like $N^{1/2}$ in the large N limit. For most operators the limit of Δ_a is infinite. Such operators are presumed to decouple in the Penrose/BMN limit and are therefore not considered.

This duality can be tested perturbatively in three quantities

$$\lambda' \quad \leftrightarrow \quad 1/(\mu\alpha' P_-)^2, \tag{12.167}$$

$$g_2 = J^2/N \quad \leftrightarrow \quad 4\pi g_s (\mu\alpha' P_-)^2, \tag{12.168}$$

$$1/J \quad \leftrightarrow \quad 1/(\mu R^2 P_-). \tag{12.169}$$

In each case, we have written dimensionless gauge-theory quantities on the left and the corresponding string-theory quantities on the right. The λ' expansion is the loop expansion in the gauge theory (for correlation functions of BMN operators), and the g_2 expansion is the loop expansion of the string-theory description.

Since the plane-wave string theory is tractable, it is possible to obtain results that are exact in their λ' dependence. In special cases these results can be reproduced in the dual field theory. Thus, for example, Fock-space states of the form $a_n^{I\dagger} a_{-n}^{J\dagger}|0\rangle$ correspond to certain single-trace *two-impurity* operators in the gauge theory. To leading order in g_2 and $1/J$, but all orders in λ', it has been verified in the gauge theory that, for these operators,

$$\Delta_a = 2\sqrt{1 + \lambda' n^2} \tag{12.170}$$

in agreement with expectations based on Eqs (12.162) and (12.163).

Some of the first-order corrections in g_2 and $1/J$ have also been examined, and agreement with the duality predictions has been found. The g_2 corrections are obtained by using the vertex operator of the light-cone gauge string field theory. (Its construction is a long story that we won't pursue here.) The $1/J$ corrections are obtained by keeping track of the leading $1/R^2$ corrections to the Penrose limit. This is straightforward, in principle, but rather complicated in practice.

The M-theory plane-wave duality

Let us now mention the corresponding results for M-theory. The $AdS_4 \times S^7$ and the $AdS_7 \times S^4$ solutions have identical Penrose limits. This background turns out to be given by

$$ds^2 = 2dx^+ dx^- + g_{++}(x^I)(dx^+)^2 + \sum_{I=1}^{9} dx^I dx^I, \qquad (12.171)$$

where

$$g_{++}(x^I) = -\mu^2((r_3/3)^2 + (r_6/6)^2). \qquad (12.172)$$

The coordinate r_3 is the radial coordinate for three of the x^Is and r_6 is the radial coordinate for the other six of them. The transverse symmetry in this case is $SO(3) \times SO(6)$. The M theory four-form field strength takes the form

$$F_4 \sim \mu\, dx^+ \wedge dx^1 \wedge dx^2 \wedge dx^3. \qquad (12.173)$$

The dual gauge theory in this case is a version of Matrix theory. It is a massive deformation of the original Matrix-theory proposal for a dual description of M-theory in flat 11-dimensional space-time, which was discussed in Section 12.2.

EXERCISES

EXERCISE 12.11
Starting from the light-cone gauge action in Eq. (12.160), generalize the analysis given in Chapter 2 to derive the mode expansions of the bosonic fields and the quantization conditions. Also, derive the corresponding formulas for the fermions.

Solution

Varying the action gives the equation of motion

$$(-\partial_\tau^2 + \partial_\sigma^2)X^I - X^I = 0,$$

where we have set $\mu = 1$ for simplicity. Thus, the mode expansion of X^I can be written in the form

$$X^I(\sigma, \tau) = i\sum_{n=-\infty}^{\infty} \frac{1}{\sqrt{2\omega_n P_-}}\left(e^{-i\omega_n\tau}a_n^I - e^{i\omega_n\tau}a_n^{\dagger I}\right)e^{-ik_n\sigma},$$

where

$$\omega_n = \sqrt{k_n^2 + 1} \quad \text{and} \quad k_n = \frac{n}{\alpha' P_-}.$$

The canonical quantization condition

$$[X^I(\sigma), P^J(\sigma')] = i\delta^{IJ}\delta(\sigma - \sigma'),$$

where

$$P^I(\sigma, \tau) = \dot{X}^I(\sigma, \tau)/(2\pi\alpha'),$$

gives the standard bosonic oscillator commutation relations in Eq. (12.161).

Similarly, for the fermions the equation of motion is

$$i(\dot{S} + S'^\dagger) + \Gamma_* S = 0.$$

We have

$$S^a(\sigma, \tau) = \sum_{n=-\infty}^{\infty} \frac{1}{\sqrt{4\omega_n P_-}}\left([\Gamma_* + \omega_n - k_n]S_n^a e^{-i\omega_n\tau}\right.$$

$$\left. + [1 - (\omega_n - k_n)\Gamma_*]S_n^{\dagger a}e^{i\omega_n\tau}\right)e^{-ik_n\sigma}.$$

Using the canonical quantization condition

$$\{S^a(\sigma, \tau), S^{b\dagger}(\sigma', \tau)\} = 2\pi\alpha'\delta(\sigma - \sigma'),$$

one obtains standard fermionic oscillator anticommutation relations

$$\{S_m^a, S_n^{b\dagger}\} = \delta_{m,n}\delta^{ab} \qquad m, n \in \mathbb{Z}.$$

□

12.6 Geometric transitions

The gauge theory that is dual to the flux model involving the type IIB superstring theory on the deformed conifold was discussed earlier in this chapter. It was emphasized that many other models describing supergravity/gauge theory duals have been constructed. Some of these dual descriptions can be obtained by analysis of a *geometric transition*.

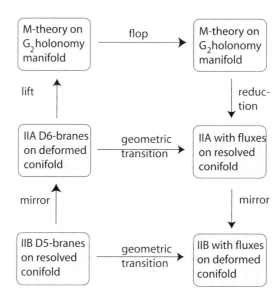

Fig. 12.10. Geometric transitions and flops in M-theory.

The basic idea of a geometric transition is that a gauge theory describing an open-string sector, that is, a gauge theory on D-branes, is dual to a flux compactification of a particular string theory in which no D-branes are present, but fluxes are present instead. In other words, as a modulus is varied, there is a transition connecting the two descriptions. Many quantities in the gauge theory, in particular the superpotential, can be computed in terms of fluxes integrated over suitable cycles. Some of the models that are related in terms of such a transition are displayed in Fig. 12.10. Let us now explain the basic idea.

Recall that, in Chapter 9, the conifold was presented as an example of a noncompact Calabi–Yau manifold that is described as a cone with an $S^2 \times S^3$ base. Two methods to blow up the singularity at the tip of the cone ($r = 0$) were discussed. They give the deformed conifold and the resolved conifold. In the former case the S^3 is blown up, while in the later case the S^2 is blown up, so that the resulting geometry is nonsingular. These geometries can be

12.6 Geometric transitions

related in terms of a conifold transition, in which the S^3 shrinks to zero size and the S^2 is blown up. Both geometries play a role in the context of geometric transitions, which link a series of gauge theory/supergravity models. These dualities can be checked by computing the corresponding topological string partition functions.[25] This is beyond the scope of this book, so here we settle for a description of the result.

The first geometric transition described in the lower two boxes of Fig. 12.10 is in the context of the type IIB theory. The precise statement is that the gauge theory resulting from a system of D5-branes wrapping the S^2 of the resolved conifold undergoes a geometric transition in the strongly coupled limit in which the S^2 shrinks to zero size, but the theory avoids the singularity as an S^3 is blowing up. The resulting model has no D-branes but fluxes that thread through the S^3. Since two different geometries are related in this process, the term geometric transition is used to describe it.

This type IIB process has a type IIA mirror dual, which is illustrated in the two boxes appearing in the middle of the figure. In the mirror picture a system of D6-branes wrapping the S^3 of the deformed conifold undergoes a geometric transition in which the S^3 shrinks to zero size and an S^2 is blowing up. The resulting closed-string theory has fluxes threading through the S^2 but no branes. Again, there is a duality between an open-string sector containing D-branes and a closed-string sector containing no branes.

An interesting result emerges once this type IIA theory is lifted to M-theory. In this process the background becomes a G_2-holonomy manifold, described in terms of a cone with an $S_3 \times \tilde{S}_3$ base, as the S^2 of the deformed conifold is lifted to an S^3. The 3-sphere S^3 has finite size while $\tilde{S}_3$ has vanishing size at the tip of the cone. A so-called *flop* interchanges S_3 and $\tilde{S}_3$ leading to M-theory compactified on another G_2-holonomy manifold. This provides an alternative (geometrical) description of the type II geometric transition, in terms of a flop in M-theory. Using dimensional reduction and mirror symmetry, the complete duality chain presented in Fig. 12.10 can be understood.

Geometric transitions provide a beautiful relation between gauge theories and flux compactifications, similar in spirit to that discussed for the type IIB theory on the deformed conifold in the previous section. Through a series of string dualities several different theories can be related in terms of the duality chain presented in Fig. 12.10. Using further string dualities, geometric transitions also can be discussed in the context of the heterotic string theory.

---

25 An alternative derivation can be obtained from the explicit string-theory backgrounds.

Homework Problems

Problem 12.1
Verify that Eqs (12.4) and (12.5) give a solution of the Killing spinor equation (12.3).

Problem 12.2
Derive the temperature and entropy per unit p-volume of the nonextremal black p-brane solution in Eq. (12.31).

Problem 12.3
Derive Eq. (12.41) for the relation between the entropy of a black string versus an array of black holes discussed in Section 12.1 in the context of the Gregory–Laflamme instability.

Problem 12.4
The superalgebra $SU(M|N)$ can be represented by supermatrices, as sketched at the end of Exercise 12.5. They can be written in block form

$$X = \begin{pmatrix} A & B \\ C & D \end{pmatrix},$$

where A is $M \times M$ hermitian, D is $N \times N$ hermitian and $B = C^\dagger$ is $M \times N$ fermionic. Also, the supertrace vanishes: $\mathrm{Str} X = \mathrm{tr} A - \mathrm{tr} D = 0$. Show that commutation of these matrices defines a closed superalgebra that satisfies the super Jacobi identities. Explain why a $U(1)$ factor decouples for $M = N$.

Problem 12.5
Consider $SU(N)$ super Yang–Mills theory for $D = 5$ with 16 supercharges. Determine the global R symmetry and the field content of this theory.

Problem 12.6
Derive equation Eq. (12.101) from the three previous equations.

Problem 12.7
Verify that the identification of coordinates in (12.99) relates the Poincaré patch metric to the global AdS metric.

PROBLEM 12.8
Derive the field equation following from the action Eq. (12.115). Show that the solution for this equation is given in terms of Bessel functions whose asymptotic behavior at $z = 0$ is of the form $\phi \sim z^\alpha$ with α given by Eq. (12.117).

PROBLEM 12.9
Consider the Born–Infeld action for a single probe D3-brane in an $AdS_5 \times S^5$ background. Show that, when the metric is expressed in terms of a radial coordinate $u = r/\alpha'$, all α' dependence cancels. What do you think is the significance of this result?

PROBLEM 12.10
The volume ratio in the formula for the AdS radius in Eq. (12.143) is determined in the dual gauge theory by certain gravitational anomalies that are analogous to the R-symmetry anomalies considered in the text. In the case of the conifold, the field theory analysis predicts that $\mathrm{Vol}(T^{1,1}) = \frac{16}{27}\mathrm{Vol}(S^5)$. By computing $\mathrm{Vol}(T^{1,1})$ show that this is correct.

PROBLEM 12.11
Consider an A_{n-1} singularity obtained by modding $\mathbb{C}^2$ by $\mathbb{Z}_n$

$$(z_1, z_2) \to (\omega z_1, \omega^{-1} z_2),$$

where $\omega = \exp(2\pi i/n)$ is an nth root of unity. Consider Nn D5-branes in the type IIB theory placed at $(z_1, z_2) = (0, 0)$ and consider the action of $\mathbb{Z}_n$ to correspond to permuting the Nn branes arranged in n groups of N D5-branes. Using orbifold techniques, determine which gauge theory lives on the branes.

PROBLEM 12.12
Consider the type IIB theory on a A_2 singularity times a T^2. Describe what the 1/4 BPS states of the $SU(3)$ $\mathcal{N} = 4$, $D = 4$ Yang–Mills theory correspond in this picture. Hint: the three-string junction introduced in Chapter 8 is relevant.

PROBLEM 12.13
Verify that the M2-brane solution in Section 12.1 is a solution of the field equations if H satisfies the eight-dimensional Laplace equation and the warp factor is that in Eq. (12.6). Repeat your calculation for the M5-brane described in Section 12.1.

PROBLEM **12.14**

Using Eqs (12.55) to (12.60) show that the explicit form of the action for Matrix theory takes the form of a sum

$$\mathcal{L} = \mathcal{L}_Y + \mathcal{L}_A + \mathcal{L}_\mathcal{G} + \mathcal{L}_{\text{fermi}},$$

as stated in Eq. (12.61). The action for the fluctuations Y is

$$\mathcal{S}_Y = i \int d\tau \left(\tfrac{1}{2} Y_1^i (\partial_\tau^2 - r^2) Y_1^i + \tfrac{1}{2} Y_2^i (\partial_\tau^2 - r^2) Y_2^i + \tfrac{1}{2} Y_3^i \partial_\tau^2 Y_3^i \right.$$

$$\left. - \sqrt{g} \epsilon^{a3d} \epsilon^{cbd} B_3^i Y_a^j Y_b^i Y_c^j - \tfrac{g}{4} \epsilon^{abe} \epsilon^{cde} Y_a^i Y_b^j Y_c^i Y_d^j \right).$$

The action for the gauge field is

$$\mathcal{S}_A = i \int d\tau \left(\tfrac{1}{2} A_1 (\partial_\tau^2 - r^2) A_1 + \tfrac{1}{2} A_2 (\partial_\tau^2 - r^2) A_2 + \tfrac{1}{2} A_3 \partial_\tau^2 A_3 \right.$$

$$+ 2 \epsilon^{ab3} \partial_\tau B_3^i A_a Y_b^i + \sqrt{g} \epsilon^{abc} \partial_\tau Y_a^i A_b Y_c^i$$

$$\left. - \sqrt{g} \epsilon^{a3d} \epsilon^{bcd} B_3^i A_a A_b Y_c^i - \tfrac{g}{2} \epsilon^{abe} \epsilon^{cde} A_a Y_b^i A_c Y_d^i \right).$$

The action for the ghost fields is

$$\mathcal{S}_\mathcal{G} = i \int d\tau \left(C_1^* (-\partial_\tau^2 + r^2) C_1 + C_2^* (-\partial_\tau^2 + r^2) C_2 - C_3^* \partial_\tau^2 C_3 \right.$$

$$\left. + \sqrt{g} \epsilon^{abc} \partial_\tau C_a^* C_b A_c - \sqrt{g} \epsilon^{a3d} \epsilon^{cbd} B_3^i C_a^* C_b Y_c^i \right).$$

Finally the action for the fermionic fields is

$$\mathcal{S}_{\text{fermi}} = i \int d\tau \left(\psi_+^T (\partial_\tau - v\tau\gamma_1 - b\gamma_2) \psi_- + \sqrt{\tfrac{g}{2}} (Y_1^i - i Y_2^i) \psi_+^T \gamma^i \psi_3 \right.$$

$$+ \tfrac{1}{2} \psi_3^T \partial_\tau \psi_3 + \sqrt{\tfrac{g}{2}} (Y_1^i + i Y_2^i) \psi_3^T \gamma^i \psi_- - i \sqrt{\tfrac{g}{2}} (A_1 - i A_2) \psi_+^T \psi_3$$

$$\left. + i \sqrt{\tfrac{g}{2}} (A_1 + i A_2) \psi_-^T \psi_3 - \sqrt{g} Y_3^i \psi_+^T \gamma^i \psi_- + i \sqrt{g} A_3 \psi_+^T \psi_- \right).$$

Derive the explicit form of $\mathcal{S}_Y$, $\mathcal{S}_A$ and $\mathcal{S}_{\text{fermi}}$. In this last action new fermionic fields were introduced

$$\psi_+ = \frac{1}{\sqrt{2}} (\psi_1 + i\psi_2) \qquad \psi_- = \frac{1}{\sqrt{2}} (\psi_1 - i\psi_2).$$

Hint: once you decompose the fields of the theory in terms of Pauli matrices according to Eq. (12.60), use the following decomposition the gamma matrices appearing in the action Eq. (12.55)

$$\Gamma^0 = \sigma^3 \otimes \mathbb{1}_{16 \times 16}, \qquad \Gamma^i = i\sigma^1 \otimes \gamma^i,$$

where σ^i are Pauli matrices and γ^i are real and symmetric.

PROBLEM **12.15**
Derive the equations of motion that follow from the Dp-brane action in Eq. (12.18), and show that these equations are solved by Eqs (12.19), (12.20), (12.21) and the flux given by Eq. (12.22).

PROBLEM **12.16**
Generalize Problem 12.15 to show that the nonextremal Dp-branes described by Eq. (12.31) are nonsupersymmetric solutions to the equations of motion following from the action Eq. (12.18).

PROBLEM **12.17**
Verify that the leading term in the large v large r expansion for Matrix theory at one-loop is given by Eq. (12.70). Also, verify the numerical coefficient.

PROBLEM **12.18**
Show that the fermionic propagator in Eq.(12.66) is a solution to Eq.(12.65).

Bibliographic discussion

In the following we briefly indicate some of the main references for each of the chapters other than the introductory one. Many more works are listed at the end than are mentioned in the discussion. The subject is so vast that it is impossible to include every important contribution. We apologize in advance for any omissions. Additional bibliographic discussions and references can be found in the previous books Green, Schwarz and Witten (1987), referred to as GSW, Polchinski (1998) and Zwiebach (2004). In addition to these, other previous string theory books include Kaku (1988, 1991, 1999, 2000), Polyakov (1987b), Lüst and Theisen (1989), Kiritsis (1998), Johnson (2003), Hori *et al.* (2003), Douglas *et al.* (2004) and Szabo (2004).

Most contributions since 1991 have been posted on the eprint archives. Thus, for example, an article whose listing includes the information *E-print hep-th/9612080* can be found on the internet at http://arxiv.org/abs/hep-th/9612080. This gives a page containing the abstract of the article as well as links to PostScript and PDF versions of the entire manuscript.

Chapter 2

While there is an important prehistory that set the scene, discussed in GSW, string theory begins with the discovery of a four-particle scattering amplitude for open strings in Veneziano (1968). This was rapidly generalized to multiparticle amplitudes and closed-string amplitudes. The recognition that these amplitudes actually describe one-dimensional extended objects (strings) was made independently in Nambu (1970a), Nielsen (1970) and Susskind (1970).

The formula for the string action as the area of the world sheet was introduced independently in Nambu (1970b), Goto (1971) and Hara (1971). The harmonic-oscillator operator description of the string spectrum and amplitude was introduced in Fubini, Gordon and Veneziano (1969) and developed in Fubini and Veneziano (1969, 1970, 1971). The Virasoro constraints first appear in Virasoro (1970). The central extension (or conformal anomaly) in the Virasoro algebra was discovered by J. Weis (unpublished). The first indication of the critical dimension $D = 26$ was obtained in Lovelace (1971). Two different proofs of the no-ghost theorem were presented in Brower (1972) and Goddard and Thorn (1972). The latter is the one described in the text. The interpretation of the relation between the critical dimension and the mass of the ground state in terms of zero-point fluctuations was given in Brink and Nielsen (1973).

Light-cone gauge quantization of the Nambu–Goto action was worked out in Goddard, Goldstone, Rebbi and Thorn (1973). The string sigma-model action with an auxiliary two-dimensional world-sheet metric tensor was constructed independently in Brink, Di Vecchia and Howe (1976) and Deser and Zumino (1976b). In fact, they also presented the generalization to the RNS string of Chapter 4.

Review articles describing the developments in the early 1970s are Alessandrini, Amati, Le Bellac and Olive (1971), Schwarz (1973), Veneziano (1974), Rebbi (1974), Mandelstam (1974) and Scherk (1975). The first five of these are reprinted in Jacob (1974), and the last one (Scherk) is reprinted in Schwarz (1985).

Chapter 3

The modern path-integral treatment of string theory was initiated for the bosonic string in Polyakov (1981a) and for the RNS string in Polyakov (1981b). This led to an appreciation of the importance of conformal symmetry and the significance of the conformal anomaly. The Polyakov approach was developed in Friedan (1984) and Alvarez (1983).

Important original papers developing the techniques of two-dimensional conformal field theory include Belavin, Polyakov and Zamolodchikov (1984) and Friedan, Qiu and Shenker (1984). Minimal models, in particular, first appear in these papers. The construction of conformal field theories associated with Lie groups was developed in Witten (1983, 1984), while the coset construction given in the text is based on Goddard, Kent and Olive (1985). Useful reviews of two-dimensional conformal field theory include Goddard and Olive (1986), Moore and Seiberg (1989), Lüst and Theisen (1989) and Ginsparg (1991).

The BRST symmetry of Becchi, Rouet and Stora (1974, 1976) and Tyutin (1975) was first applied to string theory in Kato and Ogawa (1983).

The calculation of beta functions of two-dimensional sigma models was explained in Alvarez-Gaumé, Freedman and Mukhi (1981) and Friedan (1985). This was applied to the string world-sheet action in the presence of background fields in Callan, Friedan, Martinec and Perry (1985). The subject is reviewed in Callan and Thorlacius (1989) and Tseytlin (1989). The linear dilaton theory is discussed in Chodos and Thorn (1974) and Myers (1987).

Witten's open-string field theory is presented in Witten (1986).

Chapter 4

The RNS model originated with the construction of a free wave equation for fermionic strings in Ramond (1971) and the discovery of the interacting bosonic sector in Neveu and Schwarz (1971). The formalism was developed further in Neveu, Schwarz and Thorn (1971) clarifying how the super-Virasoro constraints are implemented. The no-ghost theorem was proved in Goddard and Thorn (1972), Schwarz (1972) and Brower and Friedman (1973).

The global world-sheet supersymmetry of the gauge-fixed RNS model was first explained in Gervais and Sakita (1971). This supersymmetric theory was understood at about the same time as the discovery of the four-dimensional super-Poincaré algebra in Gol'fand and Likhtman (1971). Moreover, the Gervais–Sakita work motivated the construction of supersymmetric theories in four dimensions in Wess and Zumino (1974). Two-dimensional superspace was introduced in Fairlie and Martin (1973) and Montonen (1974), while four-dimensional superspace first appears in Salam and Strathdee (1974). Following the development of $\mathcal{N} = 1$ supergravity in four dimensions in Freedman, van Nieuwenhuizen and Ferrara (1976) and Deser and Zumino (1976a), a locally supersymmetric world-sheet action was constructed in Brink, Di Vecchia and Howe (1976) and Deser and Zumino (1976b). This action was utilized in Polyakov (1981b).

Gliozzi, Scherk and Olive (1976, 1977) discovered that, when the ten-dimensional RNS string spectrum is projected in the manner described in the text, the number of bosons and fermions agrees at every mass level, as is required for unbroken space-time supersymmetry. They also constructed ten-dimensional super Yang–Mills theory (as well as its various dimensional reductions and truncations), as did Brink, Schwarz and Scherk (1977).

The application of the BRST formalism to the construction of the fermion emission vertex operator was developed in Friedan, Martinec and Shenker (1986) and Knizhnik (1985). This was not described in this book.

Chapter 5

The formalism with manifest space-time supersymmetry was developed by Green and Schwarz in the period 1979–84. The light-cone gauge formalism was found first and utilized to prove the supersymmetry of the GSO projected theory. In particular, the type I, type IIA and type IIB superstring theories were identified and named. The spectra of these theories were analyzed and various amplitudes were computed in Green and Schwarz (1981a, 1981b, 1982). This work is reviewed in Schwarz (1982b) and Green (1984).

Brink and Schwarz (1981) found a covariant and supersymmetric action for a massless superparticle. This corresponds to the massless limit of the D0-brane action described in the text. Following the observation that this action possesses local kappa symmetry in Siegel (1983), Green and Schwarz (1984a) constructed the covariant superstring action with local kappa symmetry. The light-cone gauge results can be obtained by gauge-fixing this action, but covariant quantization of the GS action has proved elusive.

The history of anomalies in gauge theories is discussed in GSW. Gravitational anomalies in arbitrary dimensions were first systematically investigated in Alvarez–Gaumé and Witten (1984). In particular, it was proved that the gravitational anomalies cancel in type IIB supergravity and hence in type IIB superstring theory. Following this, Green and Schwarz (1985) computed the hexagon diagram contribution to the gauge anomaly in type I superstring theory and found that the cylinder and Möbius strip contributions cancel for the gauge group $SO(32)$. Using the results of Alvarez–Gaumé and Witten (1984), Green and Schwarz (1984b) found that all gauge and gravitational anomalies could cancel provided the gauge group is either $SO(32)$ or $E_8 \times E_8$. The analysis presented in the text is somewhat simpler than in the original paper, because it utilizes techniques developed later in Morales, Scrucca and Serone (1999), Stefanski (1999) and Schwarz and Witten (2001). Harvey (2005) reviews the subject of anomalies.

Chapter 6

T-duality symmetry is manifest in formulas given in Green, Schwarz and Brink (1982), but it was first discussed explicitly in Kikkawa and Yamasaki (1984). The T-duality transformations of constant background fields were derived in Buscher (1987, 1988). T-duality was reviewed in Giveon, Porrati and Rabinovici (1994) and Alvarez, Alvarez-Gaumé and Lozano (1995).

The subject of D-branes originated in works of Dai, Leigh and Polchinski (1989) and Leigh (1989). However, it did not achieve prominence until Polchinski (1995) pointed out that D-branes in superstring theories carry R-R charges. Some results were anticipated in Shenker (1991). Other insights into D-brane physics were provided in Bachas (1996) and Douglas, Kabat, Pouliot and Shenker (1997). The subject of D-branes was reviewed in Polchinski (1997). Johnson (2003) is a book about D-branes. For reviews of the properties of non-BPS D-branes see Sen (1999) and Schwarz (2001).

Chan–Paton charges were introduced to describe $U(N)$ symmetry in the very early days of string theory in Paton and Chan (1969), but it took more than another quarter century until Witten (1996a) pointed out that these rules describe coincident D-branes. In the interim Neveu and Scherk (1972) noted that the Chan–Paton symmetry is a local gauge symmetry. The Chan–Paton construction was generalized to orthogonal and symplectic groups in Schwarz (1982a) and Marcus and Sagnotti (1982).

The generalization of the Dirac quantization condition to p-branes was discovered independently in Nepomechie (1985) and Teitelboim (1986a, 1986b). The fact that D-brane charges should be understood mathematically as K-theory classes was pointed out in Minasian and Moore (1997) and elucidated in Witten (1998c) and Hořava (1999). Sen (1998c, 1998d, 1999) also contributed important insights.

The appearance of the Born–Infeld action as an effective action in string theory is due to Fradkin and Tseytlin (1985a, 1985b, 1985c). It was extended to superstrings in Callan, Lovelace, Nappi and Yost (1997, 1998). Kappa-symmetric D-brane actions were constructed by several groups: Cederwall, von Gussich, Nilsson and Westerberg (1997), Aganagic, Popescu and Schwarz (1997a, 1997b), Bergshoeff and Townsend (1997) and Cederwall, von Gussich, Nilsson, Sundell and Westerberg (1997). The study of non-abelian Born–Infeld theory in string theory was pioneered in Tseytlin (1997). The work on the nonabelian world-volume theory of coincident D-branes, and the discovery of the Myers effect, is contained in Myers (1999). Reviews of Born–Infeld theory and brane dynamics include Giveon and Kutasov (1999) and Tseytlin (2000).

Chapter 7

Shortly after Green and Schwarz (1984b) showed that an anomaly-free supersymmetric theory in ten dimensions could have $SO(32)$ or $E_8 \times E_8$

gauge symmetry, the heterotic string theory was constructed in Gross, Harvey, Martinec and Rohm (1985a, 1985b, 1986). They presented both the fermionic and the bosonic constructions. Some of the mathematical background required for the bosonic construction had been explained previously for physicists in Goddard and Olive (1985).

Toroidal compactification of the heterotic string was first studied in Narain (1986). The associated moduli space, parametrized by constant background fields, was identified in Narain, Sarmadi and Witten (1987). This was described in terms of a low-energy effective action in Maharana and Schwarz (1993).

Chapter 8

The action for eleven-dimensional supergravity was constructed in Cremmer, Julia and Scherk (1978). The type IIA supergravity action is obtained by dimensional reduction of eleven-dimensional supergravity. The formulas given in the text differ somewhat from those in the literature. The effective action for type I supergravity coupled to super Yang–Mills theory was constructed in Bergshoeff, de Roo, de Wit and Van Nieuwenhuizen (1982) and Chapline and Manton (1983). This was supplemented by some higher-dimension terms required for anomaly cancellation in Green and Schwarz (1984b). Type IIB supergravity was constructed in Schwarz and West (1983), Schwarz (1983) and Howe and West (1984). The heterotic string effective action and its S-duality relationship to the type I supergravity action was given in Witten (1995).

Electric-magnetic duality symmetry in Yang–Mills theory was first proposed in Montonen and Olive (1977). The conjecture was sharpened to $\mathcal{N} = 4$ theories in Osborn (1979). The subject is reviewed in Harvey (1997).

The notion of S-duality in string theory was first conjectured in Font, Ibañez, Lüst and Quevedo (1990) for the heterotic string compactified on T^6. This was pursued in subsequent years in Sen (1994a, 1994b) and Schwarz (1993). The duality was explained for $\mathcal{N} = 2$ gauge theories in Seiberg and Witten (1994a, 1994b). Hull and Townsend (1995) proposed the $SL(2,\mathbb{Z})$ S-duality of type IIB superstring theory as well as the $E_n(\mathbb{Z})$ U-duality generalizations. Evidence for the S-duality relationship between type I superstring theory and the $SO(32)$ heterotic string theory was given in Polchinski and Witten (1996).

The proposal that type IIA superstring theory becomes 11-dimensional at strong coupling was made in Townsend (1995) and Witten (1995). This relationship had been hinted at in Duff, Howe, Inami and Stelle (1987), which related the 11-dimensional supermembrane of Bergshoeff, Sezgin and Townsend (1987, 1988) to the ten-dimensional type IIA GS string. The term *M-theory* was introduced by Witten in 1995 lectures given at the IAS. The 11-dimensional interpretation of the strongly coupled $E_8 \times E_8$ heterotic string is due to Hořava and Witten (1996a, 1996b).

The use of the duality between M-theory on T^2 and type IIB superstring theory on S^1 as a way of understanding the S-duality of the type IIB theory was given in Aspinwall (1996) and Schwarz (1996a, 1996b). The existence of an infinite $SL(2, \mathbb{Z})$ multiplet of type IIB strings is pointed out in Schwarz (1995). An interpretation as bound states was given in Witten (1996a). Review articles discussing M-theory and superstring dualities include Townsend (1996b), Schwarz (1997), Vafa (1997), Sen (1998b) and Obers and Pioline (1999).

Chapter 9

Kaluza (1921) and Klein (1926) proposed unifying electromagnetism and Einstein's theory of gravity in four dimensions by compactifying five-dimensional Einstein gravity on a circle. The generalization and application of this idea to 11-dimensional supergravity was an active subject in the early 1980s. Reviews of Kaluza–Klein supergravity include Duff, Nilsson and Pope (1986), Townsend (1996b) and Overduin and Wesson (1997).

Compactification of string theory with an internal six-dimensional Calabi–Yau manifold was given in Candelas, Horowitz, Strominger and Witten (1985). The literature on Calabi–Yau manifolds is very large. Some basics appears in chapter 16 of GSW and Candelas (1987). A more elaborate discussion appears in Hübsch (1992). An advanced and detailed description is given in Hori *et al.* (2003). Orbifolds were first introduced in Dixon, Harvey, Vafa and Witten (1985, 1986), and their CFT description was developed in Dixon, Friedan, Martinec and Shenker (1987). Reviews of special holonomy manifolds include Joyce (2000), Acharya and Gukov (2004) and Gubser (2004).

The local constraints imposed by $\mathcal{N} = 2$, $D = 4$ supersymmetry were derived in special coordinates in De Wit, Lauwers and van Proeyen (1985). A global description of special geometry was developed in Strominger (1990). The form of the prepotential and the geometry of the moduli space of Calabi–Yau manifolds were derived in Candelas, De la Ossa, Green and Parkes (1991) using mirror symmetry. In the same paper, it was shown that conifold singularities appear in the moduli space of classical string vacua. The first evidence of mirror symmetry was found in Candelas, Lynker and Schimmrigk

(1990) and Greene and Plesser (1990). Strominger, Yau and Zaslow (1996) interpreted mirror symmetry in terms of T-duality.

Strominger (1995) showed that massless black holes coming from branes wrapped around the supersymmetric cycles introduced in Becker, Becker and Strominger (1995) give nonperturbative corrections to the low-energy effective action, and that the singularity pointed out in Becker, Becker and Strominger (1995) is lifted. This was explored further in Greene, Morrison and Strominger (1995).

The duality between M-theory on K3 and the heterotic string on T^3 was one of many dualities introduced in Witten (1995). F-theory was introduced in Vafa (1996) following related studies of cosmic strings in Greene, Shapere, Vafa and Yau (1990).

Chapter 10

Flux compactifications were introduced in Strominger (1986) and De Wit, Smith and Hari Dass (1987) as a generalization of conventional Calabi–Yau compactifications. Such compactifications include a warp factor, so that the ten-dimensional metric is no longer a direct product of the external and internal space-time. No-go theorems implied that in most cases such theories reduce to ordinary Calabi–Yau compactifications. However, with the development of nonperturbative string theory and M-theory, it became evident that the no-go theorems could be circumvented. Flux compactifications were first studied in the context of M-theory in Becker and Becker (1996) and in the context of F-theory in Dasgupta, Rajesh and Sethi (1999). Giddings, Kachru and Polchinski (2002) explained how flux compactifications can give a large hierarchy of scales. Graña (2006) reviews flux compactifications.

Gukov, Vafa and Witten (2001) made it evident that flux compactifications can lead to a solution of the moduli-space problem, since a nonvanishing potential for the moduli fields is generated. This led to the introduction of the string theory landscape, which describes a huge number of possible string theory vacua, in Susskind (2003). Their properties were analyzed in Douglas (2003) using statistical methods. Flux compactifications are dual supergravity descriptions of confining gauge theories, as was pointed out in Klebanov and Strassler (2000) and Polchinski and Strassler (2000). The idea that a brane-world scenario provides an alternative to compactification was introduced in Randall and Sundrum (1999b).

The application of flux compactifications to cosmology is an active area of research. Kachru, Kallosh, Linde and Trivedi (2003) discussed the construction of long-lived metastable de Sitter vacua, and Kachru, Kallosh, Linde,

Maldacena, McAllister and Trivedi (2003) discussed the application to inflation. Review articles on string cosmology include Linde (1999), Quevedo (2002) and Danielsson (2005).

Chapter 11

The general relativity textbooks Wald (1984) and Carroll (2004) provide useful background. Bekenstein (1973) proposed that the entropy of black holes should be proportional to the area of the event horizon. The discovery of black-hole radiation in Hawking (1975) confirmed previous indications in Bardeen, Carter and Hawking (1973) of the thermodynamic behavior of black holes. The information loss problem, which implies a possible breakdown of quantum mechanics, was pointed out in Hawking (1976). A statistical derivation of black-hole entropy using string theory techniques was given first in Strominger and Vafa (1996). Review articles describing this and other aspects of black holes in string theory are Sen (1998b), Maldacena (1998a), Peet (2001) and Mathur (2006).

The attractor mechanism was introduced in Ferrara, Kallosh and Strominger (1995). Our presentation follows Denef (2000). Black-ring solutions were found in Emparan and Reall (2002). The conjecture relating microscopic degeneracies to the topological string was proposed in Ooguri, Strominger and Vafa (2004). The subject is reviewed in Pioline (2006). Our discussion of microscopic black holes follows Dabholkar, Denef, Moore and Pioline (2005).

Chapter 12

Black p-brane solutions were constructed in Horowitz and Strominger (1996). Relevant reviews include Townsend (1996b), Duff (1996) and Stelle (1998).

Matrix theory was introduced in Banks, Fischler, Shenker and Susskind (1997). The discrete light-cone quantization interpretation for finite N was proposed in Susskind (1997). Reviews of matrix theory include Bigatti and Susskind (1997), Taylor (1998), Banks (1998) and Bilal (1999). An explanation of why Matrix theory is correct was given in Seiberg (1997) and Sen (1998a). Matrix string theory was formulated in Dijkgraaf, Verlinde and Verlinde (1997). Berenstein, Maldacena and Nastase (2002) gave a generalization of matrix theory that describes M-theory in a plane-wave background.

The large-N expansion of $U(N)$ gauge theory was given in 't Hooft (1974). The AdS/CFT correspondence was spelled out in the landmark paper Maldacena (1998). There had been earlier hints of such a connection in Maldacena and Strominger (1997a, 1997b) and Douglas, Polchinski and Strominger (1997). Some aspects of AdS/CFT also appear in Klebanov (1997), Gubser, Klebanov and Tseytlin (1997) and Gubser and Klebanov (1997). Important details were elucidated in Gubser, Klebanov and Polyakov (1998) and Witten (1998b). A detailed review of the AdS/CFT correspondence and related topics was given in Aharony, Gubser, Maldacena, Ooguri and Oz (2000). Some recent developments, not discussed in the text, include Kazakov, Marshakov, Minahan and Zarembo (2004), Lin, Lunin and Maldacena (2004), Beisert and Staudacher (2005) and Hofman and Maldacena (2006).

The field theory dual of type IIB superstring theory on $AdS_5 \times T^{1,1}$, that is, the conifold, was identified in Klebanov and Witten (1998). The duality cascade associated with the addition of fractional branes was explained in Klebanov and Strassler (2000) building on the earlier works Polchinski and Strassler (2000), Klebanov and Nekrasov (2000) and Klebanov and Tseytlin (2000).

Blau, Figueroa-O'Farrill, Hull and Papadopoulos (2002a) discovered that type IIB superstring theory admits a maximally supersymmetric plane-wave solution. Metsaev (2002) showed the world-sheet action for this background becomes a free theory in the light-cone GS formalism. The plane-wave limit of the AdS/CFT duality was introduced in Berenstein, Maldacena and Nastase (2002).

Geometric transitions were first discussed in Gopakumar and Vafa (1999). They have been used in the study of large-N limits in Vafa (2001) and Maldacena and Nuñez (2001b) among others.

Bibliography

Abouelsaood, A., Callan, C. G., Nappi, C. R., and Yost, S. A. (1987). Open strings in background gauge fields. *Nucl. Phys.*, **B280**, 599.

Acharya, B. S., and Gukov, S. (2004). M theory and singularities of exceptional holonomy manifolds. *Phys. Reports*, **392**, 121. E-print hep-th/0409191.

Aganagic, M., Popescu, C., and Schwarz, J. H. (1997a). D-brane actions with local kappa symmetry. *Phys. Lett.*, **B393**, 311. E-print hep-th/9610249.

Aganagic, M., Popescu, C., and Schwarz, J. H. (1997b). Gauge-invariant and gauge-fixed D-brane actions. *Nucl. Phys.*, **B495**, 99. E-print hep-th/9612080.

Aharony, O. (2000). A brief review of 'little string theories.' *Class. Quant. Grav.*, **17**, 929. E-print hep-th/9911147.

Aharony, O., Gubser, S. S., Maldacena, J. M., Ooguri, H., and Oz, Y. (2000). Large N field theories, string theory and gravity. *Phys. Reports*, **323**, 183. E-print hep-th/9905111.

Albrecht, A., and Steinhardt, P. J. (1982). Cosmology for grand unified theories with radiatively induced symmetry breaking. *Phys. Rev. Lett.*, **48**, 1220.

Alessandrini, V., Amati, D., Le Bellac, M., and Olive, D. (1971). The operator approach to dual multiparticle theory. *Phys. Reports*, **6**, 269–346.

Alishahiha, M., Silverstein, E., and Tong, D. (2004). DBI in the sky. *Phys. Rev.*, **D70**, 123505. E-print hep-th/0404084.

Alvarez, E., Alvarez-Gaume, L., and Lozano, Y. (1995). An introduction to T Duality in string theory. *Nucl. Phys. Proc. Suppl.*, **41**, 1. E-print hep-th/9410237.

Alvarez, O. (1983). Theory of strings with boundaries: Fluctuations, topology and quantum geometry. *Nucl. Phys.*, **B216**, 125.

Alvarez-Gaume, L., and Freedman, D. Z. (1981). Geometrical structure and ultraviolet finiteness in the supersymmetric sigma model. *Commun. Math. Phys.*, **80**, 443.

Alvarez-Gaume, L., and Vazquez-Mozo, M. A. (1992). Topics in string theory and quantum gravity. In *Les Houches Summer School 1992*, pp. 481–636. E-print hep-th/9212006.

Alvarez-Gaume, L., and Witten, E. (1984). Gravitational anomalies. *Nucl. Phys.*, **B234**, 269.

Alvarez-Gaume, L., Freedman, D. Z., and Mukhi, S. (1981). The background field method and the ultraviolet structure of the supersymmetric nonlinear sigma model. *Annals Phys.*, **134**, 85.

Angelantonj, C., and Sagnotti, A. (2002). Open strings. *Phys. Reports*, **371**, 1. Erratum – ibid. **376**, 339. E-print hep-th/0204089.

Antoniadis, I. (1990). A possible new dimension at a few TeV. *Phys. Lett.*, **B246**, 377.

Antoniadis, I., Arkani-Hamed, N., Dimopoulos, S., and Dvali, G. R. (1998). New dimensions at a millimeter to a Fermi and superstrings at a TeV. *Phys. Lett.*, **B436**, 257. E-print hep-ph/9804398.

Antoniadis, I., Bachas, C. P., and Kounnas, C. (1987). Four-dimensional superstrings. *Nucl. Phys.*, **B289**, 87.

Antoniadis, I., Gava, E., Narain, K. S., and Taylor, T. R. (1994). Topological amplitudes in string theory. *Nucl. Phys.*, **B413**, 162. E-print hep-th/9307158.

Arkani-Hamed, N., Dimopoulos, S., and Dvali, G. R. (1998). The hierarchy problem and new dimensions at a millimeter. *Phys. Lett.*, **B429**, 263. E-print hep-ph/9803315.

Arkani-Hamed, N., Dimopoulos, S., and Dvali, G. R. (1999). Phenomenology, astrophysics and cosmology of theories with sub-millimeter dimensions and TeV scale quantum gravity. *Phys. Rev.*, **D59**, 086004. E-print hep-ph/9807344.

Ashok, S., and Douglas, M. R. (2004). Counting flux vacua. *JHEP*, **0401**, 060. E-print hep-th/0307049.

Aspinwall, P. S. (1996). Some relationships between dualities in string theory. *Nucl. Phys. Proc. Suppl.*, **46**, 30. E-print hep-th/9508154.

Aspinwall, P. S. (1997). K3 surfaces and string duality. In *Fields, Strings and Duality: TASI 96*, eds. C. Efthimiou and B. Greene, pp. 421–540. Singapore: World Scientific. E-print hep-th/9611137.

Aspinwall, P. S. (2001). Compactification, geometry and duality: $N = 2$. In *Strings, Branes and Gravity: TASI 1999*, eds. J. Harvey, S. Kachru and E. Silverstein, pp. 723–805. Singapore: World Scientific. E-print hep-th/0001001.

Aspinwall, P. S. (2005). D-branes on Calabi–Yau manifolds. In *Progress in String Theory: TASI 2003*, ed. J. Maldacena, pp. 1–152. Singapore: World Scientific. E-print hep-th/0403166.

Aspinwall, P. S., Greene, B. R., and Morrison, D. R. (1994). Calabi–Yau moduli space, mirror manifolds and spacetime topology change in string theory. *Nucl. Phys.*, **B416**, 414. E-print hep-th/9309097.

Atiyah, M., and Witten, E. (2003). M-theory dynamics on a manifold of G(2) holonomy. *Adv. Theor. Math. Phys.*, **6**, 1. E-print hep-th/0107177.

Bachas, C. (1996). D-brane dynamics. *Phys. Lett.*, **B374**, 37. E-print hep-th/9511043.

Bailin, D., and Love, A. (2004). *Cosmology in Gauge Field Theory and String Theory*, Bristol: Institute of Physics Publishing.

Banados, M., Teitelboim, C., and Zanelli, J. (1992). The black hole in three-dimensional space-time. *Phys. Rev. Lett.*, **69**, 1849. E-print hep-th/9204099.

Banks, T. (1998). Matrix theory. *Nucl. Phys. Proc. Suppl.*, **67**, 180. E-print hep-th/9710231.

Banks, T. (1999). M-theory and cosmology. In *Les Houches 1999, The Primordial Universe*, eds. P. Binetruy et al., pp. 495–579. Berlin: Springer-Verlag. E-Print hep-th/9911067.

Banks, T. (2001). TASI lectures on matrix theory. In *Strings, Branes and Gravity: TASI 1999*, eds. J. Harvey, S. Kachru and E. Silverstein, pp. 495–542.

Singapore: World Scientific. E-print hep-th/9911068.

Banks, T., Fischler, W., Shenker, S. H., and Susskind, L. (1997). M theory as a matrix model: A conjecture. *Phys. Rev.*, **D55**, 5112. E-print hep-th/9610043.

Bardeen, J. M., Carter, B., and Hawking, S. W. (1973). The Four laws of black hole mechanics. *Commun. Math. Phys.*, **31**, 161.

Battefeld, T., and Watson, S. (2006). String gas cosmology. *Rev. Mod. Phys.*, **78**, 435. E-Print hep-th/0510022.

Becchi, C., Rouet, A., and Stora, R. (1974). The Abelian Higgs–Kibble model. Unitarity of the S operator. *Phys. Lett.*, **B52**, 344.

Becchi, C., Rouet, A., and Stora, R. (1976). Renormalization of gauge theories. *Annals Phys.*, **98**, 287.

Beasley, C., and Witten, E. (2002). A note on fluxes and superpotentials in M-theory compactifications on manifolds of G(2) holonomy. *JHEP*, **0207**, 046. E-print hep-th/0203061.

Becker, K., and Becker, M. (1996). M-theory on eight-manifolds. *Nucl. Phys.*, **B477**, 155. E-print hep-th/9605053.

Becker, K., and Becker, M. (1997). A two-loop test of M(atrix) theory. *Nucl. Phys.*, **B506**, 48. E-print hep-th/9705091.

Becker, K., Becker, M., Dasgupta, K., and Green, P. S. (2003). Compactifications of heterotic theory on non-Kähler complex manifolds. I. *JHEP*, **0304**, 007. E-print hep-th/0301161.

Becker, K., Becker, M., Polchinski, J., and Tseytlin, A. A. (1997). Higher order graviton scattering in M(atrix) theory. *Phys. Rev.*, **D56**, 3174. E-print hep-th/9706072.

Becker, K., Becker, M., and Strominger, A. (1995). Five-branes, membranes and nonperturbative string theory. *Nucl. Phys.*, **B456**, 130. E-print hep-th/9507158.

Beisert, N., and Staudacher, M. (2005). Long-range PSU(2,2—4) Bethe ansaetze for gauge theory and strings. *Nucl. Phys.*, **B727**, 1. E-print hep-th/0504190.

Bekenstein, J.D. (1973). Black holes and entropy. *Phys. Rev.*, **D7**, 2333.

Belavin, A. A., Polyakov, A. M., and Zamolodchikov, A. B. (1984). Infinite conformal symmetry in two-dimensional quantum field theory. *Nucl. Phys.*, **B241**, 333.

Bena, I., Polchinski, J., and Roiban, R. (2004). Hidden symmetries of the $AdS(5) \times S^5$ superstring. *Phys. Rev.*, **D69**, 046002. E-print hep-th/0305116.

Berenstein, D., and Corrado, R. (1997). M(atrix)-theory in various dimensions. *Phys. Lett.*, **B406**, 37. E-print hep-th/9702108.

Berenstein, D., Maldacena, J. M., and Nastase, H. (2002). Strings in flat space and pp waves from $N = 4$ super Yang Mills. *JHEP*, **0204**, 013. E-print hep-th/0202021.

Bergshoeff, E., and Townsend, P. K. (1997). Super D-branes. *Nucl. Phys.*, **B490**, 145. E-print hep-th/9611173.

Bergshoeff, E., de Roo, M., de Wit, B., and van Nieuwenhuizen, P. (1982). Ten-dimensional Maxwell–Einstein supergravity, its currents, and the issue of its auxiliary fields. *Nucl. Phys.*, **B195**, 97.

Bergshoeff, E., Sezgin, E., and Townsend, P. K. (1987). Supermembranes and eleven-dimensional supergravity. *Phys. Lett.*, **B189**, 75.

Bergshoeff, E., Sezgin, E., and Townsend, P. K. (1988). Properties of the eleven-dimensional supermembrane theory. *Annals Phys.*, **185**, 330.

Berkooz, M., Douglas, M. R., and Leigh, R. G. (1996). Branes intersecting at

angles. *Nucl. Phys.*, **B480**, 265. E-print hep-th/9606139.
Bershadsky, M., Cecotti, S., Ooguri, H., and Vafa, C. (1994). Kodaira–Spencer theory of gravity and exact results for quantum string amplitudes. *Commun. Math. Phys.*, **165**, 311. E-print hep-th/9309140.
Besse, A. L. (1987). *Einstein Manifolds*. Berlin: Springer-Verlag.
Bianchi, M., and Sagnotti, A. (1990). On the systematics of open string theories. *Phys. Lett.*, **B247**, 517.
Bigatti, D., and Susskind, L. (1997). Review of matrix theory. In *Cargèse 1997, Strings, branes and dualities*, pp. 277–318. E-print hep-th/9712072.
Bigatti, D., and Susskind, L. (2001). TASI lectures on the holographic principle. In *Strings, Branes and Gravity: TASI 1999*, eds. J. Harvey, S. Kachru and E. Silverstein, pp. 883–933. Singapore: World Scientific. E-print hep-th/0002044.
Bilal, A. (1999). M(atrix) theory: A pedagogical introduction. *Fortsch. Phys.*, **47**, 5. E-print hep-th/9710136.
Bilal, A. and Metzger, S. (2003). Anomaly cancellation in M-theory: A critical review. *Nucl. Phys.*, **B675**, 416. E-Print hep-th/0307152.
Birmingham, D., Blau, M., Rakowski, M., and Thompson, G. (1991). Topological field theory. *Phys. Reports*, **209**, 129.
Blau, M., Figueroa-O'Farrill, J., Hull, C., and Papadopoulos, G. (2002a). A new maximally supersymmetric background of IIB superstring theory. *JHEP*, **0201**, 047. E-print hep-th/0110242.
Blau, M., Figueroa-O'Farrill, J., Hull, C., and Papadopoulos, G. (2002b). Penrose limits and maximal supersymmetry. *Class. Quant. Grav.*, **19**, L87. E-print hep-th/0201081.
Blumenhagen, R., Cvetic, M., Langacker, P., and Shiu, G. (2005). Toward realistic intersecting D-brane models. *Ann. Rev. Nucl. Part. Sci.*, **55**,) 71. E-print hep-th/0502005.
Boucher, W., Friedan, D., and Kent, A. (1986). Determinant formulae and unitarity for the $N = 2$ superconformal algebras in two dimensions or exact results on string compactification. *Phys. Lett.*, **B172**, 316.
Bousso, R. (2002). The holographic principle. *Rev. Mod. Phys.*, **74**, 825. E-print hep-th/0203101.
Bousso, R., and Polchinski, J. (2000). Quantization of four-form fluxes and dynamical neutralization of the cosmological constant. *JHEP*, **0006**, 006. E-print hep-th/0004134.
Brandenberger, R. H. (2005a). Challenges for string gas cosmology. E-Print hep-th/0509099.
Brandenberger, R. H. (2005b). Moduli stabilization in string gas cosmology. E-Print hep-th/0509159.
Brandenberger, R. H., and Vafa, C. (1989). Superstrings in the early universe. *Nucl. Phys.*, **B316**, 391.
Breckenridge, J. C., Myers, R. C., Peet, A. W., and Vafa, C. (1997). D-branes and spinning black holes. *Phys. Lett.*, **B391**, 93. E-print hep-th/9602065.
Breitenlohner P., and Freedman, D. Z. (1982). Stability in gauged extended supergravity. *Annals Phys.*, **144**, 249.
Brézin, E., and Kazakov, V. A. (1990). Exactly solvable field theories of closed strings. *Phys. Lett.*, **B236**, 144.
Brink, L., Di Vecchia, P., and Howe, P. S. (1976). A locally supersymmetric and reparametrization invariant action for the spinning string. *Phys. Lett.*, **B65**, 471.

Brink, L., and Nielsen, H. B. (1973). A simple physical interpretation of the critical dimension of space-time in dual models. *Phys. Lett.*, **B45**, 332.

Brink, L., and Schwarz, J. H. (1981). Quantum superspace. *Phys. Lett.*, **B100**, 310.

Brink, L., Schwarz, J. H., and Scherk, J. (1977). Supersymmetric Yang–Mills theories. *Nucl. Phys.*, **B121**, 77.

Brower, R. C. (1972). Spectrum generating algebra and no-ghost theorem for the dual model. *Phys. Rev.*, **D6**, 1655.

Brower, R. C., and Friedman, K. A. (1973). Spectrum generating algebra and no-ghost theorem for the Neveu–Schwarz model. *Phys. Rev.*, **D7**, 535.

Burgess, C. P., Majumdar, M., Nolte, D., Quevedo, F., Rajesh, G., and Zhang, R. J. (2001). The inflationary brane-antibrane universe. *JHEP*, **0107**, 047. E-print hep-th/0105204.

Buscher, T. H. (1987). A symmetry of the string background field equations. *Phys. Lett.*, **B194**, 59.

Buscher, T. H. (1988). Path integral derivation of quantum duality in nonlinear sigma models. *Phys. Lett.*, **B201**, 466.

Callan, C. G., and Maldacena, J. M. (1996). D-brane approach to black hole quantum mechanics. *Nucl. Phys.*, **B472**, 591. E-print hep-th/9602043.

Callan, C. G., and Maldacena, J. M. (1998). Brane dynamics from the Born–Infeld action. *Nucl. Phys.*, **B513**, 198. E-print hep-th/9708147.

Callan, C. G., and Thorlacius, L. (1989). Sigma models and string theory. In TASI 1988 *Particles, strings and supernovae, vol. 2*, 795–878.

Callan, C. G., Giddings, S. B., Harvey, J. A., and Strominger, A. (1992). Evanescent black holes. *Phys. Rev.*, **D45**, 1005. E-print hep-th/9111056.

Callan, C. G., Harvey, J. A., and Strominger, A. (1991). World-sheet approach to heterotic instantons and solitons. *Nucl. Phys.*, **B359**, 611.

Callan, C. G., Lovelace, C., Nappi, C. R., and Yost, S. A. (1987). Adding holes and crosscaps to the superstring. *Nucl. Phys.*, **B293**, 83.

Callan, C. G., Lovelace, C., Nappi, C. R., and Yost, S. A. (1988). Loop corrections to superstring equations of motion. *Nucl. Phys.*, **B308**, 221.

Callan, C. G., Martinec, E. J., Perry, M. J., and Friedan, D. (1985). Strings in background fields. *Nucl. Phys.*, **B262**, 593.

Candelas, P. (1987). Lectures on complex manifolds. In *Superstrings '87*, eds. L. Alvarez-Gaumé *et al.*, pp. 1–88. Singapore: World Scientific.

Candelas, P., and de la Ossa, X. C. (1990). Comments on conifolds. *Nucl. Phys.*, **B342**, 246.

Candelas, P., and de la Ossa, X. C. (1991). Moduli space of Calabi–Yau manifolds. *Nucl. Phys.*, **B355**, 455.

Candelas, P., De La Ossa, X. C., Green, P. S., and Parkes, L. (1991). A pair of Calabi–Yau manifolds as an exactly soluble superconformal theory. *Nucl. Phys.*, **B359**, 21.

Candelas, P., Horowitz, G. T., Strominger, A., and Witten, E. (1985). Vacuum configurations for superstrings. *Nucl. Phys.*, **B258**, 46.

Candelas, P., Lynker, M., and Schimmrigk, R. (1990). Calabi–Yau manifolds in weighted P(4). *Nucl. Phys.*, **B341**, 383.

Carroll, S. M. (2001). TASI lectures: Cosmology for string theorists. In *Strings, Branes and Gravity: TASI 1999*, eds. J. Harvey, S. Kachru and E. Silverstein, pp. 437–492. Singapore: World Scientific. E-print hep-th/0011110.

Carroll, S. M. (2004). *Spacetime and Geometry: an Introduction to General*

Relativity. San Francisco: Addison-Wesley.
Cederwall, M., von Gussich, A., Nilsson, B. E. W., Sundell, P., and Westerberg, A. (1997). The Dirichlet super *p*-branes in ten-dimensional type IIA and IIB supergravity. *Nucl. Phys.*, **B490**, 179. E-print hep-th/9611159.
Cederwall, M., von Gussich, A., Nilsson, B. E. W., and Westerberg, A. (1997). The Dirichlet super three-brane in ten-dimensional type IIB supergravity. *Nucl. Phys.*, **B490**, 163. E-print hep-th/9610148.
Chapline, G. F., and Manton, N. S. (1983). Unification of Yang–Mills theory and supergravity in ten dimensions, PL **B120**, 105.
Chodos, A., and Thorn, C. B. (1974). Making the massless string massive. *Nucl. Phys.*, **B72**, 509.
Connes, A., Douglas, M. R., and Schwarz, A. (1998). Noncommutative geometry and matrix theory: Compactification on tori. *JHEP*, **9802**, 003. E-print hep-th/9711162.
Constable, N. R., Freedman, D. Z., Headrick, M., Minwalla, S., Motl, L., Postnikov, A., and Skiba, W. (2002). PP-wave string interactions from perturbative Yang–Mills theory. *JHEP*, **0207**, 017. E-print hep-th/0205089.
Cormier, D., and Holman, R. (2000). Spinodal decomposition and inflation: Dynamics and metric perturbations. *Phys. Rev.*, **D62**, 023520. E-print hep-ph/9912483.
Cremmer, E., and Julia, B. (1979). The SO(8) supergravity. *Nucl. Phys.*, **B159**, 141.
Cremmer, E., Ferrara, S., Girardello, L., and Van Proeyen, A. (1983). Yang–Mills theories with local supersymmetry: Lagrangian, transformation laws and superhiggs effect. *Nucl. Phys.*, **B212**, 413.
Cremmer, E., Julia, B., and Scherk, J. (1978). Supergravity theory in 11 dimensions. *Phys. Lett.*, **B76**, 409.
Cvetic, M., Shiu, G., and Uranga, A. M. (2001). Chiral four-dimensional $N=1$ supersymmetric type IIA orientifolds from intersecting D6-branes. *Nucl. Phys.*, **B615**, 3. E-print hep-th/0107166.
Dabholkar, A. (1998). Lectures on orientifolds and duality. In *Trieste 1997, High energy physics and cosmology*, pp. 128–191. E-print hep-th/9804208.
Dabholkar, A., and Harvey, J. A. (1989). Nonrenormalization of the superstring tension. *Phys. Rev. Lett.*, **63**, 478.
Dabholkar, A., Denef, F., Moore, G. W., and Pioline, B. (2005). Precision counting of small black holes. *JHEP*, **0510**, 096. E-print hep-th/0507014.
Dabholkar, A., Gibbons, G. W., Harvey, J. A., and Ruiz Ruiz, F. (1990). Superstrings and solitons. *Nucl. Phys.*, **B340**, 33.
Dai, J., Leigh, R. G., and Polchinski, J. (1989). New connections between string theories. *Mod. Phys. Lett.*, **A4**, 2073.
Danielsson, U. H. (2005). Lectures on string theory and cosmology. *Class. Quant. Grav.*, **22**, S1. E-print hep-th/0409274.
Dasgupta, K., Herdeiro, C., Hirano, S., and Kallosh, R. (2002). D3/D7 inflationary model and M-theory. *Phys. Rev.*, **D65**, 126002. E-Print hep-th/0203019.
Dasgupta, K., Rajesh, G., and Sethi, S. (1999). M theory, orientifolds and G-flux. *JHEP*, **9908**, 023. E-print hep-th/9908088.
David, F. (1988). Conformal field theories coupled to 2-D gravity in the conformal gauge. *Mod. Phys. Lett.*, **A3**, 1651.
Denef, F. (2000). Supergravity flows and D-brane stability. *JHEP*, **0008**, 050.

E-print hep-th/0005049.

Denef, F., and Douglas, M. R. (2004). Distributions of flux vacua. *JHEP*, **0405**, 072. E-print hep-th/0404116.

Derendinger, J. P., Ibañez, L. E., and Nilles, H. P. (1985). On the low-energy $D = 4$, $N = 1$ supergravity theory extracted from the $D = 10$, $N = 1$ superstring. *Phys. Lett.*, **B155**, 65.

Deser, S., and Zumino, B. (1976a). Consistent supergravity. *Phys. Lett.*, **B62**, 335.

Deser, S., and Zumino, B. (1976b). A complete action for the spinning string. *Phys. Lett.*, **B65**, 369.

Deser, S., Jackiw, R., and Templeton, S. (1982). Topologically massive gauge theories. *Annals Phys.*, **140**, 372. Erratum – ibid. **185**, 406.

de Wit, B., and Nicolai, H. (1982). $N = 8$ supergravity. *Nucl. Phys.*, **B208**, 323.

de Wit, B., Hoppe, B. J., and Nicolai, H. (1988). On the quantum mechanics of supermembranes. *Nucl. Phys.*, **B305**, 545.

de Wit, B., Lauwers, P. G., and Van Proeyen, A. (1985). Lagrangians of $N = 2$ supergravity - matter systems. *Nucl. Phys.*, **B255**, 569.

de Wit, B., Smit, D. J., and Hari Dass, N. D. (1987). Residual supersymmetry of compactified $D = 10$ supergravity. *Nucl. Phys.*, **B283**, 165.

DeWolfe, O., Freedman, D. Z., Gubser, S. S., and Karch, A. (2000). Modeling the fifth dimension with scalars and gravity. *Phys. Rev.*, **D62**, 046008. E-print hep-th/9909134.

DeWolfe, O., Giryavets, A., Kachru, S., and Taylor, W. (2005a). Enumerating flux vacua with enhanced symmetries. *JHEP*, **0502**, 037. E-print hep-th/0411061.

DeWolfe, O., Giryavets, A., Kachru, S., and Taylor, W. (2005b). Type IIA moduli stabilization. *JHEP*, **0507**, 066. E-print hep-th/0505160.

D'Hoker, E. (1993). TASI lectures on critical string theory. In *Recent Directions in Particle Theory, TASI 1992*, eds. J. Harvey and J. Polchinski, pp. 1–98. Singapore: World Scientific.

D'Hoker, E., and Freedman, D. Z. (2004). Supersymmetric gauge theories and the AdS/CFT correspondence. In *Strings, Branes and Extra Dimensions: TASI 2001*, eds. S. S. Gubser and J. D. Lykken, pp. 3–158. Singapore: World Scientific. E-print hep-th/0201253.

D'Hoker, E., and Phong, D. H. (1988). The geometry of string perturbation theory. *Rev. Mod. Phys.*, **60**, 917–1065.

Dijkgraaf, R., and Vafa, C. (2002a). Matrix models, topological strings, and supersymmetric gauge theories. *Nucl. Phys.*, **B644**, 3. E-print hep-th/0206255.

Dijkgraaf, R., and Vafa, C. (2002b). A perturbative window into non-perturbative physics. E-print hep-th/0208048.

Dijkgraaf, R., Verlinde, E. P., and Verlinde, H. L. (1997). Matrix string theory. *Nucl. Phys.*, **B500**, 43. E-print hep-th/9703030.

Dine, M. (2001). TASI lectures on M-theory phenomenology. In *Strings, Branes and Gravity: TASI 1999*, eds. J. Harvey, S. Kachru and E. Silverstein, pp. 545–612. Singapore: World Scientific. E-print hep-th/0003175.

Dine, M., Kaplunovsky, V., Mangano, M. L., Nappi, C., and Seiberg, N. (1985). Superstring model building. *Nucl. Phys.*, **B259**, 549.

Dine, M., Rohm, R., Seiberg, N., and Witten, E. (1985). Gluino condensation in superstring models. *Phys. Lett.*, **B156**, 55.

Distler, J., and Kawai, H. (1989). Conformal field theory and 2-D quantum gravity or who's afraid of Joseph Liouville?. *Nucl. Phys.*, **B321**, 509.

Dixon, L.J., Friedan, D., Martinec, E. J., and Shenker, S. H. (1987). The conformal field theory of orbifolds. *Nucl. Phys.*, **B282**, 13.

Dixon, L. J., Harvey, J. A., Vafa, C., and Witten, E. (1985). Strings on orbifolds. *Nucl. Phys.*, **B261**, 678.

Dixon, L. J., Harvey, J. A., Vafa, C., and Witten, E. (1986). Strings on orbifolds. 2. *Nucl. Phys.*, **B274**, 285.

Dixon, L. J., Kaplunovsky, V., and Louis, J. (1991). Moduli dependence of string loop corrections to gauge coupling constants. *Nucl. Phys.*, **B355**, 649.

Douglas, M. R. (1995). In *Cargèse 1997, Strings, Branes and Dualities*, pp. 267–275. E-print hep-th/9512077.

Douglas, M. R. (2003). The statistics of string / M-theory vacua. *JHEP*, **0305**, 046. E-print hep-th/0303194.

Douglas, M. R., and Hull, C. M. (1998). D-branes and the noncommutative torus. *JHEP*, **9802**, 008. E-print hep-th/9711165.

Douglas, M. R., and Moore, G. W. (1996). D-branes, quivers, and ALE instantons. E-print hep-th/9603167.

Douglas, M. R., and Nekrasov, N. A. (2001). Noncommutative field theory. *Rev. Mod. Phys.*, **73**, 977. E-print hep-th/0106048.

Douglas, M. R., and Shenker, S. H. (1990). Strings in less than one dimension. *Nucl. Phys.*, **B335**, 635.

Douglas, M. R., Kabat, D., Pouliot, P., and Shenker, S. H. (1997). D-branes and short distances in string theory. *Nucl. Phys.*, **B485**, 85. E-print hep-th/9608024.

Douglas, M. R., Polchinski, J., and Strominger, A. (1997). Probing five-dimensional black holes with D-branes. *JHEP*, **9712**, 003. E-print hep-th/9703031.

Douglas, M. R. *et al.*, eds. (2004). *Strings and Geometry: Proceedings of the 2002 Clay School*, American Mathematical Society.

Duff, M. J. (1996). Supermembranes. In *Fields, Strings and Duality: TASI 96*, eds. C. Efthimiou and B. Greene, pp. 219–289. Singapore: World Scientific. E-print hep-th/9611203.

Duff, M. J. (2001). Lectures on branes, black holes and anti-de Sitter space. In *Strings, Branes and Gravity: TASI 1999*, eds. J. Harvey, S. Kachru and E. Silverstein, pp. 3–125. Singapore: World Scientific. E-print hep-th/9912164.

Duff, M. J., Howe, P. S., Inami, T., and Stelle, K. S. (1987). Superstrings in $D=10$ from supermembranes in $D=11$. *Phys. Lett.*, **B191**, 70.

Duff, M. J., Khuri, R. R., and Lu, J. X. (1995). String solitons. *Phys. Reports*, **259**, 213. E-print hep-th/9412184.

Duff, M. J., Liu, J. T., and Minasian, R. (1995). Eleven-dimensional origin of string/string duality: A one-loop test. *Nucl. Phys.*, **B452**, 261. E-print hep-th/9506126.

Duff, M. J., Nilsson, B. E. W., and Pope, C. N. (1986). Kaluza–Klein supergravity. *Phys. Reports*, **130**, 1.

Dvali, G., and Kachru, S. (2005). New old inflation. In *From Fields to Strings: Circumnavigating Theoretical Physics, vol. 2*, eds. M. Shifman, A. Vainshtein and J. Wheater, pp. 1131–1155. Singapore: World Scientific. E-Print hep-th/0309095.

Dvali, G. R., and Tye, S. H. H. (1999). Brane inflation. *Phys. Lett.*, **B450**, 72. E-print hep-ph/9812483.

Dvali, G. R., Shafi, Q., and Solganik, S. (2001). D-brane inflation. E-print hep-th/0105203.

Eguchi, T., Gilkey, P. B., and Hanson, A. J. (1980). Gravitation, gauge theories and differential geometry. *Phys. Reports*, **66**, 213.

Elvang, H., Emparan, R., Mateos, D., and Reall, H. S. (2004). A supersymmetric black ring. *Phys. Rev. Lett.*, **93**, 211302. E-print hep-th/0407065.

Emparan, R., and Reall, H. S. (2002). A rotating black ring in five dimensions. *Phys. Rev. Lett.*, **88**, 101101 E-print hep-th/0110260.

Fairlie, D. B., and Martin, D. (1973). New light on the Neveu–Schwarz model. *Nuovo Cim.*, **18A**, 373.

Fairlie, D. B., and Nielsen, H. B. (1970). An analogue model for KSV theory. *Nucl. Phys.*, **B20**, 637.

Ferrara, S., and Kallosh, R. (1996). Supersymmetry and attractors. *Phys. Rev.*, **D54**, 1514. E-print hep-th/9602136.

Ferrara, S., Kallosh, R., and Strominger, A. (1995). $N = 2$ extremal black holes. *Phys. Rev.*, **D52**, 5412. E-print hep-th/9508072.

Font, A., and Theisen, S. (2005). Introduction to string compactification. Lect. Notes Phys. **668**, 101.

Font, A., Ibañez, L. E., Lüst, D., and Quevedo, F. (1990). Strong–weak coupling duality and nonperturbative effects in string theory. *Phys. Lett.*, **B249**, 35.

Fradkin, E. S., and Tseytlin, A. A. (1985a). Effective field theory from quantized strings. *Phys. Lett.*, **B158**, 316.

Fradkin, E. S., and Tseytlin, A. A. (1985b). Nonlinear electrodynamics from quantized strings. *Phys. Lett.*, **B163**, 123.

Fradkin, E. S., and Tseytlin, A. A. (1985c). Quantum string theory effective action. *Nucl. Phys.*, **B261**, 1.

Freedman, D. Z., Gubser, S. S., Pilch, K., and Warner, N. P. (1999). Renormalization group flows from holography – supersymmetry and a c-theorem. *Adv. Theor. Math. Phys.*, **3**, 363. E-print hep-th/9904017.

Freedman, D. Z., Mathur, S. D., Matusis, A., and Rastelli, L. (1999). Correlation functions in the CFT(d)/AdS$(d+1)$ correspondence. *Nucl. Phys.*, **B546**, 96. E-print hep-th/9804058.

Freedman, D. Z., Van Nieuwenhuizen, P., and Ferrara, S. (1976). Progress toward a theory of supergravity. *Phys. Rev.*, **D13**, 3214.

Freund, P. G. O., and Rubin, M. A. (1980). Dynamics of dimensional reduction. *Phys. Lett.*, **B97**, 233.

Frey, A. R. (2003). Warped strings: Self-dual flux and contemporary compactifications. E-print hep-th/0308156.

Friedan, D. (1984). Introduction to Polyakov's string theory. p. 839 in *Recent Advances in Field Theory and Statistical Mechanics*, Proc. of 1982 Les Houches Summer School, eds. J. B. Zuber and R. Stora. Amsterdam: Elsevier.

Friedan, D. (1985). Nonlinear models in $2 + \varepsilon$ dimensions. 1980 Ph.D. thesis, *Annals Phys.*, **163**, 318.

Friedan, D., Martinec, E. J., and Shenker, S. H. (1986). Conformal invariance, supersymmetry and string theory. *Nucl. Phys.*, **B271**, 93.

Friedan, D., Qiu, A., and Shenker, S. H. (1984). Conformal invariance, unitarity and two-dimensional critical exponents. *Phys. Rev. Lett.*, **52**, 1575.

Friedan, D., Qiu, A., and Shenker, S. H. (1985). Superconformal invariance in two dimensions and the tricritical Ising model. *Phys. Lett.*, **B151**, 37.

Friedman, R., Morgan, J., and Witten, E. (1997). Vector bundles and F theory.

Commun. Math. Phys., **187**, 679. E-print hep-th/9701162.
Fubini, S., and Veneziano, G. (1969). Level structure of dual resonance models. *Nuovo Cim.*, **A64**, 811.
Fubini, S., and Veneziano, G. (1970). Duality in operator formalism. *Nuovo Cim.*, **A67**, 29.
Fubini, S., and Veneziano, G. (1971). Algebraic treatment of subsidiary conditions in dual resonance models. *Annals Phys.*, **63**, 12.
Fubini, S., Gordon, D., and Veneziano, G. (1969). A general treatment of factorization in dual resonance models. *Phys. Lett.*, **B29**, 679.
Gaiotto, D., Strominger, A., and Yin, X. (2006a). New connections between 4D and 5D black holes. *JHEP*, **0602**, 024. E-print hep-th/0503217.
Gaiotto, D., Strominger, A., and Yin, X. (2006b). 5D black rings and 4D black holes. *JHEP*, **0602**, 023. E-print hep-th/0504126.
Gasperini, M., and Veneziano, G. (1993). Pre big bang in string cosmology. *Astropart. Phys.*, **1**, 317. E-print hep-th/9211021.
Gates, S. J., Hull, C. M., and Rocek, M. (1984). Twisted multiplets and new supersymmetric nonlinear sigma models. *Nucl. Phys.*, **B248**, 157.
Gauntlett, J. P., and Pakis, S. (2003). The geometry of $D = 11$ Killing spinors. *JHEP*, **0304**, 039. E-print hep-th/0212008.
Gauntlett, J. P., Gutowski, J. B., Hull, C. M., Pakis, S., and Reall, H. S. (2003). All supersymmetric solutions of minimal supergravity in five dimensions. *Class. Quant. Grav.*, **20**, 4587. E-print hep-th/0209114.
Gauntlett, J. P., Harvey, J. A., and Liu, J. T. (1993). Magnetic monopoles in string theory. *Nucl. Phys.*, **B409**, 363. E-print hep-th/9211056.
Gauntlett, J. P., Martelli, D., Sparks, J., and Waldram, D. (2004). Sasaki-Einstein metrics on S(2) x S(3). *Adv. Theor. Math. Phys.*, **8**, 711. E-print hep-th/0403002.
Gepner, D. (1988). Space-time supersymmetry in compactified string theory and superconformal models. *Nucl. Phys.*, **B296**, 757.
Gepner, D., and Witten, E. (1986). String theory on group manifolds. *Nucl. Phys.*, **B278**, 493.
Gervais, J. L., and Sakita, B. (1971). Field theory interpretation of supergauges in dual models. *Nucl. Phys.*, **B34**, 632.
Ghoshal, D., and Vafa, C. (1995). $c = 1$ string as the topological theory of the conifold. *Nucl. Phys.*, **B453**, 121. E-print hep-th/9506122.
Gibbons, G. W., and Hawking, S. W. (1997). Action integrals and partition functions in quantum gravity. *Phys. Rev.*, **D15**, 2752.
Gibbons, G. W., and Maeda, K. (1988). Black holes and membranes in higher dimensional theories with dilaton fields. *Nucl. Phys.*, **B298**, 741.
Giddings, S. B., Kachru, S., and Polchinski, J. (2002). Hierarchies from fluxes in string compactifications. *Phys. Rev.*, **D66**, 106006. E-print hep-th/0105097.
Gimon, E. G., and Polchinski, J. (1996). Consistency conditions for orientifolds and D-manifolds. *Phys. Rev.*, **D54**, 1667. E-print hep-th/9601038.
Ginsparg, P.H. (1991). Applied conformal field theory. In *Les Houches Summer School 1988*, pp. 1–168. E-print hep-th/9108028.
Ginsparg, P. H., and Moore, G. W. (1993). Lectures on 2D gravity and 2D string theory. In *Recent Directions in Particle Theory, TASI 1992*, eds. J. Harvey and J. Polchinski, pp. 277–469. Singapore: World Scientific. E-print hep-th/9304011.
Giveon, A., and Kutasov, D. (1999). Brane dynamics and gauge theory. *Rev. Mod.*

Phys., **71**, 983–1084. E-print hep-th/9802067.

Giveon, A., Porrati, M., and Rabinovici, E. (1994). Target-space duality in string theory. *Phys. Reports*, **244**, 77–202. E-print hep-th/9401139.

Gliozzi, F., Scherk, J., and Olive, D. I. (1976). Supergravity and the spinor dual model. *Phys. Lett.*, **B65**, 282.

Gliozzi, F., Scherk, J., and Olive, D. I. (1977). Supersymmetry, supergravity theories and the dual spinor model. *Nucl. Phys.*, **B122**, 253.

Goddard, P., and Olive, D. I. (1985). Algebras, lattices and strings. p. 51 in 1983 Berkeley workshop *Vertex Operators in Mathematics and Physics*, eds. J. Lepowsky, S. Mandelstam, I. M. Singer. New York: springer-Verlag.

Goddard, P., and Olive, D. I. (1986). Kac-Moody and Virasoro algebras in relation to quantum physics. *Int. J. Mod. Phys.*, **A1**, 303.

Goddard, P., and Thorn, C. B. (1972). Compatibility of the dual Pomeron with unitarity and the absence of ghosts in the dual resonance model. *Phys. Lett.*, **B40**, 235.

Goddard, P., Goldstone, J., Rebbi, C., and Thorn, C. B. (1973). Quantum dynamics of a massless relativistic string. *Nucl. Phys.*, **B56**, 109.

Goddard, P., Kent, A., and Olive, D. I. (1985). Virasoro algebras and coset-space models. *Phys. Lett.*, **B152**, 88.

Goddard, P., Kent, A., and Olive, D. I. (1986). Unitary representations of the Virasoro and super-Virasoro algebras. *Commun. Math. Phys.*, **103**, 105.

Gol'fand, Y. A., and Likhtman, E. P. (1971). Extension of the algebra of Poincaré group generators and violation of P invariance. *JETP Lett.*, **13**, 323.

Gopakumar, R., and Vafa, C. (1999). On the gauge theory/geometry correspondence. *Adv. Theor. Math. Phys.*, **3**, 1415. E-print hep-th/9811131.

Gopakumar, R., Minwalla, S., and Strominger, A. (2000). Noncommutative solitons. *JHEP*, **0005**, 020. E-print hep-th/0003160.

Goto, T. (1971). Relativistic quantum mechanics of one-dimensional mechanical continuum and subsidiary condition of dual resonance model. *Prog. Theor. Phys.*, **46**, 1560.

Graña, M. (2006). Flux compactifications in string theory: A comprehensive review. *Phys. Reports*, **423**, 91. E-print hep-th/0509003.

Graña, M., and Polchinski, J. (2001). Supersymmetric three-form flux perturbations on AdS(5). *Phys. Rev.*, **D63**, 026001. E-print hep-th/0009211.

Graña, M., and Polchinski, J. (2002). Gauge/gravity duals with holomorphic dilaton. *Phys. Rev.*, **D65**, 126005. E-print hep-th/0106014.

Green, M. B. (1984). Supersymmetrical dual string theories and their field theory limits: a review. *Surveys High Energ. Phys.*, **3**, 127–160.

Green, M. B. (1999). Interconnections between type II superstrings, M theory and $N = 4$ Yang–Mills. In *Corfu 1998, Quantum Aspects of Gauge Theories, Supersymmetry and Unification*, ed. A. Ceresole, pp. 22–996. Berlin: Springer-Verlag. E-Print hep-th/9903124.

Green, M. B., and Gutperle, M. (1997). Effects of D-instantons. *Nucl. Phys.*, **B498**, 195. E-print hep-th/9701093.

Green, M. B., and Schwarz, J. H. (1981a). Supersymmetrical dual string theory. *Nucl. Phys.*, **B181**, 502.

Green, M. B., and Schwarz, J. H. (1981b). Supersymmetrical dual string theory. 2. Vertices and trees. *Nucl. Phys.*, **B198**, 252.

Green, M. B., and Schwarz, J. H. (1982). Supersymmetrical string theories. *Phys. Lett.*, **B109**, 444.

Green, M. B., and Schwarz, J. H. (1984a). Covariant description of superstrings. *Phys. Lett.*, **B136**, 367.

Green, M. B., and Schwarz, J. H. (1984b). Anomaly cancellation in supersymmetric $D = 10$ gauge theory and superstring theory. *Phys. Lett.*, **B149**, 117.

Green, M. B., and Schwarz, J. H. (1985). The hexagon gauge anomaly in type I superstring theory. *Nucl. Phys.*, **B255**, 93.

Green, M. B., Harvey, J. A., and Moore, G. W. (1997). I-brane inflow and anomalous couplings on D-branes. *Class. Quant. Grav.*, **14**, 47. E-print hep-th/9605033.

Green, M. B., Schwarz, J. H., and Brink, L. (1982). $N = 4$ Yang–Mills and $N = 8$ supergravity as limits of string theories. *Nucl. Phys.*, **B198**, 474.

Green, M. B., Schwarz, J. H., and Witten, E. (1987). *Superstring Theory* in two volumes. Cambridge: Cambridge University Press.

Greene, B. R. (1995). Lectures on quantum geometry. *Nucl. Phys. Proc. Suppl.*, **41**, 92.

Greene, B. R. (1997). String theory on Calabi–Yau manifolds. In *Fields, Strings and Duality: TASI 96*, eds. C. Efthimiou and B. Greene, pp. 543–726. Singapore: World Scientific. E-print hep-th/9702155.

Greene, B. R., and Plesser, M. R. (1990). Duality in Calabi–Yau moduli space. *Nucl. Phys.*, **B338**, 15.

Greene, B. R., Morrison, D. R., and Strominger, A. (1995). Black hole condensation and the unification of string vacua. *Nucl. Phys.*, **B451**, 109. E-print hep-th/9504145.

Greene, B. R., Shapere, A. D., Vafa, C., and Yau, S. T. (1990). Stringy cosmic strings and noncompact Calabi–Yau manifolds. *Nucl. Phys.*, **B337**, 1.

Gregory, R., and Laflamme, R. (1993). Black strings and p-branes are unstable. *Phys. Rev. Lett.*, **70**, 2837. E-print hep-th/9301052.

Gross, D. J., and Migdal, A. A. (1990). Nonperturbative two-dimensional quantum gravity. *Phys. Rev. Lett.*, **64**, 127.

Gross, D. J., and Perry, M. J. (1983). Magnetic monopoles in Kaluza–Klein theories. *Nucl. Phys.*, **B226**, 29.

Gross, D. J., and Witten, E. (1986). Superstring modifications of Einstein's equations. *Nucl. Phys.*, **B277**, 1.

Gross, D. J., Harvey, J. A., Martinec, E. J., and Rohm, R. (1985a). The heterotic string. *Phys. Rev. Lett.*, **54**, 502.

Gross, D. J., Harvey, J. A., Martinec, E. J., and Rohm, R. (1985b). Heterotic string theory. 1. The free heterotic string. *Nucl. Phys.*, **B256**, 253.

Gross, D. J., Harvey, J. A., Martinec, E. J., and Rohm, R. (1986). Heterotic string theory. 2. The interacting heterotic string. *Nucl. Phys.*, **B267**, 75.

Gross, D. J., Neveu, A., Scherk, J., and Schwarz, J. H. (1970). Renormalization and unitarity in the dual resonance model. *Phys. Rev.*, **D2**, 697.

Gubser, S. S. (2004). Special holonomy in string theory and M-theory. In *Strings, Branes and Extra Dimensions: TASI 2001*, eds. S. S. Gubser and J. D. Lykken, pp. 197–233. Singapore: World Scientific. E-print hep-th/0201114.

Gubser, S. S., and Klebanov, I. R. (1997). Absorption by branes and Schwinger terms in the world volume theory. *Phys. Lett.*, **B413**, 41. E-print hep-th/9708005.

Gubser, S. S., Klebanov, I. R., and Polyakov, A. M. (1998). Gauge theory correlators from non-critical string theory. *Phys. Lett.*, **B428**, 105. E-print hep-th/9802109.

Gubser, S. S., Klebanov, I. R., and Polyakov, A. M. (2002). A semi-classical limit of the gauge/string correspondence. *Nucl. Phys.*, **B636**, 99. E-print hep-th/0204051.

Gubser, S. S., Klebanov, I. R., and Tseytlin, A. A. (1997). String theory and classical absorption by three-branes. *Nucl. Phys.*, **B499**, 217. E-print hep-th/9703040.

Gukov, S. (2000). Solitons, superpotentials and calibrations. *Nucl. Phys.*, **B574**, 169. E-print hep-th/9911011.

Gukov, S., and Sparks, J. (2002). M-theory on Spin(7) manifolds. I. *Nucl. Phys.*, **B625**, 3. E-Print hep-th/0109025.

Gukov, S., Vafa, C., and Witten, E. (2001). CFT's from Calabi–Yau four-folds. *Nucl. Phys.*, **B584**, 69. Erratum – ibid. **B608**, 477. E-print hep-th/9906070.

Günaydin, M., and Marcus, N. (1985). The spectrum of the S^5 compactification of the chiral $N = 2$, $D = 10$ supergravity and the unitary supermultiplets of U(2, 2/4). *Class. Quant. Grav.*, **2**, L11.

Günaydin, M., Romans, L. J., and Warner, N. P. (1985). Gauged $N = 8$ supergravity in five dimensions. *Phys. Lett.*, **B154**, 268.

Guth, A. H. (1981). The inflationary universe: A possible solution to the horizon and flatness problems. *Phys. Rev.*, **D23**, 347.

Gutperle, M., and Strominger, A. (2002). Spacelike branes. *JHEP*, **0204**, 018. E-print hep-th/0202210.

Haack, M., and Louis, J. (2001). M-theory compactified on Calabi–Yau fourfolds with background flux. *Phys. Lett.*, **B507**, 296. E-print hep-th/0103068.

Hanany, A., and Witten, E. (1997). Type IIB superstrings, BPS monopoles, and three-dimensional gauge dynamics. *Nucl. Phys.*, **B492**, 152. E-print hep-th/9611230.

Hara, O. (1971). On origin and physical meaning of Ward-like identity in dual-resonance model. *Prog. Theor. Phys.*, **46** 1549.

Harvey, J. A. (1997). Magnetic monopoles, duality, and supersymmetry. In *Fields, Strings and Duality: TASI 96*, eds. C. Efthimiou and B. Greene, pp. 157–216. Singapore: World Scientific. E-print hep-th/9603086.

Harvey, J. A. (2001). Komaba lectures on noncommutative solitons and D-branes. E-print hep-th/0102076.

Harvey, J. A. (2005). TASI 2003 lectures on anomalies. E-print hep-th/0509097.

Harvey, J. A., and Moore, G. W. (1998). Fivebrane instantons and R^2 couplings in $N = 4$ string theory. *Phys. Rev.*, **D57**, 2323. E-Print hep-th/9610237.

Hatfield, B. (1992). *Quantum Field Theory of Point Particles and Strings*. Redwood City: Addison-Wesley.

Hawking, S. W. (1975). Particle creation by black holes. *Commun. Math. Phys.*, **43**, 199. Erratum – ibid. **46**, 206.

Hawking, S. W. (1976). Breakdown of predictability in gravitational collapse. *Phys. Rev.*, **D14**, 2460.

Hawking, S. W., and Page, D. N. (1983). Thermodynamics of black holes in Anti de Sitter space. *Commun. Math. Phys.*, **87**, 577.

Henningson, M., and Skenderis, K. (1998). The holographic Weyl anomaly. *JHEP*, **9807**, 023. E-print hep-th/9806087.

Herzog, C. P., Klebanov, I. R., and Ouyang, P. (2002). D-branes on the conifold and $N = 1$ gauge / gravity dualities. *Cargèse 2002, Progress in string, field and particle theory*, 189–223. Also in *Les Houches 2001, Gravity, gauge theories and strings*, 383–422. E-print hep-th/0205100.

Hewett, J. L., and Rizzo, T. G. (1989). Low-energy phenomenology of superstring inspired E(6) models. *Phys. Reports*, **183**, 193.

Hofman, D. M., and Maldacena, J. M. (2006). Giant magnons. E-print hep-th/0604135.

Hořava, P. (1999). Type IIA D-branes, K-theory, and matrix theory. *Adv. Theor. Math. Phys.*, **2**, 1373. E-print hep-th/9812135.

Hořava, P., and Witten, E. (1996a). Heterotic and type I string dynamics from eleven dimensions. *Nucl. Phys.*, **B460**, 506. E-print hep-th/9510209.

Hořava, P., and Witten, E. (1996b). Eleven-dimensional supergravity on a manifold with boundary. *Nucl. Phys.*, **B475**, 94. E-print hep-th/9603142.

Hori, K., Katz, S., Klemm, A., Pandharipande, R., Thomas, R., Vafa, C., Vakil, R., and Zaslow, E. (2003). *Mirror Symmetry*. Clay Mathematics Monographs, V. 1, Providence: American Mathematical Society.

Horowitz, G. T., and Polchinski, J. (1997). A correspondence principle for black holes and strings. *Phys. Rev.*, **D55**, 6189. E-print hep-th/9612146.

Horowitz, G. T., and Strominger, A. (1991). Black strings and p-branes. *Nucl. Phys.*, **B360**, 197.

Horowitz, G. T., and Strominger, A. (1996). Counting states of near-extremal black holes. *Phys. Rev. Lett.*, **77**, 2368. E-print hep-th/9602051.

Howe, P. S., and West, P. C. (1984). The complete $N = 2$, $D = 10$ supergravity. *Nucl. Phys.*, **B238**, 181.

Hübsch, T. (1992). *Calabi–Yau Manifolds: A Bestiary for Physicists*. Singapore: World Scientific.

Hull, C. M., and Townsend, P. K. (1995). Unity of superstring dualities. *Nucl. Phys.*, **B438**, 109. E-print hep-th/9410167.

Ibañez, L. E., and Lüst, D. (1992). Duality anomaly cancellation, minimal string unification and the effective low-energy Lagrangian of 4-D strings. *Nucl. Phys.*, **B382**, 305. E-print hep-th/9202046.

Intriligator, K. A., and Seiberg, N. (1996). Lectures on supersymmetric gauge theories and electric-magnetic duality. *Nucl. Phys. Proc. Suppl.*, **45BC**, 1. E-print hep-th/9509066.

Isham, C. J., and Pope, C. N. (1988). Nowhere vanishing spinors and topological obstructions to the equivalence of the NSR and GS superstrings. *Class. Quant. Grav.*, **5**, 257.

Isham, C. J., Pope, C. N., and Warner, N. P. (1988). Nowhere vanishing spinors and triality rotations in eight manifolds. *Class. Quant. Grav.*, **5**, 1297.

Ishibashi, N., Kawai, H., Kitazawa, Y., and Tsuchiya, A. (1997). A large-N reduced model as superstring. *Nucl. Phys.*, **B498**, 467. E-print hep-th/9612115.

Itzhaki, N., Maldacena, J. M., Sonnenschein, J., and Yankielowicz, S. (1998). Supergravity and the large N limit of theories with sixteen supercharges. *Phys. Rev.*, **D58**, 046004. E-print hep-th/9802042.

Jackson, M. G., Jones, N. T., and Polchinski, J. (2004). Collisions of cosmic F- and D-strings. E-print hep-th/0405229.

Jacob, M. (1974). *Dual Theory* (Physics Reports Reprint Book Series Vol. 1.). Amsterdam: North-Holland.

Johnson, C. V. (2001). D-brane primer. In *Strings, Branes and Gravity: TASI 1999*, eds. J. Harvey, S. Kachru and E. Silverstein, pp. 129–350. Singapore: World Scientific. E-print hep-th/0007170.

Johnson, C. V. (2003). *D-Branes.* Cambridge: Cambridge University Press.

Joyce, D. D. (2000). *Compact Manifolds with Special Holonomy.* Oxford: Oxford University Press.

Kabat, D., and Pouliot, P. (1996). A Comment on zero-brane quantum mechanics. *Phys. Rev. Lett.*, **77**, 1004. E-print hep-th/9603127.

Kachru, S. (2001). Lectures on warped compactifications and stringy brane constructions. In *Strings, Branes and Gravity: TASI 1999*, eds. J. Harvey, S. Kachru and E. Silverstein, pp. 849–880. Singapore: World Scientific. E-print hep-th/0009247.

Kachru, S., and Silverstein, E. (1998). 4d conformal theories and strings on orbifolds. *Phys. Rev. Lett.*, **80**, 4855. E-print hep-th/9802183.

Kachru, S., and Vafa, C. (1995). Exact results for $N = 2$ compactifications of heterotic strings. *Nucl. Phys.*, **B450**, 69. E-print hep-th/9505105.

Kachru, S., Kalloh, R., Linde, A., Maldacena, J., McAllister, L., and Trivedi, S. P. (2003). Towards inflation in string theory. *JCAP*, **0310**, 013. E-print hep-th/0308055.

Kachru, S., Kalloh, R., Linde, A., and Trivedi, S. P. (2003). De Sitter vacua in string theory. *Phys. Rev.*, **D68**, 046005. E-print hep-th/0301240.

Kachru, S., Schulz, M. B., Tripathy, P. K., and Trivedi, S. P. (2003). New supersymmetric string compactifications. *JHEP*, **0303**, 061. E-print hep-th/0211182.

Kachru, S., Schulz, M. B., and Trivedi, S. P. (2003). Moduli stabilization from fluxes in a simple IIB orientifold. *JHEP*, **0310**, 007. E-print hep-th/0201028.

Kaku, M. (1988). *Introduction to Superstrings.* New York: Springer.

Kaku, M. (1991). *Strings, Conformal Fields, and Topology: An Introduction.* New York: Springer.

Kaku, M. (1997). *Introduction to Superstrings and M-theory.* New York: Springer.

Kaku, M. (2000). *Strings, Conformal Fields, and M-theory.* New York: Springer.

Kalb, M., and Ramond, P. (1974). Classical direct interstring action. *Phys. Rev.*, **D9**, 2273.

Kalloh, R., Kofman, L., and Linde, A.D. (2001). Pyrotechnic universe. *Phys. Rev.*, **D64**, 123523. E-print hep-th/0104073.

Kaluza, T. (1921). On the problem of unity in physics. *Sitzungsber. Preuss. Akad. Wiss. Berlin (Math.Phys.)*, 966.

Kaplunovsky, V. S. (1988). One loop threshold effects in string unification. *Nucl. Phys.*, **B307**, 145. Erratum – ibid. **B382**, 436. E-print hep-th/9205068.

Karch, A., and Randall, L. (2005). Relaxing to three dimensions. E-print hep-th/0506053.

Kato, M., and Ogawa, K. (1983). Covariant quantization of string based on BRS invariance. *Nucl. Phys.*, **B212** 443.

Kawai, H., Lewellen, D. C., and Tye, S. H. H. (1987). Construction of fermionic string models in four dimensions. *Nucl. Phys.*, **B288**, 1.

Kazakov, V. A., Marshakov, A., Minahan, J. A., and Zarembo, K. (2004). Classical / quantum integrability in AdS/CFT. *JHEP*, **0405**, 024. E-print hep-th/0402207.

Khoury, J., Ovrut, B. A., Seiberg, N., Steinhardt, P. J., and Turok, N. (2002). From big crunch to big bang. *Phys. Rev.*, **D65**, 086007. E-print hep-th/0108187.

Khoury, J., Ovrut, B. A., Steinhardt, P. J., and Turok, N. (2001). The ekpyrotic universe: Colliding branes and the origin of the hot big bang. *Phys. Rev.*,

D64, 123522. E-print hep-th/0103239.

Kikkawa, K., and Yamasaki, M. (1984). Casimir effects in superstring theories. *Phys. Lett.*, **B149**, 357.

Kikkawa, K., Sakita, B., and Virasoro, M. A. (1969). Feynman-like diagrams compatible with duality. I: Planar diagrams. *Phys. Rev.*, **184**, 1701.

Kim, H. J., Romans, L. J., and Van Nieuwenhuizen, P. (1985). The mass spectrum of chiral $N = 2$ $D = 10$ supergravity on S^5. *Phys. Rev.*, **D32**, 389.

Kiritsis, E. (1998). *Introduction to Superstring Theory.* Leuven: Leuven University Press. E-print hep-th/9709062.

Klebanov, I. R. (1992). String theory in two dimensions. In *String Theory and Quantum Gravity, Trieste 1991*, eds. J. Harvey, et al., pp. 30–101. Singapore: World Scientific. E-print hep-th/9108019.

Klebanov, I. R. (1997). World-volume approach to absorption by non-dilatonic branes. *Nucl. Phys.*, **B496**, 231. E-print hep-th/9702076.

Klebanov, I. R. (2001). TASI lectures: Introduction to the AdS/CFT correspondence. In *Strings, Branes and Gravity: TASI 1999*, eds. J. Harvey, S. Kachru and E. Silverstein, pp. 615–650. Singapore: World Scientific. E-print hep-th/0009139.

Klebanov, I. R., and Nekrasov, N. A. (2000). Gravity duals of fractional branes and logarithmic RG flow. *Nucl. Phys.*, **B574**, 263. E-print hep-th/9911096.

Klebanov, I. R., and Strassler, M. J. (2000). Supergravity and a confining gauge theory: Duality cascades and χSB resolution of naked singularities. *JHEP*, **0008**, 052. E-print hep-th/0007191.

Klebanov, I. R., and Tseytlin, A. A. (1996). Entropy of near-extremal black p-branes. *Nucl. Phys.*, **B475**, 164. E-print hep-th/9604089.

Klebanov, I. R., and Tseytlin, A. A. (2000). Gravity duals of supersymmetric $SU(N) \times SU(N + M)$ gauge theories. *Nucl. Phys.*, **B578**, 123. E-print hep-th/0002159.

Klebanov, I. R., and Witten, E. (1998). Superconformal field theory on threebranes at a Calabi–Yau singularity. *Nucl. Phys.*, **B536**, 199. E-print hep-th/9807080.

Klein, O. (1926). Quantum theory and five-dimensional theory of relativity. *Z. Phys.*, **37**, 895. Reprinted in *Surveys High Energ. Phys.*, **5**, 241 (1986).

Klemm, A., Lian, B., Roan, S. S., and Yau, S. T. (1998). Calabi–Yau fourfolds for M- and F-theory compactifications. *Nucl. Phys.*, **B518**, 515. E-print hep-th/9701023.

Knizhnik, V. G. (1985). Covariant fermionic vertex in superstrings. *Phys. Lett.*, **B160**, 403.

Knizhnik, V. G., and Zamolodchikov, A. B. (1984). Current algebra and Wess–Zumino model in two dimensions. *Nucl. Phys.*, **B247**, 83.

Knizhnik, V. G., Polyakov, A. M., and Zamolodchikov, A. B. (1988). Fractal structure of 2d quantum gravity. *Mod. Phys. Lett.*, **A3**, 819.

Kolb, E. W., and Michael S. Turner, M. S. (1990). *The Early Universe* (Frontiers in physics, 69). Redwood City: Addison-Wesley.

Kumar, J. (2006). A review of distributions on the string landscape. E-print hep-th/0601053.

Leigh, R. G. (1989). Dirac–Born–Infeld action from Dirichlet sigma model. *Mod. Phys. Lett.*, **A4**, 2767.

Lerche, W., Lüst, D., and Schellekens, A. N. (1987). Chiral four-dimensional heterotic strings from self-dual lattices. *Nucl. Phys.*, **B287**, 477.

Lerche, W., Schellekens, A. N., and Warner, N. P. (1989). Lattices and strings. *Phys. Reports*, **177**, 1–140.

Lerche, W., Vafa, C., and Warner, N. P. (1989). Chiral rings in $N=2$ superconformal theories. *Nucl. Phys.*, **B324**, 427.

Liddle, A. R. (1999). An introduction to cosmological inflation. In *Trieste 1998, High Energy Physics and Cosmology*, pp. 260-295. E-print astro-ph/9901124.

Liddle, A. R., and Lyth, D. H. (2000). *Cosmological Inflation and Large-Scale Structure*. Cambridge: Cambridge University Press.

Lidsey, J. E., Wands, D., and Copeland, E. J. (2000). Superstring cosmology. *Phys. Reports*, **337**, 343. E-print hep-th/9909061.

Lifschytz, G., and Mathur, S. D. (1997). Supersymmetry and membrane interactions in M(atrix) theory. *Nucl. Phys.*, **B507**, 621. E-print hep-th/9612087.

Lin, H., Lunin, O., and Maldacena, J. M. (2004). Bubbling AdS space and 1/2 BPS geometries. *JHEP*, **0410**, 025. E-print hep-th/0409174.

Linde, A. D. (1982). A new inflationary universe scenario: A possible solution of the horizon, flatness, homogeneity, isotropy and primordial monopole problems. *Phys. Lett.*, **B108** 389.

Linde, A. D. (1990). *Particle Physics and Inflationary Cosmology*. Chur, Switzerland: Harwood Press. E-print hep-th/0503203.

Linde, A. D. (1994a). Hybrid inflation. *Phys. Rev.*, **D49**, 748. E-Print astro-ph/9307002.

Linde, A. D. (1994b). Lectures on inflationary cosmology. In *Proceedings of the Nineteenth Lake Louise Winter Institute*, pp. 72–109. Singapore: World Scientific. E-Print hep-th/9410082.

Linde, A. (1999). Inflation and string cosmology. eConf **C040802**, L024. E-print hep-th/0503195.

Lopes Cardoso, G., Curio, G., Dall'Agata, G., Lüst, D., Manousselis, P., and G. Zoupanos, G. (2003). Non-Kähler string backgrounds and their five torsion classes. *Nucl. Phys.*, **B652**, 5. E-print hep-th/0211118.

Lovelace, C. (1971). Pomeron form-factors and dual Regge cuts. *Phys. Lett.*, **B34**, 500.

Lukas, A., Ovrut, B. A., Stelle, K. S., and Waldram, D. (1999a). The universe as a domain wall. *Phys. Rev.*, **D59**, 086001. E-print hep-th/9803235.

Lukas, A., Ovrut, B. A., Stelle, K. S., and Waldram, D. (1999b). Heterotic M-theory in five dimensions. *Nucl. Phys.*, **B552**, 246. E-print hep-th/9806051.

Lüst, D., and Theisen, S. (1989). *Lectures on String Theory*. Berlin: Springer-Verlag.

Lykken, J. D. (1996). Weak scale superstrings. *Phys. Rev.*, **D54**, 3693. E-print hep-th/9603133.

Lykken, J. D. (1996). Introduction to supersymmetry. In *Boulder 1996, Fields, strings and duality*, pp. 85–153. E-print hep-th/9612114.

Maharana, J., and Schwarz, J. H. (1993). Noncompact symmetries in string theory. *Nucl. Phys.*, **B390**, 3. E-print hep-th/9207016.

Maldacena, J. M. (1998a). Black holes and D-branes. *Nucl. Phys. Proc. Suppl.*, **61A**, 111. E-print hep-th/9705078.

Maldacena, J. M. (1998b). The large N limit of superconformal field theories and supergravity. *Adv. Theor. Math. Phys.*, **2**, 231. E-print hep-th/9711200.

Maldacena, J. M. (1998c). Wilson loops in large N field theories. *Phys. Rev. Lett.*, **80**, 4859. E-print hep-th/9803002.

Maldacena, J. M. (2001). Large N field theories, string theory and gravity. In *Trieste 2001, Superstrings and related matters*, pp. 2–75.
Maldacena, J. M. (2005). TASI 2003 lectures on AdS/CFT. In *Progress in String Theory: TASI 2003*, ed. J. Maldacena, pp. 155–203. Singapore: World Scientific. E-print hep-th/0309246.
Maldacena, J. M., and Nuñez, C. (2001a). Supergravity description of field theories on curved manifolds and a no-go theorem. *Int. J. Mod. Phys.*, **A16**, 822. E-print hep-th/0007018.
Maldacena, J. M., and Nuñez, C. (2001b). Towards the large N limit of pure $N=1$ super Yang Mills. *Phys. Rev. Lett.*, **86**, 588. E-print hep-th/0008001.
Maldacena, J. M., and Strominger, A. (1997a). Black hole greybody factors and D-brane spectroscopy. *Phys. Rev.*, **D55**, 861. E-print hep-th/9609026.
Maldacena, J. M., and Strominger, A. (1997b). Universal low-energy dynamics for rotating black holes. *Phys. Rev.*, **D56**, 4975. E-print hep-th/9702015.
Maldacena, J., Strominger, A., and Witten, E. (1997). Black hole entropy in M-theory. *JHEP*, **9712**, 002. E-Print hep-th/9711053.
Mandal, G., Sengupta, A. M., and Wadia, S. R. (1991). Classical solutions of two-dimensional string theory. *Mod. Phys. Lett.*, **A6**, 1685.
Mandelstam, S. (1973). Interacting string picture of dual resonance models. *Nucl. Phys.*, **B64**, 205.
Mandelstam, S. (1974). Dual resonance models. *Phys. Reports*, **13**, 259–353.
Mandelstam, S. (1983). Light-cone superspace and the ultraviolet finiteness of the $N=4$ model. *Nucl. Phys.*, **B213**, 149.
Marcus, N., and Sagnotti, A. (1982). Tree-level constraints on gauge groups for type I superstrings. *Phys. Lett.*, **B119**, 97.
Martelli, D., and Sparks, J. (2003). G-structures, fluxes and calibrations in M-theory. *Phys. Rev.*, **D68**, 085014. E-Print hep-th/0306225.
Mateos, D., and Townsend, P. K. (2001). Supertubes. *Phys. Rev. Lett.*, **87**, 011602. E-print hep-th/0103030.
Mathur, S. D. (2005). The fuzzball proposal for black holes: An elementary review. *Fortsch. Phys.*, **53**, 793. E-print hep-th/0502050.
Mathur, S. D. (2006). The quantum structure of black holes. *Class. Quant. Grav.*, **23**, R115. E-print hep-th/0510180.
Mavromatos, N. E. (2002). String cosmology. *Lect. Notes Phys.*, **592**, 392. E-Print hep-th/0111275.
Metsaev, R. R. (2002). Type IIB Green–Schwarz superstring in plane-wave Ramond–Ramond background. *Nucl. Phys.*, **B625**, 70. E-print hep-th/0112044.
Metsaev, R. R., and Tseytlin, A. A. (2002). Exactly solvable model of superstring in plane wave Ramond–Ramond background. *Phys. Rev.*, **D65**, 126004. E-print hep-th/0202109.
Michelson, J. (1997). Compactifications of type IIB strings to four dimensions with non-trivial classical potential. *Nucl. Phys.*, **B495**, 127. E-print hep-th/9610151.
Minahan, J. A., and Zarembo, K. (2003). The Bethe-ansatz for $N=4$ super Yang-Mills. *JHEP*, **0303**, 013. E-print hep-th/0212208.
Minasian, R., and Moore, G. W. (1997). K-theory and Ramond–Ramond charge. *JHEP*, **9711**, 002. E-print hep-th/9710230.
Minwalla, S., Van Raamsdonk, M., and Seiberg, N. (2000). Noncommutative perturbative dynamics. *JHEP*, **0002**, 020. E-print hep-th/9912072.

Montonen, C. (1974). Multiloop amplitudes in additive dual resonance models. *Nuovo Cim.*, **19A**, 69.

Montonen, C., and Olive, D. I. (1977). Magnetic monopoles as gauge particles?. *Phys. Lett.*, **B72**, 117.

Moore, G. W. (2004). Les Houches lectures on strings and arithmetic. E-print hep-th/0401049.

Moore, G. W., and Seiberg, N. (1989). Classical and quantum conformal field theory. *Commun. Math. Phys.*, **123**, 177.

Morales, J. F., Scrucca, C. A., and Serone, M. (1999). Anomalous couplings for D-branes and O-planes. *Nucl. Phys.*, **B552**, 291. E-print hep-th/9812071.

Morrison, D. R., and Plesser, M. R. (1999). Non-spherical horizons. I. *Adv. Theor. Math. Phys.*, **3**, 1. E-print hep-th/9810201.

Morrison, D. R., and Vafa, C. (1996a). Compactifications of F-theory on Calabi–Yau threefolds – I. *Nucl. Phys.*, **B473**, 74. E-print hep-th/9602114.

Morrison, D. R., and Vafa, C. (1996b). Compactifications of F-theory on Calabi–Yau threefolds – II. *Nucl. Phys.*, **B476**, 437. E-print hep-th/9603161.

Myers, R. C. (1987). New dimensions for old strings. *Phys. Lett.*, **B199**, 371.

Myers, R. C. (1999). Dielectric branes. *JHEP*, **9912**, 022. E-print hep-th/9910053.

Myers, R. C., and Perry, M. J. (1986). Black holes in higher dimensional space-times. *Annals Phys.*, **172**, 304.

Nakahara, M. (2003). *Geometry, Topology and Physics*, 2nd ed. IOP Graduate Series in Physics. London: Taylor and Francis.

Nambu, Y. (1970a). Quark model and the factorization of the Veneziano model, p. 269 in *Proc. Intern. Conf. on Symmetries and Quark Models*, ed. R. Chand, New York: Gordon and Breach. [Reprinted in *Broken Symmetry: Selected Papers of Y. Nambu*, eds. T. Eguchi and K. Nishijima, Singapore: World Scientific, 1995].

Nambu, Y. (1970b). Duality and hadrodynamics, Notes prepared for the Copenhagen High Energy Symposium. [Reprinted in *Broken Symmetry: Selected Papers of Y. Nambu*, eds. T. Eguchi and K. Nishijima, Singapore: World Scientific, 1995].

Narain, K. S. (1986). New heterotic string theories in uncompactified dimensions ¡ 10. *Phys. Lett.*, **B169**, 41.

Narain, K. S., Sarmadi, M. H., and Vafa, C. (1987). Asymmetric orbifolds. *Nucl. Phys.*, **B288**, 551.

Narain, K. S., Sarmadi, M. H., and Witten, E. (1987). A note on toroidal compactification of heterotic string theory. *Nucl. Phys.*, **B279**, 369.

Neitzke, A., and Vafa, C. (2004). Topological strings and their physical applications. E-print hep-th/0410178.

Nekrasov, N., and Schwarz, A. (1998). Instantons on noncommutative R^*4 and $(2,0)$ superconformal six-dimensional theory. *Commun. Math. Phys.*, **198**, 689. E-print hep-th/9802068.

Nepomechie, R. I. (1985). Magnetic monopoles from antisymmetric tensor gauge fields. *Phys. Rev.*, **D31**, 1921.

Neveu, A., and Scherk, J. (1972). Connection between Yang–Mills fields and dual models. *Nucl. Phys.*, **B36**, 155.

Neveu, A., and Schwarz, J. H. (1971). Factorizable dual model of pions. *Nucl. Phys.*, **B31**, 86.

Neveu, A., Schwarz, J. H., and Thorn, C. B. (1971). Reformulation of the dual pion model. *Phys. Lett.*, **B35**, 529.

Nielsen, H. B. (1970). An almost physical interpretation of the integrand of the
  n-point Veneziano amplitude. Submitted to the 15th International
  Conference on High Energy Physics (Kiev).
Nilles, H. P. (1984). Supersymmetry, supergravity and particle physics. *Phys.
  Reports*, **110**, 1.
Obers, N. A., and Pioline, B. (1999). U-duality and M-theory. *Phys. Reports*, **318**,
  113–225. E-print hep-th/9809039.
Ooguri, H., and Yin, Z. (1997). Lectures on perturbative string theories. In *Fields,
  Strings and Duality: TASI 96*, eds. C. Efthimiou and B. Greene, pp. 5–81.
  Singapore: World Scientific. E-print hep-th/9612254.
Ooguri, H., Strominger, A., and Vafa, C. (2004). Black hole attractors and the
  topological string. *Phys. Rev.*, **D70**, 106007. E-print hep-th/0405146.
Osborn, H. (1979). Topological charges for $N = 4$ supersymmetric gauge theories
  and monopoles of spin 1. *Phys. Lett.*, **B83**, 321.
Overduin, J. M., and P. S. Wesson, P. S. (1997). Kaluza–Klein gravity. *Phys.
  Reports*, **283**, 303. E-print gr-qc/9805018.
Ovrut, B. A. (2004). Lectures on heterotic M-theory. In *Strings, Branes and Extra
  Dimensions: TASI 2001*, eds. S. S. Gubser and J. D. Lykken, pp. 359–406.
  Singapore: World Scientific. E-print hep-th/0201032.
Paton, J. E., and Chan, H. (1969). Generalized Veneziano model with isospin.
  Nucl. Phys., **B10**, 516.
Peebles, P. J. E. (1993). *Principles of Physical Cosmology*. Princeton: Princeton
  University Press.
Peet, A. W. (2001). TASI lectures on black holes in string theory. In *Strings,
  Branes and Gravity: TASI 1999*, eds. J. Harvey, S. Kachru and E. Silverstein,
  pp. 353–433. Singapore: World Scientific. E-print hep-th/0008241.
Peskin, M. (1987). Introduction to string and superstring theory. In *From the
  Planck Scale to the Weak Scale, TASI 1986*, ed. H. Haber, pp. 277–408.
  Singapore: World Scientific.
Pioline, B. (2006). Lectures on on black holes, topological strings and quantum
  attractors. E-print hep-th/0607227.
Polchinski, J. (1995). Dirichlet-branes and Ramond-Ramond charges. *Phys. Rev.
  Lett.*, **75**, 4724. E-print hep-th/9510017.
Polchinski, J. (1996a). What is string theory? In *Fluctuating Geometries in
  Statistical Mechanics and Field Theory, Les Houches 1994*, eds. F. David, P.
  Ginsparg, and J. Zinn-Justin, pp. 287–422. Amsterdam: North-Holland.
  E-print hep-th/9411028.
Polchinski, J. (1996b). String duality: a colloquium. *Rev. Mod. Phys.*, **68**, 1245.
  E-print hep-th/9607050.
Polchinski, J. (1997). Lectures on D-branes. In *Fields, Strings and Duality: TASI
  96*, eds. C. Efthimiou and B. Greene, pp. 293–356. Singapore: World
  Scientific. E-print hep-th/9611050.
Polchinski, J. (1998). *String Theory* in two volumes. Cambridge: Cambridge
  University Press.
Polchinski, J. (1999). Quantum gravity at the Planck length. *Int. J. Mod. Phys.*,
  A14, 2633. E-print hep-th/9812104.
Polchinski, J. (2004). Introduction to cosmic F- and D-strings. E-print
  hep-th/0412244.
Polchinski, J., and Strassler, M. J. (2000). The string dual of a confining
  four-dimensional gauge theory. E-print hep-th/0003136.

Polchinski, J., and Strominger, A. (1996). New vacua for type II string theory. *Phys. Lett.*, **B388**, 736. E-print hep-th/9510227.

Polchinski, J., and Witten, E. (1996). Evidence for heterotic – type I string duality. *Nucl. Phys.*, **B460**, 525. E-print hep-th/9510169.

Polyakov, A. M. (1981a). Quantum geometry of bosonic strings. *Phys. Lett.*, **B103**, 207.

Polyakov, A. M. (1981b). Quantum geometry of fermionic strings. *Phys. Lett.*, **B103**, 211.

Polyakov, A. M. (1987a). Quantum gravity in two dimensions. *Mod. Phys. Lett.*, **A2**, 893.

Polyakov, A. M. (1987b). *Gauge Fields and Strings*. Chur, Switzerland: Harwood.

Pope, C. N., and Van Nieuwenhuizen, P. (1989). Compactifications of $d = 11$ supergravity on Kähler manifolds. *Commun. Math. Phys.*, **122**, 281.

Quevedo, F. (1996). Lectures on superstring phenomenology. E-print hep-th/9603074.

Quevedo, F. (2002). Lectures on string/brane cosmology. *Class. Quant. Grav.*, **19**, 5721. E-print hep-th/0210292.

Ramond, P. (1971). Dual theory for free fermions. *Phys. Rev.*, **D3**, 2415.

Randall, L., and Sundrum, R. (1999a). A large mass hierarchy from a small extra dimension. *Phys. Rev. Lett.*, **83**, 3370. E-print hep-ph/9905221.

Randall, L., and Sundrum, R. (1999b). An alternative to compactification. *Phys. Rev. Lett.*, **83**, 4690. E-print hep-th/9906064.

Rebbi, C. (1974). Dual models and relativistic quantum strings. *Phys. Reports*, **12**, 1–73.

Romans, L. J. (1985). New compactifications of chiral $N = 2$, $D = 10$ supergravity. *Phys. Lett.*, **B153**, 392.

Romans, L. J. (1986). Massive $N = 2a$ supergravity in ten dimensions. *Phys. Lett.*, **B169**, 374.

Salam, A., and Sezgin, E. (1989). *Supergravities in Diverse Dimensions* in two volumes. Amsterdam: North-Holland.

Salam, A., and Strathdee, J. A. (1974). Supergauge transformations. *Nucl. Phys.*, **B76**, 477.

Salam, A., and Strathdee, J. A. (1982). On Kaluza–Klein theory. *Annals Phys.*, **141**, 316.

Scherk, J. (1975). An introduction to the theory of dual models and strings. *Rev. Mod. Phys.*, **47**, 123.

Scherk, J., and Schwarz, J. H. (1974). Dual models for nonhadrons. *Nucl. Phys.*, **B81**, 118.

Scherk, J., and Schwarz, J. H. (1979a). Spontaneous breaking of supersymmetry through dimensional reduction. *Phys. Lett.*, **B82**, 60.

Scherk, J., and Schwarz, J. H. (1979b). How to get masses from extra dimensions. *Nucl. Phys.*, **B153**, 61.

Schnabl, M. (2005). Analytic solution for tachyon condensation in open string field theory. E-Print hep-th/0511286.

Schwarz, J. H. (1972). Physical states and Pomeron poles in the dual pion model. *Nucl. Phys.*, **B46**, 61.

Schwarz, J. H. (1973). Dual resonance theory. *Phys. Reports*, **4**, 269–335.

Schwarz, J. H. (1982a). Gauge groups for type I superstrings. In *Proceedings of the Johns Hopkins Workshop on Current Problems in Particle Theory 6*, 233.

Schwarz, J. H. (1982b). Superstring theory. *Phys. Reports*, **89**, 223–322.

Schwarz, J. H. (1983). Covariant field equations of chiral $N=2$, $D=10$ supergravity. *Nucl. Phys.*, **B226**, 269.
Schwarz, J. H. (1985). *Superstrings: The first 15 years of superstring theory* in two volumes. Singapore: World Scientific.
Schwarz, J. H. (1993). Does string theory have a duality symmetry relating weak and strong coupling? In *Strings 1993*, pp. 339-352. E-print hep-th/9307121.
Schwarz, J. H. (1995). An SL(2,Z) multiplet of type IIB superstrings. *Phys. Lett.*, **B360**, 13. Erratum – ibid. **B364**, 252. E-print hep-th/9508143.
Schwarz, J. H. (1996a). Superstring dualities. *Nucl. Phys. Proc. Suppl.*, **49**, 183. E-print hep-th/9509148.
Schwarz, J. H. (1996b). The power of M theory. *Phys. Lett.*, **B367**, 97. E-print hep-th/9510086.
Schwarz, J. H. (1997). Lectures on superstring and M-theory dualities. In *Fields, Strings and Duality: TASI 96*, eds. C. Efthimiou and B. Greene, pp. 359–418. Singapore: World Scientific. E-print hep-th/9607201.
Schwarz, J. H. (2001). TASI lectures on non-BPS D-brane systems. In *Strings, Branes and Gravity: TASI 1999*, eds. J. Harvey, S. Kachru and E. Silverstein. Singapore: World Scientific. E-print hep-th/9908144.
Schwarz, J. H., and West, P. C. (1983). Symmetries and transformations of chiral $N=2$, $D=10$ supergravity. *Phys. Lett.*, **B126**, 301.
Schwarz, J. H., and Witten, E. (2001). Anomaly analysis of brane-antibrane systems. *JHEP*, **0103**, 032. E-print hep-th/0103099.
Seiberg, N. (1995). Electric-magnetic duality in supersymmetric non-Abelian gauge theories. *Nucl. Phys.*, **B435**, 129. E-print hep-th/9411149.
Seiberg, N. (1997). Why is the matrix model correct? *Phys. Rev. Lett.*, **79**, 3577. E-print hep-th/9710009.
Seiberg, N., and Witten, E. (1994a). Electric-magnetic duality, monopole condensation, and confinement in $N=2$ supersymmetric Yang–Mills theory. *Nucl. Phys.*, **B426**, 19. Erratum – ibid. **B430**, 485. E-print hep-th/9407087.
Seiberg, N., and Witten, E. (1994b). Monopoles, duality and chiral symmetry breaking in $N=2$ supersymmetric QCD. *Nucl. Phys.*, **B431**, 484. E-print hep-th/9408099.
Seiberg, N., and Witten, E. (1999). String theory and noncommutative geometry. *JHEP*, **9909**, 032. E-print hep-th/9908142.
Sen, A. (1994a). Strong - weak coupling duality in four-dimensional string theory. *Int. J. Mod. Phys.*, **A9**, 3707. E-print hep-th/9402002.
Sen, A. (1994b). Dyon - monopole bound states, self-dual harmonic forms on the multi-monopole moduli space, and SL(2,Z) invariance in string theory. *Phys. Lett.*, **B329**, 217. E-print hep-th/9402032.
Sen, A. (1996). F-theory and orientifolds. *Nucl. Phys.*, **B475**, 562. E-print hep-th/9605150.
Sen, A. (1998a). D0 branes on T(n) and matrix theory. *Adv. Theor. Math. Phys.*, **2**, 51. E-print hep-th/9709220.
Sen, A. (1998b). An introduction to non-perturbative string theory. In *Cambridge 1997, Duality and Supersymmetric Theories*, 297–413. E-print hep-th/9802051.
Sen, A. (1998c). Stable non-BPS bound states of BPS D-branes. *JHEP*, **9808**, 010. E-print hep-th/9805019.
Sen, A. (1998d). Tachyon condensation on the brane-antibrane system. *JHEP*, **9808**, 012. E-print hep-th/9805170.

Sen, A. (1999). Non-BPS states and branes in string theory. In *Cargèse 1999, Progress in String Theory and M-Theory*, 187–234. E-print hep-th/9904207.

Sen, A. (2002a). Rolling tachyon. *JHEP*, **0204**, 048. E-print hep-th/0203211.

Sen, A. (2002b). Tachyon matter. *JHEP*, **0207**, 065. E-print hep-th/0203265.

Sen, A. (2005). Tachyon dynamics in open string theory. In *Progress in String Theory: TASI 2003*, ed. J. Maldacena, pp. 207–378. Singapore: World Scientific. E-print hep-th/0410103.

Sethi, S., Vafa, C., and Witten, E. (1996). Constraints on low-dimensional string compactifications. *Nucl. Phys.*, **B480**, 213. E-print hep-th/9606122.

Shandera, S. E., and Tye, S. H. (2006). Observing brane inflation. *JCAP*, **0605**, 007. E-Print hep-th/0601099.

Shapiro, J. A. (1970). Electrostatic analog for the Virasoro model. *Phys. Lett.*, **B33**, 361.

Shenker, S. H. (1991). In *Random Surfaces, Quantum Gravity and Strings*, eds. O. Alvarez, E. Marinari and P. Windey, pp. 191–200. New York: Plenum.

Shenker, S. H. (1995). Another length scale in string theory? E-print hep-th/9509132.

Shih, D., and Yin, X. (2005). Exact black hole degeneracies and the topological string. E-print hep-th/0508174.

Siegel, W. (1983). Hidden local supersymmetry in the supersymmetric particle action. *Phys. Lett.*, **B128**, 397.

Silverstein, E. (2005). TASI/PITP/ISS lectures on moduli and microphysics. In *Progress in String Theory: TASI 2003*, ed. J. Maldacena, pp. 381–415. Singapore: World Scientific. E-print hep-th/0405068.

Silverstein, E., and Tong, D. (2004). Scalar speed limits and cosmology: Acceleration from D-cceleration. *Phys. Rev.*, **D70**, 103505. E-Print hep-th/0310221.

Sorkin, R. D. (1983). Kaluza–Klein monopole. *Phys. Rev. Lett.*, **51**, 87.

Stefanski, B. J. (1999). Gravitational couplings of D-branes and O-planes. *Nucl. Phys.*, **B548**, 275. E-print hep-th/9812088.

Stelle, K. S. (1998). BPS branes in supergravity. E-print hep-th/9803116. In *Trieste 1997, High Energy Physics and Cosmology*, pp. 29–127.

Strassler, M. J. (2005). The duality cascade. In *Progress in String Theory: TASI 2003*, ed. J. Maldacena, pp. 419–510. Singapore: World Scientific. E-print hep-th/0505153.

Strominger, A. (1986). Superstrings with torsion. *Nucl. Phys.*, **B274**, 253.

Strominger, A. (1990). Special geometry. *Commun. Math. Phys.*, **133**, 163.

Strominger, A. (1995). Massless black holes and conifolds in string theory. *Nucl. Phys.*, **B451**, 96. E-print hep-th/9504090.

Strominger, A. (1996). Open p-branes. *Phys. Lett.*, **B383**, 44. E-print hep-th/9512059.

Strominger, A. (2001). The dS/CFT correspondence. *JHEP*, **0110**, 034. E-print hep-th/0106113.

Strominger, A., and Vafa, C. (1996). Microscopic origin of the Bekenstein–Hawking entropy. *Phys. Lett.*, **B379**, 99. E-print hep-th/9601029.

Strominger, A., Yau, S. T., and Zaslow, E. (1996). Mirror symmetry is T-duality. *Nucl. Phys.*, **B479**, 243. E-print hep-th/9606040.

Sundrum, R. (2005). To the fifth dimension and back. (TASI 2004). E-print hep-th/0508134.

Susskind, L. (1969). Harmonic oscillator analogy for the Veneziano amplitude.

Phys. Rev. Lett., **23**, 545.

Susskind, L. (1970). Dual-symmetric theory of hadrons I. *Nuovo Cim.*, **69A**, 457.

Susskind, L. (1995). The world as a hologram. *J. Math. Phys.*, **36**, 6377. E-print hep-th/9409089.

Susskind, L. (1997). Another conjecture about M(atrix) theory. E-print hep-th/9704080.

Susskind, L. (2003). The anthropic landscape of string theory. E-print hep-th/0302219.

Szabo, R. J. (2003). Quantum field theory on noncommutative spaces. *Phys. Reports*, **378**, 207. E-print hep-th/0109162.

Szabo, R. J. (2004). *An Introduction to String theory and D-brane Dynamics*. London: Imperial College Press.

Taylor, T. R., and Vafa, C. (2000). RR flux on Calabi–Yau and partial supersymmetry breaking. *Phys. Lett.*, **B474**, 130. E-print hep-th/9912152.

Taylor, W. I. (1998). Lectures on D-branes, gauge theory and M(atrices). In *Trieste 1997, High Energy Physics and Cosmology*, pp. 192-271. E-print hep-th/9801182.

Taylor, W. I. (2001). M(atrix) theory: Matrix quantum mechanics as a fundamental theory. *Rev. Mod. Phys.*, **73**, 419. E-print hep-th/0101126.

Taylor, W. I. (2003). Lectures on D-branes, tachyon condensation, and string field theory. In *Valdivia 2002, Lectures on Quantum Gravity*, pp. 151–206. E-print hep-th/0301094.

Taylor, W. I. (2005). Perturbative computations in string field theory. In *Progress in String Theory: TASI 2003*, ed. J. Maldacena, pp. 513–536. Singapore: World Scientific. E-print hep-th/0404102.

Taylor, W. I., and Zwiebach, B. (2004). D-branes, tachyons, and string field theory. In *Strings, Branes and Extra Dimensions: TASI 2001*, eds. S.S. Gubser and J.D. Lykken, pp. 641–759. Singapore: World Scientific. E-print hep-th/0311017.

Teitelboim, C. (1986a). Gauge invariance for extended objects. *Phys. Lett.*, **B167**, 63.

Teitelboim, C. (1986b). Monopoles of higher rank. *Phys. Lett.*, **B167**, 69.

't Hooft, G. (1974). A planar diagram theory for strong interactions. *Nucl. Phys.*, **B72**, 461.

't Hooft, G. (1993). Dimensional reduction in quantum gravity. In *Salamfest* pp. 284–296. Singapore: World Scientific. E-print gr-qc/9310026.

Tian, G., and Akveld, M. (2000). *Canonical Metrics in Kaehler Geometry*. Basel: Birkhauser.

Townsend, P. K. (1984). A new anomaly-free chiral supergravity theory from compactification on K3. *Phys. Lett.*, **B139**, 283.

Townsend, P. K. (1995). The eleven-dimensional supermembrane revisited. *Phys. Lett.*, **B350**, 184. E-print hep-th/9501068.

Townsend, P. K. (1996a). D-branes from M-branes. *Phys. Lett.*, **B373**, 68. E-print hep-th/9512062.

Townsend, P. K. (1996b). Four lectures on M theory. In *High Energy Physics and Cosmology, Trieste 1996*, eds. E. Gava *et al.*, pp. 385–438. Singapore: World Scientific. E-print hep-th/9612121.

Trodden. M., and Carroll, S. M. (2004). Introduction to cosmology. In *Progress in String Theory: TASI 2003*, ed. J. Maldacena, pp. 703–793. Singapore: World Scientific. E-print astro-ph/0401547.

Tseytlin, A. A. (1989). Sigma model approach to string theory. *Int. J. Mod. Phys.*, **A4**, 1257.

Tseytlin, A. A. (1997). On nonabelian generalization of the Born–Infeld action in string theory. *Nucl. Phys.*, **B501**, 41. E-print hep-th/9701125.

Tseytlin, A. A. (2000). Born–Infeld action, supersymmetry and string theory. In *The Many Faces of the Superworld*, ed. M. Shifman, pp. 417-452. Singapore: World Scientific. E-print hep-th/9908105.

Tyutin, I. V. (1975). Gauge invariance in field theory and in statistical physics in the operator formulation. Lebedev preprint FIAN No. 39 (in Russian), unpublished.

Vafa, C. (1996). Evidence for F-theory. *Nucl. Phys.*, **B469**, 403. E-print hep-th/9602022.

Vafa, C. (1997). Lectures on strings and dualities. In *High Energy Physics and Cosmology, Trieste 1996*, eds. E. Gava *et al.*, pp. 66–119. Singapore: World Scientific. E-print hep-th/9702201.

Vafa, C. (2001). Superstrings and topological strings at large N. *J. Math. Phys.*, **42**, 2798. E-print hep-th/0008142.

Vafa, C. (2005). The string landscape and the swampland. E-print hep-th/0509212.

Vafa, C., and Witten, E. (1995). A one loop test of string duality. *Nucl. Phys.*, **B447**, 261. E-print hep-th/9505053.

Veneziano, G. (1968). Construction of a crossing-symmetric, Regge-behaved amplitude for linearly rising trajectories. *Nuovo Cim.*, **A57**, 190.

Veneziano, G. (1974). An introduction to dual models of strong interactions and their physical motivations. *Phys. Reports*, **9**, 199–242.

Veneziano, G. (1991). Scale factor duality for classical and quantum strings. *Phys. Lett.*, **B265**, 287.

Verlinde, E. P. (1998). Fusion rules and modular transformations in 2D conformal field theory. *Nucl. Phys.*, **B300**, 360.

Verlinde, E. P., and Verlinde, H. L. (1987). Multiloop calculations in covariant superstring theory. *Phys. Lett.*, **B192**, 95.

Vilenkin, A., and Shellard, E. P. S. (1999). *Cosmic Strings and Other Topological Defects*. Cambridge: Cambridge University Press.

Virasoro, M. (1969). Alternative constructions of crossing-symmetric amplitudes with Regge behavior. *Phys. Rev.*, **177**, 2309.

Virasoro, M. (1970). Subsidiary conditions and ghosts in dual resonance models. *Phys. Rev.*, **D1**, 2933.

Wald, R. (1984). *General Relativity*. Chicago: University of Chicago Press.

Wald, R. M. (1993). Black hole entropy in the Noether charge. *Phys. Rev.*, **D48**, 3427. E-print gr-qc/9307038.

Weinberg, S. (1989). The cosmological constant problem. *Rev. Mod. Phys.*, **61**, 1.

Wess, J., and Bagger, J. (1992). *Supersymmetry and Supergravity*. Princeton: Princeton University Press.

Wess, J., and Zumino, B. (1974). Supergauge transformations in four dimensions. *Nucl. Phys.*, **B70**, 39.

Witten, E. (1983). Global aspects of current algebra. *Nucl. Phys.*, **B223**, 422.

Witten, E. (1984). Nonabelian bosonization in two dimensions. *Commun. Math. Phys.*, **92**, 455.

Witten, E. (1985a). Cosmic Superstrings. *Phys. Lett.*, **B153**, 243.

Witten, E. (1985b). Dimensional reduction of superstring models. *Phys. Lett.*,

B155, 151.

Witten, E. (1985c). Symmetry breaking patterns in superstring models. *Nucl. Phys.*, **B258**, 75.

Witten, E. (1986). Noncommutative geometry and string field theory. *Nucl. Phys.*, **B268**, 253.

Witten, E. (1988). Topological quantum field theory. *Commun. Math. Phys.*, **117**, 353.

Witten, E.(1991). On string theory and black holes. *Phys. Rev.*, **D44**, 314.

Witten, E. (1993). Phases of $N = 2$ theories in two dimensions. *Nucl. Phys.*, **B403**, 159. E-print hep-th/9301042.

Witten, E. (1995). String theory dynamics in various dimensions. *Nucl. Phys.*, **B443**, 85. E-print hep-th/9503124.

Witten, E. (1996a). Bound states of strings and p-branes. *Nucl. Phys.*, **B460**, 335. E-print hep-th/9510135.

Witten, E. (1996b). Small instantons in string theory. *Nucl. Phys.*, **B460**, 541. E-print hep-th/9511030.

Witten, E. (1996c). Strong coupling expansion of Calabi–Yau compactification. *Nucl. Phys.*, **B471**, 135. E-print hep-th/9602070.

Witten, E. (1997a). On flux quantization in M-theory and the effective action. *J. Geom. Phys.*, **22**, 1. E-print hep-th/9609122.

Witten, E. (1997b). Solutions of four-dimensional field theories via M-theory. *Nucl. Phys.*, **B500**, 3. E-print hep-th/9703166.

Witten, E. (1998a). Anti-de Sitter space and holography. *Adv. Theor. Math. Phys.*, **2**, 253. E-print hep-th/9802150.

Witten, E. (1998b). Anti-de Sitter space, thermal phase transition, and confinement in gauge theories. *Adv. Theor. Math. Phys.*, **2**, 505. E-print hep-th/9803131.

Witten, E. (1998c). D-branes and K-theory. *JHEP*, **9812**, 019. E-print hep-th/9810188.

Witten, E. (2004). Perturbative gauge theory as a string theory in twistor space. *Commun. Math. Phys.*, **252**, 189. E-print hep-th/0312171.

Witten, E., and Olive, D. I. (1978). Supersymmetry algebras that include topological charges. *Phys. Lett.*, **B78**, 97.

Yoneya, T. (1974). Connection of dual models to electrodynamics and gravidynamics. *Prog. Theor. Phys.*, **51**, 1907.

Zwiebach, B. (1985). Curvature squared terms and string theories. *Phys. Lett.*, **B156**, 315.

Zwiebach, B. (2004). *A First Course in String Theory*. Cambridge: Cambridge University Press.

Index

acceleration equation, 528
action
   p-brane, 7, 19, 23
      sigma-model form, 29
   D0-brane, 149, 185
   Einstein–Hilbert, 301
      in D dimensions, 550
   Einstein–Maxwell, 560
   Nambu–Goto, 24, 26, 27, 155
   point-particle, 18
      nonrelativistic limit, 20
   Polyakov, 26
   string sigma-model, 26, 30, 37
      cosmological constant term, 28
   supergravity in 11 dimensions, 304
   type I supergravity, 318
   type IIA supergravity, 311
   type IIB supergravity, 314, 317
ADE
   classification of singularities, 360, 436
   Dynkin diagrams, 423
   gauge group, 437
   groups, 422, 423
   subgroups of $SU(2)$, 437
adjunction formula, 370
Adler–Bardeen theorem, 170
AdS/CFT duality, 15, 330, 612, 638
   for D3-branes, 638, 642
   for half-BPS states, 654
   for M2-branes, 643
   for M5-branes, 644
affine connection, 445
affine Lie algebra, 68
Aharonov–Bohm effect, 198
Aichelburg–Sexl metric, 633
almost complex structure, 448, 466
amplitude
   N tachyon, 91
   absence of UV divergence, 83
   D0-brane scattering, 630
   in GS formalism, 148
   large-N gauge-theory, 641
   M-theory, 632
   multiloop, 91
   off-shell, 105
   one-loop, 94
   Shapiro–Virasoro, 91
anomalous dimension, 653, 672
   BMN operator, 680
anomaly
   characteristic class, 174, 176, 177
   conformal, 62, 66, 76, 658
   form, 174, 178
   hexagon diagram, 170
   inflow mechanism, 183
   M5-brane, 184
   NS5-brane, 184
   R symmetry, 657, 673, 676
   superconformal, 143
   triangle diagram, 170
anomaly cancellation, 9, 421
   $E_8 \times E_8$, 179
   $SO(16) \times SO(16)$, 292
   11-dimensional supergravity, 181
   at M-theory boundary, 182
   heterotic M-theory, 470
   in six dimensions, 182, 426, 454
   R symmetry, 672
   type I superstring theory, 176
   type IIB superstring theory, 175
anthropic principle, 522
anthropic reasoning, 15, 522
anti-de Sitter space, 15, 375, 377, 612
   $D = 4$, 351
   $D = 5$, 494
   boundary-operator renormalization, 653
   conformal equivalence class, 652
   covering space, 646
   Euclideanized, 647, 651
   Feynman rules, 652
   global coordinates, 647, 686
   in $(d+1)$ dimensions, 616
   nonnormalizable modes, 654
   normalizable modes, 654
   path integral, 651
   Penrose diagram, 646

Poincaré coordinates, 645, 646, 686
antibrane, 211
anticommutation relations
   fermionic oscillators, 124, 164
   heterotic string fermions, 256
   world-sheet fermion fields, 120
area law, 661, 662
attractor equation, 590, 593
attractor mechanism, 587
auxiliary field
   in point-particle action, 19, 22, 145, 185
   in world-sheet supermultiplet, 113–115
   PST, 314
   world-sheet metric, 7, 26, 27, 31
axion, 316, 401, 492, 498, 541
axion–dilaton field, 316, 476, 481, 484, 485, 524

background fields, 81, 89, 227, 228, 230, 237, 267, 275, 281, 295, 628, 629
   NS–NS, 227, 237, 245
   R–R, 229, 237
background-field gauge condition, 627
Bekenstein–Hawking entropy formula, 14, 563, 603, 609, 648
   leading correction to, 585
Beltrami differential, 92
Bertotti–Robinson metric, 562
Betti numbers, 363, 442
   of S^N, 443
Bianchi identity
   for antisymmetric tensor field, 245
   for four-form in heterotic M-theory, 521
   for Maxwell field, 560
   for Riemann tensor, 384
   with H flux, 511, 516
   with D3-brane source, 483, 492
   with M5-brane source, 353
Big Bang, 16
biholomorphic function, 61
Birkhoff's theorem, 552
black p-branes, 14, 551, 613
   nonextremal, 625
black D-branes
   nonextremal, 686
black holes, 14, 408, 549
   challenges posed by, 550
   dyonic, 590
   entropy of, 14, 563, 570, 604
   extremal, 558
      with four charges for $D = 4$, 574
      with three charges for $D = 5$, 567
   inner and outer horizons, 557
   multi-center solutions, 594
   nonextremal, 572
      $D = 4$, 581
      $D = 5$, 586
   production at accelerators, 552
   Reissner–Nordström, 557
      entropy of, 564
      extremal, 558

      extremal for $D = 5$, 559
      temperature of, 565
   rotating and BPS for $D = 5$, 573
   Schwarzschild, 552, 554
      in D dimensions, 553
   supermassive, 549
   temperature of, 562
   thermodynamic properties, 14, 562
black ring, 596, 608
blow-up, 367, 368, 373, 435, 455
BMN limit, 680
BMN operator, 680
   anomalous dimension, 680
   correlation function, 681
Born–Infeld action, 231, 246, 687
bosonic ghost fields, 142
bosonic string, 17
   number operator, 43
bosonization, 67
   of ghosts, 80
bound state
   of D0-branes, 331, 626
   of strings, 328
   threshold, 331
boundary condition
   Dirichlet, 33, 164, 187, 193–195, 201, 202, 210, 622, 623
   Neumann, 164, 202, 209
boundary state, 171
BPS
   (p, q) string, 327
   bound, 151, 297, 300, 353, 408, 485, 655
      D-string, 325
   branes, 296, 421
      tensions, 307, 341
   D-branes, 208, 209, 213, 228, 230, 621, 638
      type IIA, 405
      type IIB, 328, 405
   F-string, 221
   M-branes, 307, 332, 614
   M-theory solutions, 617
   states, 298, 328, 408
      in nine dimensions, 340
      wrapped D-branes, 414
   tensions, 427
brane–antibrane annihilation, 212, 587
brane–antibrane inflation, 537, 540
brane–brane inflation, 539
brane-world scenario, 11, 354, 459, 493
Brink–Schwarz superparticle, 185
BRST
   charge, 77
      for RNS string, 143
      mode expansion, 77
      nilpotency, 78, 107
   cohomology, 79, 108
   quantization, 75, 108
   symmetry, 77
      of RNS string, 142
bubbling AdS, 655

Calabi–Yau four-fold, 458, 460–463, 466, 467, 469, 471, 477, 478, 547, 548
  elliptically fibered, 475
  examples, 476
  Hodge numbers, 473
  holomorphich four-form, 474
  Kähler form, 473
Calabi–Yau manifold, 5, 10, 356
  n-folds, 363
  mirror symmetry, 10
  one-folds, 366
  two-folds, 366
Calabi–Yau three-fold, 588
  conifold singularity, 488
  elliptic fibration, 455
  Euler characteristic, 365
  heterotic string compactification, 415
  Hodge diamond, 364
  intersection numbers
    of three-cycles, 392
  Kähler form, 381
  M-theory compactification, 401, 410
  metric deformations, 388
  product structure of moduli space, 391
  quintic, 370
    supersymmetric three-cycle, 409
  special Lagrangian submanifold, 405
  three-torus fibration, 414
  type IIA compactification, 400, 402
  type IIB compactification, 400, 402, 403, 408, 588
  volume form, 384
calibration, 435, 439
canonical homology basis, 590
canonical momentum, 35
Cartan matrix
  E_8, 294
  $SU(3)$, 285
Casimir invariants
  of $SU(N)$, 657, 669
central charge, 42, 62, 297, 300
central extension, 42, 57, 62
Chan–Paton
  charges, 101, 196, 197, 222, 244, 250
  factors, 196
  index, 197
  matrix, 197, 198
chaotic inflation, 534
characteristic class, 157, 172–174
  factorization, 180
charge-conjugation matrix, 153
Chern character, 174
  factorization property, 177
Chern class, 471
  first, 363, 370, 371, 382, 451, 453
  second, 186
Chern–Simons form, 157, 174, 179, 186
Chern–Simons term
  D7-brane, 483
  in five dimensions, 658
chiral algebra, 71

chiral fields, 170, 174
  of type I theory, 176
  of type IIB theory, 175
  on an M5-brane, 184
  self-dual tensor, 172, 175
  Weyl fermion, 174
  Weyl gravitino, 175
chiral matter, 417
chiral primary operator, 656, 672
chiral superfield, 666
chiral supermultiplet, 417, 473
chiral-symmetry breaking, 611
chirality, 134, 302
  in ten dimensions, 136
  of dilatino, 137
  of gravitino, 137
  of R sector, 136
  of type IIB theory, 136
  of Weyl spinor, 134
  projection operators, 134
Christoffel connection, 23, 384, 509, 510, 513, 517, 547
Clifford algebra, 110, 166, 299
closed form, 442
closed string, 32
closed time-like curve, 359, 551, 645
CMB anisotropy, 531
cocycle, 67
cohomology, 79, 441
  BRST, 79, 87, 105, 108, 144
  classes, 79, 211, 321, 395, 589
  constraint, 416
  de Rham, 442, 467
  Dolbeault, 449
  of Calabi–Yau four-fold, 474, 500
  of Calabi–Yau three-fold, 392
  of K3, 419
  of Riemann surface, 95
  primitive, 467–469
cold dark matter, 531
commutation relations, 55
  Kac–Moody algebra, 69
  of bosonic oscillators, 124
  Poincaré algebra, 56
compactification, 5
  F-theory
    on K3, 427, 433
  heterotic string
    on a Calabi–Yau three-fold, 357, 374, 387, 415, 418
    on a three-torus, 420
    with flux, 508
  M-theory
    on a G_2 manifold, 433, 434, 436
    on a $Spin(7)$ manifold, 438
    on a Calabi–Yau four-fold, 461, 499
    on a Calabi–Yau three-fold, 401, 405, 410
    on K3, 419, 423
    with conical singularity, 436
  on a circle, 188, 190, 193, 198, 199, 202, 209,

211, 214, 234, 244, 271, 329, 330, 333, 337, 339
toroidal, 265, 266, 268, 270, 274, 275, 278, 280, 287, 288, 291, 340, 345, 397
  type IIA superstring
    on a Calabi–Yau three-fold, 400, 402
    on K3, 424–426
  type IIB superstring
    on a Calabi–Yau three-fold, 400, 402, 403, 408, 498, 502
    on K3, 426, 454
    with flux, 480
  warped, 355, 456
  with branes, 376
  with flux, 13, 458
complex geometry, 449
complex projective space, 369, 453
  Fubini–Study metric, 369
  Kähler potential, 369
complex structure, 90, 448, 513
  deformations, 370, 388, 391
  moduli, 590
  moduli space, 391, 589
    Kähler potential, 391, 397
    prepotential, 393
complexified Kähler form, 390, 401
confinement, 611, 663
  of $\mathcal{N}=1$ gauge theory, 675
conformal anomaly, 7, 49, 62, 66
  cancellation, 76
  ghost contribution, 76
conformal compactification, 60
conformal dimension, 64, 106
  of bosonic ghosts, 142
  of conserved currents, 70
  of free fermion, 141
  of ghost fields, 75
  of minimal model fields, 72
  of primary field, 107
  of vertex operator, 85, 86, 88, 108
conformal field, 64
conformal field theory, 15, 58
  anomalous dimensions, 653
  correlation functions, 651
  energy–momentum tensor, 652
  Euclideanized, 651
  in three dimensions, 643
  scaling dimension, 653
  state–operator correspondence, 67, 85, 655
conformal fixed point, 650
conformal flatness, 59
conformal group
  D dimensions, 59, 61, 612
  conformal weights, 668
  inversion, 60
  restricted, 62
  two dimensions, 61, 62
conformal isometry group, 91
conformal symmetry
  anomaly, 658
conformal transformation, 60

infinitesimal, 61, 74
special, 60, 62
conical singularity, 360, 670
conifold, 354, 487–489, 669, 671, 675, 684, 687
  deformed, 488, 490, 496, 498, 673, 675, 684
    with wrapped D6-branes, 685
  geometry, 489, 491
  isometry group, 489
  Kähler form, 548
  resolved, 488, 490, 684
    with wrapped D5-branes, 685
  singularity, 403, 404, 454, 487, 488, 490, 497
  transition, 357, 385, 403, 487, 685
  warped, 492, 674
  with fluxes, 491
coordinate singularity, 554, 559
correlation functions, 73
  generating function, 652
coset-space theory, 69
cosmic censorship conjecture, 550, 551, 558
cosmic fluids, 528
cosmic microwave background (CMB), 527, 530
  polarization, 538
  scalar perturbations, 538
  spectral indices, 539
  tensor perturbations, 538
cosmic strings, 537, 543
cosmological constant, 15, 84, 474, 482, 528, 547
  in brane-world scenario, 494
  positive, 504, 505
  problem, 377, 439, 503, 522
coupling constant
  $\mathcal{N}=4$ super Yang–Mills, 642
  $\mathcal{N}=4$ super Yang–Mills theory, 323
  't Hooft, 639, 650
  11-dimensional, 304
  effective, 639
  gravitational, 301, 310
    in four dimensions, 499
  heterotic string, 335, 344, 420
    in seven dimensions, 422
  of gauge unification, 519, 520
  open string, 233, 639
  QED, 8
  S-duality transformation, 323
  string, 8, 11, 85, 90, 227, 311
  super Yang–Mills in $D=10$, 318
  type I superstring, 325
  type IIA superstring, 307, 310, 331, 643
  type IIB superstring, 327, 427, 492
  Yang–Mills in five dimensions, 636
covariantly constant spinor, 363, 376
critical density, 529
critical dimension, 7, 47
  of RNS string, 129
critical string, 47, 76
cross-cap, 82, 178
cross-cap state, 171
cubic invariant

with $E_{6,6}$ symmetry, 570
current algebra, 274
    $SO(32)$, 292
    bosonic representation, 286
    fermionic representation, 263
    level-one, 292
curvature tensor, 446
curvature two-form, 173
cycle, 388, 443
    ADE classification on K3, 423
    collapsing, 485, 497
    degenerate, 415, 420, 423, 497
    holomorphic, 408
    on a Riemann surface, 95
    on a torus, 341
    special Lagrangian, 404, 407, 408, 414, 415
    supersymmetric, 357, 404, 405, 407–409, 414, 435, 436, 464
    three-dimensional, 392
    two-dimensional, 373
    vanishing, 404
    with flux, 457
    wrapped by a D-brane, 483, 485
cylinder, 31, 170, 171, 178, 361

Dp-brane, 188
D-branes, 8, 11, 14, 83, 109, 164, 171, 187, 194, 250
    bosonic actions with background fields, 237
    extremal black, 617
    half-BPS, 208, 228, 230
        T-duality transformation, 209, 210
    non-BPS, 208
    nonabelian actions, 239
    nonextremal black, 619
    space-time filling, 8, 195, 221
    tension of, 233
    type I, 223
    world-volume actions, 229
        with κ symmetry, 230
D-instanton, 207, 405
D0-branes, 149, 150, 206, 611, 626
    action, 149, 185
        kappa symmetry, 152
    coincident, 241
    mass, 235
    non-BPS, 223
D3-branes
    charge, 483
    extremal black, 619
    fractional, 672
    near-extremal black, 658
        entropy of, 659
        temperature of, 659
    nonextremal black, 620
D6-branes, 206, 332, 437
D8-branes, 207, 225, 246
Dabholkar–Harvey states, 600
dark energy, 15, 84, 532
dark matter, 417, 531
DBI action, 233, 246

in static gauge, 236, 247
de Rham cohomology, 363, 442
de Sitter space, 375, 480, 503, 613, 664
Dedekind η function, 431, 601, 608
deficit angle, 360, 428
deformed conifold, 490, 673, 675, 684
density perturbations, 538
descendant operator, 656
descendant state, 68
descent equations, 174, 178, 184, 186
diffeomorphism, 30
dilatino, 137, 175
dilaton, 11, 53, 81, 137, 618
dimensional reduction, 307
dimensional regularization, 631
Dirac algebra
    for a Kähler manifold, 466
    in D dimensions, 125
    in two dimensions, 110
Dirac matrices
    $3 + 8$ decomposition, 462
    $4 + 6$ decomposition, 512
    in six dimensions, 379
    in two dimensions, 110, 164
Dirac quantization condition, 205, 207, 214, 215, 343
    p-brane generalization, 207, 215
    for M-branes, 353
    modified, 521
Dirac roof genus, 174
Dirac string, 215
Dirac–Ramond equation, 128
Dirichlet boundary condition, 8, 33, 164, 187, 193, 622
discrete light-cone quantization (DLCQ), 626
Dolbeault cohomology groups, 449
Dolbeault operators, 449
domain wall, 673
dS/CFT duality, 613, 664
dual Coxeter number, 69
dual lattice, 277
duality
    F-theory/heterotic string, 427
    heterotic/type IIA superstring, 424
    heterotic/type IIB superstring, 426
    M-theory/$SO(32)$ superstring, 343
    M-theory/heterotic string, 420
    M-theory/type IIA superstring, 330
    M-theory/type IIB superstring, 339
    type I/$SO(32)$ heterotic, 324
duality cascade, 674
Dynkin diagrams, 423, 424
    G_2, 434
    ADE, 423
    ADE groups, 423
    Spin(8), 163
dyon, 205
dyonic black holes, 590

effective action
    low-energy, 170, 229, 296, 300, 301, 317, 386

Index

anomaly analysis, 171
  breakdown of, 403
  five-dimensional, 405, 407, 658
  for D-branes, 231, 638
  four-dimensional, 354, 416, 456, 499, 519
  heterotic, 320, 321, 456, 519
  M-theory, 329, 617
  one-loop, 629
  two-loop, 630, 631
  type I, 319
  type IIA, 184, 211
  type IIB, 327, 480
  with background fields, 628
quantum, 173
effective coupling constant, 639
  scale dependence of, 639
effective potential, 84, 195, 386, 504, 534
effective supergravity theories, 301
Eguchi–Hanson space, 367, 435, 453
  Hodge numbers, 368
  metric, 367
Einstein space, 670
Einstein tensor, 550
Einstein–Hilbert action
  corrections to, 603
  in D dimensions, 301, 550
electric flux, 457, 615
electric–magnetic duality
  $\mathcal{N} = 2$ super Yang–Mills theory, 324
  $\mathcal{N} = 4$ super Yang–Mills theory, 323
electromagnetism, 5
electroweak scale, 4
eleventh dimension, 11, 181, 643
  size of, 11
elliptic fibration
  Calabi–Yau four-fold, 433
  Calabi–Yau three-fold, 455
  K3 manifold, 432, 433
elliptic modular function, 429
end-of-the-world 9-branes, 302, 335
energy–momentum tensor
  for ghosts, 76, 80, 107
  for Kac–Moody algebra, 69
  holomorphic component, 63
  of RNS string, 119
  OPE, 66, 73
  open-string, 39
  world-sheet, 27, 30, 32, 34, 39
    central charge, 69
    conformal transformation of, 66
    modes of, 62, 63, 73
    of coset theory, 70
    vanishing of, 41
enhanced gauge symmetry, 246, 273, 274, 284, 285, 287, 288, 295, 422
  E_8, 261
  $SO(2N)$, 223
  $SU(2) \times SU(2)$, 271, 273
  $SU(2)^4$, 285
  $SU(3) \times SU(3)$, 286
  $USp(2N)$, 223
  at singularities of K3, 436
enhanced supersymmetry, 151
equation of state, 527
Euclideanized time, 562
Euler characteristic, 31, 82, 83, 442, 667
  of Calabi–Yau four-fold, 484
  of Riemann surface, 83
Euler's theorem, 641
event horizon, 550, 553, 556
  area of, 563
  topology of, 596, 621
exact form, 442
exceptional-holonomy manifolds, 357, 433
exterior derivative, 78, 102, 441

F-theory, 427
  K3 compactification, 433
  on Calabi–Yau four-fold, 475
Faddeev–Popov ghosts, 75, 142
fermi field
  free, 66
fermionization, 67
Feynman diagrams
  anomalous, 170
  in anti-de Sitter space, 652
  in large-N gauge theory, 641, 667
  in Matrix theory, 628, 630, 632
  in quantum field theory, 6
  in string theory, 8, 83, 171
  in Witten's string field theory, 105
Feynman path integral, 7
fibration
  T^3, 414, 415, 425, 437
  circle, 413
  elliptic, 432, 455, 483
  K3, 437
Fierz transformation, 380, 513
  in any dimension, 545
  in ten dimensions, 157, 185
  in two dimensions, 114
fine-structure constant, 8
first quantization, 36
first superstring revolution, 8
fixed point, 97, 221, 271, 273, 335, 360, 361, 367, 372, 435, 455
flat potential, 84, 198, 273
flatness problem, 531, 534
flop, 685
flow to the infrared, 674
flux
  five-form, 618, 640
  four-form, 615
  imaginary self-dual three-form, 484
  quantization, 457, 524
    in heterotic M-theory, 521
  self-dual five-form, 481
flux compactification, 13, 458
  tadpole-cancellation constraint, 524
  type IIB theory, 480
flux quantization, 457
fractional D3-branes, 491, 672

Friedmann equation, 528
Friedmann–Robertson–Walker (FRW) metric, 528
Fubini–Study metric, 453
fundamental form, 513
fundamental region, 93
fuzzy two-sphere, 241

G-parity, 134–136, 138
gauge anomalies, 169
gauge field
   n-form, 204
gauge hierarchy problem, 459, 493, 496
gauge symmetry
   $E_8 \times E_8$, 263
   E_n, 226
   $SO(2N)$, 225
   $SO(32)$, 222
   $SO(N)$, 251
   $SU(N)$, 251, 638
   $U(1)^N$, 251
   $U(N)$, 196, 225
   $USp(32)$, 222
   enhancement of, 271
gaugino condensation, 417, 675
general relativity, 1, 3
generations, 418, 456
geodesic equation, 23
geometric transition, 613, 684
ghost fields, 75
   OPE of, 75
ghost number, 77, 78
ghost-number operator, 79, 80
giant graviton, 660
global time coordinate, 646
gluino condensation, 504, 521
Goldstone boson, 201
Goldstone fermion, 236
grand unification, 259, 356, 416
   gauge group, 374
Grassmann algebra, 117
Grassmann coordinates, 113
Grassmann integration, 115
Grassmann numbers, 110
graviphoton, 589, 603
gravitational anomalies, 173
gravitational collapse, 599
gravitational waves, 538
gravitino, 137, 175
   in 11 dimensions, 303
graviton, 8, 53, 137
   in 11 dimensions, 302
Green–Schwarz counterterm, 179
Gregory–Laflamme instability, 621, 686
ground state
   of Ramond sector, 125
GS superstring
   action with world-sheet metric, 161
   equations of motion, 161
   light-cone gauge, 160
      action, 162, 163
      equations of motion, 162
      quantization, 164
   quantization, 160
GSO projection, 133, 135, 136

Hagedorn temperature, 52
half hypermultiplet, 300
half-BPS states, 298
harmonic forms, 364, 444, 450
harmonic oscillators, 36
Hawking radiation, 564, 586
Hawking–Page phase transition, 663
heavy Z bosons, 418
hermitian metric, 449
heterotic M-theory, 337, 518
   anomaly cancellation, 470
heterotic string theory
   $E_8 \times E_8$, 259
      strong coupling limit, 334, 518
   $SO(16) \times SO(16)$, 292
   $SO(32)$, 254
   $Spin(32)/\mathbb{Z}_2$, 259
   as M5-brane wrapped on K3, 421
   Calabi–Yau compactification, 374
   fermionic construction, 252
   in eight dimensions, 433
   in five dimensions, 427
   in seven dimensions, 420
   in six dimensions, 424
   massless spectrum, 257
   supergravity approximation, 376
heterotic three-form field strength
   Bianchi identity, 511
hexagon diagram anomaly, 170
hidden sector, 416
Higgs mechanism, 298
highest-weight state, 67, 107
Hirzebruch L-function, 175
Hodge $\star$-operator, 445
Hodge diamond, 364
Hodge dual, 445
Hodge numbers, 363, 364, 449
holographic duality, 15, 612, 648
   energy/radius correspondence, 651
holomorphic n-form, 363
holomorphic cycles, 408
holomorphic three-form, 371, 589
holonomy group, 215, 378, 447
   G_2, 357, 434
   $SU(3)$, 378, 379
   $SU(4)$, 463
   $SU(n)$, 356, 363
   $Spin(7)$, 357, 438
   $USp(2n)$, 366
homology, 441
horizon problem, 531, 534
horizon size, 534
Hubble length, 532
Hubble parameter, 528
Hubble redshift law, 530
hybrid inflation, 535, 536, 543

hypermultiplet, 300, 400, 670

infinite-momentum frame, 626
inflation, 16, 531
  brane–brane, 539
  brane-antibrane, 540
  chaotic, 534
  exit from, 536, 543
  hybrid, 535, 543
  in string theory, 539, 542
  number of e-foldings, 537
  power-law, 535
  slow-roll, 540
inflationary cosmology, 532
inflaton, 533, 535
  potential, 539
information loss, 564
inner horizon, 558
instanton contributions, 405, 407, 606
intersection numbers
  of three-cycles, 392
  of two-forms, 396
isometry of compact dimensions, 251
isotopic spin, 196

K-theory, 211, 214, 223
K3 manifold, 366
  elliptic fibration, 432
  Euler characteristic, 368
  harmonic two-forms, 453
  Hodge diamond, 368
  Hodge numbers, 368
  in complex projective space, 370
  intersections of two-cycles, 423
  Kähler-structure deformations, 419
  moduli space, 419
  orbifold limit, 367
  singularities, 423
    ADE classification, 423
Kähler cone, 390
Kähler form, 372, 449
  complexified, 390
Kähler manifold, 364, 449
  volume form, 371
Kähler potential, 450, 453, 589
  for type IIB flux compactification, 498
  for type IIB moduli, 485
  of complex-structure moduli space, 392
Kähler structure
  deformations, 388, 395
  moduli space, 395
    Kähler potential, 395
    prepotential, 396
Kähler transformation, 394, 502
Kac–Moody algebra, 68
  $SU(2)$, 281
  level, 69
  representation by free fermi fields, 264
Kaluza–Klein 5-brane, 334, 337
Kaluza–Klein compactification, 5, 251, 354
  excitation number, 188, 189, 279

  fractional, 199
  excitations, 190, 267, 330, 331, 333, 626, 646, 651, 657, 661
    on a five-sphere, 656
    on a five-sphere, 653
Kaluza–Klein monopole, 333
kappa symmetry
  D-brane actions, 230, 234
  gauge fixing, 236
  M2-brane, 406
  of D0-brane action, 152, 185
  of GS world-sheet action, 156, 158
Kerr black hole, 573
Killing spinor, 306, 614, 642
Killing spinor equation, 376, 461, 477, 493, 516, 614, 615, 686
Killing vector, 306, 553, 642
Klebanov–Strassler model, 459
Klein bottle, 82, 170, 171, 178
Kruskal diagram, 555
Kruskal–Szekeres coordinates, 554

Laplace operator, 444
Large Hadron Collider (LHC), 4, 459, 496, 552
large-N expansion, 639, 641, 667
large-N limit, 639, 641, 657
  entropy density, 660
lattice, 278
  $E_8 \times E_8$, 286
  E_8, 286, 294
  $Spin(32)/\mathbb{Z}_2$, 287, 294
  dual, 277
  even, 278
  even self-dual, 277, 288
    Euclidean, 286
  integral, 278
  self-dual, 278
  unimodular, 278
Lefschetz decomposition, 467, 469, 486, 500
left-movers, 34
Legendre transform, 604
leptons, 4
level-matching condition, 43, 190
Levi–Civita connection, 23, 446
Lichnerowicz equation, 389, 453
Lie algebra
  E_6, 416, 417
  $E_8 \times E_8$, 9, 94, 179–181, 186, 222, 250, 254, 259, 263, 274, 286, 320, 338, 416
    Cartan matrix, 294
  E_8, 181, 186, 259, 335, 336
    $E_6 \times SU(3)$ embedding, 417
    $SO(10) \times SU(4)$ embedding, 418
    $SU(5) \times SU(5)$ embedding, 418
  E_n, 226
  $E_{n,n}$, 345
  G_2, 434
  $SL(2,\mathbb{R})$, 42
  $SO(16) \times SO(16)$, 261
  $SO(32)$, 9, 94, 222, 250, 254, 274, 320
  $SU(3) \times SU(2) \times U(1)$, 4

dual Coxeter number, 69
exceptional, 434
light-cone coordinates, 48
light-cone gauge, 49
    for RNS string, 130
    quantization, 48
linear-dilaton vacuum, 98, 100
Liouville field theory, 99
local counterterm, 178
local Lorentz symmetry, 172
local Lorentz transformation
    infinitesimal, 173
local supersymmetry
    of superparticle action, 145
localized gravity, 493
Lorentz anomaly, 49
Lorentz group
    in D dimensions, 110
    in four dimensions, 300, 379
    in ten dimensions, 254
    in two dimensions, 111
Lorentz transformations, 38
    generators, 44, 51, 56
        in NS sector, 146
    infinitesimal, 30
loss of quantum coherence, 564

M-theory, 12, 184, 296, 329, 625
    higher-dimension terms, 421
    K3 compactification, 419
        dual heterotic description, 420
    nongeometric duality, 347, 636
    on a Calabi–Yau four-fold, 461
M2-brane, 307, 332
    as instanton, 521
    as source of flux, 472
    Euclidean, 405
    extremal black, 615
        near-horizon geometry, 616
    kappa symmetry, 406
M5-brane, 184, 307, 332, 353
    extremal black, 615
        near-horizon geometry, 617
    wrapped on K3, 421
magnetic flux, 457
    four-form, 616
magnetic monopole, 205, 214, 531
Majorana representation, 136, 153
    in two dimensions, 111
Majorana spinor, 153, 185
    in two dimensions, 110
Majorana–Weyl spinor
    in ten dimensions, 136, 137
    in two dimensions, 111
manifold
    G_2 holonomy, 433, 685
        singularities, 436
    $Spin(7)$ holonomy, 438
    Calabi–Yau, 356, 447
    complex, 448
    hyper-Kähler, 366, 447

    Kähler, 364, 447, 449
    pseudo-Riemannian, 444
    quaternionic Kähler, 447
    real, 440
    Riemannian, 444
    special holonomy, 447
mass gap, 663
    of $\mathcal{N}=1$ gauge theory, 675
mass-energy density, 527
mass-shell condition, 39, 50
massless superparticle, 144
Matrix theory, 13, 330, 611, 625
    action, 627, 688
    bosonic propagator, 629
    fermionic propagator, 629, 637
    Feynman rules, 628
    in plane-wave space-time, 682
    one-loop effective action, 629, 689
    toroidal compactification, 635
    two-loop effective action, 630
matter, 527
maximally symmetric space-time, 374
    Riemann tensor, 374
Maxwell field, 81
Maxwell theory, 204
    field equations, 205
membrane, 17, 19, 307, 335, 405–407, 438
metric tensor, 444
minimal model, 71
mirror symmetry, 357, 411
    circle, 413
    torus, 413
    type IIA and type IIB, 412, 685
mixed ensemble, 605
modified Bessel function, 602
modular form
    of weight eight, 286
modular invariance, 94, 286
modular transformation, 94, 431
moduli fields, 84, 387, 456
moduli space
    metric, 391
    of Calabi–Yau three-fold, 386
    of complex structures, 91, 391
    of heterotic string vacua, 287, 420
    of K3 manifolds, 419
        elliptically fibered, 433
    of Kähler structures, 395
    of M-theory vacua, 346
    of Riemann surfaces, 90
    of string compactifications, 526
    of torus, 96, 274, 284, 651
    of vacua, 84
    product structure, 391
moduli stabilization, 499
    by nonperturbative effects, 503
    in heterotic M-theory, 521
moduli-space problem, 13, 456
Moebius strip, 82, 170, 171, 178, 415
Myers effect, 241, 660

naked singularity, 551, 558
Nambu–Goto action, 7, 24, 26, 27, 155
Narain lattice, 287
naturalness, 523
near-horizon geometry, 561
near-horizon limit, 559
negative pressure, 532
negative-norm states, 37, 120
Neumann boundary condition, 8, 32, 164, 192
Neveu–Schwarz boundary condition, 123
Newton's constant, 6
   for weakly coupled heterotic string, 519
   in D dimensions, 554
   in four dimensions, 2
   in heterotic M-theory, 520
Nijenhuis tensor, 380, 448, 513, 517
no-go theorem, 480, 482, 503
Noether current, 38
Noether method, 37, 119, 121
noncritical string, 47, 98
nonrelativistic limit, 20, 27
nonrenormalizability, 2, 301
nonrenormalization theorem, 474
normal ordering, 41, 66
NS–NS fields, 227, 245
   type IIB, 313
NS–NS sector, 124, 137, 168, 184, 227, 237
   type I, 221
   type IIB, 220
NS–NS two-form, 327, 390, 470
nucleosynthesis, 530
null hypersurface, 556
number operator
   bosonic string, 43
   NS sector, 126
   R sector, 125

open string, 32
   coupling constant, 639
open-string spectrum
   NS sector, 131
   R sector, 132
operator product expansion (OPE), 64–66, 72
   bosonic ghosts, 142
   contour-integral evaluation, 65
   free fermi fields, 66, 141, 264
   ghost fields, 75
   Kac–Moody algebra, 68, 264, 281
   super-Virasoro, 141
orbifold, 358
   blow-up, 455
   circle, 359
   complex plane, 360
   heterotic string, 454
   of S^5, 670
   supersymmetry breaking, 362
   torus, 362, 367, 372
   twisted states, 361
   untwisted states, 361
orientifold plane, 171, 178, 222, 250
orientifold projection, 9, 220, 226

outer horizon, 558

p-brane, 7, 17, 19
   tension of, 7, 19
parallel transport, 510
parity
   violation, 169, 170, 175
   world-sheet, 220–222, 224, 226
partition function, 91, 275, 287–289, 293, 652, 663, 685
partons, 626
Peccei–Quinn symmetry, 396, 397
Penrose diagram, 646
Penrose limit, 678
   $1/R^2$ corrections, 682
perfect fluid, 527
perimeter law, 661
period matrix, 96
perturbation theory, 8
Pfaffian, 577
phase transition
   confinement/deconfinement, 661, 664
   Hawking–Page, 663
physical-state conditions
   bosonic string, 44
   in the NS sector, 127
   in the R sector, 127
picture-changing operators, 144
Planck brane, 495
Planck length, 6
   in 11 dimensions, 304, 310
Planck mass, 6
plane-wave space-time, 677
   M-theory, 682
      BMN Matrix theory, 682
   type IIB, 677, 678
      light-cone gauge Hamiltonian, 680
      light-cone gauge quantization, 682
      light-cone gauge world-sheet action, 679
      two-impurity states, 681
Poincaré coordinates, 646
Poincaré duality, 443, 451
Poincaré transformations, 30, 38
Poincaré–Hopf index theorem, 464
point particle, 18, 21
   massive, 149
   supersymmetric, 150, 152
Poisson bracket, 35
Poisson resummation formula, 276, 288, 290, 608
Polyakov action, 26
Polyakov path integral, 90
Pontryagin class, 183, 471
potential
   quark-antiquark, 661
power-law inflation, 535
pp-wave, 678
prepotential, 393, 396, 602, 606
   for complex structure moduli space, 393, 394
   for Kähler structure moduli space, 396
   holomorphic, 400, 603

homogeneity equation, 603, 604
  tree-level, 605
pressure, 527
primary field, 64, 107
primary operator, 656
primitive form, 467
primitivity condition, 467, 475, 479
probe graviton, 632
projective plane, 82
pseudo-Riemannian manifold, 444
PST auxiliary field, 314
puncture, 91
pure gravity
  $D = 3$, 301
pure spinor, 160

Q-bit, 648
quadratic Casimir number, 69
quantum chromodynamics (QCD), 2, 610
quantum effective action, 173
quantum electrodynamics (QED), 8
quantum field theory, 1
quantum mechanics, 1
  breakdown of, 550
quark-gluon plasma, 664
quarks, 4
quarter-BPS states, 298, 687
quartic invariant
  with $E_{7,7}$-symmetry, 576
quaternionic-Kähler manifold, 400, 401

R symmetry, 641
  $SO(8)$, 644
  $SU(4)$, 641, 657, 666, 669
  $U(1)$, 665, 671
  $USp(4)$, 426
  $\mathbb{Z}_2$, 675
  $\mathbb{Z}_{2M}$, 674
  anomaly, 657, 673, 676
  breaking of, 673
  current, 673
R–R sector, 124, 137, 168, 179, 187, 221, 229, 313
  charge, 208, 211, 225, 228, 405, 524, 620
  9-brane, 212
  fields, 168, 211, 229, 237, 238, 241, 245, 311, 314, 318, 425, 617, 618, 677
    type I, 223
  five-form, 642, 679
    flux, 677
  tadpoles, 222
  two-form, 327, 674
  type I, 221
  type IIA, 207
  type IIB, 207
  zero-form, 427, 481
radial modulus, 458, 502
radial ordering, 64
radiation, 527
Ramond boundary condition, 123
Randall–Sundrum construction, 495

red-shift factor, 649
reduced Planck mass, 528
Regge slope parameter, 34
Reissner–Nordström black hole, 557
  entropy of, 564
  extremal for $D = 4$, 558
  extremal for $D = 5$, 559
    entropy of, 570
  temperature of, 565
relevant perturbation, 655
renormalizability, 301
renormalization group equation, 649
renormalization-group flow, 673, 675
resolved conifold, 490, 684
restricted conformal group, 62
RHIC collider, 664
Ricci form, 382, 451
Ricci tensor, 351, 446, 607, 670
Ricci-flat metric, 333, 363, 366, 378, 382, 384
Riemann curvature tensor, 446
Riemann surface, 8, 83, 89
  canonical homology basis, 95
  first homology group, 95
  genus one, 92
  genus zero, 91
  higher-genus, 95
  holomorphic one-forms, 95
Riemann zeta function, 50
Riemann–Roch theorem, 91
Riemann–Schottky problem, 96
Riemannian geometry, 444
Riemannian manifold, 444
right-movers, 34
Rindler space, 647
RNS string, 110
  BRST charge, 143
    nilpotency, 143, 147
  BRST symmetry, 142
  closed-string spectrum, 136
  energy–momentum tensor, 119
  Euclideanized world-sheet action, 140
  NS boundary condition, 123
  NS–NS sector, 137
  open-string spectrum, 131
  R–R sector, 137
  supercurrent, 119
Routhian, 634

S-branes, 220, 613, 622
  in field theory, 622
  in string theory, 623
S-duality, 10, 323
  $D = 4$ heterotic, 606
  $\mathcal{N} = 4$ super Yang–Mills, 636
  type I/$SO(32)$ heterotic, 324, 352
  type IIA/heterotic, 426
  type IIB/type IIB, 327, 571, 576, 650
Sasaki–Einstein space, 671, 676
  $L^{p,q,r}$, 676
  $T^{1,1}$, 671, 676
  $Y^{p,q}$, 676

scalar curvature, 447
scalar potential
   flux compactification
      M-theory on a Calabi–Yau four-fold, 502
      type IIB on a Calabi–Yau three-fold, 502
   no-scale type, 503
   with nonperturbative terms, 504
scale factor, 528, 535
scale invariance, 649
scale transformation, 60
Schwarzian derivative, 66
Schwarzschild black hole, 552
   in D dimensions, 553
   in Kruskal–Szekeres coordinates, 555
   temperature of, 563
Schwarzschild radius, 553
second law of thermodynamics, 563
second superstring revolution, 10, 403
Seiberg duality, 674
self-dual charge, 208
self-dual field strength, 313
self-dual radius, 209
short supermultiplet, 297
Siegel upper half plane, 96
sigma-model action
   for p-brane, 29
   for string, 26, 27, 30
simplicial homology, 443
simply-connected manifold, 364
simply-laced Lie groups, 69, 422
singularities
   ADE classification, 360
   A_N, 360, 436
   D_N, 373
   blow-up, 367, 368, 373, 435
slow-roll
   approximation, 535
   conditions, 535
   inflation, 540
   parameters, 535, 539, 540
small black holes, 599
source graviton, 632
special conformal transformation, 60
special coordinates, 392
special geometry, 391, 394, 589
   homogeneous coordinates, 393
   inhomogeneous coordinates, 393
special Lagrangian cycle, 404, 589
special Lagrangian submanifold, 407
special-Kähler manifold, 400, 401, 589
spin connection, 172, 305, 445, 446
   embedding in the gauge group, 417
spin manifold, 378, 447
Spin(1,1), 111
Spin(8), 162, 169
   Clebsch–Gordon coefficients, 166
   Dynkin diagram, 163
   spinors, 133, 136, 162
   triality symmetry, 163
   vector, 136
spurious states, 44

standard big-bang model (SBB), 530
standard model, 4, 355
standard-model brane, 495
state–operator correspondence, 67, 85, 655
static gauge, 27, 233, 236
stationary metric, 553
Stokes' theorem, 215, 217, 443, 451
string charge, 227
string coupling constant, 11, 85, 90, 227, 324
   in six dimensions, 426
   nonperturbative effects, 357
   T-duality mapping, 210
   type IIA, 310
string field theory, 100
   Chern–Simons action, 104
   cubic string interaction, 105
   gauge transformation, 103
   in light-cone gauge, 682
string geometry, 411
string length scale, 6, 34, 310
string perturbation theory, 89
string sigma-model action, 26, 27, 30
string tension, 34
string theory landscape, 459, 522
string world sheet, 6
   topology, 8
stringy exclusion principle, 660
strong nuclear force, 2
submanifold
   special Lagrangian, 408
super Yang–Mills theory
   $\mathcal{N}=2$ for $D=4$, 670
   $\mathcal{N}=4$ for $D=4$, 636, 638, 640
      θ term, 650
      in terms of $\mathcal{N}=1$ superfields, 666
      multi-trace operators, 656
      planar approximation, 641
      S-duality, 650
      single-trace operators, 656
   in $0+1$ dimensions, 627
   in ten dimensions, 157, 176, 626
super-Poincaré algebra, 112, 149
super-Virasoro algebra
   constraints, 118
   generators, 126
   in the NS sector, 127
   in the R sector, 127
   operator-product expansions, 141
   superspace formulation, 141
superalgebra
   $OSp(1\,|\,2)$, 128
   $OSp(6,2|4)$, 644
   $OSp(8|4)$, 644
   $PSU(2,2|4)$, 640, 667
   $SU(1,1\,|\,1)$, 128
supercharges, 114
superconformal algebra
   $SU(2,2|1)$, 671
superconformal field theory, 612
   chiral primary operator, 655
superconformal gauge theory, 670, 671

superconformal symmetry, 76, 118, 612
    representations, 655
    supercharges, 668
superconnection, 211
supercovariant derivative, 114
supercurrent
    holomorphic, 141
    of RNS world-sheet theory, 119, 121
superfields, 113
supergraviton, 330, 626
supergravity
    $\mathcal{N} = 1$ for $D = 10$, 176, 179
    $\mathcal{N} = 1$ for $D = 6$, 182
    $\mathcal{N} = 2$ for $D = 10$, 137
    $\mathcal{N} = 2$ for $D = 4$, 400, 588
    $\mathcal{N} = 8$ for $D = 4$, 400, 576
    $\mathcal{N} = 8$ for $D = 5$, 567
    heterotic, 320
        action, 320
        supersymmetry transformations, 321
    in 11 dimensions, 15, 139, 159, 167, 302, 461, 614
        action, 304
        Freund–Rubin solution, 351
        quantum correction, 469
        supersymmetry transformations, 305
    type I, 317
        action, 318
        supersymmetry transformations, 319
    type IIA, 307
        action, 311
        supersymmetry transformations, 311
    type IIB, 313, 480
        $SL(2, \mathbb{R})$ symmetry, 315
        action, 314, 317
        self-dual five-form, 313
        supersymmetry transformations, 315
supermatrices, 668
    supertrace, 668
supermembrane, 159
superparticle
    massive, 149, 150, 152
    massless, 144, 185
    quantization, 145
superpotential, 665, 666, 672
    for complex-structure moduli, 474, 485
    for type IIB flux compactification, 497
superspace
    for world sheet, 113
    supersymmetry transformations, 149
    ten-dimensional, 148
superstring cosmology, 15, 460, 526
superstring theory, 4, 109
    p-branes, 7
    Green–Schwarz formulation, 148
    RNS formulation, 109
    types of, 9
supersymmetric cycle, 357, 404, 435
supersymmetric grand unification, 356, 418
supersymmetric state, 382
supersymmetry, 4, 109
    $D = 4, \mathcal{N} = 1$, 355
    algebra
        $\mathcal{N}$-extended, 297
        $\mathcal{N} = 1$, 298
        $\mathcal{N} = 2$, 300
    breaking, 4
    of world sheet, 112
    space-time, 4
    transformations, 112, 114
supersymmetry-breaking solutions, 501
supertrace, 668
symplectic coordinates, 589
symplectic modular group, 96, 392, 589

T-duality, 10, 187, 190
    $O(16 + n, n; \mathbb{Z})$, 287
    $O(n, n; \mathbb{Z})$, 269
    closed bosonic string, 188
    for NS–NS background fields, 227, 281
    for R–R background fields, 229
    heterotic string, 295
    in light-cone GS formulation, 218
    open bosonic string, 193
    relation to mirror symmetry, 413
tachyon, 51, 53, 109
tachyon condensation, 587
tadpole-cancellation condition, 212, 469, 470, 483
Taub–NUT metric, 333, 352, 579
    multi-center, 334, 437, 455
tensor supermultiplet, 421
three-form gauge field
    in 11 dimensions, 304
three-string junction, 329, 687
threshold bound states, 331, 626
topological string theory, 603, 613, 685
topology change, 462
toric geometry, 371
toroidal compactification, 265
    bosonic string, 266
        level-matching condition, 267
        modular invariance, 274
        moduli space, 270
        with background fields, 267
    heterotic string, 287
    type II superstrings, 273, 291
torsion, 312, 321, 508, 509, 547
torus, 31, 92, 170
    complex-structure moduli space, 397
    modular parameter, 92
    periods of, 92
translation symmetry, 38
triangle diagram anomaly, 170
twisted sector, 221, 318, 359, 361, 584
two-dimensional Ising model, 72
two-form gauge field, 81, 137
type I superstring theory, 83, 165, 170, 176, 220
    anomaly cancellation, 176
    closed, 165
    gauge groups, 176

open, 164, 166
  world-sheet topologies, 178
type I′ theory, 224, 246
type II superstring theory
  GS world-sheet action, 155
    kappa symmetry, 156
  spectrum, 167
type IIA superstring theory
  Calabi–Yau compactification, 400
  D0-branes, 330
  D6-brane, 332
  K3 compactification, 424
    mirror symmetry, 425
    S-duality, 426
  strong coupling limit, 330
type IIB superstring theory, 175
  (p, q) 5-branes, 329
  (p, q) strings, 327
  7-branes, 329, 428, 429
  anomaly cancellation, 175
  Calabi–Yau compactification, 400
  D3-brane, 329
  K3 compactification, 426
  string webs, 329
  three-string junction, 329, 687

U-duality, 345, 636
  $E_3(\mathbb{Z})$, 347
  $E_n(\mathbb{Z})$, 346
  $E_{6,6}$, 567, 570
  $E_{7,7}$, 576
ultraviolet divergences, 2, 7
  absence of, 83
  cancellation, 83
unitarity bound, 672
universal hypermultiplet, 401
unoriented string, 83, 196, 220, 221

vacuum energy density, 15, 84, 527
vacuum selection
  problem of, 14
vanishing cycle, 404
vector supermultiplet, 204
  $D = 10$, 257, 261
  $D = 3$, 473
  $D = 4, \mathcal{N} = 4$, 400
  $D = 4, \mathcal{N} = 2$, 400
    massive, 300
  E_8, 181
  $SO(32)$, 258
  $U(N)$, 239
Verma module, 68
vertex operator, 85
  integrated, 87
  of tachyon, 85
  unintegrated, 87
  with ghosts, 86
vielbein, 444
Virasoro algebra, 40, 57
  classical, 61
  generators, 39, 40

operator product expansions, 140
Volkov–Akulov theory, 236
volume of unit n-sphere, 554

Wald's formula, 603, 609
warp factor, 462
  of warped conifold, 493
warped compactification, 355, 456
  of the heterotic theory, 508
  of the type IIB theory, 480
warped conifold
  geometry of, 492
warped geometry, 457, 461, 612
warped metric, 461
  type IIB compactification, 481
wave equation, 32
web of dualities, 12, 296
weighted complex projective space, 371
Wess–Zumino consistency conditions, 173
Wess–Zumino–Witten model, 69
Weyl symmetry, 89
Weyl transformations, 30
white hole, 555
Wigner–Inönü contraction, 679
Wilson line, 198, 288, 295, 418
Wilson-loop operator, 661
winding number, 188
  fractional, 199
WMAP, 538
world line, 6
world-sheet action
  with ghost fields, 76
  cosmological constant term, 28, 31, 99
  light-cone coordinates, 33
  metric, 7, 26
    conformally flat, 89
  with fermions, 110
  with local supersymmetry, 120
world-sheet parity, 220, 226
  projection, 221
world-sheet topology, 6, 8

Yang–Mills theory, 4, 11
  anomalies, 173
  field strength, 102, 172
  gauge field, 172
  gauge transformation, 103, 172
  in three dimensions, 662
    finite temperature, 663
  large-N expansion, 639
  multi-trace operators, 656
  planar approximation, 641
  single-trace operators, 656
  with $\mathcal{N} = 4$ supersymmetry, 15, 298, 323, 636, 640, 641, 657, 687

zero-norm spurious states, 45, 129
zero-point energy, 133, 146, 257
zeta-function regularization, 50, 146